Forms Containing $a + bx$

9. $\displaystyle\int (a + bx)^n \, dx = \begin{cases} \dfrac{(a+bx)^{n+1}}{(n+1)b} + C, & n \neq -1 \\[2ex] \dfrac{1}{b}\ln|a+bx| + C, & n = -1 \end{cases}$

10. $\displaystyle\int x(a + bx)^n \, dx = \begin{cases} \dfrac{1}{b^2(n+2)}(a+bx)^{n+2} - \dfrac{a}{b^2(n+1)}(a+bx)^{n+1} + C, & n \neq -1, -2 \\[2ex] \dfrac{1}{b^2}\left[a + bx - a\ln|a+bx|\right] + C, & n = -1 \\[2ex] \dfrac{1}{b^2}\left[\ln|a+bx| + \dfrac{a}{a+bx}\right] + C, & n = -2 \end{cases}$

11. $\displaystyle\int x^2(a + bx)^n \, dx = \dfrac{1}{b^3}\left[\dfrac{(a+bx)^{n+3}}{n+3} - 2a\dfrac{(a+bx)^{n+2}}{n+2} + a^2\dfrac{(a+bx)^{n+1}}{n+1}\right] + C, \quad n \neq -1, n \neq -2, n \neq -3$

12. $\displaystyle\int x^m(a + bx)^n \, dx, \quad m > 0, m + n + 1 \neq 0$

In terms of a lower power of n:

$$\frac{x^{m+1}(a+bx)^n}{m+n+1} + \frac{an}{m+n+1}\int x^m(a+bx)^{n-1}\, dx$$

In terms of a lower power of m:

$$\frac{1}{b(m+n+1)}\left[x^m(a+bx)^{n+1} - ma\int x^{m-1}(a+bx)^n\, dx\right]$$

13. $\displaystyle\int x^m\sqrt{a + bx} \, dx = \dfrac{2}{b(2m+3)}\left[x^m\sqrt{(a+bx)^3} - ma\int x^{m-1}\sqrt{a+bx}\, dx\right]$

If $m = 1$: $\quad \dfrac{-2(2a - 3bx)\sqrt{(a+bx)^3}}{15b^2} + C$

If $m = 2$: $\quad \dfrac{2(8a^2 - 12abx + 15b^2x^2)\sqrt{(a+bx)^3}}{105b^3}$

14. $\displaystyle\int \dfrac{\sqrt{a + bx}}{x^m} \, dx = \begin{cases} -\dfrac{1}{(m-1)a}\left[\dfrac{\sqrt{(a+bx)^3}}{x^{m-1}} + \dfrac{(2m-5)b}{2}\int\dfrac{\sqrt{a+bx}\, dx}{x^{m-1}}\right], & m \neq 1 \\[3ex] 2\sqrt{a+bx} + \sqrt{a}\ln\left|\dfrac{\sqrt{a+bx} - \sqrt{a}}{\sqrt{a+bx} + \sqrt{a}}\right|, & m = 1, a > 0 \end{cases}$

College Mathematics and Calculus

WITH APPLICATIONS TO MANAGEMENT, LIFE AND SOCIAL SCIENCES

The Smith Business Series

This book is part of the Smith Business Series of textbooks, which includes:

Business Mathematics, Second Edition
This is an arithmetic based business mathematics textbook. Published by Wm. C. Brown.

Mathematics for Business Students
This is an algebra based business mathematics textbook. Published by Wm. C. Brown.

Finite Mathematics, Second Edition
This is a standard finite mathematics textbook. Published by Brooks/Cole.

Calculus with Applications
This is a business calculus textbook for students in management, life science, and social science. Published by Brooks/Cole.

College Mathematics and Calculus with Applications to Management, Life and Social Sciences
This book combines the material in the finite mathematics and business calculus books. Published by Brooks/Cole.

Computer Aided Finite Mathematics and Calculus
Written by Chris Avery and Charles Barker to accompany books in this mathematics series. Published by Brooks/Cole.

Other Brooks/Cole Titles by Karl J. Smith

Mathematics: Its Power and Utility, Second Edition

The Nature of Mathematics, Fifth Edition

Essentials of Trigonometry, Second Edition

Trigonometry for College Students, Fourth Edition

Precalculus Mathematics, Third Edition

Algebra and Trigonometry

The Smith and Boyle Precalculus Series Published by Brooks/Cole

Beginning Algebra for College Students, Third Edition

Intermediate Algebra for College Students, Third Edition

Study Guide for Intermediate Algebra for College Students, Third Edition

College Algebra, Third Edition

College Mathematics and Calculus

WITH APPLICATIONS TO MANAGEMENT, LIFE AND SOCIAL SCIENCES

Karl J. Smith

 Brooks/Cole Publishing Company
Pacific Grove, California

Brooks/Cole Publishing Company   A Division of Wadsworth, Inc.

© 1988 by Wadsworth, Inc., Belmont, California 94002. All rights reserved. No part of this book may be reproduced, stored in a retrieval system, or transcribed, in any form or by any means—electronic, mechanical, photocopying, recording, or otherwise—without the prior written permission of the publisher, Brooks/Cole Publishing Company, Pacific Grove, California 93950, a division of Wadsworth, Inc.

Printed in the United States of America

10   9   8   7   6   5   4   3   2   1

Library of Congress Cataloging-in-Publication Data

Smith, Karl J.
 College mathematics and calculus with applications to management, life and social sciences/Karl J. Smith.
       p.   cm.
 Includes index.
 ISBN 0-534-08910-0
 1. Mathematics—1961–   I. Title.
QA37.2.S576   1988   87-19403   CIP

510—dc19

Sponsoring Editor: Jeremy Hayhurst
Editorial Assistant: Maxine Westby
Production Services Coordinator: Joan Marsh
Production: Cece Munson, The Cooper Company
Manuscript Editor: Betty Berenson
Interior Design: Jamie Sue Brooks
Cover Design: Katherine Minerva
Cover Photo: Lee Hocker
Interior Illustration: Carl Brown
Typesetting: Polyglot Pte Ltd, Singapore
Cover Printing: Phoenix Color Corp.
Printing and Binding: R. R. Donnelley & Sons, Harrisonburg, Virginia

This book is dedicated, with love, to my wife,
Linda Ann Smith

Preface

College Mathematics and Calculus is a textbook for a two-semester or three-quarter course to provide the mathematics background necessary for students in business, management, life sciences, or social sciences. The text is divided into two parts, Finite Mathematics in Chapters 3–8 and Calculus in Chapters 9–16.

FINITE MATHEMATICS | OPTIONAL REVIEW MATERIAL | CALCULUS

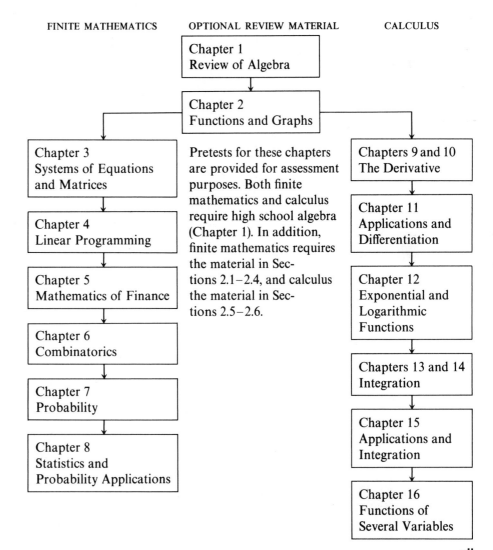

Chapter 1
Review of Algebra

Chapter 2
Functions and Graphs

Chapter 3
Systems of Equations
and Matrices

Chapter 4
Linear Programming

Chapter 5
Mathematics of Finance

Chapter 6
Combinatorics

Chapter 7
Probability

Chapter 8
Statistics and
Probability Applications

Pretests for these chapters are provided for assessment purposes. Both finite mathematics and calculus require high school algebra (Chapter 1). In addition, finite mathematics requires the material in Sections 2.1–2.4, and calculus the material in Sections 2.5–2.6.

Chapters 9 and 10
The Derivative

Chapter 11
Applications and
Differentiation

Chapter 12
Exponential and
Logarithmic
Functions

Chapters 13 and 14
Integration

Chapter 15
Applications and
Integration

Chapter 16
Functions of
Several Variables

The emphasis in this course is not on algebraic manipulation but rather on an understanding of the modeling process and using mathematics to make statements about real world applications. The prerequisite for this course is intermediate algebra.

It seems as if every new book claims innovation, state-of-the-art production, supplementary materials, readability, abundant problems, and relevant applications. How, then, is one book chosen over another? More specifically, how does this book differ from other books for this type of course, and what factors were taken into consideration as it was being written?

Content

First, every book must cover the appropriate topics, and hopefully in the right order. Finite mathematics has evolved and changed considerably since it was first introduced in the 1960s. Today's course focuses on matrices to solve both linear systems of equations and inequalities (linear programming), sets, combinatorics, and probability, and some supplementary topics such as Markov chains, game theory, and mathematics of finance. New recommendations regarding discrete mathematics are also influencing finite mathematics courses, so I have included appendixes on logic and mathematical induction.

The calculus in Chapters 9–16 is presented by using a variety of real world applications. In 1986, the Sloan Foundation sponsored a four-day conference to study the calculus curriculum. The conference's work was summarized in the following statement: "We propose a calculus syllabus which emphasizes intuition and conceptual understanding by giving priority to numerical and geometrical methods and approximation techniques." The development of integration that I used in this book follows the recommendations of this conference. The use of tables and numerical integration is emphasized and we do not spend a great deal of time on integration techniques.

Mathematical Models

When a real life situation is analyzed, it usually does not easily lend itself to mathematical analysis because the real world is far too complicated to be precisely and mathematically defined. It is therefore necessary to develop what is known as a **mathematical model**. This is a body of mathematics based on certain assumptions about the real world, which is modified by experimentation and accumulation of data, and is then used to predict some future occurrence in the real world. A mathematical model is not static or unchanging but is continually being revised and modified as additional relevant information becomes known.

Some mathematical models are quite accurate, particularly in the physical sciences. For example, the path of a projectile, the distance that an object falls in a vacuum, or the time of sunrise tomorrow all have mathematical models that provide very accurate predictions about future occurrences. On the other hand, in the fields of management, life sciences, and social sciences, the models provide much less accurate predictions because they deal with situations that involve many unknowns or undefined variables that are often random in character. How can we go about constructing a model in management, life sciences, or social sciences? You need to observe a real world problem and make assumptions about influencing factors, a process called *abstraction*.

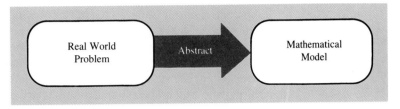

You must know enough about the mechanics of mathematics to *derive results* from the model.

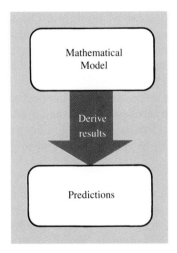

The next step is to gather data. Does the prediction fit all the known data? If not, you will use this data to *modify* the assumptions used in the model. This is an ongoing process.

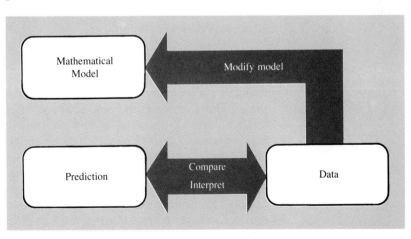

The primary skill addressed in this book is that of problem solving and building a mathematical model. Most books pay lip service to building models but rarely *develop* the skill. True model building cannot be learned in a single lesson, or even in a single course. Learning this skill must be a gradual process, and that is how I approach it in this book—in small steps with realistic applications. I devote an entire section to the nature of linear programming and how to go about the *formulation* of an appropriate model (Section 4.2). In addition, the modeling applications at the end of each chapter are included to illustrate the *model-building process* in real life situations. These applications are presented as suggestions for term papers, which are assignments that are open-ended and require a model-building approach for their development. Even though they are designed to be used as enrichment or for supplementary study, they are also designed to illustrate mathematical modeling. One essay written in response to each of these modeling applications is given in its entirety in the *Student Solutions Manual* in order to show *how* the model building can be developed. These model-building applications, even if not assigned, serve to illustrate, in a very real way, how the material developed in the rest of the chapter can be used to answer some nontrivial questions (see, for example, the Modeling Application on the Cobb-Douglas Production Function at the end of Chapter 16).

The text is divided into sections of nearly equal size that each take about one class day to develop. Many sections are marked optional in the Table of Contents. This does not mean that I feel these topics are not important; it simply means that the material in these sections is not used in subsequent sections.

Style

The author's writing style is another factor that distinguishes one textbook from another. My writing style is informal, and I always write with the student in mind. I offer study hints along the way and let the students know what is important. Frequent and abundant examples are provided so that the student can understand each step before proceeding to the next. A second color is used to highlight the important steps or particular parts of an equation or formula. The chapter reviews list important terms and provide a sample test. A listing of the chapter objectives as well as from 50 to 100 additional practice problems keyed to the stated objectives are provided in the *Student Solutions Manual* so that the student has a clear grasp of all important ideas in the course.

Problems

The third factor, which is one of the most important in deciding upon a textbook, is the number *and quality* of the problems presented. This is where I have spent a great deal of effort in writing this book. The problems should help to *develop* the students' understanding of the material, and not inhibit or thwart that understanding by being obscure. The problems start with the simpler ideas, presenting the problems in matched pairs so that for each new problem at least one is provided with an answer in the back and one without the answer. There are about three times more problems than are needed for assignments so that students will have the opportunity to go back and practice additional problems, both for the midterm and for the final. I have provided almost 7,000 problems in this text. After the manipulative skills have been developed, each problem set presents a large number of applications to show how the material can be used in business, management, and life and social sciences.

The types of problems include:

Drill. I have included a large number of drill problems to provide adequate practice so that the student can develop a clear understanding of each topic.

Applications. Self-contained applied problems are provided in almost every section to give relevance and practicality to the topic at hand. A list showing the extent and variety of the applications is part of each chapter opening and gives a preview of the material.

Modeling Applications. Each chapter concludes with an optional real life application that allows the material to be applied to a *real* (rather than a textbook) problem, but at a level of difficulty that is manageable for the student. The answers to these questions are open-ended, and a sample essay written in answer to each modeling application is included in the *Student Solutions Manual.*

CPA, CMA, and Actuary Exam Questions. Actual questions from Actuary, CPA, and CMA exams are scattered throughout the textbook. These test questions provide a link between textbook and profession.

Historical Questions. Historical notes provide insight into the humanness of mathematics, but instead of being superfluous commentaries, these are integrated into the problem sets to give students a taste of some of the ways the topics were originally developed as solutions to mathematical problems.

Supplementary Materials

The final factor often used in selecting a textbook is the type and quality of available supplements. In addition to the answers in the back of the book, I have provided several supplements:

Student Solutions Manual This provides complete solutions to every odd-numbered problem in the book. It also includes essays to illustrate modeling and lists of objectives that define the necessary skills studied in each chapter. It also provides additional practice problems to supplement those in the textbook.

Instructor's Supplement This includes answers to all of the problems in the book and additional questions to accompany the modeling applications.

Testing Program This is a computerized test bank with text editing capabilities that allows you to create an almost unlimited number of tests or retests of the material. This test bank is also available in printed form for those instructors not using a computer.

Computer Aided Finite Mathematics This is a computer supplement prepared by Chris Avery and Charles Barker. Appendix D lists the programs available in this supplement. An extensive list of other commercially available computer software is also provided in Appendix D.

As the author, I am also available to help you create any other set of supplementary materials that you feel are necessary or worthwhile for your course.

Acknowledgments

Many persons have reviewed either the entire manuscript or parts of this manuscript and have offered many valuable suggestions, and I would like to offer each and every one of them my sincere thanks:

John Alberghini
  Manchester Community College
Patricia Bannantine
  Marquette University

Craig Benham
  University of Kentucky
Thomas Covington
  Northwestern State University

Joe S. Evans
  Middle Tennessee State University
James L. Forde
  Concordia College
Marjorie S. Freeman
  University of Houston
Matthew Gould
  Vanderbilt University
Kevin Hastings
  University of Delaware
Judith Hector
  Walters State Community College
Lou Hoelzle
  Bucks County Community College
Edwin Klein
  University of Wisconsin
Ann Barber Megaw
  University of Texas, Austin
Norman Mittman
  Northeastern Illinois University
Wendell L. Motter
  Florida A & M University

Jacqueline Payton
  Virginia State University
William Ramaley
  Fort Lewis College
Robert Sharpton
  Miami-Dade Community College
Gordon Shilling
  University of Texas, Arlington
Roland Sink
  Pasadena City College
Harriette J. Stephens
  State University of New York, Canton
Mary T. Teegarden
  Mesa Junior College
William T. Watkins
  Pan American University
Jan Wynn
  Brigham Young University
Donald Zalewski
  Northern Michigan University

The problem solvers were John Spellman, Ernest Ratliff, Ricardo Torrejon, and Gregory Passty of Southwest Texas State University, Donna Szott of Allegheny County Community College—South, and Pat Bannantine of Marquette University, and I thank them not only for reviewing the manuscript, but for the long hours spent working all the examples and problems. In addition, Terry Shell of Santa Rosa Junior College worked all the problems (at least twice) and offered many helpful suggestions, and I give him my heartfelt thanks. Chris Avery and Charles Barker from De Anza College were most cooperative and helpful in modifying their computer material to match the development in this book.

The production of a textbook is a team effort. I would like to thank Jeremy Hayhurst, Maxine Westby, Joan Marsh, Katherine Minerva, Cece Munson, Jamie Sue Brooks, and Carl Brown.

And last, but not least, my thanks go to my family, Linda, Missy, and Shannon, for their love, support, and continued understanding of my involvement in writing this book.

Karl J. Smith
Sebastopol, California

Contents

CHAPTER 1 **Review of Algebra** (Optional)   **1**

1.1   Algebra Pretest   2
1.2   Real Numbers   4
1.3   Algebraic Expressions   8
1.4   Factoring   14
1.5   Linear Equations and Inequalities   18
1.6   Quadratic Equations and Inequalities   22
1.7   Rational Expressions   27
*1.8   Summary and Review   31

CHAPTER 2 **Functions and Graphs**   **33**

2.1   Functions and Graphs Pretest   34
2.2   Functions   35
2.3   Graphs of Functions   42
2.4   Linear Functions   47
2.5   Quadratic and Polynomial Functions   58
2.6   Rational Functions   65
*2.7   Summary and Review   71

FINITE MATHEMATICS   73

CHAPTER 3 **Systems of Equations and Matrices**   **75**

3.1   Systems of Equations   76
3.2   Introduction to Matrices   84

* Optional sections

3.3   Gauss–Jordan Elimination   99
3.4   Inverse Matrices   113
*3.5   Leontief Models   122
*3.6   Summary and Review   127
Modeling Application 1: Gaining a Competitive Edge in Business   130

CHAPTER 4   **Linear Programming**   **131**

4.1   Systems of Linear Inequalities   132
4.2   Formulating Linear Programming Models   137
4.3   Graphical Solution of Linear Programming Problems   147
4.4   Slack Variables and the Pivot   157
4.5   Maximization by the Simplex Method   164
4.6   Nonstandard Linear Programming Problems   174
*4.7   Duality   182
*4.8   Summary and Review   191
Modeling Application 2: Air Pollution Control   194

CHAPTER 5   **Mathematics of Finance**   **195**

5.1   Difference Equations   196
5.2   Interest   204
5.3   Annuities   214
5.4   Amortization   217
*5.5   Sinking Funds   222
*5.6   Summary and Review   226
Modeling Application 3: Investment Opportunities   229

CHAPTER 6   **Combinatorics**   **230**

6.1   Sets and Set Operations   231
6.2   Permutations   242
6.3   Combinations   250
6.4   Binomial Theorem   259
*6.5   Summary and Review   264
Modeling Application 4: Earthquake Epicenter   266

CHAPTER 7   **Probability**   **267**

7.1   Probability Models   268
7.2   Probability of Equally Likely Events   273
7.3   Calculated Probabilities   280
7.4   Conditional Probability   285
7.5   Probability of Intersections and Unions   293

7.6   Mathematical Expectation   296
7.7   Stochastic Processes and Bayes' Theorem   302
*7.8   Summary and Review   308
Modeling Application 5: Medical Diagnosis   311

CHAPTER 8   **Statistics and Probability Applications   312**

8.1   Random Variables   313
8.2   Analysis of Data   317
8.3   The Binomial Distribution   327
8.4   The Normal Distribution   333
8.5   Normal Approximation to the Binomial   338
*8.6   Correlation and Regression   343
*8.7   Summary and Review   351
Cumulative Review Chapters 3–8   354

CALCULUS   359

CHAPTER 9   **The Derivative   360**

9.1   Limits   361
9.2   Continuity   371
9.3   Rates of Change   379
9.4   Definition of Derivative   390
9.5   Differentiation Techniques, Part I   395
9.6   Differentiation Techniques, Part II   400
9.7   The Chain Rule   405
*9.8   Summary and Review   408
Modeling Application 6: Instantaneous Acceleration: A Case Study of the Mazda 626   410

CHAPTER 10   **Additional Derivative Topics   411**

10.1   Implicit Differentiation   412
10.2   Differentials   415
10.3   Business Models Using Differentiation   420
*10.4   Related Rates   426
*10.5   Summary and Review   431
Modeling Application 7: Publishing: An Economic Model   432

CHAPTER 11   **Applications and Differentiation   433**

11.1   First Derivatives and Graphs   434
11.2   Second Derivatives and Graphs   443

11.3   Curve Sketching—Relative Maximums and Minimums   450

11.4   Absolute Maximum and Minimum   454

*11.5   Summary and Review   461

Cumulative Review for Chapters 9–11   463

CHAPTER 12   **Exponential and Logarithmic Functions   464**

12.1   Exponential Functions   465

12.2   Logarithmic Functions   471

12.3   Logarithmic and Exponential Equations   476

12.4   Derivatives of Logarithmic and Exponential Functions   482

*12.5   Summary and Review   487

Modeling Application 8: World Running Records   489

CHAPTER 13   **Integration   490**

13.1   The Antiderivative   491

13.2   Integration by Substitution   497

13.3   The Fundamental Theorem of Calculus   503

13.4   Area between Curves   513

*13.5   Summary and Review   522

Modeling Application 9: Computers in Mathematics   524

CHAPTER 14   **Additional Integration Topics   525**

14.1   Integration by Parts   526

14.2   Integration by Table   530

14.3   Numerical Integration   535

*14.4   Summary and Review   544

Modeling Application 10: Modeling the Nervous System   546

CHAPTER 15   **Applications and Integration   547**

15.1   Business Models   548

15.2   Probability Density Functions   555

15.3   Improper Integrals   567

*15.4   Differential Equations   570

*15.5   Summary and Review   574

Cumulative Review for Chapters 12–15   576

Cumulative Review for Chapters 9–15   577

CHAPTER 16   **Functions of Several Variables   579**

16.1   Three Dimensional Coordinate System   580

16.2   Partial Derivatives   587

16.3   Maximum–Minimum Applications   592

16.4   Lagrange Multipliers   598

*16.5   Multiple Integrals   604

*16.6   Summary and Review   615

Modeling Application 11: The Cobb-Douglas Production Function   617

Appendixes   618

A   Logic   619

B   Mathematical Induction   628

C   Calculators   633

D   Computers   637

E   Tables   641

  1.   Pascal's Triangle—Combinatorics   642

  2.   Squares and Square Roots   643

  3.   Binomial Probabilities   644

  4.   Standard Normal Cumulative Distribution   648

  5.   Compound Interest—Future Value and Present Value   650

  6.   Ordinary Annuity   652

  7.   Annuity Due   654

  8.   Present Value of an Annuity   656

  9.   Mortality Table   658

  10.   Powers of e 659

  11.   Common Logarithms   661

F   Markov Chains   663

G   Answers to Odd-Numbered Problems   681

Index   719

CHAPTER 1
Review of Algebra
(Optional)

CHAPTER CONTENTS

1.1 **Algebra Pretest**

1.2 **Real Numbers**

1.3 **Algebraic Expressions**

1.4 **Factoring**

1.5 **Linear Equations and Inequalities**

1.6 **Quadratic Equations and Inequalities**

1.7 **Rational Expressions**

1.8 **Summary and Review**
   Important Terms
   Sample Test

The prerequisite for the material in this book is a course in basic algebra. Since it has probably been a while since you studied algebra, this chapter includes a review of the algebraic topics you need to understand the mathematical models presented in this book. The chapter begins with an algebra pretest. Part I of the pretest reviews general algebraic concepts. If you have difficulty with these questions, take an algebra course before attempting the material in this book. Part II of the pretest reviews more advanced algebraic topics that are discussed in this chapter. If you have difficulty with these questions, review the appropriate section of the chapter. The answers to all the questions on the pretest are given in the answer section of this book.

1.1 Algebra Pretest

Part I. Basic Algebraic Concepts

Choose the best answer in Problems 1–15.

1. In $5x^2y$, the 5 is

  A. a term   **B.** a binomial   **C.** an exponent

  D. a literal factor   **E.** a numerical coefficient

2. In $6x^2y + 3z$, the 6 and 3 are

  A. terms   **B.** exponents   **C.** binomials

  D. coefficients   **E.** literal factors

3. If $(-2)(-2)(-2) = x$, the value of x is

  A. -8 **B.** -6 **C.** 6

  D. 8   **E.** none of these

4. $21 - (-5)$ equals

  A. -26 **B.** -16 **C.** 16

  D. 26   **E.** none of these

5. If $a = -3$ and $b = 5$, then $a^2 - 2ab + b^2$ equals

  A. -4 **B.** 2   **C.** 64

  D. -2 **E.** none of these

6. The expression $8 + 2 \cdot 3 - 8 \div 2$ equals

  A. 11   **B.** 10   **C.** 3

  D. -3 **E.** none of these

7. $3x - 5x$ equals

  A. $15x^2$ **B.** $8x$ **C.** $15x$

  D. $-2x$ **E.** none of these

8. -5^2 equals

  A. -10 **B.** 25   **C.** -25

  D. 10   **E.** none of these

9. $(-3y^2)^3$ equals

  A. $27y^6$ **B.** $-27y^5$ **C.** $-3y^6$

  D. $-27y^6$ **E.** none of these

10. $\dfrac{a + b}{2}$ means

  A. $a + b \cdot \frac{1}{2}$ **B.** $a + b \div 2$ **C.** $a + (b \div 2)$

  D. $(a + b) \cdot \frac{1}{2}$ **E.** all of these

11. If $3x + 12 = 6$, x equals

  A. 6   **B.** 2   **C.** -2

  D. -6 **E.** none of these

12. If $6 + 3x = x - 4$, x equals

  A. 5   **B.** $2\frac{1}{2}$ **C.** 1

  D. 2   **E.** none of these

13. If $2x - 16 = 3x - 9$, x equals

  A. $-\frac{7}{5}$ **B.** -7 **C.** 5

  D. 25   **E.** none of these

14. $(32x^8) \div (-2x)$ equals

  A. $30x^7$ **B.** $16x^8$ **C.** $34x^8$

  D. $-30x^7$ **E.** none of these

15. Simplify $3(x + 2) - (x - 4y) + (x + y)$.

  A. $3x + 11y$ **B.** $3x - 3y + 6$

  C. $x + 5y + 6$ **D.** $3x + 5y + 6$

  E. none of these

Indicate if Problems 16–30 are true or false.

16. If $x \neq 0$, then $-x$ always indicates a number that is less than zero.

17. $2(xy) = (2x)(2y)$ **18.** $-2^4 = 16$

19. x^2 is always positive.   **20.** $(a + b)^2 = a^2 + b^2$

21. $\dfrac{A + C}{B + C} = \dfrac{A}{B}$ **22.** $\dfrac{2 + x}{6} = \dfrac{x}{3}$

23. $\frac{0}{5}$ is not defined.

24. A natural number is both rational and real.

25. $\dfrac{A}{B} + \dfrac{C}{B} = \dfrac{1}{B}(A + C)$ **26.** $\dfrac{-x}{-y} = -\dfrac{x}{y}$

27. $(x - y)^2 = (y - x)^2$ **28.** $\dfrac{x}{y} + \dfrac{y}{x} = 1$

29. $\dfrac{1}{x} y \dfrac{1}{z} = \dfrac{y}{xz}$ **30.** $\sqrt{x^2 + y^2} = x + y$

Part II. Algebra Pretest

Choose the best answer for each of the following problems.

[1.2]
REAL NUMBERS

1. $|2\pi - 7| =$

  A. $2\pi - 7$ **B.** $7 - 2\pi$ **C.** -0.7168 **D.** 0.7168

  E. The expression may have more than one value.

2. $6.23\overline{4}$ is an example of

  A. a natural number   **B.** a real number   **C.** an irrational number

  D. all the answers in parts A, B, and C   **E.** none of these

3. The distance between the points (5) and ($\sqrt{10}$) on a real number line is
   - A. -1.834
   - B. 1.834
   - C. $5 - \sqrt{10}$
   - D. $\sqrt{10} - 5$
   - E. none of these

[1.3]
ALGEBRAIC EXPRESSIONS

4. If $a = 2$ and $b = -3$, then $2a(a + 2b)$ equals
   - A. 32
   - B. 16
   - C. -16
   - D. -48
   - E. none of these

5. $(x + y)^2$ equals
   - A. $x^2 + y^2$
   - B. $x^2 + xy + x$
   - C. $x^2 + 2xy + y^2$
   - D. $x^2 y^2$
   - E. none of these

6. The result of simplifying $42x - [10 - 2(3x - 4) - 2]$ is
   - A. $48x + 16$
   - B. $48x - 16$
   - C. $36x - 20$
   - D. $36x - 16$
   - E. none of these

[1.4]
FACTORING

7. The greatest common factor of $10x^3$, $5x^2$, and $25x$ is
   - A. $5x$
   - B. x
   - C. $5x^2$
   - D. $10x^2$
   - E. none of these

8. One of the factors of $x^2 - 18x + 80$ is
   - A. $x + 4$
   - B. $x + 5$
   - C. $x - 10$
   - D. $x - 16$
   - E. none of these

9. The complete factorization of $15x^3 - 15xy^2$ is
   - A. $x(15x^2 - 15y^2)$
   - B. $15x(x + y)(x - y)$
   - C. $x(15x + y)(x - y)$
   - D. $3xy(5x^2 - 5xy)$
   - E. none of these

[1.5]
LINEAR EQUATIONS AND INEQUALITIES

10. Solve $\dfrac{(x + 3)(x - 1)}{x - 1} = 4$.
    - A. 1
    - B. -1
    - C. 3
    - D. -3
    - E. none of these

11. The solution for $-2x \leq 8$ is
    - A. $x \leq -4$
    - B. $\{-4, -3, -2, -1, 0, 1, 2, 3, 4, \ldots\}$
    - C. $x > -4$
    - D. $x \geq -4$
    - E. none of these

12. If $3 \leq x + 4 \leq 5$, then
    - A. $-1 \geq x$ or $x \geq 1$
    - B. $-1 \leq x$ or $x \leq 1$
    - C. $-1 \leq x \leq 1$
    - D. $7 \leq x \leq 9$
    - E. none of these

[1.6]
QUADRATIC EQUATIONS AND INEQUALITIES

13. Solve $x^2 - 7x + 12 = 0$.
    - A. $\{-3, -4\}$
    - B. $\{-3, 4\}$
    - C. $\{3, 4\}$
    - D. $\{2, -6\}$
    - E. none of these

14. Solve $(x + 1)(2x - 3) = 25$.
    - A. $\{-1, 3\}$
    - B. $\{4, 8\}$
    - C. $\{4, -\frac{7}{2}\}$
    - D. $\{6, 2\}$
    - E. none of these

15. If $(x + 1)(2 - x) < 0$, then
    - A. $x < -1$ or $x > 2$
    - B. $x < -2$ or $x > 1$
    - C. $x < -2$ or $x < 1$
    - D. $-1 < x < 2$
    - E. none of these

16. Solve $x^2 - 2x - 2 = 0$.
    - A. $-2 \pm \sqrt{3}$
    - B. $4 \pm \sqrt{3}$
    - C. $2 \pm \sqrt{3}$
    - D. $4 \pm 4\sqrt{3}$
    - E. none of these

[1.7]
RATIONAL EXPRESSIONS

17. Simplify $\dfrac{x^2 - 5x + 4}{x + 3} \cdot \dfrac{x^2 + 2x - 3}{x - 4}$.

  A. 1   **B.** $x - 1$ **C.** $x^2 - 1$
  D. $(x - 1)^2$ **E.** none of these

18. Simplify $\dfrac{x}{x^2 - 4} - \dfrac{2}{4 - x^2}$.

  A. $x - 2$ **B.** $\dfrac{1}{x - 2}$ **C.** $\dfrac{1}{2 - x}$

  D. $\dfrac{x}{x - 2}$ **E.** none of these

1.2 Real Numbers

In mathematics, we need to be clear about the types of numbers under consideration. For example, if someone asked you to "pick a number," you probably wouldn't choose $\sqrt{5}$ or $\pi/2$, but in this course we want to include such choices—the set of numbers called **real numbers**. The various **sets** of numbers include:

Natural numbers: $N = \{1, 2, 3, 4, \ldots\}$
Whole numbers: $W = \{0, 1, 2, 3, 4, \ldots\}$
Integers: $I = \{\ldots, -3, -2, -1, 0, 1, 2, 3, \ldots\}$
Rationals: $Q = \{$all Quotients p/q where p is an integer and q is a nonzero integer$\}$
  Rationals are numbers whose decimal representations either terminate or repeat, such as $\frac{1}{3}$, 0.5, $\frac{1}{8}$, or $0.\overline{1}$, but the set also includes integers such as 2, 6, $-19, \ldots$ since $2 = 2.0$, $6 = 6.0$, and so on.
Irrationals: $Q' = \{$numbers whose decimal representation does not terminate or repeat$\}$
  For example, π, $\sqrt{5}$, $\sqrt[3]{2}$
Real numbers: $R = \{$numbers in Q or $Q'\}$*

A capital letter is usually used to name a set. Thus the set of natural numbers (also called counting numbers) is often referred to by the letter N, the set of integers by I, and the set of rationals by Q (for quotients). One method of designating a set is to enclose the list of its **members**, or **elements**, in braces, $\{\ \}$, and to use three dots, if needed, to indicate that the numbers continue in the shown pattern. The set is said to be **finite** if the number of elements in it is a natural number or less than some natural number. A set that is not finite is said to be **infinite**. A set with no members is called the **empty set**, or **null set**, and is labeled $\{\ \}$ or $\varnothing$.

This course is focused on the set of real numbers, which is easily visualized by using a **coordinate system** called a **number line**, as shown in Figure 1.1. A **one-to-one correspondence** exists between all real numbers and all points on such a number line:

1. Every point on the line corresponds to precisely one real number.
2. For each real number, there corresponds one and only one point.

---

* There is another set of numbers used in mathematics, the set of **complex numbers**. In this book, however, we limit ourselves to the set of real numbers.

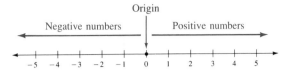

Figure 1.1 A real number line

A point associated with a particular number is called the **graph** of that number, and the number is called the **coordinate** of the point. Numbers associated with points to the right of the origin are called **positive real numbers**, and those associated with points to the left are called **negative real numbers**. Thus a number line is also a convenient way of ordering any two real numbers. If a point whose coordinate is a lies to the right of a point whose coordinate is b on a number line, then **a is greater than b**, written $a > b$. For example, $6 > 3$ and $-4 > -10$. And if a point's coordinate b is to the left of a point's coordinate a, then we say that **b is less than a**, or $b < a$. For example, $3 < 6$ and $-10 < -4$. The other symbols of comparison are:

$a = b$ a is equal to b

$a \neq b$ a is not equal to b

$a > b$ a is greater than b

$a \geq b$ a is greater than or equal to b

$a < b$ a is less than b

$a \leq b$ a is less than or equal to b

The symbols $>$, $\geq$, $<$, and $\leq$ are referred to as the **inequality symbols**.

A concept called **absolute value** is also associated with points on a number line. The symbol $|a|$ is read "the absolute value of a" and means the distance between the graph of a and the origin. For example:

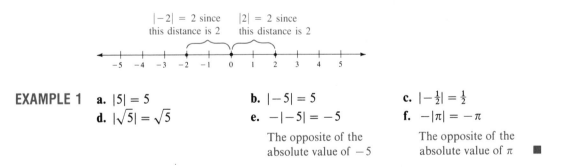

EXAMPLE 1

a. $|5| = 5$ **b.** $|-5| = 5$ **c.** $|-\frac{1}{2}| = \frac{1}{2}$

d. $|\sqrt{5}| = \sqrt{5}$ **e.** $-|-5| = -5$ **f.** $-|\pi| = -\pi$

                                       The opposite of the   The opposite of the
                                       absolute value of -5 absolute value of π ■

Since we will use the notion of absolute value in a variety of contexts, we need a more formal definition:

Absolute Value

The absolute value of a real number a is defined by

$$|a| = \begin{cases} a & \text{if } a \geq 0 \\ -a & \text{if } a < 0 \end{cases}$$

EXAMPLE 2

a. $|5| = 5$ since $5 \geq 0$

b. $|-5| = -(-5)$ since $-5 < 0$
$= 5$

c. $-|5| = -5$ since $5 \geq 0$

d. $|-\pi| = -(-\pi)$ since $-\pi < 0$
$= \pi$

e. $|\pi - 3| = \pi - 3$ since $\pi - 3 \geq 0$

f. $|3 - \pi| = -(3 - \pi)$ since $3 - \pi < 0$
$= -3 + \pi$
$= \pi - 3$

g. $|\sqrt{30} - 5| = \sqrt{30} - 5$ since $\sqrt{30} - 5 \geq 0$

h. $|\sqrt{30} - 6| = -(\sqrt{30} - 6)$ since $\sqrt{30} - 6 < 0$
$= 6 - \sqrt{30}$ ◾

In algebra, it is necessary to assume certain properties of equality and inequality. The first of these is called the **trichotomy property**, or **property of comparison**:

Property of Comparison

Given any two real numbers a and b, exactly one of the following holds:

1. $a = b$ 2. $a < b$ 3. $a > b$

This property tells us that if we are given *any* two real numbers, either they are equal or one of them is greater than the other. It establishes order on the number line. We need this property to derive a formula for the distance between two points on a number line. For example, let P_1 and P_2 be any points on a number line with coordinates x_1 and x_2, usually denoted as $P_1(x_1)$ and $P_2(x_2)$. Then, by the property of comparison, we know that $x_1 = x_2$, $x_1 < x_2$, or $x_1 > x_2$. Consider these possibilities one at a time.

$x_1 = x_2$ That is, the distance between x_1 and x_2 is 0.

$x_1 < x_2$

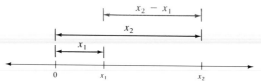

The distance is $x_2 - x_1$.

$x_1 > x_2$

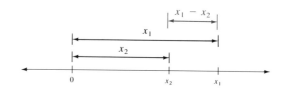

The distance is $x_1 - x_2$.

We can combine these different possibilities into one distance formula by using the idea of absolute value.

Distance on a Number Line

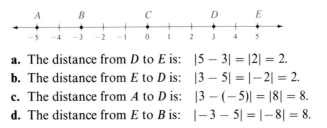

The distance d between points $P_1(x_1)$ and $P_2(x_2)$ is

$$d = |x_2 - x_1|$$

EXAMPLE 3   Let A, B, C, D, and E be points with coordinates as shown on the number line below:

a. The distance from D to E is: $|5 - 3| = |2| = 2$.
b. The distance from E to D is: $|3 - 5| = |-2| = 2$.
c. The distance from A to D is: $|3 - (-5)| = |8| = 8$.
d. The distance from E to B is: $|-3 - 5| = |-8| = 8$.   ■

Problem Set 1.2

Classify each example in Problems 1–10 as a natural number (N), whole number (W), integer (I), rational number (Q), irrational number (Q'), or real number (R). Examples may be in more than one set.

1. **a.** -9 **b.** $\sqrt{30}$
2. **a.** $\sqrt{49}$ **b.** $\frac{17}{2}$
3. **a.** 5 **b.** $\sqrt{4}$
4. **a.** $\frac{0}{5}$ **b.** $\frac{5}{0}$
5. **a.** $0.\overline{4}$ **b.** $0.5252\ldots$
6. **a.** 3.1416 **b.** π
7. **a.** 0.381 **b.** $\pi/6$
8. **a.** $\sqrt{121}$ **b.** $\sqrt{125}$
9. **a.** $16/0$ **b.** $0/16$
10. **a.** $\pi/3$ **b.** $\sqrt{4/25}$

Certain relationships among these sets are assumed from your previous courses and are illustrated below:

For Problems 11–40, let

$P = \{\text{primes}\}$
$\quad = \{2, 3, 5, 7, 11, 13, \ldots\}$
Primes are natural numbers with exactly two divisors (itself and 1)

$C = \{\text{composites}\}$
$\quad = \{4, 6, 8, 9, 10, \ldots\}$
Composites are natural numbers greater than 1 that are *not* primes

$E = \{\text{evens}\}$
$\quad = \{0, 2, 4, 6, 8, 10, \ldots\}$
Even numbers are whole numbers divisible by 2

$D = \{\text{odds}\}$
$\quad = \{1, 3, 5, 7, 9, 11, \ldots\}$
Odd numbers are natural numbers that can be written as an even number plus 1

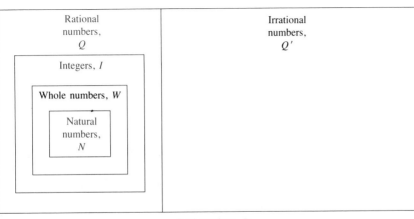

Real numbers, R

Figure 1.2 Relationships within the set of real numbers

11. Draw Figure 1.2 and show the set P.

12. Draw Figure 1.2 and show the set C.

13. Draw Figure 1.2 and show the set E.

14. Draw Figure 1.2 and show the set D.

Indicate if Problems 15–40 are true or false. Test your understanding of these sets, both from this section and from previous courses.

15. 0 is a natural number.

16. $\sqrt{3}$ is a real number.

17. $0.333\ldots$ is an irrational number.

18. $-\frac{1}{4}$ is an integer.

19. 2 is a real number.

20. 1 is a prime number

21. $|5|$ is a prime number.

22. The sum of two odd numbers is an odd number.

23. The sum of two even numbers is an even number.

24. The sum of two primes is a prime.

25. The sum of two composites is a composite.

26. The product of two even numbers is an even number.

27. The product of two odd numbers is an odd number.

28. The product of two primes is a prime.

29. The product of two rational numbers is a rational number.

30. The product of two irrational numbers is an irrational number.

31. All natural numbers are integers.

32. All irrational numbers are real.

33. All repeating decimals are positive.

34. No even integers are primes.

35. All odd integers are primes.

36. All numbers are either even or odd.

37. Some negative numbers are primes.

38. a/b is a rational number for all numbers a and b.

39. Some real numbers are integers.

40. Some irrational numbers are integers.

Rewrite Problems 41–61 without the absolute value symbols.

41. $|9|$

42. $-|19|$

43. $|-19|$

44. $-|-8|$

45. $-|\pi|$

46. $|\pi - 2|$

47. $|2 - \pi|$

48. $|\pi - 5|$

49. $|\pi - 10|$

50. $|\sqrt{20} - 5|$

51. $|\sqrt{20} - 4|$

52. $|\sqrt{50} - 7|$

53. $|\sqrt{50} - 8|$

54. $|2\pi - 5|$

55. $|2\pi - 7|$

56. $|x + 3|$ if $x \geq 3$

57. $|x + 3|$ if $x \leq -3$

58. $|y - 5|$ if $y < 5$

59. $|y - 5|$ if $y \geq 5$

60. $|5 - 2s|$ if $s > 10$

61. $|4 + 3t|$ if $t < -10$

Find the distance between the pairs of points whose coordinates are given in Problems 62–70.

62. $A(5)$ and $B(17)$

63. $C(-5)$ and $D(-15)$

64. $E(22)$ and $F(-8)$

65. $G(-2)$ and $H(9)$

66. $I(\pi)$ and $J(5)$

67. $K(\pi)$ and $L(2)$

68. $M(\sqrt{3})$ and $N(1)$

69. $P(\sqrt{3})$ and $Q(2)$

70. $R(6)$ and $S(2\pi)$

1.3 Algebraic Expressions

An **algebraic expression** is a grouping of constants and variables obtained by applying a finite number of operations (such as addition, subtraction, multiplication, nonzero division, extraction of roots). **Constants** are symbols with just one possible value (such as 5, 11, 0, 0.5, or π), and **variables** are symbols (such as x, y, or z) used to represent unspecified numbers selected from some set called the **domain**. In this book, if a domain for a variable is not specified, the domain is the set of real numbers.

Numbers or expressions that are added are called **terms** while those that are multiplied are called **factors**. If a number and a variable are multiplied together, then the factor consisting of the number alone is called the **numerical coefficient**. It is very common to have repeated factors, or **exponents**:

Exponent

If b is any real number and n is any natural number, then

$$b^n = \underbrace{b \cdot b \cdot b \cdots b}_{n \text{ factors}}$$

Furthermore, if $b \neq 0$, then

$$b^0 = 1 \quad \text{and} \quad b^{-n} = \frac{1}{b^n}$$

The number b is called the **base**, n is called the **exponent**, and b^n is called a **power**.

If an algebraic expression is a finite sum of terms with variables containing whole number exponents, it is called a **polynomial**. For example,

$$3x, \ 5, \ 2x^2 - 3x + 5, \ \frac{2}{3}x, \ 0, \ 5.5, \text{ and } \frac{x-2}{4}$$

are polynomials, but

$$\sqrt{x}, \ \frac{2}{x}, \ \frac{x}{0}, \ x^{1/2}$$

are not. The general form of a polynomial with a single variable is:

General Form of an nth Degree Polynomial in x

$$a_n x^n + a_{n-1} x^{n-1} + \cdots + a_2 x^2 + a_1 x + a_0 \qquad a_n \neq 0$$

Polynomials are frequently classified according to the number of terms they contain. A **monomial** is a polynomial with one term, a **binomial** has two terms, and a **trinomial** has three. The **degree** of a monomial is the number of variable factors in that monomial. For example, the degree of $3x^2$ is two because $3x^2 = 3xx$ (two variable factors); 3^2x is of degree one because there is one variable factor; and $3x^2y^3$ is of degree five because there are five variable factors ($xxyyy$). The expressions $5, 6^2$, and π are all of degree zero since none contains a variable factor. A special case is the monomial 0, to which no degree is assigned. The degree of a polynomial is the degree of the highest degree term.

From beginning algebra, recall the proper order of operations given in the following box.

Order of Operations Agreement

When simplifying an algebraic expression:

1. Carry out all operations within parentheses (begin with the innermost parentheses).
2. Do exponents next.
3. Complete multiplications and divisions, working from left to right.
4. Finally, do additions and subtractions, working from left to right.

EXAMPLE 1 $2 + 3 \cdot 4 = 2 + 12 = 14$ ■

EXAMPLE 2 $9 + 12 \div 3 + 4 \div 2 + 1 \cdot 2 = 9 + 4 + 2 + 2 = 17$ ■

In algebra we rarely use the symbol $\div$ but, instead, write the expression in Example 2 as

$$9 + \tfrac{12}{3} + \tfrac{4}{2} + 1 \cdot 2$$

The fractional bar used for division is also used as a grouping symbol:

$$6 + 4 \div 2 \text{ is written as } 6 + \tfrac{4}{2}$$

whereas

$$(6 + 4) \div 2 \text{ is written as } \frac{6 + 4}{2}$$

EXAMPLE 3   In algebra, $6 + 4 \div 2 + (6 + 4) \div 2 - (6 + 4 \div 2)$ is written as

$$6 + \frac{4}{2} + \frac{6 + 4}{2} - \left(6 + \frac{4}{2}\right) = 6 + \frac{4}{2} + \frac{10}{2} - (6 + 2)$$
$$= 6 + \tfrac{4}{2} + \tfrac{10}{2} - 8$$
$$= 6 + 2 + 5 - 8$$
$$= 5$$ ■

EXAMPLE 4   Simplify: $2 + 4[1 - 3(3 - 5) - \tfrac{12}{4}]$
Do the innermost parentheses first:

$$2 + 4[1 - 3(3 - 5) - \tfrac{12}{4}] = 2 + 4[1 - 3(-2) - \tfrac{12}{4}] \quad \text{Multiply and divide inside the brackets.}$$
$$= 2 + 4[1 + 6 - 3] \quad \text{Add and subtract inside the parentheses.}$$
$$= 2 + 4[4] \quad \text{Multiply before adding.}$$
$$= 2 + 16$$
$$= 18$$ ■

To **evaluate** an expression means to replace the variables with a given numerical value and then to carry out the order of operations to simplify the resulting numerical expression. This process is illustrated in Example 5.

EXAMPLE 5   Find $-(x + y)[-(s - t)]$ where $x = -2$, $y = -3$, $s = 4$, and $t = 7$.

Solution $-[(-2) + (-3)][-(4 - 7)] = -[-5][-(-3)] = -[-5][3] = -[-15] = 15$

■

Addition and Subtraction of Polynomials

Terms that are identical, except for their numerical coefficients, are called **similar terms**. That is, similar terms contain the same exponent (or exponents) on the variables. Since the variable parts are identical, they can be simplified as follows:

$$5x + 4x = (5 + 4)x = 9x$$

If you add five apples to four apples, you obtain nine apples; so, too, if you add $5x$'s to $4x$'s, you get $9x$'s. This is a statement of an algebraic property called the **distributive property**:

Distributive Property

> For real numbers a, b, and c,
>
> $$a(b + c) = ab + ac$$

You will, of course, use this property and combine similar terms mentally as shown by the following more complicated examples.

EXAMPLE 6

$$(5x^2 + 2x + 1) + (3x^3 - 4x^2 + 3x - 2) = 3x^3 + \underbrace{5x^2 + (-4)x^2}_{\text{Second-degree terms}} + \underbrace{2x + 3x}_{\text{First-degree terms}} + \underbrace{1 + (-2)}_{\text{Zero-degree terms}}$$

Similar terms, Similar terms, Similar terms

Third-degree term

$$= 3x^3 + x^2 + 5x - 1 \qquad \blacksquare$$

EXAMPLE 7
$$(4x - 5) + (5x^2 + 2x + 1) = 5x^2 + (4x + 2x) + (-5 + 1)$$
$$= 5x^2 + 6x - 4 \qquad \blacksquare$$

EXAMPLE 8
$$(4x - 5) - (5x^2 + 2x + 1) = 4x - 5 - 5x^2 - 2x - 1$$
$$= -5x^2 + 2x - 6 \qquad \blacksquare$$

Note that in Example 7 we added similar terms. In Example 8, notice that the procedure for subtracting a polynomial is to subtract *each* term.

EXAMPLE 9
$$(5x^2 + 2x + 1) - (3x^3 - 4x^2 + 3x - 2) = 5x^2 + 2x + 1 - 3x^3 + 4x^2 - 3x + 2$$
$$= -3x^3 + 9x^2 - x + 3 \qquad \blacksquare$$

Multiplication of Polynomials

To understand multiplication of polynomials, it is necessary to first understand multiplication of monomials.

EXAMPLE 10
$$x^2(x^3) = xx(xxx) = x^5 \qquad \text{There are five } x\text{'s used as a factor.}$$
$$(x^2)^3 = (x^2)(x^2)(x^2) = (xx)(xx)(xx) = x^6 \qquad \text{There are six } x\text{'s.}$$
$$(x^2 y^3)^4 = (x^2 y^3)(x^2 y^3)(x^2 y^3)(x^2 y^3) = (x^2 x^2 x^2 x^2)(y^3 y^3 y^3 y^3) = x^8 y^{12} \qquad \blacksquare$$

Example 10 illustrates three laws of exponents that are used to simplify algebraic expressions.

Laws of Exponents

Let a and b be any real numbers, m and n be any integers, and assume that each expression is defined.

FIRST LAW: $\quad b^m \cdot b^n = b^{m+n}$

SECOND LAW: $\quad (b^n)^m = b^{mn}$

THIRD LAW: $\quad (ab)^m = a^m b^m$

These laws of exponents are used along with the distributive property when multiplying polynomials.

EXAMPLE 11
$$(4x - 5)(5x^2 + 2x + 1) = (4x - 5)5x^2 + (4x - 5)2x + (4x - 5)1$$
$$= 20x^3 - 25x^2 + 8x^2 - 10x + 4x - 5$$
$$= 20x^3 - 17x^2 - 6x - 5$$ ∎

EXAMPLE 12
$$(2x - 3)(x + 4) = (2x - 3)x + (2x - 3)4$$
$$= 2x^2 - 3x + 8x - 12$$
$$= 2x^2 + 5x - 12$$ ∎

Notice that one term, $P = 4x - 5$ in Example 11 and $P = 2x - 3$ in Example 12, is simply distributed to each term in the other factor. We call this **distributive multiplication** to remind us to use the distributive property to do multiplication of polynomials.

Distributive Multiplication

$$P(A_1 + A_2 + \cdots + A_n) = PA_1 + PA_2 + \cdots + PA_n$$

EXAMPLE 13
$$\mathbf{(2x + 3y + 1)(5x - 2y - 3)}$$
$$= \mathbf{(2x + 3y + 1)5x + (2x + 3y + 1)(-2y) + (2x + 3y + 1)(-3)}$$
$$= 10x^2 + 15xy + 5x - 4xy - 6y^2 - 2y - 6x - 9y - 3$$
$$= 10x^2 + 11xy - 6y^2 - x - 11y - 3$$ ∎

It is frequently necessary to multiply binomials, and even though we use distributive multiplication, we want to be able to carry out the process quickly and efficiently in our heads. Consider Example 14.

EXAMPLE 14 **a.** $(2x + 3)(4x - 5) = (2x + 3)(4x) + (2x + 3)(-5)$
$$= \underset{\substack{\uparrow \\ \text{Product of} \\ \text{first terms}}}{8x^2} + \underset{\substack{\uparrow \\ \text{Sum of inner and} \\ \text{outer terms}}}{\underbrace{12x + (-10x)}} + \underset{\substack{\uparrow \\ \text{Product} \\ \text{of last terms}}}{3(-15)}$$

$$= 8x^2 + 2x - 15$$

b. $(5x - 3)(2x + 3) = 10x^2 \ + \ \underbrace{(15x - 6x)} \ + \ (-9)$

	$\uparrow$		$\uparrow$		$\uparrow$
Mentally	Product of first terms		Sum of outer and inner terms		Product of last terms

$$= 10x^2 + 9x - 9$$

c. $(4x - 3)(3x - 2) = 12x^2 \ - \ 17x \ + \ 6$

	$\uparrow$	$\uparrow$	$\uparrow$
	Product of first terms	Sum of outer and inner terms	Product of last terms

It is frequently necessary to multiply binomials such as those shown in Example 12; however the process illustrated in that example is too lengthy, so a shortened version, called **FOIL**, is often used instead. Four pairs of terms are multiplied.

FOIL

$$(ax + b)(cx + d) = acx^2 + \underbrace{(ad + bc)}x + bd$$

① First terms
② Outer terms
③ Inner terms
④ Last terms

Do this mentally

EXAMPLE 15 $(2x - 3)(x + 3) = 2x^2 + 3x - 9$

$$② + ③: \ 6x + (-3x)$$

$$(x + 3)(3x - 4) = 3x^2 + 5x - 12$$
$$(5x - 2)(3x + 4) = 15x^2 + 14x - 8$$

Problem Set 1.3

Simplify the expressions in Problems 1–14.

1. $3 + 2 \cdot 5$ **2.** $4 + 2 \cdot 3$ **3.** $5 + 3 \cdot 2$

4. $(-5)^2$ **5.** $(-6)^2$ **6.** $(-7)^2$

7. -5^2 **8.** -6^2 **9.** -7^2

10. $10 + 6(3 - 5)$ **11.** $8 - 5(2 - 7)$

12. $3 \cdot 5 - (-2)(-4)$

13. $\dfrac{(-2)6 + (-4)}{-4} + \dfrac{(-3)(-4)}{6}$

14. $(-2) + \dfrac{6}{-3} - 4 + \dfrac{-8}{-4}$

Let $x = -3$, $y = 2$, $z = -1$, and $w = -4$. Evaluate the expressions in Problems 15–27.

15. x^2 **16.** $-x^2$ **17.** y^2

18. $-y^2$ **19.** z^2 **20.** $-z^2$

21. $x + y - z$ **22.** $x - (y - z)$

23. $(xy)^2 + xy^2 + x^2y$ **24.** $5x - (4x + 3w)$

25. $x^2 - w^2(y^2 + z^2)$ **26.** $\dfrac{x - w^2}{z}$

27. $\dfrac{x - z}{y}$

Mentally multiply the expressions in Problems 28–41.

28. a. $(x + 3)(x + 2)$ **b.** $(x + 1)(x + 5)$

29. a. $(x - 2)(x + 6)$ **b.** $(x + 5)(x - 4)$

30. a. $(x + 1)(x - 2)$ **b.** $(x - 3)(x + 2)$

31. a. $(x - 5)(x - 3)$ **b.** $(x + 3)(x - 4)$

32. a. $(y + 1)(y - 7)$ **b.** $(y - 3)(y + 5)$

33. a. $(2y + 1)(y - 1)$ **b.** $(2y - 3)(y - 1)$

34. a. $(y + 1)(3y + 1)$ **b.** $(y + 1)(3y + 2)$

35. a. $(2y + 3)(3y - 2)$ **b.** $(2y + 3)(3y + 2)$

36. a. $(x + y)(x + y)$ **b.** $(x - y)(x - y)$

37. a. $(x + y)(x - y)$ **b.** $(a + b)(a - b)$

38. a. $(5x - 4)(5x + 4)$ **b.** $(3y - 2)(3y + 2)$

39. a. $(x + 2)^2$ **b.** $(x - 2)^2$

40. a. $(x + 4)^2$ **b.** $(x - 3)^2$

41. a. $(a + b)^2$ **b.** $(a - b)^2$

Simplify the expressions in Problems 42–68.

42. $(x + y - z) + (2x - 3y + z)$

43. $(x - y - z) + (2x + y - 3z)$

44. $(x + 2y - 3z) + (x - 3y + 5z)$

45. $(x + 3y - 2z) + (3x - 5y + 3z)$

46. $(x + 2y) - (2x - y)$

47. $(2x - y) - (2x + y)$

48. $(x - 3y) - (5x + y)$

49. $(6x - 4y) - (4x - 6y)$

50. $(2x + y + 3) - (x - y + 4)$

51. $(x + y - 5) - (2x - 3y + 4)$

52. $(5x^2 + 3x - 5) - (3x^2 + 2x + 3)$

53. $(6x^2 - 3x + 2) - (2x^2 + 5x + 3)$

54. $(3x^2 + 2x + 6) - (2x^2 + 5x + 3)$

55. $(2x^2 - 3x - 5) - (5x^2 - 6x + 4)$

56. $(x^2 - 1) - (2 - x) + (x^2 - x)$

57. $(x^2 - x) + (x - 3) - (x - x^2)$

58. $(3x - x^2) - (5 - x) - (3x - 2)$

59. $(x^2 - 7) - (3x + 4) - (x - 2x^2)$

60. $(x^2 - 5) - (3x^2 + 2x + 5)$

61. $(x + 1)^3$

62. $(x - 1)^3$

63. $(3x^2 - 5x + 2) + (x^3 - 4x^2 + x - 4)$

64. $(5x + 1) + (x^3 - 4x^2 + x - 4)$

65. $(3x^2 - 5x + 2) - (5x + 1)$

66. $(x^3 - 4x^2 + x - 4) - (3x^2 - 5x + 2)$

67. $(5x + 1)(3x^2 - 5x + 2)$

68. $(3x - 1)(x^2 + 32x - 2)$

1.4 Factoring

A **factor** of a given algebraic expression is an algebraic expression that divides evenly into the given expression. The process of **factoring** involves resolving a given expression into its factors. The procedure we will use is to carry out a series of "tests" for different types of factors. Table 1.1 lists these types in the order in which we should check them when factoring an expression.

TABLE 1.1
Factoring procedure

Type	Form	Comments
1. Common factors	$ax + ay + az = a(x + y + z)$	Use the distributive property. It can be applied with any number of terms.
2. Difference of squares	$x^2 - y^2 = (x - y)(x + y)$	Remember that the *sum* of two squares cannot be factored in the set of real numbers.
3. FOIL	$x^2 + (c + d)x + cd = (x + c)(x + d)$ $acx^2 + (ad + bc)xy + bdy^2 = (ax + by)(cx + dy)$	Use this trial-and-error procedure with trinomials, after checking for common factors and difference of squares. See the examples in this section.

Common Factoring

The distributive property leads to a very important type of factoring called **common factoring**. When simplifying expressions, we read the distributive property from left to right:

$$a(b + c) = ab + ac$$

When factoring common factors, we read it from right to left, as shown in the following box.

Common Factoring

$$ab + ac = a(b + c)$$
$$ab - ac = a(b - c)$$

EXAMPLE 1
$$5x^2 - 25x = (5x)x - (5x)5$$
$$= 5x(x - 5)$$
∎

Common factoring extends to any number of terms.

EXAMPLE 2
$$2x^3 - 20x^2 + 6x = 2x(x^2) - 2x(10x) + 2x(3)$$
$$= 2x(x^2 - 10x + 3)$$
∎

If we factor out the *greatest factor* common to each term, or if no other factor can be found, we say the polynomial is **completely factored**.

EXAMPLE 3
$$10x^2y + 25xy^2 - 15x^2y^3 = 5xy(2x) + 5xy(5y) + 5xy(-3xy^2)$$
$$= 5xy(2x + 5y - 3xy^2)$$
∎

Common factors refer to the base numerals, not to the exponents, but notice that the smallest exponent (namely, one on the x and y in Example 3) on a common base number leads to the appropriate common factor. For example, $3x^3y^5 + 5x^4y^2$ has a common factor x^3y^2, since the smallest exponent on the common factor x is 3 and on the common factor y is 2.

We usually factor **over the set of integers**, which means that all the numerical coefficients are integers. If the original polynomial has fractional coefficients, however, we factor out the fractional part first.

EXAMPLE 4
$$\tfrac{1}{36}x^2 - 5x + 1 = \tfrac{1}{36}x^2 - \tfrac{36}{36}5x + \tfrac{36}{36}(1) = \tfrac{1}{36}(x^2 - 180x + 36)$$
∎

Common factors do not have to be monomials; they can be any algebraic expression.

EXAMPLE 5
$$5x(3a - 2b) + y(3a - 2b) = (5x + y)(3a - 2b) \qquad \text{The common factor is } (3a - 2b)$$
∎

Difference of Squares

The second type of factoring involves determining whether the expression is a difference of squares.

Difference of Squares

$$x^2 - y^2 = (x - y)(x + y)$$

EXAMPLE 6 **a.** $9x^2 - 25y^2 = (3x)^2 - (5y)^2 = (3x - 5y)(3x + 5y)$

b. $16x^4 - 1 = (4x^2)^2 - (1)^2$ This can be a mental step
$$= (4x^2 - 1)(4x^2 + 1)$$ Difference of squares
$$= (2x - 1)(2x + 1)(4x^2 + 1)$$ Difference of squares again

c. $x^2 - 3$ is irreducible over the set of integers (3 is not a perfect square). Factoring over the set of integers rules out factoring this expression as

$$x^2 - 3 = (x - \sqrt{3})(x + \sqrt{3})$$

since the factors do not have integer coefficients. In this book, an expression is called completely factored if all fractions are eliminated by common factoring and if no further factoring is possible *over the set of integers*.

d. $x^2 + 4$ is irreducible over the set of integers (it is not a difference but a sum of two squares). ■

FOIL—Factoring a Trinomial

Consider the product of two binomials:

$$(2x - 3)(x + 1) = 2x^2 - 3x + 2x - 3 = 2x^2 - x - 3$$

If you understand where the terms of the product came from, you will find it easier to reverse the process. In the following discussion we will concentrate on a particular product, a second-degree polynomial in a single variable, like the one above. You can then see how this case applies to similar products.

First, examine the second-degree (or leading) term and the last term (the constant):

These terms are the products of the variable terms and the constants of the binomial factors.

Now, recall the origin of the first-degree (or middle) term of the product:

$$(2x - 3)(x + 1)$$

$$-3x$$

$$2x$$

This term is the sum of the products of the variable and constant terms in the binomial factors.

Now consider a product and reverse the multiplication procedure to determine the factors:

$$3x^2 + 13x - 10$$

1. First, find two factors whose product is $3x^2$. These determine the variable terms of the factors and hence the form of the factors:

$$(x \quad)(3x \quad)$$

2. Next, factor the constant term. These factors will yield all possible pairs of factors:

$$(x + 2)(3x - 5) \qquad (x - 2)(3x + 5) \qquad (x + 5)(3x - 2)$$
$$(x - 5)(3x - 2) \qquad (x + 1)(3x - 10) \qquad (x - 1)(3x + 10)$$
$$(x + 10)(3x - 1) \qquad (x - 10)(3x + 1)$$

3. Then check each of the possibilities to see which give the correct middle term:

$$(x + 5)(3x - 2) = 3x^2 + 13x - 10$$

Thus we factor a polynomial by reversing our knowledge of multiplication. Not all examples are this easy, and, indeed, not all can be factored over the integers. If no possibility yields the correct middle term, the polynomial is not factorable over the set of integers. For example, $x^2 + x + 1$ must be of the form

$$(x \quad 1)(x \quad 1)$$

but the only possibilities for the constant terms are

$$(x - 1)(x - 1) \qquad \text{or} \qquad (x + 1)(x + 1)$$

and no possibility yields the correct middle term. Thus we say that $x^2 + x + 1$ is not factorable over the set of integers.

Procedure for Factoring a
Trinomial

1. Find the factors of the leading term and set up the binomials.
2. Find the factors of the constant term, and consider all possible binomials.
3. Determine the factors that yield the correct middle term.
4. If no pair of factors produces the correct full product, then the trinomial is not factorable over the set of integers.

EXAMPLE 7 $x^2 - 8x + 15 = (x - 5)(x - 3)$ ∎

EXAMPLE 8 $6x^2 + x - 12 = (2x + 3)(3x - 4)$ ∎

EXAMPLE 9 $x^2 - 2xy + y^2 = (x - y)(x - y) = (x - y)^2$

If both the binomial factors are the same, then the expression is called a **perfect square**. ∎

Factoring problems are generally not divided into categories as in Examples 1–9. We should always begin by looking for a common factor and then a difference of squares, and finally trying FOIL, as shown in Examples 10–12.

EXAMPLE 10 $3x^2 - 75 = 3(x^2 - 25)$ Common factor
$$= 3(x - 5)(x + 5) \qquad \text{Difference of squares}$$ ∎

EXAMPLE 11 $(x + 3y)^2 - 1 = [(x + 3y) - 1][(x + 3y) + 1]$
$$= (x + 3y - 1)(x + 3y + 1) \qquad \blacksquare$$

EXAMPLE 12 $6ax^2 - 21ax - 12a = 3a(2x^2 - 7x - 4)$
$$= 3a(2x + 1)(x - 4) \qquad \blacksquare$$

Problem Set 1.4

Factor the expressions, if possible, in Problems 1–48.

1. $20xy - 12x$

2. $8xy - 6x$

3. $6x - 2$

4. $5y + 5$

5. $xy + xz^2 + 3x$

6. $a^2 - b^2$

7. $a^2 + b^2$

8. $s^2 + 2st + t^2$

9. $m^2 - 2mn + n^2$

10. $u^2 + 2uv + v^2$

11. $(4x - 1)x + (4x - 1)3$

12. $(a + b)x + (a + b)y$

13. $2x^2 + 7x - 15$

14. $x^2 - 2x - 35$

15. $3x^2 - 5x - 2$

16. $6y^2 - 7y + 2$

17. $2x^2 - 10x - 48$

18. $4x^3 + x^2 - 21x$

19. $12x^4 + 11x^3 - 15x^2$

20. $8x^2 + 10xy - 3y^2$

21. $4x^4 - 17x^2 + 4$

22. $4x^4 - 45x^2 + 81$

23. $(x - y)^2 - 1$

24. $(2x + 3)^2 - 1$

25. $(5a - 2)^2 - 9$

26. $(3p - 2)^2 - 16$

27. $\dfrac{4}{25}x^2 - (x + 2)^2$

28. $\dfrac{4x^2}{9} - (x + y)^2$

29. $(a + b)^2 - (x + y)^2$

30. $(m - 2)^2 - (m + 1)^2$

31. $2x^2 + x - 6$

32. $3x^2 - 11x - 4$

33. $6x^2 + 47x - 8$

34. $6x^2 - 47x - 8$

35. $6x^2 + 49x + 8$

36. $6x^2 - 49x + 8$

37. $4x^2 + 13x - 12$

38. $9x^2 - 43x - 10$

39. $9x^2 - 56x + 12$

40. $12x^2 + 12x - 25$

41. $10x^2 - 9 - x^4$

42. $5x^2 - 4 - x^4$

43. $(x^2 - \frac{1}{4})(x^2 - \frac{1}{9})$

44. $(x^2 - \frac{1}{4})(x^2 - \frac{1}{16})$

45. $(x^2 - 3x - 6)^2 - 4$

46. $(x + y + 2z)^2 - (x - y + 2z)^2$

47. $2(x + y)^2 - 5(x + y)(a + b) - 3(a + b)^2$

48. $2(s + t)^2 + 3(s + t)(s + 2t) - 2(s + 2t)^2$

1.5 Linear Equations and Inequalities

A **linear equation in one variable** is an equation that can be written in the form

$$ax + b = 0 \qquad a \neq 0$$

where x is a variable and a and b are any real numbers. An **open** or **conditional equation** is an equation containing a variable that may be either true or false, depending on the replacement for the variable. A **root** or a **solution** is a replacement for the variable that makes the equation true. We also say that the root **satisfies** the equation. The **solution set** of an open equation is the set of all solutions of the equation. To **solve an equation** means to find its solution set. If there are no values for the variable that satisfy the equation, then the solution set is said to be **empty** and is denoted by ∅. If every replacement of the variable makes the equation true, then the equation is called an **identity**. Two equations with the same solution set are called **equivalent equations**. The process of solving an equation involves finding a sequence of equivalent equations; it ends when the solution or solutions are obvious. There are a few operations that produce equivalent equations, some of which are summarized on page 19.

Properties of Equations

> If P and Q are algebraic expressions, and k is a real number, then each of the following is equivalent to $P = Q$:
>
> ADDITION PROPERTY: $P + k = Q + k$
>
> SUBTRACTION PROPERTY: $P - k = Q - k$
>
> MULTIPLICATION PROPERTY: $kP = kQ \qquad k \neq 0$
>
> DIVISION PROPERTY: $\dfrac{P}{k} = \dfrac{Q}{k} \qquad k \neq 0$

EXAMPLE 1   Solve $3x + 5 = x - 3$.

Solution

$$3x + 5 = x - 3$$
$$3x + 5 - x = x - x - 3 \qquad \text{Subtract } x \text{ from both sides (Do this step in your head)}$$
$$2x + 5 = -3 \qquad \text{Simplify both sides}$$
$$2x + 5 - 5 = -3 - 5 \qquad \text{Subtract 5 from both sides (mentally)}$$
$$2x = -8$$
$$x = -4 \qquad \text{Divide both sides by 2 (mental step—not shown)}$$

Check: $3(-4) + 5 = -4 - 3$
$$-7 = -7 \qquad \text{This is true, so the solution checks} \qquad \blacksquare$$

EXAMPLE 2   Solve $2x - 5(x - 2) = 3(3 - x)$.

Solution $2x - 5(x - 2) = 3(3 - x)$
$$2x - 5x + 10 = 9 - 3x \qquad \text{Eliminate the parentheses, then simplify}$$
$$-3x + 10 = 9 - 3x$$
$$10 = 9 \qquad \text{Add } 3x \text{ to both sides. This results in a false equation}$$

This is a contradiction, and, since it is equivalent to the original equation, the solution set is empty. $\qquad \blacksquare$

EXAMPLE 3   Solve $2x - (7 - x) = x + 1 - 2(4 - x)$.

Solution $2x - (7 - x) = x + 1 - 2(4 - x)$
$$2x - 7 + x = x + 1 - 8 + 2x \qquad \text{Remove the parentheses first}$$
$$3x - 7 = 3x - 7 \qquad \text{Combine similar terms}$$
$$-7 = -7 \qquad \text{True equation}$$

This is an identity. Because it is equivalent to the original equation, the solution set is the set of all real numbers so that x can be replaced by any real number and the equation will be true. $\qquad \blacksquare$

Another type of statement is an **inequality**. Inequalities can be sentences that are always true (for example, $x - 1 < x$ or $5 < 7$), called **absolute inequalities**; always false (for example, $x > x$), called **contradictions**; or sometimes true and sometimes false (for example, $x > 2$), called **conditional inequalities**. The latter can be solved by using a set of properties similar to those for equations.

Properties of Inequalities

If P and Q are algebraic expressions, and k is a real number, then each of the following is equivalent to $P < Q$:

ADDITION PROPERTY: $\qquad P + k < Q + k$

SUBTRACTION PROPERTY: $\qquad P - k < Q - k$

MULTIPLICATION PROPERTY:
  Positive number k: $kP < kQ$
  Negative number k: $kP > kQ$ $\qquad$ Note that the inequality is reversed

DIVISION PROPERTY:

  Positive number k: $\dfrac{P}{k} < \dfrac{Q}{k}$

  Negative number k: $\dfrac{P}{k} > \dfrac{Q}{k}$ $\qquad$ Note that the inequality is reversed

These properties also hold for $\leq$, $>$, and $\geq$.

Essentially, properties of inequalities allow any operation that is allowed for equations, *except that multiplying or dividing by a negative number reverses the sense of the inequality.*

EXAMPLE 4   Solve $5 - 2(3x - 4) < 4x - 7$.

Solution

$$5 - 2(3x - 4) < 4x - 7$$
$$5 - 6x + 8 < 4x - 7$$
$$-6x + 13 < 4x - 7$$
$$-10x + 13 < -7 \qquad \text{Subtract } 4x \text{ from both sides}$$
$$-10x < -20 \qquad \text{Subtract 13 from both sides}$$
$$x > 2 \qquad \text{Divide both sides by } -10 \text{; note that the order of the inequality has changed}$$

The solution set for Example 4 is graphed here.

Notice that the open point indicates that $x = 2$ is excluded. ■

EXAMPLE 5 $5(2x - 3) - 4(x - 2) \leq 3(x - 3)$
$$10x - 15 - 4x + 8 \leq 3x - 9$$
$$6x - 7 \leq 3x - 9$$
$$3x - 7 \leq -9$$
$$3x \leq -2$$
$$x \leq -\tfrac{2}{3}$$

The solution set for Example 5 is graphed here.

Notice that the closed point indicates that $x = -\frac{2}{3}$ is included.   ■

A "string of inequalities" may be used to show the order of three or more quantities. For example, $2 < x < 5$ states that x is a number *between* 2 and 5. The statement is a *compound inequality*, equivalent to $x > 2$ and $x < 5$ at the *same* time, and is graphed as the interval on the number line between 2 and 5:

Such inequalities may be solved in a similar way to that used for other inequalities. Note that what is done to one member of the string in each of the following inequalities is done to *all* three members. We try to isolate the x in the middle part of the string.

EXAMPLE 6

$$-3 \le 2x - 5 \le 7$$
$$-3 + 5 \le 2x - 5 + 5 \le 7 + 5 \qquad \text{Add 5 to all three members}$$
$$2 \le 2x \le 12$$
$$1 \le x \le 6 \qquad \text{Divide all three members by 2}$$
$$\text{(the inequality stays the same)}$$

This says the solution is between 1 and 6 (inclusive).   ■

EXAMPLE 7

$$-2 < 1 - 3x < 7$$
$$-3 < \quad -3x < 6 \qquad \text{Subtract 1}$$
$$1 > x > -2 \qquad \text{Divide by } -3; \text{ reverse the order of the inequalities}$$
$$-2 < x < 1$$

The string of inequalities $1 < 2 < 3$ is equivalent to $3 > 2 > 1$. The first states the order of the three integers from smaller to larger, and the second from the larger to smaller. It is standard practice to use the ascending order from smallest to largest, so inequalities are stated with the less than ($<$) relation whenever possible. In Example 7, notice that $1 > x > -2$ is rewritten as $-2 < x < 1$.   ■

Problem Set 1.5

Solve each equation in Problems 1–16.

1. $3x = 5x - 4$
2. $2x = 9 - x$
3. $7x + 10 = 5x$
4. $9x - 8 = 5x$
5. $3x + 22 = 1 - 4x$
6. $5x - 4 = 3x + 8$
7. $2x - 13 = 7x + 2$
8. $3x + 32 = 18 - 4x$
9. $7x + 18 = -2x$
10. $6x - 1 = 13 - x$
11. $8x - 3 = 15 - x$
12. $-5x = 9 - 2x$
13. $2(x - 1) = 1 + 3(x - 2)$
14. $11 - x = 2 - 3(x - 1)$
15. $5 - 2x = 1 + 3(x - 2)$
16. $1 - 2(x - 3) = 2x - 1$

Solve each inequality in Problems 17–24.

17. $3x - 2 \le 7$
18. $2x - 1 \ge 9$
19. $4 - 5x > 29$
20. $7 - 4x < 3$
21. $9x + 7 \ge 5x - 9$
22. $3x + 7 > 7x - 5$
23. $2(3 - 4x) < 30$
24. $3(2x - 5) \le 9$

Solve the compound inequalities in Problems 25–30.

25. $7 < x + 2 < 11$
26. $12 < 5 + x < 14$
27. $-2 < x - 1 < 3$
28. $-5 \le x - 2 \le 4$
29. $9 < 1 - 2x < 15$
30. $-3 \le 1 - 2x \le 7$

Solve each equation or inequality in Problems 31–47.

31. $2(x - 3) - 5x = 3(1 - 2x)$

32. $3(2x + 7) + 11 = 5(2 - x)$

33. $5(x - 1) + 3(2 - 4x) > 8$

34. $4(1 - x) - 7(2x - 5) < 3$

35. $4(x - 2) + 1 \le 3(x + 1)$

36. $6(2x - 1) \ge 3(x + 4)$

37. $2(4 - 3x) > 4(3 - x)$

38. $3(2 - 5x) \ge 5(x + 2) + 36$

39. $2(3 - 7x) \ge -4 - (5 - x)$

40. $3(7 - x) < 2(x - 2)$

41. $3(1 - x) - 5(x - 2) = 5$

42. $5(5 - 3x) - (x - 8) = 1$

43. $6(2x + 5) = 4(3x + 1)$

44. $9(2 - 3x) = 6(5x + 3)$

45. $3(2x - 5) = 5(4x - 3)$

46. $2(4 - 3x) = 3(x - 2) - (9x - 14)$

47. $5(1 - 2x) = 3(x - 4) - (13x - 17)$

1.6 Quadratic Equations and Inequalities

Quadratic Equations

A *quadratic equation in one variable* is an equation that can be written in the form

$$ax^2 + bx + c = 0 \qquad a \ne 0$$

where x is a variable, and a, b, and c are real numbers. We will consider three different methods for solving quadratic equations—factoring, square root, and the quadratic formula.

The simplest method can be used if the quadratic expression $ax^2 + bx + c$ is factorable over the integers. The solution then depends on the following property of zero.

Property of Zero

$$AB = 0 \qquad \text{if and only if} \qquad A = 0 \quad \text{or} \quad B = 0 \quad \text{(or both)}$$

Thus, if the product of two factors is zero, then at least one of the factors is zero. If a quadratic equation is factorable, this property provides a method of solution.

EXAMPLE 1　Solve $x^2 = 2x + 15$.

Solution

$x^2 - 2x - 15 = 0$	Rewrite with a zero on one side (subtract $2x$ and 15 from both sides)
$(x + 3)(x - 5) = 0$	Factor
$x + 3 = 0 \quad \text{or} \quad x - 5 = 0$	Since the product is zero, one of the factors must be zero.
$x = -3 \qquad\qquad x = 5$	

Solution: $\{-3, 5\}$　∎

You check a quadratic equation just as you check a linear equation, by substitution to see if you obtain a true or a false equation. For Example 1,

check $x = -3$:　$(-3)^2 = 2(-3) + 15$　　check $x = 5$:　$5^2 = 2(5) + 15$

$9 = -6 + 15$　　　　　　　　　　　　$25 = 10 + 15$

$9 = 9 \checkmark$　　　　　　　　　　　　　$25 = 25 \checkmark$

Solution of Quadratic Equations by Factoring

To solve a quadratic equation that can be expressed as a product of linear factors:

1. Rewrite all nonzero terms on one side of the equation.
2. Factor the expression.
3. Set each of the factors equal to zero.
4. Solve each of the linear equations.
5. Write the solution set that is the union of the solution sets of the linear equations.

EXAMPLE 2   Solve $x^2 - 6x = 0$.

Solution   Factor, if possible: $x(x - 6) = 0$
Set each factor equal to zero: $x = 0 \qquad x - 6 = 0$
$$x = 6$$

Solution is $\{0, 6\}$. However, $x = 0$, $x = 6$ or $x = 0, 6$ is a sufficient answer.   ∎

When the quadratic equation is not factorable, other methods must be employed. One such method depends on the **square root property**.

Square Root Property

If $P^2 = Q$, then $P = \pm\sqrt{Q}$.

For example, the equation $x^2 = 4$ can be rewritten as $x^2 - 4 = 0$, factored, and solved. However, the square root property can also be used:

Square root property	Factoring
$x^2 = 4$	$x^2 - 4 = 0$
$x = \pm\sqrt{4}$	$(x - 2)(x + 2) = 0$
$x = \pm 2$	$x = 2, -2$

The square root property can be derived by using the following property of square roots:

Square Root of a Real Number

For all real numbers x,
$$\sqrt{x^2} = |x|$$

This means that, if $x \geq 0$, then $\sqrt{x^2} = x$ and if $x < 0$, then $\sqrt{x^2} = -x$. For example, $\sqrt{2^2} = |2| = 2$ and $\sqrt{(-2)^2} = |-2| = 2$.

The importance of the square root property is that it can be applied to *any* quadratic of the form $ax^2 + bx + c = 0$, $a \neq 0$. The result (which is derived by completing the square and is shown in most algebra textbooks) is called the **quadratic formula**. This method for solving quadratic equations will solve any quadratic, so in practice we will usually try to solve a quadratic equation by factoring, and if that does not easily work we will go directly to the quadratic formula.

Quadratic Formula

If $ax^2 + bx + c = 0$ with $a \neq 0$, then

$$x = \frac{-b \pm \sqrt{b^2 - 4ac}}{2a}$$

EXAMPLE 3   Solve $5x^2 + 2x - 2 = 0$.

Solution   Note that $a = 5$, $b = 2$, and $c = -2$. Thus

$$x = \frac{-2 \pm \sqrt{4 - 4(5)(-2)}}{2(5)}$$

$$= \frac{-2 \pm 2\sqrt{1 + 10}}{2(5)}$$

$$= \frac{-1 \pm \sqrt{11}}{5}$$ ∎

EXAMPLE 4   Solve $5x^2 + 2x + 2 = 0$.

Solution $$x = \frac{-2 \pm \sqrt{4 - 4(5)(2)}}{2(5)}$$

$$= \frac{-2 \pm 2\sqrt{-9}}{10}$$

Since the square root of a negative number is not a real number, the solution set is *empty over the reals.** That is, we say the solution set is $\varnothing$. ∎

Since the quadratic formula contains a radical, the sign of the radicand (the number under the radical) will determine whether the roots will be real or nonreal. This radicand is called the **discriminant** of the quadratic, and its properties are summarized below:

Discriminant

If $ax^2 + bx + c = 0$ $(a \neq 0)$, then $D = b^2 - 4ac$ is called the *discriminant*.

If $D < 0$, then there are *no real solutions*.
If $D = 0$, then there is *one real solution*.
If $D > 0$, then there are *two real solutions*.

---

* In algebra, you may have solved this equation by using complex numbers. However, since the domain for variables in this course is the set of real numbers, and since there is no real number whose square is negative, we see that there is no real number that satisfies this equation.

Quadratic Inequalities

A *quadratic inequality in one variable* is an inequality that can be written in the form

$$ax^2 + bx + c < 0 \qquad a \neq 0$$

where x is a variable, and a, b, and c are any real numbers. The symbol $<$ can be replaced by $\leq$, $>$, or $\geq$ and the inequality will still be a quadratic inequality in one variable.

The procedure for solving a quadratic inequality is similar to that for solving a quadratic equality. First, we use the properties of inequality to obtain a zero on one side of the inequality. Then we factor, if possible, the quadratic expression on the left. For example,

$$x^2 - x - 6 \geq 0$$
$$(x + 2)(x - 3) \geq 0$$

A value that causes one of the factors to be zero is called a **critical value** for the inequality; the critical values for the example above are $x = -2$ and $x = 3$. It follows, then, that the inequality must be either positive or negative for every real number that is not a critical value. We therefore examine the intervals defined by the critical values.

EXAMPLE 5   Solve $x^2 - x < 6$.

Solution   First obtain a zero on one side, and then factor, if possible:

$$x^2 - x - 6 < 0$$
$$(x + 2)(x - 3) < 0$$

Next, find the critical values (one at a time), plot them on a number line, and label the parts of the number line to the left and right of the critical value "$+$" or "$-$" depending on whether the factor is positive or negative in that region.

Factor: $x + 2$ *Critical value:* $x + 2 = 0$
$$x = -2$$

Plot:

$$
\begin{array}{c|c}
----------- & +++++++++++++++++ \\
x+2 \text{ is negative here} & x+2 \text{ is positive for values of } x \\
 & \text{on this side of the critical value}
\end{array}
$$

```
_____o_____
                           -2
                            ↑
```

Critical value (neither $+$ nor $-$)

Factor: $x - 3$ *Critical value:* $x - 3 = 0$
$$x = 3$$

Plot (on the same number line):

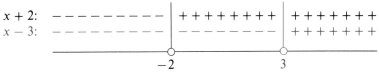

Finally, use the property of products to label the regions of the number line "positive" or "negative" as shown:

$x + 2$: – – – – – – – – – | + + + + + + + + | + + + + + + +
$x - 3$: – – – – – – – – – | – – – – – – – | + + + + + + +

Positive   −2   Negative   3   Positive

$\overbrace{\text{Neg} \cdot \text{Neg} = \text{Pos}}$ $\overbrace{\text{Pos} \cdot \text{Neg} = \text{Neg}}$ $\overbrace{\text{Pos} \cdot \text{Pos} = \text{Pos}}$

Shade in the appropriate portion of the number line indicated by the original inequality:

$$x^2 - x - 6 < 0$$ "<0" means you are looking for the portion of the number line labeled "negative"

Answer: $-2 < x < 3$ That is, x is between -2 and 3 ■

The previous example seems lengthy because all the steps are shown, but the process is quite easy and will be greatly simplified when you are doing the work. The process is summarized below, while Example 6 shows how your paper should look when doing this type of problem.

Solution of Polynomial Inequalities by Factoring

> To solve an inequality that can be expressed as a product less than or greater than zero:
>
> 1. Rewrite all nonzero terms on one side of the inequality.
> 2. Factor the expression.
> 3. Determine the critical values.
> 4. Determine the signs of the factors on the intervals between critical values.
> 5. Select the interval or intervals on which the product has the desired sign.

EXAMPLE 6   Solve $5 + 4x - x^2 \geq 0$.

Solution $(5 - x)(1 + x) \geq 0$ The critical values are 5 and −1. Notice that in this example the critical values are included in the solution since equality is included

$5 - x$ is positive for $x < 5$

$5 - x$: + + + + + + + + + + | + + + + + + + + + + + + | – – – – – – – – – – – – –

$1 + x$ is positive for $x > -1$

$1 + x$: – – – – – – – – – | + + + + + + + + + + + + | + + + + + + + + + + + + +

$(5 - x)(1 + x)$:   Negative   −1   Positive   5   Negative

$\overbrace{\text{Pos} \cdot \text{Neg} = \text{Neg}}$ $\overbrace{\text{Pos} \cdot \text{Pos} = \text{Pos}}$ $\overbrace{\text{Neg} \cdot \text{Pos} = \text{Neg}}$

Answer: $-1 \leq x \leq 5$ ■

EXAMPLE 7   Solve $x^2 + 2x - 4 < 0$.

Solution   The term on the left is in simplified form and cannot be factored using the techniques we have studied. Therefore, proceed by considering $x^2 + 2x - 4$ as a single factor.

To find the critical values, find the values for which the factor $x^2 + 2x - 4$ is zero.

$$x^2 + 2x - 4 = 0$$

$$x = \frac{-2 \pm \sqrt{4 - 4(1)(-4)}}{2} = \frac{-2 \pm 2\sqrt{5}}{2} = -1 \pm \sqrt{5}$$

Plot the critical values and check the sign of the factor in each interval:

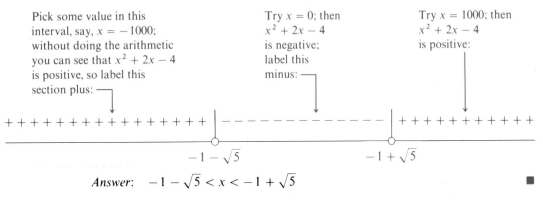

Pick some value in this interval, say, $x = -1000$; without doing the arithmetic you can see that $x^2 + 2x - 4$ is positive, so label this section plus:

Try $x = 0$; then $x^2 + 2x - 4$ is negative; label this minus:

Try $x = 1000$; then $x^2 + 2x - 4$ is positive:

$-1 - \sqrt{5}$

$-1 + \sqrt{5}$

Answer: $-1 - \sqrt{5} < x < -1 + \sqrt{5}$

Problem Set 1.6

Solve each equation in Problems 1–20 over the set of real numbers.

1. $x^2 + 2x - 15 = 0$

2. $x^2 - 8x + 12 = 0$

3. $x^2 + 7x - 18 = 0$

4. $2x^2 + 5x - 12 = 0$

5. $10x^2 - 3x - 4 = 0$

6. $6x^2 + 7x - 10 = 0$

7. $x^2 + 5x - 6 = 0$

8. $x^2 + 5x + 6 = 0$

9. $x^2 - 10x + 25 = 0$

10. $x^2 + 6x + 9 = 0$

11. $12x^2 + 5x - 2 = 0$

12. $2x^2 - 6x + 5 = 0$

13. $5x^2 - 4x + 1 = 0$

14. $2x^2 + x - 15 = 0$

15. $4x^2 - 5 = 0$

16. $3x^2 - 1 = 0$

17. $3x^2 = 7x$

18. $7x^2 = 3$

19. $3x^2 = 5x + 2$

20. $3x^2 - 2 = -5x$

Solve the inequalities in Problems 21–40.

21. $(x - 6)(x - 2) \geq 0$

22. $(x + 2)(x - 8) \leq 0$

23. $x(x + 3) < 0$

24. $x(x - 3) \geq 0$

25. $(x + 2)(8 - x) \leq 0$

26. $(2 - x)(x + 8) \geq 0$

27. $(1 - 3x)(x - 4) < 0$

28. $(2x + 1)(3 - x) > 0$

29. $x^2 \geq 9$

30. $x^2 > 4$

31. $x^2 + 9 \leq 0$

32. $x^2 + 2x - 3 < 0$

33. $x^2 - x - 6 > 0$

34. $x^2 - 7x + 12 > 0$

35. $5x - 6 \geq x^2$

36. $4 \geq x^2 + 3x$

37. $5 - 4x \geq x^2$

38. $x^2 + 2x - 1 < 0$

39. $x^2 - 2x - 2 < 0$

40. $x^2 - 8x + 13 > 0$

1.7 Rational Expressions

If a variable is used in a denominator, then the expression is not called a polynomial. Instead, it is called a **rational expression**.

Rational Expressions

A *rational expression* is an expression that can be written as a polynomial divided by a polynomial. Any values that cause division by zero are excluded from the domain.

The fundamental property used to simplify rational expressions involves factoring both the numerator and denominator and then eliminating common factors according to the property

$$\frac{PK}{QK} = \frac{P}{Q} \qquad Q, K \neq 0$$

Some rational expressions are simplified in Examples 1–5. Do not forget that all values for variables that cause division by zero are excluded from the domain.

EXAMPLE 1
$$\frac{3xyz}{x^2 + 3x} = \frac{x(3yz)}{x(x + 3)}$$
$$= \frac{3yz}{x + 3}$$
∎

EXAMPLE 2
$$\frac{x - 2}{x^2 - 4} = \frac{1(x - 2)}{(x + 2)(x - 2)}$$
$$= \frac{1}{x + 2}$$
∎

Sometimes the factors that are eliminated (as shown in color in Examples 1 and 2) are marked off in pairs as shown in Example 3. The slashes should be viewed as replacing the factor K by the number 1 since

$$\frac{PK}{QK} = \frac{P\cancel{K}}{Q\cancel{K}} = \frac{P \cdot 1}{Q \cdot 1} = \frac{P}{Q} \cdot 1 = \frac{P}{Q}$$

EXAMPLE 3
$$\frac{(x - 5)(x + 2)(x + 1)}{(x + 1)(x - 2)(x - 5)} = \frac{\cancel{(x - 5)}(x + 2)\cancel{(x + 1)}}{\cancel{(x + 1)}(x - 2)\cancel{(x - 5)}}$$
$$= \frac{x + 2}{x - 2}$$

This is reduced since the x's are terms and *not* factors. You cancel factors, not terms ∎

EXAMPLE 4
$$\frac{6x^2 + 2x - 20}{30x^2 - 68x + 30} = \frac{2(3x^2 + x - 10)}{2(15x^2 - 34x + 15)}$$

Common factor first

$$= \frac{2\cancel{(3x - 5)}(x + 2)}{2\cancel{(3x - 5)}(5x - 3)}$$

Complete the factoring; then reduce

$$= \frac{x + 2}{5x - 3}$$
∎

The laws of exponents can be extended to include rational expressions and negative exponents.

EXAMPLE 5 **a.** $\left(\dfrac{x}{y}\right)^3 = \left(\dfrac{x}{y}\right)\left(\dfrac{x}{y}\right)\left(\dfrac{x}{y}\right) = \dfrac{xxx}{yyy} = \dfrac{x^3}{y^3}$

b. $\dfrac{x^5}{x^3} = \dfrac{xxxxx}{xxx} = xx = x^2$

c. $\dfrac{x^3}{x^5} = \dfrac{xxx}{xxxxx} = \dfrac{1}{x^2} = x^{-2}$ ∎

Laws of Exponents

Let a and b be any real numbers, m and n be any integers, and assume that each expression is defined.*

FOURTH LAW: $\left(\dfrac{a}{b}\right)^m = \dfrac{a^m}{b^m}$

FIFTH LAW: $\dfrac{b^m}{b^n} = b^{m-n}$

The procedures for operating on rational expressions are identical to those for operations with fractions. However, with rational expressions, the numerators and denominators are any polynomial (division by zero is excluded) rather than constants. The procedures are summarized below.

Properties of Rational Expressions

Let $P, Q, R, S,$ and K be any polynomials such that all values of the variable that cause division by zero are excluded from the domain.

EQUALITY: $\dfrac{P}{Q} = \dfrac{R}{S}$ if and only if $PS = QR$

FUNDAMENTAL PROPERTY: $\dfrac{PK}{QK} = \dfrac{P}{Q}$

ADDITION: $\dfrac{P}{Q} + \dfrac{R}{S} = \dfrac{PS + QR}{QS}$

SUBTRACTION: $\dfrac{P}{Q} - \dfrac{R}{S} = \dfrac{PS - QR}{QS}$

MULTIPLICATION: $\dfrac{P}{Q} \cdot \dfrac{R}{S} = \dfrac{PR}{QS}$

DIVISION: $\dfrac{P}{Q} \div \dfrac{R}{S} = \dfrac{PS}{QR}$

Some operations on rational expressions are performed in Examples 6–9. All values of variables that could cause division by zero are excluded.

EXAMPLE 6 $\dfrac{13}{x - y} + \dfrac{2}{x - y} = \dfrac{15}{x - y}$ Common denominator   ∎

---

* The first three laws of exponents are in Section 1.3, page 12.

EXAMPLE 7

$$\frac{13}{x-y} - \frac{2}{y-x} = \frac{13}{x-y} + \frac{-2}{y-x}$$

It is helpful to write subtractions as additions

$$= \frac{13}{x-y} + \frac{-2}{y-x} \cdot \frac{-1}{-1}$$

Multiply by 1 (written as $-1/-1$) in order to obtain a common denominator

$$= \frac{13}{x-y} + \frac{2}{x-y}$$

$$= \frac{15}{x-y}$$

■

EXAMPLE 8

$$\frac{x+y}{x-y} + \frac{x-2y}{2x+y} = \frac{(x+y)(2x+y) + (x-y)(x-2y)}{(x-y)(2x+y)}$$

$$= \frac{2x^2 + 3xy + y^2 + x^2 - 3xy + 2y^2}{(x-y)(2x+y)}$$

$$= \frac{3x^2 + 3y^2}{(x-y)(2x+y)}$$

$$= \frac{3(x^2+y^2)}{(x-y)(2x+y)}$$

■

EXAMPLE 9

$$\left(\frac{x^2+5x+6}{2x^2-x-1} \cdot \frac{2x^2-9x-5}{x^2+7x+12}\right) \div \frac{2x^2-13x+15}{x^2+3x-4}$$

$$= \left[\frac{(x+2)(x+3)}{(x-1)(2x+1)} \cdot \frac{(2x+1)(x-5)}{(x+3)(x+4)}\right] \div \frac{(x-5)(2x-3)}{(x-1)(x+4)}$$

$$= \frac{(x+2)\cancel{(x+3)}\cancel{(2x+1)}\cancel{(x-5)}\cancel{(x-1)}\cancel{(x+4)}}{\cancel{(x-1)}\cancel{(2x+1)}\cancel{(x+3)}\cancel{(x+4)}\cancel{(x-5)}(2x-3)}$$

$$= \frac{x+2}{2x-3}$$

■

Problem Set 1.7

Simplify the expressions in Problems 1–30. Values that cause division by zero are excluded. All expressions should be reduced and negative exponents eliminated.

1. $\dfrac{2}{x+y} + 3$

2. $\dfrac{3}{x+y} - 2$

3. $\dfrac{x^2-y^2}{2x+2y}$

4. $\dfrac{x^2-y^2}{3x-3y}$

5. $\dfrac{3x^2-4x-4}{x^2-4}$

6. $\dfrac{x^2+3x-18}{x^2-9}$

7. $\dfrac{3}{x+y} + \dfrac{5}{2x+2y}$

8. $\dfrac{4}{x-y} - \dfrac{3}{2x-2y}$

9. $[7^3 + 2^5(3^3 + 4^4)]^0$

10. $[9^3 + 3^6(5^3 + 8^3)]^0$

11. $(x^2 - 36)\left(\dfrac{3x+1}{x+6}\right)$

12. $x^2 - 9 \div \dfrac{x+3}{x-3}$

13. $\dfrac{x+3}{x} + \dfrac{3-x}{x^2}$

14. $\dfrac{2}{x-y} + \dfrac{5}{y-x}$

15. $\dfrac{x}{x-1} + \dfrac{x-3}{1-x}$

16. $\dfrac{x+1}{x} + \dfrac{2-x}{x^2}$

17. $\dfrac{2x+3}{x^2} + \dfrac{3-x}{x}$

18. $\dfrac{1}{x^3} + 2xy + \dfrac{x^2}{y^2}$

19. $\dfrac{x}{y} + 2 + \dfrac{y}{x}$

20. $\dfrac{1}{x^3y^2} + \dfrac{y}{x} + 2xy$

21. $\dfrac{x}{y} - 2 + \dfrac{y}{x}$

22. $\dfrac{1}{2y} + \dfrac{1}{x} + \dfrac{y}{2x^2}$

23. $\dfrac{1}{3xy^2} + xy + \dfrac{1}{x^3y}$

24. $\dfrac{2x+y}{(x+y)^2} + \dfrac{x^2-2y^2}{(x+y)^3}$

25. $\dfrac{4x - 12}{x^2 - 49} \div \dfrac{18 - 2x^2}{x^2 - 4x - 21}$

26. $\dfrac{36 - 9x}{3x^2 - 48} \div \dfrac{15 + 13x + 2x^2}{12 + 11x + 2x^2}$

27. $\dfrac{1}{x^2 + 1} - \dfrac{x^2}{x^2 + 1}$

28. $(x^2 + 1) + \dfrac{x^2}{(x^2 + 1)^2}$

29. $\dfrac{4x^2}{x^4 - 2x^3} + \dfrac{8}{4x - x^3} - \dfrac{-4}{x + 2}$

30. $\dfrac{6x}{2x + 1} - \dfrac{2x}{x - 3} + \dfrac{4x^2}{2x^2 - 5x - 3}$

1.8 Summary and Review

IMPORTANT TERMS

Absolute value   [1.2]
Algebraic expression   [1.3]
Base   [1.3]
Binomial   [1.3]
Composite   [1.2]
Compound inequality   [1.5]
Constant   [1.3]
Coordinate of a point   [1.2]
Coordinate system   [1.2]
Critical value   [1.6]
Degree   [1.3]
Discriminant   [1.6]
Distance between points   [1.2]
Distributive property   [1.3]
Domain   [1.3]
Element   [1.2]
Empty set   [1.2, 1.5]
Equivalent equations   [1.5]
Evaluate   [1.3]
Exponent   [1.3]
Factor   [1.3, 1.4]
Factoring   [1.4]
Finite set   [1.2]
FOIL   [1.3, 1.4]
Graph of a number   [1.2]
Greater than   [1.2]
Identity   [1.5]
Inequality   [1.2, 1.5]
Infinite set   [1.2]
Integers   [1.2]
Irrationals   [1.2]
Laws of exponents   [1.3, 1.7]
Less than   [1.2]
Linear equation   [1.5]
Member   [1.2]

Monomial   [1.3]
Natural numbers   [1.2]
Negative real number   [1.2]
Null set   [1.2]
Number line   [1.2]
Numerical coefficient   [1.3]
One-to-one correspondence   [1.2]
Order of operations   [1.3]
Polynomial   [1.3]
Positive real number   [1.2]
Power   [1.3]
Prime   [1.2]
Properties of equations   [1.5]
Properties of inequalities   [1.5]
Properties of rational
  expressions   [1.7]
Property of comparison   [1.2]
Quadratic equation   [1.6]
Quadratic formula   [1.6]
Quadratic inequality   [1.6]
Rational expression   [1.7]
Rationals   [1.2]
Real numbers   [1.2]
Root   [1.5]
Set   [1.2]
Similar terms   [1.3]
Solution   [1.5]
Solution set   [1.5]
Square root of a real number   [1.6]
Square root property   [1.6]
Term   [1.3]
Trichotomy property   [1.2]
Trinomial   [1.3]
Variable   [1.3]
Whole numbers   [1.2]

SAMPLE TEST   *For additional practice there are a large number of review problems categorized by objective in the Student Solutions Manual. The following sample test (40 minutes) is intended to review the main ideas of this chapter.*

1. Draw a diagram showing how the natural numbers, whole numbers, integers, rationals, and real numbers are related.

2. Write $|2\pi - 10|$ without absolute value symbols (do not approximate).

3. Find the distance between $A(\sqrt{10})$ and $B(5)$.

4. Simplify $\dfrac{1 + 3 \cdot 5}{-3^2}$.

5. Evaluate $-x^2 + 5x(3 - 2y)$ where $x = -6$ and $y = -3$.

6. Simplify $(3x + 1)(4x - 5)$.

7. Simplify $(x + y)^2$.

8. Simplify $(x + 3)(2x^2 - 5x - 4)$.

9. Factor $6x^2 - 29x - 5$.

10. Factor $1 - 9x^2$.

11. Solve $4 - 3x = 2(5 - 2x)$.

12. Solve $15 < -x$.

13. Solve $2(x + 1) + 5 < 3(1 - x)$.

14. Solve $-3 \le x - 5 < 0$.

15. Solve $x = 5x^2$.

16. Solve $x^2 + 2x - 2 = 0$.

17. Solve $(1 - x)(3x - 2) > 0$.

18. Solve $x^2 - 5 \ge 4x$.

19. Simplify $\dfrac{7x}{30} - \dfrac{y}{12}$.

20. Simplify $\dfrac{x^2 - 25}{x + 1} \div \dfrac{x^2 - 4x - 5}{x^2 + 2x + 1}$.

CHAPTER 2
Functions and Graphs

CHAPTER CONTENTS

2.1 **Functions and Graphs Pretest**

2.2 **Functions**

2.3 **Graphs of Functions**

2.4 **Linear Functions**

2.5 **Quadratic and Polynominal Functions**

2.6 **Rational Functions**

2.7 **Summary and Review Important Terms Sample Test**

APPLICATIONS

Management (*Business, Economics, Finance, and Investments*)

Purchasing power of the dollar (2.2, Problems 45–54)
Sales graph (2.3, Problem 44)
Supply and demand equations (2.4, Problems 70–71)
1985 Tax Rate Schedule (2.4, Problems 77–81)
Demand, revenue, and break-even points (2.5, Problems 49–50)
Maximum profit (2.1, Problem 15; 2.5, Problems 51–53)
Minimum cost (2.5, Problem 54)
Cost and revenue functions (2.5, Problems 54–55)
Equilibrium point for supply and demand (2.6, Problem 40)

Life Sciences (*Biology, Ecology, Health, and Medicine*)

Alcohol content in bloodstream (2.2, Problems 55–56)
Atmospheric pressure (2.3, Problem 43)
Response of a nerve (2.3, Problem 45)
Cost of removing a pollutant (2.3, Problem 46)
Cost-benefit model for removing pollutants (2.6, Problems 28–35; 2.7, Problems 19–20)
Radiology (2.6, Problems 36–37)
Pressure-volume relationship (2.6, Problems 38–39)

Social Sciences (*Demography, Political Science, Population, Psychology, Society, and Sociology*)

U.S. population (2.2, Problems 57–58)
Marriage rate (2.2, Problem 59)
Divorce rate (2.2, Problem 60)
Population as predicted by an equation (2.4, Problems 72–73)

General

Admission price for a performance (2.3, Problem 41)
Taxable income (2.3, Problem 42)
Cost analysis for a car rental (2.4, Problems 74–76)
Skycycle ride across Snake River Canyon (2.5, Problem 57)
Stopping distance for a car (2.5, Problem 58)

CHAPTER
OVERVIEW   Chapter 2 introduces the building blocks of finite mathematics: the concepts of mathematical models and of graphs.

PREVIEW   We begin by reviewing the concept of a function and evaluating functions and functional notation. Then we discuss graphing functions, and in the remainder of the chapter focus on three very important types of functions: linear, quadratic, and rational.

PERSPECTIVE   This book is divided into two parts, finite mathematics and calculus. Calculus is concerned with infinite processes, and, as the name implies, finite mathematics deals with finite, or countable, ideas. This chapter is a prerequisite for both parts of the course.

2.1 Functions and Graphs Pretest

Choose the best answer in Problems 1–20.

1. Which of the following is *false*?
    A. A function is a number.
    B. A function can be defined by a verbal rule.
    C. A function can be defined by a table.
    D. A function can be defined by a graph.
    E. A function can be defined by an algebraic formula.

2. If $f(x) = 3x^2 + 2$, then $f(-2)$ is
    A. y B. -10 C. $-6x^2 - 4$
    D. 14 E. none of these

3. If $f(x) = 3x^2 + x - 7$, then $f(a - 1) - f(a)$ is
    A. $6x + 2$ B. 2 C. $2 - 6a$
    D. $-4(a + 3)$ E. none of these

4. If $f(x) = 3x^2 + 2$, then $\dfrac{f(x + h) - f(x)}{h}$ is
    A. $6x + 3h$ B. $\dfrac{6xh + 3h^2 + 4}{h}$

    C. $\dfrac{6x^2 + 6xh + h^2 + 2}{h}$ D. $\dfrac{3x^3 + 3x^2h + 2x + 2h}{h}$

    E. none of these

5. In (h, k), k is the
    A. slope   B. independent variable
    C. dependent variable   D. abscissa
    E. domain

6. Which of the following has *no* y-intercept?
    A. $y = 0$ B. $x + y = 0$ C. $y = x$
    D. $x = 2$ E. $x = y + 2$

7. Consider the function f defined by the figure below. The coordinates of P are
    A. (x, y) B. $(x_1, f(x_1))$ C. $(x_1, x_1 + h)$
    D. about $(1, 2)$ E. none of these

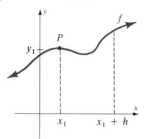

8. What is the slope of the line containing $(5, 2)$ and $(-1, 6)$?
    A. $-\frac{2}{7}$ B. $-\frac{4}{7}$ C. $-\frac{2}{3}$
    D. $-\frac{4}{3}$ E. none of these

9. If f is a function and $P(x_1, y_1)$ and $Q(x_2, y_2)$ are points on the graph of f, then the slope of the line passing through P and Q is
    A. $\dfrac{x_2 - x_1}{y_2 - y_1}$ B. $f(m)$

    C. $\dfrac{f(x_2) - f(x_1)}{x_2 - x_1}$ D. $\dfrac{f(y_2) - f(y_1)}{x_2 - x_1}$

    E. none of these

10. The equation of the line with y-intercept -2 passing through $(-3, -4)$ is
    A. $2x + y - 2 = 0$ B. $2x + y + 10 = 0$
    C. $2x + y + 11 = 0$ D. $2x - 3y - 6 = 0$
    E. none of these

11. Which of the following is the equation of the line with no slope that passes through $(2, -3)$?
    A. $x = 2$ B. $x = -3$ C. $y = 2$
    D. $y = -3$ E. none of these

12. The graph of the equation $3x - 2y^2 - 4y = 0$ is
    A. a line   B. a circle   C. a parabola
    D. a rational function with at least one asymptote
    E. none of these

13. The graph of the equations $x = 2 + 5t$ and $y = 1 + 2t$ is
   A. a line   B. a circle   C. a parabola
   D. a rational function with at least one asymptote
   E. none of these

14. Any number x such that $f(x) = 0$ is called
   A. an x-intercept   B. a y-intercept
   C. a zero of the function   D. $x = 0$
   E. none of these

15. A manufacturer selling a product produces x items per day and prices them at $440 - x$ dollars each. The overhead is found to be $9x^2 - 7560x + 1,584,000$ dollars. Then

$$\text{Profit} = \text{revenue} - \text{cost}$$
$$= (\text{number of items})(\text{price}) - \text{cost}$$
$$P = x(440 - x)$$
$$- (9x^2 - 7560x + 1,584,000)$$
$$P - 16,000 = -10(x - 400)^2$$

   What is the maximum profit (in dollars)?
   A. 80   B. 16,000   C. 400
   D. 1,658,000   E. none of these

16. The domain for the curve $y^2 + x - 5 = 0$ is
   A. $0 < x < 5$ B. $0 \le x \le 5$
   C. $x < 5$ or $x > 5$
   D. $-\sqrt{5} < x < \sqrt{5}$
   E. none of these

17. If $y = \frac{2}{3}$ is a horizontal asymptote for a curve, then
   A. $x = 0$ is a vertical asymptote
   B. the degree of the numerator is the same as the degree of the denominator
   C. the degree of the numerator is less than the degree of the denominator
   D. the degree of the numerator is greater than the degree of the denominator
   E. none of these

18. The rational function $y = \dfrac{x^3}{(x - 1)(2x + 3)}$ has
   A. vertical asymptotes at $x = 1$ and $x = -\frac{2}{3}$
   B. horizontal asymptotes at $y = 1$ and $y = -\frac{2}{3}$
   C. a horizontal asymptote at $y = 0$
   D. vertical asymptote at $x = 0$
   E. none of these

19. If $R(x)$ is the revenue function and $D(x)$ is the demand function, then
   A. $R(x) > D(x)$ B. $R(x) = xD(x)$
   C. $D(x) = xR(x)$ D. $R(x) = D(x)$
   E. none of these

20. If $R(x)$ is the revenue function and $C(x)$ is the cost function, then the break-even point is found by solving
   A. $R(x) - C(x) = 0$ B. $R(x) + C(x) = 0$
   C. $R(x) = xD(x)$ D. $xR(x) = D(x)$
   E. none of these

2.2 Functions

We begin by reviewing an idea from algebra that is fundamental for the study of calculus, namely, the idea of a **function**.

Function

> A *function* of a variable x is a rule f that assigns to each value of x a unique* number $f(x)$, called the *value of the function at x*. A function can be defined by a verbal rule, a table, a graph, or an algebraic formula.

Remember:

1. $f(x)$ is pronounced "f of x."
2. f is a *function* while $f(x)$ is a *number*.
3. The set of replacements for x is called the **domain** of the function.
4. The set of all values $f(x)$ is called the **range** of f.

---

* By a unique number, we mean exactly one number.

EXAMPLE 1   *A function defined as a verbal rule.*
Let *x* be a year from 1980 to 1985, inclusive. We define the function *p* by the rule that *p(x)* is the closing price of Xerox stock on January 4 of year *x*. The domain is the set $\{1980, 1981, 1982, 1983, 1984, 1985\}$, and the range is the set of possible prices for Xerox stock.* For example, $p(1984) = 49\frac{1}{4}$. ∎

EXAMPLE 2   *A function defined by a table.*
Let *g* be a function defined so that *g(x)* is the price of gasoline in the year *x* as given by the following table:

Year	Average price per gallon of gasoline on January 4
1944	$ .21
1954	$ .29
1964	$ .30
1974	$ .53
1984	$1.24

The domain is $\{1944, 1954, 1964, 1974, 1984\}$

The range is $\{\$.21, \$.29, \$.30, \$.53, \$1.24\}$

For example, $g(1984) = \$1.24$. ∎

EXAMPLE 3   *A function defined by a graph.*

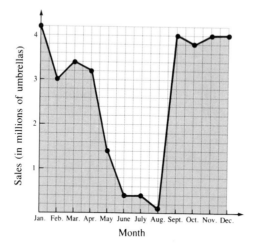

Let *s(x)* be the sales of umbrellas in millions of units for *x*, a month in 1988. The domain is $\{\text{Jan., Feb.}, \ldots, \text{Nov., Dec.}\}$, and the range is $0.1 \le y \le 4.2$. (Remember, the units on the *y*-axis are in millions, so this means $100{,}000 \le y \le 4{,}200{,}000$.) For example, $s(\text{Jan.}) \approx 4.2$, $s(\text{Aug.}) \approx 0.1$. ∎

---

* If you are familiar with the stock market, you know that Xerox stock prices are quoted in eighths of a dollar. The actual prices of Xerox stock over the years 1980–1985 provide the elements of the range.

EXAMPLE 4 *A function defined by an algebraic formula.*
Let f be a function defined by

$$f(x) = 2x - 5$$

This rule says take a number x from the domain, multiply it by 2, and then subtract 5. For example,

$$f(3) = 2(3) - 5$$
$$= 1$$

    ■

In this book, unless otherwise specified, the *domain is the set of real numbers for which the given function is meaningful.* This means x can be replaced by *any real number.* For example,

$$f(x) = 2x - 5$$

has a domain consisting of all real numbers, while

$$f(x) = \frac{1}{x - 1} \quad \text{and} \quad g(x) = \frac{3}{5x - 5}$$

have domains that exclude $x = 1$ because division by zero is not defined ($x - 1 = 0$ and $5x - 5 = 0$ if $x = 1$). We say $f(1)$ does not exist or we say $f(x)$ is not defined at $x = 1$. Finally, if

$$f(x) = \sqrt{x}$$

then $x \geq 0$ is implied since the square root of a number x is real if and only if $x \geq 0$, so $f(x)$ is defined only if $x \geq 0$.

EXAMPLE 5 Give the domain of the following functions.

  a. $f(x) = \dfrac{x - 8}{3}$ **b.** $f(x) = \dfrac{3}{x - 8}$ **c.** $f(x) = \sqrt{x + 2}$ **d.** $f(x) = \dfrac{1}{\sqrt{x + 2}}$

Solution **a.** The domain is all real numbers.
  b. The domain is all real numbers except where $x - 8 = 0$ or $x = 8$. We state this domain by simply writing $x \neq 8$.
  c. The domain is all real numbers for which

$$x + 2 \geq 0$$
$$x \geq -2$$

This can be stated in **set-builder notation** $\{x \mid x \geq -2\}$, which is read

"The set of all x, such that $x \geq -2$"
" $\{$ x $\mid$ $x \geq -2\}$"

However, set-builder notation is too formal for our work in this course, so we will just write $x \geq -2$ for the domain.
  d. This is similar to part e, but division by zero must also be excluded, so we see that the domain is $x > -2$.

    ■

Example 4 illustrates a very important process called **evaluating a function**. To find $f(3)$ you substitute 3 for every occurrence of x in the formula for $f(x)$. This process is illustrated in Examples 6 and 7.

EXAMPLE 6 Given f and g defined by $f(x) = 2x - 5$ and $g(x) = x^2 + 4x + 3$, find the indicated values:

   a. $f(1)$ **b.** $g(2)$ **c.** $g(-3)$ **d.** $f(-3)$

Solution   **a.** The symbol $f(1)$ is found by replacing x by 1 in the expression

$$f(x) = \quad 2x - 5$$
$$\updownarrow \qquad \updownarrow$$
$$f(1) = 2(1) - 5$$
$$= -3$$

Warning: This Is a Very Important Example   **b.** $g(2)$: $g(x) = x^2 + 4x + 3$
$$\updownarrow \quad \updownarrow \quad \updownarrow$$
$$g(2) = (2)^2 + 4(2) + 3$$
$$= 4 + 8 + 3$$
$$= 15$$

   c. $g(-3)$: $g(-3) = (-3)^2 + 4(-3) + 3$
$$= 9 - 12 + 3$$
$$= 0$$

   d. $f(-3) = 2(-3) - 5$
$$= -11$$

A function may be evaluated by using a variable.

EXAMPLE 7 Let F and G be defined by $F(x) = x^2 + 1$ and $G(x) = (x + 1)^2$. Then:

   a. $F(w) = w^2 + 1$

   b. $G(t) = (t + 1)^2$
$$= t^2 + 2t + 1$$

   c. $F(-a) = (-a)^2 + 1$
$$= a^2 + 1$$

   d. $G(-a) = (-a + 1)^2$
$$= a^2 - 2a + 1$$

   e. $F(w + h) = (w + h)^2 + 1$
$$= w^2 + 2wh + h^2 + 1$$

   f. $F(w^3) = (w^3)^2 + 1$
$$= w^6 + 1$$

   g. $[G(a)]^2 = [(a + 1)^2]^2$ since $G(a) = (a + 1)^2$
$$= (a + 1)^4$$

   h. $F[G(x)] = F[(x + 1)^2]$
$$= [(x + 1)^2]^2 + 1$$
$$= (x + 1)^4 + 1$$

   i. $F(\sqrt{t}) = (\sqrt{t})^2 + 1 = t + 1$

   j. $G(x + h) = [(x + h) + 1]^2$
$$= (x + h)^2 + 2(x + h) + 1$$
$$= x^2 + 2xh + h^2 + 2x + 2h + 1$$

In calculus, functional notation is used to carry out manipulations such as those shown below in Example 8. WARNING: Because of the extensive use of this manipulation, be sure you thoroughly understand Example 8.

EXAMPLE 8   Find $\dfrac{f(x + h) - f(x)}{h}$ for each function given below.

a. $f(x) = x^2$, where $x = 5$. We substitute 5 for every occurrence of x in the formula, but do not substitute for h.

$$\frac{f(5 + h) - f(5)}{h} = \frac{(5 + h)^2 - 5^2}{h}$$

$$= \frac{25 + 10h + h^2 - 25}{h}$$

$$= \frac{(10 + h)h}{h} = 10 + h$$

b. $f(x) = 2x^2 + 1$, where $x = 1$:

$$\frac{f(1 + h) - f(1)}{h} = \frac{[2(1 + h)^2 + 1] - [2(1)^2 + 1]}{h}$$

$$= \frac{[2(1 + 2h + h^2) + 1] - (2 + 1)}{h}$$

$$= \frac{2h^2 + 4h + 3 - 3}{h}$$

$$= 2h + 4$$

c. $f(x) = x^2 + 3x - 2$:

$$\frac{f(x + h) - f(x)}{h} = \frac{[(x + h)^2 + 3(x + h) - 2] - (x^2 + 3x - 2)}{h}$$

$$= \frac{x^2 + 2xh + h^2 + 3x + 3h - 2 - x^2 - 3x + 2}{h}$$

$$= \frac{2xh + h^2 + 3h}{h}$$

$$= 2x + 3 + h$$ ■

Functional notation can be used to work a wide variety of applied problems, as shown in Examples 9–12.

EXAMPLE 9   What is the average change per year in the price per gallon of gasoline from 1974 ($.53) to 1984 ($1.24)?

Solution   The price of gasoline changed by $.71 per gallon during this 10-year period ($1.24 − $.53 = $.71). Thus the average per year change in the price is:

$$\text{Change in the price of gas} = \frac{\$1.24 - \$.53}{1984 - 1974} = \frac{\$.71}{10} \quad \leftarrow 10 \text{ years}$$

$$= \$.071 \quad \blacksquare$$

EXAMPLE 10 What is the average change per year in the price per gallon of gasoline from 1974 to a year h years later?

Solution Notice that h years after 1974 is $1974 + h$, so the average is

$$\frac{g(1974 + h) - g(1974)}{h}$$ ∎

EXAMPLE 11 The alcohol concentration (as a percent) in an average person's bloodstream x hours after drinking 140 ml (8 oz) of 100 proof whiskey is given by the formula

$$A(x) = 0.106x - 0.0015x^3$$

How much alcohol is in the bloodstream after one hour?

Solution Evaluate the function A when $x = 1$:

$$A(1) = 0.106(1) - 0.0015(1)^3$$
$$= 0.1045$$

The alcohol content in the bloodstream is about 0.1%. ∎

EXAMPLE 12 In 1981 the U.S. population was 230 million. If we assume a growth rate of 2%, the population (in millions) t years after 1981 can be approximated by the formula

$$P(t) = 230(1.02)^t$$

What is the expected population for 1990?

Solution Since $1990 - 1981 = 9$, $t = 9$. Thus:

$$P(9) = 230(1.02)^9 \quad \text{BY CALCULATOR:} \quad \boxed{1.02}\;\boxed{y^x}\;\boxed{9}\;\boxed{\times}\;\boxed{230}\;\boxed{=}$$
$$\approx 275 \qquad\qquad\qquad \text{DISPLAY:} \quad 274.87129$$

The 1990 U.S. population will be about 275 million. ∎

Problem Set 2.2

1. What is a mathematical model?*

2. Why are mathematical models necessary or useful?*

3. What is a function (use your own words)?

4. If $y = f(x)$, what is the difference between the symbols f and $f(x)$, if any?

5. One of the following examples expresses y as a function of x and the other does not. Which is a function? Explain your answer.
   a. y is the closing price of IBM stock on March 3 of year x
   b. x is the closing price of Tandy stock on July 1 of year y

6. One of the following examples is a function of x and the other is not. Which is a function? Explain your answer.
   a. $y = x^2$ **b.** $x = y^2$

7. Use the table in Example 2 on page 36 to find and interpret each of the following expressions.
   a. $g(1954)$ **b.** $g(1974)$

8. Use the graph in Example 3 on page 36 to find and interpret the following expressions.
   a. $s(\text{March})$ **b.** $s(\text{June})$

9. Use $f(x) = 2x - 5$ (see Example 4) to find
   a. $f(8)$ **b.** $f(-3)$

10. Use $f(x) = 2x - 5$ (see Example 4) to find
    a. $f(-5)$ **b.** $f(7)$

Evaluate the functions in Problems 11–16.

11. Let $f(x) = 5x - 3$. Find
    a. $f(2)$ **b.** $f(10)$ **c.** $f(-15)$ **d.** $f(100)$

---

* See the Preface for a discussion of a mathematical model.

12. Let $g(x) = 4x - 10$. Find
    a. $g(-2)$ **b.** $g(5)$ **c.** $g(-10)$ **d.** $g(50)$

13. Let $h(x) = 6 - 4x$. Find
    a. $h(0)$ **b.** $h(8)$ **c.** $h(-7)$ **d.** $h(100)$

14. Let $k(x) = 10 - 3x$. Find
    a. $k(-4)$ **b.** $k(-5)$ **c.** $k(6)$ **d.** $k(10)$

15. Let $m(x) = x^2 - 3x + 1$. Find
    a. $m(0)$ **b.** $m(1)$ **c.** $m(2)$ **d.** $m(3)$

16. Let $n(x) = x^2 + x - 3$. Find
    a. $n(0)$ **b.** $n(1)$ **c.** $n(2)$ **d.** $n(3)$

State the domain of the functions in Problems 17–20.

17. a. $f(x) = 5x^2 - 3x + 2$ **b.** $f(x) = 6x^2 + 5x - \sqrt{17}$

18. a. $f(x) = \dfrac{2x - 5}{3}$ **b.** $f(x) = \dfrac{3}{2x - 5}$

19. a. $f(x) = \dfrac{3x + 2}{x + 5}$ **b.** $f(x) = \dfrac{\sqrt{2x - 1}}{2x + 1}$

20. a. $f(x) = \sqrt{2x + 6}$ **b.** $f(x) = \dfrac{1}{\sqrt{2x + 6}}$

In Problems 21–36, let $f(x) = 5x - 2$ and $g(x) = 2x^2 - 4x - 5$. Evaluate and simplify:

21. a. $f(t)$ **b.** $f(w)$ **22. a.** $g(s)$ **b.** $g(t)$

23. a. $f(t + h)$ **b.** $f(s + t)$ **24. a.** $g(t + h)$ **b.** $g(s + t)$

25. $f(t + h + 8)$ **26.** $g(t - h - 3)$

27. $g(3 + h)$ **28.** $g(t - 2)$

29. $f(2x^2)$ **30.** $f(2x^2 - 4x)$

31. $g(2x^2 - 4x)$ **32.** $f[g(x)]$

33. $g(5x)$ **34.** $g(5x - 2)$

35. $g[f(x)]$ **36.** $f[f(x)]$

Find $[f(x + h) - f(x)]/h$ for each function in Problems 37–44.

37. $f(x) = 2x$ **38.** $f(x) = 5x$

39. $f(x) = 2x^2$ **40.** $f(x) = 5x^2$

41. $f(x) = 2x^2 - 3$ **42.** $f(x) = 5x^2 - 3x$

43. $f(x) = x^2 - 2x + 1$ **44.** $f(x) = 3x^2 - 2x + 4$

APPLICATIONS

For Problems 45–54, use the following table, which reflects the purchasing power of the dollar from October 1944 to October 1984 (Source: U.S. Bureau of Labor Statistics, Consumer Division). Let x represent the year; let the domain be the set $\{1944, 1954, 1964, 1974, 1984\}$.

Year	Round steak (1 lb)	Sugar (5 lb)	Bread (loaf)	Coffee (1 lb)	Eggs (1 doz)	Milk ($\frac{1}{2}$ gal)	Gasoline (1 gal)
1944	$ .45	$ .34	$ .09	$ .30	$ .64	$ .29	$ .21
1954	.92	.52	.17	1.10	.60	.45	.29
1964	1.07	.59	.21	.82	.57	.48	.30
1974	1.78	1.08	.36	1.31	.84	.78	.53
1984	2.15	1.48	1.29	2.69	1.15	1.08	1.52

Let $r(x)$ = price of 1 lb of round steak $s(x)$ = price of 5 lb of sugar
     $b(x)$ = price of a loaf of bread $c(x)$ = price of 1 lb of coffee
     $e(x)$ = price of a dozen eggs $m(x)$ = price of $\frac{1}{2}$ gal of milk
     $g(x)$ = price of 1 gal of gasoline

45. Find: **a.** $r(1954)$ **b.** $m(1954)$

46. Find: **a.** $g(1944)$ **b.** $c(1984)$

47. Find $s(1984) - s(1944)$.

48. Find $b(1984) - b(1944)$.

49. a. Find the change in the price of eggs from 1944 to 1984.
    b. Use functional notation to write the change in the price of eggs.

50. a. Find the change in the price of round steak from 1944 to 1984.
    b. Use functional notation to write the change in the price of round steak.

51. a. Find $\dfrac{g(1944 + 40) - g(1944)}{40}$.
    b. What does the expression in part a mean?

52. a. Find $\dfrac{m(1944 + 40) - m(1944)}{40}$

    b. What does the expression in part a mean?

53. a. What is the average increase in the price of sugar per year from 1944 to 1954? Use functional notation.

    b. What is the average increase in the price of sugar per year from 1944 to 1964? Use functional notation.

    c. What is the average increase in the price of sugar per year from 1944 to 1974? Use functional notation.

    d. What is the average increase in the price of sugar per year from 1944 to 1984? Use functional notation.

    e. What is the average increase in the price of sugar per year from 1944 to $1944 + h$, where h is an unspecified number of years? Use functional notation.

54. Repeat Problem 53 for coffee instead of sugar.

55. Use $A(x) = 0.106x - 0.0015x^3$ from Example 11 to estimate the amount of alcohol in the bloodstream after 2 hours.

56. Use $A(x) = 0.106x - 0.0015x^3$ from Example 11 to estimate the amount of alcohol in the bloodstream after $\frac{1}{2}$ hour.

57. Use $P(t) = 230(1.02)^t$ from Example 12 to estimate the U.S. population in the year 2001.

58. Use $P(t) = 230(1.02)^t$ from Example 12 to estimate the U.S. population at the turn of the next century.

59. According to the U.S. Public Health Service, the number of marriages in the United States in 1977 was about 2,176,000 and in 1982 it was about 2,495,000. If $M(x)$ represents the number of marriages in year x,

    a. Find $\dfrac{M(1982) - M(1977)}{5}$.

    b. Verbally describe $\dfrac{M(1977 + h) - M(1977)}{h}$.

60. According to the U.S. Public Health Service, the number of divorces in the United States in 1977 was about 1,097,000 and in 1982 it was about 1,180,000. If $D(x)$ represents the number of divorces in year x,

    a. Find $\dfrac{D(1982) - D(1977)}{5}$.

    b. Verbally describe $\dfrac{D(1977 + h) - D(1977)}{h}$.

2.3 Graphs of Functions

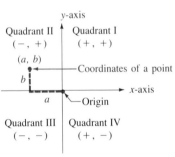

Figure 2.1 Cartesian coordinate system

A **two-dimensional coordinate system** consists of two perpendicular coordinate lines in a plane. Usually one of the coordinate lines is horizontal with the positive direction to the right; the other is vertical with the positive direction upward. These coordinate lines are called **coordinate axes**, and the point of intersection is called the **origin**. Note in Figure 2.1 that the axes divide the plane into four parts called the **first, second, third**, and **fourth quadrants**. This two-dimensional coordinate system is also called a **Cartesian coordinate system** in honor of René Descartes, who was the first to describe it in mathematical detail.

Points in a plane are denoted by ordered pairs. The term *ordered pair* refers to two real numbers represented by (a, b), where a is the **first component** and b is the **second component**. The order in which the components are listed is important since $(a, b) \neq (b, a)$ if $a \neq b$.

The horizontal number line is called the **x-axis** (or the *axis of abscissas*), and x represents the first component of the ordered pair. The set of values for x is the **domain**. The vertical number line is called the **y-axis** (or the *axis of ordinates*), and y represents the second component of the ordered pair. The set of values for y is called the **range**. The plane determined by the x- and y-axes is called the *coordinate plane*, *Cartesian plane*, or *xy-plane*. When we refer to a point (x, y), we mean a point in the coordinate plane whose abscissa is x and whose ordinate is y. To **plot a point (x, y)** means to locate the point with coordinates (x, y) in the plane and represent its location by a dot. (In this book, if variables other than x and y are used, you will be told which represents the first and which the second component.)

To **graph a relation** means to draw a picture of the ordered pairs that *satisfy* the equation in a one-to-one fashion. This process of graphing is shown for three different types of models.

EXAMPLE 1 Graph $f(x) = 2x - 5$.

Solution Let $y = f(x)$ so that we can speak about the ordered pair (x, y) instead of the more notationally cumbersome form $(x, f(x))$. One method for graphing a function, or a relation, is to plot the ordered pairs that satisfy the equation. That is, *you* choose values for x (the first component) and calculate, or find, the corresponding values for y (the second component): Let

$x_1 = 0$:	$f(0) = y_1 = 2 \cdot 0 - 5 = -5$	The ordered pair is $(0, -5)$
$x_2 = 2$:	$f(2) = y_2 = 2 \cdot 2 - 5 = -1$	The ordered pair is $(2, -1)$
$x_3 = 4$:	$f(4) = y_3 = 2 \cdot 4 - 5 = 3$	The ordered pair is $(4, 3)$
$x_4 = -2$:	$f(-2) = y_4 = 2 \cdot (-2) - 5 = -9$	The ordered pair is $(-2, -9)$

Plot the points as shown in Figure 2.2. After you have plotted enough points to see the curve's general shape, connect the points to obtain the curve.

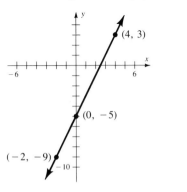

These points seem to lie on a line. In Section 2.4 we will graph lines much more efficiently than we did here.

Figure 2.2 Graph of $f(x) = 2x - 5$

EXAMPLE 2 Graph $g(x) = x^2 + 4x + 3$.

Solution Let $y = g(x)$ and

$x_1 = 0$:	$g(0) = y_1 = 0^2 + 4 \cdot 0 + 3 = 3$	Ordered pair $(0, 3)$
$x_2 = 1$:	$g(1) = y_2 = 1^2 + 4 \cdot 1 + 3 = 8$	$(1, 8)$
$x_3 = 2$:	$g(2) = y_3 = 2^2 + 4 \cdot 2 + 3 = 15$	$(2, 15)$
$x_4 = -1$:	$g(-1) = y_4 = (-1)^2 + 4(-1) + 3 = 0$	$(-1, 0)$
$x_5 = -2$:	$g(-2) = y_5 = (-2)^2 + 4(-2) + 3 = -1$	$(-2, -1)$
$x_6 = -3$:	$g(-3) = y_6 = (-3)^2 + 4(-3) + 3 = 0$	$(-3, 0)$
$x_7 = -4$:	$g(-4) = y_7 = (-4)^2 + 4(-4) + 3 = 3$	$(-4, 3)$

Plot the points and draw the curve as shown in Figure 2.3.

These points do not lie on a line. If the points are connected they form a curve called a **parabola**. We will learn how to efficiently graph parabolas in Section 2.5.

Figure 2.3 Graph of $g(x) = x^2 + 4x + 3$

Sketch the graph of $h(x) = 1/x$.

Solution Let $y = h(x)$ and

$x_1 = 0$: $y_1 = \frac{1}{0}$ But $y_1 = \frac{1}{0}$ is not defined, so $x = 0$ is not in the domain—*there is no point on the graph corresponding to $x = 0$.*

$x_2 = 1$: $y_2 = \frac{1}{1} = 1$ The ordered pair is $(1, 1)$

$x_3 = 2$: $y_3 = \frac{1}{2}$ $(2, \frac{1}{2})$

$x_4 = 3$: $y_4 = \frac{1}{3}$ $(3, \frac{1}{3})$

$x_5 = -1$: $y_5 = 1/(-1) = -1$ $(-1, -1)$

$x_6 = -2$: $y_6 = 1/(-2) = -\frac{1}{2}$ $(-2, -\frac{1}{2})$

$x_7 = -3$: $y_7 = 1/(-3) = -\frac{1}{3}$ $(-3, -\frac{1}{3})$

If you plot the ordered pairs as shown in Figure 2.4a, you see that they cannot be connected with either a line or a parabola. In fact, you cannot connect all of these points with any smooth curve because no point corresponds to $x = 0$. Consider some additional points close to zero:

$x = \frac{1}{2}$: $y = 1/\frac{1}{2} = 2$ $(\frac{1}{2}, 2)$

$x = \frac{1}{3}$: $y = 1/\frac{1}{3} = 3$ $(\frac{1}{3}, 3)$

$\vdots$ $\vdots$

We can now draw a smooth curve as shown in Figure 2.4b.

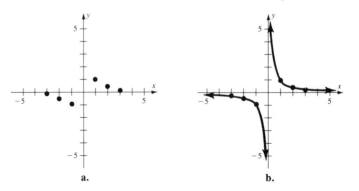

Figure 2.4 Graph of $h(x) = \frac{1}{x}$. This curve is a type of curve called a **rational function**. These curves are discussed in Section 2.6.

a.   b.   ■

To summarize, the **graph of a function** f is the set of all points $(x, f(x))$ in a coordinate plane, where x is in the domain of f. That is, the graph of f can be described as the set of all points with coordinates (x, y) such that $y = f(x)$. The graph of a typical function f is shown in Figure 2.5.

Note that the graph of a function is such that for each a in the domain there is only *one point* $(a, f(a))$ on the graph. This means that every vertical line passes through the graph of a function in at most one point. This is the so-called **vertical line test** for the graphs of functions as illustrated in Example 4.

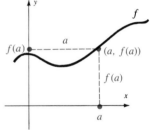

Figure 2.5 Graph of a function

EXAMPLE 4 Which of the following graphs are functions?

Solution

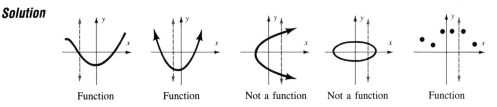

| Function | Function | Not a function | Not a function | Function |

Imagine a vertical line sweeping from left to right across the plane—if it passes through more than one point of the graph at one time, then the graph does not represent a function. ∎

There are some points on a graph that are of particular importance to us—the **intercepts**.

Intercepts

> If the number zero is in the domain of f, then $f(0)$ is called the **y-intercept** of the graph of f and is the point $(0, f(0))$. This is the point where the graph crosses the y-axis.
> If a is a real number in the domain such that $f(a) = 0$, then a is called an **x-intercept** and is the point $(a, 0)$. This is the point where the graph crosses the x-axis. Any number x such that $f(x) = 0$ is called a **zero of the function**.

EXAMPLE 5 Find the domain, range, and intercepts for f defined by the graph below.

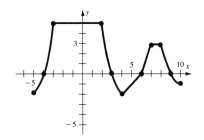

Solution Domain: $-5 \le x \le 10$

Range: $-2 \le y \le 5$
The zeros of the function
are -4, 3, 6, and 9.

y-intercept: $(0, 5)$; we usually simply say that the *y*-intercept is 5. A function will not have more than one *y*-intercept.

x-intercepts: $(-4, 0)$, $(3, 0)$, $(6, 0)$, and $(9, 0)$. ∎

Problem Set 2.3

1. Define the graph of a function f.
2. What does it mean when we say the ordered pair (a, b) satisfies the equation $y = f(x)$?
3. What are the x- and y-intercepts of the graph of a function f?
4. What is the zero of a function f?

Use Figure 2.6a to find the coordinates of the points in Problems 5–7.

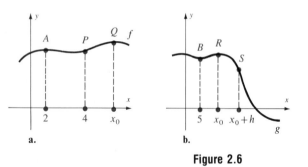

a.

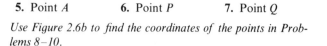

b.

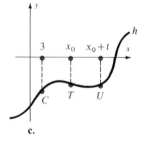

c.

Figure 2.6

5. Point A **6.** Point P **7.** Point Q

Use Figure 2.6b to find the coordinates of the points in Problems 8–10.

8. Point B **9.** Point R **10.** Point S

Use Figure 2.6c to find the coordinates of the points in Problems 11–13.

11. Point C **12.** Point T **13.** Point U

Find the domain, range, and intercepts of the relations defined by the graphs in Problems 14–19. Also state whether the graph defines a function.

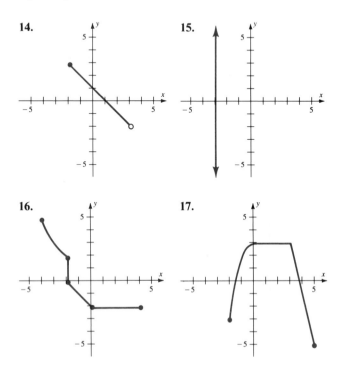

14.

15.

16.

17.

18.

19.

Graph the functions in Problems 20–40 by plotting points.

20. $f(x) = 3x + 1$ **21.** $f(x) = 2x + 3$
22. $f(x) = x - 4$ **23.** $f(x) = 6 - 2x$
24. $f(x) = 1 - x$ **25.** $f(x) = -3x - 1$
26. $g(x) = 2x^2$ **27.** $g(x) = \frac{1}{2}x^2$
28. $g(x) = \frac{1}{10}x^2$ **29.** $g(x) = x^2 + 4x + 4$
30. $g(x) = x^2 + 6x + 9$ **31.** $g(x) = x^2 - 6x + 9$
32. $g(x) = x^2 + 2x - 3$ **33.** $g(x) = 2x^2 - 4x + 5$
34. $g(x) = 2x^2 - 4x + 4$ **35.** $h(x) = -1/x$
36. $h(x) = 3/x$ **37.** $h(x) = -2/x$
38. $h(x) = 1/(x - 1)$ **39.** $h(x) = 2/(x - 2)$
40. $h(x) = -3/(x + 2)$

APPLICATIONS

41. A theater has a capacity of 1500 seats. The formula relating the number of adult tickets sold, a, and the number of children's tickets sold, c, is $a + c = 1500$. Graph (a, c) satisfying this relationship. Assume that both a and c are positive.

42. The taxable income for certain taxpayers is given by the formula $T = 0.85I - 4000$. Graph (I, T) for $18{,}200 \le I \le 23{,}500$.

43. The pressure, P, in centimeters of mercury, is given as a function of the depth in meters, d, under water, by using the formula $P = 0.4d + 7.6$. Graph (d, P) for $0 \le d \le 10$.

44. A supply company finds that the number of computer disks sold in year x is given by the function $s(x) = 5000 + x^2$, where $x = 0$ corresponds to 1980. Graph the sales for the years 1980 to 1990 (inclusive).

45. The number of responses n (per milliseconds) of a nerve is a function of the length of time t (in milliseconds) since the nerve was stimulated. For a certain nerve, this relationship is $n = 150 - (t - 10)^2$. Graph (t, n) for $0 \le t \le 22$.

46. The cost C (in thousands of dollars) of removing p percent of a certain pollutant is given by the formula $C = 20p/(105 - p)$. Graph (p, C) for $0 \le p \le 100$.

2.4 Linear Functions

This section reviews material from previous courses and also provides the notation and concepts required in the remainder of this book. Recall that a first-degree equation with two variables is called a *linear equation*.

Linear Function

> A function f is a *linear function* if it can be written in the form
>
> $$f(x) = mx + b \quad \text{or} \quad y = mx + b$$
>
> where m and b are real numbers.

If $m = 0$, then $f(x) = b$, and the graph of $f(x) = b$ is a **horizontal line** as shown in Figure 2.7a. Functions whose graphs are horizontal lines are called **constant functions**. Now let (x_1, y_1) and (x_2, y_2) be any points on a line so that $x_1 = x_2$. Then this line is parallel to the y-axis and is called a **vertical line**, as shown in Figure 2.7b. Note that vertical lines are not the graphs of functions. Vertical lines always have the form $x = c$, where c is some constant ($x = 3$, for example).

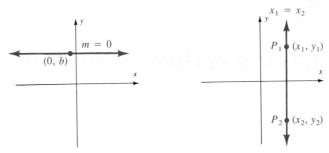

Figure 2.7a Horizontal line   **Figure 2.7b** Vertical line

A line is determined by two points, so if we know *any* two points on a line we can draw the line by using a straightedge and the given points. In the last section we evaluated formulas to find those points. However, two points that are usually easy to find are the intercepts:

To find the y-intercept: Let $x = 0$ and solve for y.

To find the x-intercept: Let $y = 0$ and solve for x.

EXAMPLE 1 Graph $f(x) = -\frac{3}{2}x + 3$ by plotting the x- and y-intercepts.

Solution If $f(x) = -\frac{3}{2}x + 3$, write $y = -\frac{3}{2}x + 3$.

Let $x = 0$: $\quad y = -\frac{3}{2}(0) + 3$

$\qquad\qquad\quad y = 3 \qquad$ The point $(0, 3)$ is the y-intercept

Let $y = 0$: $\quad 0 = -\frac{3}{2}x + 3$

$\qquad\qquad\quad \frac{3}{2}x = 3$

$\qquad\qquad\quad x = 2 \qquad$ The point $(2, 0)$ is the x-intercept

Draw the line passing through the plotted intercepts as shown in Figure 2.8. ■

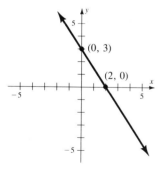

Figure 2.8 Graph of $f(x) = \frac{-3}{2}x + 3$

The constants b and m in the form $f(x) = mx + b$ give us important information about the line we wish to graph. For example, if we let $y = f(x)$, we can write $y = mx + b$. Now we can find the y-intercept. Let $x = 0$:

$$y = m \cdot x + b$$
$$y = b$$

This means that the y-intercept is $(0, b)$. We generally shorten the notation and simply say the y-intercept is b to mean the line passes through the y-axis at the point $(0, b)$.

The second constant m, which is the coefficient of x when the equation is solved for y, tells us the steepness or **slope** of a line. But, first, we need to define what we mean by the slope of a line. This definition requires that we know what is meant by the vertical change (*rise*) relative to a horizontal change (*run*). These ideas are illustrated in Figure 2.9.

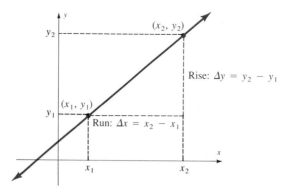

Figure 2.9 Slope of a line

Let Δx represent the horizontal change and Δy represent the vertical change.*
Note that $\Delta x = x_2 - x_1$ and $\Delta y = y_2 - y_1$.

---

* Δx is one symbol (not the multiplication of two variables) and is pronounced "delta x" (Δy is pronounced "delta y").

Slope

> Let (x_1, y_1) and (x_2, y_2) be points on a line such that $x_1 \neq x_2$. Then
>
> $$\text{slope} = \frac{\text{vertical change}}{\text{horizontal change}} = \frac{\Delta y}{\Delta x} \quad \text{or} \quad \frac{\text{rise}}{\text{run}} = \frac{y_2 - y_1}{x_2 - x_1} = \frac{\Delta y}{\Delta x}$$
>
> If $\Delta x = 0$, then the line is vertical and has *no* slope ($\Delta y / 0$ is undefined).
> If $\Delta y = 0$, then the line is horizontal and has *zero* slope ($0 / \Delta x = 0$).

To show that m in the equation $y = mx + b$ is the slope, consider the line specified by the equation $y = mx + b$, which passes through (x_1, y_1) and (x_2, y_2), with $x_1 \neq x_2$. This means that $y_1 = mx_1 + b$ and $y_2 = mx_2 + b$, so that

$$\text{slope} = \frac{\Delta y}{\Delta x} = \frac{y_2 - y_1}{x_2 - y_1}$$

$$= \frac{(mx_2 + b) - (mx_1 + b)}{x_2 - x_1} \qquad \text{Substitution}$$

$$= \frac{mx_2 - mx_1}{x_2 - x_1}$$

$$= \frac{m(x_2 - x_1)}{x_2 - x_1}$$

$$= m$$

This discussion tells us that we can find the y-intercept and slope of linear equations of the form $y = mx + b$ by inspection, as shown by Example 2.

EXAMPLE 2   Find the slope and y-intercept.

	Slope	*y-intercept*	
a. $y = \frac{1}{2}x + 3$	$m = \frac{1}{2}$	$b = 3$	By inspection
b. $y = x - 3$	$m = 1$	$b = -3$	By inspection
c. $y = -\frac{2}{3}x + \frac{5}{2}$	$m = -\frac{2}{3}$	$b = \frac{5}{2}$	By inspection
d. $3x + 4y + 8 = 0$			

Solve for y: $4y = -3x - 8$

$$y = -\tfrac{3}{4}x - \tfrac{8}{4}$$

Thus the slope $m = -\frac{3}{4}$, and the y-intercept $b = -\frac{8}{4} = -2$ ∎

Since the slope and y-intercept are easy to find after the linear equation is solved for y, we give a special name to this form of the equation.

Slope-Intercept Form of the
Equation of a Line

> The **slope-intercept** form of a linear equation is
>
> $$y = mx + b$$
>
> and the graph of this equation is the line having slope m and y-intercept b.

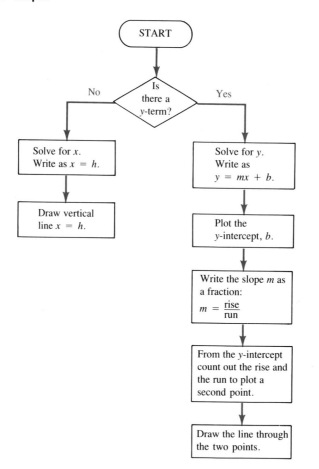

Figure 2.10 Procedure for graphing a line by the slope-intercept method

This form of the equation of a line can be used for graphing certain lines when it is not convenient to plot points. The procedure is summarized in Figure 2.10. Carefully study this procedure.

EXAMPLE 3   Graph $y = \frac{1}{2}x + 3$.

Solution   By inspection, the y-intercept is 3 and the slope is $\frac{1}{2}$; the line is graphed by first plotting the y-intercept $(0, 3)$ and then finding a second point by counting out the slope: over 2 and up 1.

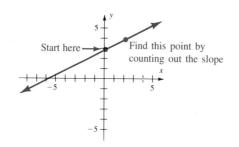

EXAMPLE 4 Graph $3x - 2y - 6 = 0$.

Solution Solve for y:

$$2y = 3x - 6$$
$$y = \tfrac{3}{2}x - 3$$

The y-intercept is -3 and the slope is $\tfrac{3}{2}$; the line is graphed by first plotting the y-intercept $(0, -3)$ and then finding a second point by counting out the slope (over 2, up 3).

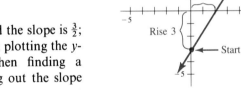

EXAMPLE 5 Graph $4x + 2y - 5 = 0$ for $-1 \le x \le 3$.

Solution Solve for y:

$$2y = -4x + 5$$
$$y = -2x + \tfrac{5}{2}$$

Because of the restriction on the domain, we do not want to graph the entire line, just that part with first components as specified by the restriction namely $-1 \le x \le 3$. The usual convention is to show the entire line as a dashed line and the answer as a solid line. For this example the y-intercept is $\tfrac{5}{2}$ and the slope is -2; the line is shown as a dashed line. Because of the restriction of the domain, the part of the line with x values between -1 and 3 (inclusive) is shown as a solid line segment.

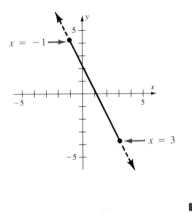

Sometimes a function will have a different equation for different parts of its domain, as shown in Example 6. If each of those equations is a line segment, then the function is called a **piecewise linear function**.

EXAMPLE 6 Graph $y = x$ if $x \ge 0$
$\qquad\qquad y = -x$ if $x < 0$

Solution Graph each of the line segments as shown in the figure.

Remember from algebra the definition of absolute value:

$$|x| = x \quad \text{if} \quad x \ge 0$$

and

$$|x| = -x \quad \text{if} \quad x < 0$$

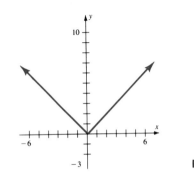

Thus the function in Example 6, the **absolute value function**, can be written $y = |x|$.

EXAMPLE 7   Graph $y = |6 - 2x|$.

Solution   We can write this function without absolute value symbols by considering the two parts of the definition of absolute value.

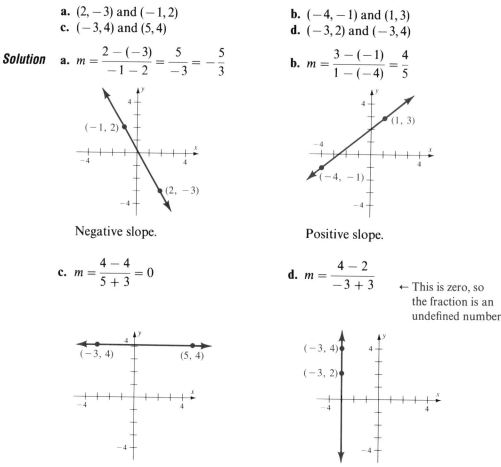

$$y = 6 - 2x \quad \text{if} \quad 6 - 2x \geq 0$$
$$-2x \geq -6$$
$$x \leq 3$$

See Section 1.5 for a review of solving linear inequalities

And

$$y = -(6 - 2x) \quad \text{if} \quad 6 - 2x < 0$$
$$x > 3$$

Note that $y = -(6 - 2x)$ is the same as $y = 2x - 6$

This function is shown at the left.   ∎

If we are given two points, it is also possible to find the slope by using $m = \Delta y / \Delta x$.

EXAMPLE 8   Sketch the line passing through the points whose coordinates are given and then find the slope of each line.

a. $(2, -3)$ and $(-1, 2)$ **b.** $(-4, -1)$ and $(1, 3)$
c. $(-3, 4)$ and $(5, 4)$ **d.** $(-3, 2)$ and $(-3, 4)$

Solution   **a.** $m = \dfrac{2 - (-3)}{-1 - 2} = \dfrac{5}{-3} = -\dfrac{5}{3}$ **b.** $m = \dfrac{3 - (-1)}{1 - (-4)} = \dfrac{4}{5}$

Negative slope.   Positive slope.

c. $m = \dfrac{4 - 4}{5 + 3} = 0$ **d.** $m = \dfrac{4 - 2}{-3 + 3}$ ← This is zero, so the fraction is an undefined number

0 slope; horizontal line.   Undefined slope; vertical line.   ∎

In constructing mathematical models it is often necessary to write a linear equation using available or given information about the line. Example 9 shows how to do this if you know the slope and y-intercept.

EXAMPLE 9   Find the equation of the line with y-intercept 5 and slope $-\frac{2}{3}$.

Solution   Use the equation $y = mx + b$, where $b = 5$ and $m = -\frac{2}{3}$.

$$y = -\frac{2}{3}x + 5$$

More often than not, unfortunately, when you need to find the equation of a line you will not know the y-intercept. You will, however, know a point and the slope or two points. In these cases it is easier to use another form of the equation of a line called the **point-slope form**. It is easy to derive this equation if we remember that

$$m = \frac{y_2 - y_1}{x_2 - x_1}$$

Since (x, y) is any point on the line passing through (x_1, y_1), we have

$$m = \frac{y - y_1}{x - x_1}$$

Thus, by multiplying both sides by $x - x_1$, we obtain

$$m(x - x_1) = y - y_1$$

Point-Slope Form of the Equation of a Line

A nonvertical line having slope m and passing through (x_1, y_1) has the equation

$$y - y_1 = m(x - x_1)$$

EXAMPLE 10   Find the equation of the line with slope 3 passing through $(-2, -5)$.

Solution   Use the equation $y - y_1 = m(x - x_1)$, where $m = 3$, $x_1 = -2$, and $y_1 = -5$:

$$y - (-5) = 3[x - (-2)]$$
$$y + 5 = 3(x + 2)$$

If you know two points and want the equation, first find the slope and *then* use the point-slope form.

EXAMPLE 11   Find the equation of the line passing through $(-2, 3)$ and $(4, -1)$.

Solution   First find the slope:

$$m = \frac{\Delta y}{\Delta x} = \frac{-1 - 3}{4 - (-2)} = \frac{-4}{6} = -\frac{2}{3}$$

Now use the point-slope form (you can use *either* of the given points):

$$(-2, 3): \quad y - 3 = -\frac{2}{3}(x + 2) \quad \text{or} \quad (4, -1): \quad y + 1 = -\frac{2}{3}(x - 4)$$

It is not easy to see that the equations in Example 11 are the same. For this reason, we are often asked to algebraically manipulate the answers into the same form. This form is called the **standard form** of the equation of a line.

Standard Form of the Equation of a Line

> The *standard form* of the equation of a line is
>
> $$Ax + By + C = 0$$
>
> where (x, y) is any point on the line, and A, B, and C are constants (A and B not both zero).

EXAMPLE 12   Change the point-slope forms given in Example 11 to standard form.

Solution

$$y - 3 = -\tfrac{2}{3}(x + 2)$$
$$3(y - 3) = -2(x + 2)$$
$$3y - 9 = -2x - 4$$
$$2x + 3y - 5 = 0$$

Eliminate fractions
(multiply by 3)
Eliminate parentheses
Obtain a 0 on the right

$$y + 1 = -\tfrac{2}{3}(x - 4)$$
$$3(y + 1) = -2(x - 4)$$
$$3y + 3 = -2x + 8$$
$$2x + 3y - 5 = 0$$

∎

Note that both equations given in Example 11 are the same in standard form.

EXAMPLE 13   Find the equation of a line passing through $(7, -2)$ with no slope.

Solution   Do not confuse "no slope" (vertical line) with "zero slope" (horizontal line). Since the line is vertical, the equation has the form $x = h$ when it passes through (h, k). Thus $x = 7$ is the equation. In standard form, $x - 7 = 0$.   ∎

A mathematical model is often constructed by assuming that the relationship between two variables is linear and then writing an equation using two known data points, as illustrated in Example 14. When relating two values in an applied setting, one value will often depend on the other, so it is customary to designate the dependent one as the **dependent variable**, and the other as the **independent variable**. When x and y are used to represent the variables, x is the independent variable and y the dependent variable.

EXAMPLE 14   Assume that the demand for a Jane Fonda video is linearly related to its price. Market research shows that at $30 about 3 million tapes will be sold, but at $60 only 1 million will be sold. Write an equation in standard form to represent this information.

Solution   Since the number sold seems to depend on the price, let $x =$ price (the independent variable) and find the corresponding y-value (the number of tapes, in millions; the dependent variable). Then the data points are $(30, 3)$ and $(60, 1)$. First find the slope:

$$m = \frac{1 - 3}{60 - 30} = \frac{-2}{30} = \frac{-1}{15}$$

Use the point-slope form to find the equation

$$y - 3 = \tfrac{-1}{15}(x - 30)$$

Finally, find the standard form of the equation:

$$15y - 45 = -x + 30$$
$$x + 15y - 75 = 0$$

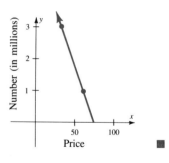

∎

Parametric Form of the Equation of a Line

Sometimes it is convenient to define x and y in terms of another variable, say, t. For example, suppose that the location of a point (x, y) is defined in terms of time, t (in seconds), as follows:

$$x = 1 + t \quad \text{and} \quad y = 2t$$

These equations must be interpreted by saying that although the first component has a 1 second "head start," the rate at which the second component is changing (with respect to time) is twice as fast as the first component. We can tabulate the values of x and y by choosing values for t, where $t \geq 0$ (since time cannot be negative).

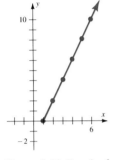

Figure 2.11 Graph of $x = 1 + t, y = 2t$

t	0	1	2	3	4	5
x	1	2	3	4	5	6
y	0	2	4	6	8	10

The variable t is called a **parameter**, and the equations $x = 1 + t$ and $y = 2t$ are called **parametric equations**. From the graph it looks like it is a line. We can prove this by *eliminating the parameter*. We solve the first equation for t ($t = x - 1$) and substitute the result into the second equation:

$$y = 2t = 2(x - 1) = 2x - 2$$

which is a linear equation.

Parametric Form of the Equation of a Line

> The graph of the parametric equations
>
> $$x = x_1 + at \quad \text{and} \quad y = y_1 + bt$$
>
> is a line passing through (x_1, y_1) with slope $m = b/a$.

It is easy to derive this result by solving one of the equations for t and substituting the result into the other equation.

$$x - x_1 = at$$

$$\frac{x - x_1}{a} = t$$

Now, by substitution,

$$y = y_1 + bt = y_1 + b\left(\frac{x - x_1}{a}\right)$$

$$y - y_1 = \frac{b}{a}(x - x_1)$$

This is the equation of a line passing through (x_1, y_1) with slope b/a.

We conclude by summarizing the various forms of linear equations.

Forms of a Linear Equation

STANDARD FORM:	$Ax + By + C = 0$	(x, y) is any point on the line; A, B, and C are constants; A and B are not both zero
SLOPE-INTERCEPT FORM:	$y = mx + b$	m is the slope; b is the y-intercept
POINT-SLOPE FORM:	$y - y_1 = m(x - x_1)$	m is the slope; (x_1, y_1) is the known point
HORIZONTAL LINE:	$y = k$	(h, k) is a point on the line
VERTICAL LINE:	$x = h$	(h, k) is a point on the line
PARAMETRIC FORM:	$x = x_1 + at$ $y = y_1 + bt$	(x_1, y_1) is a known point; $\frac{b}{a}$ is the slope

Problem Set 2.4

Find the x- and y-intercepts for the lines whose equations are given in Problems 1–8.

1. $y = 2x + 4$

2. $y = 5x - 10$

3. $4x + 3y + 4 = 0$

4. $3x + 2y - 9 = 0$

5. $100x - 250y + 500 = 0$

6. $2x - 5y - 1200 = 0$

7. $y + 2 = 0$

8. $x - 2 = 0$

Find the slope of the line passing through the points in Problems 9–14.

9. $(2, 3)$ and $(5, 4)$

10. $(4, -1)$ and $(-2, 3)$

11. $(5, 2)$ and $(-2, -3)$

12. $(-2, -3)$ and $(4, 5)$

13. $(-2, -3)$ and $(-1, -2)$

14. $(-3, -1)$ and $(-7, -10)$

Find the slope and y-intercept in Problems 15–23.

15. $y = 2x + 4$

16. $y = 5x - 3$

17. $y = 9x + 1$

18. $4x + 3y + 4 = 0$

19. $2x - 3y + 5 = 0$

20. $5x - 2y - 5 = 0$

21. $y - 5 = 0$

22. $y + 9 = 0$

23. $x - 3 = 0$

Graph the lines of the equations in Problems 24–35 by finding the slope and y-intercept.

24. $y = 3x + 4$

25. $y = 2x - 5$

26. $y = -3x + 1$

27. $y = -\frac{1}{4}x + 2$

28. $y = -\frac{2}{3}x - 4$

29. $y = \frac{3}{5}x + \frac{2}{5}$

30. $3x - y + 2 = 0$

31. $x + 3y - 9 = 0$

32. $2x - 3y + 15 = 0$

33. $x = \frac{2}{3}y$

34. $y - 3 = 0$

35. $2x + 5 = 0$

Graph the line segments or piecewise functions given in Problems 36–45.

36. $y = 4x - 2$ $-3 \le x \le 4$

37. $y = -3x + 2$ $-4 \le x \le 3$

38. $2x + 5y + 10 = 0$ $-1 \le x \le 5$

39. $3x - 2y + 8 = 0$ $-3 \le x \le 4$

40. $2x + y + 5 = 0$ if $-3 \le x \le 0$
$y + 5 = 0$ if $0 < x < 3$
$x - y - 8 = 0$ if $3 \le x \le 10$

41. $y = 3x + 2$ if $0 \le x \le 2$
$y - 8 = 0$ if $2 < x < 5$
$x - 2y + 11 = 0$ if $x \ge 5$

42. $y = 2|x|$

43. $y = |3x - 6|$

44. $y = |2x + 4|$

45. $y = -3|x|$

Graph the lines in Problems 46–57 from their parametric equations.

46. $x = 2 + 3t$
$y = 1 + t$

47. $x = 1 + 2t$
$y = 2 + t$

48. $x = 5t$
$y = 2 + t$

49. $x = 3t$
$y = 1 + 2t$

50. $x = 1 - 2t$
$y = 5 + 3t$

51. $x = 2 - 2t$
$y = 1 + t$

52. $x = -1 + t$
$y = 3 - 2t$

53. $x = -2 + t$
$y = 4 - 5t$

54. $x = -3 + 4t$
$y = -1 - 3t$

55. $x = -2 - 3t$
$y = -2 + 5t$

56. $x = 6 - 2t$
$y = -5 + 3t$

57. $x = 5 - t$
$y = -3 - 2t$

Find the standard form of the equation of the line satisfying the conditions given in Problems 58–69.

58. y-intercept 5; slope 6

59. y-intercept -3; slope -2

60. y-intercept 0; slope 0

61. y-intercept 4; slope 0

62. slope 2; passing through $(4, 3)$

63. slope -1; passing through $(-3, 5)$

64. slope $\frac{1}{2}$; passing through $(5, 3)$

65. slope $\frac{3}{5}$; passing through $(4, -3)$

66. passing through $(-3, -1)$ and $(3, 2)$

67. passing through $(5, 6)$ and $(1, -2)$

68. passing through $(4, -2)$ and $(4, 5)$

69. passing through $(5, 6)$ and $(7, 6)$

APPLICATIONS

Problems 70–76 provide some real world examples of line graphs. One way of finding the equation of the line is to write two data points from the given information (as shown in Example 14) and then to use those points to write the equation. Use the given information to write an equation in standard form of the line described by the problem.

70. The demand for a certain product is related to the price of the item. Suppose a new line of stationery is tested at two stores. At store A, 25 boxes are sold within a month at $5 each, and, at store B, 15 boxes priced at $10 each are sold during the same time. Let x be the price and y be the number of boxes sold.

71. An important factor related to the demand for a product is its supply. The amount of stationery in Problem 70 that can be supplied is also related to the price. At $5 each, 10 boxes can be supplied, and, at $10, 20 boxes can be supplied. Let x be the price and y be the number of boxes sold.

72. The population of Florida in 1970 was roughly 6.8 million, and in 1980 it was 9.7 million. Let x be the year (let 1960 be the base year; that is, $x = 0$ represents 1960, so $x = 10$

represents 1970) and y be the population. Use this equation to predict the population in 1990.

73. The population of Texas in 1970 was roughly 11.2 million, and in 1980 it was 14.2 million. Let x be the year (let 1960 be the base year; that is, $x = 0$ represents 1960, so $x = 10$ represents 1970) and y be the population. Use this equation to predict the population in 1990.

74. It costs $90 to rent a car driven 100 miles and $140 if it's driven 200 miles.

75. It costs $60 to rent a car driven 50 miles and $60 if it's driven 260 miles.

76. Suppose it costs $100 for the maintenance and repairs of a 3-year-old car driven 1000 miles and $650 for maintenance and repairs if it's driven 6500 miles.

Use the following 1985 U.S. Tax Rate Schedule to answer the questions in Problems 77–81.

Schedule X
Single Taxpayers

If the amount on Form 1040, line 37 is:		Enter on Form 1040, line 38	of the amount
Over—	But not over—		over—
$0	$2,300	—0—	
2,300	3,400	.... 11%	$2,300
3,400	4,400	$121 + 13%	3,400
4,400	8,500	251 + 15%	4,400
8,500	10,800	866 + 17%	8,500
10,800	12,900	1,257 + 19%	10,800
12,900	15,000	1,656 + 21%	12,900
15,000	18,200	2,097 + 24%	15,000
18,200	23,500	2,865 + 28%	18,200
23,500	28,800	4,349 + 32%	23,500
28,800	34,100	6,045 + 36%	28,800
34,100	41,500	7,953 + 40%	34,100
41,500	55,300	10,913 + 45%	41,500
55,300	....	17,123 + 50%	55,300

77. Graph the tax for income from $60,000 to $100,000.

78. Write the equation for the tax (T) for income (I) from $55,300 to $100,000.

79. Write the equation for the tax (T) for income (I) from $15,000 to $18,200.

80. Write the equation for the tax (T) for income (I) from $20,000 to $25,000.

81. Write the equation for the tax (T) for income (I) from $25,000 to $30,000.

2.5 Quadratic and Polynomial Functions

While many real world situations can be described by the linear models discussed, many others cannot. It is therefore important to be able to build many *different* types of models. In this section, we discuss a model that requires a second-degree equation for its description.

Quadratic Function

> A function f is a **quadratic function** if
>
> $$f(x) = ax^2 + bx + c$$
>
> where a, b, and c are real numbers and $a \neq 0$.

Note that if we write $f(x)$ as y, the quadratic function has one first-degree variable and one second-degree variable. If $b = c = 0$, however, the quadratic function has the form

$$y = ax^2$$

and has a graph called a **standard position parabola**.

EXAMPLE 1 Sketch the graph of $y = 2x^2$.

Solution Begin by finding some ordered pairs that satisfy the equation. Let

$$x_1 = 0; \quad \text{then } y_1 = 2(0)^2 = 0; \quad \text{the point is } (0,0)$$
$$x_2 = 1; \quad \text{then } y_2 = 2(1)^2 = 2; \quad \text{the point is } (1,2)$$
$$x_3 = -1; \quad \text{then } y_3 = 2(-1)^2 = -2; \quad \text{the point is } (-1,2)$$
$$x_4 = 2; \quad \text{then } y_4 = 2(2)^2 = 8; \quad \text{the point is } (2,8)$$
$$x_5 = -2; \quad \text{then } y_5 = 2(-2)^2 = -8; \quad \text{the point is } (-2,8)$$
$$\vdots$$

These points are plotted and a smooth curve is drawn through them, as shown in Figure 2.12.

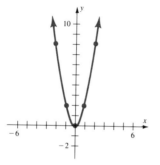

Figure 2.12 Graph of $y = 2x^2$

EXAMPLE 2   Sketch the graph of $y = -\frac{1}{2}x^2$.

Solution   Plot the points satisfying this equation. If

$$x_1 = 0; \quad \text{then } y_1 = 0; \quad \text{the point is } (0,0)$$
$$x_2 = 1; \quad \text{then } y_2 = -\tfrac{1}{2}; \quad \text{the point is } (1, -\tfrac{1}{2})$$
$$x_3 = -1; \quad \text{then } y_3 = -\tfrac{1}{2}; \quad \text{the point is } (-1, -\tfrac{1}{2})$$
$$x_4 = 2; \quad \text{then } y_4 = -2; \quad \text{the point is } (2, -2)$$
$$x_5 = -2; \quad \text{then } y_5 = -2; \quad \text{the point is } (-2, -2)$$
$$x_6 = 3; \quad \text{then } y_6 = -\tfrac{9}{2}; \quad \text{the point is } (3, -\tfrac{9}{2})$$
$$x_7 = 4; \quad \text{then } y_7 = -8; \quad \text{the point is } (4, -8)$$

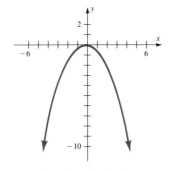

Figure 2.13 Graph of
$y = (-\frac{1}{2})x^2$

We can now make some general observations based on the special case $y = ax^2$.

1. The graph has a characteristic shape called a **parabola**.
2. If $a > 0$, the parabola opens upward; we say it is **concave upward**. If $a < 0$, the parabola opens downward and is **concave downward**.
3. The point $(0,0)$ is the lowest point if the parabola opens upward $(a > 0)$; $(0,0)$ is the highest point if the parabola opens downward $(a < 0)$. This highest or lowest point is called the **vertex**.
4. A parabola is **symmetric** with respect to the vertical line passing through the vertex. This means we can calculate points to the right of the vertex and use symmetry to plot the corresponding points to the left of the vertex.
5. Relative to a fixed scale, the magnitude of a determines the "wideness" of the parabola; small values of $|a|$ yield "wide" parabolas; large values of $|a|$ yield "narrow" parabolas.

For graphs of parabolas of the form

$$y - k = a(x - h)^2$$

the vertex is the point (h, k), and we use this point as the starting point for graphing the parabola. This means that we count out units from the point (h, k) instead of from the origin in order to find new points of the parabola, as illustrated in Example 3.

EXAMPLE 3   Sketch the graph of $y + 3 = 2(x - 2)^2$.

Solution   By inspection, $(h, k) = (2, -3)$ and is the vertex of the parabola. (Compare this equation with the equation graphed in Example 1.) If the only difference between two equations is the (h, k) point, the graphs will be the same except that they will have different vertices, as shown in Figure 2.14a.

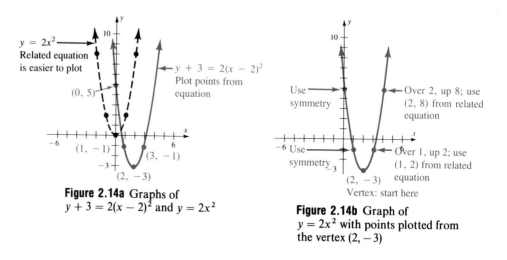

Figure 2.14a Graphs of
$y + 3 = 2(x - 2)^2$ and $y = 2x^2$

Figure 2.14b Graph of
$y = 2x^2$ with points plotted from
the vertex $(2, -3)$

Equation $y + 3 = 2(x - 2)^2$		Related equation $y = 2x^2$.
Let $x_1 = 0$:	$(0, 5)$	$(0, 0)$ vertex
$x_2 = 1$:	$(1, -1)$	$(1, 2)$
$x_3 = -1$:	$(-1, 15)$	$(-1, 2)$
$x_4 = 2$:	$(2, -3)$ vertex	$(2, 8)$
$x_5 = -2$:	$(-2, 29)$	$(-2, 8)$
$x_6 = 3$:	$(3, -1)$	$(3, 18)$

Instead of graphing both of these as in Figure 2.14a, suppose we count out the points of the related equation *from the point* (h, k) *instead of from the origin*, as shown in Figure 2.14b. Note that this graph *is the same as the graph of the more complicated equation.* This leads us to the conclusion that we can graph the simpler related equation if we remember to count out the points from the vertex rather than from the origin. ■

EXAMPLE 4   Sketch the graph of $y - \frac{1}{2} = -2(x + \frac{3}{4})^2$.

Solution   We find and plot the vertex $(-\frac{3}{4}, \frac{1}{2})$; we next plot points on the related equation $y = -2x^2$: If

$x_1 = 0$,	then $y_1 = 0$	Vertex
$x_2 = 1$,	then $y_2 = -2$	Count over 1 and down 2 *from the vertex*
$x_3 = 2$,	then $y_3 = -8$	See Figure 2.15

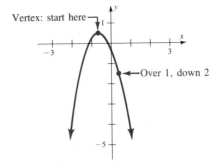

Figure 2.15 Graph of $y - \frac{1}{2} = -2(x + \frac{3}{4})^2$

Parabolas that are not of the form $y - k = a(x - h)^2$ are graphed by plotting points as shown in the last section or by using the techniques of calculus to be developed in Chapter 9. Many applications involve finding the maximum or minimum value of a function f. If f is a quadratic model, then the maximum or minimum value is at the vertex. That is, an equation of the form

$$y - k = a(x - h)^2$$

can be rewritten in functional form as

$$f(x) = a(x - h)^2 + k$$

If $a > 0$, the parabola opens upward. If $x = h$, the minimum value is at the vertex (h, k). [You can see that this is true by looking at the graph or noting that $(x - h)^2$ is necessarily nonnegative for all x and zero if and only if $x = h$.] Similarly, if $a < 0$, the parabola opens downward, and the maximum value is at the vertex (h, k).

EXAMPLE 5   A small manufacturer of citizens-band radios determines that profit is related to the number of items produced. If x items are produced per day, and the maximum number that can be produced is ten items, the profit, in dollars, is determined to be

$$P(x) = -30(x - 6)^2 + 480$$

How many radios should be produced in order to maximize profit?

Solution   The profit function has a graph that opens downward. The maximum profit is found at the vertex of this parabola, as shown in Figure 2.16. The vertex of the function

$$P(x) - 480 = -30(x - 6)^2$$

is $(6, 480)$, so the maximum profit of $480 is obtained when six radios are manufactured. Note also from the graph that if there were a strike and no radios were produced, the daily profit would be -600 (this is a $600-per-day loss).

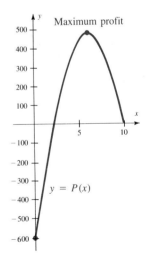

Figure 2.16 Graph of $P(x) = -30(x - 6)^2 + 480$

Break-Even Analysis

Suppose that the cost of producing the radios described in Example 5 is $5x^2 + 40x + 600$. This function,

$$C(x) = 5x^2 + 40x + 600$$

is called a **cost function** because it gives the cost of producing x items. Most cost functions are made up of two parts, a variable that depends on the number of items produced and a fixed part, which does not. The value of $C(0)$ is the **fixed cost**. For this example, $C(0) = 5(0)^2 + 40(0) + 600 = 600$. Suppose also that the price of each radio is set at $400 - 25x$ dollars. If this function represents the highest price per unit that would sell all x units, it is called the **demand function**. The demand function indicates that the price is somewhat determined by the number of radios sold. The **revenue function**, $R(x)$, is defined to be the product of the number of items and the price:

$$R(x) = (\text{number of items})(\text{price per item})$$

For this example,

$$R(x) = x(400 - 25x)$$
$$= 400x - 25x^2$$

A company will *break even* (that is, profit will balance loss) if the cost and the revenue are equal. The point (or points) at which this occurs is called a **break-even point**. For this example,

$$R(x) = C(x) \qquad \text{or} \qquad R(x) - C(x) = 0$$

This equation can be solved geometrically or algebraically. Geometrically, graph

$$y = R(x) \qquad \text{and} \qquad y = C(x)$$

as in Figure 2.17. The break-even points are the points of intersection.

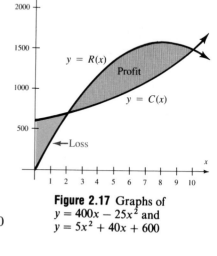

Figure 2.17 Graphs of $y = 400x - 25x^2$ and $y = 5x^2 + 40x + 600$

Algebraically,

$$R(x) = C(x)$$
$$400x - 25x^2 = 5x^2 + 40x + 600$$
$$30x^2 - 360x + 600 = 0$$
$$x^2 - 12x + 20 = 0$$
$$(x - 10)(x - 2) = 0$$
$$x = 10, 2$$

The break-even points are at $(2, 700)$ and $(10, 1500)$.

Polynomial Functions

Linear and quadratic functions are special types of more general mathematical functions called **polynomial functions**.

Polynomial Function

A function f is a *polynomial function* in x of degree n if

$$f(x) = a_n x^n + a_{n-1} x^{n-1} + \cdots + a_1 x + a_0$$

where $a_0, a_1, a_2, \ldots, a_n$ are real numbers, n is a whole number, and $a_n \neq 0$.

Examples of polynomial functions are

$$f(x) = 5x^3 + 2x^2 - 3x + 5 \qquad g(x) = x^2 - 5x + 1 \qquad h(x) = 6$$

Note that linear and quadratic functions are special types of polynomial functions of degree 1 and 2, respectively. (For a review of the terminology and algebraic simplification of polynomials, see Chapter 1.)

EXAMPLE 6   **a.** $f(x) = x$ is a polynomial.
b. $f(x) = \frac{1}{x}$ is not a polynomial since $\frac{1}{x} = x^{-1}$ and -1 is not a whole number.
c. $f(x) = \frac{1}{6}x$ is a polynomial ($n = 1, a_n = \frac{1}{6}$).
d. $f(x) = \sqrt{x} + 3x^2$ is not a polynomial since $\sqrt{x} = x^{1/2}$ and $\frac{1}{2}$ is not a whole number.
e. $f(x) = x^2 + 3x + 4x^{-2}$ is not a polynomial since the exponent of $4x^{-2}$ is not a whole number.   ∎

One method of graphing a polynomial function is to plot points as in Section 2.3, but much better and easier methods are available in calculus, so we will delay the graphing of polynomial functions until we discuss curve sketching in general in Chapter 11. In the meantime we will limit applications to the linear and quadratic functions introduced in this and the previous sections.

Problem Set 2.5

Sketch the graph of each equation in Problems 1–12.

1. $y = x^2$
2. $y = -x^2$
3. $y = -2x^2$
4. $y = 3x^2$
5. $y = -5x^2$
6. $y = 5x^2$
7. $y = \frac{1}{3}x^2$
8. $y = -\frac{1}{3}x^2$
9. $y = \frac{1}{10}x^2$
10. $y = -\frac{1}{10}x^2$
11. $y = \frac{2}{3}x^2$
12. $y = -\frac{2}{3}x^2$

Give the vertex of each parabola and tell whether it opens upward or downward.

13. $y = (x - 1)^2$
14. $y = -(x + 2)^2$
15. $y = (x + 3)^2$
16. $y = -2(x - 1)^2$
17. $y = \frac{1}{4}(x - 1)^2$
18. $y = -\frac{1}{2}(x + 1)^2$

19. $y - 2 = (x - 1)^2$
20. $y - 2 = 3(x + 2)^2$
21. $y - 2 = -\frac{3}{5}(x - 1)^2$
22. $y + 3 = \frac{2}{3}(x + 2)^2$
23. $y + \frac{1}{2} = (x - \frac{3}{4})^2$
24. $y - 0.1 = (x + 0.2)^2$

Sketch the graph of each equation in Problems 25–36.

25. $y = (x - 1)^2$
26. $y = -(x + 2)^2$
27. $y = (x + 3)^2$
28. $y = -2(x - 1)^2$
29. $y = \frac{1}{4}(x - 1)^2$
30. $y = -\frac{1}{2}(x + 1)^2$
31. $y - 2 = (x - 1)^2$
32. $y - 2 = 3(x + 2)^2$
33. $y - 2 = -\frac{3}{5}(x - 1)^2$
34. $y + 3 = \frac{2}{3}(x + 2)^2$
35. $y + \frac{2}{3} = (x + \frac{1}{3})^2$
36. $y + \frac{2}{5} = -(x - \frac{3}{5})^2$

Find the maximum or minimum value of y for each of the functions in Problems 37–48.

37. $y = -4(x + 1)^2 + 3$

38. $y = -5(x - 4)^2 + 2$

39. $y = -10(x - 450)^2 + 1250$

40. $y = 12(x + 30)^2 - 140$

41. $y = 25(x - 560)^2 - 1400$

42. $y = -150(x + 2300)^2 + 12,000$

43. $y + 14 = -3(x + 3)^2$

44. $y - 40 = -4(x + 5)^2$

45. $2(x + 3)^2 - y + 13 = 0$

46. $3(x - 2)^2 - y + 10 = 0$

47. $3y + 6(x - 1)^2 + 12 = 0$

48. $5y - 30(x + 3)^2 - 100 = 0$

APPLICATIONS

49. After extensive market research, a consulting firm has determined that the demand equation for a certain product is $1000 - 10x$ dollars, where x is the number of items produced. Since the company has fixed costs of 50,000 dollars, the cost function is found to be $C(x) = 50,000 - 500x$.
a. Find the revenue function.
b. Find the break-even point(s).

50. The demand equation for a certain type of ratchet flange is $50 - x$ dollars, where x is the number of flanges produced. The cost function is found to be $C(x) = 1200 - 20x$.
a. Find the revenue function.
b. Find the break-even point(s).

51. A manufacturer produces quality boats. The profit, in dollars, is determined to be

$$P(x) = -10(x - 375)^2 + 1,156,250$$

where x is the number of boats.
a. How many boats should be produced in order to produce a maximum profit?
b. What is the profit (or loss) if no boats are produced?
c. What is the maximum profit?

52. The profit function, P, is found by setting

$$P(x) = R(x) - C(x)$$

Using the information from Problem 49, we find

$$P(x) = -10(x - 75)^2 + 6250$$

Find the maximum profit for this function.

53. The profit function for a rachet flange is

$$P(x) = -2(x - 25)^2 + 650$$

What is the maximum profit?

54. Graph the cost and revenue functions of Problem 49 and shade the region representing the company's profit.

55. Graph the cost and revenue functions of Problem 50 and shade the region representing the company's profit.

56. The highest bridge in the world is the bridge over the Royal Gorge of the Arkansas River in Colorado. It is 1053 ft above the water. If a rock is thrown vertically upward from this bridge with an initial velocity of 64 feet per second, the height h of the rock above the river at time t is described by the function

$$h(x) = -16(t - 2)^2 + 1117$$

What is the maximum height possible for a rock thrown vertically upward from the bridge with an initial velocity of 64 feet per second? After how many seconds will it reach that height?

57. In 1974 Evel Knievel attempted a skycycle ride across the Snake River. Suppose the path of the skycycle is given by the equation

$$d(x) = -0.0005(x - 2390)^2 + 3456$$

where $d(x)$ is the height above the canyon floor for a horizontal distance of x units from the launching ramp. What was Knievel's maximum height?

58. a. In most states, drivers are required to know approximately how long it takes to stop their cars at various speeds. Suppose you estimate three car lengths for 30 mph and six car lengths for 60 mph. One car length per 10 mph assumes a linear relationship between speed and distance covered. Write a linear equation where x is the speed of the car and y is the distance traveled by the car in feet. (Assume that one car length = 20 ft.)
b. The scheme for the stopping distance of a car given in part a is convenient but not accurate. The stopping distance is more accurately approximated by the quadratic equation

$$y = 0.071x^2$$

This says that a car requires four times as many feet to stop at 60 miles per hour than at 30 mph. That is, doubling the speed quadruples the braking distance. Graph this quadratic equation and the linear equation from part a on the same coordinate axes.
c. Comment on the results from part b. At about what speed are the two measures the same?

2.6 Rational Functions

The third type of function used for mathematical models is called a **rational function**.

Rational Function

> A function f is a *rational function* if
>
> $$f(x) = \frac{P(x)}{Q(x)}$$
>
> where $P(x)$ is any polynomial function and $Q(x)$ is a polynomial function whose domain excludes values for which $Q(x) = 0$.

Functions such as

$$f(x) = \frac{1}{x} \qquad y = \frac{1}{x-2} \qquad g(x) = \frac{x^2 + 3x - 1}{x-1}$$

are examples of rational functions. The values $x = 0$, $x = 2$, and $x = 1$, respectively, are excluded from the domains of these three functions according to the definition of a rational function.

In Section 2.3, we graphed $h(x) = \frac{1}{x}$ (Example 3) by plotting points. It is important to remember the general shape of the graph of this function so we repeat it here for easy reference.

Note that the domain for the graph in Figure 2.18 consists of all real numbers except $x = 0$. That is, the graph of the function $y = \frac{1}{x}$ does not cross the vertical line $x = 0$. Now consider the graph of the rational function $y = \frac{1}{x-1}$. The definition of rational functions *excludes* from the domain all values for which $x - 1 = 0$, or, namely, $x = 1$. You can, however, view the equation $x = 1$ as the equation of a vertical line. This line is called a **vertical asymptote** for the curve $y = 1/(x - 1)$ and is illustrated in Example 1a.

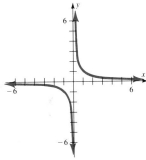

Figure 2.18 Graph of $y = \frac{1}{x}$

EXAMPLE 1 Graph the following equations by plotting points:

$$\textbf{a. } y = \frac{1}{x-1} \qquad \textbf{b. } y = \frac{1}{x+2} \qquad \textbf{c. } y = \frac{1}{2x-5}$$

Solution The vertical asymptotes are lines for which the denominators are equal to zero. To find the asymptotes, set the denominator equal to zero and solve:

$$\begin{array}{lll}
\textbf{a. } x - 1 = 0 & \textbf{b. } x + 2 = 0 & \textbf{c. } 2x - 5 = 0 \\
\qquad x = 1 & \qquad x = -2 & \qquad x = \frac{5}{2}
\end{array}$$

Plot additional points as necessary as shown at the top of the next page. ∎

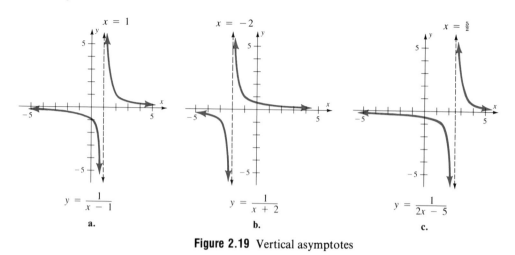

Figure 2.19 Vertical asymptotes

As you can see, a vertical asymptote is a vertical line in which as x gets closer and closer to a value that causes division by zero, the value of the function gets larger and larger or smaller and smaller. That is, a line is an asymptote for a curve if the distance between the line and the curve becomes smaller and smaller within some suitable domain. The graph of a function cannot cross its vertical asymptotes.

EXAMPLE 2   Graph

$$y = \frac{1}{3x^2 - 5x - 2}$$

Solution   Find the vertical asymptotes and compare them with those in Figure 2.19.

$$3x^2 - 5x - 2 = 0$$
$$(x - 2)(3x + 1) = 0$$
$$x = 2, -\tfrac{1}{3}$$

Draw the vertical asymptotes and plot additional points.

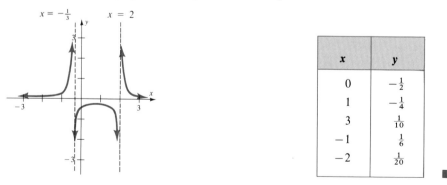

x	y
0	$-\frac{1}{2}$
1	$-\frac{1}{4}$
3	$\frac{1}{10}$
-1	$\frac{1}{6}$
-2	$\frac{1}{20}$

It would be convenient if we could say that if $f(x) = P(x)/Q(x)$, then any value for which $Q(x) = 0$ would be a vertical asymptote. Unfortunately, such is not the case, as we see in Example 3.

EXAMPLE 3   Graph

$$y = \frac{x}{x^2 - x}$$

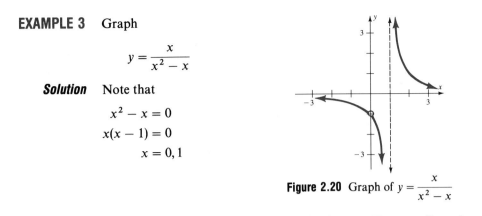

Solution   Note that

$$x^2 - x = 0$$
$$x(x - 1) = 0$$
$$x = 0, 1$$

Figure 2.20 Graph of $y = \dfrac{x}{x^2 - x}$

so there are two values that cause division by zero. However, if we plot points, we see that $x = 1$ is a vertical asymptote and that $x = 0$ is simply a point deleted from the domain.   ■

There is a way to decide if a value causing division by zero indicates a deleted point or a vertical asymptote. All we need to do is reduce the fraction and set the resulting denominator equal to zero. Thus

$$\frac{x}{x^2 - x} = \frac{x}{x(x - 1)} = \frac{1}{x - 1}$$

$$\underset{\substack{\text{Values that cancel} \\ \text{indicate deleted points} \\ \text{(that is, } x = 0)}}{x \neq 0, x \neq 1} \qquad \underset{\substack{\text{Values that do not} \\ \text{cancel indicate asymptotes}}}{\underbrace{x = 1 \text{ is an asymptote}}}$$

EXAMPLE 4   Given the following equations, find the vertical asymptotes.

a. $y = \dfrac{x + 4}{x^2 - 16}$ **b.** $y = \dfrac{3x^2 - 5x - 2}{x^2 - 5x + 6}$

Solution   **a.** $y = \dfrac{x + 4}{x^2 - 16} = \dfrac{x + 4}{(x - 4)(x + 4)} = \dfrac{1}{\underline{x - 4}}$

$$x \neq 4, \underline{x \neq -4} \quad x = 4 \text{ is an asymptote}$$
$$x = -4 \text{ is a deleted point on the graph}$$

b. $y = \dfrac{3x^2 - 5x - 2}{x^2 - 5x + 6} = \dfrac{(3x + 1)(x - 2)}{(x - 3)(x - 2)} = \dfrac{3x + 1}{x - 3}$

$$x = 3 \text{ is a vertical asymptote}$$ ■

Note that the graphs in Figures 2.18 and 2.19 and in Examples 2 and 3 not only have vertical asymptotes but also have a **horizontal asymptote**, $y = 0$. As $|x|$ gets larger, each of the named curves gets closer to the line $y = 0$. There will be a horizontal asymptote $y = 0$ whenever the degree of the numerator is less than the degree of the denominator. This result will be justified in the next chapter. For now, study the variations of asymptotes shown in Figure 2.21a–d.

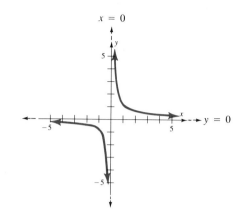

Figure 2.21a Graph of $y = \frac{1}{x}$.
Vertical asymptote, $x = 0$;
horizontal asymptote, $y = 0$.

Figure 2.21b Graph of $y = \frac{1}{x} - 2$.
Vertical asymptote, $x = 0$;
horizontal asymptote, $y = -2$.

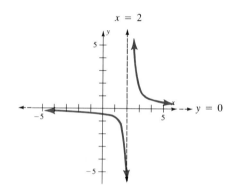

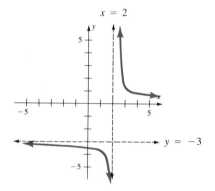

Figure 2.21c Graph of $y = \frac{1}{(x-2)}$.
Vertical asymptote, $x = 2$;
horizontal asymptote, $y = 0$.

Figure 2.21d Graph of
$y = \frac{1}{(x-2)} - 3$.
Vertical asymptote, $x = 2$;
horizontal asymptote,
$y = -3$.

Apparently, horizontal asymptotes depend on the relative degrees of the numerator and denominator polynomials. The following summary is useful when graphing rational functions.

Asymptotes

If $f(x) = P(x)/D(x)$, where $P(x)$ and $D(x)$ are polynomial functions with no common factors, then

the line $x = r$ is a *vertical asymptote* if $D(r) = 0$;
the line $y = 0$ is a *horizontal asymptote* if the degree of $P(x)$ is less than the degree of $D(x)$;

the line $y = \dfrac{a_n}{b_n}$ is a *horizontal asymptote* if the degree of $P(x)$ is the same as the degree of $D(x)$ and
$$P(x) = a_n x^n + \cdots + a_0 \text{ and}$$
$$D(x) = b_n x^n + \cdots + b_0$$

EXAMPLE 5 Graph $y = \dfrac{3x - 1}{2x + 5}$.

Solution First find the vertical asymptotes. Set the denominator equal to 0 and solve.

$$2x + 5 = 0$$
$$2x = -5$$
$$x = -\tfrac{5}{2}$$

Second, find the horizontal asymptotes. The degree of $3x - 1$ and $2x + 5$ is the same, so there is one horizontal asymptote: $y = \tfrac{3}{2}$. The remaining points are found by calculation.

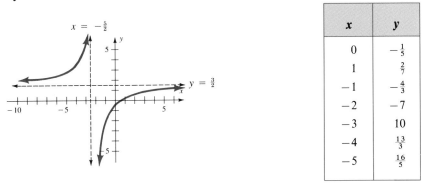

x	y
0	$-\tfrac{1}{5}$
1	$\tfrac{2}{7}$
-1	$-\tfrac{4}{3}$
-2	-7
-3	10
-4	$\tfrac{13}{3}$
-5	$\tfrac{16}{5}$

Cost-Benefit Models

Many real life situations involve some desired result, or benefit. However, to achieve this result involves a certain cost and it is often necessary to compare the benefit with the cost. An equation that compares costs and benefits is called a **cost-benefit model**. Suppose that the cost, C, of installing five-field electrostatic precipitators for the removal of p percent of pollutants released into the atmosphere from the emissions of a cement factory is given by the formula

$$C = \frac{200,000p}{105 - p}$$

Example 6 demonstrates the use of this cost-benefit model.

EXAMPLE 6 Use the cost-benefit function given above to find:

a. How much would it cost to remove 95% of the pollutants (as required by the Environmental Protection Agency)?
b. How much does it cost to remove 80% of the pollutants (which the company is presently doing)?
c. How much would it cost to remove 100% of the pollutants?
d. Graph the cost-benefit relationship for $0 \le p \le 100$.

Solution **a.** For $p = 95$, $C = \dfrac{200,000(95)}{105 - 95} = 1,900,000$

b. For $p = 80$, $C = \dfrac{200,000(80)}{105 - 80} = 640,000$

c. For $p = 100$, $C = \dfrac{200,000(100)}{105 - 100} = 4,000,000$

d. Use the points found above (along with additional ones as needed) to graph this function for $0 \le p \le 100$. Note that there is both a horizontal and vertical asymptote.

$$C = \frac{200,000p}{105 - p} \qquad \leftarrow \text{Vertical asymptote: } 105 - p = 0$$
$$= 105$$

Horizontal asymptote since the degree of the numerator and denominator is the same; it is

$$C = \frac{200,000}{-1} \qquad \text{From the coefficients of } p$$
$$= -200,000$$

Note: You will not see the graph approach this asymptote because C approaches $-200,000$ as $|p|$ gets larger, but we are asked to graph this curve only for $0 \le p \le 100$.

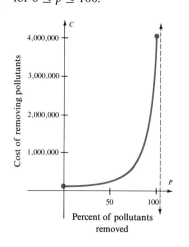

Problem Set 2.6

1. What is a rational function?
2. What is a vertical asymptote, and how do you find it?
3. What is a horizontal asymptote, and how do you find it?

Name the horizontal and vertical asymptotes for the curves in Problems 4–15.

4. $f(x) = \dfrac{1}{x + 3}$

5. $f(x) = \dfrac{5}{2x - 3}$

6. $f(x) = \dfrac{2}{3x + 2}$

7. $g(x) = \dfrac{1}{x^2 + 2x - 3}$

8. $g(x) = \dfrac{6}{2x^2 + 3x - 2}$

9. $g(x) = \dfrac{5}{6x^2 + 5x - 6}$

10. $h(x) = \dfrac{2x + 5}{3x - 4}$

11. $h(x) = \dfrac{5x^2 + 3x - 2}{2x^2 + 4x + 1}$

12. $h(x) = \dfrac{6x^2 - 3x + 1}{2x^2 - 5}$

13. $y = \dfrac{3x + 2}{x^2 + x + 2}$

14. $y = \dfrac{x - 5}{x^2 - 3x + 5}$

15. $y = \dfrac{2x + 1}{x^2 + x - 2}$

Graph the functions in Problems 16–27.

16. $y = \dfrac{2}{x}$

17. $y = -\dfrac{1}{x}$

18. $y = -\dfrac{3}{x}$

19. $y = \dfrac{1}{x - 4}$

20. $y = \dfrac{1}{x + 3}$

21. $y = \dfrac{1}{3x - 2}$

22. $y = \dfrac{1}{x} + 4$

23. $y = \dfrac{1}{x} + 3$

24. $y = \dfrac{1}{x - 1} + 3$

25. $y = \dfrac{2x^2 - 3x + 1}{x^2 + 2x - 3}$

26. $y = \dfrac{2x^2 + 5x + 3}{x^2 - x - 2}$

27. $y = \dfrac{2x^2 - x - 1}{4x^2 + 4x + 1}$

APPLICATIONS

In Problems 28–31, use the cost-benefit model

$$C = \dfrac{40{,}000p}{110 - p}$$

where C is the cost of removing p% of the pollutants.

28. What is the cost of removing 95% of the pollutants?
29. What is the cost of removing 80% of the pollutants?
30. What is the cost of removing 100% of the pollutants?
31. Graph these relationships for $0 \leq p \leq 100$.

In Problems 32–35, use the cost-benefit model

$$C = \dfrac{18{,}000p}{100 - p}$$

where C is the cost of removing p% of the pollutants.

32. What is the cost of removing 95% of the pollutants?
33. What is the cost of removing 80% of the pollutants?
34. Is it possible to remove 100% of the pollutants?
35. Graph these relationships for $0 \leq p < 100$.

Radiologists must deal with three quantities each time an X ray is taken:

t = time in seconds that the X-ray machine is on

mA = the current, measured in millirems

FFD = distance from the X-ray machine to the film

Use this information in Problems 36 and 37.

36. If FFD is held constant, the relationship between mA and t is given by

$$mA = \dfrac{400}{t}$$

Graph this relationship for $0 < t \leq 10$.

37. If time is held constant, the relationship between mA and FFD is given by

$$mA = \dfrac{256}{(FFD)^2}$$

Graph this relationship for $0 < FFD \leq 16$.

The pressure-volume relationship is

$$\dfrac{\text{Original pressure}}{\text{New pressure}} = \dfrac{\text{new volume}}{\text{original volume}}$$

Use this proportion in Problems 38 and 39.

38. If the new pressure is 840 millimeters (mm) of mercury and the new volume is 100 milliliters (mℓ), write a rational function relating the original pressure and the original volume. Graph this relationship using the pressure as the independent variable.

39. If 400 mℓ of oxygen is under a pressure of 2800 mm of mercury, write a rational function relating the new pressure and the new volume. Graph this relationship using the pressure as the independent variable.

40. Suppose that the supply and demand functions for a product are

$$S(p) = p - 20 \quad \text{and} \quad D(p) = \dfrac{800}{p}$$

The point(s) where $S(p) = D(p)$ are called the **equilibrium point(s)**.

a. Graph S and D on the same coordinate axes and estimate the equilibrium point(s).
b. Algebraically solve $S(p) = D(p)$.
c. Where does the supply function cross the p-axis? What is the economic significance of this point?

2.7 Summary and Review

IMPORTANT TERMS

Absolute value function   [2.4]
Asymptote   [2.6]
Break-even point   [2.5]
Cartesian coordinate system   [2.3]
Concave   [2.5]
Constant function   [2.4]
Coordinate axes   [2.3]
Cost function   [2.5]
Cost-benefit model   [2.6]
Demand function   [2.5]

Dependent variable   [2.4]
Domain   [2.2, 2.3]
Evaluating a function   [2.2]
First component   [2.3]
Fixed cost   [2.5]
Function   [2.2]
Graph of a relation or
   function   [2.3]
Horizontal asymptote   [2.6]
Horizontal line   [2.4]

Independent variable   [2.4]

Intercepts   [2.3]

Linear function   [2.4]

Parameter   [2.4]

Parametric equations   [2.4]

Parabola   [2.3, 2.5]

Piecewise linear function   [2.4]

Plot a point   [2.3]

Point-slope form   [2.4]

Polynomial function   [2.5]

Quadrants   [2.3]

Quadratic function   [2.5]

Range   [2.2]

Rational function   [2.6]

Revenue function   [2.5]

Second component   [2.3]

Slope   [2.4]

Slope-intercept form   [2.4]

Standard form   [2.4]

Standard position parabola   [2.5]

Symmetry   [2.5]

Two-dimensional coordinate
system   [2.3]

Vertex   [2.5]

Vertical asymptote   [2.6]

Vertical line   [2.4]

Vertical line test   [2.3]

x-axis   [2.3]

y-axis   [2.3]

Zero of a function   [2.3]

SAMPLE TEST   *For additional practice there are a large number of review problems categorized by objective in the Student Solutions Manual. The following sample test (40 minutes) is intended to review the main ideas of this chapter.*

In Problems 1–8, let $f(x) = 2x^2 - 3x + 5$ and $g(x) = 1 - 4x$. Find the values requested in Problems 1–6.

1. $f(3)$ **2.** $g(10)$ **3.** $g(t + 1)$ **4.** $f(t - h)$

5. $\dfrac{g(x + h) - g(x)}{h}$ **6.** $\dfrac{f(x + h) - f(x)}{h}$ **7.** Graph g.

8. What are the x- and y-intercepts for f?

9. Find the slope of the line passing through $(-5, 9)$ and $(-3, -7)$.

10. Graph $2x + 3y - 12 = 0$ by finding the slope and y-intercept.

11. Graph $y = |x + 3|$.

12. Graph

$$\begin{cases} x + y + 1 = 0 & \text{for} \quad -3 \le x \le 1 \\ y + 2 = 0 & \text{for} \quad 1 < x < 5 \end{cases}$$

13. Find the standard form of the equation of a line passing through $(30, 20)$ with slope -5.

14. Find the equation of a line with zero slope and y-intercept -3.

15. Find the equation of a line with no slope passing through $(-3, -2)$.

16. Sketch $y + 3 = \frac{1}{2}(x - 2)^2$.

17. What is the maximum value of the function $y - 550 = -\frac{2}{3}(x - 40)^2$?

18. What are the asymptotes for the function

$$F(x) = \frac{2x + 1}{x^2 + x - 6}$$

19. Use the cost-benefit model

$$C = \frac{500p}{100 - p}$$

to find the cost of removing 90% of the pollutants.

20. Graph C from Problem 19 for $0 \le p < 100$.

FINITE MATHEMATICS

Finite mathematics is a recent course in mathematics. It was first proposed in the early 1960s in the now classic book by Kemeny, Snell, and Thompson, *Introduction to Finite Mathematics* (Englewood Cliffs, N.J.: Prentice-Hall, 1956). The topics in that book originally defined the course. Over the years the course has evolved to include the topics that are now included in this part of the book (see list of chapters below), but the topic itself, finite mathematics, is not well defined in the sense that algebra or calculus is well defined. Very loosely defined, finite mathematics is mathematics that does not use calculus. More specifically, finite mathematics means *applying* the elementary mathematics of sets, matrices, linear programming, probability, and statistics to real life problems.

The prerequisites for this part of the book are at least 2 years of high school algebra (reviewed in Chapter 1) and the concepts of functions, graphs, and linear functions (Sections 2.1–2.4).

Chapter 3   Systems of Equations and Matrices
Chapter 4   Linear Programming
Chapter 5   Mathematics of Finance
Chapter 6   Combinatorics
Chapter 7   Probability
Chapter 8   Statistics and Probability Distributions

CHAPTER 3

Systems of Equations and Matrices

CHAPTER CONTENTS

3.1 Systems of Equations
3.2 Introduction to Matrices
3.3 Gauss–Jordan Elimination
3.4 Inverse Matrices
3.5 Leontief Models
3.6 Summary and Review
    Important Terms
    Sample Test

APPLICATIONS

Management (*Business, Economics, Finance, and Investments*)

Supply and demand (3.1, Problems 41–46)
Break-even analysis (3.1, Problems 47–53)
Value of a stock portfolio (3.1, Problems 54–55)
Inventory control (3.2, Problem 5)
Delivery routes (3.2, Problem 64)
Airline routes (3.2, Problem 65)
Manufacturing construction costs (3.2, Problem 66)
Mixing an alloy in a production process (3.3, Problem 50)
Manufacturing a candy (3.3, Problem 51)
Leontief models (3.5, Problems 7–17)
Study of the Israeli economy (3.5, Problems 18–20)

Life sciences (*Biology, Ecology, Health, and Medicine*)

Mixture of a chemical fertilizer (3.1, Problem 57)
Preparing a proper cattle-feed mix (3.3, Problem 48; 3.4, Problems 55–57)
Preparing a commercial pesticide (3.3, Problem 49)
Preparing a diet of three basic foods (3.4, Problems 58–60)

Social sciences (*Demography, Political Science, Population, Psychology, Society, and Sociology*)

Voter demographics (3.1, Problem 56; 3.3, Problem 52)
Louvre Tablet from the Babylonian civilization (3.1, Problem 58)
Diplomatic communication channels (3.2, Problem 67)

General interest

Best price for a car rental (3.1, Problems 37–38, 40)
Matrix of car rental costs (3.2, Problem 6)
Development of matrix theory (3.2, Problems 68–73)
Ice cream cone with the best value (3.3, Problem 53)

Modeling application—Gaining a Competitive Edge in Business

CHAPTER
OVERVIEW

Here we solve *systems* of equations and explore an extremely powerful concept in mathematical modeling—the *matrix*.

PREVIEW

We first learn what matrices are, then how to manipulate them, and, finally, how to use them to solve systems of equations. Wassily Leontief (1906–) received the Nobel Prize in Economics in 1973 for describing the economy in terms of an input–output model. In Section 3.5, we look at two input–output models: a *closed model*, in which we try to find the relative income of each participant within a system, and an *open model*, in which we try to find the amount of production needed to achieve an anticipated demand. The modeling application involves writing a paper based on the economic model Leontief used.

PERSPECTIVE

The most important idea of this chapter—one that must be mastered in order to succeed with the material that follows later in the book—is that of *Gauss–Jordan elimination*. It is used over and over again in this course, so work hard to understand the procedure.

3.1 Systems of Equations

Many situations involve two variables or unknowns related in some specific fashion, as in Example 1.

EXAMPLE 1

Cost Analysis   Linda is on a business trip and needs to rent a car for a day. The prices at two agencies are different, and she wants to rent the least expensive car.

Agency A: $10 per day plus 50¢ per mile
Agency B: $40 per day plus 25¢ per mile

Which agency should she rent from?

Solution

The answer to this question depends on the number of miles Linda intends to drive. Let c be the cost (in dollars) of the rental and let m be the number of miles driven. These variables are related as follows:

COST = BASIC CHARGE + MILEAGE CHARGE

where the mileage charge is the cost per mile times the number of miles driven. Thus

Agency A: $c = 10 + 0.5m$
Agency B: $c = 40 + 0.25m$

We use a Cartesian graph as a model for this problem and graph the equations for agencies A and B. The point of intersection looks like the point $(120, 70)$. What does this mean? It means that if Linda plans on driving less than 120 miles, she should rent from agency A, and if she plans on driving more than 120 miles, she should rent from agency B.

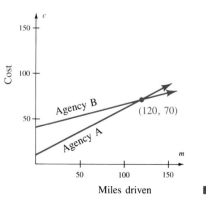

Miles driven

The model in Example 1 is called a **system of equations**. Solving systems of equations is a procedure that is used throughout mathematics. In this text we will solve a variety of systems of equations and inequalities, so we now introduce a notation that will be easy to generalize in later applications. We begin our study of this topic by considering an arbitrary system of two equations with two variables:

$$\begin{cases} a_{11}x_1 + a_{12}x_2 = b_1 \\ a_{21}x_2 + a_{22}x_2 = b_2 \end{cases}$$

The notation may seem strange, but it will prove useful. The variables are x_1 and x_2, and the constants are $a_{11}, a_{12}, a_{21}, a_{22}, b_1$, and b_2. By a *system of two equations with two variables*, we mean any two equations in those variables. The **simultaneous solution** of a system is the intersection of the solution sets of the individual equations. The brace in front of the equations indicates that this intersection is the desired solution. If all the equations in a system are linear, it is called a **linear system**. We limit our study in this section to linear systems of equations with two unknowns.

Since the graph of each equation in a system of linear equations in two variables is a line, the solution set for the system is the intersection of two lines. In two dimensions, two lines must be related to each other in one of three possible ways:

1. The lines intersect at a single point. They are called **consistent**.
2. The graphs are parallel lines. In this case, the solution set is empty, and the equations are called **inconsistent**.
3. The graphs are the same line. In this case, there are infinitely many points in the solution set, and any solution of one equation is also a solution of the other. The equations of such a system are said to be **dependent**. If the equations of a system are not dependent, then they are called **independent**.

EXAMPLE 2   Relate the system $\begin{cases} 2x - 3y = -8 \\ x + y = 6 \end{cases}$ to the general system $\begin{cases} a_{11}x_1 + a_{12}x_2 = b_1 \\ a_{21}x_1 + a_{22}x_2 = b_2 \end{cases}$ and solve by graphing.

Solution   The variables are $x_1 = x$ and $x_2 = y$. The subscripts in $a_{11}, a_{12}, a_{21},$ and a_{22} are called *double subscripts* and indicate *position*. That is, a_{12} should not be read "*a* sub twelve," but rather as "*a* sub one-two." This means that it represents the second constant in the first equation:

$$a_{12}$$

Equation number ⎯⎤ ⎣⎯ Coefficient number in equation

This effort in using notation may seem rather complicated when dealing with only two equations and two variables, but we want to develop a notation that can be used with many variables and many equations.

For this example, the constants are

$$a_{11} = 2 \qquad a_{12} = -3 \qquad b_1 = -8$$
$$a_{21} = 1 \qquad a_{22} = 1 \qquad b_2 = 6$$

The solution is $x = 2$, $y = 4$, or $(2, 4)$, as shown in the figure.

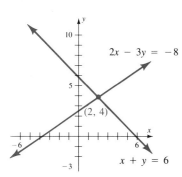

Graph of the system
$\begin{cases} 2x - 3y = -8 \\ x + y = 6 \end{cases}$

EXAMPLE 3   Relate the system $\begin{cases} 2x - 3y = -8 \\ 4x - 6y = -2 \end{cases}$ to the general system and solve by graphing.

Solution   The variables are $x_1 = x$, $x_2 = y$, and the constants are

$$a_{11} = 2 \qquad a_{12} = -3 \qquad b_1 = -8$$
$$a_{21} = 4 \qquad a_{22} = -6 \qquad b_2 = -2$$

Since the graphs of the lines are distinct and parallel, as shown in the figure, there is no point of intersection. These equations are inconsistent.

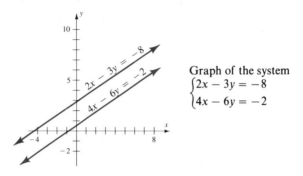

Graph of the system
$\begin{cases} 2x - 3y = -8 \\ 4x - 6y = -2 \end{cases}$

∎

EXAMPLE 4   Relate the system $\begin{cases} 2x - 3y = -8 \\ y = \frac{2}{3}x + \frac{8}{3} \end{cases}$ to the general system and solve by graphing.

Solution   To relate to the general system, both equations must be arranged in the usual form; the second equation needs to be rewritten:

$$y = \tfrac{2}{3}x + \tfrac{8}{3}$$
$$3y = 2x + 8 \qquad \text{Multiply by 3}$$
$$-2x + 3y = 8$$

Thus, in standard form, the system is

$$\begin{cases} 2x - 3y = -8 \\ -2x + 3y = 8 \end{cases}$$

This means that the variables are $x_1 = x$, $x_2 = y$, and the constants are

$$a_{11} = 2 \qquad a_{12} = -3 \qquad b_1 = -8$$
$$a_{21} = -2 \qquad a_{22} = 3 \qquad b_2 = 8$$

The equations represent the same line, as shown in the figure, and are dependent.

∎

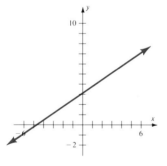

Graph of the system
$\begin{cases} 2x - 3y = -8 \\ y = \frac{2}{3}x + \frac{8}{3} \end{cases}$

The **graphing method** can give solutions only as accurate as the graphs we can draw, and, consequently, it is inadequate for most applications. We therefore need more efficient methods.

In general, given a system, the procedure is to write a simpler equivalent system. Two systems are said to be **equivalent** if they have the same solution set. In this chapter we limit ourselves to finding only real roots. There are several ways to go about writing equivalent systems. The first nongraphical method we consider comes from the substitution property of real numbers and leads to a **substitution method** for solving systems.

Substitution Method for Solving Systems of Equations

1. *Solve* one of the equations for one of the variables.
2. *Substitute* the expression obtained into the other equation.
3. *Solve* the resulting equation in a single variable for the value of that variable.
4. *Substitute* that value into either of the original equations to determine the value of the other variable.
5. *State* the solution.

EXAMPLE 5   Solve $\begin{cases} 2p + 3q = 5 \\ \quad\quad q = -2p + 7 \end{cases}$ by substitution.

Solution   Since $q = -2p + 7$, substitute $-2p + 7$ for q in the other equation:

$$2p + 3(-2p + 7) = 5$$
$$2p - 6p + 21 = 5$$
$$-4p = -16$$
$$p = 4$$

Substitute 4 for p in either of the given equations:

$$q = -2p + 7$$
$$= -2(4) + 7$$
$$q = -1$$

The solution is $(p, q) = (4, -1)$.   ■

A third method for solving systems is called the **linear combination method**. It involves substitution and the idea that if equal quantities are added to equal quantities, the resulting equation is equivalent to the original system. In general, such addition will not simplify matters unless the numerical coefficients of one or more terms are opposites. However, we can often force them to be opposites by multiplying one or both of the given equations by appropriate nonzero constants.

Linear Combination Method for Solving Systems of Equations

1. *Multiply* one or both of the equations by a constant or constants, so that the coefficients of one of the variables become opposites.
2. *Add* corresponding members of the equations to obtain a new equation in a single variable.
3. *Solve* the derived equation for that variable.
4. *Substitute* the value of the found variable into either of the original equations and solve for the second variable.
5. *State* the solution.

EXAMPLE 6   Solve: $\begin{cases} 3x + 5y = -2 \\ 2x + 3y = 0 \end{cases}$.

Solution   Multiply both sides of the first equation by 2 and both sides of the second equation by -3. This procedure, shown below, forces the coefficients of x to be opposites:

$$2 \begin{cases} 3x + 5y = -2 \\ -3 \begin{cases} 2x + 3y = 0 \end{cases} \end{cases}$$

This means you should add the equations of the system

$$+ \begin{cases} 6x + 10y = -4 \\ -6x - 9y = 0 \end{cases}$$

$$[6x + (-6x)] + [10y + (-9y)] = -4 + 0 \qquad \text{This step should be done mentally}$$

$$y = -4$$

If $y = -4$, then

$$2x + 3y = 0$$
$$2x + 3(-4) = 0$$
$$x = 6$$

The solution is $(6, -4)$. ∎

Later we will need to find the corner points for certain regions in the plane and the methods of this section, as illustrated in Example 7, will be helpful.

EXAMPLE 7   Find the corner points labeled A, B, C, and D in the illustrated region.

Solution   Point A is obvious, it is the origin $(0, 0)$. Points B and C are relatively easy to find because they are the intercepts of the lines.

For point B, the x-intercept is found when $y = 0$ in the equation $y = -2x + 12$.

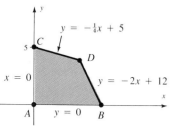
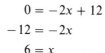

Thus

$$0 = -2x + 12$$
$$-12 = -2x$$
$$6 = x$$

Point B is $(6, 0)$.

For point C, the y-intercept is found by inspection; it is the point $(0, 5)$.

Point D is the intersection of the lines whose equations are given, namely:

$$\begin{cases} y = -\tfrac{1}{4}x + 5 \\ y = -2x + 12 \end{cases}$$

Solve these by substitution,

$$-2x + 12 = -\tfrac{1}{4}x + 5$$
$$-8x + 48 = -x + 20$$
$$-7x = -28$$
$$x = 4$$

Finally, $y = -2(4) + 12 = 4$, so D is the point $(4, 4)$. ∎

There are two very important business applications of systems: *supply and demand* and *break-even analysis*.

Supply and Demand

Let p be the price of an item and n the number of items available at price p; that is, consider the ordered pair (p, n). There are two curves relating p and n. The first, called the **supply curve**, expresses the relationship between p and n from a manufacturer's point of view. For every value of p, the supply curve gives the number n that the manufacturer is willing to produce at the price p. The higher the price, the more the manufacturer is willing to supply. Therefore, the supply curve rises when viewed from left to right. The other curve, the **demand curve**, expresses the relationship between p and n from the consumer's point of view. For every value of p, the demand curve gives the number n that the consumer is willing to buy at price p. The higher the price, the less consumers will buy, so demand curves fall when viewed from left to right.

If supply and demand curves are drawn on the same coordinate system, the intersection point (if it exists) is called the **equilibrium point** and represents values for which supply equals demand.

In this section, the supply and demand curves are linear, but in most real life situations they are more general curves, as shown by the illustrations below.

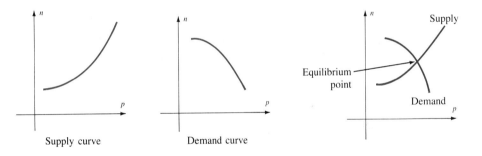

Supply curve   Demand curve

EXAMPLE 8   A product has a supply curve given by the equation $n = 500p - 8100$ and a demand curve given by the equation $n = -100p + 15{,}000$. Find the equilibrium point.

Solution   Find the intersection point by using substitution:

$$-100p + 15{,}000 = 500p - 8100$$
$$600p = 23{,}100$$
$$p = \frac{23{,}100}{600}$$
$$= 38.5$$

Now, if $p = 38.5$, then

$$n = -100(38.5) + 15{,}000$$
$$= 11{,}150$$

Thus equilibrium is reached when the price is $38.50 and the number is 11,150 items. ∎

Break-Even Analysis

In order to make a profit, revenue must exceed cost. The point at which revenue equals the cost is called the **break-even point**.

EXAMPLE 9　A company producing bicycle reflectors has fixed costs of $2,100 and variable costs of 80¢ per reflector. The selling price of each reflector is $1.50. What is the break-even point?

Solution　Let x be the number of items produced, C be the cost, and R be the revenue. Then

$$C = 0.8x + 2100$$

and

$$R = 1.5x$$

Although the graph in the figure illustrates what is meant by break-even analysis and profit and loss, it is often not accurate enough to find the break-even point by inspection. Therefore we will also solve this system of equations by substitution:

$1.5x = 0.8x + 2100$　　Since $R = C$ at the break-even point

$0.7x = 2100$

$x = 3000$　　If $x = 3000$, then $R = 1.5(3000) = 4500$

If 3000 reflectors are produced, the cost and revenue will both be $4,500.　■

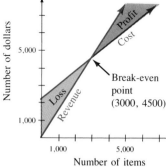

Number of dollars — Number of items. Profit, Cost, Loss, Revenue. Break-even point (3000, 4500).

Problem Set 3.1

Solve the systems in Problems 1–6 by graphing.

1. $\begin{cases} y = 3x - 7 \\ y = -2x + 8 \end{cases}$

2. $\begin{cases} x + y = 1 \\ 3x + y = -5 \end{cases}$

3. $\begin{cases} y = \frac{2}{3}x - 7 \\ 2x + 3y = 3 \end{cases}$

4. $\begin{cases} y = \frac{3}{5}x + 2 \\ 3x - 5y = 10 \end{cases}$

5. $\begin{cases} 2x - 3y = 9 \\ y = \frac{2}{3}x - 3 \end{cases}$

6. $\begin{cases} 2x - 3y = 0 \\ y = \frac{2}{3}x - 2 \end{cases}$

Solve the systems in Problems 7–12 by substitution. Relate each system to the general system $\begin{cases} a_{11}x_1 + a_{12}x_2 = b_1 \\ a_{21}x_1 + a_{22}x_2 = b_2 \end{cases}$.

7. $\begin{cases} a = 3b - 7 \\ a = -2b + 8 \end{cases}$

8. $\begin{cases} s + t = 1 \\ 3s + t = -5 \end{cases}$

9. $\begin{cases} m = \frac{2}{3}n - 7 \\ 2n + 3m = 3 \end{cases}$

10. $\begin{cases} v = \frac{3}{5}u + 2 \\ 3u - 5v = 10 \end{cases}$

11. $\begin{cases} 2p - 3q = 9 \\ q = \frac{2}{3}p - 3 \end{cases}$

12. $\begin{cases} 3t_1 + 5t_2 = 1541 \\ t_2 = 2t_1 + 160 \end{cases}$

Solve the systems in Problems 13–18 by linear combinations. Relate each system to the general system $\begin{cases} a_{11}x_1 + a_{12}x_2 = b_1 \\ a_{21}x_1 + a_{22}x_2 = b_2 \end{cases}$.

13. $\begin{cases} c + d = 2 \\ 2c - d = 1 \end{cases}$

14. $\begin{cases} 2s_1 + s_2 = 10 \\ 5s_1 - 2s_2 = 16 \end{cases}$

15. $\begin{cases} 3q_1 - 4q_2 = 3 \\ 5q_1 + 3q_2 = 5 \end{cases}$

16. $\begin{cases} 9x + 3y = 5 \\ 3x + 2y = 2 \end{cases}$

17. $\begin{cases} 7x + y = 5 \\ 14x - 2y = -2 \end{cases}$

18. $\begin{cases} 2x + 3y = 1 \\ 3x - 2y = 0 \end{cases}$

Solve the systems in Problems 19–30 by any method.

19. $\begin{cases} 5x + 4y = 5 \\ 15x - 2y = 8 \end{cases}$

20. $\begin{cases} 3x + 2y = 1 \\ 6x + 4y = 2 \end{cases}$

21. $\begin{cases} 4x - 2y = -28 \\ y = \frac{1}{2}x + 5 \end{cases}$

22. $\begin{cases} 12x - 5y = -39 \\ y = 2x + 9 \end{cases}$

23. $\begin{cases} y = 2x - 1 \\ y = -3x - 9 \end{cases}$

24. $\begin{cases} y = \frac{2}{3}x - 5 \\ y = -\frac{4}{3}x + 7 \end{cases}$

25. $\begin{cases} 2x - 3y - 5 = 0 \\ 3x - 5y + 2 = 0 \end{cases}$

26. $\begin{cases} 2x + 3y = a \\ x - 5y = b \end{cases}$

27. $\begin{cases} ax + by = 1 \\ bx + ay = 0 \end{cases}$

28. $\begin{cases} 0.12x + 0.06y = 210 \\ x + y = 2000 \end{cases}$

29. $\begin{cases} 0.12x + 0.06y = 228 \\ x + y = 2000 \end{cases}$

30. $\begin{cases} 0.12x + 0.06y = 1140 \\ x + y = 10,000 \end{cases}$

Find the coordinates of the corner points for each of the regions in Problems 31–36.

31.

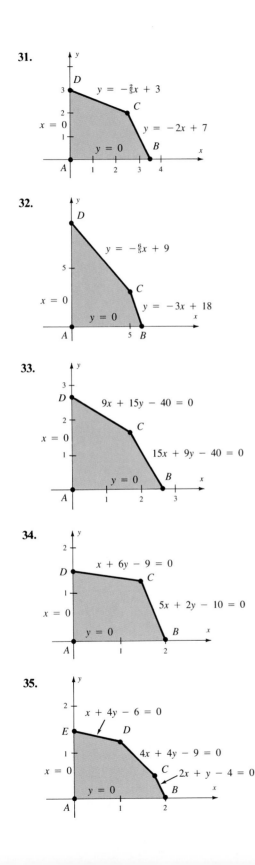

$y = -\frac{2}{3}x + 3$

$x = 0$

$y = -2x + 7$

$y = 0$

32.

$y = -\frac{6}{5}x + 9$

$x = 0$

$y = -3x + 18$

$y = 0$

33.

$9x + 15y - 40 = 0$

$x = 0$

$15x + 9y - 40 = 0$

$y = 0$

34.

$x + 6y - 9 = 0$

$5x + 2y - 10 = 0$

$x = 0$

$y = 0$

35.

$x + 4y - 6 = 0$

$4x + 4y - 9 = 0$

$x = 0$

$2x + y - 4 = 0$

$y = 0$

36.

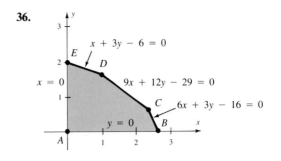

$x + 3y - 6 = 0$

$x = 0$

$9x + 12y - 29 = 0$

$6x + 3y - 16 = 0$

$y = 0$

APPLICATIONS

37. Suppose a car rental agency gives the following choices:

Option A: $30 per day plus 40¢ per mile

Option B: Flat $50 per day with unlimited mileage

At what mileage are both rates the same?

38. Suppose a car rental agency gives the following options:

Option A: $40 per day plus 50¢ per mile

Option B: Flat $60 per day with unlimited mileage

At what mileage are both rates the same?

39. A paint sprayer rents for $4 per hour or $24 per day. At how many hours are the rates the same?

40. Two car rental agencies have the following rates:

Agency A: $15 per day plus 20¢ per mile

Agency B: $25 per day plus 15¢ per mile

At what mileage are both rates the same?

41. The supply curve for a certain commodity is $n = 250p - 2500$, and the demand curve for the same commodity is $n = 3500 - 150p$. What should the price of the commodity be in order for the market to be stable?

42. The supply curve for a new product is $n = 2000p + 6000$, and the demand curve for the same product is $n = 13,000 - 1500p$. What is the equilibrium point for this product?

43. The supply curve for a new software product is $n = 2.5p - 500$, and the demand curve for the same product is $n = 200 - 0.5p$.

a. At $250 for the product, how many items would be supplied? How many demanded?

b. At what price would no items be supplied?

c. At what price would no items be demanded?

d. What is the equilibrium price for this product?

e. How many units will be produced at the equilibrium price?

44. The supply curve for a new commodity is $n = 2500p - 500$, and the demand curve for the same product is $n = 31,500 - 1500p$.
    a. At \$15 per commodity, how many items would be supplied? How many demanded?
    b. At what price would no items be supplied?
    c. At what price would no items be demanded?
    d. What is the equilibrium price for this product?
    e. How many units will be produced at the equilibrium price?

45. Suppose you want to sell T-shirts on your campus. You are trying to decide on a price between \$1 and \$7, so you conduct a market-research survey and find that 800 students will buy the shirt for \$1, but none will buy it for \$7. You also find a supplier who can supply 200 shirts if they are sold for \$1 and 600 if they are sold for \$7. At what price will supply equal demand, and how many could you expect to sell if both supply and demand are linear?

46. A new greeting card is introduced in a test market. Within the month, 3000 cards are sold priced at 50¢ and 1500 are sold priced at \$2. On the other hand, 2500 cards could be supplied at a cost of \$2 but only 1500 could be supplied at a cost of \$1. At what price does the supply equal the demand, assuming that both are linear?

47. A firm producing sunglasses has fixed costs of \$1,200 and variable costs of 80¢ per pair. Find the break-even point if the sunglasses sell for \$2 per pair.

48. Repeat Problem 47 if the sunglasses sell for \$3 per pair.

49. Microdrop Corporation produces floppy computer disks at a cost of \$1.20 per disk with total fixed costs of \$18,000. If the disks sell for \$3 each, what is the break-even point?

50. Repeat Problem 49 if the disks sell for \$2.50 each.

51. Hallmark introduces a new greeting card that has fixed costs of \$2,800 and variable costs of 75¢ per card. If the cards sell for \$1.50 each, what is the break-even point?

52. Repeat Problem 51 if the cards sell for \$1.25 each.

53. Repeat Problem 51 if the cards sell for \$2 each.

54. An investor owns Xerox and Standard Oil stock. The closing prices of each stock for the week are given in the table. If the value of the portfolio was \$19,500 on Monday and \$20,300 on Friday, how many shares of each stock does the investor own?

	Standard Oil	Xerox
Monday	45	60
Tuesday	$45\frac{1}{2}$	$62\frac{1}{8}$
Wednesday	46	63
Thursday	$47\frac{1}{4}$	$61\frac{1}{2}$
Friday	47	62

55. Another investor owns the same stocks as the investor described in Problem 54. This investor's portfolio was valued at \$34,500 on Monday and at \$35,600 on Wednesday. How many shares of each stock are owned by this investor?

56. The total number of registered Democrats and Republicans in a certain community is 100,005. Voter turnout in a recent election showed that 50% of the Democrats voted, 60% of the Republicans voted, and a total of 55,200 votes were cast, which also included Independents and write-in votes. If the registered Democrats outnumber the registered Republicans by 2 to 1, how many registered Democrats and Republicans are there in the community and how many independents or write-in votes were there in the election?

57. Two chemicals are combined to form two grades of fertilizer. One unit of grade A fertilizer requires 10 pounds of potassium and 3 pounds of calcium, while a unit of grade B fertilizer requires 9 pounds of potassium and 4 pounds of calcium. If 780 pounds of potassium and 260 pounds of calcium are available, how many units of each grade of fertilizer can be mixed?

58. *Historical Question* The Louvre Tablet from the Babylonian civilization is dated at about 1500 B.C. It shows a system equivalent to

$$\begin{cases} xy = 1 \\ x + y = a \end{cases}$$

Even though this is not a linear system, you can solve it by substitution if you let $a = 2$. Find x and y.

3.2 Introduction to Matrices

One of the biggest problems in applying mathematics to the real world is developing a means of systematizing and handling the great number of variables and large amounts of data that are inherent in real life situations. One step in handling large amounts of data is the creation of a mathematical model involving what are called **matrices**.

Definition of Matrix | A *matrix* is a rectangular array of numbers. (The plural is *matrices*.)

You are already familiar with matrices from everyday experiences. For example, Table 3.1 shows a car rental chart in the form of a matrix.

TABLE 3.1
Example of a Matrix: Costs of
Weekly Car Rentals in Europe

Country	Fiat Panda	Ford Fiesta	Opel Kadett	Renault R14	Volkswagen Microbus
Austria	US $149	US $179	US $219	US $269	US $289
Belgium	130	143	222	273	425
Denmark	164	212	269	408	480
France	189	214	257	326	386
Great Britain	160	174	206	215	243
Holland	156	183	213	259	305
Ireland	176	185	198	222	233
Italy	179	213	259	353	408
Luxembourg	130	143	222	273	425
Spain	156	188	206	247	306
Sweden	178	205	246	281	315

In mathematics we enclose the array of numbers in brackets and denote the entire array by using a capital letter. The size of the matrix is called the **order** and is specified by naming the number of **rows** (rows are horizontal) first and then the number of **columns** (columns are vertical). A matrix with order $m \times n$ (read "m by n") means that the matrix has m rows and n columns. The order of the matrix in Table 3.1 is 11×5 (eleven by five). The price of the Renault in Denmark is found in row 3, column 4.

EXAMPLE 1 Name the order of each given matrix.

$$A = \begin{bmatrix} 1 & 6 & 3 \\ 4 & 8 & -2 \end{bmatrix} \qquad B = \begin{bmatrix} 6 & 1 \\ 4 & -3 \\ 2 & 0 \\ -1 & 4 \end{bmatrix} \qquad C = \begin{bmatrix} 5 \\ 8 \\ 1 \end{bmatrix}$$

$$D = \begin{bmatrix} 4 & 8 & 6 & 2 \end{bmatrix} \qquad E = \begin{bmatrix} 6 & 8 & -2 \\ 5 & 4 & -1 \\ 0 & -5 & 7 \end{bmatrix}$$

Solution The order of A is 2×3; B is 4×2; C is 3×1; D is 1×4; and E is 3×3. ∎

If the matrix has only one column (for example, matrix C), then it is called a **column matrix**; if it has only one row (matrix D, for example), then it is called a **row matrix**; and if the number of rows and columns are the same, it is called a **square matrix**. The numbers in a matrix are called the **entries**, or **components**, of a matrix. If all the entries of a matrix are zero, it is called a **zero matrix** and is represented by **0**.

Two matrices are said to be **equal** if they have the same order and if the corresponding entries are equal. (Corresponding entries are numbers in the same position—that is, same row and column location.)

EXAMPLE 2 Find x and y, if possible, so that the pairs of matrices are equal.

a. $\begin{bmatrix} 2 & x \\ y & -3 \end{bmatrix} = \begin{bmatrix} 2 & 5 \\ -1 & -3 \end{bmatrix}$

By inspection, $x = 5$ and $y = -1$.

b. $\begin{bmatrix} 6 & 4 \\ x & y \end{bmatrix} = \begin{bmatrix} 6 & 0 \\ 4 & -7 \end{bmatrix}$

No values of x and y will make these matrices equal since $4 \neq 0$ in the upper right entries.

c. $\begin{bmatrix} x & 8 \\ y & -4 \end{bmatrix} = \begin{bmatrix} 2 & 8 & -1 \\ 3 & -4 & 0 \end{bmatrix}$

No values of x and y will make these matrices equal since the matrices must have the same order to be equal. ■

EXAMPLE 3 **Inventory Control** Suppose a company produces two models of wireless telephones and manufactures them at four factories. The number of units produced at each factory can be summarized by the following matrix:

	Factory			
	A	B	C	D
Standard model	16	25	15	8
Deluxe model	10	0	14	3

If the entries represent thousands of units, answer the following questions by reading the matrix:

a. How many standard models are produced at factory D?
b. How many standard models are produced at factory A?
c. How many deluxe models are produced at factory C?
d. At which factory are no deluxe models produced?

Solution **a.** 8000   **b.** 16,000   **c.** 14,000   **d.** Factory B ■

Sometimes entries 1 and 0 are used to summarize information in matrix form, as illustrated by Example 4.

EXAMPLE 4 **Communication Theory** Use matrix notation to summarize the following:

The United States has diplomatic relations with the Soviet Union and Mexico, but not with Cuba.

Mexico has diplomatic relations with the United States and the Soviet Union, but not with Cuba.

The Soviet Union has diplomatic relations with the United States, Mexico, and Cuba.

Cuba has diplomatic relations with the Soviet Union, but not with the United States and Mexico.

Solution Use 1 if the row entry and the corresponding column entry have diplomatic relations and 0 if they do not. Then:

	U.S.	U.S.S.R.	Cuba	Mexico
U.S.	0	1	0	1
U.S.S.R.	1	0	1	1
Cuba	0	1	0	0
Mexico	1	1	0	0

∎

The solution to Example 4 is called a **communication matrix** or *incidence matrix*.

If two matrices have the same order, we define the **sum of the matrices** as the matrix whose components result from the sum of the corresponding components of the given matrices. We also say the original matrices are **conformable**. If the matrices do not have the same order, then we say the matrices are **not conformable** for addition.

EXAMPLE 5 Let

$$\mathbf{0} = [0 \quad 0 \quad 0] \qquad A = [4 \quad 2 \quad 6]$$

$$B = \begin{bmatrix} 6 \\ 1 \end{bmatrix} \qquad C = [1 \quad 7 \quad 2] \qquad D = \begin{bmatrix} 1 \\ 2 \end{bmatrix}$$

$$E = \begin{bmatrix} 2 & 1 & 0 \\ 4 & 7 & 3 \\ -2 & 0 & 1 \end{bmatrix} \qquad F = \begin{bmatrix} 6 & 1 & 2 \\ 3 & -10 & 4 \\ 1 & 3 & -2 \end{bmatrix}$$

Find (wherever possible):

a. $A + C$ **b.** $B + D$ **c.** $A + \mathbf{0}$ **d.** $A + B$ **e.** $E + F$

Solution **a.** $A + C = [4 + 1 \quad 2 + 7 \quad 6 + 2] = [5 \quad 9 \quad 8]$

b. $B + D = \begin{bmatrix} 6 + 1 \\ 1 + 2 \end{bmatrix} = \begin{bmatrix} 7 \\ 3 \end{bmatrix}$

c. $A + \mathbf{0} = [4 + 0 \quad 2 + 0 \quad 6 + 0] = [4 \quad 2 \quad 6]$

d. Not conformable

e. $E + F = \begin{bmatrix} 2 + 6 & 1 + 1 & 0 + 2 \\ 4 + 3 & 7 + (-10) & 3 + 4 \\ -2 + 1 & 0 + 3 & 1 + (-2) \end{bmatrix} = \begin{bmatrix} 8 & 2 & 2 \\ 7 & -3 & 7 \\ -1 & 3 & -1 \end{bmatrix}$

∎

EXAMPLE 6   Let S_{1987} and S_{1988} represent the sales (in thousands of dollars) for a company in 1987 and 1988, respectively:

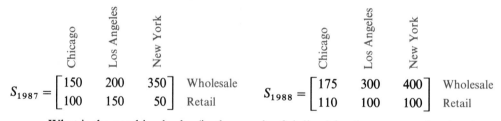

$$S_{1987} = \begin{bmatrix} 150 & 200 & 350 \\ 100 & 150 & 50 \end{bmatrix} \begin{matrix} \text{Wholesale} \\ \text{Retail} \end{matrix} \qquad S_{1988} = \begin{bmatrix} 175 & 300 & 400 \\ 110 & 100 & 100 \end{bmatrix} \begin{matrix} \text{Wholesale} \\ \text{Retail} \end{matrix}$$

What is the combined sales (in thousands of dollars) for the company for the given years?

Solution   The answer is found by finding the sum of the matrices:

$$S_{1987} + S_{1988} = \begin{bmatrix} 150 + 175 & 200 + 300 & 350 + 400 \\ 100 + 110 & 150 + 100 & 50 + 100 \end{bmatrix}$$

$$= \begin{bmatrix} 325 & 500 & 750 \\ 210 & 250 & 150 \end{bmatrix}$$

For the company in Example 6, we might be interested in finding the increase (or decrease) in sales between 1987 and 1988. We see that the difference in wholesale sales in Chicago is found by subtracting:

$$175 - 150 = 25$$

The difference for the other components is found similarly. The **difference of matrices** A and B of the same order, $A - B$, is found by subtracting the entries of B from the corresponding entries of A.

EXAMPLE 7   Find the change in sales between 1987 and 1988 for the company with sales S_{1987} and S_{1988} in Example 6.

Solution   The answer is found by finding the difference of the matrices:

$$S_{1988} - S_{1987} = \begin{bmatrix} 175 - 150 & 300 - 200 & 400 - 350 \\ 110 - 100 & 100 - 150 & 100 - 50 \end{bmatrix}$$

$$= \begin{bmatrix} 25 & 100 & 50 \\ 10 & -50 & 50 \end{bmatrix}$$

The negative entry indicates a decrease in sales.

Continuing with the same example, suppose we want to find the profit generated by the sales in each category and have found that 3% of the gross sales for a given year is actual profit. Then, we find the 1988 profit from wholesale sales in Chicago by multiplying 0.03 (3%) by 175 (the wholesale sales in thousands of dollars in Chicago in 1988):

$$0.03 \cdot 175 = 5.25$$

To find the profit in each category, we multiply the matrix

$$S_{1985} = \begin{bmatrix} 175 & 300 & 400 \\ 110 & 100 & 100 \end{bmatrix}$$

by 0.03. This operation of multiplying each entry of a matrix A by a number c is denoted by cA and is called **scalar multiplication**.

EXAMPLE 8   Find the profit (in thousands of dollars) for S_{1988}.

Solution   Calculate $(0.03)S_{1988}$.

$$(0.03)S_{1988} = \begin{bmatrix} (0.03)(175) & (0.03)(300) & (0.03)(400) \\ (0.03)(110) & (0.03)(100) & (0.03)(100) \end{bmatrix}$$

$$= \begin{bmatrix} 5.25 & 9 & 12 \\ 3.3 & 3 & 3 \end{bmatrix} \qquad ∎$$

You now know how to represent information in matrix form, how to add and subtract matrices, and how to multiply a matrix and a number (scalar multiplication). We now turn our attention to multiplying two matrices.

It might seem natural to multiply two matrices by multiplying corresponding entries. However, such a definition has not proved to be useful, and it turns out that the definition that is useful is rather unusual, so we will consider several examples before formally defining matrix multiplication.

Example 8 illustrates scalar multiplication by considering a 3% profit in three sales locations, as given by

$$(0.03)S_{1988} = (0.3) \begin{matrix} & \text{Chicago} & \text{Los Angeles} & \text{New York} \\ \begin{bmatrix} 175 & 300 & 400 \\ 110 & 100 & 100 \end{bmatrix} & & \end{matrix} \begin{matrix} \text{Wholesale} \\ \text{Retail} \end{matrix}$$

Now, however, suppose an inventory tax of 3% is imposed on wholesale sales and a 6% tax is imposed on retail sales. What is the total tax (in thousands of dollars) at each location? First, the necessary calculations to answer this question must be summarized:

	Wholesale tax	+	Retail tax		Total tax
Chicago tax:	$0.03(175)$	+	$0.06(110)$	$= 5.25 + 6.6$	$= 11.85$
Los Angeles tax:	$0.03(300)$	+	$0.06(100)$	$= 9 + 6$	$= 15$
New York tax:	$0.03(400)$	+	$0.06(100)$	$= 12 + 6$	$= 18$

These calculations summarize the operation of matrix multiplication.

COMPUTER APPLICATION
See Program 2 on the computer disk.

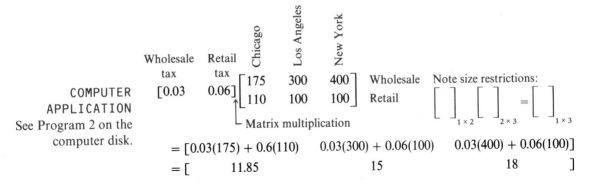

$$= [0.03(175) + 0.6(110) \qquad 0.03(300) + 0.06(100) \qquad 0.03(400) + 0.06(100)]$$

$$= [\qquad 11.85 \qquad\qquad\qquad 15 \qquad\qquad\qquad 18 \qquad]$$

The operation of matrix multiplication has widespread applications, but it is difficult to understand, so we will proceed slowly through several examples, each progressing from the previous one.

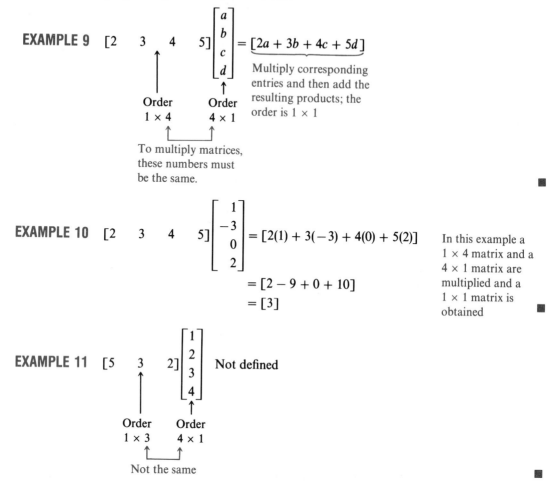

EXAMPLE 9
$$[2 \quad 3 \quad 4 \quad 5]\begin{bmatrix} a \\ b \\ c \\ d \end{bmatrix} = \underbrace{[2a + 3b + 4c + 5d]}$$

Multiply corresponding entries and then add the resulting products; the order is 1×1

Order 1×4 Order 4×1

To multiply matrices, these numbers must be the same.

EXAMPLE 10
$$[2 \quad 3 \quad 4 \quad 5]\begin{bmatrix} 1 \\ -3 \\ 0 \\ 2 \end{bmatrix} = [2(1) + 3(-3) + 4(0) + 5(2)]$$
$$= [2 - 9 + 0 + 10]$$
$$= [3]$$

In this example a 1×4 matrix and a 4×1 matrix are multiplied and a 1×1 matrix is obtained

EXAMPLE 11
$$[5 \quad 3 \quad 2]\begin{bmatrix} 1 \\ 2 \\ 3 \\ 4 \end{bmatrix} \quad \text{Not defined}$$

Order 1×3 Order 4×1

Not the same

The process illustrated in Examples 9–11 is repeated over and over for larger-order matrices. If two matrices cannot be multiplied because of their order (as in Example 11), then they are said to be **nonconformable** for multiplication.

You can multiply matrices that are not both row or column matrices. For example,

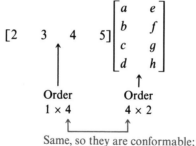

$$[2 \quad 3 \quad 4 \quad 5]\begin{bmatrix} a & e \\ b & f \\ c & g \\ d & h \end{bmatrix}$$

Order 1×4 Order 4×2

Same, so they are conformable; the answer is a 1×2 matrix

If there are two columns, then the product is found in two steps:

$$[2 \quad 3 \quad 4 \quad 5] \begin{bmatrix} a & e \\ b & f \\ c & g \\ d & h \end{bmatrix} = [2a + 3b + 4c + 5d \qquad]$$

Row 1, column 1 ⎯⎯⎯⎯⎯⎯⎯

Answer for position 1, 1

$$[2 \quad 3 \quad 4 \quad 5] \begin{bmatrix} a & e \\ b & f \\ c & g \\ d & h \end{bmatrix} = [2a + 3b + 4c + 5d \qquad 2e + 3f + 4g + 5h]$$

Row 1, column 2 ⎯⎯⎯⎯⎯⎯⎯

Answer for position 1, 2

EXAMPLE 12
$$[2 \quad 3 \quad 4 \quad 5] \begin{bmatrix} 1 & 2 \\ -3 & 1 \\ 0 & -2 \\ 2 & 3 \end{bmatrix} = [2(1) + 3(-3) + 4(0) + 5(2) \qquad]$$

$$[2 \quad 3 \quad 4 \quad 5] \begin{bmatrix} 1 & 2 \\ -3 & 1 \\ 0 & -2 \\ 2 & 3 \end{bmatrix} = [\mathbf{3} \qquad]$$

$$[2 \quad 3 \quad 4 \quad 5] \begin{bmatrix} 1 & 2 \\ -3 & 1 \\ 0 & -2 \\ 2 & 3 \end{bmatrix} = [3 \qquad 2(2) + 3(1) + 4(-2) + 5(3)]$$

$$[2 \quad 3 \quad 4 \quad 5] \begin{bmatrix} 1 & 2 \\ -3 & 1 \\ 0 & -2 \\ 2 & 3 \end{bmatrix} = [3 \qquad \mathbf{14}]$$ ■

EXAMPLE 13
$$[2 \quad -1 \quad 2] \begin{bmatrix} 4 & 2 & -1 \\ 1 & 0 & 2 \\ 3 & -1 & 3 \end{bmatrix} = [2(4) + (-1)1 + 2(3) \qquad]$$

$$[2 \quad -1 \quad 2] \begin{bmatrix} 4 & 2 & -1 \\ 1 & 0 & 2 \\ 3 & -1 & 3 \end{bmatrix} = [13 \qquad 2(2) + (-1)(0) + 2(-1) \qquad]$$

$$[2 \quad -1 \quad 2] \begin{bmatrix} 4 & 2 & -1 \\ 1 & 0 & 2 \\ 3 & -1 & 3 \end{bmatrix} = [13 \qquad 2 \qquad 2(-1) + (-1)(2) + 2(3)]$$

$$= [13 \qquad 2 \qquad 2]$$ ■

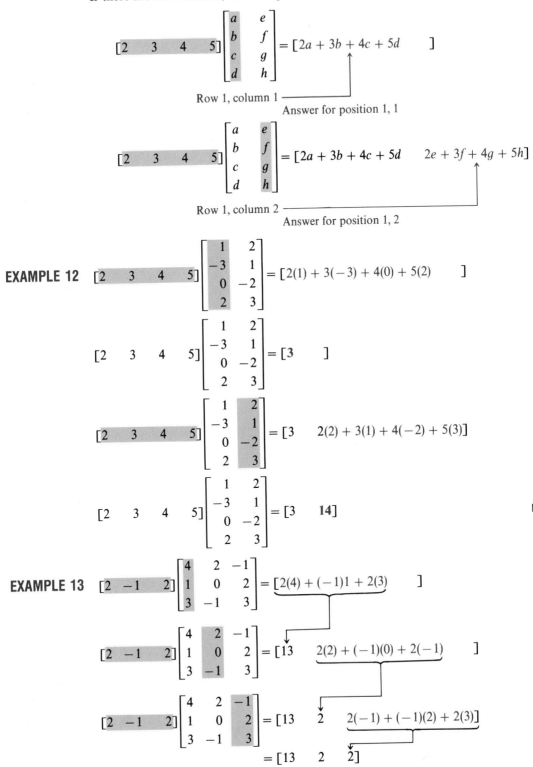

Example 13 is written out to clearly demonstrate what is happening. Your work, however, would probably look like this:

Be sure you know where these numbers came from; this is the way work will be shown in subsequent examples

$$[2 \quad -1 \quad 2]\begin{bmatrix} 4 & 2 & -1 \\ 1 & 0 & 2 \\ 3 & -1 & 3 \end{bmatrix} = [8-1+6 \quad 4+0-2 \quad -2-2+6]$$

$$= [13 \quad 2 \quad 2]$$

EXAMPLE 14 $\begin{bmatrix} 1 & 3 & 2 \\ -1 & 4 & 3 \end{bmatrix}\begin{bmatrix} 4 & 2 \\ -1 & 3 \\ 2 & -3 \end{bmatrix} = \begin{bmatrix} 4-3+4 & 2+9-6 \\ -4-4+6 & -2+12-9 \end{bmatrix} = \begin{bmatrix} 5 & 5 \\ -2 & 1 \end{bmatrix}$ ■

Now, to define matrix multiplication, let the symbol

$$A = [a_{ij}]_{m,n}$$

mean A is a matrix of order m by n with $1 \le i \le m$ and $1 \le j \le n$. That is, a_{23} is the entry in the second row, third column, or, more generally, a_{ij} is an element of A in the ith row, jth column. Now let $A = [a_{ij}]_{m,r}$, $B = [b_{ij}]_{p,n}$, and $C = [c_{ij}]_{m,n}$, and write $AB = C$, where C is called the **product** of A and B. If $p = r$, then matrix multiplication is possible, so we can define multiplication for

$$A = [a_{ij}]_{m,r} \quad \text{and} \quad B = [b_{ij}]_{r,n}$$

These numbers tell the size of the product matrix

Note that these must be the same for multiplication to be defined

Matrix
Multiplication

$AB = C$, where

$$[c_{ij}]_{m,n} = [a_{i1} \quad a_{i2} \quad \cdots \quad a_{ir}]\begin{bmatrix} b_{1j} \\ b_{2j} \\ \vdots \\ b_{rj} \end{bmatrix}$$

$$= [a_{i1}b_{1j} + a_{i2}b_{2j} + \cdots + a_{ir}b_{rj}]$$

EXAMPLE 15 Give the *order* of the following products and give the *position* of the entry found by the highlighted product. If multiplication is not possible, say so.

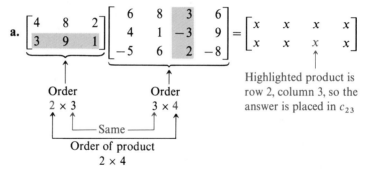

a. $\begin{bmatrix} 4 & 8 & 2 \\ 3 & 9 & 1 \end{bmatrix}\begin{bmatrix} 6 & 8 & 3 & 6 \\ 4 & 1 & -3 & 9 \\ -5 & 6 & 2 & -8 \end{bmatrix} = \begin{bmatrix} x & x & x & x \\ x & x & x & x \end{bmatrix}$

Order
2 × 3

Order
3 × 4

Same

Order of product
2 × 4

Highlighted product is row 2, column 3, so the answer is placed in c_{23}

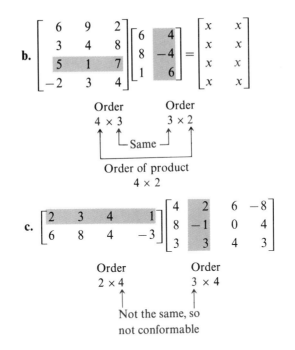

b.

$$
\begin{bmatrix} 6 & 9 & 2 \\ 3 & 4 & 8 \\ 5 & 1 & 7 \\ -2 & 3 & 4 \end{bmatrix}
\begin{bmatrix} 6 & 4 \\ 8 & -4 \\ 1 & 6 \end{bmatrix}
=
\begin{bmatrix} x & x \\ x & x \\ x & x \\ x & x \end{bmatrix}
$$

Order   Order
4 × 3   3 × 2
└── Same ──┘
Order of product
4 × 2

c.

$$
\begin{bmatrix} 2 & 3 & 4 & 1 \\ 6 & 8 & 4 & -3 \end{bmatrix}
\begin{bmatrix} 4 & 2 & 6 & -8 \\ 8 & -1 & 0 & 4 \\ 3 & 3 & 4 & 3 \end{bmatrix}
$$

Order   Order
2 × 4   3 × 4

Not the same, so
not conformable

For Examples 16 and 17, let

$$
A = \begin{bmatrix} 1 & 2 \\ 3 & 4 \end{bmatrix} \quad \text{and} \quad B = \begin{bmatrix} -1 & 2 \\ 1 & 3 \end{bmatrix}
$$

EXAMPLE 16 $AB = \begin{bmatrix} 1 & 2 \\ 3 & 4 \end{bmatrix} \begin{bmatrix} -1 & 2 \\ 1 & 3 \end{bmatrix} = \begin{bmatrix} -1+2 & 2+6 \\ -3+4 & 6+12 \end{bmatrix} = \begin{bmatrix} 1 & 8 \\ 1 & 18 \end{bmatrix}$

EXAMPLE 17 $BA = \begin{bmatrix} -1 & 2 \\ 1 & 3 \end{bmatrix} \begin{bmatrix} 1 & 2 \\ 3 & 4 \end{bmatrix} = \begin{bmatrix} -1+6 & -2+8 \\ 1+9 & 2+12 \end{bmatrix} = \begin{bmatrix} 5 & 6 \\ 10 & 14 \end{bmatrix}$

Note from Examples 16 and 17 that $AB \neq BA$. That is, matrix multiplication is *not commutative*. This means you must be careful not to switch the order of the matrix factors when working with matrix products.

This section concludes with some models that use matrix multiplication.

EXAMPLE 18   A developer builds a housing complex featuring two-, three-, and four-bedroom units. Each unit comes in two different floor plans. The matrix P (for production) tells the number of each type of unit for this development.

Plan I   Plan II

$$
P = \begin{bmatrix} 10 & 5 \\ 25 & 10 \\ 15 & 10 \end{bmatrix}
\begin{matrix} \text{2 bedrooms} \\ \text{3 bedrooms} \\ \text{4 bedrooms} \end{matrix}
$$

Many materials are used in building these homes, but this model will be simplified to include only lumber, concrete, fixtures, and labor. The matrix M (for materials) gives the amounts of these materials used (in appropriate units of each).*

$$M = \begin{bmatrix} 7 & 8 & 9 & 20 \\ 8 & 9 & 9 & 22 \end{bmatrix} \begin{array}{l} \text{Plan I} \\ \text{Plan II} \end{array}$$

with column labels Lumber, Concrete, Fixtures, Labor.

The matrix product PM gives the amount of material needed for the development.

$$PM = \begin{array}{l} 2 \\ 3 \\ 4 \end{array} \begin{bmatrix} 10 & 5 \\ 25 & 10 \\ 15 & 10 \end{bmatrix} \begin{bmatrix} 7 & 8 & 9 & 20 \\ 8 & 9 & 9 & 22 \end{bmatrix}$$

with column labels I, II over the first matrix and Lumber, Concrete, Fixtures, Labor over the second.

$$= \begin{bmatrix} 70 + 40 & 80 + 45 & 90 + 45 & 200 + 110 \\ 175 + 80 & 200 + 90 & 225 + 90 & 500 + 220 \\ 105 + 80 & 120 + 90 & 135 + 90 & 300 + 220 \end{bmatrix} \begin{array}{l} \text{2 bedrooms} \\ \text{3 bedrooms} \\ \text{4 bedrooms} \end{array}$$

with column labels Lumber, Concrete, Fixtures, Labor.

$$= \begin{bmatrix} 110 & 125 & 135 & 310 \\ 255 & 290 & 315 & 720 \\ 185 & 210 & 225 & 520 \end{bmatrix} \begin{array}{l} \text{2 bedrooms} \\ \text{3 bedrooms} \\ \text{4 bedrooms} \end{array}$$

with column labels Lumber, Concrete, Fixtures, Labor.

If the cost of each unit of material is given by the matrix C (for cost), then the total cost of each model for this development is $(PM)C$.

Cost per unit

$$C = \begin{bmatrix} 800 \\ 90 \\ 1000 \\ 1000 \end{bmatrix} \begin{array}{l} \text{Lumber} \\ \text{Concrete} \\ \text{Fixtures} \\ \text{Labor} \end{array}$$

$$(PM)C = \begin{bmatrix} 110 & 125 & 135 & 310 \\ 255 & 290 & 315 & 720 \\ 185 & 210 & 225 & 520 \end{bmatrix} \begin{bmatrix} 800 \\ 90 \\ 1000 \\ 1000 \end{bmatrix} = \begin{bmatrix} 544,250 \\ 1,265,100 \\ 911,900 \end{bmatrix} \begin{array}{l} \text{2 bedrooms} \\ \text{3 bedrooms} \\ \text{4 bedrooms} \end{array}$$

Since 15 of the units have two bedrooms, the average cost per unit is $36,283 ($544,250 ÷ 15). For three-bedroom units the average cost per unit is $36,146, and for four-bedroom units the average cost is $36,476. ■

EXAMPLE 19 In Example 4 a matrix was developed to represent the diplomatic relations of the United States, the Soviet Union, Cuba, and Mexico. Let A represent the communi-

---

* For example, lumber in 1000 board feet, concrete in cubic yards, etc.

cation matrix from that example:

$$
A = \begin{array}{c c c c c} & \text{U.S.} & \text{U.S.S.R.} & \text{Cuba} & \text{Mexico} \\ \left[\begin{array}{cccc} 0 & 1 & 0 & 1 \\ 1 & 0 & 1 & 1 \\ 0 & 1 & 0 & 0 \\ 1 & 1 & 0 & 0 \end{array}\right] & \begin{array}{l} \text{U.S.} \\ \text{U.S.S.R.} \\ \text{Cuba} \\ \text{Mexico} \end{array} \end{array}
$$

How many channels of communication are open to the various countries if they are willing to speak through an intermediary?

Solution Consider a particular example, say, Mexico sending a message to Cuba through a single intermediary. Mexico can send a message to the United States, but the United States cannot pass the message along to Cuba; Mexico can send a message to the Soviet Union and then the Soviet Union can forward it to Cuba; Mexico to Cuba direct does not qualify as sending the message through an intermediary. Therefore, there is only one way that Mexico can send a message to Cuba via an intermediary. How about the United States sending a message to the United States through an intermediary? (Perhaps to test for leaks in the communications network!) The United States can send a message to the Soviet Union and then back to the United States, as well as to Mexico and then back. Thus there are two ways that the United States can send a message to itself through an intermediary. Do you see that there are three ways that the Soviet Union can send a round-trip message through one intermediary? All of this information can be summarized in matrix form:

$$
\begin{array}{c c c c c} & \text{U.S.} & \text{U.S.S.R.} & \text{Cuba} & \text{Mexico} \\ \begin{array}{l} \text{U.S.} \\ \text{U.S.S.R.} \\ \text{Cuba} \\ \text{Mexico} \end{array} & \left[\begin{array}{cccc} 2 & 1 & 1 & 1 \\ 1 & 3 & 0 & 1 \\ 1 & 0 & 1 & 1 \\ 1 & 1 & 1 & 2 \end{array}\right] \end{array}
$$

This matrix is the same as $AA = A^2$:

$$
A^2 = \begin{bmatrix} 0 & 1 & 0 & 1 \\ 1 & 0 & 1 & 1 \\ 0 & 1 & 0 & 0 \\ 1 & 1 & 0 & 0 \end{bmatrix} \begin{bmatrix} 0 & 1 & 0 & 1 \\ 1 & 0 & 1 & 1 \\ 0 & 1 & 0 & 0 \\ 1 & 1 & 0 & 0 \end{bmatrix}
$$

$$
= \begin{bmatrix} 0+1+0+1 & 0+0+0+1 & 0+1+0+0 & 0+1+0+0 \\ 0+0+0+1 & 1+0+1+1 & 0+0+0+0 & 1+0+0+0 \\ 0+1+0+0 & 0+0+0+0 & 0+1+0+0 & 0+1+0+0 \\ 0+1+0+0 & 1+0+0+0 & 0+1+0+0 & 1+1+0+0 \end{bmatrix}
$$

$$
= \begin{array}{c c c c c} & \text{U.S.} & \text{U.S.S.R.} & \text{Cuba} & \text{Mexico} \\ \left[\begin{array}{cccc} 2 & 1 & 1 & 1 \\ 1 & 3 & 0 & 1 \\ 1 & 0 & 1 & 1 \\ 1 & 1 & 1 & 2 \end{array}\right] & \begin{array}{l} \text{U.S.} \\ \text{U.S.S.R.} \\ \text{Cuba} \\ \text{Mexico} \end{array} \end{array}
$$

■

In general, given a communication matrix A:

One intermediary message—one multiplication: $AA = A^2$

Two intermediary messages—two multiplications: $AAA = A^3$

Three intermediary messages—three multiplications: $AAAA = A^4$

And so on.

Another common application of communication matrices is the network of airline routes in the United States. For example, there are several different ways to fly from Kansas City to San Francisco: direct, through one intermediate city, two intermediate cities, and so on. Problem 67 in Problem Set 3.2 involves this type of application.

Problem Set 3.2

State the order of each matrix in Problems 1–2 and if it is a square matrix, a column matrix, or a row matrix.

1. a. $A = \begin{bmatrix} 6 & 1 & 4 \\ 7 & 9 & 2 \\ 1 & 5 & 3 \end{bmatrix}$ **b.** $B = \begin{bmatrix} 2 \\ 1 \\ 5 \end{bmatrix}$

c. $C = \begin{bmatrix} 4 & 9 \\ 1 & 6 \\ 7 & 5 \end{bmatrix}$ **d.** $D = \begin{bmatrix} 4 \\ 0 \\ 1 \\ 3 \end{bmatrix}$

e. $E = \begin{bmatrix} 4 & 1 & 7 \\ 9 & 6 & 5 \end{bmatrix}$

f. $F = \begin{bmatrix} 5 & 0 & 1 & 2 \end{bmatrix}$

g. $G = \begin{bmatrix} 6 & 9 \end{bmatrix}$

2. a. $H = \begin{bmatrix} 6 & 5 \\ 9 & 2 \end{bmatrix}$ **b.** $I = \begin{bmatrix} 1 & 0 & 0 \\ 0 & 1 & 0 \\ 0 & 0 & 1 \end{bmatrix}$

c. $J = \begin{bmatrix} 0 & 1 \\ 1 & 0 \end{bmatrix}$ **d.** $K = \begin{bmatrix} 1 & 0 \\ 0 & 1 \\ 0 & 0 \\ 0 & 0 \\ 0 & 0 \end{bmatrix}$

e. $L = \begin{bmatrix} 1 & 0 & 0 & 0 \\ 0 & 1 & 0 & 0 \end{bmatrix}$

f. $M = \begin{bmatrix} 5 \end{bmatrix}$

In Problems 3–4 find replacements of the variables that make the matrices equal. If it is not possible to make the matrices equal, state that.

3. a. $\begin{bmatrix} 4 & 8 \\ -1 & x \end{bmatrix} = \begin{bmatrix} y & 8 \\ -1 & 2 \end{bmatrix}$

b. $\begin{bmatrix} 6 \\ x \\ y \end{bmatrix} = \begin{bmatrix} z \\ 4 \\ -9 \end{bmatrix}$

c. $\begin{bmatrix} 1 & 0 \\ 0 & 1 \end{bmatrix} = \begin{bmatrix} a & b \\ c & d \end{bmatrix}$

4. a. $\begin{bmatrix} 0 & x \\ 0 & 1 \end{bmatrix} = \begin{bmatrix} 1 & 2 \\ 0 & y \end{bmatrix}$

b. $\begin{bmatrix} 2 & 4 & 6 \\ 5 & 9 & 3 \end{bmatrix} = \begin{bmatrix} 2 & x & y \\ z & 9 & 3 \end{bmatrix}$

c. $\begin{bmatrix} -1 & 2 & 3 \\ 4 & x & 0 \end{bmatrix} = \begin{bmatrix} -1 & 2 & y \\ z & 2 & 1 \end{bmatrix}$

5. A company supplies four parts for General Motors. The parts are manufactured at four factories. Summarize the following production information using a matrix: Factory A produces 25 units of part one, 42 units of part two, and 193 units of part three; factory B produces 16 units of part one, 39 units of part two, and 150 units of part three; factory C produces 50 units of each part; and factory D produces 320 units of part four only.

6. a. Write Table 3.1 in matrix form.
   b. What is the entry in row 2, column 3?
   c. What is the entry in row 3, column 2?

d. In which row and column is the price of a Volkswagen Microbus in Sweden?

e. In which row and column is the price of an Opel Kadett in Holland?

f. Would the prices of a Volkswagen Microbus in various countries be represented as a row matrix or a column matrix?

In Problems 7–8 give the order of the product and tell the position of the entry found by the highlighted product. If multiplication is not possible, state that.

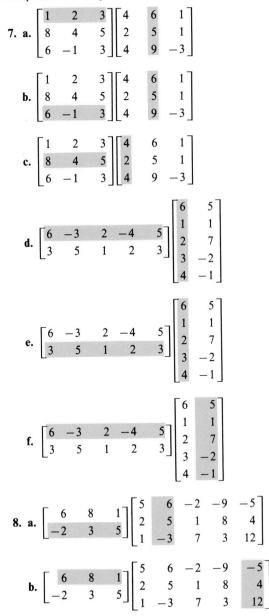

7. a.
$$\begin{bmatrix} 1 & 2 & 3 \\ 8 & 4 & 5 \\ 6 & -1 & 3 \end{bmatrix} \begin{bmatrix} 4 & 6 & 1 \\ 2 & 5 & 1 \\ 4 & 9 & -3 \end{bmatrix}$$

b.
$$\begin{bmatrix} 1 & 2 & 3 \\ 8 & 4 & 5 \\ 6 & -1 & 3 \end{bmatrix} \begin{bmatrix} 4 & 6 & 1 \\ 2 & 5 & 1 \\ 4 & 9 & -3 \end{bmatrix}$$

c.
$$\begin{bmatrix} 1 & 2 & 3 \\ 8 & 4 & 5 \\ 6 & -1 & 3 \end{bmatrix} \begin{bmatrix} 4 & 6 & 1 \\ 2 & 5 & 1 \\ 4 & 9 & -3 \end{bmatrix}$$

d.
$$\begin{bmatrix} 6 & -3 & 2 & -4 & 5 \\ 3 & 5 & 1 & 2 & 3 \end{bmatrix} \begin{bmatrix} 6 & 5 \\ 1 & 1 \\ 2 & 7 \\ 3 & -2 \\ 4 & -1 \end{bmatrix}$$

e.
$$\begin{bmatrix} 6 & -3 & 2 & -4 & 5 \\ 3 & 5 & 1 & 2 & 3 \end{bmatrix} \begin{bmatrix} 6 & 5 \\ 1 & 1 \\ 2 & 7 \\ 3 & -2 \\ 4 & -1 \end{bmatrix}$$

f.
$$\begin{bmatrix} 6 & -3 & 2 & -4 & 5 \\ 3 & 5 & 1 & 2 & 3 \end{bmatrix} \begin{bmatrix} 6 & 5 \\ 1 & 1 \\ 2 & 7 \\ 3 & -2 \\ 4 & -1 \end{bmatrix}$$

8. a.
$$\begin{bmatrix} 6 & 8 & 1 \\ -2 & 3 & 5 \end{bmatrix} \begin{bmatrix} 5 & 6 & -2 & -9 & -5 \\ 2 & 5 & 1 & 8 & 4 \\ 1 & -3 & 7 & 3 & 12 \end{bmatrix}$$

b.
$$\begin{bmatrix} 6 & 8 & 1 \\ -2 & 3 & 5 \end{bmatrix} \begin{bmatrix} 5 & 6 & -2 & -9 & -5 \\ 2 & 5 & 1 & 8 & 4 \\ 1 & -3 & 7 & 3 & 12 \end{bmatrix}$$

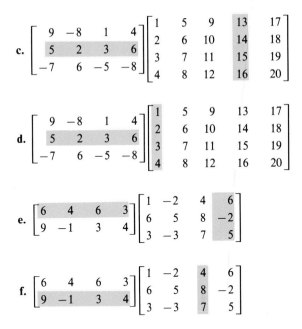

c.
$$\begin{bmatrix} 9 & -8 & 1 & 4 \\ 5 & 2 & 3 & 6 \\ -7 & 6 & -5 & -8 \end{bmatrix} \begin{bmatrix} 1 & 5 & 9 & 13 & 17 \\ 2 & 6 & 10 & 14 & 18 \\ 3 & 7 & 11 & 15 & 19 \\ 4 & 8 & 12 & 16 & 20 \end{bmatrix}$$

d.
$$\begin{bmatrix} 9 & -8 & 1 & 4 \\ 5 & 2 & 3 & 6 \\ -7 & 6 & -5 & -8 \end{bmatrix} \begin{bmatrix} 1 & 5 & 9 & 13 & 17 \\ 2 & 6 & 10 & 14 & 18 \\ 3 & 7 & 11 & 15 & 19 \\ 4 & 8 & 12 & 16 & 20 \end{bmatrix}$$

e.
$$\begin{bmatrix} 6 & 4 & 6 & 3 \\ 9 & -1 & 3 & 4 \end{bmatrix} \begin{bmatrix} 1 & -2 & 4 & 6 \\ 6 & 5 & 8 & -2 \\ 3 & -3 & 7 & 5 \end{bmatrix}$$

f.
$$\begin{bmatrix} 6 & 4 & 6 & 3 \\ 9 & -1 & 3 & 4 \end{bmatrix} \begin{bmatrix} 1 & -2 & 4 & 6 \\ 6 & 5 & 8 & -2 \\ 3 & -3 & 7 & 5 \end{bmatrix}$$

Find the indicated matrices in Problems 9–11, if possible. Give the order of your answer.

9. a.
$$\begin{bmatrix} 3 & 5 & 8 & 9 \end{bmatrix} \begin{bmatrix} w \\ x \\ y \\ z \end{bmatrix}$$

b.
$$\begin{bmatrix} 2 & -3 & 4 & 5 \end{bmatrix} \begin{bmatrix} w \\ x \\ y \\ z \end{bmatrix}$$

c.
$$\begin{bmatrix} 2 & 3 & 5 \end{bmatrix} \begin{bmatrix} a & d \\ b & e \\ c & f \end{bmatrix}$$

d.
$$\begin{bmatrix} 3 & -2 & 1 \end{bmatrix} \begin{bmatrix} a & b \\ c & d \\ e & f \end{bmatrix}$$

10. a.
$$\begin{bmatrix} 1 & -2 & 2 \end{bmatrix} \begin{bmatrix} 1 & 2 & 3 \\ 4 & -1 & 3 \\ -3 & 2 & 1 \end{bmatrix}$$

b.
$$\begin{bmatrix} 3 & 2 & -4 \end{bmatrix} \begin{bmatrix} 1 & 2 & -3 \\ -4 & 3 & 2 \\ 2 & 3 & 1 \end{bmatrix}$$

c. $\begin{bmatrix} 1 & 3 & 2 \\ -1 & 1 & 2 \end{bmatrix} \begin{bmatrix} 6 & 3 & 2 & -3 \\ 1 & 1 & -4 & -1 \\ 2 & 1 & 1 & 2 \end{bmatrix}$

d. $\begin{bmatrix} 6 & 1 & 3 \\ 2 & -3 & 5 \end{bmatrix} \begin{bmatrix} 1 & 0 & 2 & 1 & 1 \\ 2 & 0 & 1 & 1 & 0 \\ 3 & 1 & 4 & 0 & 0 \end{bmatrix}$

11. a. $\begin{bmatrix} w \\ x \\ y \\ z \end{bmatrix} \begin{bmatrix} 8 & -1 & 3 & -2 \end{bmatrix}$

b. $\begin{bmatrix} w \\ x \\ y \\ z \end{bmatrix} \begin{bmatrix} 6 & -1 & 3 & 2 \end{bmatrix}$

c. $\begin{bmatrix} 6 \\ 3 \\ -1 \end{bmatrix} \begin{bmatrix} 2 & -4 & 3 \end{bmatrix}$

d. $\begin{bmatrix} 5 \\ -2 \\ -1 \end{bmatrix} \begin{bmatrix} 2 & 6 & 4 \end{bmatrix}$

Find the indicated matrices in Problems 12–51. Let

$$A = \begin{bmatrix} 1 & 0 & 2 \\ 3 & -1 & 2 \\ 4 & 1 & 0 \end{bmatrix}$$

$$B = \begin{bmatrix} 1 & 4 & 0 \\ 3 & -1 & 2 \\ -2 & 1 & 5 \end{bmatrix}$$

$$C = \begin{bmatrix} 8 & 1 & 6 \\ 3 & 5 & 7 \\ 4 & 9 & 2 \end{bmatrix}$$

$$D = \begin{bmatrix} -2 \\ 1 \\ 3 \end{bmatrix} \qquad E = \begin{bmatrix} 4 \\ 1 \\ 6 \end{bmatrix}$$

$$F = \begin{bmatrix} 2 & 4 & 7 \end{bmatrix} \qquad G = \begin{bmatrix} 6 & -1 & 0 \end{bmatrix}$$

12. $A + B$
13. $B + A$
14. $A - B$
15. $B - A$
16. $C + D$
17. $D + C$
18. $C - D$
19. $D - C$
20. $E + F$
21. $F + E$
22. $E - F$
23. $F - E$

24. $2A + B$
25. $3A - 4B$
26. $2C + 3D$
27. $C - 2D$
28. $3E - 2D$
29. $3E - 2F$
30. $E + 3F$
31. $3B + 2C$
32. $A + (B + C)$
33. $(A + B) + C$
34. $(B + C) + D$
35. $B + (C + D)$
36. AB
37. BA
38. AC
39. CA
40. BC
41. CB
42. $A(BC)$
43. $(AB)C$
44. $B(AC)$
45. $(BA)C$
46. $C(AB)$
47. $(CA)B$
48. $B + C$
49. $A(B + C)$
50. $(B + C)A$
51. $AB + AC$

Find the indicated matrices in Problems 52–63. Let

$$A = \begin{bmatrix} 1 & 2 \\ 4 & 0 \\ -1 & 3 \\ 2 & 1 \end{bmatrix}$$

$$B = \begin{bmatrix} 4 & 2 \\ -1 & 3 \end{bmatrix}$$

$$C = \begin{bmatrix} 1 & 0 & 0 & 0 \\ 0 & 1 & 0 & 0 \\ 0 & 0 & 1 & 0 \\ 0 & 0 & 0 & 1 \end{bmatrix}$$

$$D = \begin{bmatrix} 4 & 1 & 3 & 6 \\ -1 & 0 & -2 & 3 \end{bmatrix}$$

52. AB
53. BA
54. B^2
55. CA
56. BD
57. DB
58. $(B + C)A$
59. $BA + CA$
60. CD
61. A^2
62. C^3
63. B^3

APPLICATIONS

64. A company with five offices in San Francisco operates its own delivery service for office mail. Arrows represent the direction of communication in the figure. Note that A can send mail directly to offices B, D, and E but not to C.

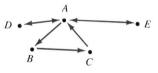

a. Express this delivery network with a communication matrix.

b. Develop a matrix showing the possible two-stop deliveries.

c. Develop a matrix showing the possible three-stop deliveries.

65. Consider the map of airline routes shown below.

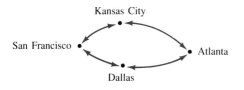

a. Fill in the blanks in the following communication matrix representing the airline routes:

b. In how many ways can you travel from Kansas City to San Francisco making exactly one stop?
c. In how many ways can you travel from San Francisco to Kansas City making two stops?
d. Write the communication matrix showing the number of routes among these cities if you make exactly one stop.
e. Write the communication matrix showing the number of routes among these cities if you make exactly two stops.

66. In Example 18, add 40 units of overhead for the plan I homes and 45 units of overhead for the plan II models. If the cost for overhead is 1000 per unit, find the total cost for the housing complex as well as the average cost for two-, three-, and four-bedroom homes assuming all the other information in Example 18 remains unchanged.

67. If the countries in Example 19 are willing to speak through two intermediaries, find the communication matrix and tell how many different ways the United States can communicate with each of the other countries in this fashion.

Historical Question The English mathematician Arthur Cayley (1821–1895) was responsible for matrix theory. Cayley began his career as a lawyer in 1849, but he was always a mathematician at heart, publishing more than 200 papers in mathematics while practicing law. In 1863 he gave up law to accept a chair in mathematics at Cambridge. He originated the notion of matrices in 1858 and worked on matrix theory over the next several years. His work was entirely theoretical and was not used for any practical purpose until 1925 when matrices were used in quantum mechanics. Since 1950, matrices have played an important role in the social sciences and business. Much of the early work with matrices focused upon their properties. In Problems 68–73 we look at some of these properties. Let A, B, and C be square matrices of order 3 and let a and b be real numbers. Give an example to show that each of the following is false or else explain why you think the property is true.

68. $A + B = B + A$
69. $A - B = B - A$
70. $(A - B) - C = A - (B - C)$
71. $(A + B) + C = A + (B + C)$
72. $a(B + C) = aB + aC$
73. $(a + b)C = aC + bC$

3.3 Gauss–Jordan Elimination

In this section we develop a procedure to solve more general linear systems. Consider a system of m equations and n variables:

$$\begin{cases} a_{11}x_1 + a_{12}x_2 + \cdots + a_{1n}x_n = b_1 \\ a_{21}x_1 + a_{22}x_2 + \cdots + a_{2n}x_n = b_2 \\ \qquad \vdots \qquad \vdots \qquad \vdots \qquad \vdots \\ a_{m1}x_1 + a_{m2}x_2 + \cdots + a_{mn}x_n = b_m \end{cases}$$

Now consider what is called the **augmented matrix** for this system:

$$\begin{bmatrix} a_{11} & a_{12} & \cdots & a_{1n} & \vdots & b_1 \\ a_{21} & a_{22} & \cdots & a_{2n} & \vdots & b_2 \\ \vdots & \vdots & & \vdots & \vdots & \vdots \\ a_{m1} & a_{m2} & \cdots & a_{mn} & \vdots & b_m \end{bmatrix}$$

EXAMPLE 1   Write the given systems in augmented matrix form.

a. $\begin{cases} 2x + 3y = 8 \\ 3x + 2y = 7 \end{cases}$

b. $\begin{cases} 2x + y = 3 \\ 3x - y = 2 \\ 4x + 3y = 7 \end{cases}$

c. $\begin{array}{l} 3x - 2y + z = -2 \\ 4x - 5y + 3z = -9 \\ 2x - y + 5z = -5 \end{array}$

d. $\begin{cases} 5x - 3y + z = -3 \\ 2x + 5z = 14 \end{cases}$

e. $\begin{cases} x_1 - 3x_3 + x_5 = -3 \\ x_2 + x_4 = -1 \\ x_3 + x_5 = 7 \\ x_1 + x_2 - x_3 + 4x_4 = -8 \\ x_1 + x_2 + x_3 + x_4 + x_5 = 8 \end{cases}$

Solution   Note that some coefficients are negative and some are zeros.

a. $\left[\begin{array}{cc|c} 2 & 3 & 8 \\ 3 & 2 & 7 \end{array}\right]$

b. $\left[\begin{array}{cc|c} 2 & 1 & 3 \\ 3 & -1 & 2 \\ 4 & 3 & 7 \end{array}\right]$

c. $\left[\begin{array}{ccc|c} 3 & -2 & 1 & -2 \\ 4 & -5 & 3 & -9 \\ 2 & -1 & 5 & -5 \end{array}\right]$

d. $\left[\begin{array}{ccc|c} 5 & -3 & 1 & -3 \\ 2 & 0 & 5 & 14 \end{array}\right]$

e. $\left[\begin{array}{ccccc|c} 1 & 0 & -3 & 0 & 1 & -3 \\ 0 & 1 & 0 & 1 & 0 & -1 \\ 0 & 0 & 1 & 0 & 1 & 7 \\ 1 & 1 & -1 & 4 & 0 & -8 \\ 1 & 1 & 1 & 1 & 1 & 8 \end{array}\right]$ ∎

EXAMPLE 2   Write a system of equations (use $x_1, x_2, \ldots, x_n$ for the variables) that has the given augmented matrix.

a. $\left[\begin{array}{ccc|c} 2 & 1 & -1 & -3 \\ 3 & -2 & 1 & 9 \\ 1 & -4 & 3 & 17 \end{array}\right]$

b. $\left[\begin{array}{ccc|c} 1 & 0 & 0 & 3 \\ 0 & 1 & 0 & -2 \\ 0 & 0 & 1 & -5 \end{array}\right]$

c. $\left[\begin{array}{cc|c} 1 & 0 & 4 \\ 0 & 1 & 3 \end{array}\right]$

d. $\left[\begin{array}{cc|c} 1 & 0 & -5 \\ 0 & 1 & 2 \\ 2 & 1 & -3 \end{array}\right]$

e. $\left[\begin{array}{ccc|c} 1 & 0 & 0 & 3 \\ 0 & 1 & 0 & -3 \\ 0 & 0 & 1 & 6 \\ 0 & 0 & 0 & 1 \end{array}\right]$

f. $\left[\begin{array}{ccc|c} 1 & 0 & 0 & 1 \\ 0 & 1 & 0 & 5 \\ 0 & 0 & 1 & -2 \\ 0 & 0 & 0 & 0 \end{array}\right]$

Solution   a. $\begin{cases} 2x_1 + x_2 - x_3 = -3 \\ 3x_1 - 2x_2 + x_3 = 9 \\ x_1 - 4x_2 + 3x_3 = 17 \end{cases}$

b. $\begin{cases} x_1 = 3 \\ x_2 = -2 \\ x_3 = -5 \end{cases}$

c. $\begin{cases} x_1 = 4 \\ x_2 = 3 \end{cases}$

d. $\begin{cases} x_1 = -5 \\ x_2 = 2 \\ 2x_1 + x_2 = -3 \end{cases}$

e. $\begin{cases} x_1 = 3 \\ x_2 = -3 \\ x_3 = 6 \\ 0 = 1 \end{cases}$

f. $\begin{cases} x_1 = 1 \\ x_2 = 5 \\ x_3 = -2 \\ 0 = 0 \end{cases}$ ∎

The goal of this section is to solve a system of m equations with n unknowns. We have already looked at systems of two equations with two unknowns. In high school you may have solved three equations with three unknowns. Now, however, we want to be able to solve problems with two equations and five unknowns, or three equations and two unknowns, or any mixture of linear equations or unknowns. The procedure of this section—**Gauss–Jordan elimination**—is a general method for solving all of these types of systems. We write the system in augmented matrix form (as in Example 1), then carry out a process that transforms the matrix until the solution is obvious. Look back at Example 2—the solutions to parts b and c are obvious. Part e shows $0 = 1$ in the last equation, so this system has no solution (0 cannot equal 1), and part f shows $0 = 0$ (which is true for all replacements of the variable), which means that the solution is found by looking at the other equations (namely, $x_1 = 1$, $x_2 = 5$, and $x_3 = -2$). The terms with nonzero coefficients in these examples are arranged on a diagonal, and such a system is said to be in *diagonal form*.

But what process will allow us to transform a matrix into diagonal form? We begin with some steps called **elementary row operations**. Elementary row operations change the *form* of a matrix, but the new form represents an equivalent system. Matrices which represent equivalent systems are called **equivalent matrices** and we introduce these elementary row operations in order to write equivalent matrices. Let us work with a system with three equations and three unknowns (any size will work the same way).

System format	Matrix format

$$\begin{cases} 3x - 2y + 4z = 11 \\ x - y - 2z = -7 \\ 2x - 3y + z = -1 \end{cases} \qquad \begin{bmatrix} 3 & -2 & 4 & \vdots & 11 \\ 1 & -1 & -2 & \vdots & -7 \\ 2 & -3 & 1 & \vdots & -1 \end{bmatrix}$$

Interchanging two equations is equivalent to interchanging two rows in matrix format, and, certainly, if we do this, the solution to the system will be the same:

System format	Matrix format

$$\begin{cases} x - y - 2z = -7 \\ 3x - 2y + 4z = 11 \\ 2x - 3y + z = -1 \end{cases} \qquad \begin{bmatrix} 1 & -1 & -2 & \vdots & -7 \\ 3 & -2 & 4 & \vdots & 11 \\ 2 & -3 & 1 & \vdots & -1 \end{bmatrix}$$

The first and second equations (rows) are interchanged

Elementary Row Operation 1: Interchange any two rows. Since multiplying or dividing both sides of any equation by any nonzero number does not change the solution, then, in matrix format, the solution will not be changed if any row is multiplied or divided by a nonzero constant. For example, multiply both sides of the first equation by -3:

System format	Matrix format

$$\begin{cases} -3x + 3y + 6z = 21 \\ 3x - 2y + 4z = 11 \\ 2x - 3y + z = -1 \end{cases} \qquad \begin{bmatrix} -3 & 3 & 6 & \vdots & 21 \\ 3 & -2 & 4 & \vdots & 11 \\ 2 & -3 & 1 & \vdots & -1 \end{bmatrix}$$

Both sides of the first equation are multiplied by -3; the first row is multiplied by -3.

Because it is easier to program software by dividing multiplication and division into two separate row operations, we have done the same.

Elementary Row Operation 2: Multiply all the elements of a row by the same nonzero real number.

Elementary Row Operation 3: Divide all the elements of a row by the same nonzero real number.

The last property we need to carry out Gauss–Jordan elimination rests on the property we used with the linear combination method, namely, adding equations to eliminate a variable. In terms of the system format this means that one equation can be replaced by its sum with another equation in the system. For example, if we add the first equation to the second equation we have:

System format *Matrix format*

$$\begin{cases} -3x + 3y + 6z = 21 \\ \phantom{-3x +{}} y + 10z = 32 \\ 2x - 3y + z = -1 \end{cases} \qquad \begin{bmatrix} -3 & 3 & 6 & \vdots & 21 \\ 0 & 1 & 10 & \vdots & 32 \\ 2 & -3 & 1 & \vdots & -1 \end{bmatrix}$$

In terms of matrix format, any row can be replaced by its sum with some other row.

This process is usually used in conjunction with Elementary Row Operation 2. That is, an equation is changed by adding to it a nonzero multiple of another equation in the system. Go back to the original system:

System format *Matrix format*

$$\begin{cases} 3x - 2y + 4z = 11 \\ x - y - 2z = -7 \\ 2x - 3y + z = -1 \end{cases} \qquad \begin{bmatrix} 3 & -2 & 4 & \vdots & 11 \\ 1 & -1 & -2 & \vdots & -7 \\ 2 & -3 & 1 & \vdots & -1 \end{bmatrix}$$

We can change this system by multiplying the second equation by -3 and adding the result to the first equation. In matrix terminology we would say multiply the second row by -3 and add it to the first row:

System format *Matrix format*

$$\begin{cases} \phantom{x -{}} y + 10z = 32 \\ x - y - 2z = -7 \\ 2x - 3y + z = -1 \end{cases} \qquad \begin{bmatrix} 0 & 1 & 10 & \vdots & 32 \\ 1 & -1 & -2 & \vdots & -7 \\ 2 & -3 & 1 & \vdots & -1 \end{bmatrix}$$

Once again, multiply the second row, this time by -2, and add it to the third row. Wait! Why -2? Where did that come from? The idea is the same one we used in the linear combination method—we use a number that will give a zero coefficient to the x in the third equation.

System format *Matrix format*

$$\begin{cases} \phantom{x -{}} y + 10z = 32 \\ x - y - 2z = -7 \\ \phantom{x -{}} -y + 5z = 13 \end{cases} \qquad \begin{bmatrix} 0 & 1 & 10 & \vdots & 32 \\ 1 & -1 & -2 & \vdots & -7 \\ 0 & -1 & 5 & \vdots & 13 \end{bmatrix}$$

Note that the multiplied row is not changed, that, instead, the changed row is the one to which the multiplied row is added. We call the original row the **pivot row** and the changed row the **target row**.

Elementary Row Operation 4: Multiply all the entries of a row (the pivot row) by a nonzero real number and add each resulting product to the corresponding entry of another specified row (the target row). Note that this operation changes only the target row.

There you have it! You should carry out these four elementary row operations until you have a system for which the solution is obvious, as illustrated in Example 3.

EXAMPLE 3   Solve $\begin{cases} 2x - 5y = 5 \\ x - 2y = 1 \end{cases}$.

Solution

System notation	*Matrix notation*
$\begin{cases} 2x - 5y = 5 \\ x - 2y = 1 \end{cases}$	$\left[\begin{array}{cc:c} 2 & -5 & 5 \\ 1 & -2 & 1 \end{array}\right]$

Elementary Row Operation 1

Interchange the first and second equations

Interchange the first and second rows

$\begin{cases} x - 2y = 1 \\ 2x - 5y = 5 \end{cases}$
$\left[\begin{array}{cc:c} 1 & -2 & 1 \\ 2 & -5 & 5 \end{array}\right]$ ⟵ Pivot row
⟵ Target row

Elementary Row Operation 4

Add -2 times the first equation to the second

Add -2 times the first row to the second row

$\begin{cases} x - 2y = 1 \\ -\ y = 3 \end{cases}$
$\left[\begin{array}{cc:c} 1 & -2 & 1 \\ 0 & -1 & 3 \end{array}\right]$ ⟵ Pivot row remains unchanged
⟵ Target row changes

Elementary Row Operation 2

Multiply both sides of the second equation by -1

Multiply row 2 by -1

$\begin{cases} x - 2y = 1 \\ -\ y = -3 \end{cases}$
$\left[\begin{array}{cc:c} 1 & -2 & 1 \\ 0 & 1 & -3 \end{array}\right]$ ⟵ New pivot row

Elementary Row Operation 4

Add 2 times the second equation to the first

Add 2 times the second row to the first row

$\begin{cases} x = -5 \\ y = -3 \end{cases}$
$\left[\begin{array}{cc:c} 1 & 0 & -5 \\ 0 & 1 & -3 \end{array}\right]$ This is called the **reduced row-echelon form**

The solution, $(-5, -3)$, is now obvious.   ∎

As you study Example 3, first look at how the elementary row operations led to a system equivalent to the first—but one for which the solution is obvious. Next, try to decide *why* a particular row operation was chosen when it was. Most students quickly learn the elementary row operations, but then use a series of (almost random) steps until the obvious solution results. This often works, but is not very efficient. The steps chosen in Example 3 illustrate a very efficient method of using the elementary row operations to determine a system whose solution is obvious. The method was discovered independently by two mathematicians, Karl Friedrich Gauss (1777–1855) and Camille Jordan (1838–1922), so today the process is known as the Gauss–Jordan method. (Since the process was only recently attributed to Jordan, many books refer to the method simply as *Gaussian elimination*.) Before stating the process, however, we will consider a procedure called **pivoting**.

Pivoting

1. Divide all entries in the row in which the pivot appears (called the *pivot row*) by the pivot element so that the pivot entry becomes a 1. This uses Elementary Row Operation 3.
2. Obtain zeros above and below the pivot element by using Elementary Row Operation 4.

EXAMPLE 4 Pivot the given matrix about the circled element.

Solution
$$\begin{bmatrix} 15 & ⑤ & | & 35 \\ 5 & 2 & | & -3 \end{bmatrix} \xrightarrow{\text{R1} \div 5} \begin{bmatrix} 3 & 1 & | & 7 \\ 5 & 2 & | & -3 \end{bmatrix} \xrightarrow{-2\text{R1} + \text{R2}} \begin{bmatrix} 3 & 1 & | & 7 \\ -1 & 0 & | & -17 \end{bmatrix}$$

This shorthand notation shows what we did: divide row 1 by 5

Multiply row 1 (pivot row) by -2 and add it to row 2 (target row)

Consider another matrix.

$$\begin{bmatrix} 2 & 3 & 4 & | & 1 \\ -1 & ② & 3 & | & -2 \\ 0 & 1 & -1 & | & 3 \end{bmatrix} \xrightarrow{\frac{1}{2}\text{R2}} \begin{bmatrix} 2 & 3 & 4 & | & 1 \\ -\frac{1}{2} & 1 & \frac{3}{2} & | & -1 \\ 0 & 1 & -1 & | & 3 \end{bmatrix}$$

$$\xrightarrow{-3\text{R2} + \text{R1}} \begin{bmatrix} \frac{7}{2} & 0 & -\frac{1}{2} & | & 4 \\ -\frac{1}{2} & 1 & \frac{3}{2} & | & -1 \\ 0 & 1 & -1 & | & 3 \end{bmatrix} \xrightarrow{-1\text{R2} + \text{R3}} \begin{bmatrix} \frac{7}{2} & 0 & -\frac{1}{2} & | & 4 \\ -\frac{1}{2} & 1 & \frac{3}{2} & | & -1 \\ \frac{1}{2} & 0 & -\frac{5}{2} & | & 4 \end{bmatrix} \quad ∎$$

The four elementary row operations are listed below for easy reference.

Elementary Row Operations

There are *four elementary row operations* for producing equivalent matrices:

1. Interchange any two rows.
2. Multiply all the elements of a row by the same nonzero real number.
3. Divide all the elements of a row by the same nonzero real number.
4. Multiply all the entries of a row (*pivot row*) by a nonzero real number and add each resulting product to the corresponding entry of another specified row (*target row*). (Note that this operation changes only the target row.)

COMPUTER APPLICATION The disk accompanying this book has a program called Row Reduction. It lists the four elementary row operations as options. The program allows the operator input by asking what to do and then carrying out the arithmetic (in decimal form), or else it automatically carries out the pivoting process.

You are now ready to see the method worked out by Gauss and Jordan. It efficiently uses the elementary row operations to diagonalize the matrix. That is, the first pivot is the first entry in the first row, first column; the second is the entry in the second row, second column; and so on until the solution is obvious.

Gauss–Jordan Elimination

1. Select the element in the first row, first column, as a pivot.
2. Pivot.
3. Select the element in the second row, second column, as a pivot.
4. Pivot.
5. Repeat the process until you arrive at the last row, or until the pivot element is a zero. If it is a zero and you can interchange that row with a row below it, so that the pivot element is no longer a zero, do so and continue. If it is a zero and you cannot interchange rows so that it is not a zero, the process is complete.

EXAMPLE 5 Solve $\begin{cases} 3x - 2y + z = -2 \\ 4x - 5y + 3z = -9. \\ 2x - y + 5z = -5 \end{cases}$

Solution We will solve this system by choosing the steps according to the Gauss–Jordan method.

$$\begin{bmatrix} ③ & -2 & 1 & | & -2 \\ 4 & -5 & 3 & | & -9 \\ 2 & -1 & 5 & | & -5 \end{bmatrix} \xrightarrow{R1 \div 3} \begin{bmatrix} 1 & -\frac{2}{3} & \frac{1}{3} & | & -\frac{2}{3} \\ 4 & -5 & 3 & | & -9 \\ 2 & -1 & 5 & | & -5 \end{bmatrix} \xrightarrow{-4R1 + R2} \begin{bmatrix} 1 & -\frac{2}{3} & \frac{1}{3} & | & -\frac{2}{3} \\ 0 & -\frac{7}{3} & \frac{5}{3} & | & -\frac{19}{3} \\ 2 & -1 & 5 & | & -5 \end{bmatrix}$$

3 is the pivot · Obtain a 1 in the pivot position · Pivot row is 1; target row is 2; -4 is the opposite of the corresponding element in the target row so that a zero is obtained

$$\xrightarrow{-2R1 + R3} \begin{bmatrix} 1 & -\frac{2}{3} & \frac{1}{3} & | & -\frac{2}{3} \\ 0 & ⑤{-\frac{7}{3}} & \frac{5}{3} & | & -\frac{19}{3} \\ 0 & \frac{1}{3} & \frac{13}{3} & | & -\frac{11}{3} \end{bmatrix} \xrightarrow{R2 \div -\frac{7}{3}} \begin{bmatrix} 1 & -\frac{2}{3} & \frac{1}{3} & | & -\frac{2}{3} \\ 0 & 1 & -\frac{5}{7} & | & \frac{19}{7} \\ 0 & \frac{1}{3} & \frac{13}{3} & | & -\frac{11}{3} \end{bmatrix} \xrightarrow{\frac{2}{3}R2 + R1}$$

Pivot row is 1; target row is 3 · Select a new pivot $\left(-\frac{7}{3}\right)$ · Pivot

$$\begin{bmatrix} 1 & 0 & -\frac{1}{7} & | & \frac{8}{7} \\ 0 & 1 & -\frac{5}{7} & | & \frac{19}{7} \\ 0 & \frac{1}{3} & \frac{13}{3} & | & -\frac{11}{3} \end{bmatrix} \xrightarrow{-\frac{1}{3}R2 + R3} \begin{bmatrix} 1 & 0 & -\frac{1}{7} & | & \frac{8}{7} \\ 0 & 1 & -\frac{5}{7} & | & \frac{19}{7} \\ 0 & 0 & ⑨{\frac{96}{21}} & | & -\frac{96}{21} \end{bmatrix} \xrightarrow{R3 \div \frac{96}{21}}$$

Select a new pivot

$$\begin{bmatrix} 1 & 0 & -\frac{1}{7} & | & \frac{8}{7} \\ 0 & 1 & -\frac{5}{7} & | & \frac{19}{7} \\ 0 & 0 & 1 & | & -1 \end{bmatrix} \xrightarrow{\frac{5}{7}R3 + R2} \begin{bmatrix} 1 & 0 & -\frac{1}{7} & | & \frac{8}{7} \\ 0 & 1 & 0 & | & 2 \\ 0 & 0 & 1 & | & -1 \end{bmatrix} \xrightarrow{\frac{1}{7}R3 + R1} \begin{bmatrix} 1 & 0 & 0 & | & 1 \\ 0 & 1 & 0 & | & 2 \\ 0 & 0 & 1 & | & -1 \end{bmatrix}$$

Pivot

The solution, $(1, 2, -1)$, is now obvious. ■

Note that the Gauss–Jordan method usually introduces (often ugly) fractions. For this reason, many people are turning to readily available computer programs to carry out the drudgery of this method. However, if you do not have access to a computer, you can often reduce the amount of arithmetic by forcing the pivot elements to be 1 by using elementary row operations other than division. Also, you can combine the pivoting steps. In this example we could have forced the first pivot to be 1 by multiplying row 3 by -1 and adding it to the first row instead of dividing by 3. Let us look at the arithmetic in this simplified version:

$$\begin{bmatrix} 3 & -2 & 1 & \vdots & -2 \\ 4 & -5 & 3 & \vdots & -9 \\ 2 & -1 & 5 & \vdots & -5 \end{bmatrix} \longrightarrow \begin{bmatrix} ① & -1 & -4 & \vdots & 3 \\ 4 & -5 & 3 & \vdots & -9 \\ 2 & -1 & 5 & \vdots & -5 \end{bmatrix} \longrightarrow \begin{bmatrix} 1 & -1 & -4 & \vdots & 3 \\ 0 & -1 & 19 & \vdots & -21 \\ 0 & 1 & 13 & \vdots & -11 \end{bmatrix} \longrightarrow$$

$-1R3 + R1$ Pivot; combine two steps in one:
$-4R1 + R2; -2R1 + R3$

$$\begin{bmatrix} 1 & -1 & -4 & \vdots & 3 \\ 0 & ① & -19 & \vdots & 21 \\ 0 & 1 & 13 & \vdots & -11 \end{bmatrix} \longrightarrow \begin{bmatrix} 1 & 0 & -23 & \vdots & 24 \\ 0 & 1 & -19 & \vdots & 21 \\ 0 & 0 & ㉜ & \vdots & -32 \end{bmatrix} \longrightarrow \begin{bmatrix} 1 & 0 & -23 & \vdots & 24 \\ 0 & 1 & -19 & \vdots & 21 \\ 0 & 0 & 1 & \vdots & -1 \end{bmatrix} \longrightarrow \begin{bmatrix} 1 & 0 & 0 & \vdots & 1 \\ 0 & 1 & 0 & \vdots & 2 \\ 0 & 0 & 1 & \vdots & -1 \end{bmatrix}$$

New pivot is   Pivot second row   New pivot is   Pivot third row:
$R2 \div (-1)$ (combine two steps): $R3/32$ $19R3 + R2;$
             $R2 + R1; -1R2 + R3$ $23R3 + R1$

The real beauty of Gauss–Jordan elimination is that it works with all sizes of linear systems. Consider the following example consisting of five equations and five unknowns.

EXAMPLE 6   Solve $\begin{cases} x_1 - 3x_3 + x_5 = -3 \\ x_2 + x_4 = -1 \\ x_3 + x_5 = 7 \\ x_1 + x_2 - x_3 + 4x_4 = -8 \\ x_1 + x_2 + x_3 + x_4 + x_5 = 8 \end{cases}$

Solution

$$\begin{bmatrix} ① & 0 & -3 & 0 & 1 & \vdots & -3 \\ 0 & 1 & 0 & 1 & 0 & \vdots & -1 \\ 0 & 0 & 1 & 0 & 1 & \vdots & 7 \\ 1 & 1 & -1 & 4 & 0 & \vdots & -8 \\ 1 & 1 & 1 & 1 & 1 & \vdots & 8 \end{bmatrix} \xrightarrow[\substack{-R1 + R4 \\ -R1 + R5}]{} \begin{bmatrix} 1 & 0 & -3 & 0 & 1 & \vdots & -3 \\ 0 & ① & 0 & 1 & 0 & \vdots & -1 \\ 0 & 0 & 1 & 0 & 1 & \vdots & 7 \\ 0 & 1 & 2 & 4 & -1 & \vdots & -5 \\ 0 & 1 & 4 & 1 & 0 & \vdots & 11 \end{bmatrix}$$

$$\xrightarrow[\substack{-R2 + R4 \\ -R2 + R5}]{} \begin{bmatrix} 1 & 0 & -3 & 0 & 1 & \vdots & -3 \\ 0 & 1 & 0 & 1 & 0 & \vdots & -1 \\ 0 & 0 & ① & 0 & 1 & \vdots & 7 \\ 0 & 0 & 2 & 3 & -1 & \vdots & -4 \\ 0 & 0 & 4 & 0 & 0 & \vdots & 12 \end{bmatrix} \xrightarrow[\substack{3R3 + R1 \\ -2R3 + R4 \\ -4R3 + R5}]{} \begin{bmatrix} 1 & 0 & 0 & 0 & 4 & \vdots & 18 \\ 0 & 1 & 0 & 1 & 0 & \vdots & -1 \\ 0 & 0 & 1 & 0 & 1 & \vdots & 7 \\ 0 & 0 & 0 & ③ & -3 & \vdots & -18 \\ 0 & 0 & 0 & 0 & -4 & \vdots & -16 \end{bmatrix}$$

$$\xrightarrow{R4 \div 3}
\begin{bmatrix}
1 & 0 & 0 & 0 & 4 & 18 \\
0 & 1 & 0 & 1 & 0 & -1 \\
0 & 0 & 1 & 0 & 1 & 7 \\
0 & 0 & 0 & \textcircled{1} & -1 & -6 \\
0 & 0 & 0 & 0 & -4 & -16
\end{bmatrix}
\xrightarrow{(-1)R4 + R2}
\begin{bmatrix}
1 & 0 & 0 & 0 & 4 & 18 \\
0 & 1 & 0 & 0 & 1 & 5 \\
0 & 0 & 1 & 0 & 1 & 7 \\
0 & 0 & 0 & 1 & -1 & -6 \\
0 & 0 & 0 & 0 & \boxed{-4} & -16
\end{bmatrix}$$

$$\xrightarrow{R5 \div (-4)}
\begin{bmatrix}
1 & 0 & 0 & 0 & 4 & 18 \\
0 & 1 & 0 & 0 & 1 & 5 \\
0 & 0 & 1 & 0 & 1 & 7 \\
0 & 0 & 0 & 1 & -1 & -6 \\
0 & 0 & 0 & 0 & 1 & 4
\end{bmatrix}
\xrightarrow[\substack{-R5 + R3 \\ -R5 + R2 \\ -4R5 + R1}]{R5 + R4}
\begin{bmatrix}
1 & 0 & 0 & 0 & 0 & 2 \\
0 & 1 & 0 & 0 & 0 & 1 \\
0 & 0 & 1 & 0 & 0 & 3 \\
0 & 0 & 0 & 1 & 0 & -2 \\
0 & 0 & 0 & 0 & 1 & 4
\end{bmatrix}$$

The solution is $(x_1, x_2, x_3, x_4, x_5) = (2, 1, 3, -2, 4)$. ∎

Example 7 is an example of a system of three equations with two unknowns that has a solution. Example 8 shows what Gauss–Jordan elimination looks like when there are three equations with two unknowns and no solution. (Remember, if there is no solution, the system is inconsistent.)

EXAMPLE 7 Solve $\begin{cases} 2x + y = 3 \\ 3x - y = 2. \\ 4x + 3y = 7 \end{cases}$

Solution
$$\begin{bmatrix}
2 & 1 & 3 \\
3 & -1 & 2 \\
4 & 3 & 7
\end{bmatrix}
\xrightarrow{-1R1 + R2}
\begin{bmatrix}
2 & 1 & 3 \\
1 & -2 & -1 \\
4 & 3 & 7
\end{bmatrix}
\xrightarrow[\substack{\text{rows 1 and 2} \\ R_1 \leftrightarrow R_2}]{\text{Interchange}}
\begin{bmatrix}
1 & -2 & -1 \\
2 & 1 & 3 \\
4 & 3 & 7
\end{bmatrix}$$

$$\xrightarrow[-4R1 + R3]{-2R1 + R2}
\begin{bmatrix}
\textcircled{1} & -2 & -1 \\
0 & 5 & 5 \\
0 & 11 & 11
\end{bmatrix}
\xrightarrow{R2 \div 5}
\begin{bmatrix}
1 & -2 & -1 \\
0 & \textcircled{1} & 1 \\
0 & 11 & 11
\end{bmatrix}$$

$$\xrightarrow[-11R2 + R3]{2R2 + R1}
\begin{bmatrix}
1 & 0 & 1 \\
0 & 1 & 1 \\
0 & 0 & 0
\end{bmatrix}$$

This final matrix is equivalent to the system: $\begin{cases} x = 1 \\ y = 1. \\ 0 = 0 \end{cases}$

The solution is $(1, 1)$.

Check:
$$2x + y = 2(1) + 1 = 3 \quad \checkmark$$
$$3x - y = 3(1) - 1 = 2 \quad \checkmark$$
$$4x + 3y = 4(1) + 3(1) = 7 \quad \checkmark$$

∎

EXAMPLE 8   Solve $\begin{cases} 2x + y = 3 \\ 3x - y = 2 \\ x - 2y = 4 \end{cases}$

Solution

$$\begin{bmatrix} 2 & 1 & \vdots & 3 \\ 3 & -1 & \vdots & 2 \\ 1 & -2 & \vdots & 4 \end{bmatrix} \xrightarrow{R1 \leftrightarrow R3} \begin{bmatrix} 1 & -2 & \vdots & 4 \\ 3 & -1 & \vdots & 2 \\ 2 & 1 & \vdots & 3 \end{bmatrix} \xrightarrow[-2R1 + R3]{-3R1 + R2} \begin{bmatrix} 1 & -2 & \vdots & 4 \\ 0 & 5 & \vdots & -10 \\ 0 & 5 & \vdots & -5 \end{bmatrix}$$

$$\xrightarrow{R2 \div 5} \begin{bmatrix} 1 & -2 & \vdots & 4 \\ 0 & 1 & \vdots & -2 \\ 0 & 5 & \vdots & -5 \end{bmatrix} \xrightarrow[-5R2 + R3]{2R2 + R1} \begin{bmatrix} 1 & 0 & \vdots & 0 \\ 0 & 1 & \vdots & -2 \\ 0 & 0 & \vdots & 5 \end{bmatrix}$$

This is equivalent to $\begin{cases} x = 0 \\ y = -2 \\ 0 = 5 \end{cases}$. But, since $0 \neq 5$ regardless of the values of x and y, this is an inconsistent system.   ∎

A dependent system has infinitely many solutions, but just because a system has infinitely many solutions does not mean that it is satisfied by any set of values. For example, $x + y = 5$ has infinitely many solutions, but not just *any* replacements for x and y will satisfy that equation. In fact, we might say if we choose any value, say, t, for x, then y is fixed to be $5 - t$. In the last chapter we called t a *parameter*, and we could say that any ordered pair of the form $(t, 5 - t)$ will satisfy the equation $x + y = 5$. Example 9 illustrates a dependent system and the use of a parameter in specifying the solution.

EXAMPLE 9   Solve $\begin{cases} 3x - 2y + 4z = 8 \\ x - y - 2z = 5 \\ 4x - 3y + 2z = 13 \end{cases}$.

Solution

$$\begin{bmatrix} 3 & -2 & 4 & \vdots & 8 \\ 1 & -1 & -2 & \vdots & 5 \\ 4 & -3 & 2 & \vdots & 13 \end{bmatrix} \xrightarrow{R_1 \leftrightarrow R2} \begin{bmatrix} 1 & -1 & -2 & \vdots & 5 \\ 3 & -2 & 4 & \vdots & 8 \\ 4 & -3 & 2 & \vdots & 13 \end{bmatrix}$$

$$\xrightarrow[-4R1 + R3]{-3R1 + R2} \begin{bmatrix} 1 & -1 & -2 & \vdots & 5 \\ 0 & 1 & 10 & \vdots & -7 \\ 0 & 1 & 10 & \vdots & -7 \end{bmatrix} \xrightarrow[-R2 + R3]{R2 + R1} \begin{bmatrix} 1 & 0 & 8 & \vdots & -2 \\ 0 & 1 & 10 & \vdots & -7 \\ 0 & 0 & 0 & \vdots & 0 \end{bmatrix}$$

This is equivalent to the system $\begin{cases} x + 8z = -2 \\ y + 10z = -7 \\ 0 = 0 \end{cases}$. If we pick any value for z, say, t, then x and y are, in turn, determined: $x = -2 - 8t$ and $y = -7 - 10t$. This is called a *parametric solution*: $(-2 - 8t, -7 - 10t, t)$. Such a solution means that there are infinitely many ordered triplets satisfying the system (a dependent system), but just *not any* ordered triplet. Let us list a few of the solutions:

if $t = 0$: $(-2, -7, 0)$;

if $t = 1$: $(-10, -17, 1)$;

if $t = \frac{1}{2}$: $(-6, -12, \frac{1}{2})$; and so on.   ∎

You may also need a parameter when there are more unknowns than equations, as illustrated by Example 10.

EXAMPLE 10 Solve $\begin{cases} 5x - 3y + z = -3 \\ 2x + 5z = 14 \end{cases}$

Solution
$$\begin{bmatrix} 5 & -3 & 1 & \vdots & -3 \\ 2 & 0 & 5 & \vdots & 14 \end{bmatrix} \xrightarrow{-2R2 + R1} \begin{bmatrix} 1 & -3 & -9 & \vdots & -31 \\ 2 & 0 & 5 & \vdots & 14 \end{bmatrix}$$

$$\xrightarrow{-2R1 + R2} \begin{bmatrix} 1 & -3 & -9 & \vdots & -31 \\ 0 & 6 & 23 & \vdots & 76 \end{bmatrix} \xrightarrow{R2 \div 6} \begin{bmatrix} 1 & -3 & -9 & \vdots & -31 \\ 0 & 1 & \frac{23}{6} & \vdots & \frac{38}{3} \end{bmatrix}$$

$$\xrightarrow{3R2 + R1} \begin{bmatrix} 1 & 0 & \frac{5}{2} & \vdots & 7 \\ 0 & 1 & \frac{23}{6} & \vdots & \frac{38}{3} \end{bmatrix}$$

This is equivalent to the system $\begin{cases} x + \frac{5}{2}z = 7 \\ y + \frac{23}{6}z = \frac{38}{3} \end{cases}$.

Here, again, we can choose any z, but then x and y are determined. We could call our choice t (as in the previous examples), but common practice is to let $z = kt$, where k is the least common multiple of the denominators of the fractions. By making this choice, many fractions can be avoided (for those who are not particularly fond of fractions). In this example, we can let $z = 6t$, then

$$x = 7 - \tfrac{5}{2}(6t) = 7 - 15t$$
$$y = \tfrac{38}{3} - \tfrac{23}{6}(6t) = \tfrac{38}{3} - 23t$$

This gives a parametric solution: $(7 - 15t, \frac{38}{3} - 23t, 6t)$. Remember, *every* different choice of t gives another solution. ∎

It is possible that you will need more than one parameter for a problem. Suppose you have 2 equations with 4 unknowns, then you will need at least $4 - 2 = 2$ parameters, as illustrated by Example 11.

EXAMPLE 11 Solve $\begin{cases} -w + 2x - 3y + z = 3 \\ 2w - 3x + y - 2z = 4 \end{cases}$.

Solution
$$\begin{bmatrix} -1 & 2 & -3 & 1 & \vdots & 3 \\ 2 & -3 & 1 & -2 & \vdots & 4 \end{bmatrix} \longrightarrow \begin{bmatrix} 1 & -2 & 3 & -1 & \vdots & -3 \\ 2 & -3 & 1 & -2 & \vdots & 4 \end{bmatrix} \longrightarrow$$

$$\begin{bmatrix} 1 & -2 & 3 & -1 & \vdots & -3 \\ 0 & 1 & -5 & 0 & \vdots & 10 \end{bmatrix} \longrightarrow \begin{bmatrix} 1 & 0 & -7 & -1 & \vdots & 17 \\ 0 & 1 & -5 & 0 & \vdots & 10 \end{bmatrix}$$

This is equivalent to the system

$$\begin{cases} w - 7y - z = 17 \quad \text{or} \quad w = 17 + 7y + z \\ x - 5y = 10 \quad \text{or} \quad x = 10 + 5y \end{cases}$$

You can choose any values for both y and z, and then w and x will be determined; for example, if we use s and t as the parameters so that $y = s$ and $z = t$, then

$$w = 17 + 7s + t$$
$$x = 10 + 5s$$

and the solution is $(17 + 7s + t, 10 + 5s, s, t)$. For example,

if $s = 0$ and $t = 0$, then $(17, 10, 0, 0)$ is a solution;
if $s = 0$ and $t = 1$, then $(18, 10, 0, 1)$ is a solution;
if $s = 1$ and $t = 0$, then $(24, 15, 1, 0)$ is a solution;
if $s = 1$ and $t = 1$, then $(25, 15, 1, 1)$ is a solution; and so on.

Each of these solutions can be checked in the original system to verify that each, in turn, satisfies the system.   ∎

EXAMPLE 12   A rancher has to mix three types of feed for her cattle. The analysis shown in the table gives the amounts per bag (100 lb) of grain.

Grain	Protein	Carbohydrates	Sodium
A	7 lb	88 lb	1 lb
B	6 lb	90 lb	1 lb
C	10 lb	70 lb	2 lb

How many bags of each type of grain should the rancher mix to provide 71 pounds of protein, 854 pounds of carbohydrates, and 12 pounds of sodium?

Solution   Let a, b, and c be the number of bags of grains A, B, and C, respectively, needed for the mixture. Then:

Grain	Protein	Carbohydrates	Sodium
A	$7a$	$88a$	a
B	$6b$	$90b$	b
C	$10c$	$70c$	$2c$
Total	71	854	12

Thus

$$\begin{cases} 7a + 6b + 10c = 71 \\ 88a + 90b + 70c = 854 \\ a + b + 2c = 12 \end{cases}$$

$$\begin{bmatrix} 7 & 6 & 10 & \vdots & 71 \\ 88 & 90 & 70 & \vdots & 854 \\ 1 & 1 & 2 & \vdots & 12 \end{bmatrix} \longrightarrow \begin{bmatrix} 1 & 1 & 2 & \vdots & 12 \\ 88 & 90 & 70 & \vdots & 854 \\ 7 & 6 & 10 & \vdots & 71 \end{bmatrix}$$

$$\longrightarrow \begin{bmatrix} 1 & 1 & 2 & \vdots & 12 \\ 0 & 2 & -106 & \vdots & -202 \\ 0 & -1 & -4 & \vdots & -13 \end{bmatrix} \longrightarrow \begin{bmatrix} 1 & 1 & 2 & \vdots & 12 \\ 0 & 1 & -53 & \vdots & -101 \\ 0 & -1 & -4 & \vdots & -13 \end{bmatrix}$$

$$\longrightarrow \begin{bmatrix} 1 & 0 & 55 & 113 \\ 0 & 1 & -53 & -101 \\ 0 & 0 & -57 & -114 \end{bmatrix} \longrightarrow \begin{bmatrix} 1 & 0 & 55 & 113 \\ 0 & 1 & -53 & -101 \\ 0 & 0 & 1 & 2 \end{bmatrix}$$

$$\longrightarrow \begin{bmatrix} 1 & 0 & 0 & 3 \\ 0 & 1 & 0 & 5 \\ 0 & 0 & 1 & 2 \end{bmatrix}$$

The rancher should mix three bags of grain A, five bags of grain B, and two bags of grain C. ∎

Problem Set 3.3

Write the systems in Problems 1–2 in augmented matrix form.

1. a. $\begin{cases} 4x + 5y = -16 \\ 3x + 2y = 5 \end{cases}$

b. $\begin{cases} x + y + z = 4 \\ 3x + 2y + z = 7 \\ x - 3y + 2z = 0 \end{cases}$

c. $\begin{cases} x_1 + 3x_2 + x_3 + x_4 = 3 \\ x_1 - 2x_3 + 2x_4 = 0 \\ x_3 + 5x_4 = -14 \\ x_2 - 3x_3 - x_4 = 2 \end{cases}$

2. a. $\begin{cases} 2x + y = 0 \\ 3x - 2y = -7 \\ x - 3y = -1 \end{cases}$

b. $\begin{cases} 2x - y + z = 2 \\ 2x - 4z = 32 \end{cases}$

c. $\begin{cases} x_1 - 2x_2 - x_3 - x_4 = 4 \\ 5x_1 + 2x_2 - x_3 + 2x_4 = 23 \\ 3x_1 + 4x_2 - 3x_3 + x_4 = 2 \\ 2x_1 - 2x_2 + x_3 - x_4 = 10 \end{cases}$

Write a system of equations (use $x_1, x_2, x_3, \ldots, x_n$ for the variables) that has the given augmented matrix in Problems 3–4.

3. a. $\begin{bmatrix} 2 & 1 & 4 & 3 \\ 6 & 2 & -1 & -4 \\ -3 & -1 & 0 & 1 \end{bmatrix}$

b. $\begin{bmatrix} 1 & 0 & 0 & 5 \\ 0 & 1 & 0 & -3 \\ 0 & 0 & 1 & 4 \end{bmatrix}$

c. $\begin{bmatrix} 1 & 0 & 0 & 3 \\ 0 & 1 & 0 & 2 \\ 0 & 0 & 1 & -8 \\ 0 & 0 & 0 & 1 \end{bmatrix}$

4. a. $\begin{bmatrix} 1 & 0 & 0 & 5 \\ 0 & 1 & 2 & 4 \end{bmatrix}$

b. $\begin{bmatrix} 1 & 0 & 0 & 0 & 3 \\ 0 & 1 & 0 & 0 & -6 \\ 0 & 0 & 1 & 5 & 4 \end{bmatrix}$

c. $\begin{bmatrix} 1 & 0 & 0 & 0 & 6 \\ 0 & 1 & 0 & 0 & -3 \\ 0 & 0 & 1 & 0 & 5 \\ 0 & 0 & 0 & 1 & 4 \end{bmatrix}$

5. Let $A = \begin{bmatrix} -2 & 2 & 4 & 8 \\ -1 & 3 & -2 & 3 \\ 1 & 5 & 3 & -2 \end{bmatrix}$

Work each part separately: use this matrix for each part and not the results from another part.

a. Interchange two rows to obtain a 1 in the first position of the first row.
b. Multiply each member of row 2 by 2.
c. Multiply each member of row 1 by $\frac{1}{2}$.
d. Multiply row 3 by 2 and add the result to row 1.
e. Multiply row 3 by -1 and add the result to row 2.

6. Let $B = \begin{bmatrix} 2 & -1 & 4 & 8 \\ 3 & -2 & 1 & 4 \\ -5 & 1 & 2 & -7 \end{bmatrix}$

Work each part separately: use this matrix for each part and not the results from another part.

a. Interchange two rows to obtain a 1 in the second position of the second row.
b. Multiply each member of row 1 by $\frac{1}{2}$.
c. Multiply each member of row 3 by -2.
d. Multiply row 1 by 3 and add the result to row 3.
e. Multiply row 2 by 2 and add the result to row 3.

Solve the systems in Problems 7–47 by using Gauss–Jordan elimination.

7. $\begin{cases} x + y = 3 \\ 2x + 3y = 8 \end{cases}$

8. $\begin{cases} x - y = 2 \\ 3x + 2y = 11 \end{cases}$

9. $\begin{cases} 3x - 4y = -2 \\ 2x + 5y = -32 \end{cases}$

10. $\begin{cases} 5x + 2y = 4 \\ 3x + y = -1 \end{cases}$

11. $\begin{cases} 3x - 2y = 10 \\ 6x - 4y = 20 \end{cases}$

12. $\begin{cases} 4x + 5y = 9 \\ 7x + 7y = 14 \end{cases}$

13. $\begin{cases} 2x + y = 0 \\ 3x - 2y = -7 \\ x - 3y = -1 \end{cases}$

14. $\begin{cases} 3x - 5y = 9 \\ 2x - 4y = 8 \\ x + 3y = -11 \end{cases}$

15. $\begin{cases} 2x - 3y = -8 \\ 9x - 7y = -10 \\ 7x + 5y = -6 \end{cases}$

16. $\begin{cases} 3x - 2y = 13 \\ 4x + 5y = 2 \\ -2x - 3y = 0 \\ x + y = 1 \end{cases}$

17. $\begin{cases} 4x + 3y = -19 \\ -x - 5y = 24 \\ 2x + 3y = -17 \\ 5x - 2y = 5 \end{cases}$

18. $\begin{cases} 3x - 2y = 6 \\ 5x + 4y = 2 \\ -2x + 3y = 1 \\ x - 4y = -3 \end{cases}$

19. $\begin{cases} x + y + z = 6 \\ 2x - y + z = 3 \\ x - 2y - 3z = -12 \end{cases}$

20. $\begin{cases} 2x - y + z = 3 \\ x - 3y + 2z = 7 \\ x - y - z = -1 \end{cases}$

21. $\begin{cases} x + y + z = 4 \\ x + 3y + 2z = 4 \\ x - 2y + z = 7 \end{cases}$

22. $\begin{cases} x + 2z = 7 \\ x + y = 11 \\ -2y + 9z = -3 \end{cases}$

23. $\begin{cases} x + 2z = 13 \\ 2x + y = 8 \\ -2y + 9z = 41 \end{cases}$

24. $\begin{cases} 6x + y + 20z = 27 \\ x - y = 0 \\ y + z = 2 \end{cases}$

25. $\begin{cases} 4x + y + 2z = 7 \\ x + 2y = 0 \\ 3x - y - z = 7 \end{cases}$

26. $\begin{cases} 2x - y + 4z = 13 \\ 3x + 6y = 0 \\ 2y - 3z = 3 + 3x \end{cases}$

27. $\begin{cases} 3x - 2y + z = 5 \\ 5x - 3y = 24 \\ 2y + z = -5 \end{cases}$

28. $\begin{cases} x + y + z = 4 \\ x - 2y - z = 1 \\ 3x + y - 2z = -1 \end{cases}$

29. $\begin{cases} x + y + z = 4 \\ 3x + 2y + z = 7 \\ x - 3y + 2z = 2 \end{cases}$

30. $\begin{cases} x + y + z = 6 \\ x - 2y - z = 2 \\ 3x - y - 2z = 1 \end{cases}$

31. $\begin{cases} 2x - y + 3z = 7 \\ -x + 3y - 2z = -13 \\ 3x - 4y + 5z = 20 \end{cases}$

32. $\begin{cases} 3x - 2y + z = 13 \\ x - 5y + 2z = 24 \\ 2x + 3y - z = -11 \end{cases}$

33. $\begin{cases} 6x - y - 2z = 7 \\ 5x - 4y - 5z = 5 \\ x + 3y + 3z = 4 \end{cases}$

34. $\begin{cases} 2x - y + z = 2 \\ 2x - 4z = 32 \end{cases}$

35. $\begin{cases} x - 2y + z = -3 \\ 3y - 7z = -6 \end{cases}$

36. $\begin{cases} 2x - 3y = 1 \\ 4y + 6z = -8 \end{cases}$

37. $\begin{cases} w + x - 2y + 3z = 5 \\ 3w - 2x + y - 5z = 8 \end{cases}$

38. $\begin{cases} -2w + 3x - y - z = 6 \\ 3w - 2x + y + 2z = 5 \end{cases}$

39. $\begin{cases} 2w - 3x + z = 5 \\ 3x + 4y - z = 6 \end{cases}$

40. $\begin{cases} x_1 + 2x_2 + x_4 = 3 \\ 3x_1 + x_2 - 2x_4 = -1 \\ x_1 + 3x_3 - x_4 = 2 \end{cases}$

41. $\begin{cases} 6x_1 - 3x_2 + x_4 = -12 \\ 2x_2 + 4x_3 - x_4 = 1 \\ 3x_1 + 2x_2 + 2x_3 = -3 \end{cases}$

42. $\begin{cases} 3x_2 - x_3 - 4x_4 = 1 \\ 2x_1 + x_3 + 3x_4 = -1 \\ 5x_1 - 3x_2 - 5x_3 = -4 \end{cases}$

43. $\begin{cases} x_1 + 3x_2 + x_3 + x_4 = -1 \\ x_1 - 2x_3 + 2x_4 = -4 \\ x_3 + 5x_4 = -14 \\ x_2 - 3x_3 - x_4 = -1 \end{cases}$

44. $\begin{cases} x_1 - 2x_2 - x_3 - x_4 = -1 \\ 5x_1 + 2x_2 - x_3 + 2x_4 = 10 \\ 3x_1 + 4x_2 - 3x_3 + x_4 = -13 \\ 2x_1 - 2x_2 + x_3 - x_4 = 9 \end{cases}$

45. $\begin{cases} x_1 - 3x_2 + x_3 - x_4 = 3 \\ -x_1 + 2x_2 - 2x_3 + x_4 = -3 \\ 3x_1 - 5x_2 - 6x_3 + 4x_4 = 10 \\ 2x_1 + 3x_2 + 4x_3 + 2x_4 = 7 \end{cases}$

46. $\begin{cases} 2x_1 + 3x_2 + 2x_3 - x_4 = 10 \\ x_1 + 2x_2 - 4x_3 + 3x_4 = -3 \\ x_1 + 5x_2 - 2x_3 + 4x_4 = 1 \\ 3x_1 + 3x_2 + 2x_3 - 3x_4 = 14 \\ 5x_1 - 6x_2 - x_3 + 4x_4 = -1 \end{cases}$

47. $\begin{cases} x_1 - x_2 + 2x_3 - x_4 + 2x_5 = 2 \\ 2x_1 + x_2 + x_3 + 2x_4 - 2x_5 = 0 \\ -x_1 - x_2 - x_3 - 3x_4 - x_5 = 3 \\ 3x_1 + 2x_2 - x_3 - x_4 + x_5 = 7 \end{cases}$

APPLICATIONS

48. A farmer must mix three types of cattle feed. The table on page 113 gives the amounts per bag (100 lb) of grain. How many bags of each type of grain should be mixed to provide 58 pounds of protein, 655 pounds of carbohydrates, and 11 pounds of sodium?

Grain	Protein	Carbohydrates	Sodium
A	9 lb	75 lb	2 lb
B	5 lb	90 lb	1 lb
C	8 lb	80 lb	1 lb

49. In order to control a certain type of disease, it is necessary to use 23 liters of pesticide A and 34 liters of pesticide B. The dealer can order commercial spray I, each container of which holds 5 liters of pesticide A and 2 liters of pesticide B, and commercial spray II, each container of which holds 2 liters of pesticide A and 7 liters of pesticide B. How many containers of each type of commercial spray should be used to attain exactly the right proportion of pesticides needed?

50. In order to manufacture a certain alloy it is necessary to use 33 kg of metal A and 56 kg of metal B. The manufacturer mixes alloy I, each bar of which contains 3 kg of metal A and 5 kg of metal B, with alloy II, each bar of which contains 4 kg of metal A and 7 kg of metal B, to make the desired alloy. How much of the two alloys should be used in order to produce the alloy desired?

51. A candy maker mixes chocolate, milk, and coconut to produce three kinds of candy (I, II, and III) with the following proportions:

Candy I: 7 lb chocolate, 5 gal milk, and 1 oz coconut

Candy II: 3 lb chocolate, 2 gal milk, and 2 oz coconut

Candy III: 4 lb chocolate, 3 gal milk, and 3 oz coconut

If 67 pounds of chocolate, 48 gallons of milk, and 32 ounces of coconut are available, how much of each kind of candy can be produced?

52. The total number of registered Democrats, Republicans, and Independents in a certain community is 100,000. Voter turnout in a recent election was tabulated as follows:

50% of the Democrats voted
60% of the Republicans voted
70% of the Independents voted
55,200 votes were cast (assume that there were no write-in votes; everyone voted Democratic, Republican, or Independent)

The ratio of registered Democrats to registered Independents is 9 to 1. How many registered Democrats, Republicans, and Independents are there in the community?

53. Baskin-Robbins stores recently began offering three sizes of ice cream cones:

Small scoop: 2.5 oz for $.70
Medium scoop: 4 oz for $.95
Large scoop: 6 oz for $1.40

Are these sizes consistently priced? If not, which size is the best bargain? [Hint: you must consider not only the price per ounce for ice cream, but also the price of the cone.]

3.4 Inverse Matrices

In the set of real numbers there are two properties that we now extend to matrices.

Identity Elements

ADDITION: $a + 0 = 0 + a = a$ — 0 is the identity for addition in the set of real numbers.

MULTIPLICATION: $a \cdot 1 = 1 \cdot a = a$ — 1 is the identity for multiplication in the set of real numbers.

Inverse Elements

ADDITION: $a + (-a) = (-a) + a = 0$ — For each number a there exists an opposite (*additive inverse*) so that the sum of a and its opposite is 0.

MULTIPLICATION: $a \cdot a^{-1} = a^{-1} \cdot a = 1$ — For each nonzero number a there exists a reciprocal (*multiplicative inverse*) so that the product of a and its reciprocal is 1.

For matrices, the identity for addition is the zero matrix $\mathbf{0}$ so that $A + 0 = 0 + A = A$ for conformable matrices $\mathbf{0}$ and A. For example,

$$\begin{bmatrix} 1 & 2 & 3 \\ 4 & 5 & 6 \\ 7 & 8 & 9 \end{bmatrix} + \begin{bmatrix} 0 & 0 & 0 \\ 0 & 0 & 0 \\ 0 & 0 & 0 \end{bmatrix} = \begin{bmatrix} 1 & 2 & 3 \\ 4 & 5 & 6 \\ 7 & 8 & 9 \end{bmatrix}$$

Identity matrix
for 3 × 3 matrices
and addition

Identical

For matrix multiplication, the square matrix I of order $n \times n$ consisting of 1s on the main diagonal and zeros elsewhere is called the **identity matrix** of order n, since $IA = AI = A$ for every conformable matrix A. For example,

$$\begin{bmatrix} 1 & 2 & 3 \\ 4 & 5 & 6 \\ 7 & 8 & 9 \end{bmatrix} \begin{bmatrix} 1 & 0 & 0 \\ 0 & 1 & 0 \\ 0 & 0 & 1 \end{bmatrix} = \begin{bmatrix} 1 & 2 & 3 \\ 4 & 5 & 6 \\ 7 & 8 & 9 \end{bmatrix}$$

Identity matrix
for 3 × 3 matrices
and multiplication

Identical

The inverse matrix for addition is simply the matrix whose entries are opposites of the corresponding entries of the original matrix. However, it is the inverse for multiplication that is of particular interest to us.

Inverse of a Matrix

If A is a square matrix and if there exists a matrix A^{-1} such that
$$A^{-1}A = AA^{-1} = I$$
then A^{-1} is called the *inverse of* A for multiplication.

Usually, in the context of matrices, when we talk simply of the inverse of A we mean the inverse of A for multiplication.

EXAMPLE 1   Verify that the inverse of $A = \begin{bmatrix} 2 & 1 \\ 3 & 2 \end{bmatrix}$ is $B = \begin{bmatrix} 2 & -1 \\ -3 & 2 \end{bmatrix}$.

Solution $AB = \begin{bmatrix} 2 & 1 \\ 3 & 2 \end{bmatrix} \begin{bmatrix} 2 & -1 \\ -3 & 2 \end{bmatrix}$ $BA = \begin{bmatrix} 2 & -1 \\ -3 & 2 \end{bmatrix} \begin{bmatrix} 2 & 1 \\ 3 & 2 \end{bmatrix}$

$\qquad\qquad = \begin{bmatrix} 4-3 & -2+2 \\ 6-6 & -3+4 \end{bmatrix}$ $= \begin{bmatrix} 4-3 & 2-2 \\ -6+6 & -3+4 \end{bmatrix}$

$\qquad\qquad = \begin{bmatrix} 1 & 0 \\ 0 & 1 \end{bmatrix} = I$ $= \begin{bmatrix} 1 & 0 \\ 0 & 1 \end{bmatrix} = I$

Thus $B = A^{-1}$.   ∎

EXAMPLE 2 Show that A and B are inverses of each other, where

$$A = \begin{bmatrix} 0 & 1 & 2 \\ -1 & 1 & 2 \\ 1 & -2 & -5 \end{bmatrix} \quad \text{and} \quad B = \begin{bmatrix} 1 & -1 & 0 \\ 3 & 2 & 2 \\ -1 & -1 & -1 \end{bmatrix}$$

Solution $AB = \begin{bmatrix} 0 & 1 & 2 \\ -1 & 1 & 2 \\ 1 & -2 & -5 \end{bmatrix} \begin{bmatrix} 1 & -1 & 0 \\ 3 & 2 & 2 \\ -1 & -1 & -1 \end{bmatrix} = \begin{bmatrix} 0+3-2 & 0+2-2 & 0+2-2 \\ -1+3-2 & 1+2-2 & 0+2-2 \\ 1-6+5 & -1-4+5 & 0-4+5 \end{bmatrix}$

$= \begin{bmatrix} 1 & 0 & 0 \\ 0 & 1 & 0 \\ 0 & 0 & 1 \end{bmatrix} = I$

$BA = \begin{bmatrix} 1 & -1 & 0 \\ 3 & 2 & 2 \\ -1 & -1 & -1 \end{bmatrix} \begin{bmatrix} 0 & 1 & 2 \\ -1 & 1 & 2 \\ 1 & -2 & -5 \end{bmatrix} = \begin{bmatrix} 0+1+0 & 1-1+0 & 2-2+0 \\ 0-2+2 & 3+2-4 & 6+4-10 \\ 0+1-1 & -1-1+2 & -2-2+5 \end{bmatrix}$

$= \begin{bmatrix} 1 & 0 & 0 \\ 0 & 1 & 0 \\ 0 & 0 & 1 \end{bmatrix} = I$

Since $AB = I = BA$, then $B = A^{-1}$. ∎

If a given matrix has an inverse, we say that it is **nonsingular**. The unanswered question, however, is how to *find* an inverse matrix. Consider the following two examples.

EXAMPLE 3 Find the inverse of $A = \begin{bmatrix} 1 & 2 \\ 1 & 4 \end{bmatrix}$.

Solution Find a matrix B (if it exists) so that $AB = I$; since we do not know B, let its entries be variables. That is, let

$$B = \begin{bmatrix} x_1 & x_2 \\ y_1 & y_2 \end{bmatrix}$$

Then

$AB = \begin{bmatrix} 1 & 2 \\ 1 & 4 \end{bmatrix} \begin{bmatrix} x_1 & x_2 \\ y_1 & y_2 \end{bmatrix}$

$= \begin{bmatrix} x_1 + 2y_1 & x_2 + 2y_2 \\ x_1 + 4y_1 & x_2 + 4y_2 \end{bmatrix}$

$= \begin{bmatrix} 1 & 0 \\ 0 & 1 \end{bmatrix}$

By the definition of the equality of matrices, we see that

$$\begin{cases} x_1 + 2y_1 = 1 \\ x_1 + 4y_1 = 0 \end{cases} \qquad \begin{cases} x_2 + 2y_2 = 0 \\ x_2 + 4y_2 = 1 \end{cases}$$

Solve these systems by using Gauss–Jordan elimination:

$$\begin{bmatrix} 1 & 2 & | & 1 \\ 1 & 4 & | & 0 \end{bmatrix} \longrightarrow \begin{bmatrix} 1 & 2 & | & 1 \\ 0 & 2 & | & -1 \end{bmatrix} \qquad \begin{bmatrix} 1 & 2 & | & 0 \\ 1 & 4 & | & 1 \end{bmatrix} \longrightarrow \begin{bmatrix} 1 & 2 & | & 0 \\ 0 & 2 & | & 1 \end{bmatrix}$$

$$\longrightarrow \begin{bmatrix} 1 & 2 & | & 1 \\ 0 & 1 & | & -\frac{1}{2} \end{bmatrix} \qquad\qquad \longrightarrow \begin{bmatrix} 1 & 2 & | & 0 \\ 0 & 1 & | & \frac{1}{2} \end{bmatrix}$$

$$\longrightarrow \begin{bmatrix} 1 & 0 & | & 2 \\ 0 & 1 & | & -\frac{1}{2} \end{bmatrix} \qquad\qquad \longrightarrow \begin{bmatrix} 1 & 0 & | & -1 \\ 0 & 1 & | & \frac{1}{2} \end{bmatrix}$$

$$\begin{cases} x_1 = 2 \\ y_1 = -\frac{1}{2} \end{cases} \qquad\qquad\qquad \begin{cases} x_2 = -1 \\ y_2 = \frac{1}{2} \end{cases}$$

Therefore the inverse is

$$B = \begin{bmatrix} x_1 & x_2 \\ y_1 & y_2 \end{bmatrix} = \begin{bmatrix} 2 & -1 \\ -\frac{1}{2} & \frac{1}{2} \end{bmatrix}$$ ■

The solution of the two systems is shown side by side so you can easily see that the steps are identical since the two rows on the left are the same; the third columns are

$$\begin{bmatrix} 1 \\ 0 \end{bmatrix} \begin{bmatrix} 0 \\ 1 \end{bmatrix}$$

respectively. It would seem that we might combine steps, but, before we do, consider a three-by-three example.

EXAMPLE 4 Find the inverse for the matrix

$$\begin{bmatrix} 1 & -1 & 0 \\ 3 & 2 & 2 \\ -1 & -1 & -1 \end{bmatrix}$$

(This is matrix B from Example 2.)

Solution We need to find a matrix

$$\begin{bmatrix} x_1 & x_2 & x_3 \\ y_1 & y_2 & x_3 \\ z_1 & z_2 & z_3 \end{bmatrix}$$

so that

$$\begin{bmatrix} 1 & -1 & 0 \\ 3 & 2 & 2 \\ -1 & -1 & -1 \end{bmatrix}\begin{bmatrix} x_1 & x_2 & x_3 \\ y_1 & y_2 & y_3 \\ z_1 & z_2 & z_3 \end{bmatrix} = \begin{bmatrix} 1 & 0 & 0 \\ 0 & 1 & 0 \\ 0 & 0 & 1 \end{bmatrix}$$

$$\begin{cases} x_1 - y_1 + 0z_1 = 1 \\ 3x_1 + 2y_1 + 2z_1 = 0 \\ -x_1 - y_1 - z_1 = 0 \end{cases} \quad \begin{cases} x_2 - y_2 + 0z_2 = 0 \\ 3x_2 + 2y_2 + 2z_2 = 1 \\ -x_2 - y_2 - z_2 = 0 \end{cases} \quad \begin{array}{l} x_3 - y_3 + 0z_3 = 0 \\ 3x_3 + 2y_3 + 2z_3 = 0 \\ -x_3 - y_3 - z_3 = 1 \end{array}$$

We could solve these as three separate systems using Gauss–Jordan elimination; however, all the steps would be identical since the variables on the left of the equal signs are the same in each system. Therefore suppose we augment the matrix of the coefficients by the *three* rows and do all three at once.

$$\left[\begin{array}{rrr|rrr} 1 & -1 & 0 & 1 & 0 & 0 \\ 3 & 2 & 2 & 0 & 1 & 0 \\ -1 & -1 & -1 & 0 & 0 & 1 \end{array}\right] \longrightarrow \left[\begin{array}{rrr|rrr} 1 & -1 & 0 & 1 & 0 & 0 \\ 0 & 5 & 2 & -3 & 1 & 0 \\ 0 & -2 & -1 & 1 & 0 & 1 \end{array}\right] \xrightarrow{2R3 + R2}$$

$$\left[\begin{array}{rrr|rrr} 1 & -1 & 0 & 1 & 0 & 0 \\ 0 & 1 & 0 & -1 & 1 & 2 \\ 0 & -2 & -1 & 1 & 0 & 1 \end{array}\right] \longrightarrow \left[\begin{array}{rrr|rrr} 1 & 0 & 0 & 0 & 1 & 2 \\ 0 & 1 & 0 & -1 & 1 & 2 \\ 0 & 0 & -1 & -1 & 2 & 5 \end{array}\right] \longrightarrow$$

$$\left[\begin{array}{rrr|rrr} 1 & 0 & 0 & 0 & 1 & 2 \\ 0 & 1 & 0 & -1 & 1 & 2 \\ 0 & 0 & 1 & 1 & -2 & -5 \end{array}\right]$$

Now, if we relate this back to the original three systems, we see that the inverse is

$$\left[\begin{array}{rrr} 0 & 1 & 2 \\ -1 & 1 & 2 \\ 1 & -2 & -5 \end{array}\right]$$ ∎

By studying Examples 3 and 4, we are led to a procedure for finding the inverse of a nonsingular matrix:

Procedure for Finding the Inverse of a Matrix

1. Augment the given matrix with I; that is, write $[A \mid I]$, where I is the identity matrix of the same order as the given square matrix A.
2. Perform elementary row operations using Gauss–Jordan elimination in order to change the matrix A into the identity matrix (if possible).
3. If steps 1 and 2 can be performed, the result in the augmented part is the inverse of A.

COMPUTER APPLICATION Most inverses involve (ugly) fractions, and it is often not only convenient, but necessary, to have numerical assistance. Computer programs for finding the inverse of a matrix are fairly easy to obtain and use. Program 4 on the computer disk accompanying this book is helpful if you have access to a computer. There are two options: manual and automatic reduction. With manual reduction, the computer asks what you want to do and then simply carries out the arithmetic. With automatic reduction, the computer not only does the calculations but also makes the decisions as to what step comes next. This program will also generate additional problems so that you can practice the procedure. You should use the automatic reduction option only after you understand the process.

EXAMPLE 5   Find the inverse of the matrix $A = \begin{bmatrix} 1 & 2 \\ 0 & 0 \end{bmatrix}$.

Solution   Write the augmented matrix $[A \mid I]$:

$$\begin{bmatrix} 1 & 2 & \vdots & 1 & 0 \\ 0 & 0 & \vdots & 0 & 1 \end{bmatrix}$$

We want to make the left-hand side look like the corresponding identity matrix. This is impossible since there are no elementary row operations that will put it into the required form. Thus there is no inverse. ∎

EXAMPLE 6   Find the inverse of the matrix $A = \begin{bmatrix} 0 & 1 & 2 \\ 2 & -1 & 1 \\ -1 & 1 & 0 \end{bmatrix}$.

Solution   Write the augmented matrix $[A \mid I]$ and make the left-hand side look like the corresponding identity matrix (if possible):

$$\begin{bmatrix} 0 & 1 & 2 & \vdots & 1 & 0 & 0 \\ 2 & -1 & 1 & \vdots & 0 & 1 & 0 \\ -1 & 1 & 0 & \vdots & 0 & 0 & 1 \end{bmatrix} \xrightarrow{R1 \leftrightarrow R3} \begin{bmatrix} -1 & 1 & 0 & \vdots & 0 & 0 & 1 \\ 2 & -1 & 1 & \vdots & 0 & 1 & 0 \\ 0 & 1 & 2 & \vdots & 1 & 0 & 0 \end{bmatrix}$$

$$\longrightarrow \begin{bmatrix} 1 & -1 & 0 & \vdots & 0 & 0 & -1 \\ 2 & -1 & 1 & \vdots & 0 & 1 & 0 \\ 0 & 1 & 2 & \vdots & 1 & 0 & 0 \end{bmatrix} \longrightarrow \begin{bmatrix} 1 & -1 & 0 & \vdots & 0 & 0 & -1 \\ 0 & 1 & 1 & \vdots & 0 & 1 & 2 \\ 0 & 1 & 2 & \vdots & 1 & 0 & 0 \end{bmatrix}$$

$$\longrightarrow \begin{bmatrix} 1 & 0 & 1 & \vdots & 0 & 1 & 1 \\ 0 & 1 & 1 & \vdots & 0 & 1 & 2 \\ 0 & 0 & 1 & \vdots & 1 & -1 & -2 \end{bmatrix} \longrightarrow \begin{bmatrix} 1 & 0 & 0 & \vdots & -1 & 2 & 3 \\ 0 & 1 & 0 & \vdots & -1 & 2 & 4 \\ 0 & 0 & 1 & \vdots & 1 & -1 & -2 \end{bmatrix}$$

Thus $A^{-1} = \begin{bmatrix} -1 & 2 & 3 \\ -1 & 2 & 4 \\ 1 & -1 & -2 \end{bmatrix}$. ∎

Inverses, when they are known, can be used to solve matrix equations. For example, suppose

$$A = \begin{bmatrix} 0 & 1 & 2 \\ 2 & -1 & 1 \\ -1 & 1 & 0 \end{bmatrix} \qquad X = \begin{bmatrix} x \\ y \\ z \end{bmatrix} \qquad B = \begin{bmatrix} 0 \\ -1 \\ 1 \end{bmatrix}$$

Then

$$AX = \begin{bmatrix} 0 & 1 & 2 \\ 2 & -1 & 1 \\ -1 & 1 & 0 \end{bmatrix} \begin{bmatrix} x \\ y \\ z \end{bmatrix} = \begin{bmatrix} y + 2z \\ 2x - y + z \\ -x + y \end{bmatrix}$$

and $AX = B$ means

$$\begin{cases} y + 2z = 0 \\ 2x - y + z = -1 \\ -x + y = 1 \end{cases}$$

since matrices are equal if and only if corresponding entries are equal. If you can write a system in matrix form, then you can solve the system $AX = B$ provided A^{-1} exists:

$$AX = B \qquad \text{Given system}$$
$$(A^{-1})AX = A^{-1}B \qquad \text{Multiply both sides by } A^{-1} \text{ on the left}$$
$$(A^{-1}A)X = A^{-1}B \qquad \text{Associative property}$$
$$IX = A^{-1}B \qquad \text{Inverse property}$$
$$X = A^{-1}B \qquad \text{Identity property}$$

This means that to solve a system, all we have to do is multiply A^{-1} and B to find X. Since

$$A = \begin{bmatrix} 0 & 1 & 2 \\ 2 & -1 & 1 \\ -1 & 1 & 0 \end{bmatrix}$$

we see from Example 6,

$$A^{-1} = \begin{bmatrix} -1 & 2 & 3 \\ -1 & 2 & 4 \\ 1 & -1 & -2 \end{bmatrix}$$

so we can solve for X as follows:

$$X = A^{-1}B$$
$$= \begin{bmatrix} -1 & 2 & 3 \\ -1 & 2 & 4 \\ 1 & -1 & -2 \end{bmatrix} \begin{bmatrix} 0 \\ -1 \\ 1 \end{bmatrix} = \begin{bmatrix} 0 - 2 + 3 \\ 0 - 2 + 4 \\ 0 + 1 - 2 \end{bmatrix} = \begin{bmatrix} 1 \\ 2 \\ -1 \end{bmatrix}$$

Thus $x = 1$, $y = 2$, and $z = -1$.

The method of solving a system by using the inverse matrix is very efficient if you know the inverse. Unfortunately, *finding* the inverse for only one system is usually more work than using another method to solve the system. However, there are certain applications that yield the same coefficient matrix over and over. In these cases the inverse method is worthwhile. And, finally, computers can often find approximations for inverse matrices quite easily.

COMPUTER APPLICATION Program 5 on the accompanying computer disk is designed to solve systems by using inverses. If the given matrix is singular, you can also use this software to decide if the system has no solutions or an infinite number of solutions.

Problem Set 3.4

Use multiplication in Problems 1–12 to determine whether the matrices are inverses.

1. $\begin{bmatrix} 1 & 2 \\ 2 & 3 \end{bmatrix}$, $\begin{bmatrix} -3 & 2 \\ 2 & -1 \end{bmatrix}$

2. $\begin{bmatrix} 2 & -5 \\ -1 & 2 \end{bmatrix}$, $\begin{bmatrix} -2 & -5 \\ -1 & -2 \end{bmatrix}$

3. $\begin{bmatrix} 3 & 5 \\ 4 & 7 \end{bmatrix}$, $\begin{bmatrix} 7 & -5 \\ -4 & 3 \end{bmatrix}$

4. $\begin{bmatrix} 4 & 7 \\ 5 & 9 \end{bmatrix}$, $\begin{bmatrix} 9 & -7 \\ -5 & 4 \end{bmatrix}$

5. $\begin{bmatrix} 4 & 3 \\ 2 & 2 \end{bmatrix}$, $\begin{bmatrix} 1 & -\frac{3}{2} \\ -1 & 2 \end{bmatrix}$

6. $\begin{bmatrix} 2 & 3 \\ 2 & 1 \end{bmatrix}$, $\begin{bmatrix} -\frac{1}{4} & \frac{3}{4} \\ \frac{1}{2} & -\frac{1}{2} \end{bmatrix}$

7. $\begin{bmatrix} 0 & 1 & 0 \\ 1 & -1 & 0 \\ -1 & 2 & 1 \end{bmatrix}$, $\begin{bmatrix} -1 & -1 & 0 \\ -1 & 0 & 0 \\ 1 & -1 & -1 \end{bmatrix}$

8. $\begin{bmatrix} 1 & 0 & 0 \\ 0 & 1 & 1 \\ 2 & 0 & 1 \end{bmatrix}$, $\begin{bmatrix} 1 & 0 & 0 \\ 2 & 1 & -1 \\ -2 & 0 & 1 \end{bmatrix}$

9. $\begin{bmatrix} 3 & -2 & 4 \\ 2 & 1 & 2 \\ 5 & 3 & 5 \end{bmatrix}$, $\begin{bmatrix} -1 & 22 & -8 \\ 0 & -5 & 2 \\ 1 & -19 & 7 \end{bmatrix}$

10. $\begin{bmatrix} 6 & -1 & -5 \\ -7 & 1 & 5 \\ -10 & 2 & 11 \end{bmatrix}$, $\begin{bmatrix} -1 & -1 & 0 \\ -27 & -16 & -5 \\ 4 & 2 & 1 \end{bmatrix}$

11. $\begin{bmatrix} 1 & 1 & 0 & 1 \\ 2 & 0 & -5 & 8 \\ 0 & 1 & 3 & 2 \\ 0 & 0 & 3 & 31 \end{bmatrix}$, $\begin{bmatrix} 1 & 95 & -1 & -29 \\ 0 & -92 & 1 & 28 \\ 0 & 31 & 0 & -10 \\ 0 & -3 & 0 & 1 \end{bmatrix}$

12. $\begin{bmatrix} 1 & 0 & 0 & 2 \\ 3 & 1 & -1 & 0 \\ 0 & 2 & 1 & 3 \\ 1 & 0 & 0 & 1 \end{bmatrix}$, $\begin{bmatrix} -2 & 0 & 0 & 2 \\ 0 & \frac{1}{3} & \frac{1}{3} & -1 \\ -3 & -\frac{2}{3} & \frac{1}{3} & 5 \\ 1 & 0 & 0 & -1 \end{bmatrix}$

Find the inverses of the matrices in Problems 13–30, if possible.

13. $\begin{bmatrix} 4 & -7 \\ -1 & 2 \end{bmatrix}$

14. $\begin{bmatrix} 4 & 0 \\ 0 & 5 \end{bmatrix}$

15. $\begin{bmatrix} 3 & 5 \\ 1 & 2 \end{bmatrix}$

16. $\begin{bmatrix} 3 & -1 \\ -4 & 2 \end{bmatrix}$

17. $\begin{bmatrix} 1 & 3 \\ 2 & 0 \end{bmatrix}$

18. $\begin{bmatrix} 2 & 3 \\ 1 & -6 \end{bmatrix}$

19. $\begin{bmatrix} 8 & 6 \\ -2 & 4 \end{bmatrix}$

20. $\begin{bmatrix} 2 & 1 \\ 4 & 3 \end{bmatrix}$

21. $\begin{bmatrix} 1 & -\frac{3}{2} \\ -1 & 2 \end{bmatrix}$

22. $\begin{bmatrix} 1 & 0 & 2 \\ 2 & 1 & 0 \\ 0 & -2 & 9 \end{bmatrix}$

23. $\begin{bmatrix} 6 & 1 & 20 \\ 1 & -1 & 0 \\ 0 & 1 & 3 \end{bmatrix}$

24. $\begin{bmatrix} 4 & 1 & 0 \\ 2 & -1 & 4 \\ -3 & 2 & 1 \end{bmatrix}$

25. $\begin{bmatrix} 1 & -1 & 1 \\ 0 & 2 & -1 \\ 2 & 3 & 0 \end{bmatrix}$

26. $\begin{bmatrix} 15 & 4 & -5 \\ -12 & -3 & 4 \\ -4 & -1 & 1 \end{bmatrix}$

27. $\begin{bmatrix} 1 & 0 & 2 \\ 3 & -1 & 2 \\ 4 & 1 & 0 \end{bmatrix}$

28. $\begin{bmatrix} 1 & 0 & 0 & 1 \\ 0 & 2 & 0 & 0 \\ 0 & 0 & 0 & 1 \\ 2 & 0 & 1 & 0 \end{bmatrix}$

29. $\begin{bmatrix} 0 & 1 & 2 & 0 \\ 0 & 0 & 0 & 1 \\ 1 & 1 & 3 & 0 \\ 2 & 4 & 0 & 0 \end{bmatrix}$

30. $\begin{bmatrix} 1 & 2 & 0 & 0 \\ 0 & 0 & 1 & 0 \\ 1 & 3 & 0 & 1 \\ 2 & 4 & 0 & 0 \end{bmatrix}$

Solve the systems in Problems 31–54 by solving the corresponding matrix equation with an inverse, if possible.

Problems 31–33 use the inverse found in Problem 13.

31. $\begin{cases} 4x - 7y = -2 \\ -x + 2y = 1 \end{cases}$

32. $\begin{cases} 4x - 7y = -65 \\ -x + 2y = 18 \end{cases}$

33. $\begin{cases} 4x - 7y = 48 \\ -x + 2y = -13 \end{cases}$

Problems 34–36 use the inverse found in Problem 19.

34. $\begin{cases} 8x + 6y = 12 \\ -2x + 4y = -14 \end{cases}$ 35. $\begin{cases} 8x + 6y = 16 \\ -2x + 4y = 18 \end{cases}$

36. $\begin{cases} 8x + 6y = -6 \\ -2x + 4y = -26 \end{cases}$

Problems 37–39 use the inverse found in Problem 18.

37. $\begin{cases} 2x + 3y = 9 \\ x - 6y = -3 \end{cases}$ 38. $\begin{cases} 2x + 3y = 2 \\ x - 6y = 16 \end{cases}$

39. $\begin{cases} 2x + 3y = 2 \\ x - 6y = -14 \end{cases}$

Problems 40–42 use the inverse found in Problem 20.

40. $\begin{cases} 2x + y = 5 \\ 4x + 3y = 9 \end{cases}$ 41. $\begin{cases} 2x + y = 16 \\ 4x + 3y = 2 \end{cases}$

42. $\begin{cases} 2x + y = -3 \\ 4x + 3y = 1 \end{cases}$

Problems 43–45 use the inverse found in Problem 22.

43. $\begin{cases} x + 2z = 7 \\ 2x + y = 16 \\ -2y + 9z = -3 \end{cases}$ 44. $\begin{cases} x + 2z = 4 \\ 2x + y = 0 \\ -2y + 9z = 19 \end{cases}$

45. $\begin{cases} x + 2z = 7 \\ 2x + y = 0 \\ -2y + 9z = 31 \end{cases}$

Problems 46–48 use the inverse found in Problem 23.

46. $\begin{cases} 6x + y + 20z = 27 \\ x - y = 0 \\ y + 3z = 4 \end{cases}$

47. $\begin{cases} 6x + y + 20z = 14 \\ x - y = 1 \\ y + 3z = 1 \end{cases}$

48. $\begin{cases} 6x + y + 20z = 11 \\ x - y = 5 \\ y + 3z = -3 \end{cases}$

Problems 49–51 use the inverse found in Problem 24.

49. $\begin{cases} 4x + y = 6 \\ 2x - y + 4z = 12 \\ -3x + 2y + z = 4 \end{cases}$

50. $\begin{cases} 4x + y = 7 \\ 2x - y + 4z = -11 \\ -3x + 2y + z = -12 \end{cases}$

51. $\begin{cases} 4x + y = -10 \\ 2x - y + 4z = 20 \\ -3x + 2y + z = 20 \end{cases}$

Problems 52–54 use the inverse found in Problem 29.

52. $\begin{cases} x + 2y = 5 \\ z = 3 \\ w + x + 3y = 9 \\ 2w + 4x = 8 \end{cases}$

53. $\begin{cases} x + 2y = 0 \\ z = -4 \\ w + x + 3y = 4 \\ 2w + 4x = -2 \end{cases}$

54. $\begin{cases} x + 2y = 7 \\ z = -7 \\ w + x + 3y = 16 \\ 2w + 4x = -4 \end{cases}$

APPLICATIONS

55. A rancher has to mix three types of feed. The table gives the amounts per 100-pound bag of grain.

Grain	Protein	Carbohydrates	Sodium
I	10 lb	75 lb	2 lb
II	20 lb	70 lb	3 lb
III	15 lb	80 lb	1 lb

How many bags of each type of grain should be mixed to provide 135 pounds of protein, 740 pounds of carbohydrates, and 22 pounds of sodium?

56. Repeat Problem 55 to provide 350 pounds of protein, 1885 pounds of carbohydrates, and 48 pounds of sodium.

57. Repeat Problem 55 to provide 545 pounds of protein, 3090 pounds of carbohydrates, and 79 pounds of sodium.

58. A dietitian is to arrange a diet of three basic foods. The diet is to include 1800 units of vitamin A, 9800 units of vitamin C, and 1420 units of calcium. The number of units per ounce of each of the foods is given in the table.

Food	Vitamin A	Vitamin C	Calcium
I	50 units	300 units	20 units
II	30 units	200 units	40 units
III	40 units	100 units	30 units

How many ounces of each food should be supplied?

59. Repeat Problem 58 to supply 1200 units of vitamin A, 6000 units of vitamin C, and 900 units of calcium.

60. Repeat Problem 58 to supply 1140 units of vitamin A, 7100 units of vitamin C, and 1030 units of calcium.

3.5 Leontief Models*

Wassily Leontief received the Nobel Prize in Economics in 1973 for describing the economy in terms of an **input–output model**. We now look at two extended applications of the Leontief model: a **closed model** in which we try to find the relative income of each participant within a system, and an **open model** in which we try to find the amount of production needed to achieve a forecast demand.

A Closed Model

Suppose a simplified economy consists of three individuals: a farmer, a tailor, and a builder, who provide the essentials of food, clothing, and shelter for each other. Of the food produced by the farmer, 50% is used by the farmer, 20% by the tailor, and 30% by the builder. The builder's production is utilized 40% by the builder, 40% by the farmer, and 20% by the tailor. The tailor's production is divided among them as 45% for the builder, 20% for the farmer, and 35% for the tailor. Suppose that each person's wages are to be about $1,000 (it could be shells, sheep, or any other convenient unit used as money). How much should each person pay the others?

We begin by writing the given information in matrix form. This is called an **input–output matrix**:

$$
\begin{array}{c}
\\
\\
A = \\
\\
\end{array}
\begin{array}{ccc}
& \text{PRODUCER} & \\
\text{Farmer} & \text{Builder} & \text{Tailor} \\
\end{array}
$$

$$
A = \begin{bmatrix} 0.5 & 0.4 & 0.2 \\ 0.3 & 0.4 & 0.45 \\ 0.2 & 0.2 & 0.35 \end{bmatrix}
\begin{array}{l} \text{Farmer} \\ \text{Builder} \quad \text{USER} \\ \text{Tailor} \end{array}
$$

In general, an input–output matrix for a closed model consists of n rows and n columns where the sum of each column is 1 and each entry a_{ij} is a fraction between 0 and 1, inclusive: $0 \leq a_{ij} \leq 1$. The a_{ij} entry is the fraction used by i and produced by j.

Let X be the column matrix representing the price of each output in the system. That is, let

$$
X = \begin{bmatrix} x_1 \\ x_2 \\ x_3 \end{bmatrix}
$$

where

$$x_1 = \text{Farmer's wages}$$
$$x_2 = \text{Tailor's wages}$$
$$x_3 = \text{Builder's wages}$$

The wages include those wages paid to themselves as part of the economic system. For the wages to come out even, the total amount paid out by each must equal the

---

* This section is optional.

total amount received by each. That is,

$$X = AX$$

$$\begin{bmatrix} x_1 \\ x_2 \\ x_3 \end{bmatrix} = \begin{bmatrix} 0.5 & 0.4 & 0.2 \\ 0.3 & 0.4 & 0.45 \\ 0.2 & 0.2 & 0.35 \end{bmatrix} \begin{bmatrix} x_1 \\ x_2 \\ x_3 \end{bmatrix}$$

This can be written as the system

$$\begin{bmatrix} x_1 \\ x_2 \\ x_3 \end{bmatrix} = \begin{bmatrix} 0.5x_1 + 0.4x_2 + 0.2x_3 \\ 0.3x_1 + 0.4x_2 + 0.45x_3 \\ 0.2x_1 + 0.2x_2 + 0.35x_3 \end{bmatrix} \quad \text{or} \quad \begin{cases} x_1 = 0.5x_1 + 0.4x_2 + 0.2x_3 \\ x_2 = 0.3x_1 + 0.4x_2 + 0.45x_3 \\ x_3 = 0.2x_1 + 0.2x_2 + 0.35x_3 \end{cases}$$

By subtracting to obtain zeros on the right, we have

$$\begin{cases} 0.5x_1 - 0.4x_2 - 0.2x_3 = 0 \\ -0.3x_1 + 0.6x_2 - 0.45x_3 = 0 \\ -0.2x_1 - 0.2x_2 + 0.65x_3 = 0 \end{cases}$$

A system of equations in which the right-hand side is zero is called a **homogeneous system** of equations.

There is, of course, a trivial solution for which $x_1 = x_2 = x_3 = 0$, but that is not the only solution. For a closed model, the system will have a one-parameter solution. This parameter will serve as a scaling factor.

The solution using Gauss–Jordan elimination is shown below:

$$\begin{bmatrix} 0.5 & -0.4 & -0.2 & \vdots & 0 \\ -0.3 & 0.6 & -0.45 & \vdots & 0 \\ -0.2 & -0.2 & 0.65 & \vdots & 0 \end{bmatrix} \longrightarrow \begin{bmatrix} 5 & -4 & -2 & \vdots & 0 \\ -2 & 4 & -3 & \vdots & 0 \\ -4 & -4 & 13 & \vdots & 0 \end{bmatrix}$$

$$\longrightarrow \begin{bmatrix} 1 & -8 & 11 & \vdots & 0 \\ -2 & 4 & -3 & \vdots & 0 \\ -4 & -4 & 13 & \vdots & 0 \end{bmatrix} \longrightarrow \begin{bmatrix} 1 & -8 & 11 & \vdots & 0 \\ 0 & -12 & 19 & \vdots & 0 \\ 0 & -36 & 57 & \vdots & 0 \end{bmatrix}$$

$$\longrightarrow \begin{bmatrix} 1 & -8 & 11 & \vdots & 0 \\ 0 & 1 & -\frac{19}{12} & \vdots & 0 \\ 0 & 0 & 0 & \vdots & 0 \end{bmatrix} \longrightarrow \begin{bmatrix} 1 & 0 & -\frac{5}{3} & \vdots & 0 \\ 0 & 1 & -\frac{19}{12} & \vdots & 0 \\ 0 & 0 & 0 & \vdots & 0 \end{bmatrix}$$

Thus $x_1 = \frac{5}{3}x_3$ and $x_2 = \frac{19}{12}x_3$. Since the choice of x_3 is arbitrary, choose x_3 so that it is about \$1,000 and also so that it is divisible by 12 (to give an integer solution). Let $x_3 = 1008$:

$$x_1 = \frac{5}{3}(1008) = 1680$$
$$x_2 = \frac{19}{12}(1008) = 1596$$
$$x_3 = 1008$$

The wages to be paid are \$1,680 to the farmer, \$1,596 to the builder, and \$1,008 to the tailor. Note that if you let $x_3 = 636$, then the farmer's wages are \$1,060, the builder's wages \$1,007, and the tailor's wages \$636. No matter what choice is made for x_3, each other participant's share is determined.

An Open Model

Suppose a simplified economy consists of three industries: farming, clothing, and construction, which provide the essentials of food, clothing, and shelter not only for each other, but also for the consumer. Suppose we take $100 to be the total value of the goods produced by the farming industry. The clothing and construction industries require $10 and $15, respectively, of those goods, and the farmers consume $25 worth of their own output. The consumer requires the remaining $50 worth of the goods produced by the farming industry. The construction industry consumes $5 of its own production, the farming industry consumes $15, and the clothing industry consumes $10. The construction industry produces $110 of goods, so the remaining $80 is used by the consumer. The third industry, the clothing industry, uses $5 worth of its own production, while the farming segment uses $10 and the construction industry uses $20. The consumer requires $85 of the total clothing industry's output of $120. Construct a model to analyze how the output of any one of the industries is affected by changes in the other two and how production can be changed to anticipate changes in consumer demand.

Once again, begin by writing the input–output matrix:

$$
\begin{array}{c}
\\
\text{PRODUCER}\\
\begin{array}{ccc}
\text{Farming} & \text{Construction} & \text{Clothing}
\end{array}
\end{array}
\qquad
\begin{array}{l}
\text{USER}\\
\text{(numerators are the}\\
\text{amounts used by each of}\\
\text{these industries)}\\
\text{Consumer}
\end{array}
$$

$$
A = \begin{bmatrix} \frac{25}{100} & \frac{15}{110} & \frac{10}{120} \\ \frac{15}{100} & \frac{5}{110} & \frac{10}{120} \\ \frac{10}{100} & \frac{20}{110} & \frac{5}{120} \end{bmatrix}
\qquad
\begin{array}{ll}
\text{Farming} & 25 + 15 + 10 + 50 = 100 \\
\text{Construction} & 15 + \ 5 + 10 + 80 = 110 \\
\text{Clothing} & 10 + 20 + \ 5 + 85 = 120
\end{array}
$$

denominators are the production of each industry

$$
A \approx \begin{bmatrix} 0.25 & 0.14 & 0.08 \\ 0.15 & 0.05 & 0.08 \\ 0.10 & 0.18 & 0.04 \end{bmatrix}
$$

This is the input–output matrix correct to the nearest hundredth

The matrix A for an open model consists of n rows and n columns where the entries a_{ij} represent the output of industry j required for one unit of output of industry i. This concept is shown by arrows on the upper matrix. Note that the sum of the columns in an open model are each less than or equal to 1.

Let D_i be the column matrix representing the consumer demand in the year i. That is, $i = 0$ represents the *present* consumer demand:

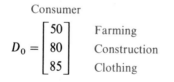

$$
\begin{array}{c}
\text{Consumer}\\
D_0 = \begin{bmatrix} 50 \\ 80 \\ 85 \end{bmatrix}
\end{array}
\quad
\begin{array}{l}
\text{Farming}\\
\text{Construction}\\
\text{Clothing}
\end{array}
$$

Now, to anticipate future demand, let

x_i = Output of the farming industry in year i

y_i = Output of the construction industry in year i

z_i = Output of the clothing industry in year i

so that

$$X_i = \begin{bmatrix} x_i \\ y_i \\ z_i \end{bmatrix}$$

In particular,

$$X_0 = \begin{bmatrix} 100 \\ 110 \\ 120 \end{bmatrix}$$

The product AX_0 represents the amounts required for internal consumption, and $AX_0 + D_0$ represents the total output. That is,

INTERNAL CONSUMPTION + DEMAND = TOTAL OUTPUT

Suppose marketing forecasts predict a 5-year plan setting demand at

$$D_5 = \begin{bmatrix} 110 \\ 200 \\ 200 \end{bmatrix}$$

We then need to solve

$$AX_5 + D_5 = X_5$$

The solution of this type of forecasting problem is called an *open Leontief model*. For the closed model we found the solution for the particular matrix equation, but with the open model we can anticipate finding $X_1, X_2, X_3, \ldots$ for any number of years. For this reason we solve this system using an inverse matrix (if it exists). First, we solve for X_5:

$$AX_5 + D_5 = X_5$$
$$X_5 - AX_5 = D_5$$
$$IX_5 - AX_5 = D_5$$
$$(I - A)X_5 = D_5$$
$$X_5 = (I - A)^{-1}D_5:$$

Now, find the difference

$$I - A = \begin{bmatrix} 1 & 0 & 0 \\ 0 & 1 & 0 \\ 0 & 0 & 1 \end{bmatrix} - \begin{bmatrix} 0.25 & 0.14 & 0.08 \\ 0.15 & 0.05 & 0.08 \\ 0.10 & 0.18 & 0.04 \end{bmatrix} = \begin{bmatrix} 0.75 & -0.14 & -0.08 \\ -0.15 & 0.95 & -0.08 \\ -0.10 & -0.18 & 0.96 \end{bmatrix}$$

and form the augmented matrix to find the inverse $(I - A)^{-1}$:

$$
\begin{bmatrix}
0.75 & -0.14 & -0.08 & \vdots & 1 & 0 & 0 \\
-0.15 & 0.95 & -0.08 & \vdots & 0 & 1 & 0 \\
-0.10 & -0.18 & 0.96 & \vdots & 0 & 0 & 1
\end{bmatrix}
\longrightarrow
\begin{bmatrix}
1 & -0.19 & -0.11 & \vdots & 1.33 & 0 & 0 \\
0 & 0.92 & -0.10 & \vdots & 0.20 & 1 & 0 \\
0 & -0.20 & 0.95 & \vdots & 0.13 & 0 & 1
\end{bmatrix}
$$

$$
\longrightarrow
\begin{bmatrix}
1 & -0.19 & -0.11 & \vdots & 1.33 & 0 & 0 \\
0 & 1 & -0.10 & \vdots & 0.22 & 1.09 & 0 \\
0 & 0 & 0.93 & \vdots & 0.18 & 0.22 & 1
\end{bmatrix}
\longrightarrow
\begin{bmatrix}
1 & -0.19 & -0.11 & \vdots & 1.33 & 0 & 0 \\
0 & 1 & -0.10 & \vdots & 0.22 & 1.09 & 0 \\
0 & 0 & 1 & \vdots & 0.19 & 0.23 & 1.08
\end{bmatrix}
$$

$$
\longrightarrow
\begin{bmatrix}
1 & -.19 & 0 & \vdots & 1.35 & 0.03 & 0.12 \\
0 & 1 & 0 & \vdots & 0.24 & 1.11 & 0.11 \\
0 & 0 & 1 & \vdots & 0.19 & 0.23 & 1.08
\end{bmatrix}
\longrightarrow
\begin{bmatrix}
1 & 0 & 0 & \vdots & 1.40 & 0.23 & 0.14 \\
0 & 1 & 0 & \vdots & 0.24 & 1.11 & 0.11 \\
0 & 0 & 1 & \vdots & 0.19 & 0.23 & 1.08
\end{bmatrix}
$$

Substitute this inverse matrix $(I - A)^{-1}$ and the given matrix D_5 into the equation $X_5 = (I - A)^{-1}D_5$:

$$
X_5 =
\begin{bmatrix}
1.40 & 0.23 & 0.14 \\
0.24 & 1.11 & 0.11 \\
0.19 & 0.23 & 1.08
\end{bmatrix}
\begin{bmatrix}
110 \\
200 \\
200
\end{bmatrix}
=
\begin{bmatrix}
154.00 + 46.00 + 28.00 \\
26.40 + 222.00 + 22.00 \\
20.90 + 46.00 + 216.00
\end{bmatrix}
=
\begin{bmatrix}
228.00 \\
270.40 \\
282.90
\end{bmatrix}
$$

The results shown above include a considerable amount of round-off error and are provided so you can follow the process. If you use the computer disk and the given fractions, the last column matrix will be 227.27, 270.00, and 283.64. Thus the required output in 5 years is about $230 for farming, $270 for construction, and $280 for clothing.

Problem Set 3.5

Solve the homogeneous systems in Problems 1–6.

1. $\begin{cases} x + 3y - 5z = 0 \\ 3x + y - 3z = 0 \\ 4x + 6y - 11z = 0 \end{cases}$

2. $\begin{cases} 2x - y + 3z = 0 \\ x + 3y - 4z = 0 \\ 8x + 3y + z = 0 \end{cases}$

3. $\begin{cases} 3x + 6y + 2z = 0 \\ 2x - 3y + 6z = 0 \\ 2x + 3y + 2z = 0 \end{cases}$

4. $\begin{cases} 0.4x_1 - 0.5x_2 - 0.2x_3 = 0 \\ -0.1x_1 + 0.6x_2 - 0.3x_3 = 0 \\ -0.3x_1 - 0.1x_2 + 0.5x_3 = 0 \end{cases}$

5. $\begin{cases} 0.3x_1 - 0.3x_2 - 0.65x_3 = 0 \\ -0.25x_1 + 0.5x_2 - 0.2x_3 = 0 \\ -0.5x_1 - 0.2x_2 + 0.85x_3 = 0 \end{cases}$

6. $\begin{cases} 0.7x_1 - 0.5x_2 - 0.45x_3 = 0 \\ -0.4x_1 + 0.5x_2 - 0.35x_3 = 0 \\ -0.3x_1 - 0.25x_2 + 0.8x_3 = 0 \end{cases}$

APPLICATIONS

7. Suppose each person's wages for the closed model discussed in this section should be around $30,000. Give two possible answers to the question if all other information in the problem stays the same.

8. In the closed model discussed in this section, suppose the farmer uses 45% of his own production and gives 25% to the tailor. Answer the question if the other information stays the same.

9. Suppose a merchant joins the closed model and allocates her resources as shown by the following input–output matrix:

% of production				
Farmer	Builder	Tailor	Merchant	
0.4	0.3	0.1	0.1	Farmer
0.2	0.3	0.3	0.3	Builder
0.2	0.2	0.4	0.3	Tailor
0.2	0.2	0.2	0.3	Merchant

How much should each person pay the others if the wages are to be about $1,000?

Apply the given demand in Problems 10–12 to the open model discussed in this section to find the required output to meet the demand.

10. $D_1 = \begin{bmatrix} 50 \\ 80 \\ 90 \end{bmatrix}$ **11.** $D_3 = \begin{bmatrix} 70 \\ 100 \\ 110 \end{bmatrix}$ **12.** $D_4 = \begin{bmatrix} 100 \\ 180 \\ 170 \end{bmatrix}$

13. Solve the open model discussed in this section for the following input–output matrix, with the other information remaining the same:

$$A = \begin{bmatrix} 0.3 & 0.2 & 0.1 \\ 0.2 & 0.4 & 0.3 \\ 0.1 & 0.1 & 0.2 \end{bmatrix}$$

14. Consider a simple economy with three industries: manufacturing (M), agriculture (A), and transportation (T). Find A and $I - A$ for this system given the information in the table.

Use	Producer M	A	T	External demand	Total output
M	50	20	100	30	200
A	30	80	100	90	300
T	20	10	50	420	500

15. Find $(I - A)^{-1}$ for Problem 14.

16. Find the output in Problem 14 to meet the given demand.

17. Find the output in Problem 14 to meet the external demand if it changes to 200 for M, 90 for A, and 150 for T.

18. **Historical Question** In 1966 Leontief studied a simplified model of the 1958 Israeli economy by dividing it into three sectors—agriculture, manufacturing, and energy—as shown in the table:*

	Agriculture	Manufacturing	Energy
Agriculture	0.293	0	0
Manufacturing	0.014	0.207	0.017
Energy	0.044	0.010	0.216

The exports (in thousands of Israeli pounds) are shown in the following table:

Agriculture	138,213
Manufacturing	17,597
Energy	1786

Find A and $I - A$ for this system.

19. Find $(I - A)^{-1}$ for Problem 18.

20. Find the output in Problem 18 to meet the external demand.

3.6 Summary and Review

IMPORTANT TERMS

Augmented matrix   [3.3]
Break-even analysis   [3.1]
Closed model   [3.5]*
Column matrix   [3.2]
Communication matrix   [3.2]
Component   [3.2]
Conformable matrices   [3.2]
Consistent   [3.1]
Dependent   [3.1]
Difference of matrices   [3.2]
Elementary row operations   [3.3]
Entry   [3.2]
Equal matrices   [3.2]

Equilibrium point   [3.1]
Equivalent matrices   [3.3]
Equivalent systems   [3.1]
Gauss–Jordan elimination   [3.3]
Graphing method   [3.1]
Homogeneous system   [3.5]*
Identity matrix   [3.4]
Inconsistent   [3.1]
Independent   [3.1]
Input–output model   [3.5]*
Inverse matrix   [3.4]
Leontief model   [3.5]*
Linear combination method   [3.1]

---

* Wassily Leontief, *Input–Output Economics* (New York: Oxford University Press, 1966), pp. 54–57.

Linear system [3.1]

Matrix [3.2]

Nonconformable matrices [3.2]

Nonsingular matrix [3.4]

Open model [3.5]

Order [3.2]

Pivit row [3.3]

Pivoting [3.3]

Product of matrices [3.3]

Reduced row-echelon form [3.3]

Row matrix [3.2]

Scalar multiplication [3.2]

Simultaneous solution [3.1]

Square matrix [3.2]

Substitution method [3.1]

Sum of matrices [3.2]

Supply and demand [3.1]

System of equations [3.1]

Target row [3.3]

Zero matrix [3.2]

SAMPLE TEST *For additional practice there are a large number of review problems categorized by objective in the Student Solutions Manual. The following sample test (40 minutes) is intended to review the main ideas of the chapter.*

1. Solve the systems using each of the following methods exactly once: graphing, adding, and substitution.

a. $\begin{cases} 2x + 3y = 2 \\ 5x - 3y = -27 \end{cases}$
b. $\begin{cases} y = 2x - 4 \\ y = \frac{1}{3}x + 1 \end{cases}$
c. $\begin{cases} x + y = 13 \\ 3x - 4y = -17 \end{cases}$

Perform the indicated operations in Problems 2–5, if possible.

$$A = \begin{bmatrix} 4 & -1 & 2 \\ 2 & 1 & -1 \\ 0 & 5 & 3 \end{bmatrix} \quad B = \begin{bmatrix} 1 & 3 & -2 \\ 1 & -1 & 1 \\ 2 & -2 & 2 \end{bmatrix} \quad C = \begin{bmatrix} 1 & -2 & 3 \end{bmatrix}$$

$$D = \begin{bmatrix} 4 & -5 & 1 \end{bmatrix} \quad E = \begin{bmatrix} 2 & 0 & 1 \\ 1 & -3 & 4 \end{bmatrix} \quad I = \begin{bmatrix} 1 & 0 & 0 \\ 0 & 1 & 0 \\ 0 & 0 & 1 \end{bmatrix}$$

2. a. $2A$ b. $3C - 2D$
3. a. EA b. AE
4. a. AB b. BA
5. a. $A(B + I)$ b. $I(A + B)$

6. a. Write the information on this map of airline routes in matrix form.

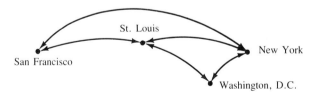

St. Louis

San Francisco

New York

Washington, D.C.

b. Use a matrix to show the number of ways a person could fly making one stop.

c. In how many ways can a person fly from San Francisco to New York making two stops?

Solve the systems in Problems 7–8 by Gauss–Jordan elimination.

7. a. $\begin{cases} 4x - 3y = 18 \\ 5x + 2y = 11 \end{cases}$
b. $\begin{cases} 2x - y = 9 \\ x + 5y = -23 \\ 3x - 4y = 26 \end{cases}$
c. $\begin{cases} 3x - y + z = 9 \\ 2x + y - 3z = -2 \end{cases}$

8. $\begin{cases} x - 2y + z = 6 \\ y - 3x + w = -2 \\ w - x - y = 2 \\ 2y - 3w + 2z = -2 \end{cases}$

9. **a.** Find the inverse of $A = \begin{bmatrix} 3 & 3 & -1 \\ -2 & -2 & 1 \\ -4 & -5 & 2 \end{bmatrix}$.

   b. Use the inverse from part a to solve

   $$\begin{cases} 3x + 3y - z = 8 \\ -2x - 2y + z = -1 \\ -4x - 5y + 2z = 3 \end{cases}$$

10. A manufacturer produces two products, I and II. The products require three ingredients: A, B, and C, to be used as follows:

	A	B	C
Product I	6	3	5
Product II	5	4	7

If the manufacturer has a supply of 3400 units of items A and C and 2000 units of item B, how many of each product can be manufactured?

Modeling Application 1

Gaining a Competitive Edge in Business

Solartex manufactures solar collector panels. During the first year of operation, rent, insurance, utilities, and other fixed costs averaged $8,500 per month. Each panel sold for $320 and cost the company $95 in materials and $55 in labor. Since company resources are limited, Solartex cannot spend more than $20,000 in any one month. Sunenergy, another company selling similar panels, competes directly with Solartex. Last month, Sunenergy manufactured 85 panels at a total cost of $20,475, but the previous month produced only 60 panels at a total cost of $17,100.

Mathematical modeling involves creating mathematical equations and procedures to make predictions about the real world. Typical textbook problems focus on limited, specific skills, but when confronted with a real life example you are often faced with a myriad of "facts" without specific clues on how to fit them together to make predictions. In this book, you will be given a modeling application and asked to write a paper using the given information. You will need to do some research to have adequate data. There are no "right answers" for these papers. In the *Student's Solution Manual*, a paper is presented for each modeling application, but your paper could certainly take a quite different direction.

For this first application, the following questions might help you get started: What is the cost equation for Solartex? How many panels can be manufactured by Solartex given the available capital? What is the minimum number of panels that should be produced? In order to make a profit, revenue must exceed cost; what is Solartex's break-even point? How would you compare Solartex and Sunenergy?

Feel free to supply information that is not given above. For example, a study of market demand could be useful. Suppose you find that at $75 per panel, Solartex (or a real company you have studied) could sell 200 panels per month, but at $450, they could sell only 20 panels per month. On the other hand, if Solartex had to sell the panels at $225 each, they could only afford material that would limit them to 30 panels per month, but at $450 each, they could afford sufficient material to supply 100 panels per month. What is the equilibrium point for this information?

Write a paper based on this modeling application.

CHAPTER 4

Linear Programming

CHAPTER CONTENTS

4.1 Systems of Linear Inequalities

4.2 Formulating Linear Programming Models

4.3 Graphical Solution of Linear Programming Problems

4.4 Slack Variables and the Pivot

4.5 Maximization by the Simplex Method

4.6 Nonstandard Linear Programming Problems

4.7 Duality

4.8 Summary and Review
Important Terms
Sample Test

APPLICATIONS

Management (*Business, Economics, Finance, and Investments*)

Allocation of resources in production (4.2, Problems 1–2, 9, 13, 16, and 17; 4.3, Problems 37–39 and 41)

Maximizing profit (4.2, Problems 8, 11; 4.5, Problems 29–31)

Allocation of resources in manufacturing (4.2, Problems 3, 4, and 18; 4.6, Problems 19–21; 4.8, Problem 10)

Maximizing yield from an investment (4.2, Problem 7; 4.3, Problems 44–45)

Office management (4.2, Problem 10)

Operating costs (4.2, Problem 12)

Staff utilization (4.2, Problems 20 and 21; 4.8, Problem 9)

Transportation problem; minimizing shipping costs (4.2, Problems 22–24)

Society of Actuaries examination (4.2, Problem 18; 4.5, Problem 31)

Minimizing costs (4.6, Problems 17–18; 4.7, Problem 23)

Management of an investment fund (4.7, Problem 26)

Staff utilization in minimizing labor costs (4.7, Problems 27–28)

Transportation problem (4.7, Problems 29–31)

CPA examination (4.8, Problems 5–10)

Life sciences (*Biology, Ecology, Health, and Medicine*)

Maximum number of animals (4.2, Problems 14–15)

Minimizing exercise time to achieve certain goals (4.2, Problem 19)

Diet problem:
   Kellogg's Corn Flakes and Post Honeycombs (4.2, Problem 5; 4.3, Problem 43)
   Kellogg's Corn Flakes and Kellogg's Raisin Bran (4.2, Problem 6)
   Meeting the minimum daily requirements at the lowest cost (4.3, Problem 42)

Determining the proper diet for horses (4.6, Problem 17)

Allocation of oil reserves (low, medium, and high grade) (4.7, Problem 24)

General interest

Allocation of time in an exercise program (4.7, Problem 25)

Modeling application—Air Pollution Control Model

This chapter introduces the topic of linear programming. Linear programming is applied when you are interested in maximizing or minimizing a linear function—for example, maximizing profits, minimizing costs, finding the most efficient shipping schedules, minimizing waste in production, securing the proper mixes of ingredients in a product, controlling inventories, or finding the most efficient assignment of personnel.

PREVIEW Graphing systems of linear inequalities is reviewed in the first section, and then formulating linear programming problems is discussed. These two ideas are then tied together to solve linear programming problems by graphing. Next, an algebraic process for solving linear programming problems called the *simplex method* is introduced. The simplex method is first applied to what is called a "standard" problem. Then we apply it to variations of standard problems. The final section, Duality, ties together, in a very nice and surprising way, maximum and minimum type problems.

PERSPECTIVE This chapter gives us a basic understanding of one of the most important topics in this course—linear programming. However, most real life applications are not limited to two variables, and since graphing is limited to two dimensions, we also need an algebraic method. The simplex method provides a step-by-step procedure for optimization. The step-by-step method can easily be adapted to computer solution. Computer solution is absolutely essential for many real world applications, which may use hundreds of variables and constraints. So, if you have a computer available, you might wish to use Program 6 on the computer disk accompanying this book.

4.1 Systems of Linear Inequalities

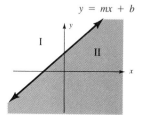

Figure 4.1 Half-planes

A **linear inequality** in two variables can be written in one of the following forms:

$$ax + by < c \qquad ax + by > c$$
$$ax + by \leq c \qquad ax + by \geq c$$

where a, b, and c are constants. The graphing of a linear inequality is similar to the graphing of a linear equation. Any line divides the plane into three regions, as shown in Figure 4.1. Regions I and II are called **half-planes**, so we can say the three sets determined by a line are the two half-planes and the set of points on the line itself. The line is called the **boundary** of the half-planes. If the boundary is included in the half-plane, the half-plane is said to be **closed**; if the boundary is not included, it is said to be **open**. Half-planes are drawn by showing the boundary as a dashed line, as in Figure 4.2c.

It is apparent in Figure 4.2 that evey linear inequality has an associated equation that is the boundary of that half-plane. For example, $2x + 3y < 12$ has the boundary line $2x + 3y = 12$. This line is graphed by solving for y: $y = -\frac{2}{3}x + 4$. After you have drawn the boundary (either solid or dashed), the remaining question is Which half-plane satisfies the inequality? The easiest way to decide is to choose *any* test point not on the boundary line. If it satisfies the inequality, then the solution is the half-plane containing the test point. If the test point does not satisfy the inequality, then the solution is the half-plane that doesn't contain the test point.

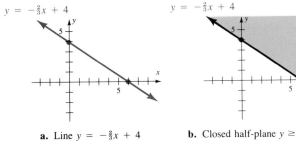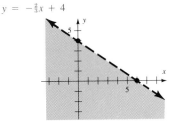

a. Line $y = -\frac{2}{3}x + 4$

b. Closed half-plane $y \geq -\frac{2}{3}x + 4$
Draw boundary as a solid line

c. Open half-plane $y < -\frac{2}{3}x + 4$
Draw boundary as a dashed line

Figure 4.2 Graphing half-planes

EXAMPLE 1   Graph $5x + 2y - 10 \leq 0$.

Solution   *Step 1*   *Graph the boundary line.*   The boundary is included if the given inequality is $\geq$ or $\leq$, so draw $5x + 2y - 10 = 0$ as a solid line. The easiest way to sketch this line is to put it into slope-intercept form: $y = -\frac{5}{2}x + 5$.

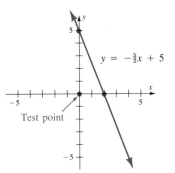

Step 2   *Choose a test point.*   The point $(0,0)$ is usually the best choice because of the simplified arithmetic:

$$5x + 2y - 10 \leq 0$$

Test $(0,0)$:

$$5(0) + 2(0) - 10 \leq 0$$
$$-10 \leq 0 \qquad \text{True}$$

Step 3   *Shade the appropriate half-plane.*   If the test point satisfies the inequality (it is true, as shown in step 2), shade in the half-plane containing the test point. Otherwise, shade in the other half-plane.

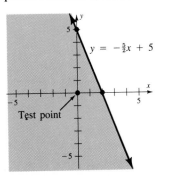

EXAMPLE 2   Graph $y > -20x + 110$.

Solution   *Step 1*   The boundary is $y = -20x + 110$; $m = -\frac{20}{1}$ and $b = 110$. Draw the boundary as a dashed line because it is not included in the half-plane ($>$ symbol).

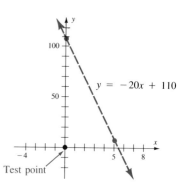

Step 2

$$y > -20x + 110$$

Test point $(0, 0)$:

$$0 > -20(0) + 110 \qquad \text{False}$$

Step 3   Graph the half-plane that does not contain the test point.

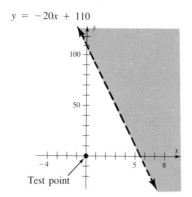

There are many applications in which inequalities play an important role. In Chapter 2 we found the *break-even point*—that is, the place where cost and revenue are the same. Sometimes, breaking even for a business takes the form of a linear equation. Suppose a company knows that it will break even if 50 items are sold at $3 each or if 10 items are sold at $5 each.* This gives a *break-even linear equation*:

$$y = -20x + 110$$

where x is the price and y is the number of items. Suppose that in order to make a profit it is necessary for

$$y > -20x + 110$$

---

* In practice, this is more likely to be 50,000 and 10,000 items. We use 50 and 10 here to keep the numbers easy to handle. To find the equation of the line where x is the price and y is the number of items, we use the points $(3, 50)$ and $(5, 10)$. The slope is

$$m = \frac{10 - 50}{5 - 3} = -20$$

and $y = mx + b$, so $50 = -20(3) + b$ or $b = 110$. Thus $y = -20x + 110$.

(This was graphed in Example 2.) However, in this application, it is also necessary that x and y both be positive:

$$x \geq 0$$
$$y \geq 0$$

Furthermore, market research shows that the price must be $8 or less:

$$x \leq 8$$

This information taken together gives a **system of linear inequalities** similar to a system of equations:

$$\begin{cases} y > -20x + 110 \\ x \geq 0 \\ y \geq 0 \\ x \leq 8 \end{cases}$$

The region determined by the system of inequalities is the solution of the system and is similar to the point of intersection for a system of equations because it is the intersection of the individual inequalities in the system.

EXAMPLE 3 Graph the system $\begin{cases} y > -20x + 110 \\ x \geq 0 \\ y \geq 0 \\ x \leq 8 \end{cases}$

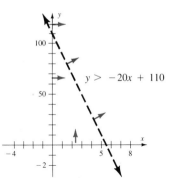

Solution Graph each half-plane, but instead of shading, mark the appropriate regions with arrows. For this example,

$$x \geq 0$$
$$y \geq 0$$

gives the first quadrant, which is marked with arrows as shown.

Next graph $y = -20x + 110$ and use a test point, say, $(0, 0)$, to determine the half-plane. Now draw $x = 8$ and mark it with arrows to show that $x \leq 8$. Finally shade the region of the plane that represents the intersection:

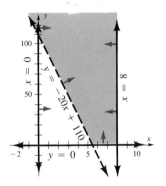

EXAMPLE 4   Graph the solution of the system $\begin{cases} 2x + y \leq 3 \\ x - y > 5 \\ x \geq 0 \\ y \geq -10 \end{cases}$

Solution   The graphs of the individual inequalities and their intersections are shown in the figure. Note the use of arrows to show the solutions of the individual inequalities. This device replaces the use of a lot of shading, which can be confusing if there are many inequalities in the system.

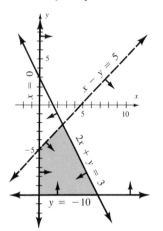

Problem Set 4.1

Graph the solution for each of the linear inequalities in Problems 1–24.

1. $y \geq 2x - 3$
2. $y \geq 3x + 2$
3. $y \geq 4x + 5$

4. $y \leq \frac{1}{2}x + 2$
5. $y \leq \frac{2}{3}x - 4$
6. $y \leq \frac{4}{5}x - 3$

7. $x - y > 3$
8. $2x + y < -3$
9. $3x + y < 4$

10. $x + 2y > 4$
11. $x - 3y > 12$
12. $3x - 4y > 8$

13. $y \leq 6$
14. $x \geq 2$
15. $y < -2$

16. $y > 0$
17. $x < 0$
18. $x < 8$

19. $260x - 1040y > 11,250$ **20.** $150x - 450y < 7200$

21. $0.06x - 0.05y < 10,000$ **22.** $0.03x - 0.05y > 10,000$

23. $0.08x + 0.10y \geq 500$ **24.** $0.10x - 0.08y \leq 1500$

Graph the solution of each system in Problems 25–51.

25. $\begin{cases} x \geq 0 \\ y \leq 0 \end{cases}$
26. $\begin{cases} x \geq 0 \\ y \geq 0 \end{cases}$
27. $\begin{cases} x \leq 0 \\ y \leq 0 \end{cases}$

28. $\begin{cases} x \geq 0 \\ y \geq 0 \\ x \leq 5 \\ y \leq 6 \end{cases}$
29. $\begin{cases} x \geq 0 \\ y \geq 0 \\ x < 8 \\ y < 5 \end{cases}$
30. $\begin{cases} x \geq 0 \\ y \geq 0 \\ x < 500 \\ y < 1000 \end{cases}$

31. $\begin{cases} y \leq 3x - 4 \\ y \geq -2x + 5 \end{cases}$
32. $\begin{cases} 2x + y > 3 \\ 3x - y < 2 \end{cases}$
33. $\begin{cases} 3x - 2y \geq 6 \\ 2x + 3y \leq 6 \end{cases}$

34. $\begin{cases} x + 2y \leq 18 \\ x + y \geq 4 \end{cases}$
35. $\begin{cases} x + y > 4 \\ x - y > -2 \end{cases}$
36. $\begin{cases} 2x + 3y \geq 3 \\ 2x + 3y \leq 9 \end{cases}$

37. $\begin{cases} x - 10 \leq 0 \\ x \geq 0 \end{cases}$
38. $\begin{cases} y - 5 \leq 0 \\ y \geq 0 \end{cases}$
39. $\begin{cases} y - 25 \leq 0 \\ y \geq 0 \end{cases}$

40. $\begin{cases} -5 < x \\ 3 \geq x \\ 5 > y \\ -2 \leq y \end{cases}$
41. $\begin{cases} -10 \leq x \\ x \leq 6 \\ 3 < y \\ y < 8 \end{cases}$
42. $\begin{cases} -5 < x \\ x \leq 2 \\ -4 \leq y \\ y < 9 \end{cases}$

43. $\begin{cases} x \geq 0 \\ y \geq 0 \\ x + y \leq 8 \\ y \leq 4 \\ x \leq 6 \end{cases}$
44. $\begin{cases} x \geq 0 \\ y \geq 0 \\ x + y \leq 9 \\ 2x - 3y \geq -6 \\ x - y \leq 3 \end{cases}$

45. $\begin{cases} x \geq 0 \\ y \geq 0 \\ 2x + y \geq 8 \\ y \leq 5 \\ x - y \leq 2 \\ 3x - y \geq 5 \end{cases}$

46. $\begin{cases} 2x + 3y \leq 30 \\ 3x + 2y \geq 20 \\ x \geq 0 \\ y \geq 0 \end{cases}$

49. $\begin{cases} 2x - 3y \leq 20 \\ 3x + 2y \leq 10 \\ x \geq 0 \\ y \leq 0 \end{cases}$

50. $\begin{cases} 6x - 3y - 15 \leq 0 \\ 4x + 2y - 30 \leq 0 \\ x \geq 0 \\ y \geq 0 \end{cases}$

47. $\begin{cases} 2x - 3y + 30 \geq 0 \\ 3x - 2y + 20 \leq 0 \\ x \leq 0 \\ y \geq 0 \end{cases}$

48. $\begin{cases} x + y - 10 \leq 0 \\ x + y + 4 \geq 0 \\ x - y \leq 6 \\ y - x \leq 4 \end{cases}$

51. $\begin{cases} 4x - 3y + 2 \geq 0 \\ x + y - 3 \leq 0 \\ x \geq 0 \\ y \geq 0 \end{cases}$

4.2 Formulating Linear Programming Models

One of the most difficult tasks in solving a real world problem is the construction of a mathematical model that simulates the situation. In the 1940s a new mathematical model called **linear programming** was found to be applicable to a wide range of situations in which the maximum or minimum value of some variables was needed. In this section, you will learn how to formulate or build a model for a variety of applications (see the examples listed in the box), and in the next section you will learn how to solve these models using graphical techniques. (In Section 4.5 an algebraic method, the *simplex method*, for solving linear programming problems is introduced.)

Examples of linear programming models

Allocation of Resources Models
    Maximize profit from a manufacturing process
    Maximize revenue from a manufacturing process
    Minimize costs from a manufacturing process
    Maximize production
    Find the optimal use of land
    Utilization of a sales force
    Utilization of office staff

Nutritional (or Diet) Models
    Minimize cost of a diet with certain nutritional requirements
    Minimize calories given certain nutritional requirements
    Manufacture a processed food to meet government regulations

Investment Strategies
    Maximize the return on an investment
    Allocate funds over several possible investments
    Minimize expenses

Transportation Model
    Minimize shipping costs

EXAMPLE 1   **Allocation of Resources in Production**   A farmer has 100 acres on which to plant two crops: corn and wheat. To produce these crops, there are certain expenses, as shown in the table.

Item	Cost per acre
Corn	
Seed	$ 12
Fertilizer	58
Planting/care/harvesting	50
Total	$120
Wheat	
Seed	$ 40
Fertilizer	80
Planting/care/harvesting	90
Total	$210

After the harvest, the farmer must usually store the crops while awaiting favorable market conditions. Each acre yields an average of 110 bushels of corn or 30 bushels of wheat. The limitations of resources are

Available capital: $15,000
Available storage facilities: 4000 bushels

If the net profit (the profit *after* all expenses have been subtracted) per bushel of corn is $1.30 and for wheat is $2.00, how should the farmer plant the 100 acres to maximize profits?

Solution   First, you might try to solve this problem by using your intuition. If you plant all 100 acres with wheat, the production is $30 \times 100 = 3000$ bushels, for a net profit of $3000 \times 2 = \$6,000$. But to plant 100 acres with wheat would cost $\$210 \times 100 = \$21,000$, and only $15,000 is available. On the other hand, if 100 acres of corn are planted, the total cost is $\$120 \times 100 = \$12,000$ and the net profit is $110 \times 100 \times \$1.30 = \$14,300$. However, the yield of 11,000 bushels (110×100) cannot be stored since there are facilities to store only 4000 bushels.

To formulate a mathematical model, begin by letting

$x =$ Number of acres to be planted in corn

$y =$ Number of acres to be planted in wheat

There are certain limitations, or **constraints**. The number of acres planted cannot be negative, so

$x \geq 0$ These constraints apply in almost every model, even though they are not
$y \geq 0$ explicitly stated as part of the given problem. Be sure, however, to state
          them when listing the constraints in the solution

The amount of available land is 100 acres, so

$x + y \leq 100$

Why not $x + y = 100$? It might be more profitable for the farmer to leave some land out of production. That is, it is not *necessary* to plant all the land.

We also know that

$120x =$ Expenses for planting corn

$210y =$ Expenses for planting wheat

The total expenses cannot exceed $15,000; this is the *available capital*:

$120x + 210y \leq 15,000$

The yields are

$110x =$ Yield of acreage planted in corn

$30y =$ Yield of acreage planted in wheat

The total yield cannot exceed the storage capacity of 4000 bushels, so

$110x + 30y \leq 4000$

We summarize the constraints (in boldface above) in the following system:

$$\begin{cases} x \geq 0 \\ y \geq 0 \\ x + y \leq 100 \\ 120x + 210y \leq 15,000 \\ 110x + 30y \leq 4000 \end{cases}$$

Now let P represent the total profit. The farmer wants to maximize this profit. A function that is to be maximized or minimized is called the **objective function**.

$$\text{PROFIT FROM CORN} = \text{VALUE} \cdot \text{AMOUNT}$$
$$= 1.30 \cdot 110x$$
$$= 143x$$
$$\text{PROFIT FROM WHEAT} = \text{VALUE} \cdot \text{AMOUNT}$$
$$= 2.00 \cdot 30y$$
$$= 60y$$
$$P = \text{PROFIT FROM CORN} + \text{PROFIT FROM WHEAT}$$
$$= 143x + 60y$$

The linear programming model is stated as follows:

Maximize: $P = 143x + 60y$

Subject to: $\begin{cases} x \geq 0 \\ y \geq 0 \\ x + y \leq 100 \\ 120x + 210y \leq 15,000 \\ 110x + 30y \leq 4000 \end{cases}$

This linear programming model will be solved in the next section.

EXAMPLE 2 **Allocation of Resources in Manufacturing** The Wadsworth Widget Company manufactures two types of widgets: regular and deluxe. Each widget is produced at a station consisting of a machine and a person who finishes the widgets by hand. The regular widget requires 3 hours of machine time and 2 hours of finishing time. The deluxe widget requires 2 hours of machine time and 4 hours of finishing time. The profit on the regular widget is $25, and on the deluxe widget the profit is $30. If the workday is 8 hours, how many of each type of widget should be produced at each station per day in order to maximize the profit?

Solution We want to maximize the profit, P. There are two types of items, regular and deluxe. Let

$$x = \text{Number of regular widgets produced}$$
$$y = \text{Number of deluxe widgets produced}$$

Then

Profit from regular widgets $= 25x$

Profit from deluxe widgets $= 30y$

$P = $ PROFIT FROM REGULAR WIDGETS $+$ PROFIT FROM DELUXE WIDGETS

$\quad = 25x + 30y$

Two constraints are

$x \geq 0$ The number of widgets must be nonnegative

$y \geq 0$

Production time is summarized in the table:

	Machine time (hr)	Manual time (hr)
Regular widget, x Deluxe widget, y	3 each; $3x$ total 2 each; $2y$ total	2 each; $2x$ total 4 each; $4y$ total
Total workday	$3x + 2y$	$2x + 4y$

Thus

$3x + 2y \leq 8$ Machine workday

$2x + 4y \leq 8$ Manual workday

The linear programming model is:

Maximize: $P = 25x + 30y$

Subject to: $\begin{cases} x \geq 0 \\ y \geq 0 \\ 3x + 2y \leq 8 \\ 2x + 4y \leq 8 \end{cases}$

  ∎

EXAMPLE 3 **Diet Problem** A convalescent hospital wishes to provide, at a minimum cost, a diet that has a minimum of 200 grams of carbohydrates, 100 grams of protein, and

120 grams of fats per day. These requirements can be met with two foods:

Food	Carbohydrates	Protein	Fats
A	10 g	2 g	3 g
B	5 g	5 g	4 g

If food A costs 29¢ per ounce and food B costs 15¢ per ounce, how many ounces of each food should be purchased for each patient per day in order to meet the minimum requirements at the lowest cost?

Solution   Let

x = Number of ounces of food A

y = Number of ounces of food B

The *minimum cost*, C, is found by:

Cost of food A = $0.29x$

Cost of food B = $0.15y$

$C = 0.29x + 0.15y$

The constraints are

$x \geq 0$ The amounts of food must be nonnegative

$y \geq 0$

The table summarizes the nutrients provided:

Food	Amount (in ounces)	Total consumption (in grams)		
		Carbohydrates	Protein	Fats
A	x	$10x$	$2x$	$3x$
B	y	$5y$	$5y$	$4y$
Total		$10x + 5y$	$2x + 5y$	$3x + 4y$

Daily requirements are:

$10x + 5y \geq 200$

$2x + 5y \geq 100$

$3x + 4y \geq 120$

The linear programming model is:

Minimize: $C = 0.29x + 0.15y$

Subject to:
$$\begin{cases} x \geq 0 \\ y \geq 0 \\ 10x + 5y \geq 200 \\ 2x + 5y \geq 100 \\ 3x + 4y \geq 120 \end{cases}$$

EXAMPLE 4   **Investment**   Brown Bros., Inc., is an investment company analyzing a pension fund for a certain company. A maximum of $10 million is available to invest in two places. No more than $8 million can be invested in stocks yielding 12%, and at least $2 million can be invested in long-term bonds yielding 8%. The stock-to-bond investment ratio cannot be more than 1 to 3. How should Brown Bros. advise their client so that the pension fund will receive the maximum yearly return on investment?

Solution   To build this model you need to use the *simple interest formula*:

$$I = Prt$$

where

$I = $ Interest	The amount paid for the use of another's money
$P = $ Principal	The amount invested
$r = $ Interest rate	Write this as a decimal. It is assumed to be an annual interest rate, unless otherwise stated
$t = $ Time	In years, unless otherwise stated

Let

$x = $ Amount invested (in millions) in stocks (12% yield)
$y = $ Amount invested (in millions) in bonds (8% yield)

Then

Stocks:	$I = 0.12x$	$P = x, r = 0.12$, and $t = 1$
Bonds:	$I = 0.08y$	$P = y, r = 0.08$, and $t = 1$

The return on investment, R (in millions), is found by

$$R = 0.12x + 0.08y$$

The constraints are:

$x \geq 0$	Investments are nonnegative
$y \geq 0$	
$x + y \leq 10$	Maximum investment is $10 million; note that the unit chosen for this problem is in millions of dollars
$x \leq 8$	No more than $8 million in stocks
$y \geq 2$	No less than $2 million in bonds
$3x \leq y$	Must invest $3 million in bonds for every $1 million invested in stocks for a stock-to-bond ratio of 1 to 3. That is

$$\frac{\text{stock}}{\text{bond}} \leq \frac{1}{3}$$
$$3 \text{ stock} \leq \text{bond}$$
$$3x \leq y$$

Thus the linear programming model for this problem is:

Maximize: $R = 0.12x + 0.08y$

Subject to:
$$\begin{cases} x \geq 0 \\ y \geq 0 \\ x \leq 8 \\ y \geq 2 \\ x + y \leq 10 \\ 3x \leq y \end{cases}$$

EXAMPLE 5   **Transportation Problem**   Sears ships a certain air-conditioning unit from factories in Portland, Oregon, and Flint, Michigan, to distribution centers in Los Angeles, California, and Atlanta, Georgia. The shipping costs are summarized in the table:

Source	Destination	Shipping cost
Portland	Los Angeles	$30
	Atlanta	$40
Flint	Los Angeles	$60
	Atlanta	$50

The supply and demand, in number of units, is shown below:

Supply	Demand
Portland, 200	Los Angeles, 300
Flint, 600	Atlanta, 400

How should shipments be made from Portland and Flint to minimize the shipping cost?

Solution   The information can be summarized by the following "map."

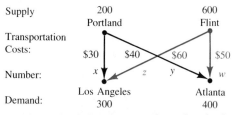

Supply: 200 Portland, 600 Flint
Transportation Costs: $30, $40, $60, $50
Number: x, z, y, w
Los Angeles, Atlanta
Demand: 300, 400

Suppose the following number of units is shipped:

Source	Destination	Number	Shipping cost
Portland	Los Angeles	x	$30x$
	Atlanta	y	$40y$
Flint	Los Angeles	z	$60z$
	Atlanta	w	$50w$

The linear programming problem is then:

Minimize: $C = 30x + 40y + 60z + 50w$

Subject to: $\begin{cases} x + y \leq 200 \\ z + w \leq 600 \\ x + z \geq 300 \\ y + w \geq 400 \\ x \geq 0, \quad y \geq 0, \quad z \geq 0, \quad w \geq 0 \end{cases}$

■

Note that as we progressed from Example 1 to Example 5, we were able to formulate the linear programming problem with fewer and fewer words so that the information in the problem was translated almost directly into a linear programming model. Also, notice in Example 5 that we stated the nonnegative constraints together for convenience. In the next section we develop graphical techniques for solving these types of problems.

Problem Set 4.2

Write a linear programming model for Problems 1–24. Do not attempt to solve the problems, but be sure to define all your variables.

APPLICATIONS

1. A farmer has 500 acres on which to plant two crops: corn and wheat. It costs $120 per acre to produce corn and $60 per acre to produce wheat, and there is $24,000 available to pay for this year's production. If the yield per acre is 100 bushels of corn or 40 bushels of wheat and the farmer has contracted to store at least 20,000 bushels, how much should the farmer plant to maximize the profits when the net profit is $1.20 per bushel for corn and $2.50 per bushel for wheat?

2. A farmer has 1500 acres on which to plant three crops: corn, wheat, and soybeans. It costs $150 per acre to produce corn, $80 per acre to produce wheat, and $60 per acre to produce soybeans, and there is $36,000 available to pay for this year's production. If the yield per acre is 100 bushels of corn, or 40 bushels of wheat, or 60 bushels of soybeans, and the farmer has contracted to store at least 20,000 bushels, how much should the farmer plant to maximize the profits when the net profit is $1.00 per bushel for corn, $2 per bushel for wheat, and $1.50 per bushel for soybeans?

3. The Thompson Company manufactures two industrial products: a standard product ($45 profit per item) and an economy product ($30 profit per item). These products are built using machine time and manual labor. The standard product requires 3 hours machine time and 2 hours manual labor. The economy model requires 3 hours machine time and no manual labor. If the week's supply of manual labor is limited to 800 hours and machine time to 1500 hours, how much of each type of product should be produced each week in order to maximize the profit?

4. A furniture manufacturer makes chairs, sofas, and sofabeds. The production process is divided into carpentry, finishing, and upholstery. Manufacture of a chair requires 3 hours of carpentry, 1 hour of finishing, and 2 hours of upholstery. Manufacture of a sofa requires 5 hours of carpentry, 2 hours of finishing, and 4 hours of upholstery. Manufacture of a sofabed requires 8 hours of carpentry, 3 hours of finishing, and 4 hours of upholstery. The manufacturer has available each day 120 hours of manual labor for carpentry, 24 hours of manual labor for finishing, and 96 hours of manual labor for upholstery. The net profit is $50 per chair, $75 per sofa, and $100 per sofabed. How many of each item should be produced each day in order to maximize profits?

5. The nutritional information in the table is found on the sides of the cereal boxes listed (for 1 ounce of cereal with $\frac{1}{2}$ cup of whole milk). What is the minimum cost in order to receive at least 322 grams of starch (and related carbohydrates) and 119 grams of sucrose (and other sugars) by consuming these two cereals if corn flakes cost 7¢ per ounce and Honeycombs cost 19¢ per ounce?

Cereal	Starch and related carbohydrates	Sucrose and other sugars
Kellogg's Corn Flakes	23 g	7 g
Post Honeycombs	14 g	17 g

6. The nutritional information in the table is found on the sides of the cereal boxes listed (for 1 ounce of cereal with $\frac{1}{2}$ cup of whole milk). What is the most cereal a person could consume and not receive more than 322 grams of starch (and related carbohydrates) or more than 126 grams of sucrose (and other sugars)?

Cereal	Starch and related carbohydrates	Sucrose and other sugars
Kellogg's Corn Flakes	23 g	7 g
Kellogg's Raisin Bran	14 g	18 g

7. Your broker tells you about two investments she thinks worthwhile. She recommends a new issue of Pertec stock, which should yield 20% over the next year, and then to balance your account she recommends Campbell Municipal Bonds with a 10% annual yield. The stock-to-bond ratio should be no less than 1 to 3. If you have no more than $100,000 to invest and do not want to invest more than $70,000 in Pertec or less than $20,000 in bonds, how much should you invest in each to maximize your return?

8. Mama's Home Bakery produces two specialties: German chocolate cake and apple strudel. Daily business requires that at least 50 cakes and 100 strudels be baked every day. The capacity of the ovens limits the total number of items that can be baked each day to no more than 250. How many of each item should be baked in order to maximize profits if the net profit on each cake is $1.50 and on each strudel is $1.20?

9. Karlin Manufacturing has discontinued production of a line of tires that has not been profitable. This has created some excess production time, and management is considering devoting this excess to their Glassbelt or Rainbelt tire lines. The amount of excess production available is 200 hours per month on the composition machine and 400 hours per month on production line #3. The number of hours required for each tire on each of these production lines is given in the table. How many of each tire should Karlin produce to maximize the return if the unit return on each Glassbelt is $52 and on each Rainbelt is $36?

Machine	Glassbelt tire	Rainbelt tire
Composition machine	3 hr	1 hr
Production line #3	2 hr	2 hr

10. An office manager decides to purchase some personal-size computers. Apple computers cost $900 each, will provide 250 K storage capability, and will take up 5 square feet of desk space. IBM/PC computers cost $1,200 each, will provide 612 K storage, and will take up only 3 square feet of desk space. Proper utilization of space requires that a total of not more than 165 square feet be used for this new equipment. If there is $36,000 available to spend on computers, how many of each type of computer should be purchased if the office manager wishes to maximize storage capacity?

11. Foley's Motel has 200 rooms and a restaurant that seats 50 people. Experience shows that 40% of the commercial guests and 20% of the other guests (i.e., noncommercial) eat in the restaurant. There is a net profit of $4.50 per day from each commercial guest and $3.50 per day from each other guest. Find the number of commercial and other guests needed in order to maximize the net profits. Assume one guest per room and that the capacity of the restaurant will not be exceeded.

12. Karlin Enterprises manufactures two games. Standing orders require that at least 24,000 space-battle games and 5000 football games be produced. The company has two factories: The Gainesville plant can produce 600 space-battle games and 100 football games per day; the Sacramento plant can produce 300 space-battle games and 100 football games per day. If the Gainesville plant costs $20,000 per day to operate and the Sacramento plant costs $15,000 per day, find the number of days per month that each factory should operate to minimize the cost. (Assume that each month has 30 days.)

13. Alco Company manufactures two products: Alpha and Beta. Each product must pass through two processing operations, and all materials are introduced at the first operation. Alco may produce either one product exclusively or various combinations of both products subject to the constraints given in the table. A shortage of technical labor has limited Alpha production to no more than 700 units per day. There are no constraints on the production of Beta other than the hour constraints shown in the table. How many of each product should be manufactured in order to maximize profits?

Product	Hours required to produce one unit		Profit per unit
	First process	Second process	
Alpha	1 hr	1 hr	$5
Beta	3 hr	2 hr	$8
Total capacity per day	1200 hr	1000 hr	

14. An island is inhabited by two species of animals, A1 and A2, which feed on three types of food, F1, F2, and F3. The first species requires 12 units of food F1, 6 units of F2, and 16 units of F3 to survive. The second species requires 15, 20, and 5 units of foods F1, F2, and F3, respectively. The island can normally supply 1200 units of F1, 1200 units of F2, and 800 units of F3. What is the maximum number of animals that this island can support?

15. An isolated geographic region is inhabited by three species of animals, A1, A2, and A3, which feed on three types of food, F1, F2, and F3. The first species requires 12 units of food F1, 6 units of F2, and 16 units of F3 to survive. The second species requires 15, 20, and 5 units of food F1, F2, and F3, respectively, and the third species requires 20, 10, and 10 of food F1, F2, and F3, respectively. The region can supply 8000 units of F1, 10,000 units of F2, and 6000 units of F3. What is the maximum number of animals that this region can support?

16. The Cosmopolitan Oil Company requires at least 8000 barrels of low-grade oil per month; it also needs at least 12,000 barrels of medium-grade oil per month and at least 2000 barrels of high-grade oil per month. Oil is produced at one of two refineries. The daily production of these refineries is summarized in the table. If it costs $17,000 per day to operate refinery I and $15,000 per day to operate refinery II, how many days per month should each refinery be operated to satisfy the requirements and at the same time minimize the costs? (Assume that each month has 30 days.)

	Oil production (barrels)		
Refinery	**Low grade**	**Medium grade**	**High grade**
I	200	400	100
II	200	300	100

17. Pell, Inc., manufactures two products: tables and shelves. Each product must be processed in each of three departments: machining, assembling, and finishing. The hours needed to produce one unit of product per department and the maximum possible hours per department are shown in the table. Standing orders require that Pell manufacture at least 30 tables and 25 shelves. Pell's net profit is $4 per table and $2 per shelf. How many tables should be manufactured to maximize the net profit?

	Production time per unit (hr)		Maximum capacity (hr)
Department	**Tables**	**Shelves**	
Machining	2	1	200
Assembling	2	2	240
Finishing	2	3	300

18. A shoe company produces three models of athletic shoes. Shoe parts are first produced in the manufacturing shop and then put together in the assembly shop. The number of hours of labor required per pair in each shop is given in the table. The company can sell as many pairs of shoes as it can produce. During the next month, no more than 1000 hours can be expended in the manufacturing shop and no more than 800 hours can be expended in the assembly shop. The expected profit from the sale of each pair of shoes is:

Model 1: $12
Model 2: $7
Model 3: $5

The company wants to determine how many pairs of each shoe model to produce to maximize total profits over the next month.*

Shop	**Model 1**	**Model 2**	**Model 3**
Manufacturing	8 hr	5 hr	3 hr
Assembly	5 hr	1 hr	3 hr

19. A woman wants to design a weekly exercise schedule that involves jogging, handball, and aerobic dance. She decides to jog at least 3 hours per week, play handball at least 2 hours per week, and dance at least 5 hours per week. She also wants to devote at least as much time to jogging as to handball because she does not like handball as much as she likes the other exercises. She also knows that jogging consumes 900 calories per hour, handball 600 calories per hour, and dance 800 calories per hour. Her physician told her that she must burn up a total of at least 9000 calories per week in this exercise program. How many hours should she devote to each exercise if she wishes to minimize her exercise time?

20. To utilize costly equipment efficiently, an assembly line with a capacity of 50 units per shift must operate 24 hours per day. This requires scheduling three shifts with varying labor costs as follows:

Day shift: Labor cost $100 per unit
Swing shift: Labor cost $150 per unit
Graveyard shift: Labor cost $180 per unit

Each unit produced uses 60 linear feet of sheeting material, and there is a total of 1800 feet available for the day shift,

---

* This problem is taken from a recent examination administered by the Society of Actuaries. Reprinted by permission of the Society of Actuaries.

1500 feet for the swing shift, and 3000 feet for the graveyard shift. The day shift must produce at least 20 units and the other shifts must produce at least 50 units together. How many units should be produced on each shift to maximize the revenue if the units are sold for $850 each?

21. Repeat Problem 20 except minimize labor costs rather than maximize revenue.

22. Suppose a computer dealer has stores in Hillsborough and Palo Alto and warehouses in San Jose and Burlingame. The cost of shipping a computer from one location to another as well as the supply and demand at each location are shown on the following "map":

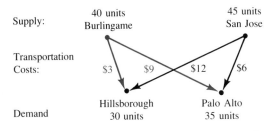

How should the dealer ship the sets to minimize the shipping costs?

23. Camstop, Inc., an importer of computer chips, can ship case lots of components to stores in Dallas and Chicago from ports in either Los Angeles or Seattle. The costs and demand are given in the table. How many cases should be shipped from each port to minimize shipping costs?

	Cost to ship (per case)		
Store	**Los Angeles port**	**Seattle port**	**Store demand**
Chicago	$9	$7	80 cases
Dallas	$7	$8	110 cases
Available inventory at each port	90 cases	130 cases	

24. A Texas appliance dealer has stores in Fort Worth and in Houston and warehouses in Dallas and San Antonio. The cost of shipping a refrigerator from Dallas to Fort Worth is $10, from Dallas to Houston it is $15, from San Antonio to Fort Worth it is $14, and from San Antonio to Houston it is $18. Suppose that the Fort Worth store has orders for 15 refrigerators and the Houston store has orders for 35. Also suppose that there are 25 refrigerators in stock at Dallas and 45 at San Antonio. What is the most economical way to supply the requested refrigerators to the two stores?

4.3 Graphical Solution of Linear Programming Problems

In the previous section we looked at some models called *linear programming problems.* In each case the model had a function called an **objective function**, which was to be maximized or minimized while satisfying several conditions, or **constraints**. If there are only two variables, we use a graphical method of solution. We begin with the set of constraints and consider them as a system of inequalities. The solution of this system of inequalities is a set of points, S. Each point of the set S is called a **feasible solution**. The objective function can be evaluated for different feasible solutions to obtain maximum or minimum values.

EXAMPLE 1   Maximize: $K = 4x + 5y$

Subject to: $\begin{cases} 2x + 5y \le 25 \\ 6x + 5y \le 45 \\ x \ge 0, \quad y \ge 0 \end{cases}$

Solution   The constraints give a set of feasible solutions as graphed in the figure. (You may want to review Section 4.1 on graphing systems of inequalities.)

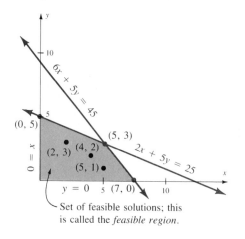

Set of feasible solutions; this is called the *feasible region*.

To solve the linear programming problem, we must now find the feasible solution that makes the objective function as large as possible. Some possible solutions are:

Feasible solution (a point in the solution set of the system)	Objective function $K = 4x + 5y$
$(2, 3)$	$4(2) + 5(3) = 8 + 15 = 23$
$(4, 2)$	$4(4) + 5(2) = 16 + 10 = 26$
$(5, 1)$	$4(5) + 5(1) = 20 + 5 = 25$
$(7, 0)$	$4(7) + 5(0) = 28 + 0 = 28$
$(0, 5)$	$4(0) + 5(5) = 0 + 25 = 25$

In this list, the point that makes the objective function the largest is $(7, 0)$. But is this the largest for all feasible solutions? How about $(6, 1)$? Or $(5, 3)$? It turns out that $(5, 3)$ provides the maximum value:

$$4(5) + 5(3) = 20 + 15 = 35$$

In Example 1, how did we know that $(5, 3)$ provides the maximum value for K? Obviously, it cannot be done by trial and error as shown in the example. Let us take a closer look. We want to find the maximum value of K. Suppose we try the value of 40. We can graph the line (see Figure 4.3)

$$4x + 5y = 40$$

or, in slope-intercept form,

$$y = -\tfrac{4}{5}x + 8$$

Notice that the slope of this line is $-\tfrac{4}{5}$ and the y-intercept is 8; also notice that it lies above the feasible set—that is, it does not intersect the feasible region, so a value of 40 is clearly unrealistic. Try again; try for the value of K to be 30. Then

$$4x + 5y = 30 \qquad \text{or} \qquad y = -\tfrac{4}{5}x + 6$$

(Again, see Figure 4.3.) Now there are many points of intersection between our line and the feasible region. Also, note that both of the lines are parallel since they have the same slope.

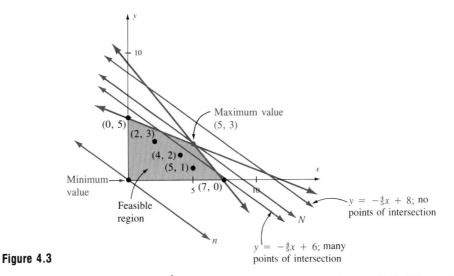

Figure 4.3

Now we begin to formulate a procedure. Think of a "family" of lines, all with slope $-\frac{4}{5}$; some of these lines will intersect the feasible region. If we are maximizing K, we want the point that gives the largest value of K and still has an intersection point (see line N); if we are minimizing K, we want the point that gives the lowest value of K and still has an intersection point (see line n).

EXAMPLE 2 Draw (or imagine) lines parallel to the given objective function in the following figures to find the points in the given feasible region that maximize and minimize that function. If there is no such point, state that.

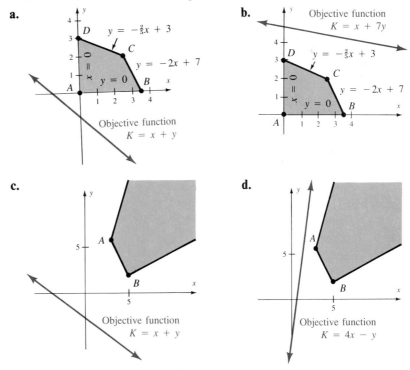

Solution **a.**

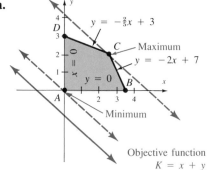

b.

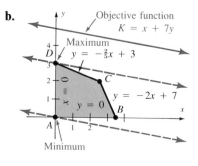

Draw lines parallel to the objective function; the maximum value from the feasible region (shaded) is point C and the minimum value is point A.

The maximum value from the feasible region is point D and the minimum value is point A.

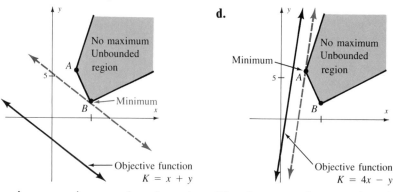

c. There's no maximum value since the region is unbounded. The minimum value is point B.

d. There's no maximum value since the region is unbounded; the minimum value is point A. ■

Note in Example 2 that all of the optimum values occur at **corner points** for the set of feasible solutions. That is, if you consider a family of parallel lines, one member of the family will either coincide with one of the sides of the feasible region or it will intersect it at one point. This follows from the fact that two lines in a plane will either coincide or will intersect at a single point. This fact leads to the **fundamental theorem of linear programming**:

Fundamental Theorem of Linear Programming

A linear expression in two variables,

$$c_1 x + c_2 y$$

defined over a convex set $S*$ whose sides are line segments, takes on its **maximum value** at a corner point of S and its **minimum value** at a corner point of S. If S is unbounded, there may or may not be an optimum value, but, if there is, then it must occur at a corner point.

_____

* A *convex set* is a set that contains the line segment joining any two of its points.

With this theorem, we can now summarize the procedure for solving a linear programming problem by graphing:

1. Find the objective expression (the quantity to be maximized or minimized).
2. Find and graph the constraints defined by a system of linear inequalities; the simultaneous solution is called the set S.
3. Find the corner points of S; this may require the solution of a system of two equations with two unknowns, one for each corner point.
4. Find the value of the objective expression for the coordinates of each corner point to find the optimum solutions.

EXAMPLE 3   Solve Example 1 by using the fundamental theorem of linear programming.

Solution   The graph from Example 1 is repeated below with the corner points labeled.

Some corner points can usually be found by inspection. In this case we can see $D = (0, 0)$ and $C = (0, 5)$.

Some corner points may require some work with the boundary *lines* (use equations of boundaries, not the inequalities giving the regions).

Point A System: $\begin{cases} y = 0 \\ 6x + 5y = 45 \end{cases}$

Solve by substitution: $6x + 5(0) = 45$

$$x = \tfrac{45}{6}$$
$$= \tfrac{15}{2}$$

Point B System: $\begin{cases} 2x + 5y = 25 \\ 6x + 5y = 45 \end{cases}$

Solve by adding: $\begin{cases} -2x - 5y = -25 \\ 6x + 5y = 45 \end{cases}$

$$4x = 20$$
$$x = 5$$

If $x = 5$, then

$$2(5) + 5y = 25$$
$$5y = 15$$
$$y = 3$$

The corner points are thus $(0, 0)$, $(0, 5)$, $(\tfrac{15}{2}, 0)$, and $(5, 3)$. Check these values.

	Objective function
Corner points	$C = 4x + 5y$
$(0, 0)$	$4(0) + 5(0) = 0$
$(0, 5)$	$4(0) + 5(5) = 25$
$(\tfrac{15}{2}, 0)$	$4(\tfrac{15}{2}) + 5(0) = 30$
$(5, 3)$	$4(5) + 5(3) = 35$

└─Look for largest value of C in this list

The maximum value for C is 35 at $(5, 3)$.

EXAMPLE 4　Maximize $P = x + 5y$ subject to the constraints of Example 1.

Solution　The problem has the same corner points as Example 3.

Corner points	Objective function $P = x + 5y$
$(0,0)$	$0 + 5(0) = 0$
$(0,5)$	$0 + 5(5) = 25$
$(\frac{15}{2}, 0)$	$\frac{15}{2} + 5(0) = \frac{15}{2}$
$(5,3)$	$5 + 5(3) = 20$

The maximum value for P is 25 at $(0,5)$.　■

EXAMPLE 5　**Allocation of Resources in Production**　A farmer has 100 acres on which to plant two crops: corn and wheat. To produce these crops, there are certain expenses, as shown in the table.

Item	Cost per acre
Corn	
Seed	$ 12
Fertilizer	58
Planting/care/harvesting	50
Total	$120
Wheat	
Seed	$ 40
Fertilizer	80
Planting/care/harvesting	90
Total	$210

After the harvest, the farmer must usually store the crops while awaiting favorable market conditions. Each acre yields an average of 110 bushels of corn or 30 bushels of wheat. The limitations of resources are

Available capital:　　　　　$15,000
Available storage facilities:　　4000 bushels

If the net profit per bushel of corn (after all expenses have been subtracted) is $1.30 and for wheat is $2.00, how should the farmer plant the 100 acres to maximize profits?

Solution　This linear programming problem was formulated in Section 4.2 (Example 1):

Maximize:　$P = 143x + 60y$　　　Where x is the number of acres
planted in corn and y is the
number of acres planted in wheat

Subject to:
$$\begin{cases} x \geq 0 \\ y \geq 0 \\ x + y \leq 100 \\ 120x + 210y \leq 15{,}000 \\ 110x + 30y \leq 4000 \end{cases}$$

First graph the set of feasible solutions by graphing the system of inequalities, as shown in the figure.

Next find the corner points. By inspection, $A = (0, 0)$.

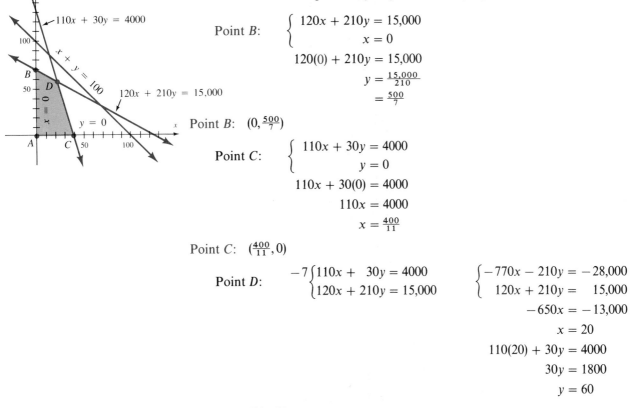

Point B: $\begin{cases} 120x + 210y = 15,000 \\ x = 0 \end{cases}$

$$120(0) + 210y = 15,000$$
$$y = \frac{15,000}{210}$$
$$= \frac{500}{7}$$

Point B: $(0, \frac{500}{7})$

Point C: $\begin{cases} 110x + 30y = 4000 \\ y = 0 \end{cases}$

$$110x + 30(0) = 4000$$
$$110x = 4000$$
$$x = \frac{400}{11}$$

Point C: $(\frac{400}{11}, 0)$

Point D: $\begin{matrix} -7 \end{matrix} \begin{cases} 110x + 30y = 4000 \\ 120x + 210y = 15,000 \end{cases}$

$\begin{cases} -770x - 210y = -28,000 \\ 120x + 210y = 15,000 \end{cases}$

$$-650x = -13,000$$
$$x = 20$$
$$110(20) + 30y = 4000$$
$$30y = 1800$$
$$y = 60$$

Point D: $(20, 60)$

Use the linear programming theorem and check the corner points:

Corner points	Objective function $P = 143x + 60y$
$(0, 0)$	$143(0) + 60(0) = 0$
$(0, \frac{500}{7})$	$143(0) + 60(\frac{500}{7}) \approx 4286$
$(\frac{400}{11}, 0)$	$143(\frac{400}{11}) + 60(0) = 5200$
$(20, 60)$	$143(20) + 60(60) = 6460$

The maximum value of P is 6460 at $(20, 60)$. This means that to maximize profits, the farmer should plant 20 acres in corn, plant 60 acres in wheat, and leave 20 acres unplanted. ∎

Note from the graph in Example 5 that some of the constraints could be eliminated and everything else would remain unchanged. For example, the boundary $x + y = 100$ was not necessary in finding the maximum value of P. Such a condition is said to be a **superfluous constraint**. It is not uncommon to have

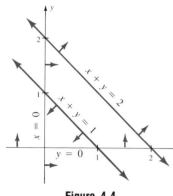

Figure 4.4

superfluous constraints in a linear programming problem. Suppose, however, that the farmer in Example 5 contracted to have the grain stored at a neighboring farm and now the contract calls for *at least* 4000 bushels to be stored. This change from $110x + 30y \leq 4000$ to $110x + 30y \geq 4000$ now makes the condition $x + y \leq 100$ important to the solution of the problem. You must be careful about superfluous constraints even though they do not affect the solution at the present time.

It is also possible to have constraints leading to an empty intersection. For example, consider the constraints

$$\begin{cases} x + y \leq 1 \\ x + y \geq 2 \\ x \geq 0, \quad y \geq 0 \end{cases}$$

The graph in Figure 4.4 shows such an empty intersection. We say that this problem has no feasible solution.

EXAMPLE 6 Solve the following linear programming problem.

$$\text{Minimize:} \quad K = 60x + 30y$$

$$\text{Subject to:} \quad \begin{cases} 2x + 3y \geq 120 \\ 2x + \ y \geq 80 \\ x \geq 0, \quad y \geq 0 \end{cases}$$

Solution Graph:

The corner points $A = (0, 80)$ and $C = (60, 0)$ are found by inspection.

$$\text{Point } B: \quad \begin{cases} 2x + 3y = 120 \\ 2x + \ y = \ 80 \end{cases}$$

$$2y = 40$$

$$y = 20$$

If $y = 20$, then

$$2x + 20 = 80$$

$$2x = 60$$

$$x = 30$$

Point B: $(30, 20)$

Extreme values:

	Objective function
Corner points	$C = 60x + 30y$
$(0, 80)$	$60(0) + 30(80) = 2400$
$(30, 20)$	$60(30) + 30(20) = 2400$
$(60, 0)$	$60(60) + 30(0) = 3600$

From the list above, there are two minimum values for the objective function: $A = (0, 80)$ and $B = (30, 20)$. In this situation, the objective function will have the same minimum value (2400) at all points along the boundary line segment joining A and B.   ∎

Problem Set 4.3

Find the points in the given feasible region that (a) maximize the given objective function and (b) minimize the given objective function. If, in either case, there is none, state that.

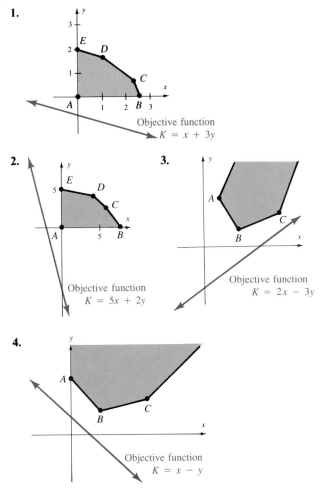

1.

Objective function
$K = x + 3y$

2.

Objective function
$K = 5x + 2y$

3.

Objective function
$K = 2x - 3y$

4.

Objective function
$K = x - y$

Find the corner points for each set of feasible solutions in Problems 5–16.

5. $\begin{cases} 2x + y \le 12 \\ x + 2y \le 9 \\ x \ge 0, \quad y \ge 0 \end{cases}$

6. $\begin{cases} 2x + 5y \le 20 \\ 2x + y \le 12 \\ x \ge 0, \quad y \ge 0 \end{cases}$

7. $\begin{cases} 3x + 2y \le 12 \\ x + 2y \le 8 \\ x \ge 0, \quad y \ge 0 \end{cases}$

8. $\begin{cases} x \le 10 \\ y \le 8 \\ 3x + 2y \ge 12 \\ x \ge 0, \quad y \ge 0 \end{cases}$

9. $\begin{cases} x + y \le 8 \\ y \le 4 \\ x \le 6 \\ x \ge 0, \quad y \ge 0 \end{cases}$

10. $\begin{cases} x + y \ge 6 \\ -2x + y \ge -16 \\ y \le 9 \\ x \ge 0, \quad y \ge 0 \end{cases}$

11. $\begin{cases} 3x + 2y \le 8 \\ x + 5y \ge 8 \\ x \ge 0, \quad y \ge 0 \end{cases}$

12. $\begin{cases} x \le 8 \\ y \ge 2 \\ x + y \le 10 \\ x \le 3y \\ x \ge 0, \quad y \ge 0 \end{cases}$

13. $\begin{cases} 10x + 5y \ge 200 \\ 2x + 5y \ge 100 \\ 3x + 4y \ge 120 \\ x \ge 0, \quad y \ge 0 \end{cases}$

14. $\begin{cases} x + y \le 9 \\ 2x - 3y \ge -6 \\ x - y \le 3 \\ x \ge 0, \quad y \ge 0 \end{cases}$

15. $\begin{cases} 2x + y \ge 8 \\ y \le 5 \\ x - y \le 2 \\ 3x - 2y \ge 5 \\ x \ge 0, \quad y \ge 0 \end{cases}$

16. $\begin{cases} 2x + y \ge 8 \\ x - 2y \le 7 \\ x - y \ge -3 \\ x \le 9 \\ x \ge 0, \quad y \ge 0 \end{cases}$

Find the optimum value for each objective function in Problems 17–36.

17. Maximize $W = 30x + 20y$ subject to the constraints of Problem 5.

18. Maximize $P = 40x + 10y$ subject to the constraints of Problem 5.

19. Maximize $T = 100x + 10y$ subject to the constraints of Problem 6.

20. Maximize $V = 20x + 30y$ subject to the constraints of Problem 6.

21. Maximize $P = 100x + 100y$ subject to the constraints of Problem 7.

22. Maximize $T = 30x + 10y$ subject to the constraints of Problem 7.

23. Minimize $C = 24x + 12y$ subject to the constraints of Problem 8.

24. Minimize $K = 6x + 18y$ subject to the constraints of Problem 8.

25. Maximize $F = 2x - 3y$ subject to the constraints of Problem 9.

26. Maximize $P = 5x + y$ subject to the constraints of Problem 9.

27. Minimize $I = 90x + 20y$ subject to the constraints of Problem 10.

28. Minimize $T = 400x + 100y$ subject to the constraints of Problem 10.

29. Maximize $P = 23x + 46y$ subject to the constraints of Problem 10.

30. Minimize $C = 12x + 15y$ subject to the constraints of Problem 10.

31. Maximize $P = 6x + 3y$ subject to the constraints of Problem 14.

32. Maximize $T = x + 6y$ subject to the constraints of Problem 14.

33. Minimize $X = 5x + 3y$ subject to the constraints of Problem 15.

34. Minimize $A = 2x - 3y$ subject to the constraints of Problem 15.

35. Minimize $K = 140x + 250y$ subject to the constraints of Problem 16.

36. Minimize $C = 640x - 130y$ subject to the constraints of Problem 16.

Solve the linear programming problems in Problems 37–45.

APPLICATIONS

37. Suppose the net profit per bushel of corn in Example 5 increases to $2.00 and the net profit for wheat drops to $1.50 per bushel. Maximize the profit if the other conditions of the example remain the same.

38. Suppose the farmer in Example 5 contracted to have the grain stored at a neighboring farm and the contract calls for at least 4000 bushels to be stored. How many acres should be planted in corn and how many in wheat to maximize profits if the other conditions of the example remain the same?

39. A farmer has 500 acres on which to plant two crops: corn and wheat. It costs $120 per acre to produce corn and $60 per acre to produce wheat, and there is $24,000 available to pay for this year's production. If the yield per acre is 100 bushels of corn or 40 bushels of wheat and the farmer has contracted to store at least 18,000 bushels, how much should the farmer plant to maximize profits when the profit is $1.20 per bushel for corn and $2.50 per bushel for wheat?

40. The Wadsworth Widget Company manufactures two types of widgets: regular and deluxe. Each widget is produced at a station consisting of a machine and a person who finishes the widgets by hand. The regular widget requires 2 hours of machine time and 1 hour of finishing time. The deluxe widget requires 3 hours of machine time and 5 hours of finishing time. The profit on the regular widget is $25, and on the deluxe widget it is $30. If the workday is 8 hours, how many of each type of widget should be produced at each station per day in order to maximize the profit?

41. The Thompson Company manufactures two industrial products: standard ($45 profit per item) and economy ($30 profit per item). These items are built using machine time and manual labor. The standard product requires 3 hours of machine time and 2 hours of manual labor. The economy model requires 3 hours of machine time and no manual labor. If the week's supply of manual labor is limited to 800 hours and machine time to 1500 hours, how much of each type of product should be produced each week in order to maximize the profit?

42. A convalescent hospital wishes to provide, at a minimum cost, a diet that has a minimum of 200 grams of carbohydrates, 100 grams of protein, and 120 grams of fats per day. These requirements can be met with two foods:

Food	Carbohydrates	Protein	Fats
A	10 g	2 g	3 g
B	5 g	5 g	4 g

If food A costs 29¢ per ounce and food B costs 15¢ per ounce, how many ounces of each food should be purchased for each patient per day in order to meet the minimum requirements at the lowest cost?

43. The nutritional information in the table below is found on the sides of the cereal boxes listed (for 1 ounce of cereal with $\frac{1}{2}$ cup of whole milk). What is the minimum cost in order to receive at least 322 grams of starch (and related

carbohydrates) and 119 grams of sucrose (and other sugars) by consuming these two cereals if corn flakes cost 7¢ per ounce and Honeycombs cost 19¢ per ounce?

Cereal	Starch and related carbohydrates	Sucrose and other sugars
Kellogg's Corn Flakes	23 g	7 g
Post Honeycombs	14 g	17 g

44. Brown Bros., Inc., is an investment company analyzing a pension fund for a certain company. A maximum of $10 million is available to invest in two places. No more than $8 million can be invested in stocks yielding 12%, and at least $2 million must be invested in long-term bonds yielding 8%. The stock-to-bond investment ratio cannot be more than 1 to 3. How should Brown Bros. advise their client so that the pension fund will receive the maximum yearly return on investment?

45. Your broker tells you of two investments she thinks are worthwhile. She recommends a new issue of Pertec stock, which should yield 20% over the next year, and then to balance your account she recommends Campbell Municipal Bonds with a 10% annual yield. The stock-to-bond ratio should be no less than 1 to 3. If you have no more than $100,000 to invest and do not want to invest more than $70,000 in Pertec or less than $20,000 in bonds, how much should you invest in each to maximize your return?

4.4 Slack Variables and the Pivot

In Sections 4.1–4.3 we solved linear programming problems using a graphical procedure. However, since graphical methods are inappropriate for more than two variables, we need an algebraic process leading to a solution of the linear programming problem. In 1946 George Dantzig developed an algebraic process called the **simplex method** for solving linear programming problems. Because of the far-reaching applications of this method, some have called it the most important development in applied mathematics of this century. The process itself is easy, but it takes a considerable amount of discussion to set up the proper terminology and notation. In this section we will:

1. Write the linear programming problem in *standard form*.
2. Introduce *slack variables*.
3. Write the linear programming problem in matrix form, called the *initial simplex tableau*.
4. Introduce a *pivoting process*.

In the next section we solve linear programming problems using the simplex method.

Throughout this section the objective function is represented by the variable z, and the linear programming variables are represented by $x_1, x_2, x_3, \ldots$. Consider the following linear programming problems using this notation:

Minimize: $z = 0.08x_1 + 0.18x_2$

Subject to: $\begin{cases} 23x_1 + 14x_2 \le 322 \\ 7x_1 + 18x_2 \le 111 \\ x_1 \ge 0, \quad x_2 \ge 0 \end{cases}$

Maximize: $z = 1.2x_1 + 2.5x_2$

Subject to: $\begin{cases} 120x_1 + 200x_2 \le 24{,}000 \\ 100x_1 + 40x_2 \ge 20{,}000 \\ x_1 \ge 0, \quad x_2 \ge 0 \end{cases}$

Maximize: $z = 35x_1 + 18x_2$

Subject to: $\begin{cases} 0.9x_1 + 0.06x_2 \le 1000 \\ 320x_1 + 118x_2 \le 10{,}000 \end{cases}$

Maximize: $z = 45x_1 + 30x_2$

Subject to: $\begin{cases} 2x_1 \le 800 \\ 2x_1 + 3x_2 \le 1500 \\ x_1 \ge 0, \quad x_2 \ge 0 \end{cases}$

Standard Linear Programming Problem

A linear programming problem is said to be a *standard linear programming problem* or, more simply, in **standard form**, if the following three conditions are met:

Condition 1.   The objective function is to be maximized.
Condition 2.   All variables are nonnegative.
Condition 3.   All constraints (except those in condition 2) are less than or equal to ($\leq$) a nonnegative constant.

EXAMPLE 1   Determine whether each of the following linear programming problems is in standard form.

a. Minimize: $z = 0.08x_1 + 0.18x_2$

Subject to: $\begin{cases} 23x_1 + 14x_2 \leq 322 \\ 7x_1 + 18x_2 \leq 111 \\ x_1 \geq 0, \quad x_2 \geq 0 \end{cases}$

This is not a standard form problem since it is a minimization problem and thus fails to meet condition 1.

b. Maximize: $z = 35x_1 + 18x_2$

Subject to: $\begin{cases} 0.9x_1 + 0.06x_2 \leq 1000 \\ 320x_1 + 118x_2 \leq 10{,}000 \end{cases}$

This is not a standard form problem since x_1 and x_2 are not necessarily nonnegative and thus condition 2 has not been met.

c. Maximize: $z = 1.2x_1 + 2.5x_2$

Subject to: $\begin{cases} 120x_1 + 200x_2 \leq 24{,}000 \\ 100x_1 + 40x_2 \geq 20{,}000 \\ x_1 \geq 0, \quad x_2 \geq 0 \end{cases}$

This is not a standard form problem since the second constraint is not less than or equal to a nonzero constant and thus condition 3 has not been met.

d. Maximize: $z = 45x_1 + 30x_2$

Subject to: $\begin{cases} 2x_1 \leq 800 \\ 2x_1 + 3x_2 \leq 1500 \\ x_1 \geq 0, \quad x_2 \geq 0 \end{cases}$

This is a standard linear programming problem.   ∎

The goal of the simplex method is to solve a standard linear programming problem by transforming it into a system of linear *equations* that can then be solved using the simplex method. To make this transformation, we introduce new variables to the problems. These variables "take up the slack" created by the inequalities and are consequently called **slack variables**. For example, when we say

$x \leq 5$

we are saying that there exists a nonnegative number y, called a *slack variable*, such that

$x + y = 5$

That is, slack variables allow us to transform linear programming problems from systems of inequalities to corresponding systems of equations.

Slack Variable	For every nonnegative constant b and each inequality of the form $$x \leq b$$ there exists a nonnegative number y, called a *slack variable*, such that $$x + y = b$$

In this book we denote slack variables by $y_1, y_2, y_3, \ldots$ to distinguish them from the linear programming variables $x_1, x_2, x_3, \ldots$. Also, a statement such as

$$z = 45x_1 + 30x_2$$

is rewritten as

$$-45x_1 - 30x_2 + z = 0$$

In Examples 2 and 3 we rewrite the given standard linear programming problems using slack variables.

EXAMPLE 2   Maximize: $z = 45x_1 + 30x_2$

Subject to: $\begin{cases} 2x_1 \leq 800 \\ 2x_1 + 3x_2 \leq 1500 \\ x \geq 0, \quad x_2 \geq 0 \end{cases}$

Solution   Using slack variables,

$$\begin{cases} -45x_1 - 30x_2 + z = 0 \\ 2x_1 + y_1 = 800 \\ 2x_1 + 3x_2 + y_2 = 1500 \\ x_1 \geq 0, \quad x_2 \geq 0 \\ y_1 \geq 0, \quad y_2 \geq 0 \end{cases}$$

Note that the nonnegative requirements are not rewritten using slack variables.   ∎

EXAMPLE 3   Maximize: $z = 3x_1 + 9x_2 + 12x_3 - 5x_4$

Subject to: $\begin{cases} x_1 + x_2 \leq 40 \\ x_3 + x_4 \leq 45 \\ x_1 + x_3 \leq 30 \\ x_2 + x_4 \leq 35 \\ x_1 \geq 0, \quad x_2 \geq 0, \quad x_3 \geq 0, \quad x_4 \geq 0 \end{cases}$

Solution   Using slack variables,

$$\begin{cases} -3x_1 - 9x_2 - 12x_3 + 5x_4 + z = 0 \\ x_1 + x_2 + y_1 = 40 \\ x_3 + x_4 + y_2 = 45 \\ x_1 + x_3 + y_3 = 30 \\ x_2 + x_4 + y_4 = 35 \\ x_1 \geq 0, \quad x_2 \geq 0, \quad x_3 \geq 0, \quad x_4 \geq 0 \\ y_1 \geq 0, \quad y_2 \geq 0, \quad y_3 \geq 0, \quad y_4 \geq 0 \end{cases}$$

∎

Note that we write the nonnegative conditions ($x_1 \geq 0, x_2 \geq 0, \ldots$) separately on the bottom lines because the variables are nonnegative for *all* standard linear programming problems. Since we know this, there is no need to include the non-negative variables in the matrix notation. We now write a matrix called the **initial simplex tableau**. When writing this matrix, it is important that you write the variables in the same order so that the proper coefficients are aligned. The standard form maximization statement at the top is generally written at the bottom in the initial simplex tableau, as shown in Examples 4 and 5.

EXAMPLE 4 Maximize: $z = 45x_1 + 30x_2$

Subject to: $\begin{cases} 2x_1 \leq 800 \\ 2x_1 + 3x_2 \leq 1500 \\ x_1 \geq 0, \quad x_2 \geq 0 \end{cases}$

Solution Align the slack variables for tableau form:

$$\begin{cases} 2x_1 \qquad\quad + y_1 \qquad\qquad = 800 \\ 2x_1 + \ 3x_2 \qquad\quad + y_2 \qquad = 1500 \\ -45x_1 - 30x_2 \qquad\qquad\quad + z = 0 \end{cases}$$

The tableau form is:

$$\begin{array}{ccccc} x_1 & x_2 & y_1 & y_2 & z \end{array} \longleftarrow \text{These are for reference only}$$

$$\left[\begin{array}{ccccc|c} 2 & 0 & 1 & 0 & 0 & 800 \\ 2 & 3 & 0 & 1 & 0 & 1500 \\ \hline -45 & -30 & 0 & 0 & 1 & 0 \end{array}\right]$$

and are not part of the tableau form

EXAMPLE 5 Maximize: $z = 3x_1 + 9x_2 + 12x_3 - 5x_4$

Subject to: $\begin{cases} x_1 + x_2 \leq 40 \\ x_3 + x_4 \leq 45 \\ x_1 + x_3 \leq 30 \\ x_2 + x_4 \leq 35 \\ x_1 \geq 0, \quad x_2 \geq 0, \quad x_3 \geq 0, \quad x_4 \geq 0 \end{cases}$

Solution Align the slack variables for tableau form:

$$\begin{cases} x_1 + \ x_2 \qquad\qquad\quad + y_1 \qquad\qquad\qquad\quad = 40 \\ \qquad\quad x_3 + \ x_4 \qquad + y_2 \qquad\qquad\quad = 45 \\ x_1 \qquad + \ x_3 \qquad\qquad\quad + y_3 \qquad\quad = 30 \\ \qquad x_2 \qquad + \ x_4 \qquad\qquad\qquad + y_4 \quad = 35 \\ -3x_1 - 9x_2 - 12x_3 + 5x_4 \qquad\qquad\qquad\quad + z = 0 \end{cases}$$

The tableau form is:

$$\begin{array}{ccccccccc} x_1 & x_2 & x_3 & x_4 & y_1 & y_2 & y_3 & y_4 & z \end{array}$$

$$\left[\begin{array}{ccccccccc|c} 1 & 1 & 0 & 0 & 1 & 0 & 0 & 0 & 0 & 40 \\ 0 & 0 & 1 & 1 & 0 & 1 & 0 & 0 & 0 & 45 \\ 1 & 0 & 1 & 0 & 0 & 0 & 1 & 0 & 0 & 30 \\ 0 & 1 & 0 & 1 & 0 & 0 & 0 & 1 & 0 & 35 \\ \hline -3 & -9 & -12 & 5 & 0 & 0 & 0 & 0 & 1 & 0 \end{array}\right]$$

Note that the initial simplex tableau looks very much like the augmented matrices we used in Chapter 3 to solve systems of equations. The procedure at that time was to follow a method called *Gauss–Jordan elimination*. For a simplex tableau the process involves a **pivoting operation**.

The *selection* of the pivot elements in the simplex method is not the same as in Gauss–Jordan elimination, as we see in the next section. Example 6, however, reviews the pivoting process. (You might also wish to review the process on page 104.)

> **COMPUTER APPLICATION**   You can practice this pivoting process without getting tied up in the arithmetic calculations by using Program 6 on the computer disk accompanying this book.

EXAMPLE 6   Pivot about the circled numbers.

a.
$$\begin{array}{ccccc} x_1 & x_2 & y_1 & y_2 & z \end{array}$$
$$\left[\begin{array}{ccccc|c} ② & 0 & 1 & 0 & 0 & 800 \\ 2 & 3 & 0 & 1 & 0 & 1500 \\ \hline -45 & -30 & 0 & 0 & 1 & 0 \end{array}\right]$$

Step 1:
$$\left[\begin{array}{ccccc|c} ① & 0 & \tfrac{1}{2} & 0 & 0 & 400 \\ 2 & 3 & 0 & 1 & 0 & 1500 \\ \hline -45 & -30 & 0 & 0 & 1 & 0 \end{array}\right]$$
Divide row 1 by 2: $R_1 \div 2$

Step 2:
$$\left[\begin{array}{ccccc|c} ① & 0 & \tfrac{1}{2} & 0 & 0 & 400 \\ 0 & 3 & -1 & 1 & 0 & 700 \\ \hline 0 & -30 & \tfrac{45}{2} & 0 & 1 & 18{,}000 \end{array}\right]$$
←Pivot row
Add $(-2)\cdot$ row 1 to row 2: $-2R_1 + R_2$
Add $45\cdot$ row 1 to row 3: $45R_1 + R_3$

b.
$$\begin{array}{cccccc} x_1 & x_2 & x_3 & y_1 & y_2 & z \end{array}$$
$$\left[\begin{array}{cccccc|c} 1 & ② & 3 & 1 & 0 & 0 & 100 \\ 2 & 4 & -1 & 0 & 1 & 0 & 200 \\ \hline -14 & -25 & -10 & 0 & 0 & 1 & 0 \end{array}\right]$$

Step 1:
$$\left[\begin{array}{cccccc|c} \tfrac{1}{2} & ① & \tfrac{3}{2} & \tfrac{1}{2} & 0 & 0 & 50 \\ 2 & 4 & -1 & 0 & 1 & 0 & 200 \\ \hline -14 & -25 & -10 & 0 & 0 & 1 & 0 \end{array}\right]$$
Divide row 1 by 2: $\tfrac{1}{2}R_1$

Step 2:
$$\left[\begin{array}{cccccc|c} \tfrac{1}{2} & ① & \tfrac{3}{2} & \tfrac{1}{2} & 0 & 0 & 50 \\ 0 & 0 & -7 & -2 & 1 & 0 & 0 \\ \hline -\tfrac{3}{2} & 0 & \tfrac{55}{2} & \tfrac{25}{2} & 0 & 1 & 1250 \end{array}\right]$$
←Pivot row
Add $(-4)\cdot$ row 1 to row 2: $-4R_1 + R_2$
Add $25\cdot$ row 1 to row 2: $25R_1 + R_2$

Problem Set 4.4

Rewrite the linear programming problems (taken from Problem Set 4.3) using the notation introduced in this section. If one of the three conditions for a standard linear programming problem is violated, tell which one.

1. Maximize: $W = 30x + 20y$

Subject to: $\begin{cases} 2x + y \le 12 \\ 5x + 8y \le 40 \\ x \ge 0, \quad y \ge 0 \end{cases}$

2. Maximize: $T = 100x + 10y$

Subject to: $\begin{cases} 2x + 5y \le 20 \\ 2x + y \le 12 \\ x \ge 0, \quad y \ge 0 \end{cases}$

3. Maximize: $P = 100x + 100y$

Subject to: $\begin{cases} 3x + 2y \le 12 \\ x + 2y \le 8 \\ x \ge 0, \quad y \ge 0 \end{cases}$

4. Minimize: $I = 90x + 20y$

Subject to: $\begin{cases} x + y \ge 6 \\ -2x + y \ge -16 \\ y \le 9 \\ x \ge 0, \quad y \ge 0 \end{cases}$

5. Minimize: $X = 5x + 3y$

Subject to: $\begin{cases} 2x + y \ge 8 \\ y \le 5 \\ x - y \le 2 \\ 3x - 2y \ge 5 \\ x \ge 0, \quad y \ge 0 \end{cases}$

6. Maximize: $P = 23x + 46y$

Subject to: $\begin{cases} x + y \le 6 \\ 2x + y \le -16 \\ y \le 9 \\ x \ge 0, \quad y \ge 0 \end{cases}$

7. Maximize: $I = 90x + 20y$

Subject to: $\begin{cases} x + y \le 6 \\ 2x + y \le -16 \\ y \le 9 \end{cases}$

8. Maximize: $P = 6x + 3y$

Subject to: $\begin{cases} x + y \le 9 \\ 2x - 3y \ge -6 \\ x - y \le 3 \\ x \ge 0, \quad y \ge 0 \end{cases}$

9. Maximize: $T = x + 6y$

Subject to: $\begin{cases} 2x + y \ge 8 \\ y \le 5 \\ x - y \le 2 \\ 3x - 2y \ge 5 \\ x \ge 0, \quad y \ge 0 \end{cases}$

10. Maximize: $P = 160x + 130y$

Subject to: $\begin{cases} 2x + 3y \le 8 \\ 4x + 5y \le 8 \end{cases}$

11. Maximize: $K = 140x + 250y$

Subject to: $\begin{cases} 2x + 5y \le 100 \\ 3x + 4y \le 120 \end{cases}$

12. Minimize: $A = 2x - 3y$

Subject to: $\begin{cases} 3x + 2y \le 12 \\ x + 2y \le 8 \\ x \ge 0, \quad y \ge 0 \end{cases}$

Write the initial simplex tableau for the standard programming problems in Problems 13–22.

13. Maximize: $z = 2x_1 + 3x_2$

Subject to: $\begin{cases} 3x_1 + x_2 \le 300 \\ 2x_1 + 2x_2 \le 400 \\ x_1 \ge 0, \quad x_2 \ge 0 \end{cases}$

14. Maximize: $z = 8x_1 + 16x_2$

Subject to: $\begin{cases} 5x_1 + 3x_2 \le 165 \\ 900x_1 + 1200x_2 \le 36,000 \\ x_1 \ge 0, \quad x_2 \ge 0 \end{cases}$

15. Maximize: $z = 45x_1 + 35x_2$

Subject to: $\begin{cases} x_1 + x_2 \le 200 \\ 4x_1 + 2x_2 \le 500 \\ x_1 \ge 0, \quad x_2 \ge 0 \end{cases}$

16. Maximize: $z = 5x_1 + 8x_2$

Subject to: $\begin{cases} x_1 + 3x_2 \le 1200 \\ x_1 + 2x_2 \le 1000 \\ x_1 \le 700 \\ x_1 \ge 0, \quad x_2 \ge 0 \end{cases}$

17. Maximize: $z = x_1 + x_2$

Subject to: $\begin{cases} 12x_1 + 150x_2 \le 1200 \\ 6x_1 + 200x_2 \le 1200 \\ 16x_1 + 50x_2 \le 800 \\ x_1 \ge 0, \quad x_2 \ge 0 \end{cases}$

18. Maximize: $z = 4x_1 + 2x_2$

Subject to: $\begin{cases} 2x_1 + x_2 \le 200 \\ 2x_1 + 2x_2 \le 240 \\ 2x_1 + 3x_2 \le 300 \\ x_1 \ge 0, \quad x_2 \ge 0 \end{cases}$

19. Maximize: $z = 12x_1 + 7x_2 + 5x_3$

Subject to: $\begin{cases} 8x_1 + 5x_2 + 3x_3 \le 1000 \\ 5x_1 + x_2 + 3x_3 \le 800 \\ x_1 \ge 0, \quad x_2 \ge 0, \quad x_3 \ge 0 \end{cases}$

20. Maximize: $z = x_1 + x_2 + x_3$

Subject to: $\begin{cases} 60x_1 \le 1800 \\ 60x_2 \le 1500 \\ 60x_3 \le 3000 \\ x_2 + x_3 \le 50 \\ x_1 \ge 0, \quad x_2 \ge 0, \quad x_3 \ge 0 \end{cases}$

21. Maximize: $z = 9x_1 + 7x_2 + 7x_3 + 8x_4$

Subject to: $\begin{cases} x_1 + x_2 \le 90 \\ x_3 + x_4 \le 130 \\ x_1 + x_3 \le 80 \\ x_2 + x_4 \le 110 \\ x_1 \ge 0, \quad x_2 \ge 0, \quad x_3 \ge 0, \quad x_4 \ge 0 \end{cases}$

22. Maximize: $z = 48x_1 + 61x_2 + 39x_3 + 45x_4$

Subject to: $\begin{cases} x_1 + x_2 \le 5 \\ x_2 + x_3 \le 4 \\ x_3 + x_4 \le 8 \\ x_2 + x_4 \le 7 \\ x_1 \ge 0, \quad x_2 \ge 0, \quad x_3 \ge 0, \quad x_4 \ge 0 \end{cases}$

Perform the pivoting process for the circled pivot in Problems 23–34.

23.

x_1	x_2	y_1	y_2	z	
③	0	1	0	0	90
6	2	0	1	0	18
−12	−6	0	0	1	0

24.

x_1	x_2	y_1	y_2	z	
⑤	0	1	0	0	20
2	3	0	1	0	30
−10	−3	0	0	1	0

25.

x_1	x_2	y_1	y_2	z	
1	0	$\frac{1}{3}$	0	0	30
0	②	−2	1	0	−162
0	−6	4	0	1	360

26.

x_1	x_2	y_1	y_2	z	
1	0	$\frac{1}{5}$	0	0	4
0	③	$-\frac{2}{5}$	1	0	22
0	−3	2	0	1	40

27.

x_1	x_2	y_1	y_2	z	
8	④	1	0	0	40
12	6	0	1	0	600
−8	−10	0	0	1	0

28.

x_1	x_2	y_1	y_2	z	
②	1	$\frac{1}{4}$	0	0	10
0	0	−6	1	0	360
−12	0	$\frac{5}{2}$	0	1	100

29.

x_1	x_2	y_1	y_2	y_3	z	
2	3	1	0	0	0	200
②	6	0	1	0	0	100
3	2	0	0	1	0	300
−10	−5	0	0	0	1	0

30.

x_1	x_2	y_1	y_2	y_3	z	
20	⑩	1	0	0	0	200
2	4	0	1	0	0	100
3	2	0	0	1	0	300
−10	−30	0	0	0	1	0

31.

x_1	x_2	y_1	y_2	y_3	z	
0	−5	1	0	0	0	300
1	②	0	1	0	0	20
0	8	0	−2	1	0	180
0	−4	0	6	0	1	300

32.

x_1	x_2	y_1	y_2	y_3	z	
2	1	$\frac{1}{2}$	0	0	0	30
−3	0	0	1	0	0	50
⊖2	0	3	0	1	0	20
−4	0	2	0	0	1	140

33.

x_1	x_2	x_3	y_1	y_2	y_3	z	
1	2	4	1	0	0	0	80
1	4	3	0	1	0	0	60
8	4	②	0	0	1	0	10
−2	−3	−4	0	0	0	1	0

34.

x_1	x_2	x_3	y_1	y_2	y_3	z	
1	2	4	1	0	0	0	80
①	4	3	0	1	0	0	20
8	4	2	0	0	1	0	200
−5	−3	−4	0	0	0	1	0

4.5 Maximization by the Simplex Method

In Section 4.4 we set the stage for the **simplex method**. This method requires a standard form linear programming problem written in an initial simplex tableau with slack variables. In this section we discuss three ideas that complete the simplex method:

1. How to read a solution from the simplex tableau
2. How to select a pivot
3. How to recognize when a maximum value is found (that is, how to know when you are finished with the simplex method)

Begin by writing Example 5 of Section 4.3 in standard form:

$$\text{Maximize:} \quad z = 143x_1 + 60x_2$$

$$\text{Subject to:} \quad \begin{cases} x_1 + x_2 \leq 100 \\ 120x_1 + 210x_2 \leq 15{,}000 \\ 110x_1 + 30x_2 \leq 4000 \\ x_1 \geq 0, \quad x_2 \geq 0 \end{cases}$$

Write this using slack variables:

$$\begin{cases} x_1 + x_2 + y_1 = 100 \\ 120x_1 + 210x_2 + y_2 = 15{,}000 \\ 110x_1 + 30x_2 + y_3 = 4000 \\ -143x_1 - 60x_2 + z = 0 \end{cases}$$

Now write the initial simplex tableau:

This dashed line separates the objective function from the constraints ⟶

$$\begin{array}{cccccc|c} x_1 & x_2 & y_1 & y_2 & y_3 & z & \\ 1 & 1 & 1 & 0 & 0 & 0 & 100 \\ 120 & 210 & 0 & 1 & 0 & 0 & 15{,}000 \\ 110 & 30 & 0 & 0 & 1 & 0 & 4000 \\ \hline -143 & -60 & 0 & 0 & 0 & 1 & 0 \end{array}$$

This vertical line separates the left and right sides of the equations

Solutions can be read from this matrix tableau. It represents the set of feasible solutions. The problem asks us to maximize an objective function subject to two variables and three constraints (not counting those that restrict the variables to nonnegative values). This means that we have a system of four equations and six variables (three given variables, x_1, x_2, and z, and three slack variables, y_1, y_2, and y_3). From our work with systems of equations we know that any two of the six variables can be chosen arbitrarily and the other four can then be found by solving the remaining system. Also, from the matrix solution of systems of equations, the matrix

$$\begin{array}{cccc|c} y_1 & y_2 & y_3 & z & \\ 1 & 0 & 0 & 0 & 100 \\ 0 & 1 & 0 & 0 & 15{,}000 \\ 0 & 0 & 1 & 0 & 4000 \\ 0 & 0 & 0 & 1 & 0 \end{array}$$

gives the solution

$$\begin{cases} y_1 = 100 \\ y_2 = 15{,}000 \\ y_3 = 4000 \\ z = 0 \end{cases}$$

Thus, if we choose to assign values to x_1 and x_2, we can solve the remaining system. A **basic solution** of a system such as this is one in which we arbitrarily assign values to two variables. A **basic feasible solution** is one in which all the variables are nonnegative. For example, if we let $x_1 = 0$ and $x_2 = 0$, then a basic feasible solution is the one specified above. Since x_1 and x_2 are both zero, this is sometimes called the **initial basic solution**. Notice that $z = 0$ is hardly a maximum value for z, but nevertheless it is a solution to the system. We sometimes write y_1, y_2, y_3, and z to the right of the matrix to help remind us which value in the last column corresponds to which variable. Thus, for this example, we would write

x_1	x_2	y_1	y_2	y_3	z		
1	1	1	0	0	0	100	y_1
120	210	0	1	0	0	15,000	y_2
110	30	0	0	1	0	4000	y_3
-143	-60	0	0	0	1	0	z

EXAMPLE 1   Consider the following matrix tableau. Find the initial basic solution.

x_1	x_2	x_3	y_1	y_2	y_3	z	
2	3	4	1	0	0	0	40
4	5	3	0	1	0	0	30
6	9	9	0	0	1	0	15
-3	-10	-5	0	0	0	1	0

Solution   If $x_1 = 0, x_2 = 0$, and $x_3 = 0$, then $y_1 = 40, y_2 = 30, y_3 = 15$, and $z = 0$ (you can see this by inspection). From this information *you begin* the simplex method by labeling the right-hand column as shown:

x_1	x_2	x_3	y_1	y_2	y_3	z		
2	3	4	1	0	0	0	40	y_1
4	5	3	0	1	0	0	30	y_2
6	9	9	0	0	1	0	15	y_3
-3	-10	-5	0	0	0	1	0	z

You do not need to choose values for x_1, x_2, and x_3 as in Example 1, but you can arbitrarily choose values for any convenient variables. For example, to find a basic solution for

x_1	x_2	x_3	y_1	y_2	y_3	z	
0	8	0	1	4	6	0	30
1	4	0	0	3	9	0	30
0	6	1	0	2	3	0	90
0	9	0	0	4	12	1	4000

you can arbitrarily choose some values to be zero. But which ones and how many? Find the columns with a 1 and with all other entries 0, as shown by the shading below:

$$
\begin{array}{ccccccc}
x_1 & x_2 & x_3 & y_1 & y_2 & y_3 & z \\
\end{array}
$$

$$
\left[
\begin{array}{ccccccc|c}
0 & 8 & 0 & 1 & 4 & 6 & 0 & 30 \\
1 & 4 & 0 & 0 & 3 & 9 & 0 & 30 \\
0 & 6 & 1 & 0 & 2 & 3 & 0 & 90 \\
\hline
0 & 9 & 0 & 0 & 4 & 12 & 1 & 4000 \\
\end{array}
\right]
$$

If you now look at the remaining columns and choose $x_2 = 0$, $y_2 = 0$, and $y_3 = 0$, then the resulting matrix is

$$
\begin{array}{cccc}
x_1 & x_3 & y_1 & z \\
\end{array}
$$

$$
\left[
\begin{array}{cccc|c}
0 & 0 & 1 & 0 & 30 \\
1 & 0 & 0 & 0 & 30 \\
0 & 1 & 0 & 0 & 90 \\
\hline
0 & 0 & 0 & 1 & 4000 \\
\end{array}
\right]
$$

From this matrix you can see that $y_1 = 30$, $x_1 = 30$, $x_3 = 90$, and $z = 4000$. Now you can label the column of the original matrix at the right to reflect this information:

$$
\begin{array}{ccccccc}
x_1 & x_2 & x_3 & y_1 & y_2 & y_3 & z \\
\end{array}
$$

$$
\left[
\begin{array}{ccccccc|c}
0 & 8 & 0 & 1 & 4 & 6 & 0 & 30 \\
1 & 4 & 0 & 0 & 3 & 9 & 0 & 30 \\
0 & 6 & 1 & 0 & 2 & 3 & 0 & 90 \\
\hline
0 & 9 & 0 & 0 & 4 & 12 & 1 & 4000 \\
\end{array}
\right]
\begin{array}{l}
y_1 \\
x_1 \\
x_3 \\
z \\
\end{array}
$$

EXAMPLE 2 Find a basic feasible solution for the given simplex tableau by labeling the column at the right.

$$
\begin{array}{ccccccc}
x_1 & x_2 & x_3 & y_1 & y_2 & y_3 & z \\
\end{array}
$$

$$
\left[
\begin{array}{ccccccc|c}
8 & 0 & 0 & 9 & 1 & 3 & 0 & 90 \\
5 & 0 & 1 & 19 & 0 & 6 & 0 & 80 \\
4 & 1 & 0 & 12 & 0 & 9 & 0 & 110 \\
\hline
-12 & 0 & 0 & 5 & 0 & 6 & 1 & 850 \\
\end{array}
\right]
$$

Solution Look for the columns with a 1 and 0s:

$$
\begin{array}{ccccccc}
x_1 & x_2 & x_3 & y_1 & y_2 & y_3 & z \\
\end{array}
$$

$$
\left[
\begin{array}{ccccccc|c}
8 & 0 & 0 & 9 & 1 & 3 & 0 & 90 \\
5 & 0 & 1 & 19 & 0 & 6 & 0 & 80 \\
4 & 1 & 0 & 12 & 0 & 9 & 0 & 110 \\
\hline
-12 & 0 & 0 & 5 & 0 & 6 & 1 & 850 \\
\end{array}
\right]
\begin{array}{l}
y_2 \\
x_3 \\
x_2 \\
z \\
\end{array}
$$

A basic feasible solution is $y_2 = 90$, $x_3 = 80$, $x_2 = 110$, and $z = 850$. [This assumes that you let $x_1 = 0$, $y_1 = 0$, and $y_3 = 0$ (the nonshaded columns).] ■

Is the basic feasible solution in Example 2 an **optimal solution**? Not as long as the last row has any negative values. This can be seen by writing the last row of the tableau in equation form. If the z value has any negative coefficients it is not yet maximized.

Optimal Solution

> The z value in a simplex tableau is maximized when all the entries in the bottom row are nonnegative.

The *process* by which we obtain an optimal solution is called the **pivoting process** and is summarized below.

Pivoting Process

> **Rule 1.** The **pivot column** is the column that has the most negative number as the bottom entry.
>
> **Rule 2.** The **pivot row** is the row above the line that has the smallest positive ratio when the entry in the last column is divided by the corresponding positive entry from the same row of the pivot column.
>
> The **pivot element** is the entry in the intersection of the pivot row and pivot column. Using this pivot element, pivoting is completed in two steps:
>
> 1. Divide the pivot row by the pivot element so that the pivot element becomes a 1.
> 2. Obtain 0s above and below the pivot element.

EXAMPLE 3 Carry out the pivoting process on the given initial simplex tableau.

$$
\begin{array}{ccccc}
x_1 & x_2 & y_1 & y_2 & z \\
\end{array}
$$

$$
\left[
\begin{array}{ccccc|c}
2 & 4 & 1 & 0 & 0 & 20 \\
3 & 2 & 0 & 1 & 0 & 18 \\
\hline
-6 & -5 & 0 & 0 & 1 & 0 \\
\end{array}
\right]
\begin{array}{l}
y_1 \\
y_2 \\
z \\
\end{array}
$$

$\uparrow$

Pivot column; this is the most negative entry

Solution The variable at the top of the pivot column (x_1 in this example) is sometimes called the **entering variable** because when the pivoting process is finished, the x_1 will appear at the right and will no longer be a zero variable.

We now find the pivot row:

$$
\begin{array}{ccccc}
x_1 & x_2 & y_1 & y_2 & z \\
\end{array}
$$

$$
\left[
\begin{array}{ccccc|c}
2 & 4 & 1 & 0 & 0 & 20 \\
3 & 2 & 0 & 1 & 0 & 18 \\
\hline
-6 & -5 & 0 & 0 & 1 & 0 \\
\end{array}
\right]
\begin{array}{l}
y_1 \\
y_2 \\
z \\
\end{array}
\qquad
\begin{array}{l}
20 \div 2 = 10 \\
18 \div 3 = 6 \leftarrow
\end{array}
\left\{
\begin{array}{l}
\text{Pivot row; this} \\
\text{is the smallest} \\
\text{positive entry}
\end{array}
\right.
$$

$\uparrow$

The variable labeling the pivot row (y_2 in this example) is sometimes called the **departing variable** because when the pivoting process is finished, the y_2 will no longer appear at the right.

Circle the pivot element and carry out the pivot process:

$$
\begin{array}{c}
\begin{array}{ccccc} x_1 & x_2 & y_1 & y_2 & z \end{array} \\
\left[\begin{array}{ccccc|c}
2 & 4 & 1 & 0 & 0 & 20 \\
\textcircled{3} & 2 & 0 & 1 & 0 & 18 \\
\hdashline
-6 & -5 & 0 & 0 & 1 & 0
\end{array}\right]
\begin{array}{l} y_1 \\ y_2 \leftarrow \\ z \end{array}
\end{array}
$$

$\uparrow$

Step 1 Divide to make the pivoting element 1:

$$
\begin{array}{c}
\begin{array}{ccccc} x_1 & x_2 & y_1 & y_2 & z \end{array} \\
\left[\begin{array}{ccccc|c}
2 & 4 & 1 & 0 & 0 & 20 \\
\textcircled{1} & \frac{2}{3} & 0 & \frac{1}{3} & 0 & 6 \\
\hdashline
-6 & -5 & 0 & 0 & 1 & 0
\end{array}\right]
\end{array}
\qquad \leftarrow \text{Each entry is divided by 3}
$$

$\uparrow$

Step 2 Perform elementary row operations to obtain 0s:

$$
\begin{array}{c}
\begin{array}{ccccc} x_1 & x_2 & y_1 & y_2 & z \end{array} \\
\left[\begin{array}{ccccc|c}
0 & \frac{8}{3} & 1 & -\frac{2}{3} & 0 & 8 \\
\textcircled{1} & \frac{2}{3} & 0 & \frac{1}{3} & 0 & 6 \\
\hdashline
0 & -1 & 0 & 2 & 1 & 36
\end{array}\right]
\begin{array}{l} y_1 \\ x_1 \\ z \end{array}
\end{array}
$$

Relabel these according to the 1s and 0s in the columns after the pivoting process

Note: x_1 entered and y_2 departed

Repeat the pivoting process until there are no negative values in the last row.

$$
\begin{array}{c}
\begin{array}{ccccc} x_1 & x_2 & y_1 & y_2 & z \end{array} \\
\left[\begin{array}{ccccc|c}
0 & \frac{8}{3} & 1 & -\frac{2}{3} & 0 & 8 \\
1 & \frac{2}{3} & 0 & \frac{1}{3} & 0 & 6 \\
\hdashline
0 & -1 & 0 & 2 & 1 & 36
\end{array}\right]
\begin{array}{l} y_1 \\ x_1 \\ z \end{array}
\end{array}
\qquad
\begin{array}{l}
8 \div \frac{8}{3} = 3 \leftarrow \\
6 \div \frac{2}{3} = 9
\end{array}
$$

$\uparrow$

Step 1

$$
\begin{array}{c}
\begin{array}{ccccc} x_1 & x_2 & y_1 & y_2 & z \end{array} \\
\left[\begin{array}{ccccc|c}
0 & \textcircled{1} & \frac{3}{8} & -\frac{1}{4} & 0 & 3 \\
1 & \frac{2}{3} & 0 & \frac{1}{3} & 0 & 6 \\
\hdashline
0 & -1 & 0 & 2 & 1 & 36
\end{array}\right]
\end{array}
\qquad
\begin{array}{l}
\leftarrow \text{Divide the entries of this} \\
 \text{row by } \frac{8}{3}
\end{array}
$$

$\uparrow$

Step 2

$$
\begin{array}{c}
\begin{array}{ccccc} x_1 & x_2 & y_1 & y_2 & z \end{array} \\
\left[\begin{array}{ccccc|c}
0 & 1 & \frac{3}{8} & -\frac{1}{4} & 0 & 3 \\
1 & 0 & -\frac{1}{4} & \frac{1}{2} & 0 & 4 \\
\hdashline
0 & 0 & \frac{3}{8} & \frac{7}{4} & 1 & 39
\end{array}\right]
\begin{array}{l} x_2 \\ x_1 \\ z \end{array}
\end{array}
$$

All values in the last row are nonnegative, so the $\leftarrow$ process is complete

Thus the maximum value of z is 39 and it occurs when

$$x_2 = 3$$
$$x_1 = 4$$
$$z = 39$$

There are three equations with five unknowns, so we can arbitrarily choose two variables. We choose $y_1 = 0$ and $y_2 = 0$; the remaining variables are then found by inspection of the last matrix by looking at the columns where a single 1 and 0s elsewhere are found

We can now summarize the simplex method.

Simplex Method

In a standard linear programming problem:

Step 1. Write the initial simplex tableau using slack variables.

Step 2. **Test for maximality:** If all entries in the last row are nonnegative, then the tableau is the final tableau; interpret the solution.

Step 3. Select the pivot element.

    a. The **pivot column** is the column that has the most negative entry at the bottom.*

    b. The **pivot row** is the row that has the smallest positive ratio. If there are no positive ratios, then there is no solution.

Step 4. Carry out the pivoting process and return to step 2.

These steps are written in flowchart form in Figure 4.5.

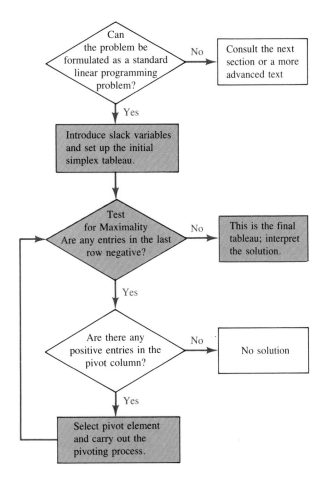

Figure 4.5 The simplex method can easily be written as a computer program. See, for example, Program 6 in the computer supplement accompanying this text. (Appendix D provides a complete listing of the programs available in the computer supplement.)

---

* If two columns have the same entry and this is the most negative entry, then either can be chosen.

EXAMPLE 4 Maximize: $z = 4x_1 + 5x_2$

Subject to: $\begin{cases} 2x_1 + 5x_2 \le 25 \\ 6x_1 + 5x_2 \le 45 \\ x_1 \ge 0, \quad x_2 \ge 0 \end{cases}$

Solution We solved this problem in Examples 1–3 of Section 4.3. The maximum value from the graphical method was 35 at $(5, 3)$. We now solve this problem using the complex method.

$$\begin{array}{ccccc|c}
x_1 & x_2 & y_1 & y_2 & z & \\
2 & ⑤ & 1 & 0 & 0 & 25 \\
6 & 5 & 0 & 1 & 0 & 45 \\
\hline
-4 & -5 & 0 & 0 & 1 & 0
\end{array}
\begin{array}{l}
y_1 \quad 25 \div 5 = 5 \leftarrow \text{Departing variable} \\
y_2 \quad\;\; 45 \div 5 = 9 \\
z
\end{array}$$

$\uparrow$
Entering variable

$$\begin{array}{ccccc|c}
\frac{2}{5} & ① & \frac{1}{5} & 0 & 0 & 5 \\
6 & 5 & 0 & 1 & 0 & 45 \\
\hline
-4 & -5 & 0 & 0 & 1 & 0
\end{array}
\quad \frac{1}{5}R1$$

After the pivoting process you obtain:

$$\begin{array}{ccccc|c}
\frac{2}{5} & 1 & \frac{1}{5} & 0 & 0 & 5 \\
4 & 0 & -1 & 1 & 0 & 20 \\
\hline
-2 & 0 & 1 & 0 & 1 & 25
\end{array}
\begin{array}{l}
\\
-5R1 + R2 \\
5R1 + R3
\end{array}$$

Select a new pivot:

$$\begin{array}{ccccc|c}
x_1 & x_2 & y_1 & y_2 & z & \\
\frac{2}{5} & 1 & \frac{1}{5} & 0 & 0 & 5 \\
④ & 0 & -1 & 1 & 0 & 20 \\
\hline
-2 & 0 & 1 & 0 & 1 & 25
\end{array}
\begin{array}{l}
x_2 \quad 5 \div \frac{2}{5} = 12.5 \\
y_2 \quad 20 \div 4 = 5 \leftarrow \text{Departing variable} \\
z
\end{array}$$

$\uparrow$
Entering variable

Pivot again:

$$\begin{array}{ccccc|c}
\frac{2}{5} & 1 & \frac{1}{5} & 0 & 0 & 5 \\
① & 0 & -\frac{1}{4} & \frac{1}{4} & 0 & 5 \\
\hline
-2 & 0 & 1 & 0 & 1 & 25
\end{array}
\quad \frac{1}{4}R2$$

$$\begin{array}{ccccc|c}
x_1 & x_2 & y_1 & y_2 & z & \\
0 & 1 & \frac{3}{10} & -\frac{1}{10} & 0 & 3 \\
1 & 0 & -\frac{1}{4} & \frac{1}{4} & 0 & 5 \\
0 & 0 & \frac{1}{2} & \frac{1}{2} & 1 & 35
\end{array}
\begin{array}{l}
x_2 \quad -\frac{2}{5}R2 + R1 \\
x_1 \\
z \quad\;\; 2R2 + R3
\end{array}$$

This gives the maximum value of 35 at $(5, 3)$. (We have chosen $y_1 = 0$ and $y_2 = 0$.)

■

EXAMPLE 5   Maximize: $z = 3x_1 + 9x_2 + 12x_3 - 5x_4$

Subject to: $\begin{cases} x_1 + x_2 \le 40 \\ x_3 + x_4 \le 45 \\ x_1 + x_3 \le 30 \\ x_2 + x_4 \le 35 \\ x_1 \ge 0, \quad x_2 \ge 0, \quad x_3 \ge 0, \quad x_4 \ge 0 \end{cases}$

Solution   Write the initial simplex tableau (this was done as Example 5 of the previous section) and determine the entering and departing variables:

x_1	x_2	x_3	x_4	y_1	y_2	y_3	y_4	z				
1	1	0	0	1	0	0	0	0	40	y_1	$40 \div 0$	Not defined
0	0	1	1	0	1	0	0	0	45	y_2	$45 \div 1 = 45$	
1	0	①	0	0	0	1	0	0	30	y_3	$30 \div 1 = 30 \leftarrow$	Departing variable
0	1	0	1	0	0	0	1	0	35	y_4	$35 \div 0$	Not defined
-3	-9	-12	5	0	0	0	0	1	0	z		

$\uparrow$ Entering variable

Next, circle the pivot element and pivot as shown below:

x_1	x_2	x_3	x_4	y_1	y_2	y_3	y_4	z				
1	1	0	0	1	0	0	0	0	40	y_1	$40 \div 1 = 40$	
-1	0	0	1	0	1	-1	0	0	15	y_2	$15 \div 0$	Not defined
1	0	1	0	0	0	1	0	0	30	x_3	$30 \div 0$	Not defined
0	①	0	1	0	0	0	1	0	35	y_4	$35 \div 1 = 35 \leftarrow$ Departing
9	-9	0	5	0	0	12	0	1	360	z		variable

$\uparrow$ Entering variable

The new entering and departing variables and pivot element are found in the above tableau. Pivot once again:

x_1	x_2	x_3	x_4	y_1	y_2	y_3	y_4	z		
1	0	0	-1	1	0	0	-1	0	5	y_1
-1	0	0	1	0	1	-1	0	0	15	y_2
1	0	1	0	0	0	1	0	0	30	x_3
0	1	0	1	0	0	0	1	0	35	x_2
9	0	0	14	0	0	12	9	1	675	z

This is the final tableau. The solution (by inspection) is

$y_1 = 5$

$y_2 = 15$

$x_3 = 30$

$x_2 = 35$

$z = 675$

This solution assumes that we let $x_1 = 0$, $x_4 = 0$, $y_3 = 0$, and $y_4 = 0$; there are five equations with nine unknowns, so four variables are arbitrarily chosen

The answer, however, is given in terms of the original variables in the problem, so we say that the objective function has a maximum value of 675 when $x_1 = 0$, $x_2 = 35$, $x_3 = 30$, and $x_4 = 0$. ∎

Problem Set 4.5

Find the initial basic solution for each matrix tableau in Problems 1–3.

1. a.
$$\begin{array}{ccccc} x_1 & x_2 & y_1 & y_2 & z \\ \end{array}$$
$$\left[\begin{array}{ccccc|c} 4 & 8 & 1 & 0 & 0 & 30 \\ 6 & 5 & 0 & 1 & 0 & 50 \\ \hline -10 & -20 & 0 & 0 & 1 & 0 \end{array}\right]$$

b.
$$\begin{array}{ccccc} x_1 & x_2 & y_1 & y_2 & z \\ \end{array}$$
$$\left[\begin{array}{ccccc|c} 9 & 12 & 1 & 0 & 0 & 120 \\ 5 & 18 & 0 & 1 & 0 & 180 \\ \hline -40 & -30 & 0 & 0 & 1 & 0 \end{array}\right]$$

2. a.
$$\begin{array}{cccccc} x_1 & x_2 & y_1 & y_2 & y_3 & z \\ \end{array}$$
$$\left[\begin{array}{cccccc|c} 5 & 9 & 1 & 0 & 0 & 0 & 18 \\ 3 & 12 & 0 & 1 & 0 & 0 & 35 \\ 7 & 21 & 0 & 0 & 1 & 0 & 49 \\ \hline -5 & -9 & 0 & 0 & 0 & 1 & 0 \end{array}\right]$$

b.
$$\begin{array}{cccccc} x_1 & x_2 & y_1 & y_2 & y_3 & z \\ \end{array}$$
$$\left[\begin{array}{cccccc|c} 9 & 5 & 1 & 0 & 0 & 0 & 19 \\ 6 & 12 & 0 & 1 & 0 & 0 & 35 \\ 12 & 1 & 0 & 0 & 1 & 0 & 48 \\ \hline -10 & -25 & 0 & 0 & 0 & 1 & 0 \end{array}\right]$$

3. a.
$$\begin{array}{ccccccc} x_1 & x_2 & x_3 & y_1 & y_2 & y_3 & z \\ \end{array}$$
$$\left[\begin{array}{ccccccc|c} 8 & 12 & 40 & 1 & 0 & 0 & 0 & 60 \\ 5 & 9 & 12 & 0 & 1 & 0 & 0 & 30 \\ 6 & 15 & 8 & 0 & 0 & 1 & 0 & 40 \\ \hline -5 & -12 & -3 & 0 & 0 & 0 & 1 & 0 \end{array}\right]$$

b.
$$\begin{array}{cccccccc} x_1 & x_2 & x_3 & y_1 & y_2 & y_3 & y_4 & z \\ \end{array}$$
$$\left[\begin{array}{cccccccc|c} 5 & 9 & 11 & 1 & 0 & 0 & 0 & 0 & 80 \\ 6 & 8 & 4 & 0 & 1 & 0 & 0 & 0 & 50 \\ 9 & 18 & 1 & 0 & 0 & 1 & 0 & 0 & 60 \\ 1 & 8 & 1 & 0 & 0 & 0 & 1 & 0 & 90 \\ \hline -8 & -12 & -5 & 0 & 0 & 0 & 0 & 1 & 0 \end{array}\right]$$

Find a basic feasible solution for each simplex tableau in Problems 4–6 and determine whether it is the final tableau.

4. a.
$$\begin{array}{ccccc} x_1 & x_2 & y_1 & y_2 & z \\ \end{array}$$
$$\left[\begin{array}{ccccc|c} 3 & 1 & 4 & 0 & 0 & 60 \\ 4 & 0 & 8 & 1 & 0 & 30 \\ \hline -3 & 0 & 12 & 0 & 1 & 80 \end{array}\right]$$

b.
$$\begin{array}{ccccc} x_1 & x_2 & y_1 & y_2 & z \\ \end{array}$$
$$\left[\begin{array}{ccccc|c} 0 & 1 & 6 & 2 & 0 & 12 \\ 1 & 0 & 1 & 4 & 0 & 80 \\ \hline 0 & 0 & -3 & 9 & 1 & 120 \end{array}\right]$$

5. a.
$$\begin{array}{cccccc} x_1 & x_2 & y_1 & y_2 & y_3 & z \\ \end{array}$$
$$\left[\begin{array}{cccccc|c} 1 & 8 & 0 & 3 & 0 & 0 & 10 \\ 0 & 4 & 1 & 1 & 0 & 0 & 12 \\ 0 & 1 & 0 & 0 & 1 & 0 & 20 \\ \hline 0 & 3 & 0 & 2 & 0 & 1 & 32 \end{array}\right]$$

b.
$$\begin{array}{ccccccc} x_1 & x_2 & x_3 & y_1 & y_2 & y_3 & z \\ \end{array}$$
$$\left[\begin{array}{ccccccc|c} 0 & 2 & 0 & 1 & 2 & 4 & 0 & 20 \\ 0 & 4 & 1 & 0 & 8 & 8 & 0 & 80 \\ 1 & 3 & 0 & 0 & 9 & 3 & 0 & 120 \\ \hline 0 & -5 & 0 & 0 & 4 & -2 & 1 & 360 \end{array}\right]$$

6. a.
$$\begin{array}{ccccccc} x_1 & x_2 & x_3 & y_1 & y_2 & y_3 & z \\ \end{array}$$
$$\left[\begin{array}{ccccccc|c} 0 & 1 & 0 & 4 & 0 & 0 & 0 & 90 \\ 1 & 0 & 0 & 6 & 1 & 0 & 0 & 70 \\ 0 & 0 & 1 & 3 & 3 & 0 & 0 & 65 \\ 0 & 0 & 0 & 5 & 2 & 1 & 0 & 20 \\ \hline 0 & 0 & 0 & 19 & 10 & 0 & 1 & 250 \end{array}\right]$$

b.
$$\begin{array}{ccccccc} x_1 & x_2 & x_3 & y_1 & y_2 & y_3 & z \\ \end{array}$$
$$\left[\begin{array}{ccccccc|c} 1 & 2 & 1 & 0 & 0 & 1 & 0 & 5 \\ 1 & -1 & 3 & 0 & 1 & 0 & 0 & 10 \\ 6 & 5 & -2 & 1 & 0 & 0 & 0 & 8 \\ \hline -3 & -5 & -6 & 0 & 0 & 0 & 1 & 0 \end{array}\right]$$

Carry out the pivoting process on each initial simplex tableau in Problems 7–12.

7.
$$\begin{array}{ccccc} x_1 & x_2 & y_1 & y_2 & z \\ \end{array}$$
$$\left[\begin{array}{ccccc|c} 6 & 3 & 1 & 0 & 0 & 20 \\ 2 & 4 & 0 & 1 & 0 & 4 \\ \hline -4 & -20 & 0 & 0 & 1 & 0 \end{array}\right]$$

8.
$$\begin{array}{ccccc} x_1 & x_2 & y_1 & y_2 & z \\ \end{array}$$
$$\left[\begin{array}{ccccc|c} 3 & 6 & 1 & 0 & 0 & 60 \\ 5 & 6 & 0 & 1 & 0 & 10 \\ \hline -1 & -30 & 0 & 0 & 1 & 0 \end{array}\right]$$

9.
$$\begin{array}{cccccc} x_1 & x_2 & y_1 & y_2 & y_3 & z \\ \end{array}$$
$$\left[\begin{array}{cccccc|c} 2 & 3 & 1 & 0 & 0 & 0 & 50 \\ 5 & 15 & 0 & 1 & 0 & 0 & 10 \\ 4 & 3 & 0 & 0 & 1 & 0 & 70 \\ \hline -5 & -20 & 0 & 0 & 0 & 1 & 0 \end{array}\right]$$

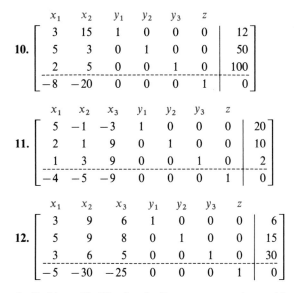

10.
$$\begin{array}{ccccccc} x_1 & x_2 & y_1 & y_2 & y_3 & z & \\ 3 & 15 & 1 & 0 & 0 & 0 & 12 \\ 5 & 3 & 0 & 1 & 0 & 0 & 50 \\ 2 & 5 & 0 & 0 & 1 & 0 & 100 \\ \hline -8 & -20 & 0 & 0 & 0 & 1 & 0 \end{array}$$

11.
$$\begin{array}{cccccccc} x_1 & x_2 & x_3 & y_1 & y_2 & y_3 & z & \\ 5 & -1 & -3 & 1 & 0 & 0 & 0 & 20 \\ 2 & 1 & 9 & 0 & 1 & 0 & 0 & 10 \\ 1 & 3 & 9 & 0 & 0 & 1 & 0 & 2 \\ \hline -4 & -5 & -9 & 0 & 0 & 0 & 1 & 0 \end{array}$$

12.
$$\begin{array}{cccccccc} x_1 & x_2 & x_3 & y_1 & y_2 & y_3 & z & \\ 3 & 9 & 6 & 1 & 0 & 0 & 0 & 6 \\ 5 & 9 & 8 & 0 & 1 & 0 & 0 & 15 \\ 3 & 6 & 5 & 0 & 0 & 1 & 0 & 30 \\ \hline -5 & -30 & -25 & 0 & 0 & 0 & 1 & 0 \end{array}$$

In Problems 13–28 solve the linear programming problems, if possible.

13. Maximize: $z = 2x_1 + 3x_2$

Subject to: $\begin{cases} 3x_1 + x_2 \le 300 \\ 2x_1 + 2x_2 \le 400 \\ x_1 \ge 0, \quad x_2 \ge 0 \end{cases}$

14. Maximize: $z = 5x_1 - 3x_2$

Subject to: $\begin{cases} 2x_1 + x_2 \le 200 \\ 5x_1 + 2x_2 \le 100 \\ x_1 \ge 0, \quad x_2 \ge 0 \end{cases}$

15. Maximize: $z = 8x_1 + 16x_2$

Subject to: $\begin{cases} 5x_1 + 3x_2 \le 165 \\ 900x_1 + 1200x_2 \le 36,000 \\ x_1 \ge 0, \quad x_2 \ge 2 \end{cases}$

16. Maximize: $z = 240x_1 + 100x_2$

Subject to: $\begin{cases} 300x_1 + 900x_2 \le 36,000 \\ 3x_1 + x_2 \le 150 \\ x_1 \ge 0, \quad x_2 \ge 0 \end{cases}$

17. Maximize: $z = 45x_1 + 35x_2$

Subject to: $\begin{cases} x_1 + x_2 \le 200 \\ 4x_1 + 2x_2 \le 500 \\ x_1 \ge 0, \quad x_2 \ge 0 \end{cases}$

18. Maximize: $z = 90x_1 + 55x_2$

Subject to: $\begin{cases} x_1 + x_2 \le 100 \\ 3x_1 + 4x_2 \le 400 \\ x_1 \ge 0, \quad x_2 \ge 0 \end{cases}$

19. Maximize: $z = 5x_1 + 8x_2$

Subject to: $\begin{cases} x_1 + 3x_2 \le 1200 \\ x_1 + 2x_2 \le 1000 \\ x_1 \le 700 \\ x_1 \ge 0, \quad x_2 \ge 0 \end{cases}$

20. Maximize: $z = 8x_1 + 10x_2$

Subject to: $\begin{cases} 2x_1 + x_2 \le 1000 \\ x_1 + 3x_2 \le 1500 \\ x_2 \le 500 \\ x_1 \ge 0, \quad x_2 \ge 0 \end{cases}$

21. Maximize: $z = x_1 + x_2$

Subject to: $\begin{cases} 12x_1 + 150x_2 \le 1200 \\ 6x_1 + 200x_2 \le 1200 \\ 16x_1 + 50x_2 \le 800 \\ x_1 \ge 0, \quad x_2 \ge 0 \end{cases}$

[*Hint:* The first pivot can be selected in the first or second column. Selecting either will give the same answer, but the arithmetic is significantly easier if you select the pivot from the first column.]

22. Maximize: $z = 140x_1 + 80x_2$

Subject to: $\begin{cases} 100x_1 + 5x_2 \le 1500 \\ 200x_1 + 8x_2 \le 1200 \\ 150x_1 + 3x_2 \le 900 \\ x_1 \ge 0, \quad x_2 \ge 0 \end{cases}$

23. Maximize: $z = 4x_1 + 2x_2$

Subject to: $\begin{cases} 2x_1 + x_2 \le 200 \\ 2x_1 + 2x_2 \le 240 \\ 2x_1 + 3x_2 \le 300 \\ x_1 \ge 0, \quad x_2 \ge 0 \end{cases}$

24. Maximize: $z = 3x_1 + 2x_2$

Subject to: $\begin{cases} 2x_1 - 3x_2 \le 80 \\ 2x_1 + x_2 \le 100 \\ 2x_1 + 2x_2 \le 120 \\ x_1 \ge 0, \quad x_2 \ge 0 \end{cases}$

25. Maximize: $z = 12x_1 + 7x_2 + 5x_3$

Subject to: $\begin{cases} 8x_1 + 5x_2 + 3x_3 \le 1000 \\ 5x_1 + x_2 + 3x_3 \le 800 \\ x_1 \ge 0, \quad x_2 \ge 0, \quad x_3 \ge 0 \end{cases}$

26. Maximize: $z = x_1 + x_2 + x_3$

Subject to: $\begin{cases} 60x_1 \le 1800 \\ 60x_2 \le 1500 \\ 60x_3 \le 3000 \\ x_2 + x_3 \le 50 \\ x_1 \ge 0, \quad x_2 \ge 0, \quad x_3 \ge 0 \end{cases}$

27. Maximize: $z = 9x_1 + 7x_2 + 7x_3 + 8x_4$

Subject to:
$$\begin{cases} x_1 + x_2 \le 90 \\ x_3 + x_4 \le 130 \\ x_1 + x_3 \le 80 \\ x_2 + x_4 \le 110 \\ x_1 \ge 0, \quad x_2 \ge 0, \quad x_3 \ge 0, \quad x_4 \ge 0 \end{cases}$$

28. Maximize: $z = 48x_1 + 61x_2 + 39x_3 + 45x_4$

Subject to:
$$\begin{cases} x_1 + x_2 \le 5 \\ x_2 + x_3 \le 4 \\ x_3 + x_4 \le 8 \\ x_2 + x_4 \le 7 \\ x_1 \ge 0, \quad x_2 \ge 0, \quad x_3 \ge 0, \quad x_4 \ge 0 \end{cases}$$

Answer Problems 29–31 by using the simplex method.

APPLICATIONS

29. Alco Company manufactures two products: Alpha and Beta. Each product must pass through two processing operations, and all materials are introduced at the first operation. Alco may produce either one product exclusively or various combinations of both products

| Product | Hours required to produce 1 unit | | |
	First process	Second process	Profit per unit
Alpha	1 hr	1 hr	$5
Beta	3 hr	2 hr	$8
Total capacity per day	1200 hr	1000 hr	

subject to the constraints given in the table. A shortage of technical labor has limited Alpha production to no more than 700 units per day. There are no constraints on the production of Beta other than the hour constraints in the table. How many of each product should be manufactured in order to maximize the profit?

30. If each Alpha costs $2 to manufacture and each Beta costs $3, rework Problem 29 with the additional restriction that only $2,000 per day is available to pay for these manufacturing costs.

31. A shoe company produces three models of athletic shoes. Shoe parts are first produced in the manufacturing shop and then put together in the assembly shop. The number of hours of labor required per pair in each shop is given in the table. The company can sell as many pairs of shoes as it can produce. However, during the next month, no more than 1000 hours can be expended in the manufacturing shop and no more than 800 hours can be expended in the assembly shop. The expected profit from the sale of each pair of shoes is $12 for model 1, $7 for model 2, and $5 for model 3. The company wants to determine how many pairs of each shoe model to produce to maximize total profits over the next month.*

Shop	Model 1	Model 2	Model 3
Manufacturing	8 hr	5 hr	3 hr
Assembly	5 hr	1 hr	3 hr

---

* This problem is taken from a recent examination administered by the Society of Actuaries. Reprinted by permission of the Society of Actuaries.

4.6 Nonstandard Linear Programming Problems*

The standard linear programming problem is subject to three conditions:

Condition 1.   The objective function is to be maximized.
Condition 2.   All variables are nonnegative.
Condition 3.   All constraints (except those in condition 2) are less than or equal to a nonnegative constant.

In this section we learn how to handle certain types of nonstandard problems.

Sometimes a linear programming problem is not standard, but can be transformed into standard form by an algebraic process or by substitution, as shown in Examples 1–4.

---

* This section is not required for subsequent textual development.

EXAMPLE 1

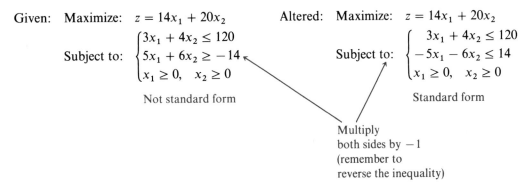

Given: Maximize: $z = 14x_1 + 20x_2$

Subject to: $\begin{cases} 3x_1 + 4x_2 \leq 120 \\ 5x_1 + 6x_2 \geq -14 \\ x_1 \geq 0, \quad x_2 \geq 0 \end{cases}$

Not standard form

Altered: Maximize: $z = 14x_1 + 20x_2$

Subject to: $\begin{cases} 3x_1 + 4x_2 \leq 120 \\ -5x_1 - 6x_2 \leq 14 \\ x_1 \geq 0, \quad x_2 \geq 0 \end{cases}$

Standard form

Multiply both sides by -1 (remember to reverse the inequality)

This is now a standard linear programming problem and you can carry out the simplex method:

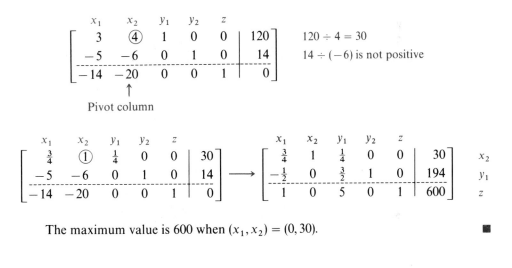

$$\begin{array}{ccccc} x_1 & x_2 & y_1 & y_2 & z \\ \left[\begin{array}{ccccc|c} 3 & ④ & 1 & 0 & 0 & 120 \\ -5 & -6 & 0 & 1 & 0 & 14 \\ \hline -14 & -20 & 0 & 0 & 1 & 0 \end{array}\right] \end{array}$$

$120 \div 4 = 30$

$14 \div (-6)$ is not positive

↑
Pivot column

$$\begin{array}{ccccc} x_1 & x_2 & y_1 & y_2 & z \\ \left[\begin{array}{ccccc|c} \frac{3}{4} & ① & \frac{1}{4} & 0 & 0 & 30 \\ -5 & -6 & 0 & 1 & 0 & 14 \\ \hline -14 & -20 & 0 & 0 & 1 & 0 \end{array}\right] \end{array} \longrightarrow \begin{array}{ccccc} x_1 & x_2 & y_1 & y_2 & z \\ \left[\begin{array}{ccccc|c} \frac{3}{4} & 1 & \frac{1}{4} & 0 & 0 & 30 \\ -\frac{1}{2} & 0 & \frac{3}{2} & 1 & 0 & 194 \\ \hline 1 & 0 & 5 & 0 & 1 & 600 \end{array}\right] \begin{array}{c} x_2 \\ y_1 \\ z \end{array} \end{array}$$

The maximum value is 600 when $(x_1, x_2) = (0, 30)$. ■

EXAMPLE 2

Given: Maximize: $z = 3x_1 + 2x_2$

Altered: Maximize: $z = 3x_1 - 2x_3$

Let $x_3 = -x_2$; then $x_2 = -x_3$ and thus $-x_3 \leq 0$ is the same as $x_3 \geq 0$

Subject to: $\begin{cases} x_1 \geq 0 \\ x_2 \leq 0 \\ 5x_1 + x_2 \leq 100 \\ 3x_1 - 4x_2 \leq 50 \end{cases}$

Not standard form

Subject to: $\begin{cases} x_1 \geq 0 \\ x_3 \geq 0 \\ 5x_1 - x_3 \leq 100 \\ 3x_1 + 4x_3 \leq 50 \end{cases}$

Standard form

Now translate this standard form into tableau form and complete the simplex process:

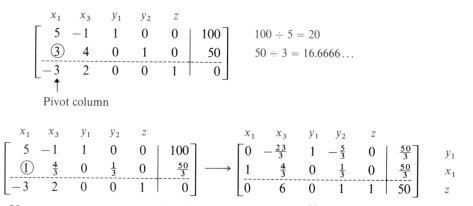

$$
\begin{array}{ccccc}
x_1 & x_3 & y_1 & y_2 & z \\
\end{array}
$$

$$
\left[\begin{array}{ccccc|c}
5 & -1 & 1 & 0 & 0 & 100 \\
\textcircled{3} & 4 & 0 & 1 & 0 & 50 \\
\hline
-3 & 2 & 0 & 0 & 1 & 0
\end{array}\right]
\qquad
\begin{array}{l}
100 \div 5 = 20 \\
50 \div 3 = 16.6666\ldots
\end{array}
$$

↑
Pivot column

$$
\begin{array}{ccccc}
x_1 & x_3 & y_1 & y_2 & z \\
\end{array}
$$

$$
\left[\begin{array}{ccccc|c}
5 & -1 & 1 & 0 & 0 & 100 \\
\textcircled{1} & \frac{4}{3} & 0 & \frac{1}{3} & 0 & \frac{50}{3} \\
\hline
-3 & 2 & 0 & 0 & 1 & 0
\end{array}\right]
\longrightarrow
\begin{array}{ccccc}
x_1 & x_3 & y_1 & y_2 & z \\
\end{array}
$$

$$
\left[\begin{array}{ccccc|c}
0 & -\frac{23}{3} & 1 & -\frac{5}{3} & 0 & \frac{50}{3} \\
1 & \frac{4}{3} & 0 & \frac{1}{3} & 0 & \frac{50}{3} \\
\hline
0 & 6 & 0 & 1 & 1 & 50
\end{array}\right]
\begin{array}{l}
y_1 \\
x_1 \\
z
\end{array}
$$

Now $x_2 = -x_3$, so $x_2 = -0 = 0$ and the solution is $(\frac{50}{3}, 0)$, giving the maximum value of z as 50. ∎

If there are mixed constraints of the type $\leq$ and $\geq$, then you can reverse the inequality by multiplying both sides by -1. This forces all of the constraints to be less than or equal to the constraints, but then the requirement that all variables be nonnegative may be violated. For example, if

$$2x_1 + 3x_2 \geq 4$$

then multiplying both sides by -1 yields

$$-2x_1 - 3x_2 \leq -4$$

This type of constraint violates condition 3:

Linear polynomial $\leq$ nonnegative constant

We will now relax this requirement by allowing

Linear polynomial $\leq$ constant

but in doing this we need to modify the simplex procedure, as illustrated in Example 3.

EXAMPLE 3   Maximize: $z = 8x_1 + 12x_2$

Subject to: $\begin{cases} x_1 + x_2 \leq 10 \\ x_1 - x_2 \geq 5 \\ x_1 \geq 0, \quad x_2 \geq 0 \end{cases}$

Solution   Multiply the second constraint by -1:

$$\begin{cases} x_1 + x_2 \leq 10 \\ -x_1 + x_2 \leq -5 \\ x_1 \geq 0, \quad x_2 \geq 0 \end{cases}$$

$$
\begin{array}{ccccc}
x_1 & x_2 & y_1 & y_2 & z \\
\end{array}
$$

$$
\left[\begin{array}{ccccc|c}
1 & 1 & 1 & 0 & 0 & 10 \\
-1 & 1 & 0 & 1 & 0 & -5 \\
\hline
-8 & -12 & 0 & 0 & 1 & 0
\end{array}\right]
$$

Everything is the same except that there is a negative number in the rightmost column, which violates the simplex method. If you were to try to carry out the simplex method on this matrix you would not arrive at the solution. Why? (For a hint, try using a computer.) All is not lost because we can put the tableau into standard form by pivoting to remove the negative entry in the rightmost column. Use the following procedure:

To Remove a Negative Element in the Rightmost Column

1. Select any negative entry in its row; this is the pivot column.
2. Find the ratios as before, using the pivot column and the entries in the rightmost column (this is the same as before; choose the smallest non-negative ratio).

Pivot:

$$
\begin{array}{ccccc}
x_1 & x_2 & y_1 & y_2 & z \\
\end{array}
$$

$$
\left[\begin{array}{ccccc|c}
1 & 1 & 1 & 0 & 0 & 10 \\
\boxed{-1} & 1 & 0 & 1 & 0 & -5 \\
\hline
-8 & -12 & 0 & 0 & 1 & 0
\end{array}\right]
$$

← Locate any negative elements in the rightmost column (-5 in this example). Next, select *any* negative element in the same row to find the pivot column (the first column in this example shown by box). Finally, consider the ratios: first row: $10 \div 1 = 10$ and second row: $-5 \div (-1) = 5$. The smallest positive ratio gives the pivot element (-1 in this example but the pivot element can be different from the boxed element)

Pivot:

$$
\begin{array}{ccccc}
x_1 & x_2 & y_1 & y_2 & z \\
\end{array}
$$

$$
\left[\begin{array}{ccccc|c}
1 & 1 & 1 & 0 & 0 & 10 \\
① & -1 & 0 & -1 & 0 & 5 \\
\hline
-8 & -12 & 0 & 0 & 1 & 0
\end{array}\right]
$$

You can now begin the simplex process

Pivot:

$$
\begin{array}{ccccc}
x_1 & x_2 & y_1 & y_2 & z \\
\end{array}
$$

$$
\left[\begin{array}{ccccc|c}
0 & 2 & 1 & 1 & 0 & 5 \\
1 & -1 & 0 & -1 & 0 & 5 \\
\hline
0 & -20 & 0 & -8 & 1 & 40
\end{array}\right]
$$

$5 \div 2 = \frac{5}{2}$

$5 \div -1$ not positive

↑
Pivot column

$$
\left[\begin{array}{ccccc|c}
0 & ① & \frac{1}{2} & \frac{1}{2} & 0 & \frac{5}{2} \\
1 & -1 & 0 & -1 & 0 & 5 \\
\hline
0 & -20 & 0 & -8 & 1 & 40
\end{array}\right]
$$

$\frac{5}{2} \div 1 = \frac{5}{2}$

$5 \div (-1)$ not positive

$$
\begin{array}{ccccc}
x_1 & x_2 & y_1 & y_2 & z \\
\end{array}
$$

$$
\left[\begin{array}{ccccc|c}
0 & 1 & \frac{1}{2} & \frac{1}{2} & 0 & \frac{5}{2} \\
1 & 0 & \frac{1}{2} & -\frac{1}{2} & 0 & \frac{15}{2} \\
\hline
0 & 0 & 10 & 2 & 1 & 90
\end{array}\right]
$$

x_2

x_1

z

The maximum value is $z = 90$, and the optimal solution is $(x_1, x_2) = (\frac{15}{2}, \frac{5}{2})$. ∎

For the rest of this section (as well as the next section), we turn our attention to *minimization* problems. The first, and easiest, type is one in which all of the constraints are of the $\leq$ type. Then we simply need to treat z as we treated x_2 in Example 2. This process is illustrated in Example 4.

EXAMPLE 4 Minimize: $z = -4x_1 - 5x_2$

Solution Subject to: $\begin{cases} 2x_1 + 5x_2 \leq 25 \\ 6x_1 + 5x_2 \leq 45 \\ x_1 \geq 0, \quad x_2 \geq 0 \end{cases}$

This is a minimization problem with $\leq$ constraints. It is *not* a standard linear programming problem. Treat z as we treated x_2 in Example 2. Let $z' = -z$. This means that the smallest value for z is the largest value for z'. To see this more clearly, write down any set of positive numbers, say,

$$S = \{3, 9, 18, 20\}$$

Opposites: $\bar{S} = \{-3, -9, -18, -20\}$

Minimize S: 3 is the smallest element in S

Maximize $\bar{S}$: -3 is the largest element in $\bar{S}$

This means that the objective function for this example can be rewritten as:

Maximize: $z' = -z = 4x_1 + 5x_2$

Subject to: $\begin{cases} 2x_1 + 5x_2 \leq 25 \\ 6x_1 + 5x_2 \leq 45 \\ x_1 \geq 0, \quad x_2 \geq 0 \end{cases}$

Solving the problem using the simplex method is the same as with the standard linear programming problem except that now the coefficient of the z is negative. For this example, the simplex tableau is

$$\begin{array}{ccccc}
x_1 & x_2 & y_1 & y_2 & z \\
\end{array}$$
$$\left[\begin{array}{ccccc|c}
2 & 5 & 1 & 0 & 0 & 25 \\
6 & 5 & 0 & 1 & 0 & 45 \\
\hline
-4 & -5 & 0 & 0 & -1 & 0
\end{array}\right] \begin{array}{c} y_1 \\ y_2 \\ z \end{array}$$

However, since this example has only two variables the most expedient method of solution is by graphing. From Example 1 in Section 4.3, the solution is $z' = 35$ when $x_1 = 5$ and $x_2 = 3$. This means that the minimum value of z is -35 when $x_1 = 5$ and $x_2 = 3$. ∎

The second type of minimization problem in this section is one that has *both* $\leq$ and $\geq$ type constraints. For $\geq$ constraints, we multiply both sides by -1 and treat the tableau like the one in Example 3. We illustrate this procedure with an applied problem from Section 4.2.

EXAMPLE 5 **Transportation Problem** Sears ships a certain air-conditioning unit from factories in Portland, Oregon, and Flint, Michigan, to distribution centers in Los Angeles, California, and Atlanta, Georgia. Shipping costs are summarized in the table:

Source	Destination	Shipping cost
Portland	Los Angeles Atlanta	$30 $40
Flint	Los Angeles Atlanta	$60 $50

Supply and demand, in number of units, are:

Supply	Demand
Portland, 200 Flint, 600	Los Angeles, 300 Atlanta, 400

How should shipments be made from Portland and Flint to minimize the shipping cost?

Solution   This model was built on page 143 and leads to the following linear programming problem. Let

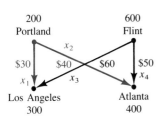

$x_1 =$ Number shipped from Portland to Los Angeles

$x_2 =$ Number shipped from Portland to Atlanta

$x_3 =$ Number shipped from Flint to Los Angeles

$x_4 =$ Number shipped from Flint to Atlanta

Thus the problem is:

Minimize: $z = 30x_1 + 40x_2 + 60x_3 + 50x_4$

Subject to: $\begin{cases} x_1 + x_2 \le 200 \\ x_3 + x_4 \le 600 \\ x_1 + x_3 \ge 300 \\ x_2 + x_4 \ge 400 \end{cases}$

These last two constraints should be rewritten as

$$-x_1 - x_3 \le -300$$
$$-x_2 - x_4 \le -400$$

Begin with the following initial tableau:

$$
\begin{array}{ccccccccc|c}
x_1 & x_2 & x_3 & x_4 & y_1 & y_2 & y_3 & y_4 & z & \\
\hline
① & 1 & 0 & 0 & 1 & 0 & 0 & 0 & 0 & 200 \\
0 & 0 & 1 & 1 & 0 & 1 & 0 & 0 & 0 & 600 \\
\boxed{-1} & 0 & -1 & 0 & 0 & 0 & 1 & 0 & 0 & -300 \\
0 & -1 & 0 & -1 & 0 & 0 & 0 & 1 & 0 & -400 \\
\hdashline
30 & 40 & 60 & 50 & 0 & 0 & 0 & 0 & -1 & 0
\end{array}
$$

First we need to work to eliminate the negative elements in the right-hand column. Select the row with right-hand entry -300 and then select one of the negative elements (see the boxed entry). Next we (after checking the ratios) select as a pivot the number circled above and carry out the pivoting process (the details are left for you).

$$
\begin{array}{cccccccccc}
x_1 & x_2 & x_3 & x_4 & y_1 & y_2 & y_3 & y_4 & z & \\
\left[\begin{array}{ccccccccc|c}
1 & 1 & 0 & 0 & 1 & 0 & 0 & 0 & 0 & 200 \\
0 & 0 & 1 & 1 & 0 & 1 & 0 & 0 & 0 & 600 \\
0 & 1 & \boxed{-1} & 0 & 1 & 0 & 1 & 0 & 0 & -100 \\
0 & -1 & 0 & -1 & 0 & 0 & 0 & 1 & 0 & -400 \\
\hline
0 & 10 & 60 & 50 & -30 & 0 & 0 & 0 & -1 & -6000
\end{array}\right]
\end{array}
$$

(circle on the -1 in row 3, x_3 column)

We still have some negative entries, so we select the row with the entry of -100 and use the only negative element in it as a pivot. After pivoting, we have:

$$
\begin{array}{cccccccccc}
x_1 & x_2 & x_3 & x_4 & y_1 & y_2 & y_3 & y_4 & z & \\
\left[\begin{array}{ccccccccc|c}
1 & 1 & 0 & 0 & 1 & 0 & 0 & 0 & 0 & 200 \\
0 & 1 & 0 & 1 & 1 & 1 & 1 & 0 & 0 & 500 \\
0 & -1 & 1 & 0 & -1 & 0 & -1 & 0 & 0 & 100 \\
0 & \boxed{-1} & 0 & -1 & 0 & 0 & 0 & 1 & 0 & -400 \\
\hline
0 & 70 & 0 & 50 & 30 & 0 & 60 & 0 & -1 & -12{,}000
\end{array}\right]
\end{array}
$$

There is one negative entry remaining. Select the negative element in the row shown by the box and use the circled element as the pivot. After pivoting:

$$
\begin{array}{cccccccccc}
x_1 & x_2 & x_3 & x_4 & y_1 & y_2 & y_3 & y_4 & z & \\
\left[\begin{array}{ccccccccc|c}
1 & 1 & 0 & 0 & 1 & 0 & 0 & 0 & 0 & 200 \\
-1 & 0 & 0 & 1 & 0 & 1 & 1 & 0 & 0 & 300 \\
1 & 0 & 1 & 0 & 0 & 0 & -1 & 0 & 0 & 300 \\
1 & 0 & 0 & \boxed{-1} & 1 & 0 & 0 & 1 & 0 & -200 \\
\hline
-70 & 0 & 0 & 50 & -40 & 0 & 60 & 0 & -1 & -26{,}000
\end{array}\right]
\end{array}
$$

Pivot again, using the boxed element:

$$
\begin{array}{cccccccccc}
x_1 & x_2 & x_3 & x_4 & y_1 & y_2 & y_3 & y_4 & z & \\
\left[\begin{array}{ccccccccc|c}
1 & 1 & 0 & 0 & 1 & 0 & 0 & 0 & 0 & 200 \\
0 & 0 & 0 & 0 & 1 & 1 & 1 & 1 & 0 & 100 \\
1 & 0 & 1 & 0 & 0 & 0 & -1 & 0 & 0 & 300 \\
-1 & 0 & 0 & 1 & -1 & 0 & 0 & -1 & 0 & 200 \\
\hline
-20 & 0 & 0 & 0 & 10 & 0 & 60 & 50 & -1 & -36{,}000
\end{array}\right]
\end{array}
$$

(circle on the 1 in row 1, x_1 column)

Now we can carry out the simplex method using the circled element above as the pivot:

$$
\begin{array}{cccccccccc}
x_1 & x_2 & x_3 & x_4 & y_1 & y_2 & y_3 & y_4 & z & \\
\left[\begin{array}{ccccccccc|c}
1 & 1 & 0 & 0 & 1 & 0 & 0 & 0 & 0 & 200 \\
0 & 0 & 0 & 0 & 1 & 1 & 1 & 1 & 0 & 100 \\
0 & -1 & 1 & 0 & -1 & 0 & -1 & 0 & 0 & 100 \\
0 & 1 & 0 & 1 & 0 & 0 & 0 & -1 & 0 & 400 \\
\hline
0 & 20 & 0 & 0 & 30 & 0 & 60 & 50 & -1 & -32{,}000
\end{array}\right]
\end{array}
$$

The solution is $x_1 = 200$, $x_2 = 0$, $x_3 = 100$, and $x_4 = 400$. This means that the maximum value is $-32,000$, so the minimum value of the original problem is $\$32,000$ and is achieved by shipping 200 air-conditioning units from Portland to Los Angeles and none from Portland to Atlanta. Flint ships 100 units to Los Angeles and 400 to Atlanta. ∎

A final note on the simplex method. There are some technical complications that were not discussed because they are beyond the scope of this course. For example, when trying certain problems (especially if you randomly choose entries on a computer) the simplex method will break down, and you may at some stage find no nonnegative ratios to consider—which means that you cannot choose a pivot element. But, when the simplex method "breaks down," it simply means that the associated linear programming problem does not have a solution.

Problem Set 4.6

Use the simplex method to solve Problems 1–16.

1. Maximize: $z = 10x_1 + 40x_2$

Subject to: $\begin{cases} 3x_1 + 5x_2 \le 50 \\ 2x_1 + 3x_2 \ge -10 \\ x_1 \ge 0, \quad x_2 \ge 0 \end{cases}$

2. Maximize: $z = 500x_1 + 300x_2$

Subject to: $\begin{cases} 9x_1 + 5x_2 \ge -5 \\ 15x_1 + 3x_2 \le 75 \\ x_1 \ge 0, \quad x_2 \ge 0 \end{cases}$

3. Maximize: $z = 30x_1 + 40x_2$

Subject to: $\begin{cases} 5x_1 + 3x_2 \ge -5 \\ 2x_1 + 3x_2 \le 40 \\ x_1 \ge 0, \quad x_2 \ge 0 \end{cases}$

4. Maximize: $z = 5x_1 + 4x_2$

Subject to: $\begin{cases} 3x_1 + x_2 \le 80 \\ 2x_1 + 5x_2 \le 100 \\ x_1 \ge 0, \quad x_2 \le 0 \end{cases}$

5. Maximize: $z = 30x_1 - 20x_2$

Subject to: $\begin{cases} 2x_1 - x_2 \le 12 \\ -2x_2 \le 9 \\ x_1 \ge 0, \quad x_2 \le 0 \end{cases}$

6. Maximize: $z = 100x_1 - 10x_2$

Subject to: $\begin{cases} 2x_1 - 5x_2 \le 20 \\ 2x_1 - x_2 \le 12 \\ x_1 \ge 0, \quad x_2 \le 0 \end{cases}$

7. Maximize: $z = 50x_1 + 40x_2$

Subject to: $\begin{cases} 3x_1 + 2x_2 \le 8 \\ x_1 + 5x_2 \le 8 \\ x_1 \ge 0, \quad x_2 \ge 0 \end{cases}$

8. Maximize: $z = 3x_1 + 2x_2$

Subject to: $\begin{cases} 2x_1 + 3x_2 \le 105 \\ 5x_1 + 10x_2 \ge 15 \\ x_1 \ge 0, \quad x_2 \ge 0 \end{cases}$

9. Maximize: $z = 5x_1 + 3x_2$

Subject to: $\begin{cases} 12x_1 + 4x_2 \le 16 \\ 2x_1 + 5x_2 \ge 10 \\ x_1 \ge 0, \quad x_2 \ge 0 \end{cases}$

10. Minimize: $z = -x_1 - 3x_2$

Subject to: $\begin{cases} 6x_1 + 2x_2 \le 5 \\ 2x_1 + 3x_2 \le 6 \\ x_1 \ge 0, \quad x_2 \ge 0 \end{cases}$

11. Minimize: $z = -20x_1 - 5x_2$

Subject to: $\begin{cases} 3x_1 + 5x_2 \le 10 \\ 3x_1 + 2x_2 \le 6 \\ x_1 \ge 0, \quad x_2 \ge 0 \end{cases}$

12. Minimize: $z = x_1 + 3x_2$

Subject to: $\begin{cases} x_1 + x_2 \le 10 \\ 5x_1 + 2x_2 \ge 20 \\ -x_1 + 2x_2 \ge 0 \\ x_1 \ge 0, \quad x_2 \ge 0 \end{cases}$

13. Minimize: $z = 35x_1 + 10x_2$

Subject to: $\begin{cases} -2x_1 + 3x_2 \ge 0 \\ 8x_1 + x_2 \le 52 \\ -2x_1 + x_2 \le 2 \\ x_1 \ge 3 \\ x_1 \ge 0, \quad x_2 \ge 0 \end{cases}$

14. Minimize: $z = 140x_1 - 60x_2$

Subject to: $\begin{cases} -2x_1 + 3x_2 \geq 0 \\ 8x_1 + x_2 \leq 52 \\ -2x_1 + x_2 \leq 2 \\ x_1 \geq 3 \\ x_1 \geq 0, \quad x_2 \geq 0 \end{cases}$

15. Maximize: $z = 2x_1 + 3x_2 + 5x_3$

Subject to: $\begin{cases} x_1 + 3x_2 + x_3 \leq 46 \\ 2x_1 + x_2 + x_3 \leq 40 \\ x_2 + x_3 = 10 \\ x_1 \geq 0, \quad x_2 \geq 0, \quad x_3 \geq 0 \end{cases}$

[*Hint:* Since $x_2 + x_3 = 10$, write $x_3 = 10 - x_2$ and eliminate the variable x_3 from the problem.]

16. Maximize: $z = x_1 + x_2 + 5x_3$

Subject to: $\begin{cases} x_1 - 3x_3 \leq 30 \\ x_2 - 2x_3 \leq 20 \\ x_1 + 5x_2 \leq 40 \\ x_1 + x_2 - x_3 = 8 \\ x_1 \geq 0, \quad x_2 \geq 0, \quad x_3 \geq 0 \end{cases}$

[*Hint:* Since $x_1 + x_2 - x_3 = 8$, write $x_3 = x_1 + x_2 - 8$ and eliminate x_3 from the problem.]

APPLICATIONS

17. Tony's veterinarian perscribes three food supplements for his horses. Tony must feed them at least 115 kilograms of oats and 50 kilograms of alfalfa, but no more than 200 kilograms of grain. The amounts of each of these in the supplements are summarized in the table.

| | Supplement | | |
	A	B	C
Oats	1 kg	2 kg	3 kg
Alfalfa	2 kg	1 kg	1 kg
Grain	1 kg	0 kg	1 kg
Cost per kg:	$1	$1	$4

How many units of A, B, and C should be mixed to minimize the cost?

18. Helmer's apple farm grows apples. Helmer's cannot grow more than 3000 bins (but can grow less). The crop is divided into two types of apples: eating and applesauce. Helmer's must supply at least 80 bins of eating apples and 800 bins of apples for sauce. The cost of producing these apples is $4 per bin of eating apples and $3.25 per bin of applesauce. How many bins of each should be produced to minimize the cost?

19. Pell, Inc., manufactures two products: tables and shelves. Each product must be processed in each of three departments: machining, assembling, and finishing. The hours needed to produce one unit of product per department and the maximum possible hours per department are shown in the table. Standing orders require that Pell manufacture at least 30 tables and 25 shelves. Pell's net profit is $4 per table and $2 per shelf. How many tables should be manufactured to maximize the profit? Solve this problem graphically.

| Department | Production time per unit (hr) | | Maximum capacity (hr) |
	Tables	Shelves	
Machining	2	1	200
Assembling	2	2	240
Finishing	2	3	300

20. Solve Problem 19 by using the simplex method.

21. The answer to Problem 19 using the graphical method gives some integral coordinates, but the answer to Problem 20 using the simplex method does not. Look at the graphical solution and reconcile the answers to Problems 19 and 20.

4.7 Duality*

The most common type of minimization problem is one in which all of the constraints are of the $\geq$ type. In this section we use a procedure for solving these problems. The process was first introduced by John von Neumann (1903–1957), who has been described as one of the greatest geniuses of this century. It involves

---

* This section is not required for subsequent textual development.

solving a related maximization problem called the **dual** of the minimization problem. We begin with a two-dimensional example so that we can work both the original problem and the dual problem graphically to illustrate the plausibility of the method. The dual can be used with any number of variables.

EXAMPLE 1 Minimize: $z = 50x_1 + 60x_2$

Subject to: $\begin{cases} 7x_1 + 2x_2 \geq 14 \\ 3x_1 + 5x_2 \geq 20 \\ x_1 \geq 0, \quad x_2 \geq 0 \end{cases}$

Solution We begin by using the graphical method:

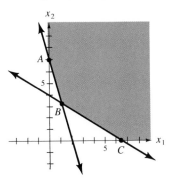

Corner points: A: $(0, 7)$
B: $(\frac{30}{29}, \frac{98}{29})$
C: $(\frac{20}{3}, 0)$

check the corner points and find:

$(0, 7)$: $z = 50(0) + 60(7) = 420$
$(\frac{30}{29}, \frac{98}{29})$: $z = 50(\frac{30}{29}) + 60(\frac{98}{29}) \approx 254.4$
$(\frac{20}{3}, 0)$: $z = 50(\frac{20}{3}) + 60(0) \approx 333.33$

The minimum value is 254.48, which occurs at $(\frac{30}{29}, \frac{98}{29})$.

Next, write the problem using an augmented matrix (the objective function is written in the last row):

$$\begin{bmatrix} 7 & 2 & \vdots & 14 \\ 3 & 5 & \vdots & 20 \\ \hdashline 50 & 60 & \vdots & 0 \end{bmatrix}$$

If we interchange the rows and columns of this matrix (it is called the **transpose** and we will discuss it below), we obtain

$$\begin{bmatrix} 7 & 3 & \vdots & 50 \\ 2 & 5 & \vdots & 60 \\ \hdashline 14 & 20 & \vdots & 0 \end{bmatrix}$$

John von Neumann discovered that if the problem specified by this matrix is maximized using $\leq$ constraints, the same answer as the corresponding minimum problem will always be obtained. Check it out; use y's instead of x's so we can keep the two problems apart.

Maximize: $z = 14y_1 + 20y_2$

Subject to: $\begin{cases} 7y_1 + 3y_2 \leq 50 \\ 2y_1 + 5y_2 \leq 60 \\ y_1 \geq 0, \quad y_2 \geq 0 \end{cases}$

Here again, solve the linear programming problem by graphing.

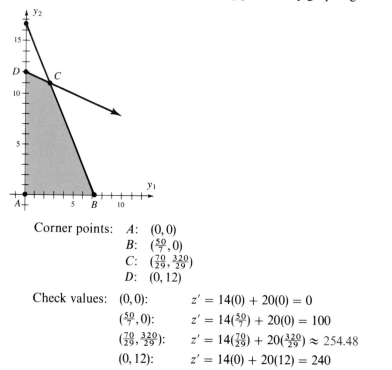

Corner points: A: $(0, 0)$
  B: $(\frac{50}{7}, 0)$
  C: $(\frac{70}{29}, \frac{320}{29})$
  D: $(0, 12)$

Check values: $(0, 0)$: $z' = 14(0) + 20(0) = 0$
  $(\frac{50}{7}, 0)$: $z' = 14(\frac{50}{7}) + 20(0) = 100$
  $(\frac{70}{29}, \frac{320}{29})$: $z' = 14(\frac{70}{29}) + 20(\frac{320}{29}) \approx 254.48$
  $(0, 12)$: $z' = 14(0) + 20(12) = 240$

The maximum value is 254.48 at the point $(\frac{70}{29}, \frac{320}{29})$.

Note that the minimum value of the original problem is the same as the maximum value of the dual: *This is always true.*   ■

The feasible regions of the two problems are different, and the corner points are different, but the *extreme values* of the objective functions are the same. You should note an even closer connection between a problem and its dual as we use the simplex method to solve the dual (the maximum problem).

For now, let us use slack variables u and v (you will soon see these turn out to be x_1 and x_2)

$50 \div 3 = 16.666\ldots$
$60 \div 5 = 12$

$$\begin{array}{ccccc} y_1 & y_2 & u & v & z' \end{array}$$
$$\left[\begin{array}{ccccc|c} 7 & 3 & 1 & 0 & 0 & 50 \\ 2 & ⑤ & 0 & 1 & 0 & 60 \\ \hline -14 & -20 & 0 & 0 & 1 & 0 \end{array}\right]$$

Pivot column

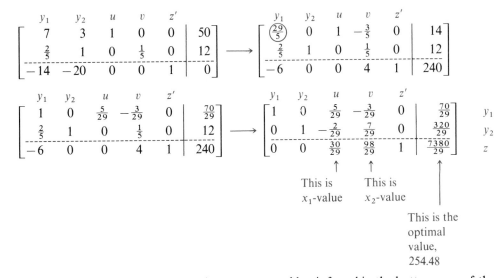

Note that the solution to the *minimum problem* is found in the bottom row of the tableau under the columns u and v (the slack variables); that is, $x_1 = \frac{30}{29}$ and $x_2 = \frac{98}{29}$. This suggests that when solving the dual, you can use the original x-values as slack variables.

The key to changing a minimization problem to its dual maximization problem is the interchanging of the rows and columns in the augmented matrix of the minimization problem. We called this the *transpose*:

Transpose of a Matrix

The *transpose*, M^T, of a matrix M is the matrix formed by interchanging the rows and columns of M.

EXAMPLE 2 Find the transpose of the given matrices.

$$A = \begin{bmatrix} 1 & 2 \\ 3 & 4 \end{bmatrix} \quad B = \begin{bmatrix} 6 & 8 & 9 \\ 4 & 1 & 7 \\ 3 & 2 & 1 \end{bmatrix} \quad C = \begin{bmatrix} 4 & 8 \end{bmatrix}$$

$$D = \begin{bmatrix} 6 \\ 1 \\ 2 \end{bmatrix} \quad E = \begin{bmatrix} 1 & 3 & 6 \\ 4 & 9 & 2 \end{bmatrix}$$

Solution

$$A^T = \begin{bmatrix} 1 & 3 \\ 2 & 4 \end{bmatrix} \quad B^T = \begin{bmatrix} 6 & 4 & 3 \\ 8 & 1 & 2 \\ 9 & 7 & 1 \end{bmatrix} \quad C^T = \begin{bmatrix} 4 \\ 8 \end{bmatrix}$$

$$D^T = \begin{bmatrix} 6 & 1 & 2 \end{bmatrix} \quad E^T = \begin{bmatrix} 1 & 4 \\ 3 & 9 \\ 6 & 2 \end{bmatrix}$$

We can now summarize this process.

Von Neumann's Duality Principle

The optimum value of a minimum linear programming problem, if a solution exists, is the same as the optimum value of its dual. That is, the maximum value of z' is the same as the minimum value of z.

To find the dual:

1. The given problem should be a minimization problem consisting only of $\geq$ constraints.* (You may have to use some of the techniques of the last section if the order of some of the constraints is not correct.)
2. Write the simplex problem in matrix form, with each constraint on a different row and with the objective function as the bottom row.
3. Find the transpose of the problem; this gives the dual problem (the dual of a minimum problem is a maximum problem). The constraints of the dual are formed from the rows of the transpose; the objective function is at the bottom; and the inequality signs are the reverse of those in the original problem.
4. The optimal solution is given by the entries in the bottom row of the columns corresponding to the slack variables, and the minimum value of the objective function of the minimization problem is the same as the maximum value of the objective function of the dual.

EXAMPLE 3   Minimize: $z = 50x_1 + 60x_2$

Subject to: $\begin{cases} 7x_1 + 2x_2 \geq 14 \\ 3x_1 + 5x_2 \geq 20 \\ 6x_1 + 10x_2 \geq 30 \\ x_1 \geq 0, \quad x_2 \geq 0 \end{cases}$

Solution   Solve this by solving the dual problem. First, write in augmented matrix form:

$$\begin{bmatrix} 7 & 2 & \vdots & 14 \\ 3 & 5 & \vdots & 20 \\ 6 & 10 & \vdots & 30 \\ \hline 50 & 60 & \vdots & 0 \end{bmatrix}$$

Next, find the transpose:

$$\begin{bmatrix} 7 & 3 & 6 & \vdots & 50 \\ 2 & 5 & 10 & \vdots & 60 \\ \hline 14 & 20 & 30 & \vdots & 0 \end{bmatrix}$$

The transpose leads to the dual problem:

Maximize: $z' = 14y_1 + 20y_2 + 30y_3$

Subject to: $\begin{cases} 7y_1 + 3y_2 + 6y_3 \leq 50 \\ 2y_1 + 5y_2 + 10y_3 \leq 60 \end{cases}$

---

* The dual can also be used to transform a maximization problem into a minimization problem. In this book, however, we will use the dual only to transform minimization problems into maximization ones.

Solve using the simplex method:

$$\begin{bmatrix} y_1 & y_2 & y_3 & x_1 & x_2 & z' & \\ 7 & 3 & 6 & 1 & 0 & 0 & 50 \\ 2 & 5 & ⑩ & 0 & 1 & 0 & 60 \\ \hline -14 & -20 & -30 & 0 & 0 & 1 & 0 \end{bmatrix}$$

$50 \div 6 = 8.333$

$60 \div 10 = 6$

↑
Pivot column

$$\begin{bmatrix} y_1 & y_2 & y_3 & x_1 & x_2 & z' & \\ 7 & 3 & 6 & 1 & 0 & 0 & 50 \\ \frac{1}{5} & \frac{1}{2} & 1 & 0 & \frac{1}{10} & 0 & 6 \\ \hline -14 & -20 & -30 & 0 & 0 & 1 & 0 \end{bmatrix}$$

$$\begin{bmatrix} y_1 & y_2 & y_3 & x_1 & x_2 & z' & \\ ㉙⁄₅ & 0 & 0 & 1 & -\frac{3}{5} & 0 & 14 \\ \frac{1}{5} & \frac{1}{2} & 1 & 0 & \frac{1}{10} & 0 & 6 \\ \hline -8 & -5 & 0 & 0 & 3 & 1 & 180 \end{bmatrix}$$

$14 \div \frac{29}{5} = 2.413\ldots$

$6 \div \frac{1}{5} = 30$

↑
Pivot column

$$\begin{bmatrix} y_1 & y_2 & y_3 & x_1 & x_2 & z' & \\ 1 & 0 & 0 & \frac{5}{29} & -\frac{3}{29} & 0 & \frac{70}{29} \\ \frac{1}{5} & \frac{1}{2} & 1 & 0 & \frac{1}{10} & 0 & 6 \\ \hline -8 & -5 & 0 & 0 & 3 & 1 & 180 \end{bmatrix}$$

$$\begin{bmatrix} y_1 & y_2 & y_3 & x_1 & x_2 & z' & \\ 1 & 0 & 0 & \frac{5}{29} & -\frac{3}{29} & 0 & \frac{70}{29} \\ 0 & ①⁄₂ & 1 & -\frac{1}{29} & \frac{35}{290} & 0 & \frac{160}{29} \\ \hline 0 & -5 & 0 & \frac{40}{29} & \frac{63}{29} & 1 & \frac{5780}{29} \end{bmatrix}$$

$$\begin{bmatrix} y_1 & y_2 & y_3 & x_1 & x_2 & z' & \\ 1 & 0 & 0 & \frac{5}{29} & -\frac{3}{29} & 0 & \frac{70}{29} \\ 0 & 1 & 2 & -\frac{2}{29} & \frac{70}{290} & 0 & \frac{320}{29} \\ \hline 0 & -5 & 0 & \frac{40}{29} & \frac{63}{29} & 1 & \frac{5780}{29} \end{bmatrix}$$

$$\begin{bmatrix} y_1 & y_2 & y_3 & x_1 & x_2 & z' & \\ 1 & 0 & 0 & \frac{5}{29} & -\frac{3}{29} & 0 & \frac{70}{29} \\ 0 & 1 & 2 & -\frac{2}{29} & \frac{70}{290} & 0 & \frac{320}{29} \\ \hline 0 & 0 & 10 & \frac{30}{29} & \frac{98}{29} & 1 & \frac{7380}{29} \end{bmatrix}$$

The maximum is $z' = \frac{7380}{29} \approx 254.5$, so von Neumann's duality principle tells us that the minimum value of z is also 254.5. In addition to giving the minimum value of z, the duality process also gives the values of x_1 and x_2. These values appear at the bottom of the columns labeled x_1 and x_2, respectively. Thus the objective function is minimized when $x_1 = \frac{30}{29}$ and $x_2 = \frac{98}{29}$. ∎

COMPUTER COMMENT   Most linear programming problems require an extensive amount of tedious arithmetic calculations and for that reason most mathematicians and financial consultants use computer programs of the simplex method to carry out the actual "work" of this method. You might wish to use Program 6 of the computer disk that accompanies this text. One of the things you will notice when using a computer program is that instead of exact results, the work is rounded to a specified number of decimal places. Example 3, for example, looks like this when Program 6 is used:

$$
\begin{array}{cccccc}
y_1 & y_2 & y_3 & x_1 & x_2 & z'
\end{array}
$$

$$
\left[
\begin{array}{cccccc|c}
\boxed{5.80} & 0 & 0 & 1 & -0.60 & 0 & 14 \\
0.20 & 0.50 & 1 & 0 & 0.10 & 0 & 6 \\
\hdashline
-8 & -5 & 0 & 0 & 3 & 1 & 180
\end{array}
\right]
$$

This is the step in Example 3 where $\frac{29}{5}$ is circled.

$\longrightarrow$

$$
\begin{array}{cccccc}
y_1 & y_2 & y_3 & x_1 & x_2 & z'
\end{array}
$$

$$
\left[
\begin{array}{cccccc|c}
1 & 0 & 0 & 0.17 & -0.10 & 0 & 2.41 \\
0 & \boxed{0.50} & 1 & -0.03 & 0.12 & 0 & 5.52 \\
\hdashline
0 & -5 & 0 & 1.38 & 2.17 & 1 & 199.31
\end{array}
\right]
$$

This is the step in Example 3 where $\frac{1}{2}$ is circled.

$$
\begin{array}{cccccc}
y_1 & y_2 & y_3 & x_1 & x_2 & z'
\end{array}
$$

$$
\left[
\begin{array}{cccccc|c}
1 & 0 & 0 & 0.17 & -0.10 & 0 & 2.41 \\
0 & 1 & 2 & -0.07 & 0.24 & 0 & 11.03 \\
\hdashline
0 & 0 & 10 & 1.03 & 3.38 & 1 & 254.48
\end{array}
\right]
$$

This is the final step.

A final note on integer solutions. In this chapter we have been requiring nonnegative variables and have not worked problems requiring integer solutions because linear programming problems that impose integer constraints involve techniques that are beyond the scope of this text. Unfortunately, you cannot simply solve such a problem by the techniques discussed here and then round your results to the nearest integer. The rounded solution *may not* be the best integer solution. However, you can use the following result: Suppose you are maximizing z and

z_0 = Maximum value for the continuous problem (the answer found using the simplex method in this chapter)

z_1 = Maximum found by rounding z_0 to the nearest integer

z_2 = Maximum integer solution (found by techniques not discussed in this text)

The best we can say is that

$$z_1 \leq z_2 \leq z_0$$

Linear programming provides extremely useful mathematical models, but there are many applied applications for which the model developed here is not sufficient. For more advanced study of linear programming, you will need to consult a linear algebra or a linear programming textbook.

Problem Set 4.7

Find the transpose of the matrices in Problems 1–4.

1. a. $A = \begin{bmatrix} 6 & 9 \\ 4 & 8 \end{bmatrix}$

b. $B = \begin{bmatrix} 5 & 6 \\ 3 & 8 \end{bmatrix}$

c. $C = \begin{bmatrix} 4 & 9 & 1 \\ 6 & 1 & 4 \end{bmatrix}$

2. a. $D = \begin{bmatrix} 8 & 1 \\ 6 & 2 \\ 4 & 5 \end{bmatrix}$ **b.** $E = \begin{bmatrix} 6 & 8 & 1 \\ 2 & 3 & 4 \\ 5 & 9 & 7 \end{bmatrix}$

c. $F = \begin{bmatrix} 1 & 0 & 3 \\ 4 & 9 & 7 \\ 8 & 6 & 5 \end{bmatrix}$

3. a. $G = \begin{bmatrix} 1 & 3 & 5 \end{bmatrix}$ **b.** $H = \begin{bmatrix} 4 \\ 9 \\ 6 \end{bmatrix}$

c. $J = \begin{bmatrix} 1 \\ 0 \\ 3 \\ 2 \end{bmatrix}$

4. a. $K = \begin{bmatrix} 1 & 8 & 7 & 4 \end{bmatrix}$

b. $L = \begin{bmatrix} 4 & 8 & 0 & 3 & 2 \\ 6 & 1 & 4 & 7 & 9 \end{bmatrix}$

c. $M = \begin{bmatrix} 1 & 4 & 3 \\ 6 & 9 & 2 \\ 4 & 7 & 1 \\ 5 & 7 & 11 \end{bmatrix}$

Write the dual of Problems 5–8.

5. Minimize: $z = 3x_1 + 4x_2$

Subject to: $\begin{cases} 2x_1 + 8x_2 \geq 10 \\ 3x_1 + 5x_2 \geq 30 \\ x_1 \geq 0, \quad x_2 \geq 0 \end{cases}$

6. Minimize: $z = 50x_1 + 30x_2$

Subject to: $\begin{cases} 5x_1 + 5x_2 \geq 25 \\ 6x_1 - 2x_2 \geq 10 \\ x_1 \geq 0, \quad x_2 \geq 0 \end{cases}$

7. Minimize: $z = 3x_1 + 2x_2 + 5x_3$

Subject to: $\begin{cases} x_1 + x_2 + x_3 \geq 10 \\ 2x_1 + 3x_3 \geq 2 \\ x_2 + 2x_3 \geq 1 \\ 3x_1 + 5x_3 \geq 15 \\ x_1 \geq 0, \quad x_2 \geq 0, \quad x_3 \geq 0 \end{cases}$

8. Minimize: $z = x_1 + 2x_2 + 3x_3$

Subject to: $\begin{cases} x_1 + x_2 + x_3 \geq 50 \\ 6x_1 + 5x_3 \geq 10 \\ 7x_2 + 5x_3 \geq 35 \\ 9x_1 + 4x_2 \geq 18 \\ x_1 \geq 0, \quad x_2 \geq 0, \quad x_3 \geq 0 \end{cases}$

Solve Problems 9–22 by using the dual.

9. Minimize: $z = 3x_1 + 4x_2$

Subject to: $\begin{cases} 2x_1 + 8x_2 \geq 10 \\ 3x_1 + 5x_2 \geq 30 \\ x_1 \geq 0, \quad x_2 \geq 0 \end{cases}$

10. Minimize: $z = 50x_1 + 30x_2$

Subject to: $\begin{cases} 5x_1 + 5x_2 \geq 25 \\ 6x_1 - 2x_2 \geq 10 \\ x_1 \geq 0, \quad x_2 \geq 0 \end{cases}$

11. Minimize: $z = 3x_1 + 2x_2 + 5x_3$

Subject to: $\begin{cases} x_1 + x_2 + x_3 \geq 10 \\ 2x_1 + 3x_3 \geq 2 \\ x_2 + 2x_3 \geq 1 \\ 3x_1 + 5x_3 \geq 15 \\ x_1 \geq 0, \quad x_2 \geq 0, \quad x_3 \geq 0 \end{cases}$

12. Minimize: $z = x_1 + 2x_2 + 3x_3$

Subject to: $\begin{cases} x_1 + x_2 + x_3 \geq 50 \\ 6x_1 + 5x_3 \geq 10 \\ 7x_2 + 5x_3 \geq 35 \\ 9x_1 + 4x_2 \geq 18 \\ x_1 \geq 0, \quad x_2 \geq 0, \quad x_3 \geq 0 \end{cases}$

13. Minimize: $z = 50x_1 + 10x_2 + 20x_3$

Subject to: $\begin{cases} x_1 + x_2 + x_3 \geq 25 \\ 2x_1 + x_2 + 3x_3 \geq 100 \\ x_1 \geq 0, \quad x_2 \geq 0, \quad x_3 \geq 0 \end{cases}$

14. Minimize: $z = 10x_1 + 20x_2 + 30x_3$

Subject to: $\begin{cases} x_1 + 3x_2 + x_3 \geq 75 \\ 2x_1 + x_2 + 5x_3 \geq 80 \\ x_1 \geq 0, \quad x_2 \geq 0, \quad x_3 \geq 0 \end{cases}$

15. Minimize: $z = 24x_1 + 12x_2$

Subject to: $\begin{cases} x_1 \leq 10 \\ x_2 \leq 8 \\ 3x_1 + 2x_2 \geq 12 \\ x_1 \geq 0, \quad x_2 \geq 0 \end{cases}$

16. Minimize: $z = 6x_1 + 18x_2$

Subject to: $\begin{cases} x_1 \leq 10 \\ x_2 \leq 8 \\ 3x_1 + 2x_2 \geq 12 \\ x_1 \geq 0, \quad x_2 \geq 0 \end{cases}$

17. Minimize: $z = 90x_1 + 20x_2$

Subject to: $\begin{cases} x_1 + x_2 \geq 6 \\ -2x_1 + x_2 \geq -16 \\ x_2 \leq 9 \\ x_1 \geq 0, \quad x_2 \geq 0 \end{cases}$

18. Minimize: $z = 400x_1 + 100x_2$

Subject to: $\begin{cases} x_1 + x_2 \geq 6 \\ -2x_1 + x_2 \geq -16 \\ x_2 \leq 9 \\ x_1 \geq 0, \quad x_2 \geq 0 \end{cases}$

19. Minimize: $z = 5x_1 + 3x_2$

Subject to: $\begin{cases} 2x_1 + x_2 \geq 8 \\ x_2 \leq 5 \\ x_1 - x_2 \leq 2 \\ 3x_1 - 2x_2 \geq 5 \\ x_1 \geq 0, \quad x_2 \geq 0 \end{cases}$

20. Minimize: $z = 2x_1 - 3x_2$

Subject to: $\begin{cases} 2x_1 + x_2 \geq 8 \\ x_2 \leq 5 \\ x_1 - x_2 \leq 2 \\ 3x_1 - 2x_2 \geq 5 \\ x_1 \geq 0, \quad x_2 \geq 0 \end{cases}$

21. Minimize: $z = 140x_1 + 250x_2$

Subject to: $\begin{cases} 2x_1 + x_2 \geq 8 \\ x_1 - x_2 \leq 7 \\ x_1 - x_2 \geq -3 \\ x_1 \leq 9 \\ x_1 \geq 0, \quad x_2 \geq 0 \end{cases}$

22. Minimize: $z = 640x_1 - 130x_2$

Subject to: $\begin{cases} 2x_1 + x_2 \geq 8 \\ x_1 - x_2 \leq 7 \\ x_1 - x_2 \geq -3 \\ x_1 \leq 9 \\ x_1 \geq 0, \quad x_2 \geq 0 \end{cases}$

APPLICATIONS

23. Karlin Enterprises manufactures two electronic games. Standing orders require that at least 24,000 space-battle games and 5000 football games be produced. The company has two factories: the Gainesville plant can produce 600 space-battle games and 100 football games per day; the Sacramento plant can produce 300 space-battle games and 100 football games per day. If the Gainesville plant costs $20,000 per day to operate and the Sacramento factory costs $15,000 per day, find the number of days per month each factory should operate to minimize the cost. (Assume that each month has 30 days.)

24. The Cosmopolitan Oil Company requires at least 8000 barrels of low-grade oil per month; it also needs at least 12,000 barrels of medium-grade oil per month and at least 2000 barrels of high-grade oil per month. Oil is produced at one of two refineries. The daily production of these refineries is summarized in the table. If it costs $17,000 per day to operate refinery I and $15,000 per day to operate refinery II, how many days per month should each refinery be operated to satisfy the requirements and at the same time minimize the costs? (Assume that each month has 30 days.)

Refinery	Oil production (barrels)		
	Low grade	Medium grade	High grade
I	200	400	100
II	200	300	100

25. A woman wants to design a weekly exercise schedule that involves jogging, handball, and aerobic dance. She decides to jog at least 3 hours per week, play handball at least 2 hours per week, and dance at least 5 hours per week. She also wants to devote at least as much time to jogging as to handball because she does not like handball as much as she likes the other exercises. She also knows that jogging consumes 900 calories per hour, handball 600 calories per hour, and dance 800 calories per hour. Her physician told her that she must burn up a total of at least 9000 calories per week in this exercise program. How many hours should she devote to each exercise if she wishes to minimize her exercise time?

26. Brown Brothers is an investment company analyzing the pension fund of a certain company. A maximum of $10 million is available to invest in two places. No more than $8 million can be invested in stocks yielding 12%, and at least $2 million can be invested in long-term bonds yielding 8%. The stock-to-bond investment ratio cannot be more than 1 to 3. How should Brown Brothers advise

their client so that the pension fund will receive the maximum yearly return on investment?

27. To utilize costly equipment efficiently, an assembly line with a capacity of 50 units per shift must operate 24 hours per day. This requires scheduling three shifts with varying labor costs as follows:

Day shift: Labor costs are $100 per unit
Swing shift: Labor costs are $150 per unit
Graveyard shift: Labor costs are $180 per unit

Each unit produced uses 60 linear feet of sheeting material, and there is a total of 1800 feet available for the day shift, 1500 feet for the swing shift, and 3000 feet for the graveyard shift. The day shift must produce at least 20 units and the other shifts must produce at least 50 units together. How many units should be produced on each shift to maximize the revenue if the units are sold for $850 each?

28. Repeat Problem 27 except minimize labor costs rather than maximize revenue.

29. Suppose a computer dealer has stores in Hillsborough and Palo Alto and warehouses in San Jose and Burlingame. The cost of shipping a computer from one location to another as well as the supply and demand at each location are shown on the following "map":

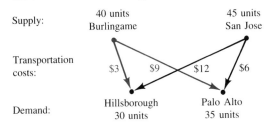

Supply:
40 units Burlingame
45 units San Jose

Transportation costs:
$3 $9 $12 $6

Demand:
Hillsborough 30 units
Palo Alto 35 units

How should the dealer ship the computers to minimize the shipping costs?

30. Camstop, an importer of computer chips, can ship case lots of components to stores in Dallas and Chicago from ports in either Los Angeles or Seattle. The costs and demand are given in the table. How many cases should be shipped from each port to minimize shipping costs?

| Store | Cost to ship (each case) | | |
	Los Angeles port	Seattle port	Store demand
Chicago	$9	$7	80 cases
Dallas	$7	$8	110 cases
Available inventory at each port	90 cases	130 cases	

31. A Texas appliance dealer has stores in Fort Worth and in Houston and warehouses in Dallas and San Antonio. The cost of shipping a refrigerator from Dallas to Fort Worth is $14, and from Dallas to Houston shipping is $15; from San Antonio to Fort Worth the cost is $10, and from San Antonio to Houston it is $18. Suppose that the Fort Worth store has orders for 15 refrigerators and the Houston store has orders for 35. Also suppose that there are 25 refrigerators in stock at Dallas and 45 at San Antonio. What is the most economical way to supply the requested refrigerators to the two stores?

4.8 Summary and Review

IMPORTANT TERMS

Basic feasible solution [4.5]
Basic solution [4.5]
Boundary of a half-plane [4.1]
Closed half-plane [4.1]
Constraints [4.2, 4.3]
Corner point [4.3]
Departing variable [4.5]
Dual [4.7]
Duality principle [4.7]
Entering variable [4.5]
Feasible solution [4.3]
Fundamental theorem of linear programming [4.3]

Half-plane [4.1]
Initial basic solution [4.5]
Initial simplex tableau [4.4]
Linear inequality [4.1]
Linear programming [4.2]
Linear programming models [4.2]
Maximization [4.5]
Maximum value [4.3]
Minimization [4.6]
Minimum value [4.3]
Nonstandard linear programming problem [4.6]
Objective function [4.2, 4.3]

Open half-plane [4.1]
Optimal solution [4.5]
Pivot [4.5]
Pivoting process [4.4, 4.5]
Simplex method [4.4, 4.5]
Simplex tableau [4.4]
Slack variable [4.4]
Standard linear programming
    problem [4.4]

Superfluous constraint [4.3]
System of linear inequalities [4.1]
Test for maximality [4.5]
Transpose of a matrix [4.7]
Von Neumann's duality
    principle [4.7]

SAMPLE TEST *For additional practice there are a large number of review problems categorized by objective in the Student Solutions Manual. The following sample test (40 minutes) is intended to review the main ideas of the chapter.*

*In Problems 1–4 set up the initial tableau for the linear programming problems and indicate the location of the first pivot. You do not need to solve.**

1. Maximize: $z = 6x_1 + 25x_2 + 3x_3$

Subject to: $\begin{cases} 2x_1 + x_3 \le 50 \\ 4x_2 + x_3 \le 90 \\ 3x_1 + 4x_2 \le 100 \\ x_1 \ge 0, \quad x_2 \ge 0, \quad x_3 \ge 0 \end{cases}$

2. Maximize: $z = 5x_1 + 3x_2$

Subject to: $\begin{cases} 7x_1 - 3x_2 \ge 3 \\ 2x_1 + x_2 \le 12 \\ x_1 \ge 0, \quad x_2 \ge 0 \end{cases}$

3. Minimize: $z = 50x_1 + 80x_2$

Subject to: $\begin{cases} 9x_1 + x_2 \ge 18 \\ 3x_1 + 12x_2 \ge 36 \\ 2x_1 + 3x_2 \ge 30 \\ x_1 \ge 0, \quad x_2 \ge 0 \end{cases}$

4. Maximize: $z = x_1 + 3x_2 + 2x_3$

Subject to: $\begin{cases} x_1 + x_2 + x_3 \le 10 \\ 4x_1 + 2x_2 + 3x_3 \le 32 \\ x_1 + 2x_2 + x_3 \le 16 \\ x_1 \ge 0, \quad x_2 \ge 0, \quad x_3 \ge 0 \end{cases}$

Problems 5–10 are reprinted with the permission of the American Institute of Certified Public Accountants, Inc. The questions are multiple-choice questions taken from recent CPA examinations. Choose the best, or most, appropriate response for each question.

CPA Exam
May 1973

The Ball Company manufactures three types of lamps: A, B, and C. Each lamp is processed in two departments—I and II. Total available hours of labor per day for departments I and II are 400 and 600, respectively. No additional labor is available. Time requirements and profit per unit for each lamp type are:

	A	B	C
Hours required in department I	2	3	1
Hours required in department II	4	2	3
Profit per unit (sales price less all variable costs)	$5	$4	$3

The company has assigned you, as the accounting member of its profit planning committee, to determine the number of types of A, B, and C lamps that it should produce in order to

---

* For extra practice you can solve Problems 1–4, but to do so will require more time.

maximize its total profit from the sale of lamps. The following questions concern the linear programming model your group has developed.

5. The coefficients of the objective function would be
   A. 4, 2, 3   B. 2, 3, 1   C. 5, 4, 3   D. 400, 600

6. The constraints in the model would be
   A. 2, 3, 1   B. 5, 4, 3   C. 4, 2, 3   D. 400, 600

7. The constraint imposed by the available hours in department I could be expressed as
   A. $4x_1 + 2x_2 + 3x_3 \leq 400$ B. $4x_1 + 2x_2 + 3x_3 \geq 400$
   C. $2x_1 + 3x_2 + 1x_3 \leq 400$ D. $2x_1 + 3x_2 + 1x_3 \geq 400$

8. The most types of lamps that would be included in the optimal solution would be
   A. 2   B. 1   C. 3   D. 0

CPA Exam May 1979

9. The Pauley Company plans to expand its sales force by opening several new branch offices. Pauley has $10,400,000 in capital available for new branch offices. Pauley will consider opening only two types of branches: 20-person branches (type A) and 10-person branches (type B). Expected initial cash outlays are $1,300,000 for a type A branch and $670,000 for a type B branch. The expected annual profit is $92,000 for a type A branch and $36,000 for a type B branch. Pauley will hire no more than 200 employees for the new branch offices and will not open more than 20 branch offices. Use linear programming to help decide how many branch offices should be opened.

   In a system of equations for a linear programming model, which of the following equations would **not** represent a constraint (restriction)?
   A. $A + B \leq 20$
   B. $20A + 10B \leq 200$
   C. $\$92,000A + \$36,000B \leq \$128,000$
   D. $\$1,300,000A + \$670,000B \leq \$10,400,000$

CPA Exam November 1975

10. Patsy, Inc., manufactures two products, X and Y. Each product must be processed in each of three departments: machining, assembling, and finishing. The hours needed to produce one unit of product per department and the maximum possible hours per department are:

Department	Production hours per unit X	Y	Maximum capacity in hours
Machining	2	1	420
Assembling	2	2	500
Finishing	2	3	600

Other restrictions are:

$$X \geq 50$$
$$Y \geq 50$$

The objective function is to maximize profits where profit $= \$4X + \$2Y$. Given the objective and constraints, what is the most profitable number of units of X and Y, respectively, to manufacture?

Modeling Application 2

Air Pollution Control

Alco Cement Company produces cement. The Environmental Protection Agency has ordered Alco to reduce the amount of emissions released into the atmosphere during production. The company wants to comply, but it also wants to do so at the least possible cost. Present production is 2.5 million barrels of cement, and 2 pounds of dust are emitted for every barrel of cement produced. The cement is produced in kilns that are presently equipped with mechanical collectors. However, in order to reduce the emissions to the required level, the mechanical collectors must be replaced by four-field electrostatic precipitators, which would reduce emissions to 0.5 pound of dust per barrel of cement, or by five-field precipitators, which would reduce emission to 0.2 pound per barrel. The capital and operating costs for the four-field precipitator are 14¢ per barrel of cement produced, and for the five-field precipitator costs are 18¢ per barrel.*

Write a paper that shows the minimum cost for the control methods that Alco should use in order to reduce particulate emissions by 4.2 million pounds. For general guidelines about writing this essay, see the commentary for Modeling Application 1 on page 130.

* This application is adapted from R. E. Kohn, "A Mathematical Programming Model for Air Pollution Control," *School Science and Mathematics*, June 1969, pp. 487–499.

CHAPTER 5
Mathematics of Finance

CHAPTER CONTENTS

5.1 Difference Equations

5.2 Interest

5.3 Annuities

5.4 Amortization

5.5 Sinking Funds

5.6 Summary and Review
      Important Terms
      Sample Test

APPLICATIONS

Management (*Business, Economics, Finance, and Investments*)

Salary increases (5.1, Problems 49–50)
Finding the future value of bank accounts (5.2, Problems 27–31; 5.6, Problem 2)
Finding the effective yield of an investment (5.2, Problems 32–35)
Effect of inflation (5.2, Problems 41–52; 5.5, Problem 23; 5.6, Problem 7)
Present value problem (5.2, Problems 36–40, 53–58; 5.6, Problem 11)
Annuity values (5.3, Problems 1–23; 5.5, Problems 22, 29–30; 5.6, Problems 9–10)
Setting up a retirement fund (5.3, Problems 26–27)
Value of Social Security deposits (5.3, Problems 28–29)
Monthly loan payment (5.4, Problems 13–24; 5.5, Problem 27; 5.6, Problems 6, 8, 16)
Present value of an annuity (5.4, Problems 1–12, 27)
Value of an insurance annuity (5.4, Problems 29–30)
Sinking funds to pay off notes (5.5, Problems 13–16, 21; 5.6, Problems 13–14)
Repayment of a bond (5.5, Problem 28; 5.6, Problem 17)
Value of a savings certificate (5.6, Problem 1)
Value of a lump-sum insurance payment (5.6, Problems 3–4)
Present value of an annuity (5.6, Problem 12)
Management accounting examination (5.6, Problem 18)
CPA examination (5.6, Problems 19–20)

Life sciences (*Biology, Ecology, Health, and Medicine*)

Cell division (5.1, Problem 53)
Price of wheat as determined by last year's production (5.1, Problem 54)

Social sciences (*Demography, Political Science, Population, Psychology, Society, and Sociology*)

Breaking news story (5.1, Problem 51)
Spread of rumors (5.1, Problem 52)

General interest

Calculating total interest paid for a consumer's loan (5.4, Problems 25–26)
Present value of a lottery prize (5.4, Problem 28; 5.6, Problem 15)
Saving for a purchase (5.5, Problems 18, 21, 24–26)
Investing an inheritance (5.5, Problem 19)
Savings account for a child (5.5, Problem 20)

Modeling application—Investment Opportunities

CHAPTER OVERVIEW

The notion of simple interest is introduced, which leads directly to the concept of compound interest. This chapter uses a unified notation, so you should learn the meanings of the variables used early in the chapter, especially the difference between present value, P, and future value, A. Pay special attention to the relationship between annual rate, r, and the rate per period, i; and to the total time, t (in years), and the number of periods, N. One of the important skills to be learned in this chapter is the ability to look at a particular situation and decide what type of financial formula applies (that is, what model to use).

PREVIEW

One of the cornerstones upon which business is built is the payment of interest for the temporary use of another's money. In this chapter we discuss simple and compound interest, present value, annuities, and sinking funds. We will be concerned with comparing the present value, P, with its value, A, at some future time. We can make payments or deposits in two ways: deposit a lump sum or make periodic payments. Consider the following simplified chapter overview:

	Present value, P	Future value, A	Method for finding unknown value
Lump-sum payment or deposit	Known	Unknown	Compound interest
	Unknown	Known	Present value
Periodic payment	Known	Unknown	Annuity
	Unknown	Known	Sinking fund

PERSPECTIVE

Intelligent handling of money goes hand in hand with financial success. While understanding compound interest, annuities, amortization, and sinking funds does not guarantee success, such understanding is necessary when handling money and investments.

5.1 Difference Equations

One of the most fundamental mathematical applications for business people, as well as for consumers, is that of finances and financial formulas. In order to derive many of the financial equations we will use in this chapter, you will first need to understand the notion of a difference equation.

An infinite set of numbers $x_0, x_1, x_2, x_3, \ldots$ is called a sequence or progression. For example, the even numbers

$$2, 4, 6, 8, 10, \ldots$$

form a sequence. The individual numbers are called *terms* of the sequence. The number x_0 is the first value of the sequence and is called the initial value. In terms of investments and interest, it is usually the amount of money you have now (initially). The number x_1 is the second value. Think of the subscript as a measurement of time—the second value of the sequence, x_1, is the value at the end of the first year.

Thus, x_5 is the value at the end of the fifth year (which is the sixth term of the sequence).

Sequences are usually defined by giving what is called the *general term*, which is a formula such as

$$x_n = 3n + 2$$

where $n = 0, 1, 2, 3, 4, \ldots$. You can then find any term by evaluation:

$$x_0 = 3(0) + 2 = 2$$
$$x_1 = 3(1) + 2 = 5$$
$$x_2 = 3(2) + 2 = 8$$
$$x_3 = 3(3) + 2 = 11$$

Thus, the sequence is $2, 5, 8, 11, \ldots$.

Although sequences are used throughout mathematics in a variety of contexts and applications, we will use them in connection with financial formulas by considering general terms of the type

$$x_{n+1} = ax_n + b$$

where a and b are constants. An equation of this type is called a **first-order linear difference equation**.

EXAMPLE 1 Find the first four terms of the sequence that satisfies the difference equation

$$x_{n+1} = 6x_n + 50$$

and has an initial value of 100.

Solution Given $x_0 = 100$; then

$$x_1 = 6x_0 + 50 = 6(100) + 50 = 650$$
$$x_2 = 6x_1 + 50 = 6(650) + 50 = 3950$$
$$x_3 = 6x_2 + 50 = 6(3950) + 50 = 23{,}750$$

The sequence is $100, 650, 3950, 23{,}750, \ldots$. ■

A **solution** of a linear difference equation is a sequence in which the successive terms will satisfy the equation. That is, a solution of $x_{n+1} = ax_n + b$ is completely determined when a, b, and x_0 are all known. The difference between finding the solution and finding the general term is that the solution allows us to calculate any term without first calculating all the preceding terms. Our goal, therefore, is to find this solution for the first-order linear difference equation. First, however, we will solve two special cases, called *arithmetic* and *geometric* sequences.

Arithmetic Sequence

If $a = 1$, then the general difference equation becomes

$$x_{n+1} = x_n + b$$

EXAMPLE 2 Let $b = 2$ in the difference equation

$$x_{n+1} = x_n + 2$$

with initial value 1. Find the first four terms of this sequence.

Solution This difference equation is $x_{n+1} = x_n + 2$ with $x_0 = 1$:

$$x_0 = 1$$
$$x_1 = x_0 + 2 = 1 + 2 = 3$$
$$x_2 = x_1 + 2 = 3 + 2 = 5$$
$$x_3 = x_2 + 2 = 5 + 2 = 7$$

The sequence is 1, 3, 5, 7,.... ∎

A difference equation with $a = 1$, namely

$$x_{n+1} = x_n + b$$

is called an **arithmetic sequence** or **progression**. It is characterized by the fact that there is a *common difference, b*. Notice other terms of the sequence are

$$x_0$$
$$x_1 = x_0 + b$$
$$x_2 = x_1 + b = x_0 + b + b = x_0 + 2b$$
$$x_3 = x_2 + b = x_0 + 2b + b = x_0 + 3b$$
$$x_4 = x_3 + b = x_0 + 3b + b = x_0 + 4b$$
$$\vdots$$
$$x_n = x_0 + nb$$

Arithmetic Sequence

> The solution of the arithmetic sequence with first term x_0 and common difference b is
>
> $$x_n = x_0 + nb$$

EXAMPLE 3 Find the x_{10} of the arithmetic sequence whose first term is -3 and whose common difference is 4.

Solution $x_0 = -3; b = 4$, so

$$x_n = x_0 + nb$$
$$x_{10} = -3 + 10(4) = 37$$

∎

Geometric Sequence

If $b = 0$, then the general difference equation becomes

$$x_{n+1} = ax_n$$

EXAMPLE 4 Find the first four terms of the sequence $x_{n+1} = ax_n$ for the initial value of 1 and $a = 2$.

Solution   The difference equation is $x_{n+1} = 2x_n$:

$$x_0 = 1$$
$$x_1 = 2x_0 = 2(1) = 2$$
$$x_2 = 2x_1 = 2(2) = 4$$
$$x_3 = 2x_2 = 2(4) = 8$$
$$\vdots$$

The sequence is 1, 2, 4, 8,.... ∎

A difference equation with $b = 0$, namely

$$x_{n+1} = ax_n$$

is called a **geometric sequence** or **progression**. It is characterized by the fact that there is a *common ratio, a*. Notice the terms of the sequence are

$$x_0$$
$$x_1 = ax_0$$
$$x_2 = ax_1 = a(ax_0) = a^2x_0$$
$$x_3 = ax_2 = a(a^2x_0) = a^3x_0$$
$$x_4 = ax_3 = a(a^3x_0) = a^4x_0$$
$$\vdots$$
$$x_n = a^nx_0$$

Geometric Sequence

The solution of the geometric sequence with first term x_0 and common ratio a is

$$x_n = a^nx_0$$

EXAMPLE 5   Find the x_5 of the geometric sequence whose first term is 100 and whose common ratio is $\frac{1}{2}$.

Solution $x_0 = 100$, $a = \frac{1}{2}$, so

$$x_n = a^nx_0$$
$$x_5 = (\tfrac{1}{2})^5(100) = 3.125$$ ∎

General Solution of a Difference Equation

We can now find the general solution of the difference equation

$$x_{n+1} = ax_n + b$$

Begin by looking for a pattern. Let x_0 be the initial value.

$$x_1 = ax_0 + b$$
$$x_2 = ax_1 + b = a(ax_0 + b) + b = a^2x_0 + ab + b$$
$$x_3 = ax_2 + b = a(a^2x_0 + ab + b) = a^3x_0 + a^2b + ab + b$$
$$x_4 = ax_3 + b = a(a^3x_0 + a^2b + ab + b) + b = a^4x_0 + a^3b + a^2b + ab + b$$
$$\vdots$$
$$x_n = a^nx_0 + a^{n-1}b + a^{n-2}b + \cdots + a^2b + ab + b$$

Even though this is the general solution, it is not particularly useful because of its algebraic form, so we now change the algebraic form so that it is easier to use.

$$x_n = a^n x_0 + a^{n-1}b + a^{n-2}b + \cdots + a^2 b + ab + b$$
$$= a^n x_0 + b(a^{n-1} + a^{n-2} + \cdots + a^2 + a + 1)$$

Now, consider the sum

$$S_n = 1 + a + a^2 + \cdots + a^n$$

and

$$aS_n = a + a^2 + a^3 + \cdots + a^{n+1} \qquad \text{Multiply both sides by } a$$
$$S_n - aS_n = 1 - a^{n+1} \qquad \text{Subtract (notice the arrows in the lines above; all of those terms are zero when you subtract)}$$

$$(1 - a)S_n = 1 - a^{n+1} \qquad \text{Factor}$$
$$S_n = \frac{1 - a^{n+1}}{1 - a} \qquad a \neq 1$$

Thus,

$$a^{n-1} + a^{n-2} + \cdots + a^2 + a + 1 = \frac{1 - a^{(n-1)+1}}{1 - a}$$
$$= \frac{1 - a^n}{1 - a}$$

The general solution is now stated (by substitution in the second step)

$$x_n = a^n x_0 + b(a^{n-1} + a^{n-2} + \cdots + a^2 + a + 1)$$
$$x_n = a^n x_0 + b\left(\frac{1 - a^n}{1 - a}\right)$$
$$= a^n x_0 + \frac{b}{1 - a}(1 - a^n)$$
$$= a^n x_0 + \frac{b}{1 - a} - \frac{b}{1 - a}a^n$$
$$= \frac{b}{1 - a} + a^n x_0 - \frac{b}{1 - a}a^n$$
$$= \frac{b}{1 - a} + \left(x_0 - \frac{b}{1 - a}\right)a^n$$

General Solution of a Difference Equation

The difference equation $x_{n+1} = ax_n + b$ with $a \neq 1$ has solution

$$x_n = \frac{b}{1 - a} + \left(x_0 - \frac{b}{1 - a}\right)a^n$$

EXAMPLE 6   Solve each difference equation and find x_4 for each.

a. $x_{n+1} = 5x_n + 8, \quad x_0 = 1$ **b.** $x_{n+1} = 3x_n, \quad x_0 = 4$
c. $x_{n+1} = x_n + 4, \quad x_0 = 3$

Solution   **a.** $a = 5$, $b = 8$, $x_0 = 1$, so using the general solution,

$$\frac{b}{1-a} = \frac{8}{1-5} = \frac{8}{-4} = -2$$

$$x_n = -2 + (1 + 2)5^n$$
$$= -2 + 3 \cdot 5^n \qquad \text{This is the solution}$$

Also, $x_4 = -2 + 3 \cdot 5^4 = 1873$

b. $a = 3$, $b = 0$, $x_0 = 4$, so using the general solution,

$$\frac{b}{1-a} = \frac{0}{1-3} = 0$$

$$x_n = 0 + (4 + 0)3 = 4 \cdot 3^n \qquad \text{This is the solution}$$

Notice, however, since $b = 0$, this is a geometric sequence, so using the theorem for geometric sequences we find (directly)

$$x_n = 4 \cdot 3^n$$

Also, $x_4 = 4 \cdot 3^4 = 324$

c. $a = 1$, $b = 4$, $x_0 = 3$. This is an arithmetic sequence. From the theorem for arithmetic sequences, we find (directly)

$$x_n = 3 + 4n$$

Also, $x_4 = 3 + 4 \cdot 4 = 19$　■

EXAMPLE 7   Suppose the average beginning salary for a recent college graduate is $25,000 and that this graduate can expect annual salary increases of $1,000 a year plus a 5% cost-of-living increase. What is the graduate's salary for the fifth year?

Solution   Let x_n be the salary in the nth year. Now,

$$x_n = \text{last year's salary} + .05 \,(\text{last year's salary}) + 1000$$
$$= x_{n-1} + .05x_{n-1} + 1000$$
$$= 1.05x_{n-1} + 1000$$

Note that the general solution is given for a difference equation of the form $x_{n+1} = ax_n + b$ and we have modeled this example in terms of x_n (instead of x_{n+1}):

$$x_n = 1.05x_{n-1} + 1000$$

These equations are the same since the relationship between x_n and x_{n+1} is the same as that between x_{n+1} and x_n (that is, one year change). Therefore, we can use the general solution formula for a difference equation where $a = 1.05$, $b = 1000$, and $x_0 = 25,000$.

$$x_n = \frac{1000}{1 - 1.05} + \left(25,000 - \frac{1000}{1 - 1.05}\right)(1.05)^n$$

In particular, we want $n = 5$ (fifth year):

$$x_5 = \frac{1000}{-.05} + \left(25,000 - \frac{1000}{-.05}\right)(1.05)^5 \approx 37,432.67$$

The graduate's salary in five years should be $37,433.　■

Sometimes models using difference equations are developed using the idea of **proportionality**. We say that two quantities are proportional if one is equal to a constant times the other. For example, if x and y are proportional then $x = ky$ for some constant k, which is called the *constant of proportionality*.

EXAMPLE 8   Suppose that a major news story breaks on the four national television networks at the same time. The number of people learning the news each hour after it broke is proportional to the number who have not heard it by the end of the preceding hour. If we assume that the population is 220 million, how many people would hear of the news 24 hours after it broke if the constant of proportionality is 0.2?

Solution   Let x_n = number of people who have heard the news after n hours. If P is the population (220 million in this example), then the number of people who have not heard the news after n hours is $P - x_n$. Thus,

$$
\begin{pmatrix} \text{number who have} \\ \text{heard the news} \\ \text{after } n+1 \text{ hours} \end{pmatrix} = \begin{pmatrix} \text{number who know} \\ \text{after } n \text{ hours} \end{pmatrix} + \begin{pmatrix} \text{number who learn} \\ \text{news during the} \\ (n+1)\text{st hour} \end{pmatrix}
$$
$$
x_{n+1} \qquad\qquad = \qquad\qquad x_n \qquad\qquad + \qquad k(P - x_n)
$$

For this example, $k = .2$ and $P = 220$:

$$
\begin{aligned}
x_{n+1} &= x_n + 0.2(220 - x_n) \\
&= x_n + 44 - 0.2x_n \\
&= 0.8x_n + 44
\end{aligned}
$$

If we assume that no one heard the news initially ($x_0 = 0$) we can find the solution where $a = 0.8$, $b = 44$, and $x_0 = 0$:

$$
\begin{aligned}
x_n &= \frac{b}{1-a} + \left(x_0 - \frac{b}{1-a} \right) a^n \\
&= \frac{44}{1-0.8} + \left(0 - \frac{44}{1-0.8} \right)(0.8)^n \\
&= 220 - 220(0.8)^n
\end{aligned}
$$

For $n = 24$,

$$
\begin{aligned}
x_{24} &= 220 - 220(0.8)^{24} \\
&\approx 219
\end{aligned}
$$

Thus, after one day (24 hours) approximately 1 million people would not have heard the news. ■

Sigma Notation

Sometimes we want to write a sum of terms of a sequence using a shorthand notation. This notation uses the uppercase Greek letter sigma, so is often referred to as **sigma notation**. For example, we wrote

$$
S_n = 1 + a + a^2 + a^3 + \cdots + a^n
$$

and using sigma notation this sum could be written as

$$S_n = \sum_{k=0}^{n} a^k \quad \text{since} \quad \sum_{k=0}^{n} a^k = 1 + a + a^2 + a^3 + \cdots + a^n$$

In other words, the sigma notation evaluates the expression immediately following the sigma (a^k in this example) first for $k = 0$, then for $k = 1$, next for $k = 2$, and so on, where k counts up (one unit at a time) until it reaches the final value of n. The value 0 in this example is called the *lower limit of summation* and n the *upper limit of summation*, while the variable k is called the *index of summation*.

EXAMPLE 9   Find $\sum_{i=1}^{4} i(i + 3)$.

Solution   We have $\sum_{i=1}^{4} i(i + 3) = \overbrace{1(1 + 3)}^{i=1} + \overbrace{2(2 + 3)}^{i=2} + \overbrace{3(3 + 3)}^{i=3} + \overbrace{4(4 + 3)}^{i=4}$

$$= 1 \cdot 4 + 2 \cdot 5 + 3 \cdot 6 + 4 \cdot 7$$
$$= 4 + 10 + 18 + 28$$
$$= 60$$

Problem Set 5.1

Compute the first four terms of the solution to the linear difference equations given in Problems 1–12.

1. $x_{n+1} = x_n + 10; x_0 = 6$
2. $x_{n+1} = x_n - 5; x_0 = 3$
3. $x_{n+1} = 4x_n; x_0 = 1$
4. $x_{n+1} = 10x_n; x_0 = 0$
5. $x_{n+1} = -5x_n; x_0 = 10$
6. $x_{n+1} = 3x_n; x_0 = 4$
7. $x_{n+1} = 2x_n + 3; x_0 = 4$
8. $x_{n+1} = -3x_n - 4; x_0 = 5$
9. $y_{n+1} = 2y_n + 3; y_0 = 1$
10. $y_{n+1} = 3y_n - 2; y_0 = 4$
11. $y_{n+1} = 5y_n - 2; y_0 = 3$
12. $y_{n+1} = 4y_n + 3; y_0 = 2$

Find x_4 for each difference equation given in Problems 13–24.

13. $x_{n+1} = x_n + 8; x_0 = 0$
14. $x_{n+1} = x_n + 100; x_0 = 100$
15. $x_{n+1} = 3x_n; x_0 = 1$
16. $x_{n+1} = 2x_n; x_0 = 1$
17. $x_n = (\tfrac{1}{2})x_{n-1} + 2; x_0 = 100$
18. $x_n = (\tfrac{1}{10})x_{n-1} + 2; x_0 = 1000$
19. $y_{n+1} = 5y_n + 2; y_0 = 0$

20. $y_{n+1} = 2y_n - 3; y_0 = 10$
21. $y_{n+1} = 2y_n + 1; y_0 = 8$
22. $y_{n+1} = 1 - 2y_n; y_0 = 0$
23. $y_{n+1} = 1 - (\tfrac{1}{2})y_n; y_0 = 0$
24. $y_{n+1} = 10 - (\tfrac{1}{10})y_n; y_0 = 0$

Solve the difference equations in Problems 25–40.

25. $x_{n+1} = x_n + 25, x_0 = 5$
26. $x_{n+1} = x_n - 20, x_0 = 20$
27. $x_{n+1} = x_n - 2, x_0 = 4$
28. $x_{n+1} = x_n + 1, x_0 = 0$
29. $x_n = x_{n-1} + 4, x_0 = 0$
30. $x_n = x_{n-1} - 5, x_0 = 5$
31. $x_n = 3x_{n-1}, x_0 = 1$
32. $x_n = (\tfrac{1}{5})x_{n-1}, x_0 = 5$
33. $x_n = 2x_{n-1}, x_0 = 1$
34. $x_n = 4x_{n-1}, x_0 = 1$
35. $x_n = 2x_{n-1} - 3, x_0 = 0$
36. $x_n = -2x_{n-1} + 15, x_0 = 0$
37. $x_n = 5x_{n-1} + 9, x_0 = 2$
38. $x_n = -3x_{n-1} + 5, x_0 = 1$
39. $x_n = 4x_{n-1} + 2, x_0 = 1$
40. $x_n = 3x_{n-1} - 4, x_0 = 10$

In Problems 41–48 evaluate the given expression.

41. $\sum_{k=2}^{6} k$

42. $\sum_{m=1}^{4} m^2$

43. $\sum_{n=0}^{6} (2n+1)$

44. $\sum_{k=2}^{5} (10-2k)$

45. $\sum_{k=1}^{5} (-2)^{k-1}$

46. $\sum_{k=0}^{4} 3(-2)^k$

47. $\sum_{k=0}^{3} 2(3^k)$

48. $\sum_{k=2}^{5} (100-5k)$

APPLICATIONS

49. Suppose a job pays a starting salary of $20,000 with a $500 per year increase and a 3% cost-of-living increase. What will the salary be in the tenth year?

50. Suppose a job pays a starting salary of $38,000 with a $2,500 per year increase and a 5% cost-of-living increase. What will the salary be in the fifth year?

51. Suppose a news story breaks in a town of 200,000 and the number of people learning of the news after the story broke is proportional to the number who have not heard it by the end of the preceding hour. How many will have heard the story in 8 hours if the constant of proportionality is 0.3?

52. Suppose the number of persons in a company of 3000 employees who have heard a particular piece of gossip is proportional to the number of persons who have not heard it by the end of the previous day. How many will have heard the gossip one week after the rumor started if the constant of proportionality is 0.4?

53. Suppose a cell divides every 20 minutes. If there is initially one cell, how many cells will there be in 8 hours?

54. According to the Department of Agriculture (Statistical Reporting Service), the U.S. wheat production was approximately 2.6 billion bushels. It is known that the current wheat crop affects next year's level of production and next year's price. Let p_n denote the price of wheat (in dollars per bushel) and q_n denote the quantity of wheat produced (in billions of bushels), and suppose that p_n and q_n are related by the equations

$$p_n = 10 - .01q_n$$
$$q_{n+1} = .75p_n - 5$$

Solve the difference equation for production.

5.2 Interest

One of the most fundamental mathematical concepts for business people and consumers is the idea of interest. Simply stated, **interest** is rent paid for the use of another's money. That is, we receive interest when we let others use our money (when we deposit money in a savings account, for example), and we pay interest when we use the money of others (for example, when we borrow from a bank).

The amount of the deposit or loan is called the **principal**, or **present value**, and the **interest rate** is stated as a percentage of the principal over a given period of **time**.

Simple Interest Formula

The *simple interest formula* is

$I = Prt$ where I = Amount of interest
P = Principal or Present value
r = Annual interest rate
t = Time (in years)

EXAMPLE 1 How much interest does a $73 deposit earn in 3 years if the interest rate is 8%?

Solution $I = Prt$
$= 73(0.08)(3)$ Notice that interest rates are written as decimals when substituted into formulas
$= 17.52$

The amount of interest is $17.52. ∎

The **future value**, A, of a deposit is the amount of money on deposit after a given amount of time. For Example 1, the future value of the account in 3 years is $A = \$73 + \$17.52 = \$90.52$.

Future Value (simple interest)

The *future value*, A, is

$$A = P + I \qquad \text{where} \quad P = \text{Principal}$$
$$I = \text{Amount of interest}$$

Example 1 is an example of simple interest, but banks and businesses pay **compound interest**. For compound interest, after some designated period of time, the earned interest is added to the account so that future calculations include this earned interest as part of the principal. For Example 1, suppose that interest is *compounded annually*. This means that at the end of the first year the value of the account is found as follows:

$$I = Prt$$
$$= 73(0.08)(1) \qquad t = 1 \text{ at the end of the first year}$$
$$= 5.84$$

Therefore

$$A = P + I$$
$$= 73 + 5.84$$
$$= 78.84$$

This amount then becomes the principal for the second year:

$$I = Prt$$
$$= 78.84(0.08)(1)$$
$$= 6.3072 \qquad \text{This is \$6.31 interest for the second year}$$

At the end of the second year,

$$A = P + I$$
$$= 78.84 + 6.31$$
$$= 85.15$$

For the third year,

$$I = 85.15(0.08)(1) \qquad \text{and} \qquad A = 85.15 + 6.81$$
$$= 6.812 \qquad\qquad\qquad\qquad\quad = 91.96$$

Note that with simple interest (Example 1) the future value in 3 years is $90.52 and with compound interest it is $91.96.

The process just described with numbers is easy to follow, but to find the general formula we need to repeat the steps algebraically.

First Year $\quad A = P + I$
$$= P + Pr \qquad I = Prt = Pr \text{ when } t = 1$$
$$= P(1 + r) \qquad \text{Factor out the common factor } P; \text{ this number}$$
$$\text{becomes the principal for the second year}$$

Second Year $A = P(1 + r) + I$ | The second year principal is
$= P(1 + r) + P(1 + r)$ | $P(1 + r)$. Also,
$= P(1 + r)[1 + r]$ | $I = \text{principal} \times \text{rate} \times \text{time}$
$= P(1 + r)^2$ | $= \underbrace{P(1 + r)}r(1)$
   | Second-year principal

This becomes the principal for the third year.

Third Year $A = P(1 + r)^2 + I$
$= P(1 + r)^2 + P(1 + r)^2 r$
$= P(1 + r)^2[1 + r]$
$= P(1 + r)^3$

If you continue in the same fashion for t years at a rate of r compounded annually, you can see the formula is

$$A = P(1 + r)^t$$

We can also derive this formula using a difference equation.

$$\underset{\text{OF NEXT YEAR}}{\text{BALANCE AT BEGINNING}} = \underset{\text{OF THIS YEAR}}{\text{BALANCE AT BEGINNING}} + \text{INTEREST}$$

$$A_{n+1} = A_n + I$$
$$A_{n+1} = A_n + A_n r \qquad \text{Since } I = Prt = A_n r(1)$$
$$A_{n+1} = A_n(1 + r)$$

This is a geometric sequence whose first term is P and whose common ratio is $(1 + r)$, so after t years the formula is

$$A_t = (1 + r)^t P \quad \text{or} \quad A = P(1 + r)^t$$

EXAMPLE 2   Find the compound interest for a $73 deposit at 8% for 3 years compounded annually.

Solution $A = 73(1 + 0.08)^3$
$= 73(1.08)^3$

On a calculator,

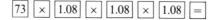

or, if your calculator has an exponent key,

$\boxed{73}$ $\boxed{\times}$ $\boxed{1.08}$ $\boxed{y^x}$ $\boxed{3}$ $\boxed{=}$

The rest of this chapter assumes that calculators have an exponent key; if yours does not, you will need to use repeated multiplication as shown on the first line. The result is **91.958976** or $91.96; the interest is $91.96 − $73 = $18.96.   ∎

If you wish, you may use a table for compound interest problems. Table 5 in Appendix E shows the compound interest for $1 for N periods. For Example 2, find the column headed 8% and look in the row $N = 3$. The entry is 1.259712. Multiply this number by the principal (73) to obtain **91.958976** or $91.96.

EXAMPLE 3   You are considering a 6-year $1,000 certificate of deposit paying 12% compounded annually. How much money will you have at the end of 5 years?

Solution   *By calculator*

$A = 1000(1.12)^5$

| 1.12 | y^x | 5 | × | 1000 | = |

The result **1762.341683** rounded to the nearest cent is $1,762.34.

By Table 5, Appendix E

Find the column headed 12% and the row labeled $N = 5$ to find the entry 1.762342. Multiply:

$1000(1.762342) \approx \$1,762.34$

∎

The disadvantages of working with tables are (1) a different table is needed for each different rate; (2) tables are not as readily available as calculators; and (3) even when using a table, a final multiplication by the principal is necessary.

If you have deposited money or seen savings and loan advertisements lately, you know that most financial institutions compound interest more frequently than annually. This is to your advantage because the shorter the compounding period, the sooner you earn interest on your interest.

Future Value (compound interest)

$$A = P(1 + i)^N \qquad \text{where} \quad A = \text{Future value}$$
$$P = \text{Present value}$$
$$r = \text{Annual interest rate}$$
$$t = \text{Number of years}$$
$$n = \text{Number of times compounded per year}$$
$$i = \text{Rate per period} = \frac{r}{n}$$
$$N = \text{Number of periods} = nt$$

EXAMPLE 4   Reconsider Example 3 for 12% compounded monthly.

Solution   By calculator:

$$A = 1000\left(1 + \frac{0.12}{12}\right)^{12(5)} = 1000(1.01)^{60}$$

$$= \textbf{1816.696699} \qquad \text{PRESS:} \quad \boxed{1.01} \ \boxed{y^x} \ \boxed{60} \ \boxed{\times} \ \boxed{1000} \ \boxed{=}$$

Answer rounded to the nearest cent: $1,816.70   ∎

We can also use Table 5 in Appendix E to work problems like Example 4. This is particularly important if your calculator does not have an exponent key. To find the rate per period (this is i in Table 5), divide the annual rate by

2   if semiannually
4   if quarterly
12   if monthly
360   if daily

Also, to find the number of periods (this is N in Table 5), multiply the time by

$$
\begin{array}{rl}
2 & \text{if semiannually} \\
4 & \text{if quarterly} \\
12 & \text{if monthly} \\
360 & \text{if daily}
\end{array}
$$

To find the amount of Example 4 using Table 5, look up a rate of 1% ($12\% \div 12$) and $N = 60$. The entry is 1.816697, so

$$A = 1000(1.816697)$$

$$\approx \$1,816.70 \qquad \text{Answer is rounded to the nearest cent}$$

Note that when paying interest, banks use 360 for the number of days in a year; this is called **ordinary interest**. Unless instructed otherwise, use 360 for the number of days in a year. If 365 days are used, it is called **exact interest**.

EXAMPLE 5 Use Table 5 in Appendix E to find the amount you would have if you invested $1,000 for 10 years at 8% interest compounded:

a. Annually   **b.** Semiannually   **c.** Quarterly

Solution **a.** $N = 10$, and the rate is 8%: $A = 1000 \times 2.158925$
$$= \$2,158.93$$
b. $N = 20$, and the rate per period is 4%: $A = 1000 \times 2.191123$
$$= \$2,191.12$$
c. $N = 40$, and the rate per period is 2%: $A = 1000 \times 2.208040$
$$= \$2,208.04 \qquad ■$$

Table 5 is not extensive enough to find monthly or daily compounding of $1,000 at an 8% annual interest; these problems require a calculator. Also, keep in mind the difference between interest and future value for both the simple and compound interest formulas. With simple interest, you find the interest first and then the future value. For compound interest, you first find the future value and then find the interest. Table 5.1 summarizes this comparison.

TABLE 5.1
Comparison of simple and compound interest

	Interest, I	Future value, A
Simple interest formula	$I = Prt$ This is found first for simple interest	$A = P + I$ This is found after using the simple interest formula
Compound interest	$I = A - P$ This is found after using Table 5	$A = P\left(1 + \dfrac{r}{n}\right)^{nt}$ or $A = P(\text{Table 5 number})$ This is found first for compound interest

You have no doubt, seen advertisements that say

$$8.00\% = 8.33\%*$$

* Effective annual yield when principal and interest are left in the account.

In order to understand the phrase **effective annual yield**, look at the earnings of $1 at 8% compounded annually and quarterly:

Annually

$$1(1 + 0.08) = 1.08$$

Quarterly

$$1\left(1 + \frac{0.08}{4}\right)^4 = (1.02)^4 = 1.08243216$$

The $1 compounded quarterly at 8% is the same as $1 compounded annually at a rate of 8.243216%. Therefore 8.24% is called the *effective annual yield* of 8% compounded quarterly.

Effective Annual Yield

The *effective annual yield*, or *effective rate*, for an account paying r percent compounded n times per year is the simple annual interest rate that would pay an equivalent amount. It is found by the following formula:

$$\left(1 + \frac{r}{n}\right)^n - 1$$

EXAMPLE 6   Verify the claims of the bank advertisement shown. (You need a calculator with an exponent key for this example.)

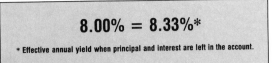

ALSO NEW, HIGHEST INTEREST EVER ON INSURED SAVINGS.

8.33% annual yield on

8% interest compounded daily

Annual yield based on daily compounding when funds and interest remain on deposit a year. Note: Federal regulations require a substantial interest penalty for early withdrawal of principal from Certificate Accounts.

Solution   The compounding is on a daily basis ($n = 360$), so:

$$\left(1 + \frac{0.08}{360}\right)^{360} \approx (1.00022222)^{360}$$ This calculation was done on a calculator with an exponent key

$$\approx 1.08327735$$

$$\text{Effective rate} \approx 1.08327735 - 1$$

$$\approx 0.0833 \quad \text{or} \quad 8.33\%$$

Businesses sometimes need to have a given amount of money on hand at some time in the future. That is, suppose a business has a $1,000 note payable due in 5 years. The business would like to know the amount of money that must be deposited *today* so that in 5 years the total amount of principal *and* interest will be $1,000. This sum to be deposited is called the **present value** and is the same variable P that appeared in the future value formulas. That is, for present value, the interest formulas are solved for P since the variable A is known.

Present Value

SIMPLE INTEREST: $P = A - I$

COMPOUND INTEREST: By formula, $P = A(1 + i)^{-N}$
By table, $P = A \div$ (Table 5, Appendix E, entry)

The variables in this formula have the same meaning they had in the compound interest formula:

P = Present value (principal)
A = Future value
n = Number of times compounded per year
r = Annual rate
t = Number of years
$i = \frac{r}{n}$ (rate per period)
$N = tn$ (number of periods)

It is easy to see where the simple interest formula comes from (subtract I from both sides of the future value formula). The compound interest formula is also easy to derive when you remember the meaning of a negative exponent:

$A = P(1 + i)^N$ Future value formula

$P = \dfrac{A}{(1 + i)^N}$ Divide both sides by $(1 + i)^N$

$= A(1 + i)^{-N}$ $(1 + i)^{-N} = \dfrac{1}{(1 + i)^N}$

EXAMPLE 7   Don's Shoe Repair needs $1,000 for a note payable in 5 years, and Don wants to make a deposit now to pay for that note. What is the present value if the money is deposited in a savings account paying 8% compounded quarterly?

Solution $i = \dfrac{r}{n} = \dfrac{0.08}{4} = 0.02$

$N = nt = 4(5) = 20$

By calculator

$A = 1000(1.02)^{-20}$

| 1.02 | y^x | 20 | +/− | × | 1000 | = |

The display shows: **672.9713331**

Don should deposit $672.97.

By Table 5, Appendix E

Find the column headed 2% and the row labeled $N = 20$ to find the entry 1.485947.

$A = 1000 \div 1.485947$

$\approx \$672.97$ ∎

EXAMPLE 8 Tom Mikalson has built a good reputation for his sporting goods store, and the business is growing at an annual rate of 21%. Tom plans on selling out and retiring in 5 years. Yesterday a developer offered Tom $240,000 for the business, and Tom is trying to decide whether he should sell now.

The business was recently appraised for $198,000, so Tom uses this as the present value of the business.

If he sells, he will invest the money with his stockbroker in an account paying 12% compounded quarterly.

He will sell if the price today is better than the one he can expect in 5 years. What should Tom do?

Solution First, find the expected value of the business in 5 years, then find the present value of that amount.

Future value

$$A = P(1 + i)^N$$

$$= 198,000(1 + 0.21)^5$$

$$= 513,561.0071$$

$n = 1$ (compounded annually)
$r =$ Growth rate (21%)
$N = nt = 1(5) = 5$
$i = r/n = \frac{0.21}{1} = 0.21$

Present value

$$P = A(1 + i)^{-N}$$

$$= 513,561.0071(1 + 0.03)^{-20}$$

$$= 284,346.2779$$

$n = 4$ (compounded quarterly)
$r =$ Interest rate (12%)
$N = nt = 4(5) = 20$
$i = \frac{0.12}{4} = 0.03$

This means that Tom should not accept less than $284,346.28 for his business (even though it is presently appraised at $198,000). ■

Inflation

Example 8 involves a very important application of present and future value—inflation. **Inflation** indicates an increase in the amount of currency in circulation, which leads to a fall in the currency's value and a consequent rise in prices. Inflation is usually specified as a percent, and for our purposes we will use the future and present value formulas with $n = 1$ so that $N =$ the number of years and $i =$ the annual rate.

EXAMPLE 9 The inflation rate in 1988 is about 2%. A person earning a $25,000 salary wants to know what salary to expect in 1994 if this rate of inflation continues and if her salary keeps pace with inflation.

Solution This is a future value problem; use Table 5 in Appendix E or a calculator.

$$A = 25,000(1 + 0.02)^6$$

$$\approx 28,154.06$$

Future value formula
$P = 25,000$
$i = 0.02$ (inflation rate)
$N = 6$ (number of years)

By table, $A = 25,000(1.126162)$ In Table 5, $N = 6$, $i = 2\% = 1.126162$

$$= 28,154.05$$

The expected wage will be about $28,000. ■

With inflation problems, the best figure we can arrive at is an estimation since in reality the inflation rate is not a constant. Therefore be careful not to try to project exact or precise answers.

EXAMPLE 10 An insurance agent wants to sell you a policy that will pay you $20,000 in 30 years. If you assume an average rate of inflation of 9% over the next 30 years, what is the value of that insurance payment in terms of today's dollars?

Solution Use Table 5 to find the present value ($N = 30, i = 9\%$):

$$P = 20,000 \div 13.267678$$
$$= 1507.42$$

The present value of the $20,000 is about $1,500. ∎

The effects of compound interest can sometimes be rather dramatic. It is not uncommon to find some investments that pay 16% (although such investments are much riskier than insured savings accounts).

EXAMPLE 11 Your first child has just been born. You want to give her a million dollars when she retires at age 65. If you invest your money at 16% compounded quarterly, how much do you need to invest today so your child will have a million dollars in 65 years?

Solution Find the present value of 1,000,000 where $n = 4, r = 16\%, t = 65, i = \frac{0.16}{4} = 0.04$, and $N = 4(65) = 260$. Use the present value formula:

$$P = 1,000,000(1 + 0.04)^{-260}$$
$$= 37.26763062$$

This means that if you invest $37.28 at 16% compounded quarterly in your child's name, she will have $1,000,000 at age 65. ∎

Problem Set 5.2

APPLICATIONS

In Problems 1–6 compare the amount of simple interest and the interest if the investment is compounded annually.

1. $1,000 at 8% for 5 years
2. $5,000 at 10% for 3 years
3. $2,000 at 12% for 3 years
4. $2,000 at 12% for 5 years
5. $5,000 at 12% for 20 years
6. $1,000 at 14% for 30 years

In Problems 7–12 compare the future amount you would have if the money were invested at simple interest or invested with interest compounded annually.

7. $1,000 at 8% for 5 years
8. $5,000 at 10% for 3 years
9. $2,000 at 12% for 3 years
10. $2,000 at 12% for 5 years
11. $5,000 at 12% for 20 years
12. $1,000 at 14% for 30 years

Fill in the blanks for Problems 13–26.

Compounding period, n	Principal, P	Yearly rate, r	Time, t	Period rate, $i = r/n$	Number of periods, $N = nt$	Entry in Table 5	Total amount	Total interest
13. Annually, 1	$1,000	9%	5 years	_____	_____	_____	_____	_____
14. Semiannually, 2	$1,000	9%	5 years	_____	_____	_____	_____	_____

Compounding period, n	Principal, P	Yearly rate, r	Time, t	Period rate, $i = r/n$	Number of periods, $N = nt$	Entry in Table 5	Total amount	Total interest
15. Annually, 1	$500	8%	3 years	_____	_____	_____	_____	_____
16. Semiannually, 2	$500	8%	3 years	_____	_____	_____	_____	_____
17. Quarterly, 4	$500	8%	3 years	_____	_____	_____	_____	_____
18. Semiannually, 2	$3,000	18%	3 years	_____	_____	_____	_____	_____
19. Quarterly, 4	$5,000	18%	10 years	_____	_____	_____	_____	_____
20. Quarterly, 4	$624	16%	5 years	_____	_____	_____	_____	_____
21. Quarterly, 4	$5,000	20%	10 years	_____	_____	_____	_____	_____
22. Monthly, 12	$350	12%	5 years	_____	_____	_____	_____	_____
23. Monthly, 12	$4,000	24%	5 years	_____	_____	_____	_____	_____
24. Quarterly, 4	$800	12%	90 days	_____	_____	_____	_____	_____
25. Quarterly, 4	$1,250	16%	450 days	_____	_____	_____	_____	_____
26. Quarterly, 4	$1,000	12%	900 days	_____	_____	_____	_____	_____

Answer the questions in Problems 27–35. A calculator with an exponent key is required for these exercises.

27. What is the future amount of $12,000 invested for 5 years at 14% compounded monthly?

28. What is the future amount of $800 invested for 1 year at 10% compounded daily?

29. What is the future amount of $9,000 invested for 4 years at 20% compounded monthly?

30. If $5,000 is compounded annually at $5\frac{1}{2}$% for 12 years, what is the total interest received at the end of that time?

31. If $10,000 is compounded annually at 8% for 18 years, what is the total interest received at the end of that time?

32. What is the effective yield of 6% compounded quarterly?

33. What is the effective yield of 6% compounded monthly?

34. What is the effective yield of 8% compounded monthly?

35. What is the effective yield of 12% compounded daily?

Find the present value in Problems 36–40.

	Amount needed	Time	Interest	Compounded
36.	$7,000	5 years	8%	Annually
37.	$25,000	5 years	8%	Semiannually
38.	$165,000	10 years	12%	Semiannually
39.	$500,000	10 years	12%	Quarterly
40.	$3,000,000	20 years	12%	Semiannually

In Problems 41–52 calculate the expected price in the year 2008 if you assume a 10% inflation rate and use the given 1988 price. Answers should be rounded to the nearest dollar.

41. Cup of coffee, $0.75

42. Small car, $6,000

43. Big Mac, $1.85

44. Movie admission, $5.00

45. Monthly rent, $400

46. Sunday paper, $1.00

47. Textbook, $28.00

48. Tuition at a private college, $14,000

49. Yearly salary, $25,000

50. Condominium, $65,000

51. House, $115,000

52. Business, $675,000

53. You owe $5,000 due in 3 years, but you would like to pay the debt today. If the present interest rate is compounded annually 11%, how much should you pay today so that the present value is equivalent to the $5,000 payment in 3 years?

54. Ricon Bowling Alley will need $20,000 in 5 years to resurface the lanes. What deposit should be made today in an account that pays 9% compounded semiannually?

55. An accounting firm agrees to purchase a computer for $150,000. It will be delivered in 270 days. How much do the owners need to deposit in an account paying 18% compounded quarterly?

56. The Fair View Market must be remodeled in 3 years. Remodeling will cost $200,000. How much should be deposited now (to the nearest dollar) in order to pay for this remodeling if the account pays 10% compounded monthly?

57. A laundromat will need seven new washing machines in $2\frac{1}{2}$ years for a total cost of $2,900. How much money (to the nearest dollar) should be deposited at the present time to pay for these machines? The interest rate is 11% compounded semiannually.

58. A computerized checkout system is planned for Able's Grocery Store. The system will be delivered in 18 months at a cost of $560,000. How much should be deposited today into an account paying 7.5% compounded daily?

59. A contest offers the winner $50,000 now or $10,000 now and $45,000 in 1 year. Which is the best choice if the current interest rate is 10% compounded monthly and the winner does not intend to use any of the money for 1 year?

5.3 Annuities

In Section 5.2 we discussed the present and future values of a lump-sum deposit bearing compound interest. However, it is far more typical to make monthly or other periodic payments into an account. Suppose you decide to give up smoking and save the $1 per day you spend on cigarettes. How much will you save in 5 years? If you save the money without earning any interest, you will have

$$\$1 \times 365 \times 5 = \$1,825$$

However, let us assume that you save $1 per day and at the end of each month you deposit the $30 (assume all months are 30 days) into a savings account earning 12% interest compounded monthly. Now, how much will you have in 5 years?

TABLE 5.2

Time	Amount saved	Interest earned	Total amount
Start	$0	$0	$0
1 month	$30	$0	$30
2 months	$60	$.30	$60.30
3 months	$90	$.60	$90.90
4 months	$120	$.90	$121.80
⋮			

If we continue to calculate the entries in Table 5.2 month by month, we will tire long before we reach 5 years (that would be 60 entries to calculate). Instead, let us shorten the problem to 6 months and try to notice a pattern so we can use a formula. At the same time, we will use the variables of the last section to mean the same thing: P = present value; A = future value; r = annual rate; t = number of years; n = number of times compounded per year; $i = \frac{r}{n}$; and $N = nt$. We will also introduce a new variable, m, as the periodic payment (monthly payment in this example).

Detail of Table 5.2   *For $n = 6$, $i = \frac{0.12}{12} = 0.01$, and $m = \$30$:*

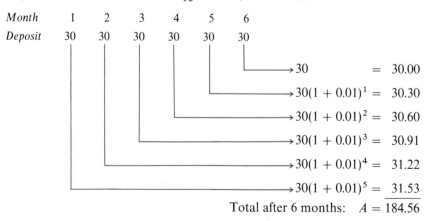

Month	1	2	3	4	5	6
Deposit	30	30	30	30	30	30

$$30 = 30.00$$
$$30(1 + 0.01)^1 = 30.30$$
$$30(1 + 0.01)^2 = 30.60$$
$$30(1 + 0.01)^3 = 30.91$$
$$30(1 + 0.01)^4 = 31.22$$
$$30(1 + 0.01)^5 = 31.53$$

Total after 6 months: $A = 184.56$

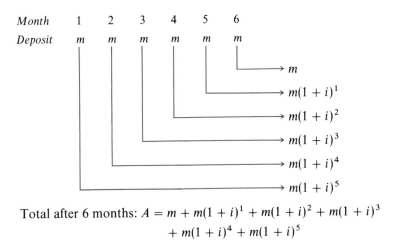

Total after 6 months: $A = m + m(1 + i)^1 + m(1 + i)^2 + m(1 + i)^3$
$$+ m(1 + i)^4 + m(1 + i)^5$$

A sequence of payments into an interest-bearing account is called an **annuity**. If the payments are made at the end of the time period, and if the frequency of payments is the same as the frequency of compounding, the annuity is called an **ordinary annuity**.

To find a general equation for an annuity, let

A_N = The amount of money in the annuity after N periods

i = Interest rate per period of time

m = Monthly deposit

Also assume that $A_0 = 0$ (that is, there is nothing in the account when the account is opened). Now,

CURRENT AMOUNT = PREVIOUS AMOUNT + INTEREST + DEPOSIT

$$A_N \qquad = \qquad A_{N-1} \qquad + \quad iA_{N-1} + \quad m$$

$$A_N = (1 + i)A_{N-1} + m$$

The solution of this difference equation is found by noticing that $a = 1 + i$, $b = m$, and $A_0 = 0$, so

$$A_N = \frac{m}{1 - (1 + i)} + \left[0 - \frac{m}{1 - (1 + i)} \right] (1 + i)^N$$

$$= \frac{m}{-i} + \frac{m}{i}(1 + i)^N = m\left[\frac{(1 + i)^N - 1}{i} \right]$$

Ordinary Annuity

By formula

Let $i = \dfrac{r}{n}$ and $N = nt$ with monthly payment m; then

$$A = m\left[\frac{(1 + i)^N - 1}{i} \right]$$

By table

Use Table 6, Appendix E:

Find this table entry as usual using i and N

$A = m$(Table 6 entry)

EXAMPLE 1 How much do you save in 5 years if you deposit $30 at the end of each month into an account paying 12% compounded monthly?

Solution The rate i is $\frac{0.12}{12} = 0.01$ and $N = 5(12) = 60$. From Table 6 in Appendix E,

$$A = 30(81.669670)$$
$$\approx \$2,450.09$$

∎

EXAMPLE 2 Repeat Example 1 for 2 years.

Solution This time $N = 2(12) = 24$. Table 6 does not have an entry for $N = 24$, so you must use the formula:

$$i = \frac{0.12}{12} = 0.01 \qquad A = 30\left[\frac{(1 + 0.01)^{24} - 1}{0.01}\right]$$

PRESS: $\boxed{1.01}\ \boxed{y^x}\ \boxed{24}\ \boxed{-}\ \boxed{1}\ \boxed{=}\ \boxed{\div}\ \boxed{.01}\ \boxed{=}\ \boxed{\times}\ \boxed{30}\ \boxed{=}$

DISPLAY: 809.203945

You would have $809.20.

∎

The annuities described thus far, in which the deposit is made at the end of each period, are ordinary annuities. Another type of annuity is one in which the payments are made at the *beginning* of each period. These are called **annuities due**. To derive a formula for an annuity due, look at the detail for Table 5.2 and note that if the periodic deposit is put into the account at the beginning of the period instead of at the end, the only difference will be that each exponent is increased by 1 (for one more period). This leads to the formula shown below:

Annuity Due

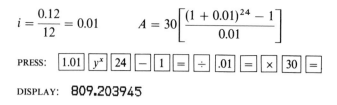

By formula

Let $i = \dfrac{r}{n}$ and $N = nt$ with monthly payment m; then

$$A = m\left[\frac{(1 + i)^{N+1} - 1}{i} - 1\right]$$

By table

Use Table 7, Appendix E:

$A = m(\text{Table 7 entry})$

EXAMPLE 3 What is the value of an annuity due for which $100 per month is deposited into an account paying 18% compounded monthly for $7\frac{1}{2}$ years?

Solution Here, $i = \frac{0.18}{12} = 0.015$ and $N = 7\frac{1}{2}(12) = 90$. Use Table 7 in Appendix E to find: 190.748849. Thus

$$A = \$100(190.748849)$$
$$= \$19,074.88$$

∎

Problem Set 5.3

Find the value of each annuity in Problems 1–23 at the end of the indicated number of years. Assume that the interest is compounded with the same frequency as the deposits. Give your answers to the nearest dollar.

	Amount of deposit	Frequency compounded	Rate	Number of years	Type of annuity
1.	$500	Annually	8%	30	Ordinary
2.	$500	Annually	6%	30	Ordinary
3.	$250	Semiannually	8%	30	Ordinary
4.	$600	Semiannually	12%	10	Ordinary
5.	$300	Quarterly	12%	10	Ordinary
6.	$100	Monthly	12%	5	Ordinary
7.	$500	Annually	8%	30	Due
8.	$500	Annually	6%	30	Due
9.	$250	Semiannually	8%	30	Due
10.	$600	Semiannually	12%	10	Due
11.	$300	Quarterly	12%	10	Due
12.	$100	Monthly	12%	5	Due
13.	$50	Monthly	18%	5	Due
14.	$200	Quarterly	18%	20	Ordinary
15.	$400	Quarterly	16%	20	Ordinary
16.	$2,500	Semiannually	18%	30	Due
17.	$30	Monthly	18%	5	Due
18.	$30	Monthly	18%	5	Ordinary
19.	$5,000	Annually	8%	10	Ordinary
20.	$5,000	Annually	8%	10	Due
21.	$100	Monthly	18%	$7\frac{1}{2}$	Due

	Amount of deposit	Frequency compounded	Rate	Number of years	Type of annuity
22.	$2,500	Semiannually	12%	20	Ordinary
23.	$1,250	Quarterly	12%	20	Due

APPLICATIONS

24. The owner of Sebastopol Tree Farm deposits $650 at the beginning of each quarter for 5 years into an account paying 8% compounded quarterly. What is the value of the account at the end of 5 years?

25. The owner of Oak Hill Squirrel Farm deposits $1,000 at the end of each quarter for 5 years into an account paying 8% compounded quarterly. What is the value at the end of 5 years?

26. John and Rosamond want to retire in 5 years and can save $150 every three months. They plan to deposit the money at the beginning of each quarter into an account paying 8% compounded quarterly. How much will they have at the end of 5 years?

27. You want to retire at age 65. You decide to make a deposit to yourself at the beginning of each year into an account paying 13%, compounded annually. Assuming you are now 25 and can spare $1,200 per year, how much will you have (to the nearest dollar) when you retire? (Alternate question: use your present age.)

28. In 1988 the maximum Social Security deposit was $3,818.10. Suppose you are 25 and make a deposit of this amount into an account at the end of each year. How much would you have (to the nearest dollar) when you retire if the account pays 8% compounded annually and you retire at age 65?

29. Repeat Problem 28 using your own age.

5.4 Amortization

Present Value of an Annuity

Suppose that instead of making monthly payments at periodic intervals, we want to deposit a lump sum today that will have the same value as an annuity at the end of some time period.

EXAMPLE 1   Chen and Mai are partners who have decided to set aside a fund for their secretary who will retire in 10 years. Chen wants to make a $50 monthly deposit into this account while Mai wants to make a lump-sum deposit today, but both want to deposit equal amounts. How much will be in the account in 10 years if the interest rate is 9% compounded monthly, and how much should Mai deposit in order to equal Chen's contribution?

Solution Chen's deposits are an ordinary annuity with $m = \$50, r = 9\%, t = 10, i = \frac{0.09}{12}$, and $N = 10(12) = 120$:

$$A = m\left[\frac{(1+i)^N - 1}{i}\right] = \$50\left[\frac{(1+\frac{0.09}{12})^{120} - 1}{\frac{0.09}{12}}\right]$$

$$= \$9,675.71$$

Mai's contribution is a present value problem. When we speak of the **present value of an annuity**, we mean the lump-sum deposit today that will equal the future value of a given annuity. Using the present value formula from Section 5.2, we find

$$P = A(1+i)^{-N} = 9675.71(1 + \tfrac{0.09}{12})^{-120}$$

$$= \$3,947.08$$

For Chen and Mai to contribute equal amounts, Chen will deposit $50 per month for 10 years and Mai will make a lump-sum contribution of $3,947.08. The total value in the account when the secretary retires will be $9,675.71   ∎

The goal of this section is to develop a formula that will allow us to calculate a lump-sum deposit (see Example 1) *without first finding the value of an annuity.* Recall that $\$P$ deposited today will amount to $A = P(1+i)^N$ (as before, $i = \frac{r}{n}$ and $N = nt$). Substitute this into the ordinary annuity formula with monthly payment m to obtain

$$P(1+i)^N = m\left[\frac{(1+i)^N - 1}{i}\right]$$

Solve this formula for P and simplify to obtain a formula for the present value of an annuity:

$$P = \frac{m}{i}\left[\frac{(1+i)^N - 1}{(1+i)^N}\right]$$

$$= \frac{m}{i}\left[1 - \frac{1}{(1+i)^N}\right] = \frac{m}{i}[1 - (1+i)^{-N}]$$

Present Value of an Annuity

By formula	*By table*
$P = m\left[\dfrac{1-(1+i)^{-N}}{i}\right]$	Use Table 8, Appendix E: $P = m$(Table 8 entry)

EXAMPLE 2 Use the formula for the present value of an annuity to calculate Mai's deposit as described in Example 1. That is, find the present value of an annuity where $m = \$50$, $r = 9\%$, and $t = 10$.

Solution $P = m\left[\dfrac{1-(1+i)^{-N}}{i}\right] = \$50\left[\dfrac{1-(1+\frac{0.09}{12})^{-120}}{\frac{0.09}{12}}\right]$

$$\approx \$3,947.08$$

PRESS: |.09| |÷| |12| |=| |STO| |+| |1| |=| |y^x| |120| |+/−| |=| |+/−| |+| |1|

|=| |÷| |RCL| |×| |50| |=| ■

The calculator steps are rather complicated, so you might want to use Table 8, Appendix E. This procedure is illustrated in Example 3.

EXAMPLE 3 What is the present value of an annuity of $30 per month for 5 years deposited into an account paying 12% compounded monthly?

Solution We have $i = \frac{0.12}{12} = 0.01$ and $N = 5(12) = 60$. From Table 8 in Appendix E,

$$P = 30(44.955038)$$
$$= \$1,348.65114$$

It would require a deposit of $1,348.65. ■

EXAMPLE 4 Teresa's Social Security benefit is $450 per month if she retires at age 62 instead of age 65. What is the present value of an annuity that would pay $450 per month for 3 years if the interest rate is 10% compounded monthly?

Solution Here, $i = \frac{0.10}{12} = 0.008\overline{3}$ and $N = 12(3) = 36$. These values are not included in Table 8, so you must use the formula:

$$P = 450\left[\dfrac{1-(1+i)^{-36}}{i}\right]$$

To use this formula you need a calculator with an exponent key. First, store i in memory:

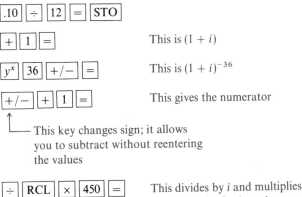

|.10| |÷| |12| |=| |STO|

|+| |1| |=| This is $(1 + i)$

|y^x| |36| |+/−| |=| This is $(1 + i)^{-36}$

|+/−| |+| |1| |=| This gives the numerator

⌐— This key changes sign; it allows you to subtract without reentering the values

|÷| |RCL| |×| |450| |=| This divides by i and multiplies by 450 for the final result

DISPLAY: **13946.056**

The present value is $13,946.06. ■

Amortization

Installment loans are one of the most common examples of the present value of an annuity. The process of paying off a debt by systematically making partial payments until the debt (principal) and the interest are repaid is called **amortization**. If the loan is paid off in regular equal installments, then we can use the formula for the present value of an annuity to find the monthly payments by algebraically solving for m:

$$P = m\left[\frac{1 - (1 + i)^{-N}}{i}\right]$$

Multiply both sides by i:

$$Pi = m[1 - (1 + i)^{-N}]$$

Divide by $[1 - (1 - i)^{-N}]$ to solve for m:

$$\frac{Pi}{1 - (1 + i)^{-N}} = m$$

Installment Payments

The monthly payment, m, for a fully amortized installment loan with annual percentage rate r is found as follows:

By formula

By table

Use the present value of an annuity table, Table 8 in Appendix E:

$$m = \frac{Pi}{1 - (1 + i)^{-N}}$$

$$m = P \div (\text{Table 8 entry})$$

where $i = \dfrac{r}{12}$.

EXAMPLE 5 In 1986 the average price of a new home was \$112,000 and the interest rate was 11%. If this amount is financed for 40 years at 11% interest, what is the monthly payment?

Solution We have $i = \frac{0.11}{12}$, $N = 12(40) = 480$, and $P = 112,000$. Thus

$$m = \frac{112,000i}{1 - (1 + i)^{-480}}$$

PRESS: $\boxed{.11}\ \boxed{\div}\ \boxed{12}\ \boxed{=}\ \boxed{\text{STO}}$ Store i for future use

$\boxed{+}\ \boxed{1}\ \boxed{=}$ Continue to find $(1 + i)$

$\boxed{y^x}\ \boxed{480}\ \boxed{+/-}\ \boxed{=}$ This gives $(1 + i)^{-480}$

$\boxed{+/-}\ \boxed{+}\ \boxed{1}\ \boxed{=}$ This gives the denominator

↑
└── Subtracts without reentering the number

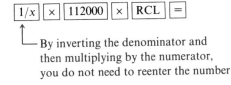

By inverting the denominator and
then multiplying by the numerator,
you do not need to reenter the number

DISPLAY: **1039.68973**

The monthly payment is $1,039.69. This is the payment for interest and principal to pay off a 40-year 11% loan of $112,000. ∎

The amount found in Example 5 is the payment for interest and principal. It does not include taxes and insurance. In passing, it might be interesting to see how much interest is paid for the home loan of Example 5. There are 480 payments of $1,039.69, so the total repaid is

$$480(1039.69) = \$499,051.20$$

Since the loan was for $112,000, the interest is

$$\$499,051.20 - \$112,000 = \$387,051.20$$

The amount of interest paid can be reduced by making a large down payment. For Example 5, if a 20% down payment was made, then $112,000(0.80) = \$89,600$ is the amount to be financed.

Problem Set 5.4

Find the present value of the ordinary annuities in Problems 1–12.

	Amount of deposit	Frequency compounded	Rate	Number of years
1.	$500	Annually	8%	30
2.	$500	Annually	6%	30
3.	$250	Semiannually	8%	30
4.	$600	Semiannually	12%	10
5.	$300	Quarterly	12%	10
6.	$100	Monthly	12%	5
7.	$200	Quarterly	18%	20
8.	$400	Quarterly	16%	20
9.	$30	Monthly	18%	5
10.	$75	Monthly	10%	10
11.	$50	Monthly	11%	30
12.	$100	Monthly	13%	40

Find the monthly payment for the loans in Problems 13–24.

13. $500 loan for 12 months at 12%

14. $100 loan for 18 months at 18%

15. $4,560 loan for 20 months at 21%

16. $3,520 loan for 30 months at 19%

17. Used-car financing of $2,300 for 24 months at 15%

18. New car financing of 2.9% on a 30-month $12,450 loan

19. Furniture financed at $3,456 for 36 months at 23%

20. A refrigerator financed for $985 at 17% for 15 months

21. A $112,000 home bought with a 20% down payment and the balance financed for 30 years at 11.5%

22. A $108,000 home bought with a 30% down payment and the balance financed for 30 years at 12.05%

23. Finance $450,000 for a warehouse with a 12.5% 30-year loan

24. Finance $859,000 for an apartment complex with a 13.2% 20-year loan

25. How much interest would be saved in Problem 21 if the home were financed for 15 rather than for 30 years?

26. How much interest would be saved in Problem 22 if the home were financed for 15 rather than for 30 years?

APPLICATIONS

27. Pat agrees to contribute $500 to the alumni fund at the end of each year for the next 5 years. Karl wants to match Pat's gift, but he wants to make a lump-sum contribution. If the current interest rate is 12.5% compounded annually, how much should Karl contribute to equal Pat's gift?

28. A $1,000,000 lottery prize pays $50,000 per year for the next 20 years. If the current rate of return is 12.25%, what is the present value of this prize?

29. Suppose you have an annuity from an insurance policy and you have the option of being paid $250 per month for 20 years or having a lump-sum payment of $25,000. Which has the most value if the current rate of return is 10% compounded monthly?

30. An insurance policy offers you the option of being paid $750 per month for 20 years or a lump sum of $50,000. Which has the most value if the current rate of return is 9% compounded monthly and you expect to live for 20 years?

5.5 Sinking Funds*

The last financial application we consider is the situation in which we need to have a lump sum of money in a certain period of time. The present value formula will tell us how much we need to have today, but we frequently do not have that amount available. Suppose your goal is $10,000 in 5 years. You can obtain 8% compounded quarterly, so the present value is $6,729.71. However, this is more than you can afford to put into the bank now. The next choice is to make a series of small equal investments to accumulate at 8% compounded quarterly, so that the end result is the same, namely, $10,000 in 5 years. The account you set up to receive those investments is called a **sinking fund**.

To find a formula for a sinking fund, we begin with the formula for an ordinary annuity and solve for m:

$$A = m\left[\frac{(1 + i)^N - 1}{i}\right]$$

$$m[(1 + i)^N - 1] = Ai$$

$$m = \frac{Ai}{(1 + i)^N - 1}$$

Sinking Fund	*By formula*	*By table*
		Use the ordinary annuity table, Table 6, Appendix E:
	$m = \dfrac{Ai}{(1 + i)^N - 1}$	$m = A \div$ (Table 6 entry)

EXAMPLE 1   A business needs to raise $1,000,000 in 20 years by making equal quarterly payments into an account paying 12% interest compounded quarterly. What is the required amount of each deposit?

---

* This section is not required for subsequent textual development.

Solution   Here, $i = \frac{0.12}{4} = 0.03$ and $N = 4(20) = 80$. From Table 6 in Appendix E, we obtain 321.363019. Thus

$$m = 1,000,000 \div 321.363019$$
$$= \$3,111.745723$$

The business needs to make quarterly deposits of $3,112 (rounded to the nearest dollar). ■

    The primary reason that businesses set up a sinking fund is to pay off bonds. A **bond** is a certificate (a written promise) of a business to repay a certain amount at some future time. It is often how a business, corporation, or government agency borrows to buy new equipment or raise money for construction. Usually, each bond has a face value of $1,000 and a specified interest rate, called the *contract rate*. This interest is paid to the bondholders at specified intervals (usually twice a year). The bonds are usually sold to an underwriter, who sells them to investors. Since the contract rate is fixed but prevailing interest rates are not, bonds are bought and sold for either more or less than the face value. This, in effect, changes the interest rate actually received (because of the changes in the principal). This interest rate is called the *market rate* or *effective rate*. Bond prices are quoted as a percent of their face amounts. A quote of 103 is 103% of the face amount, and a quote of $85\frac{1}{2}$ is $85\frac{1}{2}\%$ of the face amount.

EXAMPLE 2   The Packard-Hue Corporation issues $50,000,000 worth of bonds for a new plant. These bonds are 20-year bonds with interest payable semiannually at a contract rate of 6%. In addition to paying interest on the bonds, the company sets up a sinking fund into which they will make semiannual payments and receive 8% interest compounded semiannually. What is the amount of each semiannual payment necessary to pay the interest on the bonds and for the sinking fund?

Solution   *Bond interest*:   This is simple interest because interest is paid to bondholders and does not accumulate.

                ┌──── Semiannual payment

$$I = Prt$$
$$= \$50,000,000(0.06)(\tfrac{1}{2})$$
$$= \$1,500,000$$

Sinking fund payment: $i = \frac{0.08}{2} = 0.04$ and $N = 2(20) = 40$; thus

$$m = \$50,000,000 \div 95.025516 \longleftarrow \text{Table 6 entry}$$
$$\approx \$526,174 \qquad \text{Rounded to the nearest dollar}$$

The total semiannual cost is $1,500,000 + $526,174 = $2,026,174 ■

Summary

For many students, the most difficult part of working with business formulas is determining *which* formula or *which* table to use. To make the processes easier for you, consistent notation and consistently constructed tables are used here. Your main task, therefore, is one of classification, summarized in Table 5.3.

TABLE 5.3 Classification and formulas for financial problems

DEFINITION OF VARIABLES:

P = Present value (sometimes called principal)

A = Future value

I = Amount of interest

r = Annual interest rate

t = Number of years

n = Periods; that is, the number of times per year that the interest is compounded

m = Periodic payment

i = Rate per period $= \dfrac{r}{n}$

N = Number of periods $= nt$

FINANCIAL TABLES (Appendix E): *For all tables use i and N to find entry.*

Future value (Table 5): Lump sum known, to find future lump sum

Present value (Table 5): Future lump sum known, to find present value

Ordinary annuity (Table 6): Periodic payments at the end of each period, to find the future value

Annuity due (Table 7): Periodic payments at the beginning of each period, to find the future value

Present value of an annuity (Table 8): Find the present lump sum necessary to deposit to equal the future value of periodic payments for a given period of time

Installment payments (Table 8): Find the periodic payments to pay off both principal and interest in a given period of time

Sinking fund (Table 6): Find the amount of periodic payment in order to have a given amount at some future date

Problem Set 5.5

APPLICATIONS

Find the amount of payment necessary for each deposit to a sinking fund in Problems 1–12. Assume that the deposit is made at the end of each compounding period.

	Amount needed	Time	Interest rate	Compounded
1.	$7,000	5 years	8%	Annually
2.	$25,000	5 years	11%	Annually
3.	$25,000	5 years	12%	Semiannually
4.	$50,000	10 years	14%	Semiannually

	Amount	Time	Rate	Compounded
5.	$165,000	10 years	12%	Semiannually
6.	$3,000,000	20 years	12%	Semiannually
7.	$500,000	10 years	12%	Quarterly
8.	$55,000	5 years	16%	Quarterly
9.	$100,000	8 years	10%	Quarterly
10.	$35,000	12 years	18%	Quarterly
11.	$45,000	6 years	18%	Monthly
12.	$120,000	30 years	14%	Monthly

Classification of Types:

LUMP SUM					
Present value, P	Future value, A	Classification	Formula		Table
Known	Unknown	Future value	$A = P(1 + i)^N$		$A = P$(Table 5 entry)
Unknown	Known	Present value	$P = \dfrac{A}{(1 + i)^N}$		$P = A \div$ (Table 5 entry)

PERIODIC PAYMENTS					
Periodic payment, m	Present value, P	Future value, A	Classification	Formula	Table
Known		Unknown	Ordinary annuity (end of each period)	$A = m\left[\dfrac{(1 + i)^N - 1}{i}\right]$	$A = m$(Table 6 entry)
Known		Unknown	Annuity due (start of each period)	$A = m\left[\dfrac{(1 + i)^{N+1} - 1}{i} - 1\right]$	$A = m$(Table 7 entry)
Known	Unknown*		Present value of an annuity	$P = m\left[\dfrac{1 - (1 + i)^{-N}}{i}\right]$	$P = m$(Table 8 entry)
Unknown	Known		Installment payment	$m = \dfrac{Pi}{1 - (1 + i)^{-N}}$	$m = P \div$ (Table 8 entry)
Unknown		Known	Sinking fund	$m = \dfrac{Ai}{(1 + i)^N - 1}$	$m = A \div$ (Table 6 entry)

* Amount needed to deposit today to equal the future value of an annuity; this is the amount you can borrow with a given monthly payment on an installment loan.

13. Clearlake Optical has a $50,000 note that comes due in 4 years. The owners wish to create a sinking fund to pay this note. If the fund earns 8% compounded semi-annually, how much should each semiannual deposit be?

14. A business must raise $70,000 in 5 years. What should be the size of the owners' quarterly payment to a sinking fund paying 8% compounded quarterly?

15. Clearlake Optical has developed a new lens. The owners plan on issuing a $4,000,000 30-year bond with a contract rate of $5\frac{1}{2}\%$ paid annually to raise capital to market this new lens. To pay off the debt, they will also set up a sinking fund paying 8% interest compounded annually. What size annual payment is necessary for interest and sinking fund combined?

16. The owners of Bardoza Greeting Cards wish to introduce a new line of cards but need to raise $200,000 to do it. They decide to issue 10-year bonds with a contract rate of 6% paid semiannually. They also set up a sinking fund paying 8% interest compounded semiannually. How much money will they need to make the semiannual interest payments as well as payments to the sinking fund?

Problems 17–30 provide a mixture of financial problems. For each problem: a. Classify the type. b. Answer the questions. (Round to the nearest dollar.)

17. Rincon Bowling Alley will need $80,000 in 4 years to resurface the lanes. What lump sum would be necessary today if the owner of the business can deposit it in an account that pays 9% compounded semiannually?

18. Rita wants to save for a trip to Tahiti, so she puts $2.00 per day into a jar. After 1 year she has saved $730 and puts the money into a bank account paying 10% compounded annually. She continues to save in this manner and makes her annual $730 deposit for 15 years. How much does she have at the end of that time period?

19. Karen receives a $12,500 inheritance that she wants to save until she retires in 22 years. If she deposits the money in a fixed 11% account, compounded daily, how much will she have when she retires?

20. You want to give your child $1,000,000 when he retires at age 65. How much money do you need to deposit to an account paying 9% compounded monthly if your child is now 10 years old?

21. An accounting firm agrees to purchase a computer for $150,000 (cash on delivery) and the delivery date is in 270 days. How much do the owners need to deposit in an account paying 18% compounded quarterly?

22. For 5 years, Thompson Cleaners deposits $900 at the beginning of each quarter into an account paying 8% compounded quarterly. What is the value of the account to the nearest dollar at the end of 5 years?

23. In 1980 the inflation rate hit 16%. Suppose that the average cost of a textbook in 1980 was $15. What is the expected cost in the year 2000 if we project this rate of inflation on the cost?

24. What is the necessary amount of monthly payments to an account paying 18% compounded monthly in order to have $100,000 in $8\frac{1}{3}$ years if the deposits are made at the end of the month?

25. Thomas' Grocery Store is going to be remodeled in 5 years, and the remodeling will cost $300,000. How much should be deposited now (to the nearest dollar) in order to pay for this remodeling if the account pays 12% compounded monthly?

26. If an apartment complex will need painting in $3\frac{1}{2}$ years and the job will cost $45,000, what amount needs to be deposited into an account now in order to have the necessary funds? The account pays 12% interest compounded semiannually.

27. Teal and Associates needs to borrow $45,000. The best loan they can find is one at 12% that must be repaid in monthly installments over the next $3\frac{1}{2}$ years. How much are the monthly payments?

28. A city issues $20 million in tax-exempt 25-year bonds with 8% interest payable quarterly. In addition to paying interest on these bonds, the city sets up an account into which quarterly payments are made and 12% interest compounded quarterly is received. How much needs to be paid into this account to pay off the $20 million in 25 years?

29. Certain Concrete Company deposits $4,000 at the end of each quarter into an account paying 10% interest compounded quarterly. What is the value of the account at the end of $7\frac{1}{2}$ years?

30. Major Magic Corporation deposits $1,000 at the beginning of each month into an account paying 18% interest compounded monthly. What is the value of the account at the end of $8\frac{1}{3}$ years?

5.6 Summary and Review

IMPORTANT TERMS

Amortization [5.4]
Annuity [5.3]
Annuity due [5.3]
Arithmetic sequence [5.1]
Bond [5.5]
Compound interest formula [5.2]
Effective annual yield [5.2]
Effective rate [5.2]
Exact interest [5.2]
First-order difference equation [5.1]
Future value [5.2]
Geometric sequence [5.1]

Inflation [5.2]
Installment loan [5.4]
Interest [5.2]
Ordinary annuity [5.3]
Ordinary interest [5.2]
Present value [5.2]
Present value of an annuity [5.4]
Principal [5.2]
Rate of interest [5.2]
Sigma notation [5.1]
Simple interest formula [5.2]
Sinking fund [5.5]

SAMPLE TEST   *For additional practice there are a large number of review problems categorized by objective in the Student Solutions Manual. The following Sample Test (40 minutes) is intended to review the main ideas of this chapter.*

In Problems 1–17 classify the type of financial problem, state the appropriate formula, identify the variables and constants, and then answer the question.

1. Find the value of a $1,000 certificate in $2\frac{1}{2}$ years if the interest rate is 12% compounded monthly.

2. You deposit $300 at the end of each year into an account paying 12% compounded annually. How much is in the account in 10 years?

3. An insurance policy pays $10,000 in 5 years. What lump-sum deposit today will yield $10,000 in 5 years if it is deposited at 12% compounded quarterly?

4. A 5-year term policy has an annual premium of $300, and at the end of 5 years, all payments and interest are refunded. What lump-sum deposit is necessary to equal this amount if you assume an interest rate of 10% compounded annually?

5. What annual deposit is necessary to give $10,000 in 5 years if the money is deposited at 9% interest compounded annually?

6. A $5,000,000 apartment complex loan is to be paid off in 10 years by making 10 equal annual payments. How much is each payment if the interest rate is 14% compounded annually?

7. The prices of automobiles have increased at 6.25% per year. How much would you expect a $10,000 automobile to cost in 5 years?

8. The amount to be financed on a new car is $9,500. The terms are 17% for 4 years. What is the monthly payment?

9. At the beginning of each year, you deposit $450 into an account paying 11% compounded annually. How much is in the account in 25 years?

10. At the end of each year, you deposit $825 into an account paying 7.5% compounded annually. How much is in the account in 23 years?

11. What lump-sum deposit today will yield $1,000,000 in 37 years if it is deposited at 12% compounded annually?

12. What deposit today is equal to 33 annual deposits of $500 into an account paying 8% compounded annually?

13. What annual deposit is necessary into an account paying 11.5% to give $5,000 in 15 years?

14. You want to have $10,000 in 6 years by making a deposit at the end of each year into an account paying 12% interest compounded annually. How much should your annual deposit be?

15. A lottery offers you a choice of $1,000,000 per year for 5 years or a lump-sum payment. What lump-sum payment would equal the annual payments if the current interest rate is 14% compounded annually?

16. What is the monthly payment for a home costing $125,000 with a 20% down payment and the balance financed for 30 years at 12%?

17. Western Electric decides to raise $150,000,000 in order to develop a new geothermal energy source. They issue a 50-year bond with a contract rate of 9% paid annually. They also set up a fund to repay the debt. This fund pays 10% compounded annually. What is the size of the annual payment for both the interest and the bond repayment?

The following multiple-choice questions have appeared on recent Management accounting and CPA examinations. The questions are reprinted with permission of the National Association of Accountants and the American Institute of Certified Public Accountants, Inc.

Management Accounting Exam June 1983

18. A firm has daily cash receipts of $200,000. A commercial bank has offered to reduce the collection time by 3 days. The bank requires a monthly fee of $4,000 for providing this service. If money market rates will average 12% during the year, the additional annual income (loss) of having the service is
    A. $(24,000)   **B.** $24,000   **C.** $66,240   **D.** $68,000
    E. Some amount other than those given above

CPA Exam November 1976

19. A businesswoman wants to withdraw $3,000 (including principal) from an investment fund at the end of each year for 5 years. How should she compute her required initial investment at the beginning of the first year if the fund earns 6% compounded annually?

    A. $3,000 times the amount of an annuity of $1 at 6% at the end of each year for 5 years
    B. $3,000 divided by the amount of an annuity of $1 at 6% at the end of each year for 5 years
    C. $3,000 times the present value of an annuity of $1 at 6% at the end of each year for 5 years
    D. $3,000 divided by the present value of an annuity of $1 at 6% at the end of each year for 5 years

CPA Exam November 1976

20. A businessman wants to invest a certain sum of money at the end of each year for 5 years. The investment will earn 6% compounded annually. At the end of 5 years, he will need a total of $30,000 accumulated. How should he compute his required annual investment?

    A. $30,000 times the amount of an annuity of $1 at 6% at the end of each year for 5 years
    B. $30,000 divided by the amount of an annuity of $1 at 6% at the end of each year for 5 years
    C. $30,000 times the present value of an annuity of $1 at 6% at the end of each year for 5 years
    D. $30,000 divided by the present value of an annuity of $1 at 6% at the end of each year for 5 years

Modeling Application 3

Investment Opportunities

You have just inherited $30,000 and need to decide what to do with the money. You check around and find that the current interest rate for deposits is 13.58%, while for loans it ranges from 14% to 21%. Inflation has been around 4%, but long-range estimates are that it will average about 10%. For this problem, assume that you have no unusual debts, are 30 years old, and have a home worth $60,000 with a $30,000 mortgage at an 8% interest rate with payments of $225 per month for the next 25 years.

Write a paper discussing your investment alternatives. Omit any discussion of tax considerations. For general guidelines about writing this essay, see the commentary for Modeling Application 1 on page 130.

CHAPTER 6

Combinatorics

CHAPTER CONTENTS

6.1 Sets and Set Operations
6.2 Permutations
6.3 Combinations
6.4 Binomial Theorem
6.5 Summary and Review
    Important Terms
    Sample Test

APPLICATIONS

Management (*Business, Economics, Finance, and Investments*)

Analysis of sales representatives (6.1, Problem 55)
Comparison of inflation and unemployment rates (6.1, Problem 56)
Survey of age and owning a car (6.1, Problem 57)
Different categories of employees within the Ampex Corp. (6.1, Problem 58)
Survey of a national board (6.1, Problem 60)
Ways of rating businesses (6.3, Problem 34)
Grouping stocks (6.3, Problem 37)
Likelihood of an IRS investigation (6.3, Problem 39)

Life sciences (*Biology, Ecology, Health, and Medicine*)

Surveys (6.1, Problems 62–64)
Menu selections (6.2, Problem 43)
Human blood types (6.5, Problem 15)

Social sciences (*Demography, Political Science, Population, Psychology, Society, and Sociology*)

Surveys (6.1, Problems 65–66)
Election problem (6.2, Problems 34–36; 6.5, Problems 10–11)
Social Security numbers (6.2, Problem 44)
License plate problem (6.2, Problems 45–49, 51)
Committee problem (6.3, Problems 13, 32–33, 42–43)
Grouping individuals in a psychology experiment (6.3, Problem 35)

General interest

Likelihood of purchasing a defective tire (6.1, Problem 59)
Historical questions dealing with John Venn (6.1, Problem 71)
Selecting TV contestants (6.2, Problem 41)
Combination locks (6.2, Problem 50)
Transfinite numbers (6.2, Problem 53)
Ways of obtaining different poker hands (6.3, Problems 14–17, 22)
Ways of answering a multiple choice test (6.3, Problem 38)
Pascal's triangle (6.4, Problems 39–42)
Dealing cards to bridge players (6.5, Problem 8)

Modeling application—Earthquake Epicenter

CHAPTER OVERVIEW This chapter focuses on sets and counting techniques. In addition, you will learn about the fundamental counting principle and two very useful counting models, permutations and combinations. This chapter concludes with the binomial theorem.

PREVIEW The idea of a set is one of those very simple, and profound, concepts that is used to simplify and unify mathematics. After looking at sets, set operations, and set relationships in the first section, we move to the notion of counting and develop the ideas of permutations, combinations, and the fundamental counting principle. Make a special note of the box on page 255 because being able to tell *how* to proceed is just as important as carrying out the actual process.

PERSPECTIVE We all learned to count in a straightforward "one, two, three,..." procedure. But in mathematics, especially when working with probabilities, it is necessary to count the number of elements in involved sets, and straightforward counting is often difficult or even impossible. This chapter involves preliminary, but very necessary, work for the next chapter, on probability.

6.1 Sets and Set Operations

The concept of sets and set theory is attributed to Georg Cantor (1845–1918), a great German mathematician. Cantor's ideas were not adopted as fundamental underlying concepts in mathematics until the twentieth century, but today the idea of set is used to unify and explain nearly every other concept in mathematics. Recall that a counting number is an element of the set $\{1, 2, 3, \ldots\}$.

Sets are specified by roster or by description. In the **roster method**, sets are specified by listing the elements and enclosing those elements between braces: $\{\ \}$. In the **description method**, sets are described in words, such as

The set of subscribers to the *Wall Street Journal*

or by using **set-builder notation**, such as

$\{s \mid 0 < s < 100, s \text{ a counting number}\}$

The proper way to read the set-builder notation here is: "the set of all counting numbers, s, such that s is between 0 and 100."

If S is a set, we write $a \in S$ if a is a member of the set S and $b \notin S$ if b is not a member of S. For example, let C be the set of cities in California. If a represents the city of Anaheim and b stands for the city of Berlin, Germany, we would write

$a \in C$ and $b \notin C$

Two special sets require attention. The first is the set that contains no elements and the second is a set that contains every element under consideration:

Empty and Universal Sets

The **empty set** contains no elements and is denoted by $\{\ \}$ or $\varnothing$.

The **universal set** contains all the elements under consideration in a given discussion and is denoted by U.

For example, if we agree that $U = \{1, 2, 3, 4, 5, 6, 7, 8, 9\}$, then all sets that are considered would have elements only among the elements of U. No set could contain the number 10 since 10 is not included in what we have decided is the universe. For every problem, a universal set must be specified or implied, and it must remain fixed for that problem. However, when a new problem is begun, a new universal set can be specified.

EXAMPLE 1   Let U be the set of counting numbers. Write the given sets by roster.
$$A = \{x \mid (x - 2)(x - 3) = 0\}$$
$$B = \{y \mid 3 \le y < 7\}$$
$$C = \{y \mid y^2 = 4\}$$

Solution   By roster,
$$A = \{2, 3\}$$
$$B = \{3, 4, 5, 6\}$$
$$C = \{2\} \quad \textit{Note:} \quad \text{The number } -2 \text{ also makes the statement } y^2 = 4 \text{ true, but } -2$$
$$\text{is not in the universe for this problem} \quad \blacksquare$$

Let us now take a look at certain relationships among sets.

Subset | A set A is a **subset** of a set B, denoted by $A \subseteq B$, if every element of A is an element of B.

Consider the following sets:
$$U = \{1, 2, 3, 4, 5, 6, 7, 8, 9\}$$
$$A = \{2, 4, 6, 8\}$$
$$B = \{1, 3, 5, 7\}$$
$$C = \{5, 7\}$$
A, B, and C are subsets of the universal set U (all sets considered are subsets of the universe, by definition of the universal set). Also, $C \subseteq B$ since every element in C is also an element of B.

However, C is not a subset of A. To substantiate this claim, we must show that there is some member of C that is not a member of A. In this case we merely note that $5 \in C$ and $5 \notin A$.

Do not confuse the notions of "element" and "subset." That is, 5 is an *element* of C since it is listed in C; $\{5\}$ is a *subset* of C but is not an element since we do not find $\{5\}$ contained in C—if we did, C might look like $\{5, \{5\}, 7\}$.

To find all the possible subsets of the set $C = \{5, 7\}$ we use what is called a **tree diagram**, which is a device that helps us enumerate a list of possibilities:

Number in Subset?

5	7	Subset
Yes	Yes	$\{5, 7\}$
	No	$\{5\}$
No	Yes	$\{7\}$
	No	$\{\ \}$ or $\varnothing$

The subsets of C are: $\{5, 7\}, \{5\}, \{7\}, \varnothing$. Note that the set C has only two elements, but it has four subsets. Certainly, then, you must be careful to distinguish between a subset and an element.

The subsets of C can be classified into two categories: **proper** and **improper**.

Proper subsets of C	Improper subset of C
$\{5\}$ $\{7\}$ $\{\ \}$	$\{5, 7\}$

Every set has just one improper subset, and that is the set itself. All other subsets are proper subsets. The notation for a proper subset (as distinguished from an improper subset) does not include the small equality symbol; that is,

$A \subset B$ Used to denote the idea that A
⌐———— is a *proper subset* of B

The notion of subset can be used to define the equality of two sets. Two sets are equal if they have exactly the same elements:

Equality of Sets   Sets A and B are *equal*, denoted by $A = B$, if $A \subseteq B$ and $B \subseteq A$.

A useful way to depict relationships among sets is to represent the universal set by a rectangle, with the proper subsets in the universe represented by circular or oval-shaped regions, as in Figure 6.1.

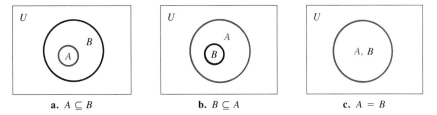

Figure 6.1 Venn diagrams showing subset and equal relationships

a. $A \subseteq B$ **b.** $B \subseteq A$ **c.** $A = B$

These figures are called **Venn diagrams**, after John Venn (1834–1923). The Swiss mathematician Leonhard Euler (1707–1783) also used circles to illustrate principles of logic, so sometimes these diagrams are called *Euler circles*. However, Venn was the first person to use them in a general way.

We can also illustrate other relationships between two sets: A and B may have no elements in common, in which case they are **disjoint**, or they may overlap and have some elements in common (Figure 6.2).

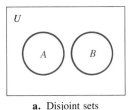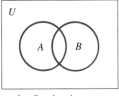

Figure 6.2 Venn diagrams for disjoint and overlapping sets

a. Disjoint sets   **b.** Overlapping sets

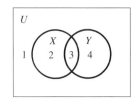

Figure 6.3 Venn diagram showing two general sets

Sometimes we are given two sets X and Y and we know nothing about how they are related. In this situation we draw a general figure, such as the one shown in Figure 6.3.

Note that these circles divide the universe into four disjoint regions. When sets are drawn in this manner it does not mean that the only possibility is overlapping sets. For example:

If $X \subseteq Y$, then region 2 is empty.
If $Y \subseteq X$, then region 4 is empty.
If $X = Y$, then regions 2 and 4 are both empty.
If X and Y are disjoint, then region 3 is empty.

Three common operations performed on sets are union, intersection, and complementation.

Union is an operation for sets A and B in which a set is formed that consists of all the elements in A or B or both. The symbol for the operation of union is $\cup$, and we write $A \cup B$.

Union of Sets

> The *union* of sets A and B, denoted by $A \cup B$, is the set consisting of all elements of A or B or both.
>
> $$A \cup B = \{x \mid x \in A \text{ or } x \in B \text{ or } x \in (\text{both } A \text{ and } B)\}$$

EXAMPLE 2 Let $U = \{1, 2, 3, 4, 5, 6, 7, 8, 9\}$, $A = \{2, 4, 6, 8\}$, $B = \{1, 3, 5, 7\}$, $C = \{5, 7\}$. Then:

a. $A \cup C = \{2, 4, 5, 6, 7, 8\}$; that is, the union of A and C is the set consisting of all elements in A or in C or in both.

b. $B \cup C = \{1, 3, 5, 7\}$; note that, even though the elements 5 and 7 appear in both sets, they are listed only once. That is, the sets $\{1, 3, 5, 7\}$ and $\{1, 3, 5, 5, 7, 7\}$ are equal (exactly the same).

c. $A \cup B = (1, 2, 3, 4, 5, 6, 7, 8\}$.

d. $(A \cup B) \cup \{9\} = \{1, 2, 3, 4, 5, 6, 7, 8, 9\} = U$; here we are considering the union of three sets; however, the parentheses indicate the operation that should be performed first. Note also that, because the solution $\{1, 2, 3, 4, 5, 6, 7, 8, 9\}$ has a name, we write down the name rather than the set. That is, $(A \cup B) \cup \{9\} = U$ is the simpler representation. ∎

We can use Venn diagrams to illustrate union. In Figure 6.4 we first shade A and then shade B. *The union is all parts that have been shaded at least once.*

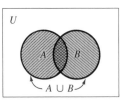

Figure 6.4 Venn diagram showing the union of two sets (color shading)

A second operation for sets is called **intersection**.

Intersection of Sets

> The *intersection* of sets A and B, denoted by $A \cap B$, is the set consisting of all elements common to A and B.
> $$A \cap B = \{x \mid x \in A \quad and \quad x \in B\}$$

EXAMPLE 3   Let $U = \{a, b, c, d, e\}$, $A = \{a, c, e\}$, $B = \{c, d, e\}$, $C = \{a\}$, and $D = \{e\}$. Then:

a. $A \cap B = \{c, e\}$; that is, the intersection of A and B is the set consisting of elements in both A and B.

b. $A \cap C = C$ since $A \cap C = \{a\}$ and $\{a\} = C$; we write down the name for the set that is the intersection.

c. $B \cap C = \emptyset$ since B and C have no elements in common (they are disjoint).

d. Parentheses tell us which operation to do first:

$$(A \cap B) \cap D = \{c, e\} \cap \{e\}$$
$$= \{e\} \qquad \text{\{e\} is the set of elements common to the sets \{c, e\}}$$
$$= D \qquad \text{and \{e\}} \qquad \blacksquare$$

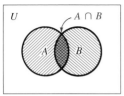

Figure 6.5 Venn diagram showing the intersection of two sets (color shading)

Intersection of sets can also be easily shown in a Venn diagram. To find the intersection of two sets A and B, first shade A and then shade B; *the intersection is all parts shaded twice* (Figure 6.5).

Complementation is an operation on a set that must be performed in relation to the universal set.

Complementation

> The *complement* of a set A, denoted by $\bar{A}$, is the set of all elements of U that are not in the set A.
> $$\bar{A} = \{x \mid x \notin A\}$$

EXAMPLE 4   Let $U = \{\text{People in California}\}$, $A = \{\text{People who are over 30}\}$, $B = \{\text{People who are 30 or under}\}$, and $C = \{\text{People who own a car}\}$. Then:

a. $\bar{C} = \{\text{Californians who do not own a car}\}$
b. $\bar{A} = B$
c. $\bar{B} = A$
d. $\bar{U} = \emptyset$
e. $\bar{\emptyset} = U$ $\blacksquare$

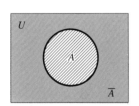

Figure 6.6 Venn diagram showing the complement of a set A (color shading)

Complementation can be shown in a Venn diagram. The color shading in Figure 6.6 shows the complement of A. In a Venn diagram, the complement is everything in U that is not in the given set (in this case, everything not in A).

The real payoff for studying these relationships comes when we have combined operations or several sets at the same time.

EXAMPLE 5   Illustrate $\overline{A \cup B}$ in a Venn diagram.

Solution   This is a combined operation. *First*, find $A \cup B$; *then* find the complement.

Step 1

Step 2

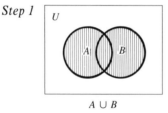

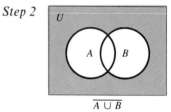

These steps are generally combined into one diagram:

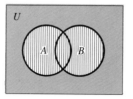

The answer is shown by color shading. The vertical lines show the preliminary step. It is generally a good idea to show the final answer (the part in color) by using a highlighter pen.   ■

EXAMPLE 6   Illustrate $\bar{A} \cup \bar{B}$ in a Venn diagram.

Solution   Compare this statement with Example 5. Here, we *first* find $\bar{A}$ and $\bar{B}$; *then* find the union.

Step 1

Step 2

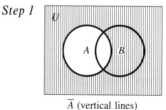

$\bar{A}$ (vertical lines)

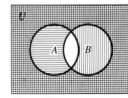

$\bar{A}$ with $\bar{B}$ (horizontal lines)

Your work will normally show the above steps with one diagram.

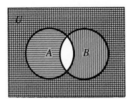

$\bar{A} \cup \bar{B}$ (color shading; the union is all parts that have horizontal or vertical lines or both)

Notice that $\overline{A \cup B} \neq \bar{A} \cup \bar{B}$. If they were equal, the final color portions of the Venn diagrams from Examples 5 and 6 would have been the same.

The *order* of operations for sets is left to right, unless parentheses indicate otherwise. Operations within parentheses are performed first, as shown by Example 7.

EXAMPLE 7   Illustrate $(A \cup \bar{C}) \cap \bar{C}$ in a Venn diagram.

Solution

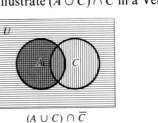

$(A \cup C) \cap \bar{C}$

Detail:

$A \cup C$: Vertical lines
$\bar{C}$: Horizontal lines
$(A \cup C) \cap \bar{C}$: The intersection of vertical and horizontal lines is the part shaded in color.

■

Sometimes you will be asked to consider relationships among three sets. The general Venn diagram for this is shown in Figure 6.7. Note that three sets divide the universe into eight regions. Can you number each?

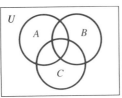

Figure 6.7 Three general sets

EXAMPLE 8   Using Figure 6.7 as a guide, shade in each of the following sets:
  a. $A \cup B$ **b.** $\overline{A \cap C}$ **c.** $B \cap C$
  d. $\bar{A}$ **e.** $\overline{A \cup B}$ **f.** $A \cap B \cap C$ **g.** $A \cup B \cup C$

Solution

a.

$A \cup B$

b.

$A \cap C$

c.

$B \cap C$

d.

$\bar{A}$

e.

$\overline{A \cup B}$

f.

$A \cap B \cap C$

g.

$A \cup B \cup C$

Venn diagrams are also used for survey problems. Suppose W. B. Blaire, a marketing research firm, conducts a survey of people buying electronic calculators during a week. It is found that 48 people bought machines with paper tape printout

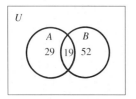

Figure 6.8 Venn diagram for sets *A* and *B*. Note that since 19 is in the intersection, then $48 - 19 = 29$ is in the remaining part of *A*; similarly, there is $71 - 19 = 52$ in the remaining part of *B*

and 71 bought machines without tape printout. How many people purchased machines?

The "obvious" answer of $48 + 71 = 119$ is not so obvious if you consider that some people who were surveyed may have purchased both types of machines. So, let

$A = \{x \mid x$ purchased an electronic calculator with paper tape$\}$

$B = \{x \mid x$ purchased an electronic calculator without paper tape$\}$

Also, **let $n(A)$ denote the number of elements** in set *A*. Thus $n(A) = 48$. Also, $n(B) = 71$. Now, $n(A \cup B) = 48 + 71 = 119$ if $A \cap B = \varnothing$. But suppose it is found that 19 persons purchased both types of machines. That is, $n(A \cap B) = 19$. Consider the Venn diagram shown in Figure 6.8.

We see that if we seek $n(A \cup B)$, we need to count the number of elements in *A* plus the number of elements in *B*. But when we do this, we have counted the number of elements in the intersection twice. Therefore

$$n(A \cup B) = n(A) + n(B) - n(A \cap B)$$
$$= 48 + 71 - 19$$
$$= 100$$

In general, for *any* sets *X* and *Y*, the number of elements in the union is the sum of the number of elements in each of the individual sets, *minus* the number of elements in the intersection, as shown in Figure 6.9.

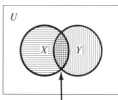

This part has been counted twice

Figure 6.9

Number of Elements in the Union of Sets

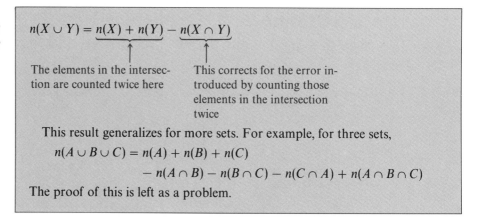

$$n(X \cup Y) = \underbrace{n(X) + n(Y)} - \underbrace{n(X \cap Y)}$$

The elements in the intersection are counted twice here

This corrects for the error introduced by counting those elements in the intersection twice

This result generalizes for more sets. For example, for three sets,

$$n(A \cup B \cup C) = n(A) + n(B) + n(C)$$
$$- n(A \cap B) - n(B \cap C) - n(C \cap A) + n(A \cap B \cap C)$$

The proof of this is left as a problem.

EXAMPLE 9   **Survey**   A survey of 470 students gives the following information:

45 students are taking finite math

41 students are taking statistics

40 students are taking computer programming
15 students are taking finite math and statistics
18 students are taking finite math and computer programming
17 students are taking statistics and computer programming
 7 students are taking all three

a. How many students are taking only finite math?
b. How many students are taking only statistics?
c. How many students are taking only computer programming?
d. How many students are not taking any of these courses?

Solution   The method of solution is to draw a Venn diagram and fill in the various regions. Let

$F = \{x \mid x \text{ is taking finite math}\}$

$S = \{x \mid x \text{ is taking statistics}\}$

$C = \{x \mid x \text{ is taking computer programming}\}$

Step 1   Fill in $F \cap S \cap C$, which is given: $n(F \cap S \cap C) = 7$.

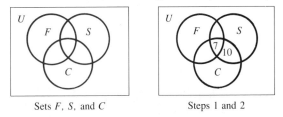

Sets F, S, and C Steps 1 and 2

Step 2   Fill in $S \cap C$, which is given: $n(S \cap C) = 17$. But 7 has previously been accounted for in this region, so we need only account for 10 additional members.

Step 3   Fill in $n(F \cap C) = 18$. Need to fill in only 11 additional members.

Step 4   Fill in $n(F \cap S) = 15$. Need to fill in only 8 additional members.

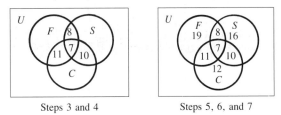

Steps 3 and 4   Steps 5, 6, and 7

Step 5   Fill in $n(C) = 40$. Note that 28 has previously been filled in, so we need an additional 12 members.

Step 6   Fill in $n(S) = 41$. Need to fill in 16 members.

Step 7   Fill in $n(F) = 45$. Need to fill in 19 members.

Step 8   Adding all the numbers in the diagram, we have accounted for 83 members. Since 470 students were surveyed, we see that 387 are not taking any of the three courses.

We now have all the answers to the questions directly from the Venn diagram:

a. 19   **b.** 16   **c.** 12   **d.** 387   ■

Problem Set 6.1

1. Explain the difference between *element of a set* and *subset of a set*. Give examples.

2. Explain the difference between *subset* and *proper subset*.

3. Explain the difference between *universal set* and *empty set*.

4. In your own words, define *union, intersection*, and *complement*. Illustrate with examples.

List all subsets of each set given in Problems 5–12.

5. $\{m, y\}$ 6. $\{4, 5\}$ 7. $\{y, o, u\}$

8. $\{6, 9, 0\}$ 9. $\{3, 6, 9\}$ 10. $\{m, a, t, h\}$

11. $\{1, 2, 3, 4\}$ 12. $\{a, b, c, d\}$

Perform the set operations in Problems 13–22. Let $U = \{1, 2, 3, 4, 5, 6, 7, 8, 9, 10\}$.

13. $\{2, 6, 8\} \cup \{6, 8, 10\}$

14. $\{2, 6, 8\} \cap \{6, 8, 10\}$

15. $\{1, 2, 3, 4, 5\} \cap \{3, 4, 5, 6, 7\}$

16. $\{1, 2, 3, 4, 5\} \cup \{3, 4, 5, 6, 7\}$

17. $\{2, 5, 8\} \cup \{3, 6, 9\}$

18. $\{2, 5, 8\} \cap \{3, 6, 9\}$

19. $\overline{\{2, 8, 9\}}$

20. $\overline{\{1, 2, 5, 7, 9\}}$

21. $\overline{\{8\}}$

22. $\overline{\{6, 7, 8, 9, 10\}}$

Let $U = \{1, 2, 3, 4, 5, 6, 7\}$, $A = \{1, 2, 3, 4\}$, $B = \{1, 2, 5, 6\}$, and $C = \{3, 5, 7\}$. List all the members of each of the sets in Problems 23–38.

23. $A \cup B$ 24. $A \cup C$ 25. $B \cup C$

26. $A \cap B$ 27. $A \cap C$ 28. $B \cap C$

29. $\bar{A}$ 30. $\bar{B}$ 31. $\bar{C}$

32. $\bar{U}$ 33. $(A \cup B) \cup C$ 34. $(A \cap B) \cap C$

35. $A \cup (B \cap C)$ 36. $\overline{A \cap (B \cup C)}$ 37. $\bar{A} \cup (B \cap C)$

38. $\overline{A \cup (B \cap C)}$

39. Determine whether each of the following is true or false:
    a. $\varnothing \in \varnothing$ b. $\varnothing \subseteq \varnothing$ c. $\varnothing \in \{\varnothing\}$
    d. $\varnothing = \varnothing$ e. $\varnothing = \{\varnothing\}$

40. Determine whether each of the following is true or false:
    a. $\varnothing \subseteq \{\varnothing\}$ b. $\varnothing \in A$ for all sets A
    c. $\varnothing = 0$ d. $\varnothing = \{0\}$ e. $\varnothing \subseteq \{0\}$

Draw a Venn diagram for each relationship in Problems 41–54.

41. $X \cup Y$ 42. $Y \cup Z$ 43. $X \cap Z$

44. $X \cap Y$ 45. $\bar{X}$ 46. $\bar{Z}$

47. $X \cup \bar{Y}$ 48. $\bar{X} \cap Z$ 49. $A \cap (B \cup C)$

50. $A \cup (B \cup C)$ 51. $\overline{(A \cup B) \cup C}$ 52. $\overline{(A \cap B) \cup C}$

53. $A \cap B \cup C$ 54. $\overline{A \cup B} \cup C$

APPLICATIONS

55. Let A represent the top sales representatives of the year 1987:

    {Bob Wisner, Joan Marsh, Craig Barth, Phyllis Niklas}

    and let B represent the top representatives of the year 1988:

    {Phyllis Niklas, Craig Barth, Shannon Smith, Christy Anton}

    a. What is the set of top sales representatives for 1987 or 1988? What set operation should be used to find this answer?
    b. What is the set of top sales representatives for the 2 years in a row? What set operation should be used to find this answer?

56. The following table gives inflation and unemployment rates for the years 1978–1985:

Year	Inflation rate	Unemployment rate
1978	7.6	6.0
1979	11.3	5.8
1980	13.6	7.1
1981	10.4	7.6
1982	6.1	9.7
1983	3.2	9.6
1984	4.2	7.5
1985	2.8	7.3

    Let $A = $ {years from 1978–1985 in which unemployment was at least 7%}
    $B = $ {years from 1978–1985 in which inflation was at least 6%}

    a. Find: $A, B, A \cup B, A \cap B$.
    b. Describe, in words, the sets $A \cup B$ and $A \cap B$.

57. Let $U = $ {All people in California}
    $A = $ {All people over 30}
    $B = $ {All people 30 or less}
    $C = $ {All people who own a car}

    a. Draw a Venn diagram showing how these sets are related.

    Determine whether each of the following is true or false:
    b. $B \subseteq A$ c. $C \subseteq U$ d. $B \subseteq U$

e. $C \in U$ **f.** $A \cap B = \varnothing$ **g.** $\bar{A} = B$
h. $A \cap C = \varnothing$ **i.** $\varnothing = U$

58. Let

$U = \{\text{All employees of Ampex Corporation}\}$
$A = \text{Set of executives} = \{\text{Employees with salaries of } \$60,000 \text{ or more}\}$
$B = \text{Set of junior executives} = \{\text{Employees with salaries between } \$40,000 \text{ and } \$60,000\}$
$C = \text{Set of nonexecutives} = \{\text{Employees with salaries of } \$20,000 \text{ and less}\}$
$D = \text{Set of white-collar workers} = \{\text{Employees with salaries of } \$12,000 \text{ or more}\}$

a. Draw a Venn diagram showing how these sets are related.

Determine whether each of the following is true or false:
b. $A \subseteq U$ **c.** $A \cup B = U$ **d.** $D \subseteq A$
e. $\varnothing \subseteq A$ **f.** $C \cap D = \varnothing$ **g.** $\varnothing \in A$
h. $B \cap C = \varnothing$ **i.** $A \cup B \cup C = D$

59. In a sample of defective tires, 72 have defects in materials, 89 have defects in workmanship, and 17 have defects of both types. How many tires are in the sample of defective tires (the sample consists of only tires with defects in materials or workmanship)?

60. In a survey of the executive board of a national charity, it is found that 15 members earn more than $100,000 per year and 9 own more than $1,000,000 in negotiable securities. If 7 earn more than $100,000 and own more than $1,000,000 in negotiable securities, how many members are on the executive board, if all the members fall into one of the two categories?

61. A survey of executives of Fortune 500 companies finds that 520 have MBA degrees, 650 have business degrees, and 450 have both degrees. How many executives with MBAs have nonbusiness undergraduate degrees? (*Note:* The MBA is a graduate degree and requires an undergraduate degree.)

62. A survey of 100 persons at Plimbo Corporation shows that 40 jog, 25 swim, and 15 both swim and jog. How many do neither?

63. A survey of 100 women finds that 59 use shampoo A, 41 use shampoo B, 35 use shampoo C, 24 use shampoos A and B, 19 use shampoos A and C, 13 use shampoos B and C, and 11 use all three. Let

$A = \{\text{Women who use shampoo A}\}$
$B = \{\text{Women who use shampoo B}\}$
$C = \{\text{Women who use shampoo C}\}$

Use a Venn diagram to show how many are in each of the eight possible categories.

64. A survey of 100 persons at Better Widgets finds that 40 jog, 25 swim, 16 cycle, 15 swim and jog, 10 swim and cycle, 8 jog and cycle, and 3 jog, swim, and cycle. Let

$J = \{\text{People who jog}\}$
$S = \{\text{People who swim}\}$
$C = \{\text{People who cycle}\}$

Use a Venn diagram to show how many are in each of the eight possible categories.

65. In an interview of 50 students: 12 liked Proposition 8 and Proposition 13, 18 liked Proposition 8 but not Proposition 2, 4 liked Proposition 8, Proposition 13, and Proposition 2, 25 liked Proposition 8, 15 liked Proposition 13, 10 liked Proposition 2 but not Proposition 8 or Proposition 13, and 1 liked Proposition 13 and Proposition 2 but not Proposition 8.

a. Of those surveyed, how many did not like any of the three propositions?
b. How many liked Proposition 8 and Proposition 2?
c. Show the completed Venn diagram.

66. To see if Shannon Smith can handle the job he has applied for, the personnel manager sends him out to poll 100 people about their favorite types of TV shows. Shannon obtains the following data: 59 prefer comedies, 38 prefer variety shows, 42 prefer serious drama, 18 prefer comedies or variety programs, 12 prefer variety or serious drama, 16 prefer comedies or serious drama, 7 prefer all types, and 2 do not like any TV shows. If you were the personnel manager, would you hire Shannon on the basis of this survey?

67. If $A \subseteq B$, describe:
a. $A \cup B$ **b.** $A \cap B$ **c.** $\bar{A} \cup B$ **d.** $A \cap \bar{B}$

68. What is the millionth number that is not a perfect square or a perfect cube?

69. What is the millionth number that is not a perfect square, a perfect cube, or a perfect fifth power?

70. Show that
$$n(A \cup B \cup C) = n(A) + n(B) + n(C) - n(A \cap B) \\ - n(A \cap C) - n(B \cap C) + n(A \cap B \cap C)$$

71. *Historical Question* John Venn (1834–1923) used diagrams to illustrate logic in his *Symbolic Logic*, published in 1881. Venn was an ordained priest, but he left the clergy in 1883 to spend all his time teaching and studying logic. In the text we noted that three intersecting circles divide the universe into eight regions. Venn also considered the case of four intersecting circles. Show what this case might look like, and tell how many disjoint regions are produced by four general intersecting circles.

6.2 Permutations

Consider the set

$$A = \{a, b, c, d, e\}$$

Remember, when set symbols are used, the order in which the elements are listed is not important. Suppose now that we wish to select elements from A by taking them *in a certain order*. For example, suppose set A is a club and the members are holding an election for president and secretary. If the first person selected is the president, and the second person selected is the secretary, then selecting members a and b is not the same as selecting members b and a for these positions. Since set notation signifies that the order of the elements is not important, another notation must be used when the order is important. This notation uses parentheses rather than braces. For this example, (a, b) or (b, a) represents the offices of (president, secretary) and $(a, b) \neq (b, a)$. If two elements are selected, then the notation (a, b) is called an **ordered pair**; if three are selected, it is called an **ordered triplet**; if four are selected, it is called an **ordered four-tuple**; and so on. These selections are called **arrangements** of the given set, A in this example. Arrangements are said to be selections from the given set *without repetitions*, since a symbol cannot be selected from the set twice.

EXAMPLE 1   List all possible arrangements of the elements a, c, and d selected from A.

Solution (a, c, d), (a, d, c), (c, a, d), (c, d, a), (d, a, c), (d, c, a) ■

Permutation

> A *permutation* of r elements of a set S with n elements is an ordered arrangement of those r elements selected without repetitions.

EXAMPLE 2   How many permutations of two elements can be selected from a set of six elements?

Solution   Let $B = \{a, b, c, d, e, f\}$ and select two elements:

$$
\left.
\begin{array}{ccccc}
(a, b) & (a, c) & (a, d) & (a, e) & (a, f) \\
(b, a) & (b, c) & (b, d) & (b, e) & (b, f) \\
(c, a) & (c, b) & (c, d) & (c, e) & (c, f) \\
(d, a) & (d, b) & (d, c) & (d, e) & (d, f) \\
(e, a) & (e, b) & (e, c) & (e, d) & (e, f) \\
(f, a) & (f, b) & (f, c) & (f, d) & (f, e)
\end{array}
\right\}
$$

There are 30 permutations of two elements selected from a set of six elements   ■

Example 2 brings up two difficulties. The first is the lack of notation for the phrase

*the number of permutations of two elements
selected from a set of six elements*

and the second is the inadequacy of relying on direct counting, especially if the sets are very large.

Notation for Permutations

$_nP_r$ is a symbol used to denote the *number of permutations* of r elements selected from a set of n elements.

Example 2 can now be shortened by writing

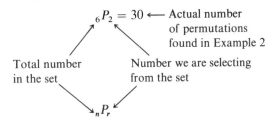

$_6P_2 = 30 \longleftarrow$ Actual number of permutations found in Example 2

Total number in the set

Number we are selecting from the set

$_nP_r$

Next, to find a formula for $_nP_r$, we turn to a general result called the *fundamental counting principle*. We begin by considering the methods of solution illustrated in the following example.

EXAMPLE 3   Consider a club with five members:

{Alfie, Bogie, Calvin, Doug, Ernie}

In how many ways can they elect a president and secretary?

Solution   There are at least two ways to solve this problem. The first, and perhaps the easiest, is by drawing a tree diagram:

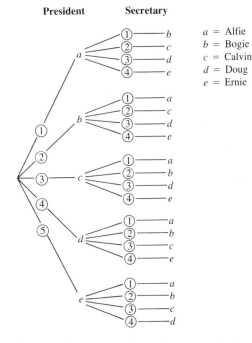

President   Secretary

a = Alfie
b = Bogie
c = Calvin
d = Doug
e = Ernie

By counting the number of branches in the tree, we see that there are 20 possibilities. This method is effective for small numbers, but the technique quickly gets out of hand. For example, if we wished to see how many ways the club could elect a president, secretary, and treasurer, this technique would be very lengthy.

A second method of solution is by using boxes or "pigeonholes" to represent each choice separately. For example:

Ways of choosing a president	Ways of choosing a secretary
5	4

↑
Since we have
chosen a president,
only four members
remain

If we multiply the numbers in the pigeonholes, we get

$$5 \cdot 4 = 20$$

and we see that the result is the same as that from the tree diagram. This pigeonhole procedure is called the **fundamental counting principle**.   ■

Fundamental Counting
Principle

> If task A can be performed in m ways, and, after task A is performed, a second task B can be performed in n ways, then task A followed by task B can be performed in $m \cdot n$ ways.

The fundamental counting principle can be used repeatedly in a particular problem so that it can be applied not only to two tasks, but also to three tasks, four tasks, or any number of tasks.

EXAMPLE 4   How many ways can the club in Example 3 select a president, secretary, and treasurer?

Solution   Continue from where we left off in Example 3 and use the fundamental counting principle a second time:

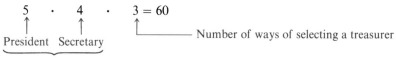

Using permutation notation, we write

$$_5P_3 = 5 \cdot 4 \cdot 3 = 60$$ ■

Can you draw a tree diagram for this example?

EXAMPLE 5 $_7P_3 = \underbrace{7 \cdot 6 \cdot 5}_{\text{Three factors}} = 210$ $_{10}P_4 = \underbrace{10 \cdot 9 \cdot 8 \cdot 7}_{\text{Four factors}} = 5040$ ■

In general, the number of permutations of n objects taken r at a time is given by the following formula:

$$_nP_r = \underbrace{n \cdot (n-1) \cdot (n-2) \cdot \cdots \cdot (n-r+1)}_{r \text{ factors}}$$

EXAMPLE 6

$_4P_4 = 4 \cdot 3 \cdot 2 \cdot 1 = 24$ $_4P_4$ is a permutation of 4 objects

$_5P_5 = 5 \cdot 4 \cdot 3 \cdot 2 \cdot 1 = 120$ $_5P_5$ is a permutation of 5 objects

$_6P_6 = 6 \cdot 5 \cdot 4 \cdot 3 \cdot 2 \cdot 1 = 720$ $_6P_6$ is a permutation of 6 objects   ■

In our work with permutations, we will frequently encounter products such as

$6 \cdot 5 \cdot 4 \cdot 3 \cdot 2 \cdot 1$

$10 \cdot 9 \cdot 8 \cdot 7 \cdot 6 \cdot 5 \cdot 4 \cdot 3 \cdot 2 \cdot 1$

$52 \cdot 51 \cdot 50 \cdot 49 \cdot \cdots \cdot 4 \cdot 3 \cdot 2 \cdot 1$

Since these are tedious to write out, **factorial notation** is used:

Definition of Factorial

$n!$ is called n *factorial* and is defined by

$$n! = n(n-1)(n-2) \cdot \cdots \cdot 3 \cdot 2 \cdot 1$$

for n a natural number. Also, $0! = 1$.

Thus, we see from Example 6 that there are $n!$ permutations of n objects.

EXAMPLE 7   Find the factorial of each whole number up to 10.

$0! = \mathbf{1}$

$1! = \mathbf{1}$

$2! = 1 \cdot 2 = \mathbf{2}$

$3! = 1 \cdot 2 \cdot 3 = \mathbf{6}$

$4! = 1 \cdot 2 \cdot 3 \cdot 4 = \mathbf{24}$

$5! = 1 \cdot 2 \cdot 3 \cdot 4 \cdot 5 = \mathbf{120}$

$6! = 1 \cdot 2 \cdot 3 \cdot 4 \cdot 5 \cdot 6 = \mathbf{720}$

$7! = 1 \cdot 2 \cdot 3 \cdot 4 \cdot 5 \cdot 6 \cdot 7 = \mathbf{5040}$

$8! = 1 \cdot 2 \cdot 3 \cdot 4 \cdot 5 \cdot 6 \cdot 7 \cdot 8 = \mathbf{40,320}$

$9! = 1 \cdot 2 \cdot 3 \cdot \cdots \cdot 8 \cdot 9 = \mathbf{362,880}$

$10! = 1 \cdot 2 \cdot 3 \cdot \cdots \cdot 9 \cdot 10 = \mathbf{3,628,800}$

■

Calculator note: Look for a $\boxed{!}$ or $\boxed{x!}$ key on your calculator. This is a factorial key. Verify one of the entries above by using your calculator.

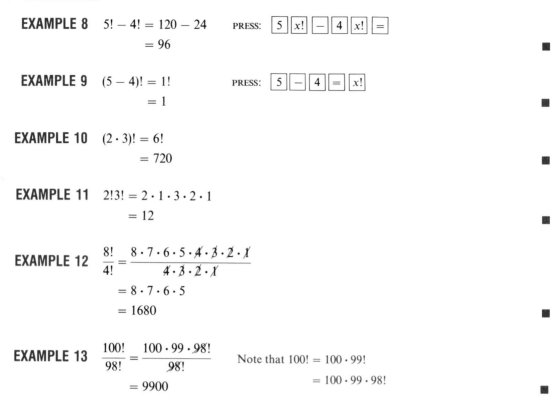

EXAMPLE 8 $5! - 4! = 120 - 24$ PRESS: $\boxed{5}\,\boxed{x!}\,\boxed{-}\,\boxed{4}\,\boxed{x!}\,\boxed{=}$
$= 96$ ∎

EXAMPLE 9 $(5 - 4)! = 1!$ PRESS: $\boxed{5}\,\boxed{-}\,\boxed{4}\,\boxed{=}\,\boxed{x!}$
$= 1$ ∎

EXAMPLE 10 $(2 \cdot 3)! = 6!$
$= 720$ ∎

EXAMPLE 11 $2!3! = 2 \cdot 1 \cdot 3 \cdot 2 \cdot 1$
$= 12$ ∎

EXAMPLE 12 $\dfrac{8!}{4!} = \dfrac{8 \cdot 7 \cdot 6 \cdot 5 \cdot \cancel{4} \cdot \cancel{3} \cdot \cancel{2} \cdot \cancel{1}}{\cancel{4} \cdot \cancel{3} \cdot \cancel{2} \cdot \cancel{1}}$
$= 8 \cdot 7 \cdot 6 \cdot 5$
$= 1680$ ∎

EXAMPLE 13 $\dfrac{100!}{98!} = \dfrac{100 \cdot 99 \cdot \cancel{98!}}{\cancel{98!}}$ Note that $100! = 100 \cdot 99!$
$= 9900$ $= 100 \cdot 99 \cdot 98!$ ∎

As the problems use larger numbers, you will find it necessary to simplify first and then use your calculator, if necessary. Example 13, for example, cannot be worked on a calculator without first simplifying.

EXAMPLE 14 $\dfrac{8!}{3!(8 - 3)!} = \dfrac{8!}{3!5!}$
$= \dfrac{8 \cdot 7 \cdot \cancel{6} \cdot \cancel{5!}}{\cancel{3} \cdot \cancel{2} \cdot 1 \cdot \cancel{5!}}$
$= 56$ ∎

Examples 13 and 14 illustrate a useful property of factorial.

Multiplication Property
of Factorial

$n! = n(n - 1)!$

This property was used in Example 13:

$100! = 100 \cdot 99!$ and $99! = 99 \cdot 98!$

Thus $100! = 100 \cdot 99 \cdot 98!$. This means that when using factorial notation, we can "count down" to any convenient number and then affix a factorial symbol. See Example 14, where we used $8! = 8 \cdot 7 \cdot 6 \cdot 5!$.

Using factorials, the formula for $_nP_r$ can be written more simply as follows:

$$_nP_r = n(n-1)(n-2)\cdot\cdots\cdot(n-r+1)$$

$$= n(n-1)(n-2)\cdot\cdots\cdot(n-r+1)\cdot\frac{(n-r)!}{(n-r)!}$$

$$= \frac{n(n-1)(n-2)\cdot\cdots\cdot(n-r+1)(n-r)(n-r-1)\cdot\cdots\cdot3\cdot2\cdot1}{(n-r)!}$$

$$= \frac{n!}{(n-r)!}$$

This is the general formula for $_nP_r$.

Permutation Formula

$$_nP_r = \frac{n!}{(n-r)!}$$

EXAMPLE 15

$$_{10}P_2 = \frac{10!}{(10-2)!}$$

$$= \frac{10\cdot9\cdot8!}{8!}$$

$$= 90$$ ∎

EXAMPLE 16

$$_nP_0 = \frac{n!}{(n-0)!}$$

$$= 1$$ ∎

EXAMPLE 17   Find the number of license plates possible in a state using only three letters, if none of the letters can be repeated.

Solution   This is a permutation of 26 objects taken 3 at a time. Thus the solution is given by

$$_{26}P_3 = \underbrace{26\cdot25\cdot24}_{\text{Three factors}} = 15{,}600$$ ∎

EXAMPLE 18   Repeat Example 17 if repetitions are allowed.

Solution   Remember, permutations do not allow repetitions, so this is *not* a permutation. However, the fundamental counting principle still applies:

$$26\cdot26\cdot26 = 17{,}576$$ ∎

EXAMPLE 19   Find the number of arrangements of letters in the word MATH.

Solution   This is a permutation of four objects taken 4 at a time:

$$_4P_4 = 4\cdot3\cdot2\cdot1 = 24$$ ∎

EXAMPLE 20 How many permutations are there of the letters in the word HATH?

Solution If you try to solve this problem as you did Example 19, you would have

$$_4P_4 = 4! = 24$$

However, if you list the possibilities, you would find only *12 different permutations*. The difficulty here is that two of the letters in the word HATH are indistinguishable. If you label them as H_1ATH_2, you would find such possibilities as

$$H_1ATH_2 \qquad H_2ATH_1 \qquad H_1AH_2T$$

If you complete this list, you would indeed find $4! = 24$ possibilities. But since there are *two indistinguishable* letters, you divide the total, 4!, by 2 to find the result:

$$\frac{4!}{2} = \frac{24}{2} = 12$$

∎

EXAMPLE 21 How many permutations are there of the letters in the word ASSIST?

Solution There are six letters, and if you consider the letters as distinguishable, as in

$$AS_1S_2IS_3T$$

there are $_6P_6 = 6! = 720$ possibilities. However,

$$AS_1S_2IS_3T \qquad AS_1S_3IS_2T \qquad AS_2S_1IS_3T$$
$$AS_2S_3IS_1T \qquad AS_3S_1IS_2T \qquad AS_3S_2IS_1T$$

are all indistinguishable, so you must divide the total by $3! = 6$.

$$\frac{6!}{3!} = \frac{6 \cdot 5 \cdot 4 \cdot 3!}{3!} = 120$$

There are 120 permutations of the letters in the word ASSIST.

∎

Examples 20 and 21 suggest a general result:

Number of Distinguishable Permutations

The number of **distinguishable permutations** of n objects, of which r are alike and the remaining objects are different from each other, is

$$\frac{n!}{r!}$$

This result generalizes to include several subcategories.

EXAMPLE 22 The number of permutations of the letters in the word ATTRACT is 7! (since there are seven letters) divided by factorials of the number of subcategories of repeated letters.

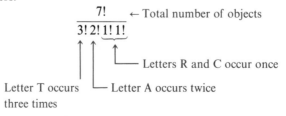

$$\frac{7!}{3!\,2!\,1!\,1!} \quad \leftarrow \text{Total number of objects}$$

Letter R and C occur once

Letter A occurs twice

Letter T occurs three times

This number can now be simplified:

$$\frac{7 \cdot 6 \cdot 5 \cdot \overset{2}{\cancel{4}} \cdot \cancel{3!}}{\cancel{3!} \cdot \cancel{2}} = 420$$

∎

General Formula for the Number of Distinguishable Permutations

The number of *distinguishable permutations* of n objects in which n_1 are of one kind, n_2 are of another kind,..., and n_k are of a further kind so that

$$n = n_1 + n_2 + \cdots + n_k$$

is given by the formula

$$\frac{n!}{n_1! n_2! \cdots n_k!}$$

Problem Set 6.2

Evaluate each expression in Problems 1–18.

1. $6! - 4!$ 2. $7! - 3!$ 3. $8! - 5!$ 4. $(6 - 4)!$

5. $(7 - 3)!$ 6. $(8 - 5)!$ 7. $4! \, 3!$ 8. $(4 \cdot 3)!$

9. $(5 \cdot 2)!$ 10. $5! \, 2!$ 11. $\dfrac{9!}{7!}$ 12. $\dfrac{10!}{6!}$

13. $\dfrac{12!}{8!}$ 14. $\dfrac{10!}{4! \, 6!}$ 15. $\dfrac{9!}{5! \, 4!}$

16. $\dfrac{12!}{3!(12 - 3)!}$ 17. $\dfrac{11!}{4!(11 - 4)!}$ 18. $\dfrac{52!}{3!(52 - 3)!}$

Evaluate each of the numbers in Problems 19–24.

19. a. $_9P_1$ b. $_9P_2$ c. $_9P_3$ d. $_9P_4$ e. $_9P_0$

20. a. $_5P_4$ b. $_{52}P_3$ c. $_7P_2$ d. $_4P_4$ e. $_{100}P_1$

21. a. $_{12}P_5$ b. $_5P_3$ c. $_8P_4$ d. $_8P_0$ e. $_gP_h$

22. a. $_{92}P_0$ b. $_{52}P_1$ c. $_7P_5$ d. $_{16}P_3$ e. $_nP_4$

23. a. $_7P_3$ b. $_5P_5$ c. $_{50}P_{48}$ d. $_{25}P_1$ e. $_mP_3$

24. a. $_8P_3$ b. $_{12}P_0$ c. $_{10}P_2$ d. $_{11}P_4$ e. $_nP_5$

How many permutations are there in the words in Problems 25–33?

25. HOLIDAY
26. ANNEX
27. ESCHEW
28. OBFUSCATION
29. MISSISSIPPI
30. CONCENTRATION
31. BOOKKEEPING
32. GRAMMATICAL
33. APOSIOPESIS

Answer the questions in Problems 34–52 by using the fundamental counting principle, the permutation formula, or both.

APPLICATIONS

34. In how many ways can a group of five people elect a president, vice president, secretary, and treasurer?

35. In how many ways can a group of 15 people elect a president and a vice president?

36. In how many ways can a group of 10 people elect a president, a vice president, and a secretary?

37. How many outfits consisting of a skirt and a blouse can a woman select if she has three skirts and five blouses?

38. How many outfits consisting of a suit and a tie can a man select if he has two suits and eight ties?

39. In how many different ways can eight books be arranged on a shelf?

40. In how many ways can you select and read three books from a shelf of eight books?

41. In how many ways can a row of 3 contestants for a TV game show be selected from an audience of 362 people?

42. How many seven-digit telephone numbers are possible if the first two digits cannot be ones or zeros?

43. Foley's Village Inn offers the following menu in its restaurant:

Main course	Dessert	Beverage
Prime rib	Ice cream	Coffee
Steak	Sherbet	Tea
Chicken	Cheesecake	Milk
Ham		Sanka
Shrimp		

In how many different ways can someone order a meal consisting of one choice from each category?

44. A typical Social Security number is 576-38-4459; the first digit cannot be zero. How many Social Security numbers are possible?

45. California license plates consist of one digit, followed by three letters, followed by three digits. How many such license plates are possible?

46. If a state issues license plates that consist of one letter followed by five digits, how many different plates are possible?

47. Repeat Problem 46 if the first letter cannot be *O*, *Q*, or *I*.

48. Repeat Problem 46 without repetition of digits.

49. New York license plates consist of three letters followed by three digits. It is also known that 245 specific arrangements of three letters are not allowed because they are considered obscene. How many license plates are possible?

50. A certain lock has five tumblers, and each tumbler can assume six positions. How many different positions are possible?

51. Texas license plates have three letters followed by three digits. If we assume that all arrangements are allowed, how many Texas license plates have a repeating letter?

52. Suppose you flip a coin and keep a record of the results. In how many ways could you obtain at least one head if you flip the coin six times?

53. Historical Question Georg Cantor (1845–1918) was the originator of set theory and the study of transfinite numbers. When he first published his paper on the theory of sets in 1874 it was very controversial because it differed from current mathematical thinking. One of Cantor's former teachers, Leopold Kronecker (1823–1891), was particularly strong in his criticism not only of Cantor's work, but also of Cantor himself. Although Cantor believed that "the essence of mathematics lies in its freedom," he tragically had a series of mental breakdowns and died in a mental hospital in 1918. Cantor proved that the set of counting numbers contains precisely as many members as its proper subset $\{2, 4, 6, \ldots\}$. This turned out to be the defining property for an infinite set. Prove that the set $\{1, 2, 3, 4, \ldots\}$ contains precisely the same number of elements as the proper subset $\{2, 4, 6, \ldots\}$.

6.3 Combinations

Consider the set

$$A = \{a, b, c, d, e\}$$

If two elements are selected from A in a certain order, they are represented by an *ordered pair* and the ordered pair is called a *permutation*. On the other hand, if two elements are selected from A *without regard to the order in which they are selected*, they are represented as a subset of A.

EXAMPLE 1   Select two elements from A:

Permutations—Order important

(a, b)	(a, c)	(a, d)	(a, e)
(b, a)	(b, c)	(b, d)	(b, e)
(c, a)	(c, b)	(c, d)	(c, e)
(d, a)	(d, b)	(d, c)	(d, e)
(e, a)	(e, b)	(e, c)	(e, d)

Notation: $_5P_2 = 20$

There are 20 permutations— order is important.

Subsets—Order not important

$\{a, b\}$	$\{a, c\}$	$\{a, d\}$	$\{a, e\}$
	$\{b, c\}$	$\{b, d\}$	$\{b, e\}$
		$\{c, d\}$	$\{c, e\}$
			$\{d, e\}$

Do not list $\{b, a\}$ since $\{b, a\} = \{a, b\}$

There are 10 subsets; note that set notation is used.   ∎

We use the following name and notation when listing different subsets of a given set.

Definition of Combination

> A **combination** of r elements of a finite set S is a subset of S that contains r distinct elements.

Remember when listing the elements of a subset that the order in which those elements are listed is not important. A notation similar to that used for permutations is used to denote the number of combinations.

Notation for Combinations

> $\begin{pmatrix} n \\ r \end{pmatrix}$ and $_nC_r$ are symbols used to denote the number of combinations of r elements selected from a set of n elements ($r \leq n$).

The notation $_nC_r$ is similar to the notation used for permutations, but since $\begin{pmatrix} n \\ r \end{pmatrix}$ is used in later work, we will use that notation for combinations.

The formula for the number of permutations leads directly to a formula for the number of combinations since each subset of r elements has $r!$ permutations of its members. Thus

$$_nP_r = \begin{pmatrix} n \\ r \end{pmatrix} \cdot r!$$

$$\begin{pmatrix} n \\ r \end{pmatrix} = \frac{_nP_r}{r!} = \frac{n!}{r!(n-r)!}$$

Combination Formula

> $$\begin{pmatrix} n \\ r \end{pmatrix} = \frac{n!}{r!(n-r)!}$$

EXAMPLE 2 $\begin{pmatrix} 10 \\ 3 \end{pmatrix} = \frac{10!}{3!\,7!}$

$$= \frac{\overset{5}{\cancel{10}} \cdot \overset{3}{\cancel{9}} \cdot 8 \cdot \cancel{7!}}{\cancel{3} \cdot \cancel{2} \cdot \cancel{7!}} = 120$$ ∎

EXAMPLE 3 $\begin{pmatrix} n \\ 0 \end{pmatrix} = \frac{n!}{0!(n-0)!} = 1$ ∎

EXAMPLE 4 $\begin{pmatrix} m-1 \\ 2 \end{pmatrix} = \frac{(m-1)!}{2!(m-1-2)!}$

$$= \frac{(m-1)(m-2)(m-3)!}{2 \cdot 1 \cdot (m-3)!}$$

$$= \frac{(m-1)(m-2)}{2}$$ ∎

EXAMPLE 5 In how many ways can a club of five members select a three-person committee?

Solution $\binom{5}{3} = \frac{5!}{3!2!} = \frac{5 \cdot 4 \cdot 3}{3!} = 10$ ∎

EXAMPLE 6 Find the number of five-card hands that can be drawn from an ordinary deck of cards.

Solution $\binom{52}{5} = \frac{52!}{5!47!} = \frac{52 \cdot 51 \cdot 50 \cdot 49 \cdot 48}{5 \cdot 4 \cdot 3 \cdot 2 \cdot 1} = 2,598,960$ ∎

EXAMPLE 7 In how many ways can a diamond flush be drawn in poker? (A diamond flush is a hand of five diamonds.)

Solution This is a combination of 13 objects (diamonds) taken 5 at a time. Thus, the solution is given by

$$\binom{13}{5} = \frac{13!}{5!8!} = \frac{13 \cdot 12 \cdot 11 \cdot 10 \cdot 9}{5!} = 1287$$ ∎

EXAMPLE 8 In how many ways can a flush be drawn in poker?*

Solution Begin with the fundamental counting principle:

$$\underbrace{4}_{\substack{\text{Number} \\ \text{of suits}}} \quad \underbrace{1287}_{\substack{\text{Number of ways of} \\ \text{drawing a flush in} \\ \text{a particular suit} \\ \text{(from Example 7)}}} = 5148$$ ∎

Ordered Partitions

As we have seen, not all counting problems can be solved as permutation or combination problems. But suppose we generalize the idea of filling pigeonholes. We wish to divide a set S into r subsets. There are several ways that we might do this. For example, suppose

$S = \{x \mid x$ is a sales representative of the Sharp Investment Company$\}$

By roster,

$S = \{$Ralph, Ron, Lim, Shannon, Mark, Todd, Tim, Greg$\}$

We divide the set S in three different ways:

1. By geographical areas covered:

North	South	East	West
Ron	Ralph	Lim	Mark
Todd		Shannon	Greg
		Tim	

---

*A flush in poker is five cards from one suit (see Figure 7.1 on page 274.)

2. By amount of sales:

Over $100,000	$50,000–$100,000	Under $50,000
Ron	Todd	Tim
Lim	Shannon	
Mark	Greg	

3. By musical instruments each plays:

Piano	Guitar	Clarinet	Accordion
Ralph	Ralph	Ron	Ralph
Todd	Mark		Todd
Greg			

If no two of the subsets have any elements in common, and if all the elements of the original set are in some subset, we call such a set of subsets a **partition** of the original set. This corresponds to our everyday usage of the word when we say a house is partitioned into several rooms. In the division by geographical area, the four sets {Ron, Todd}, {Ralph}, {Lim, Shannon, Tim}, and {Mark, Greg} form a partition of S. On the other hand, the divisions by sales and instrument are not partitions since Ralph does not appear in any of the subsets of the former, and the subsets in the latter are not disjoint.

Partition of a Set

A set S is said to be *partitioned* if S is divided into r subsets satisfying the following two conditions:
1. If A and B are any two subsets, then $A \cap B = \varnothing$; that is, the members of the subsets are pairwise disjoint.
2. The union of all the subsets is S; that is, there are no elements of S that are not included in one of the subsets.

EXAMPLE 9   Form a partition of the set $\{1, 2, 3, 4\}$.

Solution   Answers vary; some possible ones are:

$$\{1, 2\}, \{3, 4\}$$

or

$$\{1\}, \{2\}, \{3\}, \{4\}$$

or

$$\{1, 2, 3\}, \{4\}$$

However, $\{1, 2, 3\}, \{3, 4\}$ is *not* a partition.   ∎

EXAMPLE 10 Any set and its complement form a partition of a universal set. ∎

Suppose the eight sales representatives of Sharp Investment Company are to be reassigned into three geographical areas:

Area #1	Area #2	Area #3
North	East	West

We are interested in knowing how many ways we can assign two sales representatives to area #1, three sales representatives to area #2, and three sales representatives to area #3. Use the fundamental counting principle to fill the pigeonholes below:

Area #1	Area #2	Area #3
$\binom{8}{2}$	$\binom{6}{3}$	$\binom{3}{3}$

There are $\binom{8}{2}$ ways of filling area #1. This leaves six sales representatives to distribute among areas #2 and #3. For area #2, we see there are $\binom{6}{3}$ ways, and this leaves us with three sales representatives to be distributed to area #3: $\binom{3}{3}$.

Instead of calculating these numbers separately and then taking the product, let us do all the calculations at once:

$$\binom{8}{2}\binom{6}{3}\binom{3}{3} = \frac{8!}{2!\cancel{6!}} \cdot \frac{\cancel{6!}}{3!\cancel{3!}} \cdot \frac{\cancel{3!}}{3!0!}$$

$$= \frac{8!}{2!3!3!}$$

$$= \frac{8 \cdot 7 \cdot 6 \cdot 5 \cdot 4 \cdot 3!}{2 \cdot 1 \cdot 3 \cdot 2 \cdot 1 \cdot 3!}$$

$$= 560 \text{ ways}$$

This is an example of an **ordered**, or **distinguishable**, **partition** (two sales representatives for area #1 and three for area #2 is different from three sales representatives for area #1 and two for area #2—order is important).

The above example can be written in condensed form as

$$\binom{8}{2,3,3} = \frac{8!}{2!3!3!}$$

You can see that what we have done here is add a new notation to the formula for the number of distinguishable permutations that was stated in Section 6.2.

Number of Ordered Partitions

For any set of positive integers $\{n_1, n_2, \ldots, n_k\}$, where $\sum\limits_{i=1}^{k} n_i = n$, we have

$$\underbrace{\binom{n}{n_1, n_2, \ldots, n_k}}_{\substack{\text{Partition} \\ \text{notation}}} \text{for} \underbrace{= \frac{n!}{n_1! n_2! \cdots \cdot n_k!}}_{\substack{\text{Number of} \\ \text{distinguishable} \\ \text{permutations}}}$$

EXAMPLE 11 Suppose there are 15 sales representatives and they are to be divided among the three areas as follows: 4 in area #1, 5 in area #2, and 6 in area #3.

Solution Use pigeonholes to find

$$\binom{15}{4}\binom{11}{5}\binom{6}{6} = \frac{15!}{4! \, 11!} \cdot \frac{11!}{5! \, 6!} \cdot \frac{6!}{6! \, 0!}$$

$$= \frac{15!}{4! \, 5! \, 6!}$$

This can be rewritten as

$$\binom{15}{4,5,6} = \frac{15!}{4! \, 5! \, 6!}$$ ∎

Which Method?

We have now looked at several counting schemes: tree diagrams, the fundamental counting principle, permutations, combinations, and ordered partitions. In practice, you will generally not be told what type of counting problem you are dealing with — you will need to decide. Table 6.1 should help with that decision.

TABLE 6.1

Fundamental counting principle	Permutations	Combinations	Ordered partitions
Counting total of separate tasks	Number of ways of selecting r items out of n items		Partition n elements into k categories
Repetitions allowed	Repetitions are not allowed		Repetitions are not allowed
If tasks 1, 2, 3, ..., k can be performed in $n_1, n_2, n_3, \ldots, n_k$ ways, respectively, then the total number of ways the k tasks can be performed is $n_1 \cdot n_2 \cdot n_3 \cdots \cdot n_k$ ways	Order is important (select and arrange)	Order is not important (just select)	Order is not important
	Arrangements of r items from a set of n items	Subsets of r items from a set of n items	Place n items into k categories where $n = n_1 + n_2 + n_3 + \cdots + n_k$
	$_nP_r = \dfrac{n!}{(n-r)!}$	$\dbinom{n}{r} = \dfrac{n!}{r!(n-r)!}$	$\dbinom{n}{n_1, n_2, \ldots, n_k} = \dfrac{n!}{n_1! n_2! \cdots n_k!}$

> COMPUTER APPLICATION   If you have access to a computer, Program 7 on the computer disk accompanying this book involves practice with counting problems.

EXAMPLE 12   What is the number of license plates possible in Florida if each license plate consists of three letters followed by three digits, and we add the condition that repetition of letters or digits is not permitted?

Solution   This is a permutation problem since the *order* in which the elements are arranged is important. That is, CWB072 and BCW072 are different plates. The number is found by using permutations along with the fundamental counting principle:

$$_{26}P_3 \cdot {}_{10}P_3 = 26 \cdot 25 \cdot 24 \cdot 10 \cdot 9 \cdot 8 \qquad \leftarrow \text{This answer is acceptable}$$
$$= 11{,}232{,}000$$

■

EXAMPLE 13   Find the number of three-letter "words" that can be formed using three letters from $\{m, a, t, h\}$.

Solution   Finding these words is also a permutation problem since *mat* is different from *tam*.

$$_4P_3 = 4 \cdot 3 \cdot 2$$
$$= 24$$

■

EXAMPLE 14   The football team at UCLA plays 11 games each season. In how many ways can the team complete the season with 6 wins, 4 losses, and 1 tie?

Solution   This is an ordered partition $(6 + 4 + 1 = 11)$, so

$$\binom{11}{6,4,1} = \frac{11!}{6!4!1!} = \frac{11 \cdot 10 \cdot 9 \cdot 8 \cdot 7 \cdot \cancel{6!}}{\cancel{6!}4 \cdot 3 \cdot 2 \cdot 1 \cdot 1}$$
$$= 2310$$

■

EXAMPLE 15   Find the number of bridge hands (13 cards) consisting of six hearts, four spades, and three diamonds.

Solution   The order in which the cards are received is unimportant, so finding the number of bridge hands is a combination problem. Also use the fundamental counting principle:

$$\underbrace{\binom{13}{6}}_{\substack{\text{Number of ways of} \\ \text{obtaining hearts}}} \cdot \underbrace{\binom{13}{4}}_{\substack{\text{Number of ways of} \\ \text{obtaining spades}}} \cdot \underbrace{\binom{13}{3}}_{\substack{\text{Number of ways of} \\ \text{obtaining diamonds}}}$$

$$= \frac{13!}{6!(13-6)!} \cdot \frac{13!}{4!(13-4)!} \cdot \frac{13!}{3!(13-3)!}$$

$$= \frac{13!13!13!}{6!7!4!9!3!10!} \qquad \leftarrow \text{This answer is acceptable}$$

$$= 350{,}904{,}840$$

■

EXAMPLE 16 A club with 42 members wants to elect a president, a vice president, and a treasurer. From the other members, an advisory committee of 5 people is to be selected. In how many ways can this be done?

Solution This is both a permutation and a combination problem, with the final result calculated by using the fundamental counting principle:

$$\underbrace{\text{Number of ways}}_{\text{of selecting}} \quad \underbrace{\text{Number of ways}}_{\text{of selecting}}$$
$$\text{officers} \qquad \text{committee}$$

$$_{42}P_3 \quad \cdot \quad \binom{39}{5}$$

$$= 42 \cdot 41 \cdot 40 \quad \cdot \quad \frac{39!}{5!(39-5)!}$$

$$= 39{,}658{,}142{,}160 \quad \leftarrow \text{Found with a calculator} \qquad \blacksquare$$

Problem Set 6.3

Evaluate each of the numbers in Problems 1–11.

1. a. $\binom{9}{1}$ b. $\binom{9}{2}$ c. $\binom{9}{3}$ d. $\binom{9}{4}$ e. $\binom{9}{0}$

2. a. $\binom{5}{4}$ b. $\binom{52}{3}$ c. $\binom{7}{2}$ d. $\binom{4}{4}$ e. $\binom{100}{1}$

3. a. $\binom{7}{3}$ b. $\binom{5}{5}$ c. $\binom{50}{48}$ d. $\binom{25}{1}$ e. $\binom{g}{h}$

4. a. $_7P_3$ b. $_5P_5$ c. $_{52}P_2$ d. $_{25}P_1$ e. $_gP_h$

5. a. $_7P_5$ b. $\binom{8}{0}$ c. $\binom{10}{2}$ d. $_nP_4$ e. $\binom{n}{4}$

6. a. $\binom{92}{0}$ b. $_{52}P_3$ c. $_mP_n$ d. $\binom{16}{3}$ e. $\binom{m}{n}$

7. a. $_{92}P_0$ b. $\binom{53}{3}$ c. $\binom{7}{7}$ d. $_7P_7$ e. $_nP_5$

8. a. $\binom{4}{1,1,2}$ b. $\binom{6}{3,2,1}$

9. a. $\binom{5}{1,3,1}$ b. $\binom{8}{2,4,2}$

10. a. $\binom{9}{1,4,4}$ b. $\binom{7}{1,4,2}$

11. a. $\binom{10}{6,2,1,1}$ b. $\binom{8}{2,3,1,2}$

12. A bag contains 12 pieces of candy. In how many ways can 5 pieces be selected?

13. If the Senate is to form a new committee of 5 members, in how many different ways can the committee be chosen if all 100 senators are available to serve on this committee?

14. In how many ways can three aces be drawn from a deck of cards?

15. In how many ways can two kings be drawn from a deck of cards?

16. In how many ways can a heart flush be obtained? (A heart flush is a hand of five hearts.)

17. In how many ways can a spade flush be obtained? (A spade flush is a hand of five spades.)

In Problems 18–31 decide whether you would use the fundamental counting principle, a permutation, a combination, an ordered partition, or none of these. Next, write the solution using permutation, combination or ordered partition notation, if possible, and finally, answer the question asked.

18. How many different arrangements are there of letters in the word CORRECT?

19. At Artist's Dance Studio, every man must dance the last dance. If there are five men and eight women, in how many ways can dance couples be formed for the last dance?

20. Martin's Ice Cream Store sells sundaes with chocolate, strawberry, butterscotch, or marshmallow toppings, nuts, and whipped cream. If you can choose exactly three of these extras, how many possible sundaes are there?

21. Five people are to dine together at a rectangular table, but the host cannot decide on a seating arrangement. In how many ways can the guests be seated, assuming fixed chair positions?

22. In how many ways can three hearts be drawn from a deck of cards?

23. A shipment of a hundred TV sets is received. Six sets are to be chosen at random and tested for defects. In how many ways can the six sets be chosen?

24. A night watchman visits 15 offices every night. To prevent others from knowing when he will be at a particular office, he varies the order of his visits. In how many ways can this be done?

25. A certain manufacturing process calls for six chemicals to be mixed. One liquid is to be poured into the vat, and then the others are to be added in turn. All possible combinations must be tested to see which gives the best results. How many tests must be performed?

26. There are three boys and three girls at a party. In how many ways can they be seated in a row if they can sit down four at a time?

27. There are three boys and three girls at a party. In how many ways can they be seated in a row if they want to sit alternating boy, girl, boy, girl?

28. If there are ten people in a club, in how many ways can they choose a dishwasher and a bouncer?

29. How many subsets of size three can be formed from a set of five elements?

30. In how many ways can you be dealt two cards from an ordinary deck of cards?

31. In how many ways can five taxi drivers be assigned to six cars?

APPLICATIONS

32. A fraternity contains 8 sophomores, 15 juniors, and 12 seniors. In how many ways can a committee consisting of 1 sophomore, 3 juniors, and 3 seniors be chosen?

33. A sorority has 2 freshmen, 5 sophomores, 25 juniors, and 18 seniors as members. In how many ways can a committee consisting of 2 sophomores, 5 juniors, and 5 seniors be chosen?

34. A rating service rates 25 businesses as A, AA, or AAA. Suppose that 5 companies will be rated AAA, 4 companies will be rated A, and the rest rated AA. In how many ways can this be done?

35. A psychology experiment will group 50 persons into one of three categories: dominant, passive, neutral. In how many ways can 15 be dominant, 10 passive, with the rest neutral? (You can leave your answer in factorial notation.)

36. In a state lottery, 20 people will be chosen as winners. Of these 20, 5 will be awarded $3 million, 5 will win $100,000, and the rest will win $10,000. In how many ways can this be done?

37. Suppose that in a study of 50 stocks, 7 are expected to go up, 5 to go down, and the rest will stay about the same. In how many different ways (to the nearest billion) can the 50 stocks be divided up so that the conditions of this problem are satisfied?

38. A group of students found that on a certain multiple-choice test of 15 questions, the answer sheet shows response "a" occurs 3 times, "b" occurs 4 times, "c" occurs 3 times, "d" occurs 2 times, and "e" occurs 3 times. In how many ways can the questions be answered? Leave your answer in factorial notation.

39. The investigative division of the Internal Revenue Service has a list of 10 companies cited for violations of proper tax procedures. A spy for one of the companies found out that the IRS intends to bring 4 of these companies to trial, will refer 5 cases for further investigation, and the other case will be dropped. In how many ways can the disposition of the 10 cases be made according to the known information?

40. A sample of 5 watches is selected at random from a lot of 20 watches. The entire lot will be rejected if 2 or more watches are found to be defective.
   a. How many different samples can be selected?
   b. If 2 of the watches are defective, how many of the samples will lead to rejection?

41. A group of 10 Americans, 6 Russians, and 5 Finns must choose a committee of 3 persons. In how many ways can a committee be chosen that consists of:
   a. All Americans?
   b. 2 Americans?
   c. 1 American?
   d. 1 of each nationality?
   e. 2 Russians and 1 Finn?

42. How many committees of 2 men and 3 women can be chosen from 10 men and 8 women?

43. How many committees of 3 Americans and 3 Russians can be chosen from 10 Americans and 12 Russians?

44. Show that $\binom{n}{n-r} = \binom{n}{r}$.

45. Show that $\binom{n}{k-1} + \binom{n}{k} = \binom{n+1}{k}$.

46. Prove the formula for the number of ordered partitions for the case where $k = 3$.

6.4 Binomial Theorem

Combinations can be used to help us form a pattern for the expansion of a binomial $(a + b)^n$. This problem occurs frequently in mathematics, and direct calculation is too tedious for a very large exponent n. Also, we sometimes need to find only one term in the expansion, and a pattern will help us find only that term without having to find all the others.

Consider the powers of $(a + b)$ listed below, which are found by direct multiplication:

$$(a + b)^0 = 1$$
$$(a + b)^1 = 1 \cdot a + 1 \cdot b$$
$$(a + b)^2 = 1 \cdot a^2 + 2 \cdot ab + 1 \cdot b^2$$
$$(a + b)^3 = 1 \cdot a^3 + 3 \cdot a^2 b + 3 \cdot ab^2 + 1 \cdot b^3$$
$$(a + b)^4 = 1 \cdot a^4 + 4 \cdot a^3 b + 6 \cdot a^2 b^2 + 4 \cdot ab^3 + 1 \cdot b^4$$
$$(a + b)^5 = 1 \cdot a^5 + 5 \cdot a^4 b + 10 \cdot a^3 b^2 + 10 \cdot a^2 b^3 + 5 \cdot ab^4 + 1 \cdot b^5$$
$$\vdots$$

First, ignore the coefficients (shown in color) and focus on the variables:

$$(a + b)^1: \quad a \quad b$$
$$(a + b)^2: \quad a^2 \quad ab \quad b^2$$
$$(a + b)^3: \quad a^3 \quad a^2 b \quad ab^2 \quad b^3$$
$$(a + b)^4: \quad a^4 \quad a^3 b \quad a^2 b^2 \quad ab^3 \quad b^4$$
$$(a + b)^5: \quad a^5 \quad a^4 b \quad a^3 b^2 \quad a^2 b^3 \quad ab^4 \quad b^5$$
$$\vdots$$

Do you see a pattern? As you read from left to right, the powers of a decrease and the powers of b increase. Note that the sum of the exponents for each term is the same as the original exponent:

$$(a + b)^n: \quad a^n b^0 \quad a^{n-1} b^1 \quad a^{n-2} b^2 \quad \cdots \quad a^{n-r} b^r \quad \cdots \quad a^2 b^{n-2} \quad a^1 b^{n-1} \quad a^0 b^n$$

Next, consider the numerical coefficients:

$$(a + b)^0: \qquad\qquad\qquad 1$$
$$(a + b)^1: \qquad\qquad\quad 1 \quad 1$$
$$(a + b)^2: \qquad\qquad 1 \quad 2 \quad 1$$
$$(a + b)^3: \qquad\quad 1 \quad 3 \quad 3 \quad 1$$
$$(a + b)^4: \qquad 1 \quad 4 \quad 6 \quad 4 \quad 1$$
$$(a + b)^5: \quad 1 \quad 5 \quad 10 \quad 10 \quad 5 \quad 1$$
$$\vdots$$

Do you see the pattern? This arrangement of numbers is called **Pascal's triangle**. The rows and columns of Pascal's triangle are usually numbered, as shown in Figure 6.10.

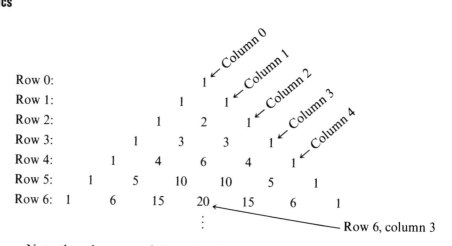

Row 0:
Row 1:
Row 2:
Row 3:
Row 4:
Row 5:
Row 6:

Figure 6.10 Pascal's triangle

Row 6, column 3

Note that the rows of Pascal's triangle are numbered to correspond to the exponent n on $(a + b)^n$. There are many relationships associated with this pattern, but, for now, we are concerned with an expression representing the entries in the pattern. Do you see how to generate additional rows of the triangle? Look at Figure 6.10 and note the following:

1. Each row begins and ends with a 1.
2. We began counting the rows with row 0. This is because after row 0, the second entry in the row is the same as the row number. Thus, row 7 begins 1   7.... .
3. The triangle is symmetric about the middle. This means that the entries of each row are the same at the beginning and the end. Thus row 7 ends with...7   1. (This property is proved in Problem 41.)
4. To find new entries, we simply add the two entries just above in the preceding row. Thus, row 7 is found by looking at row 6:

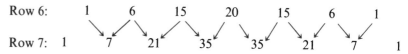

(This property is proved in Problem 40.)

Also note that the entries in Pascal's triangle are the combinations we computed in Section 6.3. That is, the entry in the third column of the fourth row is $\binom{4}{3}$, as shown below:

$$\binom{0}{0} = 1$$

$$\binom{1}{0} = 1 \qquad \binom{1}{1} = 1$$

$$\binom{2}{0} = 1 \qquad \binom{2}{1} = 2 \qquad \binom{2}{2} = 1$$

$$\binom{3}{0} = 1 \qquad \binom{3}{1} = 3 \qquad \binom{3}{2} = 3 \qquad \binom{3}{3} = 1$$

$$\binom{4}{0} = 1 \qquad \binom{4}{1} = 4 \qquad \binom{4}{2} = 6 \qquad \binom{4}{3} = 4 \qquad \binom{4}{4} = 1$$

EXAMPLE 1 Find $\binom{3}{2}$ both by Pascal's triangle and by formula.

Solution Third row, second column; entry in Pascal's triangle is 3. By formula,

$$\binom{3}{2} = \frac{3!}{2!(3-2)!}$$

$$= \frac{3!}{2!1!}$$

$$= 3$$

■

EXAMPLE 2 Find $\binom{6}{4}$ both by Pascal's triangle and by formula.

Solution Row 6, column 4; entry is 15. By formula,

$$\binom{6}{4} = \frac{6!}{4!(6-4)!}$$

$$= \frac{6!}{4!2!}$$

$$= \frac{6 \cdot 5 \cdot 4!}{2 \cdot 1 \cdot 4!}$$

$$= 15$$

■

Using this notation we can now state a very important theorem in mathematics—the **binomial theorem**.

Binomial Theorem

For any positive integer n,

$$(a + b)^n = \binom{n}{0}a^n + \binom{n}{1}a^{n-1}b + \binom{n}{2}a^{n-2}b^2 + \cdots$$

$$+ \binom{n}{r}a^{n-r}b^r + \cdots + \binom{n}{n-2}a^2b^{n-2} + \binom{n}{n-1}ab^{n-1} + \binom{n}{n}b^n$$

Pascal's triangle is efficient for finding the numerical coefficients for exponents that are relatively small, as shown in Figure 6.11. However, for larger exponents we need to use the binomial theorem to find the coefficients.

EXAMPLE 3 Find $(x + y)^8$.

Solution Use Pascal's triangle (Figure 6.11) to obtain the coefficients in the expansion. Thus

$$(x + y)^8 = x^8 + 8x^7y + 28x^6y^2 + 56x^5y^3 + 70x^4y^4$$
$$+ 56x^3y^5 + 28x^2y^6 + 8xy^7 + y^8$$

■

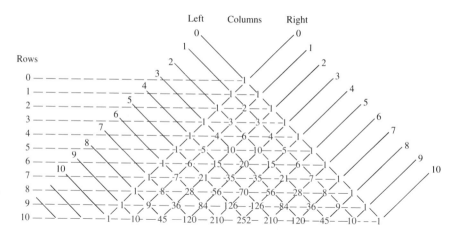

Figure 6.11 Pascal's triangle—an extended version can be found in Table 1, Appendix E

EXAMPLE 4   Find $(x - 2y)^4$.

Solution   In this example, $a = x$ and $b = -2y$ and the coefficients are shown in Figure 6.11.

$$(x - 2y)^4 = [x + (-2y)]^4$$
$$= 1 \cdot x^4 + 4 \cdot x^3(-2y) + 6 \cdot x^2(-2y)^2 + 4 \cdot x(-2y)^3 + 1 \cdot (-2y)^4$$
$$= x^4 - 8x^3y + 24x^2y^2 - 32xy^3 + 16y^4$$ ∎

EXAMPLE 5   Find $(a + b)^{15}$.

Solution   The power is rather large, so use the binomial theorem or Table 1, Appendix E, to find the coefficients:

$$(a + b)^{15} = \binom{15}{0}a^{15} + \binom{15}{1}a^{14}b + \binom{15}{2}a^{13}b^2 + \cdots + \binom{15}{14}ab^{14} + \binom{15}{15}b^{15}$$

$$= \frac{15!}{0!15!}a^{15} + \frac{15!}{1!14!}a^{14}b + \frac{15!}{2!13!}a^{13}b^2 + \cdots + \frac{15!}{14!1!}ab^{14} + \frac{15!}{15!0!}b^{15}$$

$$= a^{15} + 15a^{14}b + 105a^{13}b^2 + \cdots + 15ab^{14} + b^{15}$$ ∎

EXAMPLE 6   Find the coefficient of the term x^2y^{10} in the expansion of $(x + 2y)^{12}$.

Solution   Here, $n = 12, r = 10, a = x$, and $b = 2y$; thus

$$\binom{12}{10}x^2(2y)^{10} = \frac{12!}{10!2!}(2)^{10}x^2y^{10}$$

The coefficient is $66(1024) = 67,584$.   ∎

COMPUTER APPLICATION   If you have access to a computer, Options 1 and 2 of Program 11 on the computer disk accompanying this text will help with problems like Examples 5 and 6.

EXAMPLE 7 How many subsets of $\{1, 2, 3, 4, 5\}$ are there?

Solution In Section 6.1 we found that a set containing two elements has four subsets, and we did this by direct enumeration with a tree diagram. To follow the same procedure for this example would be too tedious. Instead, consider the following argument: There are $\binom{5}{5}$ subsets of five elements; $\binom{5}{4}$ subsets of four elements; $\binom{5}{3}$ subsets of three elements; $\binom{5}{2}$ subsets of two elements; $\binom{5}{1}$ subsets of a single element; and $\binom{5}{0}$ subsets of zero elements (the empty set). Thus the *total* number of subsets is

$$\binom{5}{0} + \binom{5}{1} + \binom{5}{2} + \binom{5}{3} + \binom{5}{4} + \binom{5}{5}$$

Now note that

$$(a + b)^5 = \binom{5}{0}a^5 + \binom{5}{1}a^4b + \binom{5}{2}a^3b^2 + \binom{5}{3}a^2b^3 + \binom{5}{4}ab^4 + \binom{5}{5}b^5$$

so that if we let $a = 1$ and $b = 1$,

$$\underbrace{(1 + 1)^5}_{2^5} = \underbrace{\binom{5}{0} + \binom{5}{1} + \binom{5}{2} + \binom{5}{3} + \binom{5}{4} + \binom{5}{5}}_{\text{Total number of subsets}}$$

We see that there are 2^5 subsets of a set containing five elements. ■

EXAMPLE 8 How many subsets are there of a set containing 10 elements?

Solution Using the ideas of Example 7, there are 2^{10} subsets. ■

Examples 7 and 8 suggest a general result:

Number of Subsets of a Finite Set | A set of n distinct elements has 2^n subsets.

Problem Set 6.4

Evaluate the expressions in Problems 1–12 using both Pascal's triangle and the binomial theorem.

1. $\binom{8}{1}$ **2.** $\binom{5}{4}$ **3.** $\binom{8}{2}$ **4.** $\binom{7}{5}$

5. $\binom{8}{3}$ **6.** $\binom{9}{5}$ **7.** $\binom{12}{1}$ **8.** $\binom{15}{0}$

9. $\binom{20}{20}$ **10.** $\binom{32}{31}$ **11.** $\binom{18}{2}$ **12.** $\binom{46}{2}$

Expand the binomial expressions in Problems 13–27 using the binomial theorem or Pascal's triangle.

13. $(x + y)^5$ **14.** $(x + y)^6$ **15.** $(x + y)^4$

16. $(x - y)^4$ **17.** $(x - y)^5$ **18.** $(x - y)^6$

19. $(x + 2)^5$ **20.** $(x - 3)^5$ **21.** $(2x + 3y)^4$

22. $(1 - x)^8$ **23.** $(1 - x)^{10}$ **24.** $(x - 2y)^8$

25. $(x - y)^{15}$ **26.** $(x + 1)^{18}$ **27.** $(x + y)^{20}$

Find the requested coefficient in Problems 28–33.

28. a^5b^6 term in $(a + b)^{11}$ **29.** a^4b^7 term in $(a - b)^{11}$

30. $a^{14}b$ term in $(a - 2b)^{15}$ **31.** $a^{10}b^4$ term in $(a + 3b)^{14}$

32. $x^{14}y^2$ term in $(2x^2 + y)^9$ **33.** $x^{14}y$ term in $(x^2 - 2y)^8$

APPLICATIONS

34. How many different subsets can be chosen from a set of seven elements?

35. How many different subsets can be chosen from the U.S. Senate? (There are 100 U.S. senators.)

36. Suppose a coin is tossed eight times. How many possible outcomes will there be of five heads and three tails? [*Hint:* Denote the possibilities of a single toss by H + T. Since there are eight tosses, we can denote the set of all possibilities by $(H + T)^8$. In this problem, you are looking for the coefficients of H^5T^3; that is, heads five times (H^5) and tails three times (T^3).]

37. Suppose a coin is tossed ten times. How many possible outcomes will there be of four heads and six tails? [See the hint in Problem 36.]

38. Suppose a coin is tossed nine times. How many possible outcomes will there be of two heads and seven tails? [See the hint in Problem 36.]

39. Show that $\binom{n}{0} + \binom{n}{1} + \binom{n}{2} + \cdots + \binom{n}{n-1} + \binom{n}{n} = 2^n$.

(This says that the sum of the entries of the nth row of Pascal's triangle is 2^n.)

40. Show that $\binom{n-1}{r-1} + \binom{n-1}{r} = \binom{n}{r}$.

[This says that to find any entry in Pascal's triangle (except the first and last), simply add the two entries directly above.]

41. Show that Pascal's triangle is symmetric.

42. Historical Question   Blaise Pascal (1623–1662) has been described as "the greatest 'might-have-been' in the history of mathematics." Pascal was frail, and, because he needed to conserve his energy, he was forbidden to study mathematics. This aroused his curiosity and forced him to acquire most of his knowledge of the subject by himself. At 16, he wrote an essay on conic sections that astounded the mathematicians of his time. At 18, he had already invented one of the first calculating machines. However, at 27, because of his health, he promised God that he would abandon mathematics and spend his time in religious study. Three years later he broke this promise and wrote *Traité du Triangle Arithmétique*, in which he investigated what we today call Pascal's triangle. The very next year he was almost killed when his runaway horse jumped an embankment. He took this to be a sign of God's displeasure with him and again gave up mathematics—this time permanently. Answer the following questions about Pascal's triangle:

a. What is the sum of the entries in the nth row?

b. How are the powers of 11 related to Pascal's triangle?

c. Find a formula for the largest number in the nth row.

6.5   Summary and Review

IMPORTANT TERMS

Arrangement   [6.2]
Binomial theorem   [6.4]
Combination   [6.3]
Complementation   [6.1]
Description method   [6.1]
Disjoint sets   [6.1]
Distinguishable permutations   [6.2]
Empty set   [6.1]
Equality of sets   [6.1]
Factorial   [6.2]
Fundamental counting principle   [6.2]
Improper subset   [6.1]
Intersection   [6.1]

Ordered pair   [6.2]
Ordered partition   [6.3]
Partition   [6.3]
Pascal's triangle   [6.4]
Permutation   [6.2]
Proper subset   [6.1]
Roster method   [6.1]
Set   [6.1]
Set-builder notation   [6.1]
Subset   [6.1]
Tree diagram   [6.1]
Union   [6.1]
Universal set   [6.1]
Venn diagram   [6.1]

SAMPLE TEST

For additional practice there are a large number of review problems categorized by objective in the Student Solutions Manual. The following sample test (40 minutes) is intended to review the main ideas of this chapter.

Let $U = \{2, 3, 4, 5, 6, 7, 8, 9, 10, 11, 12\}$, $A = \{5, 6, 7\}$, $B = \{7, 11\}$, $C = \{2, 7, 12\}$. Find the requested sets in Problems 1–4.

1. $\overline{A \cup B}$ **2.** $\bar{A} \cup \bar{B}$ **3.** $A \cap \overline{(B \cup C)}$ **4.** $\overline{A \cap (B \cup C)}$

Evaluate the expressions in Problems 5–7.

5. a. $7!$ **b.** $7! - 4!$ **c.** $(7 - 4)!$ **d.** $\dfrac{100!}{98!}$

6. a. $_6P_3$ **b.** $_6C_3$ **7. a.** $\dbinom{8}{2}$ **b.** $\dbinom{8}{5,2,1}$

8. In how many ways can a deck of 52 cards be dealt to four bridge players? Leave your answer in factorial notation.

9. In how many ways can you choose two books to read from a bookshelf containing seven books?

10. A club consists of 17 men and 19 women. In how many ways can a president, vice president, treasurer, and secretary be chosen, along with an advisory committee of 6 people, if no 2 persons can serve in more than one capacity?

11. Repeat Problem 10, but now add that exactly two of the officers must be women.

12. In how many ways can the letters in the word WORRY be arranged?

13. A mathematics test contains ten questions. In how many ways can the test be answered if the possible answers are true and false?

14. Answer Problem 13 if the possible answers are a, b, c, d, and e; that is, if the test is a multiple-choice test.

15. Human blood can be classified into types A, B, AB, or O. Furthermore, each of these types is classified as positive or negative depending on whether the antigen called Rh is present. These possibilities are summarized in the Venn diagram. Note that O blood type refers to the absence of either the A or B blood types.

Region	Blood type
1	A^-
2	B^-
3	O^+
4	A^+
5	AB^-
6	B^+
7	AB^+
8	O^-

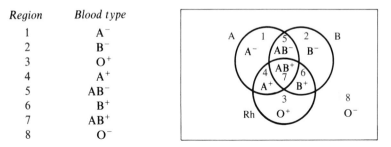

All people with the A antigen are put in circle A, those with the B antigen are put in circle B, and those with the Rh antigen are put in circle Rh. Now suppose that a hospital reports the following data:

    60 patients have the A antigen
    50 patients have the B antigen
    65 patients have the Rh antigen
    15 patients have both the A and B antigens
    20 patients have both the B and Rh antigens
    30 patients have both the A and Rh antigens
     5 patients have all three antigens
    10 patients have none of the antigens

How many patients
a. Were surveyed?
b. Have AB^+ blood?
c. Have A^+ blood?
d. Have O^- blood?
e. Have exactly one antigen?

Modeling Application 4

Earthquake Epicenter

During an earthquake, vibrations emanate in all directions from an origin called the *epicenter*. Two kinds of vibrations of importance in the study of earthquakes are called *primary* (P waves) and *secondary* (S waves). These names are due to the waves' order of arrival at a seismographic laboratory. The P wave is the fastest type of wave generated by an earthquake, with an average speed of about 8 kilometers per second. Concurrently, an S wave is produced that travels at approximately two-thirds the velocity of a P wave, so that it arrives at the recording station after the initial P wave. The greater the distance between the epicenter of the earthquake and the seismograph, the greater the time interval between the arrival of the two waves. Suppose an earthquake is recorded at three recording stations, one in Tokyo, another in Melbourne, and a third in San Francisco. The arrival times of the P and S waves are given in the following table:

	Arrival times	
Recording station	**P wave**	**S wave**
I. Tokyo	6:25:15 P.M.	6:32:19 P.M.
II. Melbourne	7:25:15 P.M.	7:34:07 P.M.
III. San Francisco	7:25:15 A.M.	7:29:15 A.M.

Write a paper that develops a mathematical model that determines the probable location of the epicenter by using Venn diagrams.* For general guidelines about writing this essay, see the commentary for Modeling Application 1 on page 130.

---

* The idea and material for this extended application are derived from Joseph Di Carlucci, "Earthquakes and Venn Diagrams," *The Mathematics Teacher*, September 1979, pp. 428–433.

CHAPTER 7
Probability

CHAPTER CONTENTS

7.1 Probability Models

7.2 Probability of Equally Likely Events

7.3 Calculated Probabilities

7.4 Conditional Probability

7.5 Probability of Intersections and Unions

7.6 Mathematical Expectation

7.7 Stochastic Processes and Bayes' Theorem

7.8 Summary and Review
Important Terms
Sample Test

APPLICATIONS

Management (*Business, Economics, Finance, and Investments*)

Expectation from life insurance policies (7.6, Problems 11–16)
Expectation from a real estate listing (7.6, Problem 26)
Distribution of cars at a car rental company (7.6, Problems 31–32)
Analysis of accident risk (7.7, Problem 28)
Probability that a shipment has a defect (7.7, Problem 29)

Life sciences (*Biology, Ecology, Health, and Medicine*)

Probabilities of family makeup—male/female (7.3, Problems 14, 24–28)
Fruit fly experiment (7.4, Problems 19–28; 7.8, Problems 6–7)
Feasibility of sinking a test oil well (7.6, Problems 27–30)
Medical diagnosis (7.7, Problem 30; 7.8, Problem 10)

Social sciences (*Demography, Political Science, Population, Psychology, Society, and Sociology*)

Public opinion polling (7.1, Problems 8, 17–19)
Survey of persons dropping a college course (7.4, Problem 36)
Qualifying for graduate school (7.7, Problem 24)
Probability that a criminal is lying (7.7, Problem 25)
Demographics of college students (7.7, Problems 26–27)

General interest

Sample space for the results of the World Series (7.1, Problem 5)
Probability of an A on an examination (7.2, Problem 17)
Dice probabilities (7.2, Problems 23–37; 7.3, Problem 33; 7.4, Problems 15–18)
Dungeons and Dragons (7.2, Problem 41)
Historical dice games (7.2, Problems 42–43)
Probability of winning a raffle (7.2, Problem 19)
Racetrack odds (7.3, Problem 32)
Poker probabilities (7.3, Problems 59–61; 7.8, Problem 4)
Historical probability problem (7.3, Problem 62)
Birthday problem (7.4, Problems 55–56)
Roulette probabilities (7.4, Problem 38; 7.6, Problems 17–25; 7.8, Problem 5)

Modeling application—Medical Diagnosis

We begin by discussing the notion of a probability model; then consider the probability of simple, equally likely events; and finally construct more complicated models by relating them back to the simpler cases.

PREVIEW This chapter introduces probabilistic models. Real world events can either be predicted with certainty or not. Of course, very little in the real world is *certain*, but events can be predicted with more or less certainty. Mathematical models are used to measure the amount of certainty, or probability. As you can imagine, it is an extremely important idea, not only in mathematics, but in almost all fields of study.

PERSPECTIVE The first major topic of finite mathematics was solving systems (both equations and inequalities) and the mathematics of matrices was used for such solutions. The second major topic of finite mathematics is probability, and is introduced in this chapter.

7.1 Probability Models

The models we have considered so far have been deterministic models. We now turn to a different type of model, called a **probabilistic model**. This model is used with situations that are random in character and attempts to predict the outcomes of events with a certain stated or known degree of accuracy. For example, if we toss a coin, it is impossible to predict in advance whether the outcome will be a head or a tail. Our intuition tells us that it is equally likely to be a head or a tail, and somehow we sense that if we repeat the experiment of tossing a coin a large number of times, heads will occur "about half the time." To check this out, I recently flipped a coin 1000 times and obtained 460 heads and 540 tails. The percentage of heads is $\frac{460}{1000} = .46 = 46\%$, which is called the **relative frequency**:

Relative Frequency of a Repeated Experiment

> If an experiment is repeated n times and an event occurs m times, then
>
> $$\frac{m}{n}$$
>
> is called the *relative frequency* of the event.

Our task in this chapter is to create a model that will assign a number p, called the *probability of an event*, which will predict the relative frequency. This means that for a *sufficiently large number of repetitions* of an experiment

$$p \approx \frac{m}{n}$$

Probabilities can be obtained in one of three ways:

1. **Theoretical probabilities** (also called *a priori models*) are obtained by logical reasoning. For example, the probability of rolling a die and obtaining a 3 is $\frac{1}{6}$

because there are six possible outcomes, each with an equal chance of occurring, so a 3 should appear $\frac{1}{6}$ of the time.

2. **Empirical probabilities** (also called *a posteriori models*) are obtained from experimental data. For example, an assembly line producing brake assemblies for General Motors produces 1500 items per day. The probability of a defective brake can be obtained by experimentation. Suppose the 1500 brakes are tested and 3 are found to be defective. Then the relative frequency, or probability, is

$$\frac{3}{1500} = .002 \quad \text{or} \quad .2\%$$

3. **Subjective probabilities** are obtained by experience and indicate a measure of "certainty." For example, a TV reporter studies the satellite maps and issues a prediction about tomorrow's weather based on past experience under similar circumstances: 80% chance of rain tomorrow.

We must define a probability measure so that it conforms to these different ways of using the word *probability*.

We begin by formalizing our terminology. We have been speaking about experiments and events. An **experiment** is the observation of any physical occurrence. A **sample space** of an experiment is the set of all possible outcomes. An **event** is a subset of the sample space. If an event is the empty set, it is called the **impossible event**; and if it has only one element, it is called a **simple event**.

EXAMPLE 1   The sample space S for the experiment of simultaneously tossing a coin and rolling a die is

$$S = \{1H, 1T, 2H, 2T, 3H, 3T, 4H, 4T, 5H, 5T, 6H, 6T\}$$

An example of an event E is "obtaining an even number or a head." That is, $E = \{1H, 2H, 2T, 3H, 4H, 4T, 5H, 6H, 6T\}$. ∎

EXAMPLE 2   A computer chip is operated until it fails, and its lifetime, t (in hours), is recorded. The sample space S is

$$S = \{t \mid t \geq 0\}$$

An example of an event E is that the chip lasts more than 1000 hours; that is, $E = \{t \mid t > 1000\}$. ∎

EXAMPLE 3   A UCLA alumnus records the results of a UCLA football game. The sample space S is

$$S = \{w, l, t\}$$

where w, l, and t denote win, lose, and tie, respectively. An example of an event E is that UCLA wins; that is, $E = \{w\}$. This is also an example of a simple event, whereas the events in Examples 1 and 2 are not simple events. ∎

Two events E and F are said to be **mutually exclusive** if $E \cap F = \varnothing$.

EXAMPLE 4   Suppose you perform an experiment of rolling a die. Then the sample space S is

$$S = \{1, 2, 3, 4, 5, 6\}$$

Let $E = \{1, 3, 5\}$, $F = \{2, 4, 6\}$, $G = \{1, 3, 6\}$, and $H = \{2, 4\}$. Then:

> E and F are mutually exclusive
> G and H are mutually exclusive

But

> F and H are *not* mutually exclusive
> E and G are *not* mutually exclusive   ■

We can now define a probability measure:

Probability Measure	Let S be a sample space associated with some experiment. With each event E we associate a real number called the *probability of E*, denoted by $P(E)$, satisfying the following properties: 1. $0 \leq P(E) \leq 1$ 2. $P(S) = 1$ 3. If E and F are mutually exclusive events, then $$P(E \cup F) = P(E) + P(F)$$

Note that this definition does not tell us how to compute $P(E)$; it simply gives some general properties of $P(E)$ from which we can build a variety of probability models. Also note that set notation is used in property 3. There is a very direct relationship between the language used to describe probabilistic situations and set notation. This relationship is shown in Table 7.1.

TABLE 7.1

$\varnothing$	Empty set	$A \cup B$	Union of sets A and B
U	Universal set	$A \cap B$	Intersection of sets A and B
S	Sample space	$\bar{A}$	Complement of a set A

Probabilistic statement	Set notation
Events A and B	A, B
A and B are mutually exclusive	$A \cap B = \varnothing$
A and B occur	$A \cap B$
A or B occurs	$A \cup B$
A does not occur	$\bar{A}$
Neither A nor B occurs	$\overline{A \cup B}$, or (equivalently) $\bar{A} \cap \bar{B}$
A and B are equally likely	$P(A) = P(B)$
A is more likely than B	$P(A) > P(B)$
A is less likely than B	$P(A) < P(B)$

In this chapter our discussion will be limited to experiments with a finite number of outcomes. Example 2 illustrates an infinite sample space. When the sample space is finite, it can be written as

$$S = \{s_1, s_2, s_3, \ldots, s_k\}$$

for some counting number k. Here $s_1, s_2, \ldots, s_k$ are the simple events or outcomes of our experiment. It follows from property 2 (above) that

$$P(s_1) + P(s_2) + P(s_3) + \cdots + P(s_k) = 1$$

which means that all the probabilities of the simple events in a probabilistic model must add up to 1.

EXAMPLE 5 Suppose a die is rolled and the number of the top face is recorded. Then $S = \{1, 2, 3, 4, 5, 6\}$. There are many possible models:

Model 1 $P(1) = \frac{1}{6}$, $P(2) = \frac{1}{6}$, $P(3) = \frac{1}{6}$, $P(4) = \frac{1}{6}$, $P(5) = \frac{1}{6}$, and $P(6) = \frac{1}{6}$. This is a model consisting of equally likely outcomes.

Model 2 $P(1) = \frac{1}{12}$, $P(2) = \frac{2}{12}$, $P(3) = \frac{3}{12}$, $P(4) = \frac{3}{12}$, $P(5) = \frac{2}{12}$, and $P(6) = \frac{1}{12}$. This is a model for a die loaded to favor the outcomes of 3 and 4.

Model 3 $P(1) = \frac{1}{10}$, $P(2) = \frac{2}{10}$, $P(3) = \frac{3}{10}$, $P(4) = \frac{3}{10}$, $P(5) = \frac{2}{10}$, and $P(6) = \frac{1}{10}$. This is *not* a probability model because the sum of the probabilities of the simple events is not 1.

Other models for this experiment are also possible. ◼

If an event is not a simple event, but has finitely many elements, then its probability can be found using the **addition principle**:

Addition Principle

If $E = \{s_1, s_2, \ldots, s_n\}$, then

$$P(E) = P(s_1) + P(s_2) + \cdots + P(s_n) \quad \text{or} \quad P(E) = \sum_{i=1}^{n} P(s_i)$$

where $P(s_1), P(s_2), \ldots, P(s_n)$ are the probabilities of the simple events.

EXAMPLE 6 Suppose a die is loaded so that the probabilities of the simple events are $P(1) = \frac{1}{12}$, $P(2) = \frac{2}{12}$, $P(3) = \frac{3}{12}$, $P(4) = \frac{3}{12}$, $P(5) = \frac{2}{12}$, and $P(6) = \frac{1}{12}$. Find the probabilities for rolling the die once and obtaining:

a. An even number   **b.** A prime number   **c.** A 2 or a 6
d. A number greater than 3   **e.** A 2 and a 6

Solution **a.** $E = \{2, 4, 6\}$ so $P(E) = P(2) + P(4) + P(6)$
$$= \tfrac{2}{12} + \tfrac{3}{12} + \tfrac{1}{12}$$
$$= \tfrac{6}{12} = \tfrac{1}{2}$$

b. $F = \{2, 3, 5\}$ so $P(F) = P(2) + P(3) + P(5)$
$$= \tfrac{2}{12} + \tfrac{3}{12} + \tfrac{2}{12}$$
$$= \tfrac{7}{12}$$

c. $G = \{2, 6\}$ so $P(G) = P(2) + P(6)$
$$= \tfrac{2}{12} + \tfrac{1}{12}$$
$$= \tfrac{3}{12} = \tfrac{1}{4}$$

d. $H = \{4, 5, 6\}$ so $P(H) = P(4) + P(5) + P(6)$
$$= \tfrac{3}{12} + \tfrac{2}{12} + \tfrac{1}{12}$$
$$= \tfrac{6}{12} = \tfrac{1}{2}$$

e. $I = \varnothing$ so $P(I) = 0$ ∎

In Example 6e, our intuition tells us that the probability of the empty event should be 0, but the following argument proves this fact by using only the definition of a probability measure.

$P(E) = P(E \cup \varnothing)$	Since $E = E \cup \varnothing$ for any set E
$P(E) = P(E) + P(\varnothing)$	Property 3 of a probability measure since E and $\varnothing$ are mutually exclusive
$0 = P(\varnothing)$	Subtract $P(E)$ from both sides

Probability of the Empty Set $P(\varnothing) = 0$

In the next section we discuss finding a model for simple events, and in Section 7.4 we find a model for the probabilities of events that are not simple.

Problem Set 7.1

Describe the sample space for the experiments in Problems 1–10.

1. A pair of dice is rolled and the sum of the top faces is recorded.

2. A coin is tossed three times and the sequence of heads and tails is recorded.

3. Ten people are asked if they graduated from college and the number of people responding "yes" is recorded.

4. A jar contains five red balls, four white balls, and three green balls. One ball is drawn and the result is recorded.

APPLICATIONS

5. The number of games necessary to complete the World Series for the years 1960–1980 is recorded. (The World Series consists of playing until one team wins four games.)

6. One hundred Eveready Long-Life® batteries are tested and the life of each battery is recorded.

7. The number of words in which an error is made on a typing test consisting of 500 words is recorded.

8. A person is randomly selected for a public opinion poll.

The person is asked two questions:

> Sex: male (m), female (f)
> Political party: Democrat (d), Republican (r), Independent (i), not registered (n)

The responses are recorded.

9. A psychologist is studying sibling relationships in families with three children. A family is asked about the sex of their children and the result is recorded. If a family has a girl, then a boy, and finally another boy, this would be recorded as {gbb}.

10. A sample of five radios is selected from an assembly line and tested. The number of defective radios is recorded.

In Problems 11–19 an experiment is described and some events are listed. Write each event using set notation and determine whether the given events are mutually exclusive.

11. A pair of dice is rolled and the sum of the numbers on the top faces is recorded.
  a. Event E is rolling an even number.
  b. Event F is rolling a prime number.

12. A pair of dice is rolled and the sum of the numbers on the top faces is recorded.
    a. Event M is rolling a 7 or an 11.
    b. Event N is rolling a 2, 3, or 12.

13. A pair of dice is rolled and the sum of the numbers on the top faces is recorded.
    a. Event G is rolling a 2 (snake eyes).
    b. Event H is rolling a sum greater than 2.

14. A coin is tossed three times and the sequence of heads and tails is recorded.
    a. Event E is tossing one head.
    b. Event F is tossing two heads.

15. A coin is tossed three times and the sequence of heads and tails is recorded.
    a. Event G is tossing a head on the first roll.
    b. Event H is tossing a head on the second roll.

16. A coin is tossed three times and the sequence of heads and tails is recorded.
    a. Event I is tossing at least one head.
    b. Event J is tossing no heads.

17. Consider the experiment described in Problem 8.
    a. Event F is the selected person is female.
    b. Event D is the selected person is a Democrat.

18. Consider the experiment described in Problem 8.
    a. Event I is the selected person is a registered Independent.
    b. Event N is the selected person is not registered.

19. Consider the experiment described in Problem 8.
    a. Event E is the selected person is either female or Republican.
    b. Event R is the selected person is a registered voter.

In Problems 20–28 consider the experiment of rolling a single die and recording the number on the top face. Let N be the event of rolling an odd number; E be the event of rolling an even number; and L be the event of rolling a number less than 3. Describe each event in words.

20. $N \cup E$ 21. $L \cup E$ 22. $N \cap E$

23. $E \cap L$ 24. $\bar{N}$ 25. $\bar{L}$

26. $\overline{N \cup L}$ 27. $\bar{E} \cap \bar{L}$ 28. $\bar{N} \cap \bar{E}$

In Problems 29–38 suppose a pair of dice are loaded so that the probabilities of the simple events are

$$P(2) = \tfrac{1}{11} \quad P(3) = \tfrac{1}{11} \quad P(4) = \tfrac{1}{11} \quad P(5) = \tfrac{1}{11}$$
$$P(6) = \tfrac{1}{11} \quad P(7) = \tfrac{1}{11} \quad P(8) = \tfrac{1}{11} \quad P(9) = \tfrac{1}{11}$$
$$P(10) = \tfrac{1}{11} \quad P(11) = \tfrac{1}{11} \quad P(12) = \tfrac{1}{11}$$

Let $C = \{2, 3, 12\}$, $E = \{7, 11\}$, and $F = \{8, 9, 10\}$.

29. Is this a probability model? If it is, explain why.

30. Find $P(E)$.   31. Find $P(C)$.

32. Find $P(F)$.   33. Find $P(E \cup F)$.

34. Find $P(E \cup C)$.   35. Find $P(E \cap C)$.

36. Find $P(\bar{F})$.   37. Find $P(\bar{C})$.

38. Find $P(\overline{E \cup F})$.

In Problems 39–48 suppose a pair of dice are loaded so that the probabilities of the simple events are

$$P(2) = \tfrac{1}{36} \quad P(3) = \tfrac{2}{36} \quad P(4) = \tfrac{3}{36} \quad P(5) = \tfrac{4}{36}$$
$$P(6) = \tfrac{5}{36} \quad P(7) = \tfrac{6}{36} \quad P(8) = \tfrac{5}{36} \quad P(9) = \tfrac{4}{36}$$
$$P(10) = \tfrac{3}{36} \quad P(11) = \tfrac{2}{36} \quad P(12) = \tfrac{1}{36}$$

Let $C = \{2, 3, 12\}$, $E = \{7, 11\}$, and $F = \{8, 9, 10\}$.

39. Is this a probability model? If it is, explain why.

40. Find $P(E)$.   41. Find $P(C)$.

42. Find $P(F)$.   43. Find $P(E \cup F)$.

44. Find $P(E \cup C)$.   45. Find $P(E \cap C)$.

46. Find $P(\bar{F})$.   47. Find $P(\bar{C})$.

48. Find $P(\overline{E \cup F})$.

7.2 Probability of Equally Likely Events

In this section we begin to calculate theoretical probabilities. Suppose a sample space is divided into simple events that are **equally likely**. For example, the experiment of flipping a coin has a sample space

$$S = \{H, T\}$$

and the simple events H = {heads}, T = {tails} are equally likely. In this text, coins are considered *fair* (equally likely heads and tails) unless otherwise noted. This means that the coin is perfectly balanced and symmetrical and the events H and T are equally likely to occur.

EXAMPLE 1   Consider two dice, each with faces labeled 1, 2, 3, 4, 5, 6.

For die A, $P(1) = \frac{1}{6}$, $P(2) = \frac{1}{6}$, $P(3) = \frac{1}{6}$, $P(4) = \frac{1}{6}$, $P(5) = \frac{1}{6}$, and $P(6) = \frac{1}{6}$.

For die B, $P(1) = \frac{1}{12}$, $P(2) = \frac{2}{12}$, $P(3) = \frac{3}{12}$, $P(4) = \frac{3}{12}$, $P(5) = \frac{2}{12}$, and $P(6) = \frac{1}{12}$. For die A, each of the six possible outcomes are equally likely. Die A is called a *fair die*. The outcomes for die B are not equally likely, so die B is called a *loaded die*. All dice used in this book are considered to be fair dice unless otherwise noted. ■

Die A in Example 1 has a sample space consisting of six equally likely simple events, so we will assign each simple event probability $\frac{1}{6}$. In so doing we are creating a probability model for rolling a single die called a **uniform probability model**.

Uniform Probability Model

> If an experiment has a sample space consisting of n mutually exclusive and equally likely simple events, then a *uniform probability model* assigns the probability of $1/n$ to each simple event.

Spades (black cards)

Hearts (red cards)

Clubs (black cards)

Diamonds (red cards)

Figure 7.1 A deck of 52 cards

EXAMPLE 2   Suppose a single card is chosen from a deck of cards. What is the probability it is an ace of spades?

Solution   In this book, when we refer to a deck of cards we are assuming the standard bridge deck shown in Figure 7.1. Since there are 52 equally likely outcomes for this experiment, we assign a probability of $\frac{1}{52}$ to the event of drawing the ace of spades. We write this as

$$P(\text{ace of spades}) = \frac{1}{52}$$ ■

EXAMPLE 3   Find $P(\text{ace})$ for the experiment in Example 2.

Solution Let $A = \{$ace of spades, ace of clubs, ace of hearts, ace of diamonds$\}$. Then

$$P(A) = P(\text{ace of spades}) + P(\text{ace of clubs}) + P(\text{ace of hearts})$$
$$+ P(\text{ace of diamonds})$$
$$= \tfrac{1}{52} + \tfrac{1}{52} + \tfrac{1}{52} + \tfrac{1}{52}$$
$$= \tfrac{4}{52}$$
$$= \tfrac{1}{13} \quad \text{or} \quad .08 \qquad \text{State probability as a reduced fraction or as a decimal rounded to the nearest hundredth} \qquad \blacksquare$$

The procedure illustrated by Example 3 is unnecessarily lengthy. Let us consider a general result. Suppose

$$E = \{k_1, k_2, k_3, \ldots, k_s\}$$

Then

$$P(E) = P(k_1) + P(k_2) + P(k_3) + \cdots + P(k_s)$$
$$= \underbrace{\frac{1}{n} + \frac{1}{n} + \frac{1}{n} + \cdots + \frac{1}{n}}_{s \text{ simple events}} \qquad \textit{Each} \text{ simple event has probability of } 1/n$$

$$= \frac{s}{n}$$

This leads us to the following important probability model:

Probability of an Event that Can Occur in Any One of n Mutually Exclusive and Equally Likely Ways

> If an experiment can occur in any of n $(n \geq 1)$ mutually exclusive and equally likely ways and if s of these ways are considered favorable, then the probability of the event E, denoted by $P(E)$, is
>
> $$P(E) = \frac{s}{n} = \frac{\text{Number of outcomes favorable to } E}{\text{Number of all possible outcomes}}$$

For Examples 4–8, suppose a single card is selected from a deck of cards (see Figure 7.1).

EXAMPLE 4 Find P(heart).

Solution There are 52 elements in the sample space and 13 of these are hearts, or successes. Therefore

$$P(\text{heart}) = \tfrac{13}{52}$$
$$= \tfrac{1}{4} \qquad\qquad \blacksquare$$

EXAMPLE 5 $P(\text{heart or an ace}) = \tfrac{16}{52}$ \qquad 13 hearts $+ 3$ *additional* aces (be careful not to count the ace of hearts twice)

$$= \tfrac{4}{13} \qquad\qquad \blacksquare$$

EXAMPLE 6 $P(\text{heart and an ace}) = \tfrac{1}{52}$ \qquad The ace of hearts is the only such card \qquad $\blacksquare$

EXAMPLE 7 P(ace or a 2) $= \frac{8}{52}$

$\qquad\qquad\qquad\qquad\quad = \frac{2}{13}$

■

EXAMPLE 8 P(ace and a 2) $= \frac{0}{52}$ There is no way of drawing a single card and obtaining an

$\qquad\qquad\qquad\qquad\quad = 0$ ace *and* a 2

■

Finding the Probability
of an Event

In summary, to find the probability of some event:

1. Describe and identify the sample space, and then count the number of elements (these should be equally likely). Call this number *n*.
2. Count the number of occurrences that are favorable to the event; call this the *number of successes* and denote it by *s*.
3. Compute the probability of the event: $P(E) = \dfrac{s}{n}$.

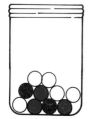

Do not forget that the simple events must be equally likely. Consider the following situation for Examples 9–11: A jar contains three red marbles, two black marbles, and five white marbles. Conduct an experiment of drawing out a single marble from the jar.

EXAMPLE 9 Find P(red).

Solution A possible sample space is {red, black, white}, but the simple events {red}, {black}, {white} are *not* equally likely. The problem, then, is to create a sample space that *has* equally likely simple events. For this example it seems that the individual marbles each have an equal chance of being chosen. Consider the sample space consisting of the individual marbles labeled as follows:

$$\{R_1, R_2, R_3, B_1, B_2, W_1, W_2, W_3, W_4, W_5\}$$

where R_1, R_2, R_3 represent the red marbles; B_1, B_2 represent the black marbles; and W_1, W_2, W_3, W_4, W_5 represent the white marbles. Then the sample space, as now described, consists of equally likely outcomes; so we use this model. Thus

$$P(\text{red}) = \frac{\text{Number of favorable outcomes}}{\text{Number of all possible outcomes}}$$

$$= \frac{3}{10}$$

■

EXAMPLE 10 P(black or white) $= \frac{7}{10}$

■

EXAMPLE 11 P(black and white) $= \frac{0}{10} = 0$ The number of favorable outcomes is 0. You cannot
draw a black *and* a white marble with only one draw

■

Sometimes probabilities are found empirically by using the model developed in this section, as shown in Example 12.

EXAMPLE 12 Suppose that in a certain study, 46 out of 155 people showed a certain kind of behavior. Assign a probability to this behavior.

Solution $p = \frac{46}{155}$

$\approx .30$ By calculator ∎

EXAMPLE 13 Suppose we conduct an experiment by rolling a pair of dice 50 times and recording the sum. This experiment is repeated 3 times for a total of 150 rolls of the dice. The results of these experiments are recorded below:

Outcome	First trial	Second trial	Third trial	Total
2	|		|	2
3	|||	||	||	7
4	⊬	||||	⊬	14
5	||	⊬ ||	⊬ |	15
6	⊬ ||||	⊬ |||	⊬ |	23
7	⊬ |||	⊬ ⊬ |	⊬ ||	26
8	⊬ |	⊬ |||	⊬ ||||	23
9	⊬ |	|||	⊬ ||	16
10	||||	||||	||	10
11	⊬	||	⊬	12
12	|	|		2
Total	50	50	50	150

The empirical probabilities can now be calculated:

$P(2) = \frac{2}{150} \approx .01$ $\quad P(3) = \frac{7}{150} \approx .05$ $\quad P(4) = \frac{14}{150} \approx .09$

$P(5) = \frac{15}{150} = .10$ $\quad P(6) = \frac{23}{150} \approx .15$ $\quad P(7) = \frac{26}{150} \approx .17$

$P(8) = \frac{23}{150} \approx .15$ $\quad P(9) = \frac{16}{150} \approx .11$ $\quad P(10) = \frac{10}{150} \approx .07$

$P(11) = \frac{12}{150} = .08$ $\quad P(12) = \frac{2}{150} \approx .01$ ∎

EXAMPLE 14 Find the theoretical probabilities for rolling a pair of dice.

Solution The sample space is $\{2, 3, 4, 5, 6, 7, 8, 9, 10, 11, 12\}$. If we assume that each simple event is equally likely, then

$$P(2) = \frac{1}{11} \approx .09$$

This same result should then hold for any of the numbers 2–12. But these theoretical probabilities do not correspond to the probabilities found in Example 13 by actually conducting an experiment of rolling a pair of dice. We know these two numbers should be approximately the same. The difficulty here is our assumption that the outcomes in this sample space are equally likely. You *must* find a sample space of *equally likely possibilities* to use the model given in this section. To find the correct sample space, let us use the fundamental counting principle. Since each of the two dice can be arranged in six ways, the total number of arrangements is $6 \cdot 6 = 36$. If you picture the dice as two different colors, you can see these arrangements in Figure 7.2.

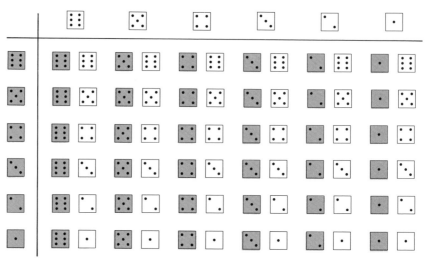

Figure 7.2 Sample space for tossing a pair of dice

Using this model, we find

$$P(2) = \frac{1}{36} \quad \leftarrow \text{Number of successful possibilities in sample space}$$
$$\leftarrow \text{Number of equally likely possibilities}$$
$$P(3) = \frac{2}{36} \quad \leftarrow \text{Look at Figure 7.2}$$
$$P(4) = \frac{3}{36}$$

After calculating the other probabilities in Example 14 (you will be asked to do some of these in Problem Set 7.2), we can compare *these* theoretical probabilities with the empirical probabilities and find consistent results, as shown in Table 7.2.

TABLE 7.2
Comparison of the empirical and theoretical probabilities for rolling a pair of dice

Outcome	Theoretical probability	Empirical probability
2	.0278	.0133
3	.0556	.0467
4	.0833	.0933
5	.1111	.1000
6	.1389	.1533
7	.1667	.1733
8	.1389	.1533
9	.1111	.1067
10	.0833	.0667
11	.0556	.0800
12	.0278	.0133

Problem Set 7.2

Use the spinner shown here for Problems 1–3 and find the requested probabilities. Assume that the pointer can never lie on a border line.

1. P(white)   **2.** P(red)   **3.** P(black)

Consider the jar containing marbles shown here. Suppose each marble has an equal chance of being picked from the jar. Find the probabilities in Problems 4–9.

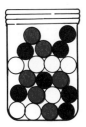

4. P(white)   **5.** P(red)

6. P(black)   **7.** P(red or black)

8. P(white or black)   **9.** P(white and black)

A single card is selected from a deck of 52 cards. Find the probabilities in Problems 10–15.

10. P(5 of clubs)   **11.** P(5)

12. P(club)   **13.** P(jack and spade)

14. P(jack)   **15.** P(jack or spade)

APPLICATIONS

Give the probabilities in Problems 16–19 in decimal form (correct to two decimal places).

16. Last year, 1485 calculators were returned to the manufacturer. If 85,000 were produced, assign a number to specify the probability that a particular calculator would be returned.

17. Last semester, Professor Math gave 13 A's out of 285 grades. If the grades were assigned randomly, what is the probability of an A?

18. Last year, it rained on 85 days in a certain city. What is the probability of rain on a day selected at random?

19. A campus club is having a raffle and needs to sell 1500 tickets. If the people on your floor of the dorm buy 285 of the tickets, what is the probability that someone on your floor will hold the winning ticket?

Perform the experiments described in Problems 20–22, tally your results, and calculate the empirical probabilities for each outcome (to the nearest hundredth).

20. Flip three coins simultaneously 50 times and note the results. The possible outcomes are (1) three heads, (2) two heads and one tail, (3) two tails and one head, and (4) three tails. Do these appear to be equally likely outcomes?

21. Simultaneously toss a coin and roll a die 50 times, and note the results. The possible outcomes are 1H, 1T, 2H, 2T, 3H, 3T, 4H, 4T, 5H, 5T, 6H, and 6T. Do these appear to be equally likely outcomes?

22. Prepare three cards that are identical except for the color. One card is black on both sides, one is white on both sides, and one is black on one side and white on the other. One card is selected at random and placed flat on the table. You will see either a black or a white card; record the color of the face. This is not the event with which we are concerned; rather, we are interested in finding the probability of the *other* side being black or white. Record the color of the underside, as shown in the table:

Color of face	Fre-quency	Outcome (color of the underside)	Fre-quency	Prob-ability
White		White		
		Black		
Black		White		
		Black		

Repeat the experiment 50 times and find the probability of occurrence with respect to the known color. Do these appear to be equally likely outcomes?

Use the sample space shown in Figure 7.2 to find the probabilities that the sums of the top faces on the dice are the numbers requested in Problems 23–34.

23. P(5)   **24.** P(6)   **25.** P(7)   **26.** P(8)

27. P(9)   **28.** P(10)   **29.** P(11)   **30.** P(12)

31. P(4 or 5)   **32.** P(4 and 5)

33. P(even number)   **34.** P(odd number)

35. Dice is a popular game in gambling casinos. Two dice are tossed, and various amounts are paid according to the outcome. If a 7 or 11 occurs on the first roll, the player wins. What is the probability of winning on the first roll?

36. In the game of dice, the player loses if the outcome of the first roll is a 2, 3, or 12. What is the probability of losing on the first roll?

37. In the game of dice, a pair of 1s is called *snake eyes*. What is the probability of losing a dice game by rolling snake eyes?

38. Consider a die with only four sides, marked 1, 2, 3, and 4. Write out a sample space similar to the one shown in Figure 7.2 for rolling a pair of these dice.

39. Using the sample space you found in Problem 38, find the probability that the sum of the dice is the given number. Assume equally likely outcomes.
 a. P(2) **b.** P(3) **c.** P(4)

40. Using the sample space you found in Problem 38, find the probability that the sum of the dice is the given number. Assume equally likely outcomes.
 a. P(5) **b.** P(6) **c.** P(7)

41. The game of Dungeons and Dragons uses nonstandard dice. Consider a die with eight sides marked 1, 2, 3, 4, 5, 6, 7, and 8. Write out a sample space similar to the one shown in Figure 7.2 for rolling a pair of these dice.

42. Historical Question The Romans played many dice games using a stone with 14 faces marked with the roman numerals I to XIV. Assuming that each face of this die has an equally likely chance of occurring, find the requested probabilities when one such stone is tossed.
 a. P(V) **b.** P(VII) **c.** P(X) **d.** P(XV)

43. Historical Question The Romans played a game of chance that required the participants to roll a pair of dice like the one described in Problem 42. Find the probability that the sum of the faces is the given number.
 a. P(20) **b.** P(2) **c.** P(15) **d.** P(25)

7.3 Calculated Probabilities

In Section 7.2 our focus was on deciding whether the events were mutually exclusive and equally likely. Then we applied the formula

$$P(E) = \frac{\text{Number of favorable outcomes}}{\text{Number of all possible outcomes}}$$

In this section we focus on different ways of counting the number of all possible outcomes and the number of favorable outcomes. You will need to use the fundamental counting principle, permutations, and combinations. It will also be helpful to have a calculator to put your answers in decimal form.

In Examples 1–3 assume that in an assortment of 20 electronic calculators there are 5 with defective switches.

EXAMPLE 1 If one machine is selected at random, what is the probability of picking one with a defective switch?

Solution The solution is given by

$$\frac{\text{Number of ways of selecting 1 defective from the 5}}{\text{Number of ways of selecting 1 machine from the 20}} = \frac{5}{20} = \frac{1}{4}$$

Now, to use the terminology of combinations in anticipation of more complicated problems, we can rework this problem:

$$\frac{\binom{5}{1}}{\binom{20}{1}} = \frac{5}{20} = .25$$

■

EXAMPLE 2   If two machines are selected at random, what is the probability that they both have defective switches?

Solution   The solution is given by

$$\frac{\left(\begin{matrix}\text{Number of ways of selecting}\\ \text{2 defectives from the 5}\end{matrix}\right)}{\left(\begin{matrix}\text{Number of ways of selecting}\\ \text{2 machines from the 20}\end{matrix}\right)} = \frac{\binom{5}{2}}{\binom{20}{2}} = \frac{\frac{5 \cdot 4}{2}}{\frac{20 \cdot 19}{2}} = \frac{1}{19} \approx .05$$ ∎

EXAMPLE 3   If three machines are selected at random, what is the probability that exactly one has a defective switch?

Solution   In this problem we use the fundamental counting principle to determine the number of successes. Picking one machine with a defective switch:

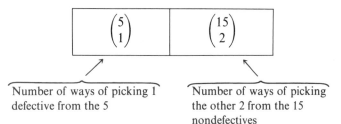

The probability we are seeking is

$$\frac{\left(\begin{matrix}\text{Number of ways of picking}\\ \text{1 defective from the 5}\end{matrix}\right) \cdot \left(\begin{matrix}\text{Number of ways of picking the}\\ \text{other 2 from the 15 nondefectives}\end{matrix}\right)}{\text{Number of ways of picking 3 from 20}}$$

$$= \frac{\binom{5}{1} \cdot \binom{15}{2}}{\binom{20}{3}} = \frac{\frac{5 \cdot 15 \cdot 14}{2}}{\frac{20 \cdot 19 \cdot 18}{3 \cdot 2 \cdot 1}}$$

$$= \frac{5 \cdot 15 \cdot 7}{20 \cdot 19 \cdot 3} = \frac{35}{76} \approx .46$$ ∎

EXAMPLE 4   Extrasensory perception (ESP) can be tested by using five colored cards. The subject is asked to arrange the red, blue, green, black, and white cards in the same order in which a person in another room has arranged them. What is the probability that the subject would arrange the cards in the same order by chance?

Solution   Let $E = \{\text{proper arrangement}\}$.

$$P(E) = \frac{\text{Number of proper arrangements}}{\text{Number of possible arrangements}} = \frac{1}{{}_5P_5}$$

$$= \frac{1}{5!} = \frac{1}{120}$$

$$\approx .008$$ ∎

EXAMPLE 5   What is the probability of being dealt a flush in poker?

Solution   A flush is 5 cards (from a deck of 52 cards) in the same suit (hearts, diamonds, clubs, or spades).

$$P(\text{flush}) = \frac{\text{Number of ways of obtaining a flush}}{\text{Number of possible poker hands}}$$

$$= \frac{(\text{Number of suits}) \cdot (\text{Number of flushes of a particular suit})}{\text{Number of ways of drawing 5 cards from 52}}$$

$$= \frac{\binom{4}{1} \cdot \binom{13}{5}}{\binom{52}{5}} \qquad \text{Fundamental counting principle}$$

$$= \frac{4 \cdot 1287}{2,598,960} \qquad \text{Detail of work:} \quad \binom{4}{1} = 4$$

$$\approx .00198$$

$$\binom{13}{5} = \frac{13!}{5!(13-5)!} = 1287$$

$$\binom{52}{5} = \frac{52!}{5!(52-5)!} = 2,598,960$$

EXAMPLE 6   What is the probability that a family that has five children has exactly two girls?

Solution   Assume that the probability of a boy or a girl is the same.

$$P(\text{exactly 2 girls}) = \frac{\text{Number of families of 5 having exactly 2 girls}}{\text{Number of all possible 5-children families}}$$

$$= \frac{\binom{5}{2}}{2^5} \qquad \leftarrow \text{Ways of selecting 2 out of 5}$$

$\leftarrow$ Fundamental counting principle;
2 possibilities for each of the
5 children

$$= \frac{10}{32}$$

$$\approx .3125$$

An easily calculated probability often involves **complementary events**. Events are complementary if they are mutually exclusive and together make up the entire sample space. The complement of any event E is denoted by $\bar{E}$. Given a sample space S and any event E,

$$E \cup \bar{E} = S \qquad \text{and} \qquad E \cap \bar{E} = \varnothing$$

so that

$$1 = P(S) = P(E \cup \bar{E}) = P(E) + P(\bar{E})$$

Therefore

$$P(E) = 1 - P(\bar{E})$$

This formula is used when it is easier to find $P(\bar{E})$ than it is to find $P(E)$.

EXAMPLE 7 Find the probability (to the nearest hundredth) that a poker hand (5 cards) has at least one ace.

Solution Direct calculation involves finding the probabilities of having one ace, two aces, three aces, or four aces in a hand. But if $E = \{$at least one ace$\}$, then $\bar{E} = \{$not at least one ace$\} = \{$no aces$\}$. It is easier to find $P(\bar{E})$ than it is to find $P(E)$:

$$P(\bar{E}) = P(\text{no aces}) = \frac{\binom{48}{5}}{\binom{52}{5}} \qquad \begin{array}{l} \leftarrow \text{5 out of 48 cards that are not aces} \\[1em] \leftarrow \text{5 out of 52 cards in deck} \end{array}$$

$$\approx .6588 \qquad \text{Using a calculator}$$

Thus

$$P(E) = 1 - P(\bar{E})$$
$$\approx 1 - .6588$$
$$\approx .34 \qquad \blacksquare$$

Probabilities are often stated in terms of **odds**. There are two ways of stating odds: *odds in favor* and *odds against*. If the odds against your winning are 2 to 5, then the odds in favor of your winning are 5 to 2. In general:

Odds in Favor The *odds in favor of an event E*, where $P(E)$ is the probability that E will occur, are

$$P(E) \quad \text{to} \quad 1 - P(E) \qquad \text{or} \qquad \frac{P(E)}{1 - P(E)} = \frac{P(E)}{P(\bar{E})}$$

Odds Against The *odds against an event E* are

$$1 - P(E) \quad \text{to} \quad P(E) \qquad \text{or} \qquad \frac{1 - P(E)}{P(E)} = \frac{P(\bar{E})}{P(E)}$$

EXAMPLE 8 If a contest has 1000 entries and you purchase 10 tickets, what are the odds against your winning?

Solution $P(\text{winning}) = \dfrac{10}{1000} = \dfrac{1}{100} \qquad 1 - P(\text{winning}) = \dfrac{99}{100}$

Odds against:

$$\frac{\frac{99}{100}}{\frac{1}{100}} = \frac{99}{100} \times \frac{100}{1}$$

$$= \frac{99}{1}$$

The odds against winning are 99 to 1. $\blacksquare$

In Example 8 we found the odds by first calculating the probability. Suppose, instead, we are given the odds and wish to calculate the probability.

Procedure for Finding Probability Given the Odds

If the odds in favor of an event E are s to b, then the probability of E is given by

$$P(E) = \frac{s}{b + s}$$

EXAMPLE 9   If the odds in favor of some event are 2 to 5, what is the probability?

Solution   In this example, $s = 2$ and $b = 5$, so $P(E) = \frac{2}{7}$.   ∎

EXAMPLE 10   If the odds against you are 100 to 1, what is the probability?

Solution   First change (mentally) to odds in favor: 1 to 100. Then, $P(E) = \frac{1}{101}$.   ∎

Problem Set 7.3

APPLICATIONS

It is known that a company has 10 pieces of machinery with no defects, 4 with minor defects, and 2 with major defects. Thus each of the 16 machines falls into one of these three categories. The inspector is about to come, and it is her policy to choose one machine and check it. Find the requested probabilities in Problems 1–6.

1. P(no defects)
2. P(major defect)
3. P(minor defect)
4. P(no major defects)
5. P(no minor defects)
6. P(defect)

Suppose a jar contains six red balls, four white balls, and three black balls. One ball is drawn at random. Find the requested probabilities in Problems 7–12.

7. P(red)
8. P(white)
9. P(black)
10. P(black or white)
11. P(red or white)
12. P(black and white)

13. What are the odds in favor of drawing a heart from an ordinary deck of 52 cards?

14. What are the odds against a family with four children having four boys?

15. Suppose the odds that a man will be bald by the time he is 60 are 9 to 1. State this as a probability.

16. Suppose the odds are 33 to 1 that someone will lie to you at least once in the next 7 days. State this as a probability.

17. What are the odds in favor of drawing an ace from an ordinary deck of 52 cards?

A certain magic trick with cards requires that five cards be randomly placed side by side in a line. Find the requested probabilities in Problems 18–23.

18. P(first card at the left is the ace of spades)
19. P(middle card is a heart)
20. P(first and last cards are diamonds)
21. P(left 3 cards are clubs and the next two are not)
22. P(second and fourth cards are kings)
23. P(middle 3 cards are red)

Suppose that a family has four children. Find the requested probabilities in Problems 24–28.

24. P(exactly 2 boys)
25. P(exactly 1 girl)
26. P(all boys)
27. P(2 girls and 2 boys)
28. P(3 girls and 1 boy)

29. What is the probability of flipping a coin five times and obtaining exactly two heads?

30. What is the probability of flipping a coin five times and obtaining three heads and two tails?

31. What is the probability of flipping a coin six times and obtaining exactly three heads?

32. Racetracks quote the approximate odds for a particular race on a large display board called a tote board. Some

examples from a particular race are:

Horse number	Odds
1	2 to 1
2	15 to 1
3	3 to 2
4	7 to 5
5	1 to 1

The odds stated are for the horse losing. Thus

$$P(\text{horse 1 losing}) = \frac{2}{2+1} = \frac{2}{3}$$

$$P(\text{horse 1 winning}) = 1 - \frac{2}{3} = \frac{1}{3}$$

What would be the probability of winning for each of these horses?

33. What are the odds in favor of rolling a 7 or 11 on a single roll of a pair of dice?

For Problems 34–39 assume that the inspector described in Problems 1–6 selects 2 machines at random. Find the requested probabilities as decimals rounded to the nearest hundredth.

34. P(both defective)
35. P(both nondefective)
36. P(both have major defects)
37. P(both have minor defects)
38. P(one with no defects and one with major defects)
39. P(one with a major defect and one with a minor defect)

For Problems 40–45 suppose two balls are drawn at random from the jar described in Problems 7–12. Find the requested probabilities as decimals rounded to the nearest hundredth.

40. P(both red)
41. P(both white)
42. P(both black)
43. P(1 red and 1 white)
44. P(1 black and 1 white)
45. P(1 red and 1 black)

For Problems 46–51 assume that the inspector described in Problems 1–6 selects 3 machines at random. Find the requested probabilities as decimals rounded to the nearest hundredth.

46. P(all have major defects)
47. P(all have minor defects)
48. P(all are nondefective)
49. P(exactly one major defect)
50. P(exactly one minor defect)
51. P(one of each type)

For Problems 52–58 suppose three balls are drawn at random from the jar described in Problems 7–12. Find the requested probabilities as decimals rounded to the nearest hundredth.

52. P(3 red)   53. P(3 white)   54. P(3 black)
55. P(2 red and 1 white)   56. P(2 white and 1 black)
57. P(2 black and 1 red)   58. P(one of each color)

Find the requested probabilities in Problems 59–61 as decimals correct to eight decimal places.

59. P(royal flush). (*Note:* A royal flush is an ace, king, queen, jack, and 10 of one suit.)
60. P(full house of three aces and a pair of 2s)
61. P(pair). [*Note:* A hand better than a pair (such as three of a kind) is not called a pair.]
62. Historical Question The mathematical theory of probability arose in France in the seventeenth century when a gambler, the Chevalier de Méré, was interested in adjusting the stakes so that he could be certain of winning if he played long enough. He was betting that he could get at least one 6 in four rolls of a die. He also bet that in 24 tosses of a pair of dice, he would get at least one 12. He found that he won more often than he lost with the first bet, but not with the second. He did not know why, so he wrote to the mathematician Blaise Pascal (1623–1662) to find out why. Pascal sent these questions to another mathematician, Pierre de Fermat (1601–1665), and together they developed the first theory of probability. What are the probabilities for winning in the two games described by de Méré?

7.4 Conditional Probability

The probability of an event depends on what information is known about that event. For example, suppose you know that a family has two children and you are interested in the probability that both the children are boys. This probability depends on additional information, as illustrated by Example 1.

EXAMPLE 1   What is the probability that a family with two children has two boys if:

a. You have no additional information.
b. You know that there is at least one boy.
c. You know that the youngest child is a boy.

Solution   Let

$$D = \{2 \text{ boys}\}$$
$$E = \{\text{at least 1 boy}\}$$
$$F = \{\text{youngest is a boy}\}$$

a. Consider the sample space:

Sample space

BB   ← Success ⎤
BG   ⎥
GB   ⎬ Four possibilities
GG   ⎦

Thus $P(D) = \frac{1}{4}$.

b. The sample space from part a is *altered* because of the additional information:

Sample space

BB   ← Success ⎤
BG   ⎬ Three possibilities
GB   ⎦
~~GG~~ ←————————— This is crossed out because we have additional information that the family has at least one boy. This is called *altering the sample space*

We write $P(D\,|\,E)$ to mean the probability of D *given the additional information* that E has occurred. Thus $P(D\,|\,E) = \frac{1}{3}$.

c. Again consider the altered sample space:

Sample space

BB
BG
~~GB~~ ⎤ These are crossed out because you know that the youngest child
~~GG~~ ⎦ is a boy

Thus $P(D\,|\,F) = \frac{1}{2}$.   ∎

In Example 1 we needed a notation to represent the probability of an event E *given that an event F has occurred*. This is the idea of **conditional probability**. We write $P(E\,|\,F)$ to denote the probability of an event E *given* that an event F has occurred. One way of finding a conditional probability is to consider an altered sample space.

EXAMPLE 2   Consider rolling a pair of dice.

a. What is the probability that the sum is 6?

b. What is the probability that the sum is 6 if you know that one of the dice shows a 5?

c. What is the probability that at least one of the numbers showing is a 6?

d. What is the probability of rolling a 2, 3, or 12 (this is known as *craps*) if you know that at least one of the dice shows a 6?

Solution The sample space is shown in Figure 7.2 (page 278).

a. Out of a sample space consisting of 36 equally likely pairs (x, y), there are five ways of obtaining a 6: (5, 1), (4, 2), (3, 3), (2, 4), and (1, 5). Thus $P(6) = \frac{5}{36}$.

b. The sample space is now reduced to 11 possibilities: (6, 5), (5, 5), (4, 5), (3, 5), (2, 5), (1, 5), (5, 6), (5, 4), (5, 3), (5, 2), and (5, 1). Of these, two are considered success. Thus $P(6 \mid \text{one is a } 5) = \frac{2}{11}$.

c. There are 11 ways to have at least one 6: (6, 6), (5, 6), (4, 6), (3, 6), (2, 6), (1, 6), (6, 5), (6, 4), (6, 3), (6, 2), and (6, 1). There are 36 equally likely pairs (x, y), so P(at least one number shown is a 6) $= \frac{11}{36}$.

d. The sample space is reduced to those pairs shown in part c, and a 2, 3, or 12 can be obtained as follows: (1, 1), (1, 2), (2, 1), and (6, 6). There is one possibility out of the sample space, so $P(2, 3, \text{ or } 12 \mid \text{at least one is a } 6) = \frac{1}{11}$. ■

EXAMPLE 3 Before an advertising campaign for a new product is launched, a survey is conducted to determine the present usage for a certain brand of hair dryer. The results are shown in the table. Assume that one respondent is chosen at random from this sample of 1400. Let

$$M = \{\text{male}\} \qquad D = \{\text{daily use}\}$$
$$F = \{\text{female}\} \qquad C = \{\text{occasional use}\}$$

	Daily use	Occasional use	Total
Male	142	258	400
Female	619	381	1000
Total	761	639	1400

Find each of the following (to the nearest hundredth) and interpret:

a. $P(M)$ **b.** $P(D)$ **c.** $P(M \mid D)$ **d.** $P(D \mid M)$

Solution **a.** $P(M)$ is the probability that the respondent is male:

$$P(M) = \frac{400}{1400} \quad \begin{array}{l} \leftarrow \text{Number of males} \\ \leftarrow \text{Total number of respondents} \end{array}$$
$$\approx .29$$

b. $P(D)$ is the probability that the respondent uses a hair dryer daily:

$$P(D) = \frac{761}{1400} \quad \begin{array}{l} \leftarrow \text{Number of daily users} \\ \leftarrow \text{Total number of respondents} \end{array}$$
$$\approx .54$$

c. $P(M|D)$ is the probability that the respondent is a male if it is known that the respondent uses a hair dryer daily:

$$P(M|D) = \frac{142}{761} \quad \begin{array}{l} \leftarrow \text{Number of males} \\ \leftarrow \text{Total number} \end{array} \Bigg\} \text{ Reduced sample space; daily users}$$

$$\approx .19$$

d. $P(D|M)$ is the probability that the respondent uses a hair dryer daily if it is known that the respondent is a male:

$$P(D|M) = \frac{142}{400} \quad \begin{array}{l} \leftarrow \text{Number of daily users} \\ \leftarrow \text{Total number} \end{array} \Bigg\} \text{ Reduced sample space; males}$$

$$\approx .36 \qquad\qquad\qquad\qquad\qquad\qquad\qquad\qquad\qquad\qquad ∎$$

Examples 2 and 3 were designed to help you understand the idea of conditional probability, but these are simple examples with simple reduced sample spaces. It is not always practical to list the sample space and then the reduced sample space, so we seek to define $P(E|F)$ in such a way as to tell us how to calculate $P(E|F)$ without listing the sample space.

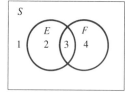

Figure 7.3

Consider the Venn diagram for two events E and F in Figure 7.3. We seek $P(E|F)$. Let $n(E)$, $n(F)$, and $n(E \cap F)$ be the number of times that events E, F, and $E \cap F$ have occurred among the n repetitions of the experiment. Now $P(E|F)$ can be found by focusing our attention on region 3. We calculate the number

$$\frac{n(E \cap F)}{n(F)}$$

which represents the number of times E has occurred among those outcomes in which F has occurred. Now

$$\frac{n(E \cap F)}{n(F)} = \frac{\dfrac{n(E \cap F)}{n}}{\dfrac{n(F)}{n}} = \frac{P(E \cap F)}{P(F)}$$

This leads to the following definition:

Conditional Probability

The *conditional probability* that an event E has occurred given that event F has occurred is denoted by $P(E|F)$ and defined by

$$P(E|F) = \frac{P(E \cap F)}{P(F)} \qquad \text{provided} \quad P(F) \neq 0$$

EXAMPLE 4 A tire manufacturer has found that 10% of the tires produced have cosmetic defects and 2% have both cosmetic and structural defects. What is the probability that one tire selected at random is structurally defective if it is known that it has a cosmetic defect?

Solution Let $C = \{\text{cosmetic defect}\}$ and $S = \{\text{structural defect}\}$; find $P(S \mid C)$:

$$P(S \mid C) = \frac{P(S \cap C)}{P(C)}$$

$$= \frac{\frac{2}{100}}{\frac{10}{100}}$$

$$= \frac{1}{5}$$
∎

The conditional probability $P(E \mid F)$ is the probability that E occurs when it is known that F occurs. Sometimes, though, the occurrence of F does not affect $P(E)$. In such circumstances we say that E and F are **independent**.

Independent Events

> The events E and F are said to be *independent* if
> $$P(E \mid F) = P(E)$$
> That is, the occurrence of E is not affected by the occurrence or nonoccurrence of F.

EXAMPLE 5 Suppose a coin is tossed twice. Let

$$H_1 = \{\text{head on first toss}\}$$
$$H_2 = \{\text{head on second toss}\}$$

These events seem to be independent since the occurrence of one event does not affect the occurrence of the other. We can verify this by using the definition.

Sample space

HH	$P(H_1) = \frac{2}{4} = \frac{1}{2}$
HT	$P(H_2) = \frac{2}{4} = \frac{1}{2}$
TH	$P(H_1 \cap H_2) = \frac{1}{4}$
TT	

These probabilities are found by looking at the sample space

Thus

$$P(H_1 \mid H_2) = \frac{P(H_1 \cap H_2)}{P(H_2)}$$

$$= \frac{\frac{1}{4}}{\frac{1}{2}}$$

$$= \frac{1}{2} = P(H_1)$$

Since $P(H_1 \mid H_2) = P(H_1)$, we know that H_1 and H_2 are independent. ∎

EXAMPLE 6 An experiment consists of drawing two consecutive cards from an ordinary deck of cards. Let $E = \{\text{first card is red}\}$ and $F = \{\text{second card is black}\}$. Are these events independent if:

a. The cards are drawn with replacement?
b. The cards are drawn without replacement?

Solution To verify independence we need to calculate $P(F \mid E)$ and compare this with $P(F)$; the events are independent if

$$P(F \mid E) = P(F)$$

a. $P(F \mid E) = \dfrac{26}{52}$ ← Number of black cards in deck
                       ← Number of cards in deck

$P(F) = \dfrac{1}{2}$ There are the same number of outcomes with the second card black as there are with the second card red

Thus $P(F \mid E) = P(F)$, so these are independent events. This should be consistent with your intuition about these events.

b. $P(F \mid E) = \dfrac{26}{51}$ ← Number of black cards in deck
                       ← Number of cards in deck (remember, this time the card is not replaced)

$P(F) = \dfrac{1}{2}$ This event, by itself, has the same probability whether or not the experiment is done with replacement

Thus $P(F \mid E) \neq P(F)$, so these are not independent events. ∎

There is an alternate way to check for independence. Suppose that E and F are independent. Then

$$P(E \mid F) = P(E)$$

Suppose we rewrite this expression:

$$P(E) = P(E \mid F)$$
$$= \frac{P(E \cap F)}{P(F)}$$

Therefore, by multiplication of both sides of the equation,

$$P(E) \cdot P(F) = P(E \cap F)$$

provided E and F are independent. We can also carry out the same calculation in reverse to arrive at the following test for independence:

Test for Independent Events

> Events E and F are independent if and only if
> $$P(E \cap F) = P(E) \cdot P(F)$$

EXAMPLE 7 Toss a single die twice. Let

$$E = \{\text{first toss is a prime}\}$$
$$F = \{\text{first toss is a 3}\}$$
$$G = \{\text{second toss is a 2}\}$$
$$H = \{\text{second toss is a 3}\}$$

Decide which of these events are independent.

Solution The sample space for both tosses has 36 possibilities and is shown in Figure 7.2 (page 278). Using this sample space, we find:

$$P(E) = \tfrac{18}{36} = \tfrac{1}{2}$$
$$P(F) = \tfrac{6}{36} = \tfrac{1}{6} \qquad P(E \cap F) = \tfrac{6}{36} = \tfrac{1}{6}$$
$$P(G) = \tfrac{6}{36} = \tfrac{1}{6} \qquad P(E \cap G) = \tfrac{3}{36} = \tfrac{1}{12} \qquad P(F \cap G) = \tfrac{1}{36}$$
$$P(H) = \tfrac{6}{36} = \tfrac{1}{6} \qquad P(E \cap H) = \tfrac{3}{36} = \tfrac{1}{12} \qquad P(F \cap H) = \tfrac{1}{36} \qquad P(G \cap H) = 0$$

Therefore:

E and F are not independent since $P(E \cap F) \neq P(E) \cdot P(F)$.
E and G are independent since $P(E \cap G) = P(E) \cdot P(G)$.
F and G are independent since $P(F \cap G) = P(F) \cdot P(G)$.
E and H are independent since $P(E \cap H) = P(E) \cdot P(H)$.
F and H are independent since $P(F \cap H) = P(F) \cdot P(H)$.
G and H are not independent since $P(G \cap H) \neq P(G) \cdot P(H)$. ∎

Problem Set 7.4

In Problems 1–8 consider a family with three children and find the requested probabilities.

1. P(exactly 2 boys)
2. P(exactly 1 girl)
3. P(exactly 3 girls)
4. P(exactly 2 boys given that at least 1 child is a boy)
5. P(exactly 1 girl given that at least 1 child is a girl)
6. P(exactly 3 girls given that at least 1 child is a girl)
7. P(at least 1 girl)
8. P(all girls given that there is at least 1 girl)

Suppose a coin is flipped four times. Find the requested probabilities in Problems 9–14.

9. P(exactly 3H)
10. P(exactly 2H)
11. P(exactly 3T)
12. P(exactly 3H given that there are at least 2H)
13. P(exactly 2H given that there is at least 1H)
14. P(exactly 3T given that there is at least 1T)

Consider rolling a pair of dice. Find the probabilities requested in Problems 15–18.

15. P(sum is 7)
16. P(at least one 5 is rolled)
17. P(sum is 7 | 2 is on one of the dice)
18. P(sum is 2 or 3 | at least one 5 is rolled)

APPLICATIONS

In an experiment it is necessary to examine fruit flies and determine their sex and whether they have mutated after exposure to radiation. For 1000 fruit flies examined, there are 643 females and 357 males. Also, 403 of the females are normal and 240 are mutated, while 190 of the males are normal and 167 are mutated. In Problems 19–28 assume a single fruit fly is chosen at random from the sample of 1000 and calculate the requested probabilities to the nearest hundredth.

19. P(male)
20. P(female)
21. P(normal male)
22. P(mutated male)
23. P(normal | male)
24. P(mutated | male)
25. P(male | normal)
26. P(male | mutated)
27. P(female | mutated)
28. P(mutated | female)

In Problems 29–34 suppose that E and F are events with the given probabilities. Use the definition of conditional probability to calculate both $P(E \mid F)$ and $P(F \mid E)$.

29. $P(E) = .5$, $P(F) = .2$, $P(E \cap F) = .1$

30. $P(E) = .85$, $P(F) = .45$, $P(E \cap F) = .3$

31. $P(E) = \frac{1}{3}$, $P(F) = \frac{1}{2}$, $P(E \cap F) = \frac{1}{6}$

32. $P(E) = \frac{3}{5}$, $P(F) = \frac{7}{10}$, $P(E \cap F) = \frac{1}{2}$

33. $P(E) = .5$, $P(F) = .8$, $P(E \cap F) = .4$

34. $P(E) = \frac{6}{7}$, $P(F) = \frac{2}{3}$, $P(E \cap F) = \frac{4}{7}$

35. The Ross Light Bulb Company has found that 5% of the bulbs it manufactures have defective filaments and 3% have both defective filaments and defective workmanship. What is the probability of defective workmanship in a particular bulb if you know it has a defective filament?

36. At Southeastern University a survey of students taking both college algebra and statistics found that 25% dropped college algebra, 30% dropped statistics, and 10% dropped both courses. If a person dropped college algebra, what is the probability that the person also dropped statistics?

37. The probability of tossing a coin four times and obtaining four heads in a row is $P(4H) = \frac{1}{2} \cdot \frac{1}{2} \cdot \frac{1}{2} \cdot \frac{1}{2} = \frac{1}{16}$. What is the probability of tossing a coin and obtaining a head if we know that heads have occurred on the previous four flips of the coin?

38. One "system" used by roulette players is to watch a game of roulette until a large number of reds occur in a row (say, 10). After 10 successive reds, they reason that black is "due" to occur and begin to bet large sums on black. Where is the fallacy in the reasoning of this "system" betting?

Use the definition in Problems 39–42 to determine whether events E and F are independent.

39. $P(E) = .5$, $P(F) = .2$, $P(E \cap F) = .1$

40. $P(E) = \frac{3}{5}$, $P(F) = \frac{7}{10}$, $P(E \cap F) = \frac{1}{2}$

41. $P(E) = \frac{6}{7}$, $P(F) = \frac{2}{3}$, $P(E \cap F) = \frac{4}{7}$

42. $P(E) = .5$, $P(F) = .8$, $P(E \cap F) = .4$

Suppose a die is rolled twice. Let

$A = \{ \text{first toss is an even} \}$

$B = \{ \text{first toss is a 6} \}$

$C = \{ \text{second toss is a 2} \}$

$D = \{ \text{second toss is a 3} \}$

Determine whether the events in Problems 43–48 are independent.

43. *A* and *B*   **44.** *A* and *C*   **45.** *A* and *D*

46. *B* and *C*   **47.** *B* and *D*   **48.** *C* and *D*

In Problems 49–54, consider a family with four children. Let

$E = \{2 \text{ boys and 2 girls}\}$

$F = \{\text{exactly 1 boy}\}$

$G = \{\text{at most 1 boy}\}$

$H = \{\text{at least 1 child of each sex}\}$

Determine whether the events are independent.

49. *E* and *F*   **50.** *E* and *G*   **51.** *E* and *H*

52. *F* and *G*   **53.** *F* and *H*   **54.** *G* and *H*

55. Birthday Problem Consider the set of people in your classroom or the set of U.S. presidents. What would you guess is the probability that two persons of the group are born on the same day of the year? As it turns out, Polk and Harding were both born on November 2. If any 24 or more people are selected at random, the probability that two or more of them will have the same birthday is greater than 50%! This is a seemingly paradoxical situation that will fool most people. The actual probabilities for the birthday problem are shown in the figure. Verify a value in this figure by experimentation. Use your class, a biographical dictionary, or other list of birth dates to see if 2 persons in a group of 24 persons have the same birthday.

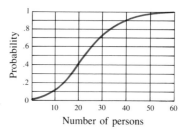

56. Birthday Problem (continued) To calculate the probabilities graphed in the figure, suppose you have a group of *n* people. Let *E* be the event that at least two of the people have the same birthday. It is very difficult to find $P(E)$, but not too difficult to find $P(\bar{E})$. If we assume that each year has 365 days (ignore leap years), then

$$P(\bar{E}) = \frac{{}_{365}P_n}{365^n} = \frac{365 \cdot 364 \cdot 363 \cdots \cdot (365 - n + 1)}{365^n}$$

Complete the following table of probabilities using the above formula:

n	5	10	20	22	23	24	30	40	50
$P(E)$	a.	b.	c.	d.	e.	f.	g.	h.	i.

7.5 Probability of Intersections and Unions

Some probability models can be built by breaking down the events being considered into simpler ones by using the words *and* (intersection) or *or* (union).

For a formula for the **probability of the intersection of two events**, recall that events E and F are independent if the occurrence of one of these events in no way affects the occurrence of the other; that is, $P(E \mid F) = P(E)$. Therefore, from the conditional probability formula,

$$P(E \mid F) = \frac{P(E \cap F)}{P(F)}$$

$$P(E \mid F) \cdot P(F) = P(E \cap F)$$

If E and F are independent, $P(E \cap F) = P(E)P(F)$. This formula is now used to derive a formula for the **probability of a union of two events**. From the definition of a probability measure (Section 7.1), we obtain a formula for

$$P(E \cup F) = P(E) + P(F) \qquad \text{if } E \text{ and } F \text{ are mutually exclusive}$$

If E and F are not mutually exclusive, then Figure 7.3 (page 288) leads us to a generalization of this formula:

$$P(E \cup F) = P(E) + P(F) - P(E \cap F)$$

We can now summarize these formulas for the probability of the intersection or union of two events.

Probability of Intersections and Unions

INTERSECTION:	$P(E \cap F) = P(E) \cdot P(F)$	Provided E and F are independent	
UNION:	$P(E \cup F) = P(E) + P(F) - P(E \cap F)$		

Suppose a coin is tossed and a die is simultaneously rolled. Let $T = \{\text{tail is tossed}\}$, $F = \{4 \text{ is rolled}\}$, and $N = \{\text{odd number is rolled}\}$. Find the probabilities in Examples 1 and 2.

EXAMPLE 1 $P(T \cap F) = P(T) \cdot P(F)$ This is the probability of a tail *and* a 4. Note that T and F

$\qquad\qquad\quad = \frac{1}{2} \cdot \frac{1}{6}$ are independent

$\qquad\qquad\quad = \frac{1}{12}$ ■

EXAMPLE 2 $P(T \cup F) = P(T) + P(F) - P(T \cap F)$ This is the probability of a tail *or* a 4

$\qquad\qquad\quad = \frac{1}{2} + \frac{1}{6} - \frac{1}{12}$

$\qquad\qquad\quad = \frac{7}{12}$ ■

In Examples 3–6 suppose a die is rolled twice and let

$A = \{\text{first toss is a prime}\}$ $B = \{\text{first toss is a 3}\}$

$C = \{\text{second toss is a 2}\}$ $D = \{\text{second toss is a 3}\}$

The sample space for this experiment has 36 possible outcomes (see Figure 7.2, page 278). By considering this sample space, we see that

$$P(A) = \tfrac{18}{36} = \tfrac{1}{2}$$
$$P(B) = P(C) = P(D) = \tfrac{6}{36} = \tfrac{1}{6}$$
$$P(A \cap B) = \tfrac{6}{36} = \tfrac{1}{6}$$

Events A and B are not independent because $P(A \cap B) \neq P(A) \cdot P(B)$

$$P(A \cap C) = P(A \cap D) = \tfrac{3}{36} = \tfrac{1}{12}$$

Events A and C *are* independent because $P(A \cap C) = P(A) \cdot P(C)$

$$P(B \cap C) = P(B \cap D) = \tfrac{1}{36}$$
$$P(C \cap D) = 0$$

This means that C and D are mutually exclusive

Can you name other pairs that are or are not independent?

EXAMPLE 3
$$\begin{aligned} P(A \cup B) &= P(A) + P(B) - P(A \cap B) \\ &= \tfrac{1}{2} + \tfrac{1}{6} - \tfrac{1}{6} \\ &= \tfrac{1}{2} \end{aligned}$$

This is the probability that a prime *or* a 3 is obtained on the first toss ∎

EXAMPLE 4
$$\begin{aligned} P(A \cup C) &= P(A) + P(C) - P(A \cap C) \\ &= \tfrac{1}{2} + \tfrac{1}{6} - \tfrac{1}{12} \\ &= \tfrac{7}{12} \end{aligned}$$

This is the probability that the first toss is a prime *or* the second toss is a 2 ∎

EXAMPLE 5
$$\begin{aligned} P(B \cup D) &= P(B) + P(D) - P(B \cap D) \\ &= \tfrac{1}{6} + \tfrac{1}{6} - \tfrac{1}{36} \\ &= \tfrac{11}{36} \end{aligned}$$

This is the probability that the first toss is a 3 *or* the second toss is a 3 ∎

EXAMPLE 6
$$\begin{aligned} P(C \cup D) &= P(C) + P(D) - P(C \cap D) \\ &= \tfrac{1}{6} + \tfrac{1}{6} - 0 \\ &= \tfrac{1}{3} \end{aligned}$$

This is the probability that the second toss is a 2 *or* a 3. Note that C and D are mutually exclusive ∎

Some experiments are said to be performed *with replacement* and others *without replacement*. Consider the following examples:

Suppose cards are drawn from a deck of 52 cards. Let

$$S_1 = \{\text{draw a spade on the first draw}\}$$
$$H_1 = \{\text{draw a heart on the first draw}\}$$
$$H_2 = \{\text{draw a heart on the second draw}\}$$

To draw **with replacement** means that the first card is drawn, the result is noted, and then the card is replaced in the deck before the second card is drawn. To draw **without replacement** means that one card is drawn, the result is noted, and then a second card is drawn without replacing the first card. Find the probabilities in Examples 7–10.

EXAMPLE 7 Find $P(S_1 \cap H_2)$, with replacement.

Solution S_1 and H_2 are independent since the cards are drawn with replacement.

$$P(S_1 \cap H_2) = P(S_1) \cdot P(H_2) = \tfrac{1}{4} \cdot \tfrac{1}{4} = \tfrac{1}{16}$$ ∎

EXAMPLE 8 Find $P(S_1 \cap H_2)$, without replacement.

Solution In this experiment the events are not independent since the probability of drawing the second card depends on what was drawn on the first card. This problem is equivalent to drawing two cards from a deck of cards. The order in which the cards are drawn is important since the question specifies that the *first* card is a spade and the *second* card is a heart; thus this is a permutation problem.

$$P(S_1 \cap H_2) = \frac{{}_{13}P_1 \cdot {}_{13}P_1}{{}_{52}P_2} = \frac{13 \cdot 13}{52 \cdot 51} = \frac{13}{204}$$ ∎

EXAMPLE 9 What is the probability of drawing two hearts with replacement?

Solution H_1 and H_2 are independent.

$$P(H_1 \cap H_2) = P(H_1) \cdot P(H_2) = \tfrac{1}{4} \cdot \tfrac{1}{4} = \tfrac{1}{16}$$ ∎

EXAMPLE 10 What is the probability of drawing two hearts without replacement?

Solution H_1 and H_2 are not independent, but the order in which the cards are drawn is not important since the question simply asks for two hearts; thus this is a combination problem.

$$P(\text{two hearts}) = \frac{{}_{13}C_2}{{}_{52}C_2} = \frac{\dfrac{13 \cdot 12}{2}}{\dfrac{52 \cdot 51}{2}} = \tfrac{1}{17} \; .$$ ∎

We conclude this section by summarizing the probability formulas:

Summary of Probability Formulas

1. $P(E) = \dfrac{s}{n}$ where event E can occur in n mutually exclusive and equally likely ways, and where s of them are considered favorable
2. $P(S) = 1$ where S is the sample space
3. $P(\emptyset) = 0$
4. $P(E) = 1 - P(\bar{E})$
5. $P(E \mid F) = \dfrac{P(E \cap F)}{P(F)}$ provided $P(F) \neq 0$
6. $P(E \text{ and } F) = P(E \cap F) = P(E) \cdot P(F)$ provided E and F are independent
7. $P(E \text{ or } F) = P(E \cup F) = P(E) + P(F) - P(E \cap F)$
8. $P(E \text{ or } F) = P(E \cup F) = P(E) + P(F)$ provided E and F are mutually exclusive

Note: When doing probability problems, assume that experiments are performed *without replacement* unless it is otherwise stated.

Problem Set 7.5

Suppose that events A, B, and C are all independent so that

$$P(A) = \tfrac{1}{2} \qquad P(B) = \tfrac{1}{3} \qquad P(C) = \tfrac{1}{6}$$

Find the probabilities in Problems 1–16.

1. $P(\bar{A})$
2. $P(\bar{B})$
3. $P(\bar{C})$
4. $P(A \cap B)$
5. $P(A \cap C)$
6. $P(B \cap C)$
7. $P(A \cup B)$
8. $P(A \cup C)$
9. $P(B \cup C)$
10. $P(\overline{A \cap B})$
11. $P(\overline{A \cap C})$
12. $P(\overline{B \cap C})$
13. $P(\overline{A \cup B})$
14. $P(\overline{A \cup C})$
15. $P(\overline{B \cup C})$
16. $P(A \cap B \cap C)$

In Problems 17–21 suppose a coin is tossed twice. Find the requested probabilities.

17. $P(2H)$
18. $P(2T)$
19. $P(3T)$
20. $P(1H \text{ and } 1T)$
21. $P(\text{match})$*

Suppose A, B, and C are independent events so that

$$P(A) = \tfrac{1}{2} \qquad P(B) = \tfrac{2}{3} \qquad P(C) = \tfrac{5}{6}$$

Find the probabilities in Problems 22–25.

22. **a.** $P(\bar{A})$ **b.** $P(\bar{B})$ **c.** $P(\bar{C})$
23. **a.** $P(A \cap B)$ **b.** $P(A \cap C)$ **c.** $P(B \cap C)$

_____

* This means P (both the same).

24. **a.** $P(A \cup B)$ **b.** $P(A \cup C)$ **c.** $P(B \cup C)$
25. **a.** $P(\overline{A \cap C})$ **b.** $P(\overline{A \cup B})$ **c.** $P(\overline{B \cup C})$

In Problems 26–38 suppose a coin is tossed and simultaneously a die is rolled. Let H = {head is tossed}, S = {6 is rolled}, and E = {even number is rolled}. Find the requested probabilities.

26. **a.** $P(H)$ **b.** $P(S)$ **c.** $P(E)$
27. $P(H \cap S)$
28. $P(H \cap E)$
29. $P(S \cap E)$
30. $P(S \cup E)$
31. $P(H \cup E)$
32. $P(H \cup S)$
33. $P(H \mid S)$
34. $P(S \mid H)$
35. $P(S \mid E)$
36. $P(E \mid S)$
37. $P(H \mid E)$
38. $P(E \mid H)$

In Problems 39–43 assume a box has five red cards and three black cards, and two cards are drawn with replacement. Find the requested probabilities.

39. P(2 red cards)
40. P(2 black cards)
41. P(1 red and 1 black card)
42. P(red on first draw and black on second draw)
43. P(red on first draw or black on second draw)

Problems 44–48 repeat the experiment of Problems 39–43 except that the cards are drawn without replacement.

44. P(2 red cards)
45. P(2 black cards)
46. P(1 red and 1 black card)
47. P(red on first draw and black on second draw)
48. P(red on first draw or black on second draw)

7.6 Mathematical Expectation

Smiles Toothpaste is giving away $10,000. All you must do to have a chance to win is send in a postcard with your name on it (the fine print says you do not need to buy a tube of toothpaste). Is it worthwhile to enter?

Suppose the contest receives one million postcards (a conservative estimate). We wish to compute your **expected value** (or your **expectation**) in entering this contest. We denote the expectation by E and compute:

$$E = (\text{Amount to win}) \cdot (\text{Probability of winning})$$

$$= \$10,000 \cdot \frac{1}{1,000,000}$$

$$= \$.01$$

A game is **fair** if the expected value equals the cost of playing the game. Is this game fair? If the toothpaste company charges you 1¢ to play, it is fair. But how much does it really cost you to enter the contest? How much is a postcard? We see that the contest is not a fair game.

EXAMPLE 1   Suppose you are going to roll two dice. You will be paid $3.60 if you roll two 1s. You do not receive anything for any other outcome. What is a fair price to pay for the privilege of rolling the dice?

Solution   A fair price is equal to the mathematical expectation of the game, which is computed as follows:

$$E = \text{(Amount of winnings)} \cdot \text{(Probability of winning)}$$

$$= \qquad \$3.60 \qquad\qquad\qquad \tfrac{1}{36}$$

$$= \$.10$$

A fair price to play this game is 10¢.   ∎

When we say that the expectation is 10¢, we certainly do *not* mean that you will win 10¢ every time you play. (Indeed, you will *never* win 10¢; you will either win $3.60 or nothing.) We mean that, if you were to play this game a very large number of times, you could expect to win an *average* of 10¢ per game.

Most games of chance will not be fair (the "house" must make a living). A realistic question would be: Which games come closest to being fair? We will consider this question in some of the problems.

Sometimes there is more than one payoff, as illustrated in Example 2.

EXAMPLE 2   A contest offers one grand prize worth $10,000, two second prizes worth $5,000 each, and ten third prizes worth $1,000 each.

Solution   The expectation can be computed by multiplying the amounts to be won by their respective probabilities and then adding the results. That is,

$$E = \text{(amount of 1st prize)} \cdot P(\text{1st prize})$$
$$+ \text{(amount of 2nd prize)} \cdot P(\text{2nd prize})$$
$$+ \text{(amount of 3rd prize)} \cdot P(\text{3rd prize})$$

Now if we assume that there are one million entries and that winners' names are replaced in the pool after being drawn, we see that:

$$P(\text{1st prize}) = \frac{1}{1,000,000}$$

$$P(\text{2nd prize}) = \frac{2}{1,000,000}$$

$$P(\text{3rd prize}) = \frac{10}{1,000,000}$$

Hence:

$$E = \$10,000 \cdot \frac{1}{1,000,000} + \$5,000 \cdot \frac{2}{1,000,000} + \$1,000 \cdot \frac{10}{1,000,000}$$

$$= \$.01 + \$.01 + \$.01$$

$$= \$.03$$ ∎

The formal definition of **mathematical expectation** follows:

Mathematical Expectation

> If an event has several possible outcomes with probabilities $p_1, p_2, p_3, \ldots,$ and for each of these outcomes the amount that can be won is $a_1, a_2, a_3, \ldots,$ the *mathematical expectation*, E, of the event is
> $$E = a_1 \cdot p_1 + a_2 \cdot p_2 + a_3 \cdot p_3 + \cdots$$

EXAMPLE 3   A contest offers the following prizes (as revealed by the fine print):

Prize	Value	Probability of winning
Grand Prize trip	$1,500 = a_1	.000026 = p_1
Weber Kettle	110 = a_2	.000032 = p_2
Magic Chef Range	279 = a_3	.000016 = p_3
Murray Bicycle	191 = a_4	.000021 = p_4
Lawn Boy Mower	140 = a_5	.000026 = p_5
Samsonite Luggage	183 = a_6	.000016 = p_6

What is the expected value for this contest?

Solution
$$E = a_1 p_1 + a_2 p_2 + a_3 p_3 + a_4 p_4 + a_5 p_5 + a_6 p_6$$
$$= 1500(.000026) + 110(.000032) + 279(.000016) + 191(.000021)$$
$$+ 140(.000026) + 183(.000016)$$
$$= .057563$$

The expected value is a little less than 6¢. Suppose we mail in our entry as specified in the rules. What is the cost of the stamp and the envelope?   ■

Sometimes there are situations in which the amount we must pay to play a game has a role in the amount we will win, so that it is not feasible to compare the expectation with the amount we must pay to play. In such situations, we calculate the expenses for playing as part of the amount to win, giving a positive, negative, or zero expectation. In this case, if the expectation is positive, then the game is in our favor; if it is negative, then the game is in our opponent's favor; and if the expectation is zero, then the game is said to be fair.

EXAMPLE 4   Suppose you are going to play a game by rolling a pair of dice. You will be paid $5.00 every time you roll a pair of 6s. You will not receive anything for any other outcome. It costs $.50 to play. What is the expected value for this game?

Solution   In this example, there is an admission price of $.50, which means that this amount is paid, win or lose. In situations for which there is an "admission price," this amount must be subtracted from the proposed winnings. Let x be the amount won. Then

Win
$x = $5.00 - $.50 = 4.50

Lose
$x = -$.50$

and
$$P(x = \$4.50) = \tfrac{1}{36} \quad \text{and} \quad P(x = -\$.50) = \tfrac{35}{36}$$

Thus
$$E(x) = 4.50(\tfrac{1}{36}) + (-.50)(\tfrac{35}{36}) \approx .125 - .486 \approx -.36 \qquad \blacksquare$$

The fact that the expectation is a loss of 36¢ per game does *not* mean that you will lose 36¢ every time you play the game. (Indeed, you will *never* lose 36¢; you will either win $5.00 or nothing—less the $.50 admission charge.) But if you were to play this game a large number of times, you could expect to lose an *average* of 36¢ per game.

EXAMPLE 5   Walt, a realtor, knows that if he takes a listing to sell a house, it will cost him $1,000. However, if he sells the house, he will receive 6% of the selling price. If another realtor sells the house, Walt will receive 3% of the selling price. If the house is unsold in 3 months, Walt will lose the listing and receive nothing. Suppose the probabilities for selling a particular $100,000 house are as follows: The probability that Walt will sell the house is .4; the probability that another agent will sell the house is .2; and the probability that the house will remain unsold is .4. What is Walt's expectation if he takes this listing?

Solution   First calculate

6% of $100,000:   .06 × $100,000 = $6,000
3% of $100,000:   .03 × $100,000 = $3,000

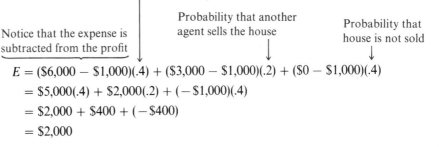

Notice that the expense is subtracted from the profit

Probability that Walt sells the house

Probability that another agent sells the house

Probability that house is not sold

$$E = (\$6,000 - \$1,000)(.4) + (\$3,000 - \$1,000)(.2) + (\$0 - \$1,000)(.4)$$
$$= \$5,000(.4) + \$2,000(.2) + (-\$1,000)(.4)$$
$$= \$2,000 + \$400 + (-\$400)$$
$$= \$2,000$$

Walt's expectation is $2000. $\blacksquare$

Operations research is the science of making optimal decisions, as illustrated in the next example.

EXAMPLE 6   Suppose that a $150,000 house has an unstable foundation and is sliding off a hillside. The only hope for the house is to drill a horizontal well to tap a reservoir of stored water that is causing the slippage. If the cost of drilling each well is $1,000, and if the probability that a particular well will hit the water reservoir is .8, how many wells should be drilled?

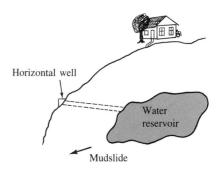

Horizontal well

Water reservoir

Mudslide

Solution It is obvious that drilling no well is too few and that drilling 150 wells is too many. Our task is to decide on the number of wells, x, to be drilled. The probability that the x wells are unsuccessful is $.2^x$ (assume that the probability of hitting water for each well is independent). Thus the probability that at least one well will hit the water reservoir is

$$1 - .2^x$$

The expected gain is $150,000(1 - .2^x)$, while the cost of drilling the wells is $1,000x$, so

$$E(x) = \$150,000(1 - .2^x) - \$1,000x$$

We want to find the maximum value of $E(x)$ for the various values of x.

$E(0) = 0$ If no wells are tried, the house will slide off the hill and be lost
$E(1) = \$150,000(1 - .2^1) - \$1,000 \cdot 1 = \$119,000$
$E(2) = \$150,000(1 - .2^2) - \$1,000 \cdot 2 = \$142,000$
$E(3) = \$150,000(1 - .2^3) - \$1,000 \cdot 3 = \$145,800$
$E(4) = \$150,000(1 - .2^4) - \$1,000 \cdot 4 = \$145,760$
$E(5) = \$150,000(1 - .2^5) - \$1,000 \cdot 5 = \$144,952$
$E(6) = \$150,000(1 - .2^6) - \$1,000 \cdot 6 = \$143,990.40$

Notice that sinking additional wells does not increase the expected gain. The expected value is greatest when three wells are drilled, so operations research tells us to drill three wells. ■

Problem Set 7.6

APPLICATIONS

1. Suppose you roll two dice. You will be paid $5 if you roll a double. You will not receive anything for any other outcomes. What is a fair price to pay for the privilege of rolling the dice?

2. A magazine subscription service is having a contest in which the prize is $80,000. If the company receives one million entries, what is the expectation of the contest?

3. You have 5 quarters, 5 dimes, 10 nickels, and 5 pennies in your pocket. You reach in and choose a coin at random so you can tip the paperboy. What is the paperboy's expectation? What is the most likely tip?

4. A game involves tossing two coins and receiving $.50 if they are both heads. What is a fair price to pay for the privilege of playing?

5. Krinkles potato chips is having a "Lucky Seven Sweepstakes." The one grand prize is $70,000; 7 second prizes each pay $7,000; 77 third prizes each pay $700; and 777 fourth prizes each pay $70. How much is the expectation of this contest if there are 10 million entries?

6. A punch-out card contains 100 spaces. One of the spaces pays $100, five of the spaces pay $10, and the others pay nothing. What is a fair price to pay to punch out one space?

7. In old gangster movies on TV, you often hear of "numbers runners" or the "numbers racket." The game, which is still played today, involves betting $1 and picking three digits, such as 245. The next day some procedure for randomly selecting numbers is used (for example, using the last three digits of the number of stocks sold on a particular day as reported in the *Wall Street Journal*). If the payoff is $500, what is the expectation for this numbers game?

8. A game involves drawing a single card from an ordinary deck. If an ace is drawn, you receive $.50; if a heart is drawn, you receive $.25; if the queen of spades is drawn, you receive $1. If the cost of playing is $.10, should you play?

9. A local company advertises a contest with the following table:

Prize	Prizes available	Approximate probability of winning
$1,000.00	13	.000005
100.00	52	.00002
10.00	520	.0002
1.00	28,900	.010989
Total	29,495	.011111*

* This is $\frac{1}{90}$.

What is the expectation for playing this game one time?

10. A club recently held a bingo contest where the following chances of winning were given:

	Playing one card 1 time	Playing one card 7 times	Playing one card 13 times
$25 prize	1 in 21,252	1 in 3036	1 in 1630
$ 3 prize	1 in 2125	1 in 304	1 in 163
$ 1 prize	1 in 886	1 in 127	1 in 68
Any prize	1 in 609	1 in 87	1 in 47

What is the expectation in dollars for playing one card 13 times?

Life insurance policies use mathematical expectation. The "winning" value is the value of the policy and the "winning" probability is the probability of dying during the life of the policy. The probability of dying is found from an actuarial table (see, for example, Table 9 in Appendix E). Note that the table gives p_x (probability of living) and q_x (probability of dying). Find the expected value for the 1-year policies described in Problems 11–16.

11. $10,000 issued at age 10
12. $10,000 issued at age 38
13. $10,000 issued at age 65
14. $25,000 issued at age 19
15. $50,000 issued at age 19
16. $125,000 issued at age 35

A U.S. roulette wheel has 38 numbered slots (1–36, 0, and 00, as shown in the figure). Some of the more common bets and payoffs are shown in the figure. If the payoff is listed as 6 to 1, you would receive $6 plus your $1 bet. Use the figure to find the expectation in Problems 17–25 for playing $1 one time. One play consists of the dealer spinning the wheel and a small white ball in opposite directions. As the ball slows to a stop it lands in one of the 38 numbered slots which are also colored black, red, or green.

17. Black
18. Odd
19. Single number bet
20. Double number bet
21. Three number bet
22. Four number bet
23. Five number bet
24. Six number bet
25. Column bet

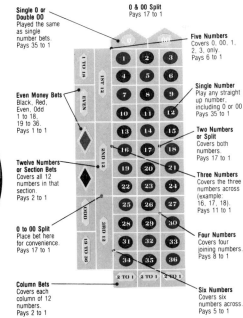

U.S. roulette wheel and board

26. A realtor who takes the listing on a house to be sold knows that she will spend $800 trying to sell the house. If she sells it herself, she will earn 6% of the selling price. If another realtor sells a house from her list, our realtor will earn only 3% of the price. If the house is unsold in 6 months, she will lose the listing. Suppose the probabilities are as follows:

	Probability
Sell by herself	.50
Sell by another realtor	.30
Not sell in 6 months	.20

What is her expected profit for listing an $85,000 house? (Be sure to subtract the $800 from the commission because the profit is 6% of the selling price *less* selling costs.)

27. An oil drilling company knows that it will cost $25,000 to sink a test well. If oil is hit, the income for the drilling company will be $425,000. If only natural gas is hit, the income will be $125,000. If nothing is hit, the company will have no income. If the probability of hitting oil is $\frac{1}{40}$ and if the probability of hitting gas is $\frac{1}{20}$, what is the expectation for the drilling company? Should the test well be sunk? (Do not forget to subtract the cost from the income in determining the payoffs.)

28. In Problem 27, suppose the income for hitting oil is changed to $825,000 and the income for gas to $225,000. Now, what is the expectation for the drilling company? Should the test well be sunk?

29. Suppose the probability of hitting the reservoir in Example 6 is .6 instead of .8. Now, how many wells should be tried?

30. Suppose each well in Example 6 costs $5,000 instead of $1,000. Now, how many wells should be tried?

31. Kingston and Associates, a consulting firm, is hired by a national car rental agency to determine the number of cars to purchase for a new outlet. After appropriate research, the consulting firm finds that the daily demand will be from 12 to 17 cars, distributed as follows:

Number of customers	12	13	14	15	16	17
Probability	.05	.1	.35	.2	.2	.1

Records also show that it costs the agency $20 per day per car whether the car is rented or not. If the rental price per car is $32, then the profit is $12 for each day the car is rented. What is the optimal number of cars that the consulting firm should recommend the agency to obtain?

32. Repeat Problem 31 for the following values:

Number of customers	10	11	12	13	14	15
Probability	.1	.2	.3	.2	.1	.1

7.7   Stochastic Processes and Bayes' Theorem

In Section 7.4 we developed the following formula for conditional probability:

$$P(E \mid F) = \frac{P(E \cap F)}{P(F)}$$

This formula can be rewritten as

$$P(E \cap F) = P(E \mid F)P(F) = P(F \mid E)P(E)$$

We now use this form to develop a formula for the probability of an event in which the possible outcomes depend on the outcomes of a preceding experiment. For example, in Section 7.3 we discussed an experiment of selecting 2 calculators from a sample space consisting of 20 calculators of which 5 had defective switches. Instead of selecting 2 at the same time, we now select them one at a time and let

$A = \{$first calculator picked has a defective switch$\}$

$B = \{$second calculator picked has a defective switch$\}$

Now, $P(A) = \frac{5}{20} = \frac{1}{4}$, but what is $P(B)$?

This sequence of experiments is called a **stochastic process**, which means that the outcome of one event depends on the outcome of the previous experiment.

EXAMPLE 1   Find $P(B)$ for the experiment just described.

Solution   To find $P(B)$ consider the following tree diagram:

Step 1   Label the paths with appropriate probabilities. The sum of these should be 1.

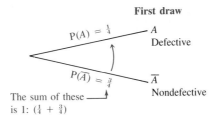

First draw

$P(A) = \frac{1}{4}$ ———— A Defective

$P(\overline{A}) = \frac{3}{4}$ ———— $\overline{A}$ Nondefective

The sum of these is 1: $(\frac{1}{4} + \frac{3}{4})$

Step 2

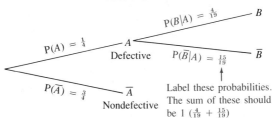

First draw   Second draw

$P(A) = \frac{1}{4}$ A Defective $P(B|A) = \frac{4}{19}$ ———— B This is $P(B|A)$; there are 4 defectives left in a sample of 19

$P(\overline{B}|A) = \frac{15}{19}$ ———— $\overline{B}$ This is $P(\overline{B}|A)$; there are 15 nondefectives left in a sample of 19.

$P(\overline{A}) = \frac{3}{4}$ $\overline{A}$ Nondefective

Label these probabilities. The sum of these should be 1 $(\frac{4}{19} + \frac{15}{19})$

Step 3

First draw   Second draw

$P(A) = \frac{1}{4}$ A Defective $P(B|A) = \frac{4}{19}$ ———— B

$P(\overline{B}|A) = \frac{15}{19}$ ———— $\overline{B}$

$P(\overline{A}) = \frac{3}{4}$ $\overline{A}$ Nondefective $P(B|\overline{A}) = \frac{5}{19}$ ———— B This is $P(B|\overline{A})$; there are 5 defectives left in a sample of 19.

$P(\overline{B}|\overline{A}) = \frac{14}{19}$ ———— $\overline{B}$ This is $P(\overline{B}|\overline{A})$; there are 14 nondefectives left in a sample of 19.

The sum of these should be 1: $(\frac{5}{19} + \frac{14}{19})$

Now, to find the probability of B you must take all the possibilities into account. Use the product rule for probability along a single path and the addition principle for different paths.

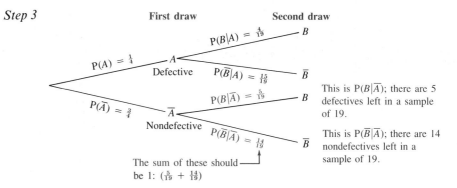

First draw   Second draw

$P(A) = \frac{1}{4}$ A Defective $P(B|A) = \frac{4}{19}$ ———— B $P(B|A)P(A) = \frac{4}{19} \cdot \frac{1}{4} = \frac{4}{76}$

$P(\overline{B}|A) = \frac{15}{19}$ ———— $\overline{B}$

$P(\overline{A}) = \frac{3}{4}$ $\overline{A}$ Nondefective $P(B|\overline{A}) = \frac{5}{19}$ ———— B $P(B|\overline{A})P(\overline{A}) = \frac{5}{19} \cdot \frac{3}{4} = \frac{15}{76}$

$P(\overline{B}|\overline{A}) = \frac{14}{19}$ ———— $\overline{B}$

Thus

$$P(B) = P(B \mid A)P(A) + P(B \mid \bar{A})P(\bar{A}) = \tfrac{4}{76} + \tfrac{15}{76}$$
$$= \tfrac{19}{76}$$
$$= \tfrac{1}{4} \qquad \blacksquare$$

To formalize the results discussed in Example 1, note that A and $\bar{A}$ form a **partition of the sample space**, as illustrated in Figure 7.4.

It can be shown that for any event E, $E = (E \cap A) \cup (E \cap \bar{A})$. Thus

$$P(E) = P[(E \cap A) \cup (E \cap \bar{A})]$$
$$= P(E \cap A) + P(E \cap \bar{A}) \qquad \text{Since } E \cap A \text{ and } E \cap \bar{A} \text{ are mutually exclusive}$$
$$P(E) = P(E \mid A)P(A) + P(E \mid \bar{A})P(\bar{A}) \qquad \text{Formula for conditional probability}$$

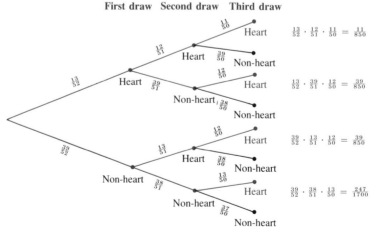

Figure 7.4

This result (shown in color in Figure 7.4) is the same result we obtained from the tree diagram in Example 1. It generalizes to any partition of the sample space.

Probability of a Partitioned Event

If $A_1, A_2, \ldots, A_n$ form a partition of a sample space and E is any event, then

$$P(E) = P(E \mid A_1)P(A_1) + P(E \mid A_2)P(A_2) + \cdots + P(E \mid A_n)P(A_n)$$

EXAMPLE 2 Three cards are drawn (without replacement) from a deck of cards. What is the probability that the third one drawn is a heart?

Solution Construct the following tree diagram. Be sure you understand how the probabilities at each step are found.

First draw   Second draw   Third draw

$\frac{13}{52}$ Heart $\frac{12}{51}$ Heart $\frac{11}{50}$ Heart $\frac{13}{52} \cdot \frac{12}{51} \cdot \frac{11}{50} = \frac{11}{850}$

  $\frac{39}{50}$ Non-heart

Heart $\frac{39}{51}$ Non-heart $\frac{38}{50}$ $\frac{12}{50}$ Heart $\frac{13}{52} \cdot \frac{39}{51} \cdot \frac{12}{50} = \frac{39}{850}$

  Non-heart

$\frac{39}{52}$ Non-heart $\frac{38}{51}$ Heart $\frac{13}{51}$ $\frac{12}{50}$ Heart $\frac{39}{52} \cdot \frac{13}{51} \cdot \frac{12}{50} = \frac{39}{850}$

  $\frac{38}{50}$ Non-heart

  Non-heart $\frac{37}{50}$ $\frac{13}{50}$ Heart $\frac{39}{52} \cdot \frac{38}{51} \cdot \frac{13}{50} = \frac{247}{1700}$

  Non-heart

Thus

$$P(\text{heart on the third draw}) = \tfrac{11}{850} + \tfrac{39}{850} + \tfrac{39}{850} + \tfrac{247}{1700}$$
$$= \tfrac{425}{1700}$$
$$= \tfrac{1}{4} \qquad \blacksquare$$

EXAMPLE 3   Three cards are drawn from a deck of cards. What is the probability that exactly two cards are hearts?

Solution   Use the tree diagram from Example 2, and trace out different paths to indicate success.

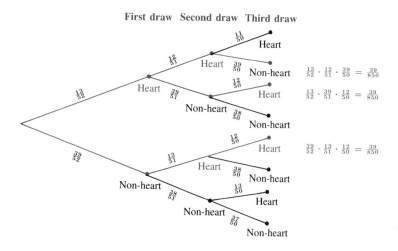

Thus

$$P(\text{exactly two hearts}) = \tfrac{39}{850} + \tfrac{39}{850} + \tfrac{39}{850} = \tfrac{117}{850} \approx .138$$ ∎

There is an important result relating $P(E\,|\,F)$ and $P(F\,|\,E)$. This formula is important because we are frequently concerned with finding $P(F\,|\,E)$ when we know (or it is easy to find) $P(E\,|\,F)$.

$$P(F\,|\,E) = \frac{P(F \cap E)}{P(E)} \qquad \text{Conditional probability formula}$$

$$= \frac{P(E \cap F)}{P(E)} \qquad P(F \cap E) = P(E \cap F)$$

$$= \frac{P(E\,|\,F)P(F)}{P(E)} \qquad \begin{array}{l}\text{Probability of intersection:}\\ P(E \cap F) = P(E\,|\,F)P(F)\end{array}$$

$$= \frac{P(E\,|\,F)P(F)}{P(E\,|\,F)P(F) + P(E\,|\,\bar{F})P(\bar{F})} \qquad \begin{array}{l}\text{Events } F \text{ and } \bar{F} \text{ form a partition}\\ \text{of the sample space}\end{array}$$

Note that $P(F\,|\,E)$ is now written in terms of $P(E\,|\,F)$. This formula was first published in 1763 by Thomas Bayes (1702–1763).

Bayes' Theorem

If $A_1, A_2, \ldots, A_n$ form a partition of the sample space and E is an event associated with S, then

$$P(A_i\,|\,E) = \frac{P(E\,|\,A_i)P(A_i)}{P(E\,|\,A_1)P(A_1) + P(E\,|\,A_2)P(A_2) + \cdots + P(E\,|\,A_n)P(A_n)}$$

$$i = 1, 2, \ldots, n$$

EXAMPLE 4 Thompson Associates, a management consulting firm, has agreed to run a quality control check on three assembly lines at Karlin Manufacturing. In checking the inventory, Thompson found a defective item. What is the probability that it came from a particular assembly line?

Analysis In checking the records, Thompson Associates gathers the following information:

Number of items in inventory	Previous testing results
50% from assembly line A	2% of items from assembly line A are defective
30% from assembly line B	3% of items from assembly line B are defective
20% from assembly line C	4% of items from assembly line C are defective

Let

$A = \{$item tested is from assembly line A$\}$

$B = \{$item tested is from assembly line B$\}$

$C = \{$item tested is from assembly line C$\}$

$D = \{$item tested is defective$\}$

Thompson is now able to calculate the following probabilities from the given information:

$$P(A) = .5 \qquad P(B) = .3 \qquad P(C) = .2$$
$$P(D\,|\,A) = .02 \qquad P(D\,|\,B) = .03 \qquad P(D\,|\,C) = .04$$

Find $P(A\,|\,D)$, $P(B\,|\,D)$, and $P(C\,|\,D)$.

Solution Find $P(A\,|\,D)$ first by formula:

$$P(A\,|\,D) = \frac{P(D\,|\,A)P(A)}{P(D\,|\,A)P(A) + P(D\,|\,B)P(B) + P(D\,|\,C)P(C)}$$

$$= \frac{(.02)(.5)}{(.02)(.5) + (.03)(.3) + (.04)(.2)} = \frac{.01}{.027} \approx .37$$

Since it can often be confusing to use the formula, a variation of a tree diagram, called a *Bayes' tree*, may be drawn:

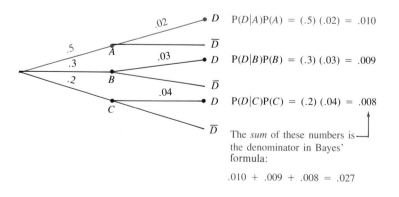

Using Bayes' tree, find the quotient of the desired path (shown in color) divided by the sum of all path possibilities:

$$P(A \mid D) = \frac{.010}{.027} \approx .37$$

Use Bayes' tree to find the other probabilities:

$$P(B \mid D) = \frac{.009}{.027} \approx .33 \qquad P(C \mid D) = \frac{.008}{.027} \approx .30$$ ■

EXAMPLE 5   Suppose a test for AIDS is given to a patient. Let

$$A = \{\text{person has AIDS}\}$$
$$R = \{\text{test shows a positive result}\}$$

If the person has AIDS, the test shows a positive result 95% of the time; if the person does not have AIDS, the test shows a positive result 5% of the time. It is also known that .1% of the population has undetected AIDS. Find the probability that if the test shows a positive result, the person actually has AIDS.

Solution   Given: $P(R \mid A) = .95$, $P(R \mid \bar{A}) = .05$, and $P(A) = .001$.
Find $P(A \mid R)$.

$$P(A \mid R) = \frac{P(R \mid A)P(A)}{P(R \mid A)P(A) + P(R \mid \bar{A})P(\bar{A})}$$

$$= \frac{(.95)(.001)}{(.95)(.001) + (.05)(.999)}$$

$$= \frac{.00095}{.00095 + .04995}$$

$$\approx .0186640472$$

This means that even though the test is highly reliable and will detect AIDS in 95% of the cases, a positive result indicates that the person actually has AIDS in only 1.8% of the cases. ■

Problem Set 7.7

In Problems 1–6 consider the experiment of selecting 2 items (without replacement) from a sample space of 100, of which 5 items are defective. Let $A = \{\text{first item selected is defective}\}$ and $B = \{\text{second item selected is defective}\}$. Find the probabilities correct to two decimal places.

1. $P(A)$ **2.** $P(\bar{A})$ **3.** $P(B \mid A)$
4. $P(B \mid \bar{A})$ **5.** $P(B)$ **6.** $P(\bar{B})$

Two cards are drawn in succession from a deck of 52 cards (without replacement). Let $D_1 = \{\text{diamond is drawn on first}$

draw$\}$ and $D_2 = \{\text{diamond is drawn on second draw}\}$. Find the probabilities in Problems 7–12.

7. $P(D_1)$ **8.** $P(\bar{D}_1)$ **9.** $P(D_2 \mid D_1)$·
10. $P(D_2 \mid \bar{D}_1)$ **11.** $P(D_2)$ **12.** $P(\bar{D}_2)$

For the experiment in Problems 1–6, draw 3 items and let $C = \{\text{third item selected is defective}\}$. Find the probabilities in Problems 13–18 correct to the nearest hundredth.

13. $P(C \mid B)$ **14.** $P(C \mid \bar{B})$ **15.** $P(C \mid A)$
16. $P(C \mid \bar{A})$ **17.** $P(C)$ **18.** $P(\bar{C})$

For the experiment in Problems 7–12, draw three cards and let $D_3 = \{$diamond is drawn on third draw$\}$. Find the probabilities in Problems 19–23 correct to the nearest hundredth.

19. $P(D_3)$

20. $P(D_3 | D_2)$

21. P(exactly 1 diamond is drawn)

22. P(3 diamonds)

23. P(at least 1 diamond)

Write your answers to Problems 24–30 correct to three decimal places.

APPLICATIONS

24. A test is given to candidates for graduate school. If the person is qualified, the probability of passing the test is 95%; if the person is not qualified, the probability of passing the test is 5%. Assume that the probability that a person taking the test is actually qualified is 70%. What is the probability that a person who passes the test is actually qualified?

25. Two persons are questioned by the police. One man tells the truth half of the time and the other always tells the truth. One man is chosen and asked one test question (a question to which the answer is known). The man answers truthfully. What is the probability that this man is the one who always tells the truth?

26. The student body of Oak Ridge University consists of 55% men and 45% women. A survey finds that 60% of the men and 30% of the women have outside jobs. What is the probability that a student who has a job is:
a. A man?   **b.** A woman?

27. At Oak Ridge University 60% of the students come from the west, 15% from the midwest, and 25% from the east. A surveys finds that 80% of the students from the west, 40% of the students from the midwest, and 60% of the students from the east had attended at least one rock concert. What is the probability that a student who has attended a rock concert comes from:
a. The east?   **b.** The midwest?   **c.** The west?

28. Evergreen Insurance Company classifies its policyholders according to the following table:

Policyholder	Percent of policyholders	Probability of an accident-free year
AA(good risk)	60%	.999
A(average)	30%	.995
B(poor risk)	10%	.99
X(bad risk)	0%	.75

If Cliff Wilson buys a policy and subsequently has an accident, what is the probability he is a class B policyholder?

29. An electronic calculator has eight essential components that a manufacturer purchases from a supplier. Three components are selected at random and are tested. One is found to be defective. A search of the records gives the following information about the supplier:

Supplier's name	Number of shipments	Number of defects per shipment			
		0	**1**	**2**	**More than 2**
Abdullah	30	21	6	3	0

Given the sample result, what is the probability that the shipment has two defects?

30. Suppose a test for cancer is given. If a person has cancer, the test will detect it in 96% of the cases; and if the person does not have cancer, the test will show a positive result 1% of the time. If we assume that 12% of the population taking the test actually has cancer, what is the probability that a person taking the test and obtaining a positive result actually has cancer?

7.8 Summary and Review

IMPORTANT TERMS

Addition principle   [7.1]
Bayes' theorem   [7.7]
Complementary events   [7.3]
Conditional probability   [7.4]
Empirical probability   [7.1]
Equally likely events   [7.2]
Event   [7.1]
Expected value   [7.6]

Experiment   [7.1]
Fair game   [7.6]
Impossible event   [7.1]
Independent events   [7.4]
Intersection of events   [7.5]
Mathematical expectation   [7.6]
Mutually exclusive   [7.1]
Odds   [7.3]

Operations research [7.6]
Partition of a sample space [7.7]
Probabilistic model [7.1]
Probability measure [7.1]
Relative frequency [7.1]
Sample space [7.1]

Simple event [7.1]
Stochastic process [7.7]
Subjective probability [7.1]
Theoretical probability [7.1]
Uniform probability model [7.2]
Union of events [7.5]

> **COMPUTER APPLICATION** If you have access to a computer, you might practice the concepts of this chapter by using Programs 8 and 9 on the computer disk accompanying this book. These programs review probability problems and probability properties.

SAMPLE TEST *For additional practice there are a large number of review problems. Categorized by objective in the Student Solutions Manual. The following sample test (40 minutes) is intended to review the main ideas of the chapter.*

1. **a.** Define a probability measure.
   b. What is meant by mutually exclusive events?
   c. What is meant by independent events?

2. Suppose a box contains five flavors of jelly beans: 10 raspberry, 20 coconut, 25 blueberry, 30 watermelon, and 15 mint. The outcomes of selecting one jelly bean at random from the box are: $A = \{coconut\}$; $B = \{raspberry, blueberry\}$; and $C = \{watermelon, mint\}$. Find:
   a. $P(A)$ **b.** $P(B)$ **c.** $P(\bar{C})$ **d.** $P(A \cup B)$ **e.** $P(\overline{A \cap B})$

3. Suppose you flip two coins simultaneously 100 times and note the results according to the following list of possibilities: $A = \{2H\}$, $B = \{2T\}$, and $C = \{1H \text{ and } 1T\}$. Will the experiment show these to be equally likely events? Why or why not?

4. Suppose two cards are drawn from a deck of cards (without replacement). Find the probabilities of the following events.
   a. P(both spaces)
   c. P(1 space and 1 heart)
   b. P(spade first, heart second)
   d. P(not both spades)

5. In the American version of roulette, there are 38 individual slots into which a steel ball has an equal chance of falling: 18 red, 18 black, and 2 green. Suppose you decide to bet $100 on red, but instead of just placing your bet you wait until black has occurred five consecutive times and then place your bet on red on the sixth play. Given these conditions, what is the probability of red on the sixth play after black occurs on the first five plays?

6. The following table shows the results of an examination of 1000 fruit flies exposed to massive amounts of radiation.

Sex of fruit fly	Normal, N	Mutated, $\bar{N}$	Total
Male, M	125	475	600
Female, $\bar{M}$	250	150	400
Total	375	625	1000

Find:
a. $P(M)$ **b.** $P(\bar{N})$ **c.** $P(M \cup N)$ **d.** $P(M \cap \bar{N})$

7. Using the information in Problem 6, find:
   a. $P(M \mid N)$ **b.** $(N \mid M)$ **c.** $P(\bar{M} \mid \bar{N})$ **d.** $P(N \mid \bar{M})$

8. A shipment of six Broncos and four Thunderbirds arrives at a dealership, and two are selected at random and tested for defects. Let

   $B_1 = \{\text{first choice is a Bronco}\}$

   $B_2 = \{\text{second choice is a Bronco}\}$

   $T_1 = \{\text{first choice is a Thunderbird}\}$

   $T_2 = \{\text{second choice is a Thunderbird}\}$

   Compute the requested probabilities by drawing with replacement.
   a. P(both Broncos)   **b.** $P(T_1 \cap B_2)$ **c.** $P(T_1 \cup B_2)$

9. Repeat Problem 8 by drawing without replacement.

10. A test for glaucoma will detect the disease in 95% of the cases and give a false positive result 2% of the time. If we assume that 10% of the population taking the test actually has glaucoma, what is the probability that a person taking the test and obtaining a positive result actually has glaucoma?

Modeling Application 5

Medical Diagnosis

After a patient with observable symptoms sees a doctor, the doctor must decide which of several possible diseases is the most probable cause of the symptoms. The doctor may order further tests before making a diagnosis. Suppose a new test for detecting renal disease is being developed. The test will be given in a double-blind study to 24 patients with confirmed renal disease and 36 patients who are free of the disease. The results are as follows: of those with renal disease, 19 test positively and 5 test negatively; for those free of the disease, 10 test positively and 26 show negative results.*

Write a paper using Bayes' theorem to develop a model to answer the question: Are the results of this test conclusive enough for a doctor to make a diagnosis? The development of this model should show that two probabilities are small. One of these is a *false positive result*, which is obtaining a positive test result when in fact the patient does not have the disease. The other is a *false negative result*, which is obtaining a negative test result when in fact the patient does have the disease. For general guidelines about writing this essay, see the commentary for Modeling Application 1 on page 130.

* This modeling application is adapted from J. S. Milton and J. J. Corbet, "Conditional Probability and Medical Tests: An Exercise in Conditional Probability," *The UMAP Journal*, Vol. III, No. 2, 1982. ©1982 Education Development Center, Inc.

CHAPTER 8

Statistics and Probability Applications

CHAPTER CONTENTS

8.1 Random Variables

8.2 Analysis of Data

8.3 The Binomial Distribution

8.4 The Normal Distribution

8.5 Normal Approximation to the Binomial

8.6 Correlation and Regression

8.7 Summary and Review
Important Terms
Sample Test
Cumulative Review for Chapters 3–8

APPLICATIONS

Management (*Business, Economics, Finance, and Investments*)

Quality control (8.1, Problems 14–15; 8.3, Problems 1–6, 38–39)
Salary comparisons (8.2, Problems 15, 34)
Analysis of telephone usage (8.2, Problem 29)
Mean life of light bulbs (8.4, Problems 31–36)
Reliability of a computer system (8.5, Problem 50)
Overbooking of airline flights (8.5, Problem 51)
Testing a computer circuit (8.6, Problem 25)
Actuary Exam questions (8.7, Problems 12–15)

Life sciences (*Biology, Ecology, Health, and Medicine*)

Probability of white-faced calves (8.3, Problems 7–12)
Physical characteristics (8.3, Problems 30–31; 8.4, Problems 17–21)
Germination rate of a seed (8.3, Problem 41)
Rainfall in Ferndale (8.4, Problems 29–30)
Death rate from cancer (8.5, Problems 48–49)
Probability of a 310-day pregnancy (8.7, Problem 9)
Radioactive contamination and cancer mortality (8.7, Problems 10–11)

Social sciences (*Demography, Political Science, Population, Psychology, Society, and Sociology*)

Analysis of test scores (8.2, Problems 17–19, 32–33; 8.3, Problem 40)
Probability of a missile reaching target (8.3, Problem 37)
Testing for ESP (8.3, Problems 42–43)
Grading on a curve (8.4, Problems 22–26)
Parapsychology experiment (8.5, Problem 46)
Correlation (8.6, Problems 25–35)

General interest

Probability that a team will win the World Series (8.3, Problems 34–35)
Game of odd person out (8.3, Problems 44–47)
Breaking strength of materials (8.4, Problems 27, 28, 38–40)
Probability of a hit in baseball (8.3, Problem 36; 8.5, Problem 47)

Cumulative review—Chapters 3–8

CHAPTER
OVERVIEW

We are now introduced to the concept of a random variable. We use random variables to discuss two important probability distributions: the binomial distribution and the normal distribution. We also discuss some statistical measures needed to analyze data. These are the measures of central tendency (often called *averages*)—the mean, median, and mode. We also discuss the measures of dispersion—the range, variance, and standard deviation. Knowledge of these statistical measures enables us to make general statements and comparisons when dealing with sets of data or large populations.

PREVIEW

When an experiment with two possible outcomes is repeated, the binomial distribution is used to analyze the results. For example, it is used to determine the failure rate on an assembly line, to determine the likelihood of certain occurrences in genetics, and in determining the reliability of predictions based on psychology experiments. The normal distribution is a continuous distribution that is used to measure variability within a population or as an approximation to certain binomial distributions. Together these distributions comprise a powerful tool in making probabilistic statements in a great variety of circumstances.

PERSPECTIVE

The central thread tying the material of this course together is that of a mathematical model. One of the difficult aspects of building a model is being able to handle the large amounts of data available in real world settings. Matrices are one type of model used for analyzing data. In this chapter other types of mathematical models are considered: those used to measure central tendency and dispersion and those used to describe binomial and normal distributions.

8.1 Random Variables

The management at Chrysler wants to introduce a "zero defects" advertising campaign on a new line of automobiles, so they decide to do extensive testing of parts coming off the assembly line. Suppose 150 parts come off the assembly line every day, and inspectors keep track of the number of items with any type of defect. If X is the number of defective items, then X can obviously assume any of the values $0, 1, 2, 3, \ldots, 149, 150$. The variable X is called a **discrete random variable**.

Random Variable

> A **random variable** X associated with a probability space S is a function that assigns a real number to each simple event in S. If S has a finite number of outcomes, then X is called a **discrete random variable**; and if X can assume any real value on an interval, then it is called a **continuous random variable**.

In finite mathematics we focus on discrete random variables, but we can often convert continuous values to discrete values by rounding off to a given number of decimal places. For example, if X represents the heights of individuals, then X is a continuous random variable. However, if we round to the nearest eighth of an inch, then it is a discrete random variable. We will look at relationships such as this and other continuous random variables in Sections 8.4 and 8.5. In the meantime, we limit our discussion to discrete random variables.

In statistics, random variables are represented by using capital letters, such as X or Y, while lowercase letters, such as x or y, are used to denote a particular value of a random variable.

EXAMPLE 1 The waiting time for a marriage license by state is given by the following 1982 list:

Alabama—None	Louisiana—72 hours	Ohio—5 days
Alaska—3 days	Maine—5 days	Oklahoma—None
Arizona—None	Maryland—48 hours	Oregon—3 days
Arkansas—3 days	Massachusetts—3 days	Pennsylvania—3 days
California—None	Michigan—3 days	Rhode Island—None
Colorado—None	Minnesota—5 days	South Carolina—24 hours
Connecticut—4 days	Mississippi—3 days	South Dakota—None
Delaware—24 hours	Missouri—3 days	Tennessee—3 days
Florida—3 days	Montana—8 days	Texas—None
Georgia—3 days	Nebraska—2 days	Utah—None
Hawaii—None	Nevada—None	Vermont—3 days
Idaho—3 days	New Hampshire—5 days	Virginia—None
Illinois—1 day	New Jersey—72 hours	Washington—3 days
Indiana—3 days	New Mexico—None	West Virginia—3 days
Iowa—3 days	New York—None	Wisconsin—5 days
Kansas—3 days	North Carolina—None	Wyoming—None
Kentucky—3 days	North Dakota—None	

These data can be summarized in a table called a **frequency distribution** as follows. Make three columns; put the number of days' wait (notice that some times are given in hours—convert these to days) into the first column, put the tally into the second column, and put the frequency into the third column:

Wait for marriage license (days)	Tally	Frequency (number of states)
0	ℍℍℍ ‖	17
1	‖‖	3
2	‖	2
3	ℍℍℍℍ ‖	21
4	‖	1
5	ℍ	5
6		0
7		0
8	‖	1
Total		50

If X is the number of days necessary to wait for a marriage license, then X is a discrete random variable where $X = 0, 1, 2, \ldots, 8$. ■

We are often interested in the probability that a random variable assumes a given value. For Example 1, we would not expect $P(X = 3)$ to be the same as $P(X = 8)$. In order to find these probabilities, we divide the frequency by the total to find the **relative frequency**.

EXAMPLE 2   Find the relative frequencies for Example 1.

Solution

Time (days)	Frequency	Relative frequency
0	17	$\frac{17}{50} = .34$
1	3	$\frac{3}{50} = .06$
2	2	$\frac{2}{50} = .04$
3	21	$\frac{21}{50} = .42$
4	1	$\frac{1}{50} = .02$
5	5	$\frac{5}{50} = .10$
6	0	$\frac{0}{50} = 0$
7	0	$\frac{0}{50} = 0$
8	1	$\frac{1}{50} = .02$
Total	50	1.00

Probability Distribution

A **probability distribution** is the collection of all values that a random variable assumes along with the probabilities that correspond to these values. Furthermore,

1. $\sum_{i=1}^{n} P(X = x_i) = 1$

2. $0 \leq P(X = x_i) \leq 1$ for every $1 \leq i \leq n$

It is often useful to display the information in a probability distribution graphically in a special kind of bar graph called a **histogram**.

EXAMPLE 3   Represent the information from Example 1 in a histogram.

Solution   Use the horizontal axis to delineate the values of the random variable and the vertical axis to represent the relative frequencies.

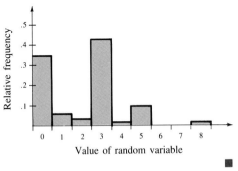

There is a very important relationship between the area of the rectangles in a histogram and the probabilities. Note that since the width of each bar is 1 unit, the **area of each bar is the probability of occurrence for that random variable.** Therefore the **total area of the rectangles is 1.** This property is very important in our study of probability distributions.

EXAMPLE 4   Draw a histogram for the probability distribution of rolling a single die and noting the number on top. Shade the portion that represents $P(X \geq 5)$.

Solution   Let $X = 1, 2, 3, 4, 5, 6$; then $P(X = x_i)$, for $1 \leq i \leq 6$, and

$$P(X \geq 5) = P(X = 5) + P(X = 6)$$

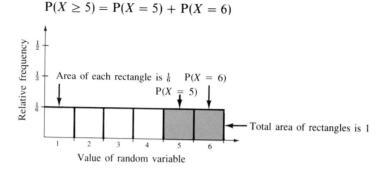

Problem Set 8.1

Give the frequency distribution and define the random variables in Problems 1–3.

1. The heights of 30 students are as follows (figures rounded to the nearest inch):

66 68 64 70 67 67 68 64 65 66
64 70 72 71 69 64 63 70 71 63
68 67 67 65 69 65 67 66 69 69

2. The number of years several leading batters played in the major leagues are as follows:

Ty Cobb	24
Rogers Hornsby	23
Ed Delehanty	16
Dan Brouthers	19
Sam Thompson	15
Al Simmons	20
Joe DiMaggio	13
Jimmy Foxx	20
Lou Gehrig	17
George Sisler	16
Nap Lajoie	21
Cap Anson	22
Eddie Collins	25
Paul Waner	20
Stan Musial	22
Henie Manush	17
Honus Wagner	21
Willie Keeler	19
Ted Williams	19
Tris Speaker	22
Billy Hamilton	14
Harry Heilmann	17
Babe Ruth	22
Jesse Burkett	16
Bill Terry	14

3. A pair of dice are rolled and the sum of the spots on the tops of the dice are recorded as follows:

3 2 6 5 3 8 8 7 10 9
7 5 12 9 6 11 8 11 11 8
7 7 7 10 11 6 4 8 8 7
6 4 10 7 9 7 9 6 6 9
4 4 6 3 4 10 6 9 6 11

Find the probability distribution (that is, the relative frequencies) and draw a histogram for Problems 4–9.

4. Use the data in Problem 1.

5. Use the data in Problem 2.

6. Use the data in Problem 3.

APPLICATIONS

7. Blane, Inc., a consulting firm, is employed to perform an efficiency study at National City Bank. As part of the study, the number of times per day that people are waiting in line is summarized.

Number waiting	Frequency
2	20
3	15
4	7
5	5
6	2
7	1
8 or more	0

8. Blane, Inc., also notes the transaction times (rounded to the nearest minute) for customers at National City Bank, as listed in the table:

Time	Frequency
1	10
2	12
3	18
4	25
5	16
6	10
7	6
8	1
9	0
10	2

9. Three coins are tossed onto a table and the following frequencies are noted:

Number of heads	Frequency
3	18
2	56
1	59
0	17

Use your knowledge of probability from Chapter 7 to find the probability distributions in Problems 10–19. Let X be the random variable.

10. Three coins are tossed onto a table and the number of tails is noted.

11. Four coins are tossed onto a table and the number of heads is noted.

12. Two dice are rolled and the total number of spots on the top faces is recorded.

13. Two cards are simultaneously drawn from a deck of 52 cards and the number of aces is noted.

14. From a lot containing 25 items, 5 of which are defective, 3 items are chosen at random and the number of defectives is noted.

15. Repeat Problem 14 except choose the 3 items one at a time, note the result, and then return the chosen item before choosing again.

16. A drug is known to effect a cure in 60% of the patients to whom it is administered. This drug is administered to three patients and the number of cures is noted.

17. Repeat Problem 16 for four patients.

18. A bag of Halloween candy contains 25 Hershey bars, 10 Rocky Roads, and 15 Almond Joys. Two pieces of candy are randomly selected and the number of Rocky Roads selected is noted.

19. Repeat Problem 18 except note the number of Almond Joys.

Draw a histogram in Problems 20–29 and shade in the region indicated for each problem.

20. P(X ≥ 2) for the experiment in Problem 10.
21. P(X is at least 3) for the experiment in Problem 11.
22. P(X = 7 or X = 11) for the experiment in Problem 12.
23. P(X = 1) for the experiment in Problem 13.
24. P(X ≥ 1) for the experiment in Problem 14.
25. P(X = 2) for the experiment in Problem 15.
26. P(X = 3) for the experiment in Problem 16.
27. P(2 ≤ X ≤ 3) for the experiment in Problem 17.
28. P(X = 2) for the experiment in Problem 18.
29. P(X at least 1) for the experiment in Problem 19.

30. Draw a histogram for $P(X = x_i)$, where $P(X = x_i) = \dfrac{x_i}{10}$ for $0 \le i \le 4$, i a counting number. For example, if $x_i = 4$, then $P(X = 4) = \frac{4}{10} = \frac{2}{5}$.

31. Draw a histogram for $P(X = x_i)$, where $P(X = x_i) = \dfrac{x_i}{20}$ for $2 \le i \le 6$, i a counting number. For example, if $x_i = 2$, then $P(X = 2) = \frac{2}{20} = \frac{1}{10}$.

8.2 Analysis of Data

The word *average* is frequently used in everyday language, but in mathematics it can take on a variety of meanings. For example, suppose we compare the "averages" of two bowlers in a tournament consisting of seven games:

Andrew: 185 average
Bob: 170 average

It would appear that Andrew did better in the tournament, but suppose the winner is found by looking at the individual games, as listed below:

	Andrew	Bob
Game 1	175	180
Game 2	150	130
Game 3	160	161
Game 4	180	185
Game 5	160	163
Game 6	183	185
Game 7	287	186
Total	1295	1190

To find each average, we divide the total by the number of games:

$$\text{Andrew: } \frac{1295}{7} = 185 \qquad \text{Bob: } \frac{1190}{7} = 170$$

However, if we consider the games separately, Bob beat Andrew in *five out of seven games*. Clearly, the "average" we calculated above does not completely describe the situation.

The first statistical measures we consider are called **measures of central tendency**, or **averages**. We already used the *summation notation* shown below:

$$\sum_{i=1}^{n} x_i = x_1 + x_2 + \cdots + x_n$$

In statistics this is sometimes abbreviated as Σx to mean *the sum of all values that x can assume*. Similarly, Σx^2 means *square each value that x can assume and then add the results*.

Averages, or Measures of Central Tendency

Given a set of data, three common *measures of central tendency* are:

1. The **mean** (or **arithmetic mean**) is found by adding the data and then dividing by the number of data items, n:

$$\text{Mean} = \frac{\Sigma x}{n}$$

The mean of a set of sample scores is denoted by $\bar{x}$.

2. The **median** is the middle number when the data are arranged in order of size. If there is an even number of data items, then the median is the mean of the two middle numbers.

3. The **mode** is the value that occurs most frequently. If there is no number that occurs more than once, there is no mode. It is possible to have more than one mode.

EXAMPLE 1   Find the mean, median, and mode for Andrew's and Bob's bowling scores.

Solution   Mean:

Andrew   Bob

$$\bar{x} = \frac{1295}{7} \qquad \bar{x} = \frac{1190}{7}$$

$$= 185 \qquad\qquad = 170$$

Median:

Andrew		Bob
150		130
160		161
160		163
175	← The middle number →	180
180	is the median	185
183		185
287		186

Mode:

Andrew		Bob
150		130
160⎤		161
160⎦	The mode is the	163
175	number that occurs	180
180	most frequently	⎰185
183		⎱185
287		186

Summary:

	Andrew	Bob
Mean	185	170
Median	175	180
Mode	160	185

Find the mean, median, and mode for the sets of data in Examples 2–4.

EXAMPLE 2   3, 5, 5, 8, 9

Solution   Mean: $\dfrac{\text{Sum of terms}}{\text{Number of terms}} = \dfrac{3+5+5+8+9}{5}$

$$= \frac{30}{5}$$

$$= 6$$

Median: Arrange in order: 3, 5, 5, 8, 9
The middle term, 5, is the median.

Mode: The term that occurs most frequently is the mode, which is 5.

EXAMPLE 3   4, 10, 9, 8, 9, 4, 5

Solution   Mean: $\dfrac{4 + 10 + 9 + 8 + 9 + 4 + 5}{7} = \dfrac{49}{7} = 7$

Median: 4, 4, 5, 8, 9, 9, 10
           The median is 8.

Mode: This set of data is *bimodal*; that is, it has two modes.
         The modes are 4 and 9. ∎

EXAMPLE 4   6, 5, 4, 7, 1, 9

Solution   Mean: $\dfrac{6 + 5 + 4 + 7 + 1 + 9}{6} = \dfrac{32}{6} \approx 5.33$

Median: 1, 4, 5, 6, 7, 9

$$\frac{5 + 6}{2} = \frac{11}{2} = 5.5$$

Mode: There is no mode. ∎

Suppose the data are presented in a frequency distribution and there are more data than can be handled conveniently by the method used in Examples 2–4. Consider Example 5, which shows how you might proceed.

EXAMPLE 5   In Section 8.1 we set up a frequency distribution for the number of days one must wait for a marriage license. This information is repeated below. Find the mean, median, and mode.

Wait for marriage license (days)	Frequency (number of states)
0	17
1	3
2	2
3	21
4	1
5	5
6	0
7	0
8	1
Total	50

Solution   Mean: To find the mean, we could, of course, add all 50 individual numbers. But, instead, notice that

       0 occurs 17 times,   so we write $0 \cdot 17$
       1 occurs 3 times,   so we write $1 \cdot 3$
       2 occurs 2 times,   so we write $2 \cdot 2$
       3 occurs 21 times,   so we write $3 \cdot 21$
               ⋮

Thus the mean is

$$\bar{x} = \frac{0 \cdot 17 + 1 \cdot 3 + 2 \cdot 2 + 3 \cdot 21 + 4 \cdot 1 + 5 \cdot 5 + 6 \cdot 0 + 7 \cdot 0 + 8 \cdot 1}{50}$$

$$= \frac{0 + 3 + 4 + 63 + 4 + 25 + 0 + 0 + 8}{50}$$

$$= \frac{107}{50}$$

$$= 2.14$$

Median: Since there are 50 values, the mean of the twenty-fifth and twenty-sixth largest values is the median. From the table, we see that the twenty-fifth term is 3 and the twenty-sixth term is 3, so the median is

$$\frac{3 + 3}{2} = \frac{6}{2}$$

$$= 3$$

Mode: The mode is the value that occurs most frequently, which is 3.   ∎

The mean from a frequency distribution is called the **weighted mean**.

Weighted Mean

If a list of scores $x_1, x_2, x_3, \ldots, x_n$ occurs $w_1, w_2, \ldots, w_n$ times, respectively, then

$$\bar{x} = \frac{\sum_{i=1}^{n} w_i \cdot x_i}{\sum_{i=1}^{n} w_i} \qquad \text{or simply} \qquad \bar{x} = \frac{\sum w \cdot x}{\sum w}$$

We have been using $\bar{x}$ to denote the mean of a set of *sample scores*. Now we let μ denote the mean of *all scores* in some population. A **population** is the complete and entire collection of elements to be studied, so a sample must be a subset of a population. In the formula for a weighted mean, if we consider the entire population, then $\sum wx / \sum w$ can be rewritten as

$$\mu = \sum x \cdot P(x)$$

if we consider $P(x)$ to be the relative frequency with which x occurs.

The measures we have been discussing can help us interpret information, but they do not give the whole story. For example, consider the following two sets of data:

Data set 1:   8, 10, 9, 9, 9
Data set 2:   9, 9, 2, 12, 13

For both these data sets the mean, median, and mode are 9. It would seem that some additional analysis is in order. Note that the data in the second set are more spread out than the data in the first set. The amount that the data are spread out is called the **dispersion**. We now consider three measures of dispersion: the **range**, **variance**, and

standard deviation. The simplest measure of dispersion is the range:

Range

> The *range* in a set of data is the difference between the largest and the smallest numbers in the set.

The ranges for the above data sets are:

Range of data set 1: $10 - 8 = 2$
Range of data set 2: $13 - 2 = 11$

Note that the range is determined only by the largest and the smallest numbers in the set; it does not give us any information about the other numbers. It thus seems reasonable to invent some other measures of dispersion that take into account all the numbers in the data. The variance and standard deviation are measures that give information about the dispersion. The **variance** uses all the numbers in the data set to measure the dispersion. When finding the variance, we must make a distinction between the variance of the entire population and the variance of a random sample from that population. When the variance is based on a set of sample scores, it is denoted by s^2; and when it is based on all scores in a population, it is denoted by σ^2 (σ is the lowercase Greek letter sigma). The variance is found by

$$s^2 = \frac{\Sigma(x - \bar{x})^2}{n - 1} \qquad \sigma^2 = \frac{\Sigma(x - \mu)^2}{n}$$

If n is large (say, greater than 30), s^2 and σ^2 will be almost the same. The formula for s^2 looks a little intimidating, but it is based on simple arithmetic procedures, which, if taken one at a time, are quite simple. Understanding the variance formula is easy if you systematically deal with the data as indicated by the following procedure:

Procedure for Finding the Variance

> *Step 1*   Determine the mean of the numbers. (Find $\bar{x}$.)
> *Step 2*   Subtract the mean from each number. (Find $x - \bar{x}$.)
> *Step 3*   Square each of these differences. [This is $(x - \bar{x})^2$.]
> *Step 4*   Find the sum of the squares of these differences. [This is $\Sigma(x - \bar{x})^2$.]
> *Step 5*   Divide this sum by 1 less than the number of pieces of data.

EXAMPLE 6   Find the variance for each set of data:

a. Data set 1:   8, 9, 9, 9, 10   **b.** Data set 2:   2, 9, 9, 12, 13

Solution   Remember that the mean, median, and mode for both these examples is the same (9). The range for data set 1 is 2, and for data set 2 it is 11. Now we want to find the second measure of dispersion, the variance. The procedure is lengthy, so make sure to follow each step carefully.

Step 1   Find the mean.
　　　　a. $\bar{x} = 9$ **b.** $\bar{x} = 9$

Step 2 Subtract each number from the mean.

a.

Data, x	Difference from the mean, $x - \bar{x}$
8	−1
9	0
9	0
9	0
10	1

b.

Data, x	Difference from the mean, $x - \bar{x}$
2	−7
9	0
9	0
12	3
13	4

Step 3 Square each of these differences. Notice that some of the differences in step 2 are positive and others are negative. Remember, we wish to find a measure of total dispersion. But if we add all these differences, we will not obtain the total variability. Indeed, if we simply add the differences for either example, the sum is zero. But we do not wish to say there is no dispersion. To resolve this difficulty with positive and negative differences, we square each difference so the result will always be nonnegative.

a.

Data, x	Difference from the mean, $x - \bar{x}$	Square of the difference, $(x - \bar{x})^2$
8	−1	1
9	0	0
9	0	0
9	0	0
10	1	1

b.

Data, x	Difference from the mean, $x - \bar{x}$	Square of the difference, $(x - \bar{x})^2$
2	−7	49
9	0	0
9	0	0
12	3	9
13	4	16

Step 4 Find the sum of these squares.
a. $1 + 0 + 0 + 0 + 1 = 2$ **b.** $49 + 0 + 0 + 9 + 16 = 74$

Step 5 Divide this sum by 1 less than the number of terms. In each of these examples there are five pieces of data, so to find the variance, divide by 4:
a. Variance $= \frac{2}{4} = .5$ **b.** Variance $= \frac{74}{4} = 18.5$ ∎

The larger the variance, the more dispersion there is in the original data, and a small variance means that the data are more closely clustered around the mean.

The **standard deviation** for a sample is s and for the entire population is σ. That is,

$$\text{Standard deviation} = \sqrt{\text{variance}}$$

EXAMPLE 7 Find the standard deviations for the data of Example 6.

a. Data set 1: 8, 9, 9, 9, 10   **b.** Data set 2: 2, 9, 9, 12, 13

Solution We did most of the work in finding the standard deviation in Example 6. Now, all we need to do is to find the square root of the answers to Example 6.

a. $\sqrt{.5} \approx .71$ **b.** $\sqrt{18.5} \approx 4.30$ ■

If the standard deviation is based on a random variable of a probability distribution, it can be written as

$$\sigma^2 = \Sigma(X - \mu)^2 \cdot P(X)$$

This formula can be manipulated into an equivalent form to help us carry out the calculations more easily:

$$\sigma^2 = [\Sigma X^2 \cdot P(X)] - \mu^2$$

EXAMPLE 8 Roll a pair of dice and let the random variable represent the total on the tops of the dice. Find the mean, variance, and standard deviation of the population.

Solution Organize the computations in table form, as shown.

X	$P(X)$	$X \cdot P(X)$	X^2	$X^2 \cdot P(X)$
2	$\frac{1}{36}$	$\frac{2}{36}$	4	$\frac{4}{36}$
3	$\frac{2}{36}$	$\frac{6}{36}$	9	$\frac{18}{36}$
4	$\frac{3}{36}$	$\frac{12}{36}$	16	$\frac{48}{36}$
5	$\frac{4}{36}$	$\frac{20}{36}$	25	$\frac{100}{36}$
6	$\frac{5}{36}$	$\frac{30}{36}$	36	$\frac{180}{36}$
7	$\frac{6}{36}$	$\frac{42}{36}$	49	$\frac{294}{36}$
8	$\frac{5}{36}$	$\frac{40}{36}$	64	$\frac{320}{36}$
9	$\frac{4}{36}$	$\frac{36}{36}$	81	$\frac{324}{36}$
10	$\frac{3}{36}$	$\frac{30}{36}$	100	$\frac{300}{36}$
11	$\frac{2}{36}$	$\frac{22}{36}$	121	$\frac{242}{36}$
12	$\frac{1}{36}$	$\frac{12}{36}$	144	$\frac{144}{36}$
Total		$\frac{252}{36} = 7$		$\frac{1974}{36}$

Mean: $\mu = \Sigma X \cdot P(X) = 7$

Variance: $\sigma^2 = [\Sigma X^2 \cdot P(X)] - \mu^2 = \dfrac{1974}{36} - 7^2 = 5.8$

Standard deviation: $\sigma = \sqrt{5.8} \approx 2.4$ ■

Many calculators have keys to calculate the mean, variance, and standard deviation. Check to see if you can carry out these operations on your calculator. If your calculator does not have these keys, you can arrange the calculations as shown in Example 8 and use the square root table given in Appendix E.

COMPUTER APPLICATION   For more real life examples, or for additional practice, use Program 10 on the computer disk accompanying this text. It gives practice with the mean, median, mode, standard deviation, and histograms.

Problem Set 8.2

Find the mean, median, mode, range, variance, and standard deviation for each set of values in Problems 1–12.

1. 1, 2, 3, 4, 5

2. 17, 18, 19, 20, 21

3. 103, 104, 105, 106, 107

4. 765, 766, 767, 768, 769

5. 4, 7, 10, 7, 5, 2, 7

6. 15, 13, 10, 7, 6, 9, 10

7. 3, 5, 8, 13, 21

8. 1, 4, 9, 16, 25

9. 79, 90, 95, 95, 96

10. 70, 81, 95, 79, 85

11. 1, 2, 3, 3, 3, 4, 5

12. 0, 1, 1, 2, 3, 4, 16

13. Compare Problems 1–4. What do you notice about the mean and standard deviation?

14. By looking at Problems 1–4 and discovering a pattern, find the mean and standard deviation of the following set of numbers:

217,850,   217,851,   217,852,   217,853,   217,854

APPLICATIONS

15. Find the mean, median, and mode of the following salaries of the employees of Green Lawn Landscaping Company:

Salary	Frequency
$10,000	4
16,000	3
20,000	2
30,000	1

16. Linda Foley, the leading salesperson for the Green Lawn Landscaping Company, turned in the following summary of sales contacts for the week of October 23–28. Find the mean, median, and mode.

Date	Number of clients contacted
Oct. 23	12
Oct. 24	9
Oct. 25	10
Oct. 26	16
Oct. 27	10
Oct. 28	21

17. Find the mean, median, and mode of the following test scores:

Test score	Frequency
90	1
80	3
70	10
60	5
50	2

18. A class obtained the following scores on a test:

Score	Frequency
90	1
80	6
70	10
60	4
50	3
40	1

Find the mean, median, mode, and range for the class.

19. A class obtained the following scores on a test:

Score	Frequency
90	2
80	4
70	9
60	5
50	3
40	1
30	2
0	4

Find the mean, median, mode, and range for the class.

Find the variance and standard deviation for Problems 20–24.

20. Problem 17   **21.** Problem 18   **22.** Problem 19

23. Problem 15   **24.** Problem 16

Find the mean, variance, and standard deviation of the probability distributions in Problems 25–28.

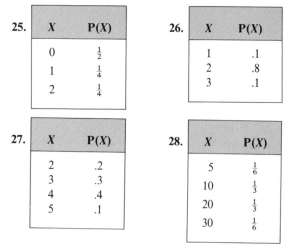

25.

X	P(X)
0	$\frac{1}{2}$
1	$\frac{1}{4}$
2	$\frac{1}{4}$

26.

X	P(X)
1	.1
2	.8
3	.1

27.

X	P(X)
2	.2
3	.3
4	.4
5	.1

28.

X	P(X)
5	$\frac{1}{6}$
10	$\frac{1}{3}$
20	$\frac{1}{3}$
30	$\frac{1}{6}$

29. A survey shows that the number of times a nonbusiness telephone rings before it is answered is:

X	P(X)
0	.00
1	.20
2	.31
3	.16
4	.18
5	.10
6	.02
7 or more	.03

Find the mean, variance, and standard deviation for the number of rings.

30. The probability for the number of customers arriving at a bank in a given period of time is summarized below:

X	P(X)
0	.25
1	.43
2	.18
3	.09
4	.03
5	.01
6 or more	.01

Find the mean, variance, and standard deviation for the number of arrivals.

31. Suppose a variance is zero. What can you say about the data?

32. A professor gives five exams. The scores for two students have the same mean, although one student seemed to do better on all the tests except one. Give an example of such scores.

33. A professor gives six exams. The scores of two students have the same mean, although one student's scores have a small standard deviation and the other student's scores have a large standard deviation. Give an example of such scores.

34. The salaries for the executives of a small company are shown below. Find the mean, median, and mode. Which measure seems to best describe the average executive salary for the company?

Position	Salary
President	$90,000
1st VP	40,000
2nd VP	40,000
Supervising manager	34,000
Accounting manager	30,000
Personnel manager	30,000

35. Roll a pair of dice 20 times. Find the mean, variance, and standard deviation for your data. Compare your results with Example 8.

36. Repeat Problem 35 for 100 rolls of the dice.

37. Roll a pair of dice until all 11 numbers occur at least once. Repeat the experiment 20 times. Find the mean, variance, and standard deviation for the number of tosses.

38. The *harmonic mean* (or H.M.) is found by dividing the number of scores, *n*, by the sum of the reciprocals of all scores:

$$\text{H.M.} = \frac{n}{\sum \frac{1}{x}}$$

Find the harmonic mean for the data in Problem 5.

39. Repeat Problem 38 for the data in Problem 6.

40. The *geometric mean* (or G.M.) is found by taking the *n*th root of the product of the *n* scores. Find the geometric mean for the data in Problem 5.

41. Repeat Problem 40 for the data in Problem 6.

42. The *quadratic mean*, or *root mean square* (R.M.S.), is found by squaring each score; adding the results; dividing by the number of scores, *n*; and then taking the square root of the result:

$$\text{R.M.S.} = \sqrt{\frac{\sum x^2}{n}}$$

Find the quadratic mean for the data in Problem 5.

43. Repeat Problem 42 for the data in Problem 6.

8.3 The Binomial Distribution

Consider now a common type of experiment—one with only two outcomes, A and $\bar{A}$. Suppose that $P(A) = p$ and $P(\bar{A}) = q = 1 - p$. We are interested in n repetitions of the experiment. If $P(A)$ remains the same for each repetition and we let X represent the number of times that event A has occurred, then we call X a **binomial random variable**.

EXAMPLE 1 Toss a coin four times. The sample space is shown below:

Number of heads	Outcomes					
4	HHHH					
3	HHHT	HHTH	HTHH	THHH		
2	HHTT	HTHT	HTTH	THTH	THHT	TTHH
1	TTTH	TTHT	THTT	HTTT		
0	TTTT					

Let the random variable X represent the number of heads that have occurred. That is, $X = 0$ if we obtain no heads; $X = 1$ means one head is obtained; $X = 2$ means two heads are obtained; $X = 3$, $X = 4$ mean three and four heads are obtained, respectively. Since there are 16 possibilities,

$$P(X = 4) = \tfrac{1}{16}$$
$$P(X = 3) = \tfrac{4}{16} = \tfrac{1}{4}$$
$$P(X = 2) = \tfrac{6}{16} = \tfrac{3}{8}$$
$$P(X = 1) = \tfrac{4}{16} = \tfrac{1}{4}$$
$$P(X = 0) = \tfrac{1}{16}$$

Note that the sum is one:

$$\tfrac{1}{16} + \tfrac{4}{16} + \tfrac{6}{16} + \tfrac{4}{16} + \tfrac{1}{16} = 1$$ ∎

This example illustrates a common discrete probability distribution called the **binomial distribution**, which is a list of outcomes and probabilities for a **binomial experiment**.

Binomial Experiment

A **binomial experiment** is an experiment that meets four conditions:
1. There must be a fixed number of trials. Denote this number by n.
2. There must be two possible mutually exclusive outcomes for each trial. Call them *success* and *failure*.
3. Each trial must be independent. That is, the outcome of a particular trial is not affected by the outcome of any other trial.
4. The probability of success and failure must remain constant for each trial.

Consider a manufacturer of transistor radios. Suppose three items are chosen at random from a day's production and are classified as defective (F) or nondefective (S). We are interested in the number of successes obtained. (A "success" is the occurrence of the event we are considering; in this case, nondefectives.) Suppose that an item has a probability of .1 of being defective and therefore a probability of .9 of being nondefective. We will assume that these probabilities remain the same throughout the experiment and that the classification of any particular item is independent of the classification of any other item. The sample space, along with the probabilities, is listed below:

Sample space	*Associated probabilities*
1. *FFF*	$(.1)(.1)(.1) = (.1)^3$
2. *FFS*	$(.1)(.1)(.9) = (.9)(.1)^2$
3. *FSF*	$(.1)(.9)(.1) = (.9)(.1)^2$
4. *SFF*	$(.9)(.1)(.1) = (.9)(.1)^2$
5. *FSS*	$(.1)(.9)(.9) = (.9)^2(.1)$
6. *SFS*	$(.9)(.1)(.9) = (.9)^2(.1)$
7. *SSF*	$(.9)(.9)(.1) = (.9)^2(.1)$
8. *SSS*	$(.9)(.9)(.9) = (.9)^3$

If we let X = the number of successes obtained, then $P(X = 0)$ is found on line 1 above:

$P(X = 0) = (.1)^3$

$P(X = 1)$ is found from lines 2, 3, and 4:

2. *FFS*	$(.9)(.1)^2$
3. *FSF*	$(.9)(.1)^2$
4. *SFF*	$(.9)(.1)^2$

Total: $\quad P(X = 1) = 3(.9)(.1)^2$

$P(X = 2)$ is found by

5. *FSS*	$(.9)^2(.1)$
6. *SFS*	$(.9)^2(.1)$
7. *SSF*	$(.9)^2(.1)$

Total: $\quad P(X = 2) = 3(.9)^2(.1)$

$P(X = 3)$ is found in line 8:

$P(X = 3) = (.9)^3$

Note that the same results can be achieved by simply considering

$$(.1 + .9)^3 = (.1)^3 + 3(.1)^2(.9) + 3(.1)(.9)^2 + (.9)^3$$

This leads us to the following theorem:

Binomial Distribution Theorem

Let X be a random variable for the number of successes in n independent and identical repetitions of an experiment with two possible outcomes, success and failure. If p is the probability of success, then

$$P(X = k) = \binom{n}{k} p^k (1 - p)^{n-k} \qquad k = 0, 1, \ldots, n$$

A sequence of independent trials for which there are only two possible outcomes is also sometimes called a sequence of **Bernoulli trials**, after Jacob Bernoulli (1654–1705).

EXAMPLE 2 Suppose a sociology teacher always gives true-false tests with 10 questions.

a. What is the probability of getting exactly 70% by guessing?
b. What is the probability of getting 70% or better by guessing?
c. If a student can be sure of getting five questions correct, but must guess at the others, what is the probability of getting 70% or better?

Solution **a.** $P(X = 7) = \binom{10}{7}\left(\frac{1}{2}\right)^7\left(\frac{1}{2}\right)^3 = \frac{10!}{7!3!} \cdot \frac{1}{2^{10}} = \frac{120}{1024} = \frac{15}{128}$

b. $P(X = 8) = \binom{10}{8}\left(\frac{1}{2}\right)^8\left(\frac{1}{2}\right)^2 = \frac{45}{1024}$

$P(X = 9) = \binom{10}{9}\left(\frac{1}{2}\right)^9\left(\frac{1}{2}\right)^1 = \frac{10}{1024} = \frac{5}{512}$

$P(X = 10) = \binom{10}{10}\left(\frac{1}{2}\right)^{10} = \frac{1}{1024}$

$P(X \geq 7) = \frac{120}{1024} + \frac{45}{1024} + \frac{10}{1024} + \frac{1}{1024} = \frac{176}{1024} = \frac{11}{64}$

c. $P(X \geq 2) = \sum_{k=2}^{5}\binom{5}{k}\left(\frac{1}{2}\right)^k\left(\frac{1}{2}\right)^{5-k}$

$= \binom{5}{2}\left(\frac{1}{2}\right)^5 + \binom{5}{3}\left(\frac{1}{2}\right)^5 + \binom{5}{4}\left(\frac{1}{2}\right)^5 + \binom{5}{5}\left(\frac{1}{2}\right)^5$

$= 10 \cdot \frac{1}{32} + 10 \cdot \frac{1}{32} + 5 \cdot \frac{1}{32} + 1 \cdot \frac{1}{32}$

$= \frac{5}{16} + \frac{5}{16} + \frac{5}{32} + \frac{1}{32} = \frac{13}{16}$

EXAMPLE 3 A missile has a probability of $\frac{1}{10}$ of penetrating enemy defenses and reaching its target. If five missiles are aimed at the same target, what is the probability that exactly one will hit its target? What is the probability that at least one will hit its target?

Solution $P(X = 1) = \binom{5}{1}\left(\frac{1}{10}\right)^1\left(\frac{9}{10}\right)^4$

$= \frac{5 \cdot 9^4}{10^5} = \frac{6561}{20,000} = .32805$

$P(X \geq 1) = 1 - P(X = 0)$

$= 1 - \binom{5}{0}\left(\frac{9}{10}\right)^5 = 1 - \frac{9^5}{10^5}$

$= 1 - .59049 = .40951$

As you can see from the above examples, the calculations for binomial probabilities can become rather tedious. For this reason, tables of binomial distributions have been compiled. If you wish to calculate the probability of X successes in n independent Bernoulli trials with the probability of success on a single trial equal to p, you can use Table 3 in Appendix E. For example, the probability

$$P(X = 1) = \binom{5}{1}\left(\frac{1}{10}\right)^1\left(\frac{9}{10}\right)^4$$

can be found in Table 3, where $n = 5$, $k = 1$, and $p = .1$:

$$P(X = 1) = .328$$

Also, we find $P(X = 0) = .590$.

EXAMPLE 4   A typist can type a page accurately (with no errors) with a probability of .25. What is the probability that there will be exactly one error in a report of five pages?

Solution   The probability of one error on one page is .75, so the solution is given by

$$P(X = 1) = \binom{5}{1}(.75)^1(.25)^4$$

But Table 3 in Appendix E only has values for $p \le .5$. However,

$$\binom{n}{k}p^k(1 - p)^{n-k} = \binom{n}{n-k}(1 - p)^{n-k}p^k$$

This means that we can interchange the second and third terms (one of these must be less than or equal to .5); p now has the value .25 and we can use the table for $X = 1$ replaced by $X = 4$ (this is $n - X$) since the exponent on the new "p term" is now 4. This means that

$$P(X = 1) = P(X = 4) = \binom{5}{4}(.25)^4(.75)^1$$

The entry in the table where $n = 5$, $k = 4$, and $p = .25$ is $P(X = 1) = .015$. ∎

COMPUTER APPLICATION   Program 11 (options 3 and 4) on the computer disk accompanying this text will help you find binomial probabilities as well as draw the accompanying histograms.

We can also find the mean, variance, and standard deviation for a binomial distribution.

EXAMPLE 5   Find the mean, variance, and standard deviation for the number of heads obtained from tossing a coin four times.

Solution   Use the information in Example 1 along with the formulas for mean and variance from Section 8.2:

$$\mu = \Sigma X \cdot P(X) \quad \text{and} \quad \sigma^2 = [\Sigma X^2 \cdot P(X)] - \mu^2$$

X	P(X)	$\overset{\mu}{X \cdot P(X)}$	X^2	$X^2 \cdot P(X)$
4	$\frac{1}{16} = .0625$	.25	16	1
3	$\frac{1}{4} = .25$	.75	9	2.25
2	$\frac{3}{8} = .375$	.75	4	1.5
1	$\frac{1}{4} = .25$	.25	1	.25
0	$\frac{1}{16} = .0625$	0	0	0
Total		2		5

From the table, $\mu = 2$, and

$$\sigma^2 = 5 - 4 = 1 \qquad \sigma = \sqrt{1} = 1$$ ■

The computations in Example 5 use the formulas for mean, variance, and standard deviation given in Section 8.2. However, if we apply some complicated algebraic manipulations to these formulas (the details are too involved to show here), we arrive at some very simple formulas for the mean, variance, and standard deviation of a binomial distribution.

Mean, Variance, and Standard Deviation of a Binomial Distribution

A binomial distribution of n independent trials, each with a probability of success p and failure $q = 1 - p$, has mean μ, variance σ^2, and standard deviation σ, given by

$$\mu = np \qquad \sigma^2 = npq \qquad \sigma = \sqrt{npq}$$

We can verify these formulas by comparison with the results found in Example 5. We had four repetitions of tossing a coin, so $n = 4$; $p = \frac{1}{2}$ since the probability of success (a head) is $\frac{1}{2}$; and $q = 1 - p = \frac{1}{2}$. Thus

$$\begin{aligned} \mu &= np & \sigma^2 &= npq & \sigma &= \sqrt{npq} \\ &= 4 \cdot \tfrac{1}{2} & &= 4 \cdot \tfrac{1}{2} \cdot \tfrac{1}{2} & &= \sqrt{1} \\ &= 2 & &= 1 & &= 1 \end{aligned}$$

EXAMPLE 6 The probability of a transmission failure in the first year on a new automobile is .005. Find the mean number of failures for 10,000 cars if we assume that the failures are independent. Also find the standard deviation.

Solution $n = 10,000$; $q = .005$; $p = 1 - .005 = .995$

Mean number of *failures*

$$\begin{aligned} \mu &= nq & \sigma &= \sqrt{npq} \\ &= 10,000(.005) & &= \sqrt{(10,000)(.995)(.005)} \\ &= 50 & &= \sqrt{49.75} \\ & & &\approx 7.05 \end{aligned}$$ ■

Problem Set 8.3

APPLICATIONS

A manufacturer of disk drives chooses three items from the day's production and classifies these as defective or nondefective. Suppose that an item has a probability of .2 of being defective. In Problems 1–5 find the requested probabilities.

1. No defective disk drives are found.

2. All three are defective.

3. Two are found to be defective.

4. At least two are defective.

5. One or more are defective.

6. Numerically, how are your answers for Problems 1 and 5 related?

All the cows in a certain herd are white-faced. The probability that a white-faced calf will be born from the mating with a certain bull is .9. Suppose four cows are bred to the same bull. Find the probabilities in Problems 7–12.

7. Four white-faced calves

8. Exactly three white-faced calves

9. No white-faced calves

10. One white-faced calf

11. At least three white-faced calves

12. Not more than two white-faced calves

Use Table 3 (Appendix E) to find the binomial probabilities in Problems 13–20.

13. $n = 5$, $X = 3$, $p = .30$

14. $n = 4$, $X = 3$, $p = .25$

15. $n = 12$, $X = 6$, $p = .65$

16. $n = 10$, $X = 4$, $p = .80$

17. $n = 6$, $X = 6$, $p = \frac{1}{2}$

18. $n = 8$, $X = 8$, $p = \frac{3}{4}$

19. $n = 7$, $X = 5$, $p = \frac{1}{10}$

20. $n = 15$, $X = 13$, $p = \frac{2}{5}$

In Problems 21–29 find the mean, variance, and standard deviation for the given values of n and p. Assume that the binomial conditions are satisfied in each case.

21. $n = 5$, $p = .3$

22. $n = 4$, $p = .25$

23. $n = 12$, $p = .65$

24. $n = 10$, $p = .8$

25. $n = 6$, $p = \frac{1}{2}$

26. $n = 8$, $p = \frac{3}{4}$

27. $n = 7$, $p = \frac{1}{10}$

28. $n = 15$, $p = \frac{2}{5}$

29. $n = 10$, $p = \frac{3}{5}$

30. The probability that children in a specific family will have blond hair is $\frac{3}{4}$. If there are four children in the family, what is the probability that two of them are blond?

31. The probability that the children in a specific family will have brown eyes is $\frac{3}{8}$. If there are four children in the family, what is the probability that at least two of them have brown eyes?

32. It is known that 85% of the graduates of Foley's School of Motel Management are placed in a job within 6 months of graduation. If a class has 20 graduates, what is the probability that 15 will be placed within 6 months?

33. If a graduating class of Foley's School in Problem 32 has 10 graduates, what is the probability that at least 9 will be placed within 6 months?

34. Suppose the National League team has a probability of $\frac{3}{5}$ of winning the World Series and the American League team has a probability of $\frac{2}{5}$. The series is over as soon as one team wins four games. Find the probability that the series is over in four games.

35. In Problem 34 what is the probability that the series will go seven games?

36. The batting average of the star baseball player Brian Thompson is .300 (that is, P(hit) = .3). What is the probability that Thompson will get at least three hits for four times at bat?

37. Suppose that research has shown that the probability that a missile penetrates enemy defenses and reaches its target is $\frac{1}{10}$. Find the smallest number of identical missiles that are necessary in order to be 80% certain of hitting the target at least once.

38. Eighty percent of the widgets of the Ampex Widget Company meet the specifications of its customers. If a sample of six widgets is tested, what is the probability that three or more of them would fail to meet the specifications?

39. Find the mean number of failures for the widgets described in Problem 38. Also find the standard deviation.

40. Ten percent of the students fail the final exam in Math 9. If the class size is 30, find the mean and standard deviation for the number of failures in each class.

41. A certain type of seed has a 60% germination rate. If the seeds are planted in rows of 50 seeds each, find the mean and standard deviation for the number of seeds that germinate in each row.

42. Suppose a person claims to have ESP and says he can read mental images 75% of the time. We set up an experiment where he must determine whether we are looking at a picture of a circle or a square. We agree that if he can correctly identify the mental image at least five out of six times we will grant his claim.
    a. What is the probability of granting his claim if he is in fact only guessing?
    b. What is the probability of denying his claim if he really does have the ability he claims?

43. Repeat Problem 42 where the person claims to read mental images 90% of the time.

44. In a certain office, three men determine who will pay for coffee by each flipping a coin. If one of them has an outcome that is different from the other two, he must pay for the coffee. What is the probability that in any play of the game there will be an odd man out?

45. In the game of Problem 44, what is the probability that there will be an odd man out in a particular play if there are five players?

46. In Problem 44, what is the probability that it will require more than two tosses to produce an odd man out?

47. Generalize Problem 44 for *n* men playing the game of odd man out.

8.4 The Normal Distribution

Suppose we survey the results of 20 children's scores on an IQ test. The scores (rounded to the nearest 5 points) are: 115, 90, 100, 95, 105, 105, 95, 105, 105, 95, 125, 120, 110, 100, 100, 90, 110, 100, 115, and 80. The mean is 103, the standard deviation is 10.93, and the frequency is shown in Figure 8.1a.

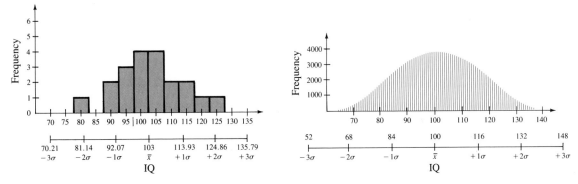

a. Frequencies of IQs for a sample of 20 children   b. Frequencies of IQs for a sample of 100,000 children

Figure 8.1 Frequencies of IQ scores

If we consider 100,000 IQ scores instead of only 20, we might obtain the frequency distribution shown in Figure 8.1b. As you can see, the frequency distribution approximates a curve. If we connect the endpoints of the bars in Figure 8.1b, we obtain a curve that is very close to a curve called the **normal distribution curve**, or simply the **normal curve**, as shown in Figure 8.2.

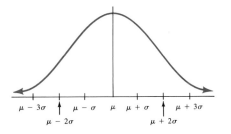

Figure 8.2 Normal distribution curve

The normal distribution is a **continuous** (rather than finite) **distribution**, and it extends indefinitely in both directions, never touching the x-axis. It is symmetric about a vertical line drawn through the mean, μ. The equation of this curve is

$$y = \frac{e^{-(x-\mu)^2/2\sigma^2}}{\sigma\sqrt{2\pi}} \qquad \text{where } \mu = \text{Mean}$$

$$\sigma = \text{Standard deviation}$$

$$\pi \approx 3.1416$$

$$e \approx 2.7183$$

Graphs of this curve for several choices of σ are shown in Figure 8.3. Using calculus, it can be shown that the "curvature" of the normal curve changes at $\mu + \sigma$ and $\mu - \sigma$. In calculus this point is called a *point of inflection*.

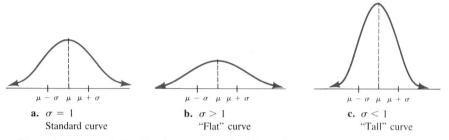

Figure 8.3 Variations of normal curves

a. $\sigma = 1$
Standard curve

b. $\sigma > 1$
"Flat" curve

c. $\sigma < 1$
"Tall" curve

Since the normal distribution is a probability distribution, we know the area under this curve is 1. Therefore we can relate the area to probabilities:

Probabilities of a Normal Probability Distribution

Let X be a random variable with a normal probability distribution. Then:

$P(a \leq X \leq b)$ is the area under the associated normal curve between a and b.

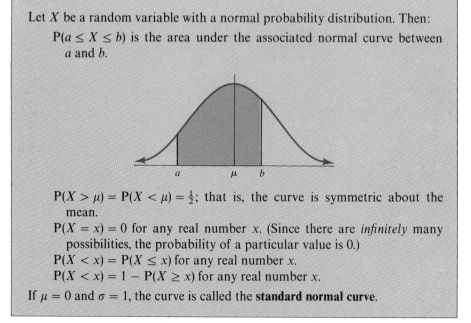

$P(X > \mu) = P(X < \mu) = \frac{1}{2}$; that is, the curve is symmetric about the mean.

$P(X = x) = 0$ for any real number x. (Since there are *infinitely* many possibilities, the probability of a particular value is 0.)

$P(X < x) = P(X \leq x)$ for any real number x.

$P(X < x) = 1 - P(X \geq x)$ for any real number x.

If $\mu = 0$ and $\sigma = 1$, the curve is called the **standard normal curve**.

Since it requires calculus to find particular areas under the standard normal curve, extensive tables have been compiled to determine the area under this curve without the necessity of going through actual computations. We know, for example, that approximately 68% (.6826) of the area lies between $\mu + \sigma$ and $\mu - \sigma$. Also, approximately 14% (.1359) is between $\mu + \sigma$ and $\mu + 2\sigma$, and 2% (.0215) between $\mu + 2\sigma$ and $\mu + 3\sigma$.

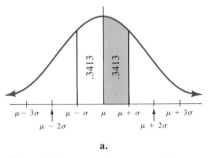
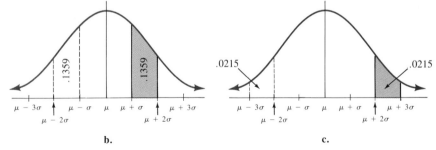

a.   b.   c.

Figure 8.4 Areas under a normal curve

For additional areas, use Table 4 in Appendix E. The table is arranged to give the area under the standard normal curve to the left of a vertical line through some number z, where $\mu = 0$ and $\sigma = 1$. Thus, to verify the area between $\mu + \sigma$ and $\mu + 2\sigma$ (shaded in color in Figures 8.4b and 8.5), you need to find $z = 2$ and then subtract $z = 1$. From the table, the area to the left of $z = 2$ is .9772 and the area to the left of $z = 1$ is .8413. Therefore the area between $z = 2$ and $z = 1$ is

$$.9772 - .8413 = .1359$$

or about 14% of the area under the curve (as stated earlier).

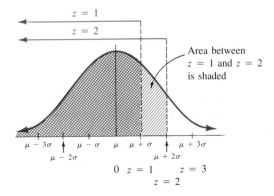

Figure 8.5

EXAMPLE 1 Find the area under the standard normal curve to the left of $z = .57$.

Solution From Table 4 in Appendix E, it is .7157. ∎

EXAMPLE 2 Find the area under the standard normal curve to the right of $z = -.13$.

Solution From Table 4, the area to the *left* of $-.13$ is .4483, so the area to the right is

$$1 - .4483 = .5517$$ ∎

EXAMPLE 3 Find the area under the standard normal curve between $z = -.05$ and $z = .93$.

Solution The area to the left of $-.05$ is .4801 and to the left of .93 it is .8238, so the area between those values is

$$.8238 - .4801 = .3437$$ ∎

What if we are not working with a standard normal curve? That is, suppose the curve is a normal curve, but it does not have a mean of 0. We can use the information in Figure 8.4, which, for convenience, is summarized in Figure 8.6.

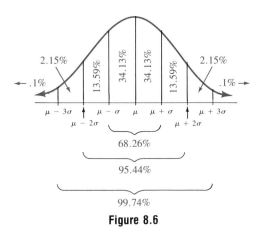

Figure 8.6

EXAMPLE 4 A teacher claims to grade "on a curve." This means the teacher believes that the scores on a given test are normally distributed. If 200 students take the exam, with mean 73 and standard deviation 9, how would the teacher grade the students?

Solution First, draw a normal curve with a mean 73 and standard deviation 9, as shown below:

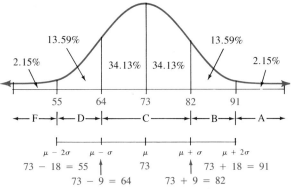

The range of 73 to 82 will contain about 34% of the class, and the range of 82 to 91 will contain about 14% of the class. Finally, about 2% of the class will score higher than 91. The teacher would therefore give grades according to the following table:

Grade on final	Letter grade	Number receiving grade	Percentage of class
Above 91	A	4	2%
83–91	B	28	14%
64–82	C	136	68%
55–63	D	28	14%
54 or below	F	4	2%

EXAMPLE 5 The Ridgemont Light Bulb Company tested a new line of light bulbs and found them to be normally distributed, with a mean life of 98 hours and a standard deviation of 13.

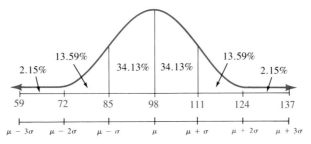

a. What percentage of bulbs will last less than 72 hours?
b. What is the probability that a bulb selected at random will last more than 111 hours?

Solution   Draw a normal curve with mean 98 and standard deviation 13.

a. About 2% will last less than 72 hours.
b. We see that about 16% of the bulbs will last longer than 111 hours, so
$$P(X > 111) \approx .16$$ ■

If you need to use a finer division of standard deviations than 1, 2, or 3, you will have to use calculus, or have a table available for that particular μ and σ, or use a number called a **z-score**. The z-score essentially translates any normal curve into a standard normal curve by use of the simple calculation given below. (This is why Table 4 in Appendix E uses z values.)

z-Score	The area to the left of a value x under a normal curve with mean μ and standard deviation σ is the same as the area under a standard normal curve to the left of the following value of z: $$z = \frac{x - \mu}{\sigma}$$

EXAMPLE 6   Find the probability that one of the light bulbs described in Example 5 will last between 110 and 120 hours.

Solution   From Example 5, $\mu = 98$ and $\sigma = 13$. Let $x = 110$; then
$$z = \frac{110 - 98}{13} \approx .92$$

Now, look up $z \approx .92$ in Table 4 (Appendix E). The area to the left of $x = 110$ is .8212.
  For $x = 120$,
$$z = \frac{120 - 98}{13} \approx 1.69$$

Again, from Table 4, the area to the left of $x = 120$ is .9545 (find $z \approx 1.69$ in the table). The area between $x = 110$ and $x = 120$ is
$$.9545 - .8212 = .1333$$

Thus $P(110 < X < 120) \approx .13$.   ■

Problem Set 8.4

Find the area under the standard normal curve satisfying the conditions in Problems 1–16.

1. Left of -2
2. Left of 0
3. Left of 1
4. Left of 1.23
5. Left of $-.61$
6. Left of $.81$
7. Right of -2
8. Right of 1
9. Right of -1
10. Right of -1.73
11. Right of 1.69
12. Right of $-.11$
13. Between $.5$ and 1.61
14. Between $-.4$ and $.4$
15. Between -1.03 and 1.59
16. Between -2.8 and $-.46$

APPLICATIONS

In Problems 17–21 suppose that people's heights (in centimeters) are normally distributed, with a mean of 170 and a standard deviation of 5. We take a sample of 50 persons.

17. How many would you expect to be between 165 and 175 centimeters tall?
18. How many would you expect to be taller than 160 centimeters?
19. How many would you expect to be taller than 175 centimeters?
20. If a person is selected at random, what is the probability that he or she is taller than 165 centimeters?
21. What is the variance (s^2) for this sample?

In Problems 22–26 suppose that, for a certain exam, a teacher grades on a curve. It is known that the mean is 50 and the standard deviation is 5. There are 45 students in the class.

22. How many students would receive a C?
23. How many students would receive an A?
24. What score would be necessary to obtain an A?
25. If an exam paper is selected at random, what is the probability that it will be a failing paper?
26. What is the variance for this exam?

27. The breaking strength of a rope (in pounds) is normally distributed, with a mean of 100 pounds and a standard deviation of 16. What is the probability that a certain rope will break with a force of 132 pounds?
28. The diameter of an electric cable is normally distributed, with a mean of .9 inch and a standard deviation of .01. What is the probability that the diameter will exceed .91 inch?
29. The annual rainfall in Ferndale, California, is known to be normally distributed, with a mean of 35.5 inches and a standard deviation of 2.5. About 2.15% of the time, how many inches will the rainfall exceed?
30. In Problem 29, what is the probability that the rainfall will exceed 30.5 inches?

A light bulb is normally distributed with a mean life of 250 hours and a standard deviation of 25 hours. Find the probabilities requested in Problems 31–36.

31. $P(X > 250)$
32. $P(X > 300)$
33. $P(X < 220)$
34. $P(200 < X < 300)$
35. $P(220 \leq X \leq 320)$
36. $P(230 \leq X \leq 240)$

37. The diameter of a pipe is normally distributed, with a mean of .4 inch and a variance of .0004. What is the probability that the diameter will exceed .44 inch?
38. The breaking strength of a certain new synthetic material is normally distributed, with a mean of 165 pounds and a variance of 9. The material is considered defective if the breaking strength is less than 159 pounds. What is the probability that a sample chosen at random will be defective?
39. Repeat Problem 37 to find the probability that the diameter will exceed .43 inch.
40. Repeat Problem 38 to find the probability that the breaking strength of a sample is between 170 and 180 pounds.

8.5 Normal Approximation to the Binomial

The normal distribution in Section 8.4 was introduced by looking at a histogram showing IQ frequencies. In this section we will look at the relationship between the binomial and normal distributions.

EXAMPLE 1   Consider an experiment of tossing a coin 16 times. What is the probability of obtaining exactly x heads for $x = 0, 1, 2, \ldots, 15, 16$?

Solution   This is a binomial distribution for which $n = 16$, $p = \frac{1}{2}$, and $q = \frac{1}{2}$. The model is

$$P(X = x) = \binom{n}{x} p^x q^{n-x}$$

$$= \binom{16}{x} \left(\frac{1}{2}\right)^x \left(\frac{1}{2}\right)^{16-x}$$

X	P(X)
0	.00002
1	.0002
2	.0018
3	.0085
4	.0278
5	.0667
6	.1222
7	.1746
8	.1964
9	.1746
10	.1222
11	.0667
12	.0278
13	.0085
14	.0018
15	.0002
16	.00002

The table in the margin shows the results of all the calculations for $x = 0, 1, 2, \ldots, 15$, and 16. Let us consider the procedure for one of these, say, $x = 9$:

$$P(X = 9) = \binom{16}{9} \left(\frac{1}{2}\right)^9 \left(\frac{1}{2}\right)^7$$

$$= \frac{16!}{(16-9)!9!} \left(\frac{1}{2}\right)^{16}$$

$$= \frac{16!}{7!9!2^{16}}$$

$$\approx .1746$$

We can construct a histogram showing the results for Example 1. In Figure 8.7, which shows this, we use $2^{16} = 65{,}536$ for the total number of possibilities.

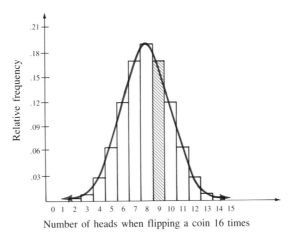

Figure 8.7 Histogram for the binomial distribution in Example 1

Number of heads when flipping a coin 16 times

Note that we have superimposed a normal curve over the histogram in Figure 8.7. Under certain conditions a normal curve can be used to approximate a binomial distribution. This relationship was first noted by the mathematician Abraham De Moivre in 1718. Suppose we use the normal curve to approximate the binomial distribution in Example 1. We first find the mean and the standard deviation:

$$\mu = np \qquad\qquad \sigma = \sqrt{npq}$$
$$= 16(\tfrac{1}{2}) \qquad\qquad = \sqrt{16(\tfrac{1}{2})(\tfrac{1}{2})}$$
$$= 8 \qquad\qquad\quad = \sqrt{4}$$
$$\qquad\qquad\qquad = 2$$

The normal curve shown in Figure 8.7 has mean 8 and standard deviation 2. The important difference between the binomial and normal distributions is that the binomial is *discrete* (or finite) and the normal is *continuous* (or infinite). If we were to find the probability of $x = 9$ for a normal distribution, we would find it to be 0. However, take a closer look at Figure 8.7, as shown in Figure 8.8.

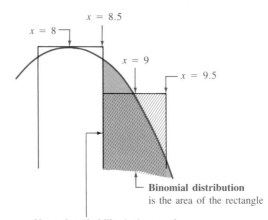

Normal probability is the area from $x = 8.5$ to $x = 9.5$ under the curve. This is used to approximate the area of the rectangle.

Binomial distribution is the area of the rectangle

Figure 8.8 Detail of Figure 8.7

The z-score for $x = 9.5$ is

$$z = \frac{x - \mu}{\sigma}$$

$$= \frac{9.5 - 8}{2}$$

$$= .75$$

The Table 4 entry for $z = .75$ is .7734

The z-score for $x = 8.5$ is

$$z = \frac{8.5 - 8}{2}$$

$$= .25$$

The Table 4 entry for $z = .25$ is .5987

The area between $x = 9.5$ and $x = 8.5$ is therefore

$$.7734 - .5987 = .1747$$

Note that this is a *very* good approximation for $P(X = 9)$ in the binomial distribution found in Example 1.

The Normal Distribution as an Approximation to the Binomial Distribution

If X is a binomial random variable of n independent trials, each with probability of success p and failure $q = 1 - p$, and if

$$np \geq 5 \quad \text{and} \quad nq \geq 5$$

then the binomial random variable is approximated by a normal distribution with mean and standard deviation given by

$$\mu = np \qquad \sigma = \sqrt{npq}$$

To show the tremendous value of this theorem, consider another, more realistic example.

EXAMPLE 2   Experimental data indicate that the cure rate of a new drug is 85%. If the drug is given to 100 patients, find the probability that at least 90 of them will be cured.

Solution   This is a binomial distribution with $p = .85$, $q = .15$, and $n = 100$. We want to find $P(X \geq 90)$.

$$P(X \geq 90) = P(X = 90) + P(X = 91) + \cdots + P(X = 99) + P(X = 100)$$

Now,

$$P(X = 90) = \binom{100}{90}(.85)^{90}(.15)^{10}$$

$$P(X = 91) = \binom{100}{91}(.85)^{91}(.15)^{9}$$

$$\vdots$$

$$P(X = 100) = \binom{100}{100}(.85)^{100}(.15)^{0}$$

These calculations are certainly out of hand (they are even out of the range of what most calculators can handle). However,

$$np = 100(.85) = 85 \geq 5 \qquad \text{and} \qquad nq = 100(.15) = 15 \geq 5$$

so we can use the normal distribution as an approximation.

$$\mu = np \qquad \sigma = \sqrt{npq}$$
$$= 85 \qquad \quad = \sqrt{100(.85)(.15)}$$
$$\approx 3.571$$

Now, $P(X \geq 90) = 1 - P(X < 90)$, so for $x = 90$, we use the z-score and Table 4 (Appendix E):

$$z = \frac{x - \mu}{\sigma}$$

$$= \frac{90 - 85}{3.571}$$

$$\approx 1.40 \qquad \text{From Table 4, } P(X < 90) \approx .9192$$

Then

$$P(X \geq 90) \approx 1 - .9192$$
$$= .0808 \qquad\qquad\qquad\qquad\qquad\qquad\qquad \blacksquare$$

In Example 2 we calculated np and nq in order to see if we could apply a normal approximation. Table 8.1 provides a quick reference as to when we may use this approximation. For Example 2, we could have used $p = .9$ and the table to find that n must be at least 50 to conclude that the normal is an acceptable approximation.

TABLE 8.1
Minimum sample required to approximate a binomial distribution by a normal distribution

p	n must be at least
.001	5000
.01	500
.1	50
.2	25
.3	17
.4	13
.5	10
.6	13
.7	17
.8	25
.9	50
.99	500
.999	5000

EXAMPLE 3   In certain grades in elementary school, students are given a vision test. Experience has shown that 18% of the students do not pass the test. Find the probability that if the test is given to 100 students, *exactly* 18 will fail the test.

Solution   This is a binomial distribution with $n = 100$ and $p = .18$.

$$P(X = 18) = \binom{100}{18}(.18)^{18}(.82)^{82}$$

This is a formidable calculation (even with a calculator). However, Table 8.1 indicates that we can use a normal approximation with

$$\mu = np \qquad\qquad \sigma = \sqrt{npq}$$
$$= (100)(.18) \qquad = \sqrt{100(.18)(.82)}$$
$$= 18 \qquad\qquad\quad \approx 3.84$$

We cannot find $x = 18$ exactly, so we find the difference of $x = 18.5$ and $x = 17.5$:

For x = 18.5:   *For x = 17.5:*

$$z = \frac{18.5 - 18}{3.84} \qquad z = \frac{17.5 - 18}{3.84}$$

$$\approx .13 \qquad\qquad \approx -.13$$

From Table 4,

$$P(X \le 18.5) \approx .5517 \qquad P(X \le 17.5) \approx .4483$$

Thus

$$P(17.5 \le X \le 18.5) \approx .5517 - .4483$$
$$= .1034$$

■

Problem Set 8.5

For the given values associated with a binomial experiment in Problems 1–9, determine whether the normal distribution is a suitable approximation.

1. $n = 1000$, $p = .01$
2. $n = 20$, $p = .5$
3. $n = 10$, $p = .6$
4. $n = 15$, $p = .6$
5. $n = 100$, $p = .4$
6. $n = 50$, $p = .1$
7. $n = 1000$, $p = .04$
8. $n = 12$, $p = .6$
9. $n = 40$, $p = .09$

A coin is tossed 15 times. Let X be a random variable representing the number of tails. Find the probabilities in Problems 10–21.

10. $P(X < 8)$
11. $P(X < 10)$
12. $P(X < 11)$
13. $P(X > 13)$
14. $P(X > 6)$
15. $P(X > 9)$
16. $P(7 < X < 9)$
17. $P(4 \leq X \leq 8)$
18. $P(5 \leq X \leq 7)$
19. $P(X = 5)$
20. $P(X = 11)$
21. $P(X = 14)$

A die is tossed 1000 times. Let X be the number of 6s tossed. Find the probabilities in Problems 22–33.

22. $P(X < 200)$
23. $P(X \leq 170)$
24. $P(X \leq 190)$
25. $P(X \geq 150)$
26. $P(X \geq 140)$
27. $P(X > 160)$
28. $P(140 \leq X \leq 150)$
29. $P(150 < X < 160)$
30. $P(140 < X < 150)$
31. $P(X = 140)$
32. $P(X = 150)$
33. $P(X = 160)$

APPLICATIONS

A new drug has a cure rate of 92%. The drug is administered to 1000 patients. Let X be the number who are cured by the drug. Find the probabilities in Problems 34–45.

34. $P(X < 900)$
35. $P(X < 950)$
36. $P(X \leq 850)$
37. $P(X \geq 920)$
38. $P(X > 900)$
39. $P(X > 910)$
40. $P(900 \leq X \leq 1000)$
41. $P(850 < X < 950)$
42. $P(800 < X < 950)$
43. $P(X = 900)$
44. $P(X = 920)$
45. $P(X = 930)$

46. An experiment in parapsychology consists of correctly identifying one of five shapes. If the test is repeated 10 times, what is the probability that the subject will correctly identify the object more than 8 times?

47. A baseball player's batting average is .250. What is the probability of more than six hits in the next ten times at bat?

48. In 1980 the yearly death rate from cancer was 186.3 per 100,000. If a person associates with roughly 500 people, estimate the probability that more than 1 out of these 500 will die of cancer in a year.

49. Repeat Problem 48 to find the probability that more than 2 of the 500 will die in a year.

50. A computer system is made up of 100 components, each with a reliability of .95 (that is, the probability that the component operates properly is .95). If these components function independently of one another, and the computer requires at least 80 components to function properly, what is the reliability of the whole system?

51. An airline is penalized for overbooking, but loses money with empty seats. Suppose the records show that for a certain flight, 8% of the advance reservations do not show up. If the plane seats 300 people, find the probability that if the airline takes 315 advance reservations, more than 300 will show up.

8.6 Correlation and Regression*

Mathematical modeling relates numerical data to assumptions about the relationship between two variables. For example:

IQ and salary
Study time and grades
Age and heart disease
Runner's speed and runner's shoes
Teacher's salaries and beer consumption

---

* The material of this section is not required for subsequent material.

All are attempts to relate two variables in some way or another. If a **correlation** is established, then the next step in the modeling process is to identify the nature of the relationship. This is called **regression analysis**. We now turn our attention to finding a *linear relationship* that is called the *best-fitting line* by a technique called the **least squares method**. The derivations of the results and the formulas given in this section are, for the most part, based on the calculus developed in Section 16.3 and will therefore not be done here. We will, however, focus on how to use the formulas as well as discuss what they mean and how they are interpreted.

We begin with correlation. We want to know if two variables are related—that is, depend on one another. Let us call one variable x and the other y. These variables can be represented as ordered pairs (x, y) in a graph called a **scatter diagram**.

EXAMPLE 1 A survey of 20 students compared the grade received on an examination with the length of time the student studied. Draw a scatter diagram to represent the data in the table.

Student Number	1	2	3	4	5	6	7	8	9	10
Length of Study Time (to nearest 5 minutes)	30	40	30	35	45	15	15	50	30	0
Grade (100 possible)	72	85	75	78	89	58	71	94	78	10
Student Number	11	12	13	14	15	16	17	18	19	20
Length of Study Time (to nearest 5 minutes)	20	10	25	25	25	30	40	35	20	15
Grade (100 possible)	75	43	68	60	70	68	82	75	65	62

Solution Let x be the study time (in minutes) and let y be the grade (in points). The graph is shown below:

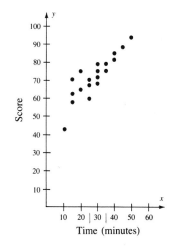

Correlation is a measure to determine whether there is a statistically significant relationship between two variables. Intuitively, it should assign a measure consistent with the scatter diagram shown in Figure 8.9.

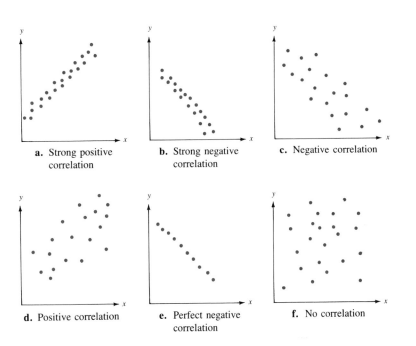

a. Strong positive correlation

b. Strong negative correlation

c. Negative correlation

d. Positive correlation

e. Perfect negative correlation

f. No correlation

Figure 8.9 Examples of correlation

Such a measure, called the **linear correlation coefficient**, r, is defined so that it has the following properties:

1. r measures the correlation between x and y.
2. r is between -1 and 1.
3. If r is close to 0, it means there is little correlation.
4. If r is close to 1, it means there is a strong positive correlation.
5. If r is close to -1, it means there is a strong negative correlation.

To write a formula for r, we let n denote the number of pairs of data present; and, as before:

Σx denotes the sum of the x values

Σx^2 means square the x values and then sum

$(\Sigma x)^2$ means sum the x values and then square

Σxy means multiply each x value by the corresponding y value and then sum

$n\Sigma xy$ means multiply n times Σxy

$(\Sigma x)(\Sigma y)$ means multiply Σx and Σy

Correlation Coefficient

The *linear correlation coefficient r* is

$$r = \frac{n \sum xy - (\sum x)(\sum y)}{\sqrt{n(\sum x^2) - (\sum x)^2} \sqrt{n(\sum y^2) - (\sum y)^2}}$$

EXAMPLE 2 Find *r* for the data in Example 1.

Solution

Study time, x	Score, y	xy	x^2	y^2
30	72	2160	900	5184
40	85	3400	1600	7225
30	75	2250	900	5625
35	78	2730	1225	6084
45	89	4005	2025	7921
15	58	870	225	3364
15	71	1065	225	5041
50	94	4700	2500	8836
30	78	2340	900	6084
0	10	0	0	100
20	75	1500	400	5625
10	43	430	100	1849
25	68	1700	625	4624
25	60	1500	625	3600
25	70	1750	625	4900
30	68	2040	900	4624
40	82	3280	1600	6724
35	75	2625	1225	5625
20	65	1300	400	4225
15	62	930	225	3844
Total 535	1378	40,575	17,225	101,104
↑ $\sum x$	↑ $\sum y$	↑ $\sum xy$	↑ $\sum x^2$	↑ $\sum y^2$

$$r = \frac{n \sum xy - (\sum x)(\sum y)}{\sqrt{n(\sum x^2) - (\sum x)^2} \sqrt{n(\sum y^2) - (\sum y)^2}}$$

$$= \frac{20(40,575) - (535)(1378)}{\sqrt{20(17,225) - (535)^2} \sqrt{20(101,104) - (1378)^2}}$$

$$= \frac{74,270}{\sqrt{58,275} \sqrt{123,196}}$$

$$\approx .8765$$

∎

Example 2 shows a very strong positive correlation. But if *r* for Example 2 had been .46, would we still be able to assume that there is a strong correlation? This question is a topic of major concern in statistics. The term **significance level** is used to denote the cutoff between results attributed to chance and the results attributed to significant differences. Table 8.2 gives **critical values** for determining whether two variables are correlated. If r is greater than the given table value, then we may assume that a correlation exists between the variables. If we use the column labeled $\alpha = .05$, then the significance level is 5%. This means that the probability is .05 that we will say the variables are correlated when, in fact, the results are attributed to chance. This is also true for a significance level of 1% ($\alpha = .01$). For Example 2, since $n = 20$, we see in Table 8.2 that $r = .46$ would show a linear correlation at a 5% significance level, but not at a 1% level.

TABLE 8.2
Correlation coefficient

n	$\alpha = .05$	$\alpha = .01$	n	$\alpha = .05$	$\alpha = .01$
4	.950	.999	18	.468	.590
5	.878	.959	19	.456	.575
6	.811	.917	20	.444	.561
7	.754	.875	25	.396	.505
8	.707	.834	30	.361	.463
9	.666	.798	35	.335	.430
10	.632	.765	40	.312	.402
11	.602	.735	45	.294	.378
12	.576	.708	50	.279	.361
13	.553	.684	60	.254	.330
14	.532	.661	70	.236	.305
15	.514	.641	80	.220	.286
16	.497	.623	90	.207	.269
17	.482	.606	100	.196	.256

Note: The derivation of this table is beyond the scope of this course. It shows the critical values of the *Pearson correlation coefficient*.

EXAMPLE 3   Find the critical value of the linear correlation coefficient for 10 pairs of data and a significance level of .05.

Solution   From Table 8.2, the critical value is $r = .632$. For $n = 10$, any value greater than $r = .632$ is considered linearly correlated. ∎

EXAMPLE 4   If $r = -.85$ and $n = 10$, are the variables correlated at a significance level of 1%?

Solution   For $n = 10$ and $\alpha = .01$, the Table 8.2 entry is .765. Since r is negative and since $|r| > .765$, we see that there is a negative linear correlation. ∎

EXAMPLE 5   The following table is a sample of some past annual mean salaries for teachers in elementary and secondary schools, along with the annual per capita beer consumption (in gallons) for Americans. Find the correlation coefficient.*

Year	1960	1965	1970	1972	1973	1983
Mean Teacher Salary	$5,000	$6,200	$8,600	$9,700	$10,200	$16,400
Per Capita Beer Consumption (gal)	24.02	25.46	28.55	29.43	29.68	35.2

Solution

$n = 6$

$\Sigma x = 56,100$

$\Sigma y = 172.34$

$\Sigma x^2 = 604,490,000$

$\Sigma y^2 = 5026.3418$

$(\Sigma x)^2 = 3,147,210,000$

$(\Sigma y)^2 = 29,701.0756$

$\Sigma xy = 1,688,969$

$$r = \frac{6(1,688,969) - (56,100)(172.34)}{\sqrt{6(604,490,000) - (56,100)^2} \ \sqrt{6(5026.3418) - (172.34)^2}}$$

$$\approx .994$$

The number $r \approx .994$ is so close to 1 that we hardly need to consult Table 8.2 to know that there is a strong positive linear correlation between teachers' salaries and per capita beer drinking. ∎

The significance of the correlation implies that teachers use their raises to buy more beer, right? Wrong. Perhaps increases in teachers' salaries precipitate higher taxes, which in turn cause taxpayers to drown their sorrows and forget their financial difficulties by drinking more beer. Or perhaps higher teachers' salaries and greater beer consumption are both manifestations of some other factor, such as a general improvement in the standard of living. In any event, the techniques in this chapter can be used only to establish a *statistical* linear relationship. *We cannot establish the existence or absence of any inherent cause-and-effect relationship.*

The final step in our discussion is to find the best-fitting line. That is, we want to find a line $y' = mx + b$ so that the sum of the distances of the data points from this line will be as small as possible. (We use y' instead of y to distinguish between the actual second component, y, and the predicted y value, y'.) Since some of these distances may be positive and some negative and since we do not want large opposites to "cancel each other out," we minimize the sum of the *squares* of these distances. Therefore, the **regression line** is sometimes called the **least squares line**.

---

* Example 5 and the paragraph following are from Mario F. Triola, *Elementary Statistics*, 2nd ed. (Menlo Park, Calif.: Benjamin/Cummings, 1983). © 1983 by The Benjamin/Cummings Publishing Company, Inc. Reprinted by permission.

Least Squares, or Regression, Line

The *least squares, or regression, line* $y' = mx + b$ is the line of best fit when

$$m = \frac{n(\Sigma\, xy) - (\Sigma\, x)(\Sigma\, y)}{n(\Sigma\, x^2) - (\Sigma\, x)^2} \qquad b = \frac{\Sigma\, y - m(\Sigma\, x)}{n}$$

EXAMPLE 6 Find the best-fitting line for the data in Example 1.

Solution From Example 2, $n = 20$, and

$$\Sigma\, xy = 40{,}575$$
$$\Sigma\, x = 535$$
$$\Sigma\, y = 1378$$
$$\Sigma\, x^2 = 17{,}225$$
$$(\Sigma\, x)^2 = (535)^2$$

Since m is used as part of the formula for b, you must first find m:

$$m = \frac{20(40{,}575) - (535)(1378)}{20(17{,}225) - (535)^2}$$

$$= \frac{74{,}270}{58{,}275}$$

$$\approx 1.27447$$

Now you can use this m when finding b:

$$b = \frac{1378 - 1.27447(535)}{20}$$

$$\approx 34.8078$$

Thus we may approximate the least squares line as $y' = 1.3x + 35$. This line is shown in the figure in the margin. ∎

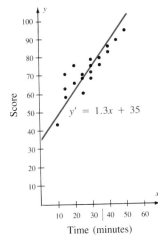

$y' = 1.3x + 35$

Regression line for the data in Example 1

EXAMPLE 7 Use the regression line of Example 6 to predict the score of a person who studied $\frac{1}{2}$ hour. ∎

Solution $x = 30$ minutes, so $y' = 1.3(30) + 35 = 74$

A final word of caution: Use the regression line only if r indicates that there is a significant linear correlation, as given in Table 8.2.

COMPUTER APPLICATION Program 12 on the computer disk accompanying this book will find the linear relationship between two variables using the least squares method, and it will also find the correlation between those variables.

Problem Set 8.6

In Problems 1–12 a sample of paired data gives a linear coefficient r. In each case use Table 8.2 to determine whether there is a significant linear correlation.

1. $n = 10$, $r = .7$, significance level 5%
2. $n = 10$, $r = .7$, significance level 1%
3. $n = 30$, $r = .4$, significance level 1%
4. $n = 30$, $r = .4$, significance level 5%
5. $n = 15$, $r = -.732$, significance level 5%
6. $n = 35$, $r = -.4127$, significance level 1%
7. $n = 50$, $r = -.3416$, significance level 1%
8. $n = 100$, $r = -.41096$, significance level 5%
9. $n = 23$, $r = .501$, significance level 1%
10. $n = 38$, $r = .416$, significance level 5%
11. $n = 28$, $r = -.214$, significance level 5%
12. $n = 55$, $r = -.14613$, significance level 1%

Draw a scatter diagram and find r for the data in each table in Problems 13–18 and determine whether there is a linear correlation at either the 5% or 1% level.

13.

x	1	2	3	4
y	1	5	8	13

14.

x	4	5	10	10
y	0	−10	−10	−20

15.

x	0	1	2	3	4
y	25	19	16	12	10

16.

x	1	3	3	5	8
y	30	22	19	15	10

17.

x	10	20	30	30	50	60
y	50	48	60	58	70	75

18.

x	85	90	100	102	105	110
y	80	40	30	28	25	15

In Problems 19–24 find the regression line for the indicated problem.

19. Problem 13   20. Problem 15   21. Problem 17
22. Problem 14   23. Problem 16   24. Problem 18

APPLICATIONS

25. A new computer circuit was tested and the times (in nanoseconds) required to carry out different subroutines were recorded as follows:

Difficulty Level	1	2	2	3	4	5	5	5
Time	10	11	13	8	15	18	21	19

Find r and determine whether it is statistically significant at the 1% level.

26. Ten people are given a standard IQ test. Their scores were then compared with their high school grades:

IQ	Grade (GPA)
117	3.1
105	2.8
111	2.5
96	2.8
135	3.4
81	1.9
103	2.1
99	3.2
107	2.9
109	2.3

Find r and determine whether it is statistically significant at the 1% level.

27. Find the regression line for the data in Problem 25.

28. Find the regression line for the data in Problem 26.

Problems 29–35 are based on the following table. Determine whether there is a correlation between the indicated variables, and, if so, is it at the 5% or the 1% significance level?

	Year					
	1965	**1970**	**1975**	**1979**	**1980**	**1983**
Birth Rate (Births per 1000)	19.4	18.3	14.8	15.8	16.2	15.5
Death Rate (Deaths per 1000)	9.4	9.4	8.9	8.7	8.9	8.6
Per Capita Income	$2,773	$3,893	$5,851	$8,757	$9,458	$11,675
Prime Interest Rate (6 months' commercial paper)	4.54%	7.72%	6.33%	10.91%	12.29%	12.15%

29. Birth rate and per capita income

30. Interest rate and birth rate

31. Interest rate and death rate

32. Birth rate and death rate

33. Per capita income and interest rate

34. Find the regression line for Problem 32.

35. Find the regression line for Problem 33.

8.7 Summary and Review

IMPORTANT TERMS

Average [8.2]
Bernoulli trial [8.3]
Binomial distribution [8.3]
Binomial experiment [8.3]
Binomial random variable [8.3]
Continuous distribution [8.4]
Continuous random variable [8.1]
Correlation [8.6]
Correlation coefficient [8.6]
Critical values [8.6]
Discrete random variable [8.1]
Dispersion [8.2]
Frequency distribution [8.1]
Histogram [8.1]
Least squares line [8.6]
Least squares method [8.6]
Linear correlation coefficient, r [8.6]
Mean [8.2]

Measure of central tendency [8.2]
Median [8.2]
Mode [8.2]
Normal distribution [8.4]
Population [8.2]
Probability distribution [8.1]
Random variable [8.1]
Range [8.2]
Regression analysis [8.6]
Regression line [8.6]
Relative frequency [8.1]
Scatter diagram [8.6]
Significance level [8.6]
Standard deviation [8.2]
Standard normal curve [8.4]
Variance [8.2]
Weighted mean [8.2]
z-Score [8.4]

SAMPLE TEST

For additional practice there are a large number of review problems categorized by objective in the Student Solutions Manual. The following sample test (40 minutes) is intended to review the main ideas of the chapter.

Use the following outcomes, obtained from rolling a pair of dice 50 times, in Problems 1–5.

4, 3, 6, 10, 8, 9, 2, 4, 7, 4, 6, 7, 11, 7, 8, 6, 4, 8, 3, 9, 7, 8, 7, 9, 5, 9, 6, 6, 10, 7, 3, 7,
10, 6, 11, 5, 9, 10, 6, 11, 8, 11, 7, 5, 6, 11, 12, 7, 8, 9

1. Prepare a frequency distribution.

2. Find the probability distribution.

3. Represent the data in a histogram.

4. Find the mean, median, and mode.

5. Find the range, variance, and standard deviation.

6. Use Table 3 in Appendix E to compute a binomial probability where $n = 10$, $k = 4$, and $p = .8$.

7. Find the mean, variance, and standard deviation for the distribution described in Problem 6.

8. Find the area under a normal curve with $\mu = 86$ and $\sigma = 5$ between 80 and 90.

9. The following item once appeared in Dear Abby's column:

> **Dear Abby:**   You wrote in your column that a woman is pregnant for 266 days. Who said so? I carried my baby for ten months and five days, and there is no doubt about it because I know the exact date my baby was conceived. My husband is in the Navy and it couldn't have possibly been conceived any other time because I saw him only once for an hour, and I didn't see him again until the day before the baby was born.
>
> I don't drink or run around, and there is no way this baby isn't his, so please print a retraction about that 266-day carrying time because otherwise I am in a lot of trouble.
>
> *San Diego Reader*

Abby's answer was consoling and gracious but not very statistical:

> **Dear Reader:**   The average gestation period is 266 days. Some babies come early. Others come late. Yours was late.

If the mean pregnancy duration is 266 with a standard deviation of 16 days, what is the probability of having a pregnancy longer than 310 days?

10. Since World War II, plutonium for use in atomic weapons has been produced at an Atomic Energy Commission facility in Hanford, Washington. One of the major safety problems encountered there has been the storage of radioactive wastes. Over the years, significant quantities of these substances have leaked from their open-pit storage areas into the nearby Columbia River, which flows through parts of Oregon and eventually empties into the Pacific Ocean. To measure the health consequences of this contamination, an index of exposure was calculated for each of the nine Oregon counties having frontage on either the Columbia River or the Pacific Ocean. The cancer mortality rate for each of these counties was also determined. The data are listed in the table, where higher index values represent higher levels of contamination.*

---

* From Richard J. Larsen and Donna Fox Stroup, *Statistics in the Real World* (New York: Macmillan, 1976).

Radioactive contamination and cancer mortality in Oregon counties

County	Index of exposure	Cancer mortality per 100,000
Clatsop	8.34	210.3
Columbia	6.41	177.9
Gilliam	3.41	129.9
Hood River	3.83	162.3
Morrow	2.57	130.1
Portland	11.64	207.5
Sherman	1.25	113.5
Umatilla	2.49	147.1
Wasco	1.62	137.5

Find the linear correlation coefficient and determine whether the variables are significantly correlated at either the 1% or 5% level.

11. Draw a scatter diagram and the regression line for the set of data in Problem 10.

The following multiple-choice questions dealing with probability and random variables are from actual actuarial exams and are reprinted with permission of the Society of Actuaries, 208 South LaSalle Street, Chicago, Illinois 60604.

Actuary Exam May 1982

12. Suppose Q and S are independent events such that the probability that at least one of them occurs is $\frac{1}{3}$ and the probability that Q occurs but S does not occur is $\frac{1}{9}$. What is $P(S)$?
   A. $\frac{4}{9}$ B. $\frac{1}{3}$ C. $\frac{2}{9}$ D. $\frac{1}{7}$ E. $\frac{1}{9}$

13. A fair coin is tossed until a head appears. Given that the first head appeared on an even-numbered toss, what is the conditional probability that the head appeared on the fourth toss?
   A. $\frac{1}{16}$ B. $\frac{1}{8}$ C. $\frac{3}{16}$ D. $\frac{1}{4}$ E. $\frac{15}{16}$

14. A card is drawn at random from an ordinary deck of 52 cards and replaced. This is done a total of 5 independent times. What is the conditional probability of drawing the ace of spades exactly 4 times, given that this ace is drawn at least 4 times?
   A. $\frac{1}{2}$ B. $\frac{12}{13}$ C. $\frac{13}{14}$ D. $\frac{60}{61}$ E. $\frac{255}{256}$

15. Suppose an experiment consists of tossing a fair coin until three heads occur. What is the probability that the experiment ends after exactly six flips of the coin with a head on the fifth toss as well as on the sixth?
   A. $\frac{1}{16}$ B. $\frac{1}{8}$ C. $\frac{5}{32}$ D. $\frac{1}{4}$ E. $\frac{10}{32}$

Cumulative Review for Chapters 3–8

1. Let

$$A = \begin{bmatrix} 4 & 0 & -1 \\ 3 & -2 & 1 \\ 0 & 3 & -2 \end{bmatrix} \qquad B = \begin{bmatrix} 4 & -1 & 3 \\ -3 & 1 & 2 \\ 2 & -3 & 0 \end{bmatrix}$$

$$I = \begin{bmatrix} 1 & 0 & 0 \\ 0 & 1 & 0 \\ 0 & 0 & 1 \end{bmatrix}$$

Perform the indicated operations.

a. AB **b.** $A(B + I)$ **c.** $I(2A + B)$ **d.** A^{-1}

Solve the systems of equations or inequalities in Problems 2–7.

2. $\begin{cases} x + y = 3 \\ 2x - y = 9 \end{cases}$

3. $\begin{cases} y = 2x + 1 \\ y = 3x - 5 \\ y = x + 7 \end{cases}$

4. $\begin{cases} x + 2y + 3z = 2 \\ 2x - y + z = -1 \end{cases}$

5. $\begin{cases} 3x + y + z = -6 \\ x + 2y - z = 9 \\ 5x + y - 3z = -4 \end{cases}$

6. $\begin{cases} 3x + 2y < -3 \\ x - y > 0 \\ x \le 0 \end{cases}$

7. $\begin{cases} 2x + 3y \le 60 \\ x + 2y \le 36 \\ 3x - 2y \le 24 \\ x \ge 0, \quad y \ge 0 \end{cases}$

8. Maximize: $23x_1 + 24x_2$

Subject to: $\begin{cases} 4x_1 + 2x_2 \le 1800 \\ 2x_1 + 3x_2 \le 1200 \\ 5x_1 + 4x_2 \le 2400 \\ x_1 \ge 0, \quad x_2 \ge 0 \end{cases}$

9. Let $U = \{1, 2, 3, \ldots, 8, 9, 10\}$, $A = \{2, 3, 5, 7\}$, $B = \{1, 3, 5, 7, 9\}$, and $C = \{2, 4, 6, 8, 10\}$. Find

a. $A \cap C$ **b.** $B \cup \bar{C}$ **c.** $U \cap C$ **d.** $\bar{U}$ **e.** $\overline{A \cap (B \cup C)}$

APPLICATIONS

10. In a survey of 600 college students:

250 had tried marijuana
350 had tried alcohol
175 had tried cocaine
110 had tried both cocaine and alcohol
140 had tried both marijuana and alcohol
100 had tried both marijuana and cocaine
70 had tried all three

a. How many had not tried any of these drugs?
b. How many had tried only one of these drugs?
c. How many had tried exactly two of these drugs?

11. a. How many subsets can be chosen from five elements?
b. In how many ways can six hats be distributed to six persons?

12. Suppose two cards are drawn from a deck of cards without replacement. Find the requested probabilities.
    a. P(both hearts)   b. P(1 heart and 1 diamond)
    c. P(heart on first draw and diamond on second draw)
    d. P(diamond on second draw given a heart on first draw)
    e. P(ace of hearts drawn both times)

13. Repeat Problem 12 with replacement.

14. A sample of 100 ball bearings is drawn from a day's production and 15 are found to be defective.
    a. What is the probability of drawing one ball bearing and finding that it is defective?
    b. Is this problem an example of empirical or theoretical probability?

15. At a fast-food outlet, 15% of the hamburgers sold were cold, 5% had a missing ingredient, and 0.5% were both cold and had a missing ingredient. What is the probability that your hamburger is cold if the pickle is missing?

16. A messenger transports money between two locations and travels along one of three routes, A Street, B Street, or C Street. One day, the route is randomly selected according to the probabilities $P(A) = .3$, $P(B) = .5$, and $P(C) = .2$. On the following day, the probabilities change so that the probability of the previously chosen route is .2, with the other two routes being equally probable. Find the indicated probabilities (subscripts are used to indicate the day on which the particular route is taken; for example, A_2 means that A was chosen on the second day).
    a. $P(A_2 | A_1)$ b. $P(A_2 | B_1)$ c. $P(A_2)$ d. $P(C_3)$

17. Consider the following test scores:

    85, 70, 75, 90, 65, 40, 70, 95, 80, 70, 55, 65, 70, 80, 95

    a. Prepare a frequency distribution.
    b. Find the probability distribution and draw a histogram.
    c. Find the mean, median, and mode.
    d. Find the range, variance, and standard deviation using the frequency distribution.

18. A psychology teacher grades on a curve. If the mean is 65 and the standard deviation is 10, how would the grades be distributed? If there are 100 students in the class, how many could expect to get As, Bs, etc.?

19. A study is conducted to test the relationship between speed (mph) and fuel consumption (mpg). The following information is obtained:

Speed	20	30	40	50	60
Fuel Consumption	35	38	40	34	29

    a. Find the linear correlation coefficient and determine whether the variables are significantly correlated at either the 1% or the 5% level.
    b. Find the regression line for this set of data.

Solve the linear programming problems in Problems 20–23.

20. A hospital wishes to provide a diet for its patients that has a minimum of 50 grams of carbohydrates, 30 grams of proteins, and 40 grams of fats per day. These requirements can be met with two foods, A and B. Food A costs $.14 per ounce and supplies 6 grams of carbohydrates, 3 grams of proteins, and 1 gram of fats per ounce. Food B costs $.06

per ounce and supplies 2 grams of carbohydrates, 2 grams of proteins, and 2 grams of fats per ounce. How many ounces of each food should be bought for each patient per day in order to meet the minimum requirements at the lowest cost?

21. An island is inhabited by three species of animals, A, B, and C, which feed on three types of food, F1, F2, and F3. The amount of each type of food (in ounces) to sustain one animal of each type is summarized below:

Animal	Food		
	F1	F2	F3
A	12	14	16
B	15	20	5
C	5	8	6

If the island contains 2000 ounces of food F1, 12,000 ounces of food F2, and 8000 ounces of food F3, what is the maximum number of animals that it can support?

22. Karlin Manufacturing produces two types of children's toys, wagons and carts. Each product must be processed in each of three departments: machining, assembling, and finishing. The hours needed to produce one toy per department and the maximum possible hours available per department are shown below:

Department	Production Time per Unit (hours)		Maximum Capacity (hours)
	Wagons	Carts	
Machining	2	1	2000
Assembling	2	2	2400
Finishing	1	2	3000

Karlin's net profit is $5 per wagon and $4 per cart. How many wagons and carts should be manufactured to maximize the profit?

23. Of the 20 or so amino acids comprising protein, 8 cannot be synthesized by humans and are known as the *essential amino acids*. A diet that provides an adequate amount of the three amino acids tryptophan (TRP), lysine (LYS), and methionine (MET) will also generally provide enough of the other essential amino acids. The table below shows three sources for these acids and the approximate number of grams of TRP, LYS, and MET per pound.

Food	Amino Acid		
	TRP	LYS	MET
Beef	1.2	8.0	2.4
Peanuts	1.5	5.0	1.2
Cashews	2.0	3.0	1.5

If peanuts cost $2.50 per pound and cashews cost $5 per pound, find the cheapest combination of peanuts and cashews that contains an amount of each of the three amino acids equal to or greater than the amount in a pound of beef.

The following two questions are from an actual CPA examination given by the American Institute of Certified Public Accountants, Inc. They are reproduced with permission.

CPA Exam
November 1974

The Golden Hawk Manufacturing Company wants to maximize the profits on products A, B, and C. The contribution margin for each product follows:

Product	Contribution margin
A	$2
B	$5
C	$4

The production requirements and departmental capacities, by departments, are:

Department	Production requirements by product (hours)			Departmental capacity (total hours)
	A	B	C	
Assembling	2	3	2	30,000
Painting	1	2	2	38,000
Finishing	2	3	1	28,000

24. What is the profit maximization formula for the Golden Hawk Company?
 A. $2A + $5B + $4C = X,$ where X = profit
 B. $5A + 8B + 5C \leq 96,000$
 C. $2A + $5B + $4C \leq X,$ where X = profit
 D. $2A + $5B + $4C = 96,000$

25. What is the constraint for the painting department of the Golden Hawk Company?
 A. $1A + 2B + 2C \geq 38,000$ **B.** $$2A + $5B + $4C \geq 38,000$
 C. $1A + 2B + 2C \leq 38,000$ **D.** $2A + 3B + 2C \leq 30,000$

CALCULUS

The ideas of calculus were first considered toward the end of the sixteenth century, but the theory was not developed until the second half of the seventeenth century when Gottfried Leibniz (1646–1716) and Isaac Newton (1642–1727) simultaneously, and independently, invented the calculus we use today. Calculus extends the range of problems we can work with. For example:

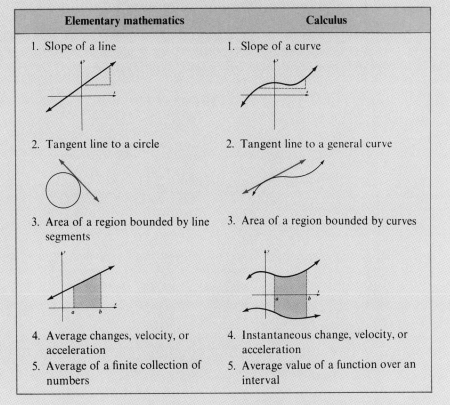

Elementary mathematics	Calculus
1. Slope of a line	1. Slope of a curve
2. Tangent line to a circle	2. Tangent line to a general curve
3. Area of a region bounded by line segments	3. Area of a region bounded by curves
4. Average changes, velocity, or acceleration	4. Instantaneous change, velocity, or acceleration
5. Average of a finite collection of numbers	5. Average value of a function over an interval

Up to now, your study of mathematics has focused on the *mechanics* of mathematics (such as solving equations, drawing graphs, and manipulating symbols), but now you will not only reinforce such skills and learn new ones but you will also apply mathematics to particular real life situations.

The prerequisites for this part of the text are at least 2 years of high school mathematics (reviewed in Chapter 1) and the concepts of functions and graphs (Chapter 2).

CHAPTER 9
The Derivative

CHAPTER CONTENTS

9.1 Limits

9.2 Continuity

9.3 Rates of Change

9.4 Definition of Derivative

9.5 Differentiation
Techniques, Part I

9.6 Differentiation
Techniques, Part II

9.7 The Chain Rule

9.8 Summary and Review
Important Terms
Sample Test

APPLICATIONS

Management (*Business, Economics, Finance, and Investments*)

Cost of manufacturing a specialized tool (9.1, Problem 57)
Rental charges for a fleet of trucks (9.2, Problem 42)
Output as a function of the number of workers (9.3, Problems 5–8)
Rate of change of the Gross National Product (9.3, Problems 13–16)
Cost of a per unit increase in production (9.3, Problems 45–47; 9.8, Problem 19)
Marginal profit (9.3, Problem 48; 9.5, Problem 32; 9.8, Problem 20)
Average rate of change of cost (9.3, Problems 49–55)
Marginal cost (9.3, Problems 56–57; 9.4, Problems 22–24; 9.5, Problem 31)
Change in the rate of earnings in a corporation (9.5, Problem 35)
Demand for a commodity in a free market (9.6, Problem 41)
Rate of change of prices in a free market (9.6, Problem 42)

Life sciences (*Biology, Ecology, Health, and Medicine*)

Temperatures at Death Valley (9.2, Problem 44)
Rate at which the number of bacteria in a culture change (9.4, Problems 25–27;
9.7, Problem 35)
Relationship between current and time on an X-ray machine (9.5,
Problems 33–34)
The relationship between a population of foxes and rabbits (9.5, Problems 38–39)
Rate of a liquid flowing into a reservoir (9.5, Problem 40)
Effect of a drug in the bloodstream (9.6, Problem 43)

Social sciences (*Demography, Political Science, Population, Psychology, Society,
and Sociology*)

The number of animals available for a psychology experiment (9.1, Problem 58)
Learning theory (9.1, Problem 59; 9.5, Problems 36–37; 9.6, Problem 44; 9.7,
Problem 36)
SAT scores of first-year college students (9.3, Problems 9–12)

General interest

Postal charges (9.2, Problem 43)
Height of a projectile after t seconds (9.3, Problems 1–4)
Average speeds in a daily commute (9.3, Problems 17–20)
Velocity of an object moving in a straight line (9.4, Problem 28)

Modeling application—Instantaneous Acceleration: A Case Study of the
Mazda 626

CHAPTER
OVERVIEW

We focus here on the concept of a derivative and are introduced to efficient ways of finding the derivative. The derivative is one of the fundamental ideas in all of calculus and is the cornerstone of more advanced mathematics.

PREVIEW

We are first introduced to the derivative as a rate of change but soon discover that it has many additional useful applications. After just a glimpse of these applications (to be continued in the next chapter), we concentrate on finding derivatives.

PERSPECTIVE

The skills learned in this chapter are used in the next chapter when we consider applications of the derivative. Then the integral is introduced as an "antiderivative"—which will tax even further your ability to work with the derivatives introduced in this chapter. Two of the single most revolutionary concepts in all of mathematics are the ideas of limits and derivatives. Remember, it took some of the greatest minds in the history of mathematics many years to formulate these ideas, so do not despair if you have trouble understanding them in one evening, or even one course. Hard work and perseverance will pay off.

9.1 Limits

Mathematical analysis can be divided into two broad categories: *continuous* and *discrete*. Let us consider a few examples to clarify the distinction. The counter on a turnstile may look very similar to the odometer on a car, but the turnstile is a discrete counting device while the odometer is a continuous device. The set of integers is a discrete set, whereas the set of real numbers is not. Calculus was being invented as mathematicians all over the world realized they needed to deal with new notions concerning the transition from discrete to continuous. Calculus is based on a continuous model, and the central key to understanding calculus is the notion of a limit.

Calculus was originally developed intuitively. Over time, every concept was subjected to the most meticulous scrutiny. It was felt that all mathematical thinking should eventually lead to the ideas of calculus and be continuous in nature. Then, in 1956, a landmark text by Kemeny, Snell, and Thompson called *Introduction to Finite Mathematics* was published. It dealt with discrete ideas not contained in calculus. For over 20 years this course has been offered at various colleges and universities but it was never meant to be a replacement for calculus. However, as computers became more and more a part of mathematics and mathematical development, the necessity of treating the continuous ideas of calculus from a discrete standpoint became increasingly important. Many mathematicians are now suggesting that continuous and discrete be accepted as fundamental classifications in mathematics and the mathematics curriculum and that courses in calculus be offered alongside courses in discrete mathematics.

So we will now turn to the notion of a limit. Consider a function f and some fixed number a in the domain of f. Next consider a sequence of values in the domain of f where the sequence of numbers gets closer and closer to a. What will happen to the values of $f(x)$? If these values also get closer and closer to some fixed number, then this number is called the **limit of f as x approaches a**.

This idea is illustrated in Example 1 by a graph, a table of values, a geometric application, and a final algebraic illustration.

EXAMPLE 1   **a.** Look at the function defined by the following graph.

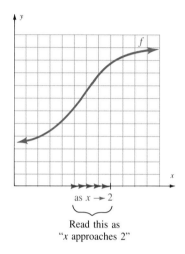

What is happening to $f(x)$ as x approaches 2? We can see that the $f(x)$ values are getting close to some value. This number is called the limit of $f(x)$ as $x \to 2$.

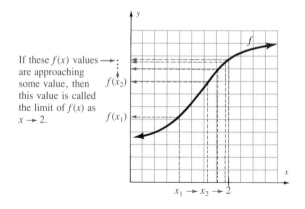

If these $f(x)$ values are approaching some value, then this value is called the limit of $f(x)$ as $x \to 2$.

b. Consider the function defined by the following table of values:

x	$\frac{1}{2}$	$\frac{1}{4}$	$\frac{1}{8}$	$\frac{1}{16}$	$\frac{1}{32}$	As $x \to 0$,
$f(x)$	$\frac{1}{2}$	$\frac{3}{4}$	$\frac{7}{8}$	$\frac{15}{16}$	$\frac{31}{32}$	what is happening to $f(x)$?

It looks like $f(x)$ is getting close to 1. This number is called the limit of $f(x)$ as $x \to 0$.

c. A few hundred years before Christ, a Greek mathematician named Archimedes approximated the value of π by calculating the limit of the perimeters of inscribed regular polygons in a circle with diameter 1.

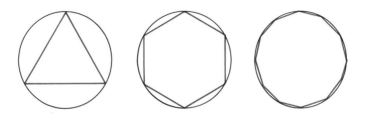

Begin by inscribing a triangle and then double the number of sides of the inscribed polygon; calculate the perimeter of each polygon. The limiting perimeter is the circumference of the circle.

d. Let $f(x) = \dfrac{x^2 + 2x - 15}{x - 3}$

What is happening to $f(x)$ as $x \to 3$? Look at

$x = 2$:	$f(2) = 7$	You can find these values on a calculator
$x = 2.5$:	$f(2.5) = 7.5$	
$x = 2.9$:	$f(2.9) = 7.9$	
$x = 2.99$:	$f(2.99) = 7.99$	

It looks like $f(x)$ is getting close to 8 as $x \to 3$. ∎

Note that there is no requirement that $f(a)$ be defined. In fact, $f(3)$ in Example 1d is not defined, but it does appear that there is a limiting value for $f(x)$. We are looking at values of x *near* a to see what happens to $f(x)$:

Limit Notation (Intuitive Notion)

> The notation
>
> $$\lim_{x \to a} f(x) = L$$
>
> is used to mean that the values of $f(x)$ get closer and closer to the number L as x gets closer and closer to (but remains different from) a.

Figure 9.1 shows the graph of a function f and the number $a = 3$. The arrowheads show possible sequences of numbers of x approaching a from both the left and the right. As x approaches $a = 3$, the $f(x)$ values get closer and closer to 5. We write this as $\lim_{x \to 3} f(x) = 5$.

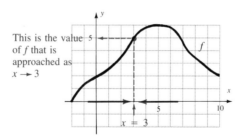

This is the value of f that is approached as $x \to 3$

Figure 9.1 Graph of f and $\lim_{x \to a} f(x)$

EXAMPLE 2 Given the function defined by the graph below, find the following limits:

$$\lim_{x \to 3} f(x) \qquad \lim_{x \to -2} f(x) \qquad \lim_{x \to 0} f(x)$$

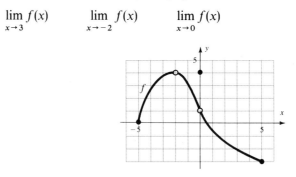

Solution Remember that $f(a)$ does not need to be defined in order to consider a limit. Also remember that an open circle on a graph indicates a deleted point. The following limits are found by inspection:

$$\lim_{x \to 3} f(x) = -2 \qquad \lim_{x \to -2} f(x) = 4 \qquad \lim_{x \to 0} f(x) = 1 \qquad ■$$

If the sequence of values is approaching a only from the right then we write $\lim_{x \to a^+} f(x)$, and if it is approaching a only from the left we write $\lim_{x \to a^-} f(x)$. These are called the right- and left-hand limits, respectively. If the right- and left-hand limits both exist and are equal, then we say the limit exists. That is, if $\lim_{x \to c^+} f(x) = L_1$ and $\lim_{x \to c^-} f(x) = L_2$ and $L_1 = L_2$, then we say that $\lim_{x \to c} f(x) = L$, where L is the common value of L_1 and L_2.

EXAMPLE 3 Find the requested limits.

a. $\lim_{x \to 3^-} f(x)$ **b.** $\lim_{x \to -2^-} f(x)$

c. $\lim_{x \to -2^+} f(x)$ **d.** $\lim_{x \to 5} f(x)$

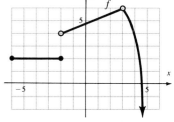

Solution **a.** $\lim_{x \to 3^-} f(x) = 6$ **b.** $\lim_{x \to -2^-} f(x) = 2$

c. $\lim_{x \to -2^+} f(x) = 4$

d. $\lim_{x \to 5} f(x)$ does not exist; that is, not all limits exist.   ■

EXAMPLE 4 Find $\lim_{x \to 2} 5$.

Solution Look at the graph of $f(x) = 5$. It is easy to see that regardless of what x is approaching, the value of $f(x)$ is 5. This means that $\lim_{x \to 2} 5 = 5$.

Limit of a Constant

If $f(x) = k$ is a constant, then

$$\lim_{x \to a} f(x) = k$$

EXAMPLE 5 $\lim\limits_{x \to 3} \dfrac{x-3}{x-3} = \lim\limits_{x \to 3} 1 = 1 \qquad x \neq 3$ ■

Next we consider limits of polynomial functions.

EXAMPLE 6 Find $\lim\limits_{x \to 4} x^2$.

Solution One method is to graph f and then use the graph to evaluate the limit as in Examples 2 and 3. You can also use a calculator or computer to construct a table of values for f as $x \to a$ (in this example $a = 4$ so $x \to 4$).

Sequence: $x \to 4^-$: 3, 3.5, 3.9, 3.99, ...
Sequence: $x \to 4^+$: 5, 4.5, 4.1, 4.01, ...

			$x \to 4^-$		
x	3	3.5	3.9	3.99	3.999
$f(x)$	9	12.25	15.21	15.9201	15.992001

$f(x) \to 16$

			$x \to 4^+$		
x	5	4.5	4.1	4.01	4.001
$f(x)$	25	20.25	16.81	16.0801	16.008001

$f(x) \to 16$

Thus: $\lim\limits_{x \to 4} x^2 = 16$. ■

After Example 6 you might be saying to yourself, why not just substitute $x = 4$ in x^2 to obtain the limit 16? Well, it is not that easy, as Example 7 illustrates.

EXAMPLE 7 The cost of removing a certain pollutant from a distillation process is given by the formula

$$f(x) = \frac{x^2 + x - 6}{x - 2}$$

Find $\lim\limits_{x \to 2} f(x)$.

Solution Construct a table of values as $x \to 2$. For the right-hand limit:

			$x \to 2^-$		
x	1	1.5	1.9	1.99	1.999	1.9999
$f(x)$	4	4.5	4.9	4.99	4.999	4.9999

$f(x) \to 5$

For the left-hand limit:

			$x \to 2^+$		
x	3	2.5	2.1	2.01	2.001	2.0001
$f(x)$	6	5.5	5.1	5.01	5.001	5.0001

$f(x) \to 5$

Since the left- and right-hand limits are the same we say the limit of $f(x)$ as x approaches 2 is 5. ■

In Example 7 you might have suggested substituting the value $x = 2$ directly into the equation. This procedure would prove to be faulty since $x = 2$ leads to division by zero. Instead, look at the *method* of graphing. This method requires us to look for common factors by factoring:

$$x^2 + x - 6 = (x - 2)(x + 3)$$

Thus

$$\frac{x^2 + x - 6}{x - 2} = x + 3, x \neq 2$$

The graph is shown in Figure 9.2. Thus $\lim_{x \to 2} f(x) = 5$.

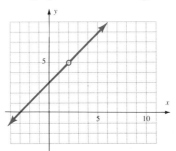

Figure 9.2 Graph of f

Note that factoring is helpful in graphing f and also that the simplified form $f(x) = x + 3, x \neq 2$ *could* have been used for direct substitution of $x = 3$:

$$\lim_{x \to 2} f(x) = \lim_{x \to 2}(x + 3) = 5$$

even though substitution was ineffective in the original unfactored form

$$f(x) = \frac{x^2 + x - 6}{x - 2}$$

What is the difference between the factored and unfactored forms of f? One is a rational function and the other is a polynomial function (with a deleted point). The graphs are identical *except* for the deleted point. Thus, if f is any polynomial function then its limit as x approaches a can be found by direct substitution of a for x in the polynomial representing f:

Limit of a Polynomial	If f is any polynomial function, then $$\lim_{x \to a} f(x) = f(a) \qquad \text{for any real number } a$$

EXAMPLE 8   Find **a.** $\lim_{x \to 1}(4x^3 - 2x^2 + x - 1)$ **b.** $\lim_{x \to 1}(x^2 - x)$

Solution   Let $f(x) = 4x^3 - 2x^2 + x - 1$ and $g(x) = x^2 - x$. Since f and g are polynomial functions, you can find the limits by substitution:

a. $\lim_{x \to 1} f(x) = f(1) = 4(1)^3 - 2(1)^2 + 1 - 1 = 2$

b. $\lim_{x \to 1} g(x) = g(1) = 1^2 - 1 = 0$ ∎

EXAMPLE 9   Find $\lim\limits_{x \to 1} \dfrac{x^2 - 1}{x - 1}$

Solution   You could proceed with a table of values but, instead, consider the following argument. Note that the function is *not defined* at $x = 1$. Since you are concerned with values *near* $x = 1$, you can say $x \neq 1$. But if $x \neq 1$, then

$$\frac{x^2 - 1}{x - 1} = \frac{(x - 1)(x + 1)}{x - 1}$$
$$= x + 1$$

and for values near 1, it is easy to see that $x + 1$ is near 2. Thus

$$\lim_{x \to 1} \frac{x^2 - 1}{x - 1} = \lim_{x \to 1}(x + 1) = 2$$ ■

EXAMPLE 10   Find $\lim\limits_{x \to 1} \dfrac{\sqrt{x} - 1}{x - 1}$.

Solution   As $x \to 1$,

$$\lim_{x \to 1} \frac{\sqrt{x} - 1}{x - 1} = \frac{\sqrt{1} - 1}{1 - 1}$$
$$= \frac{1 - 1}{1 - 1} = \frac{0}{0}$$

The expression $0/0$ is not a real number; in calculus we call this an **indeterminate form**. Do not assume, however, that when you obtain an indeterminate form that the limit does not exist. You could proceed with a table of values for this example but, instead, we will **rationalize the numerator**. Remember, from algebra, the process of *rationalizing the denominator*. The process here is the same; namely, multiply both numerator and denominator by $\sqrt{x} + 1$:

$$\frac{\sqrt{x} - 1}{x - 1} \cdot \frac{\sqrt{x} + 1}{\sqrt{x} + 1} = \frac{\sqrt{x} \cdot \sqrt{x} - \sqrt{x} + \sqrt{x} - 1}{(x - 1)(\sqrt{x} + 1)}$$
$$= \frac{x - 1}{(x - 1)(\sqrt{x} + 1)}$$
$$= \frac{1}{\sqrt{x} + 1}$$

Thus

$$\lim_{x \to 1} \frac{\sqrt{x} - 1}{x - 1} = \lim_{x \to 1} \frac{1}{\sqrt{x} + 1}$$
$$= \frac{1}{1 + 1} = \frac{1}{2}$$ ■

As long as an indeterminate form is not obtained, you may evaluate the limit as $x \to a$ by substituting a into the given expression.

EXAMPLE 11 Find $\lim\limits_{x \to 1} \dfrac{x+1}{x^2-1}$.

Solution If $x \neq 1$, then

$$\frac{x+1}{x^2-1} = \frac{x+1}{(x-1)(x+1)} = \frac{1}{x-1}$$

Consider a table of values for $\dfrac{1}{x-1}$ as x approaches 1. For the right-hand limit, $x \to 1^+$:

x	2	1.5	1.1	1.01	1.001	1.0001
$f(x)$	1	2	10	100	1000	10,000

$\lim\limits_{x \to 1^+} f(x) = ?$

For the left-hand limit, $x \to 1^-$:

x	0.5	0.9	0.99	0.999	0.9999	0.99999
$f(x)$	-2	-10	-100	-1000	$-10,000$	$-100,000$

$\lim\limits_{x \to 1^-} f(x) = ?$

It appears from the table that $f(x)$ increases without bound as x approaches 1 from the right or decreases without bound as x approaches 1 from the left. In such cases we say that the limit does not exist because the functional values do not approach a fixed number L. ∎

EXAMPLE 12 Find the value of $\frac{1}{x}$ as x increases without bound. This is called a *limit at infinity*.

x	1	2	10	100	1000	10,000
$f(x)$	1	0.5	0.1	0.01	0.001	0.0001

We say that $\frac{1}{x}$ approaches 0 as x increases without bound, and we symbolize this by $\lim\limits_{x \to \infty} \frac{1}{x} = 0$. ∎

EXAMPLE 13 Find the value of $\frac{1}{x}$ as x decreases without bound.

x	-1	-2	-10	-100	-1000	$-10,000$
$f(x)$	-1	-0.5	-0.1	-0.01	-0.001	-0.0001

We say that $\frac{1}{x}$ approaches 0 as x decreases without bound, and we symbolize this by $\lim\limits_{x \to -\infty} \frac{1}{x} = 0$.

Since $\frac{1}{x} \to 0$ for both $x \to \infty$ and $x \to -\infty$ in Examples 12 and 13, we say that $\frac{1}{x} \to 0$ as x increases or decreases without bound, and we symbolize this by

$$\lim_{|x| \to \infty} \frac{1}{x} = 0$$ ∎

We can also show that $\lim\limits_{|x| \to \infty} \dfrac{k}{x} = 0$ for any constant k. These examples lead to the following conclusion:

Limit to Infinity

$$\lim_{|x| \to \infty} \frac{1}{x^n} = 0 \quad \text{and} \quad \lim_{|x| \to \infty} \frac{k}{x^n} = 0 \quad \text{for any constant } k \text{ and } n > 0$$

EXAMPLE 14 Find $\lim\limits_{x \to \infty} \dfrac{x}{2x + 1}$.

Solution We could construct a table of values. Instead, suppose we multiply the rational

expression by 1, written as $\dfrac{\frac{1}{x}}{\frac{1}{x}}$:

$$\frac{x}{2x + 1} \cdot \frac{\frac{1}{x}}{\frac{1}{x}} = \frac{1}{2 + \frac{1}{x}}$$

Now, since $\lim\limits_{x \to \infty} \frac{1}{x} = 0$, we see that

$$\lim_{x \to \infty} \frac{1}{2 + \frac{1}{x}} = \frac{1}{2 + 0} = \frac{1}{2} \qquad \blacksquare$$

EXAMPLE 15 Find $\lim\limits_{x \to \infty} \dfrac{3x^2 - 7x + 2}{7x^2 + 2x + 5}$.

Solution Note that the largest power of x in the expression is x^2, so we multiply the numerator and denominator by $\frac{1}{x^2}$.

$$\lim_{x \to \infty} \frac{3x^2 - 7x + 2}{7x^2 + 2x + 5} \cdot \frac{\frac{1}{x^2}}{\frac{1}{x^2}}$$

Now, since k/x and k/x^2 both approach 0 as x increases without bound, we have

$$= \lim_{x \to \infty} \frac{3 - \frac{7}{x} + \frac{2}{x^2}}{7 + \frac{2}{x} + \frac{5}{x^2}} = \frac{3 - 0 + 0}{7 + 0 + 0}$$

Thus $\lim\limits_{x \to \infty} f(x) = \frac{3}{7}$. $\qquad \blacksquare$

Note that if $\lim\limits_{|x| \to \infty} f(x) = L$, then $y = L$ is a horizontal asymptote. Compare Example 15 with the result in the box on page 68.

Problem Set 9.1

Given the functions defined by the graphs in Figure 9.3, find the limits in Problems 1–12.

Figure 9.3
Functions f, g, and t

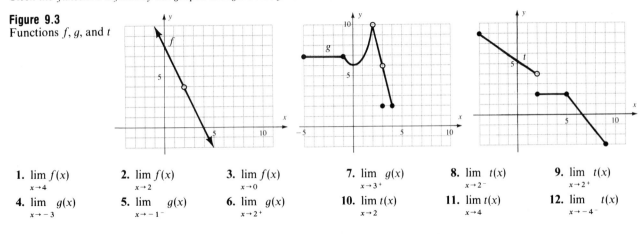

1. $\lim\limits_{x \to 4} f(x)$ **2.** $\lim\limits_{x \to 2} f(x)$ **3.** $\lim\limits_{x \to 0} f(x)$ **7.** $\lim\limits_{x \to 3^+} g(x)$ **8.** $\lim\limits_{x \to 2^-} t(x)$ **9.** $\lim\limits_{x \to 2^+} t(x)$

4. $\lim\limits_{x \to -3} g(x)$ **5.** $\lim\limits_{x \to -1^-} g(x)$ **6.** $\lim\limits_{x \to 2^+} g(x)$ **10.** $\lim\limits_{x \to 2} t(x)$ **11.** $\lim\limits_{x \to 4} t(x)$ **12.** $\lim\limits_{x \to -4^-} t(x)$

Find the limits by filling in the appropriate values in the tables in Problems 13–21.

13. $\lim\limits_{x \to 5^-} (4x - 5)$

x	2	3	4	4.5	4.9	4.99
f(x)	3					

14. $\lim\limits_{x \to 5^+} (4x - 5)$

x	6	5.5	5.1	5.01	5.001
f(x)	19				

15. $\lim\limits_{x \to -2^+} (2x^2 - 50x + 250)$

x	0	-1	-1.5	-1.9	-1.99	-1.999
f(x)	250					

16. $\lim\limits_{x \to -2^-} (2x^2 - 50x + 250)$

x	-3	-2.5	-2.1	-2.01	-2.001	-2.0001
f(x)						

17. $\lim\limits_{x \to 1} \dfrac{3x + 2}{x - 2}$

x	0	0.5	0.9	0.99	0.999	1.5	1.1	1.01	1.001
f(x)									

18. $\lim\limits_{x \to 2} \dfrac{x^3 - 8}{x^2 + 2x + 4}$

x	1	1.5	1.9	1.99	1.999	2.5	2.01	2.001
f(x)								

19. $\lim\limits_{x \to 2} \dfrac{x^2 + 2x + 4}{x^3 - 8}$

x	1	1.5	1.9	1.99	1.999	2.5	2.01	2.001
f(x)								

20. $\lim\limits_{x \to \infty} \dfrac{x^2 + 6x + 9}{x + 3}$

x	1	10	100	1000	10,000	100,000	1,000,000
f(x)							

21. $\lim\limits_{x \to -\infty} \dfrac{3x^2 - 5x + 15}{x + 3}$

x	-1	-10	-100	-1000	-10,000	-100,000	-1,000,000
f(x)							

Find the limits in Problems 22–56.

22. $\lim\limits_{x \to 0} x^8$

23. $\lim\limits_{x \to 2} (x^2 - 4)$

24. $\lim\limits_{x \to 3} (x^2 - 4)$

25. $\lim\limits_{x \to 1} \dfrac{1}{x - 3}$

26. $\lim\limits_{x \to -3} \dfrac{1}{x - 3}$

27. $\lim\limits_{x \to 3} \dfrac{1}{x - 3}$

28. $\lim\limits_{x \to 0} \dfrac{1}{x^2 + 1}$

29. $\lim\limits_{x \to \infty} \dfrac{1}{x^2 + 1}$

30. $\lim\limits_{x \to -1} \dfrac{1}{x^2 + 1}$

31. $\lim\limits_{x \to \infty} 2x$

32. $\lim\limits_{x \to \infty} (3x - 4)$

33. $\lim\limits_{x \to 2} \dfrac{x^2 - 4}{x - 2}$

34. $\lim\limits_{x \to 3} \dfrac{x^2 - 8x + 15}{x - 3}$

35. $\lim\limits_{x \to 3} \dfrac{x^2 + 3x - 10}{x - 2}$

36. $\lim\limits_{x \to -5} \dfrac{x^2 + 3x - 10}{x + 5}$

37. $\lim\limits_{x \to 4} \dfrac{\sqrt{x} - 4}{x - 16}$

38. $\lim\limits_{x \to 9} \dfrac{\sqrt{x} - 3}{x - 9}$

39. $\lim\limits_{x \to 9} \dfrac{\sqrt{x} - 3}{x - 3}$

40. $\lim\limits_{x \to 4} \dfrac{\sqrt{x} - 2}{x - 2}$

41. $\lim\limits_{x \to 2} \dfrac{x^2 - 1}{x - 2}$

42. $\lim\limits_{x \to 4} \dfrac{2x^2 - 5x - 12}{x - 4}$

43. $\lim\limits_{x \to 2} \dfrac{x^3 - 8}{x^2 + 2x + 4}$

44. $\lim\limits_{x \to 2} \dfrac{x^2 + 2x + 4}{x^3 - 8}$

45. $\lim\limits_{x \to 2} \dfrac{x + 2}{x^3 + 8}$

46. $\lim\limits_{x \to 2} \dfrac{6 - x}{2x - 15}$

47. $\lim\limits_{|x| \to \infty} \dfrac{2x^2 - 5x - 3}{x^2 - 9}$

48. $\lim\limits_{|x| \to \infty} \dfrac{x^2 + 6x + 9}{x + 3}$

49. $\lim\limits_{|x| \to \infty} \dfrac{3x - 1}{2x + 3}$

50. $\lim\limits_{|x| \to \infty} \dfrac{6x^2 - 5x + 2}{2x^2 + 5x + 1}$

51. $\lim\limits_{x \to -\infty} \dfrac{5x + 10,000}{x - 1}$

52. $\lim\limits_{x \to -\infty} \dfrac{4x + 10^6}{x + 1}$

53. $\lim\limits_{x \to \infty} \left(x + 2 + \dfrac{3}{x - 1} \right)$

54. $\lim\limits_{x \to -\infty} \left(2x - 3 + \dfrac{4}{x+2} \right)$

55. $\lim\limits_{x \to -\infty} \dfrac{4x^4 - 3x^3 + 2x + 1}{3x^4 - 9}$

56. $\lim\limits_{x \to 1} \dfrac{x^2 + x + 1}{x^3 - 1}$

APPLICATIONS

57. The cost of manufacturing a specialized machine tool is a function of the number of items manufactured. This cost is graphed below. Note that there is a jump in the cost after 10,000 items because at that point it is necessary to add a second shift.

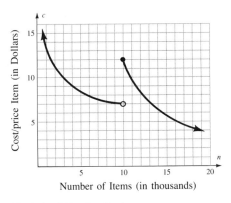

Number of Items (in thousands)

Find the following limits.

a. $\lim\limits_{n \to 10^-} c(n)$ **b.** $\lim\limits_{n \to 10^+} c(n)$

c. $\lim\limits_{n \to 13} c(n)$ **d.** $\lim\limits_{n \to 10} c(n)$

58. The number of live animals during a psychology experiment is shown in the graph below. The experiment begins with 2 animals, and after time t_4 there are 14 animals. Find the following limits.

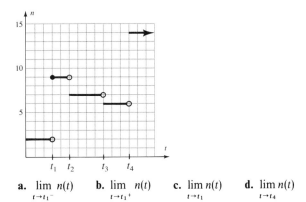

a. $\lim\limits_{t \to t_1^-} n(t)$ **b.** $\lim\limits_{t \to t_1^+} n(t)$ **c.** $\lim\limits_{t \to t_1} n(t)$ **d.** $\lim\limits_{t \to t_4} n(t)$

59. Learning theory measures the percentage of mastery of a subject as a function of time. The learning curve for a particular learning task is shown below. Note that at time t_1 there is a jump in mastery.

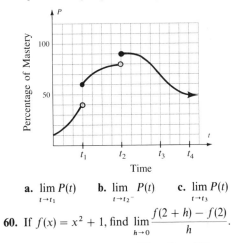

Time

a. $\lim\limits_{t \to t_1} P(t)$ **b.** $\lim\limits_{t \to t_2^-} P(t)$ **c.** $\lim\limits_{t \to t_3} P(t)$ **d.** $\lim\limits_{t \to \infty} P(t)$

60. If $f(x) = x^2 + 1$, find $\lim\limits_{h \to 0} \dfrac{f(2+h) - f(2)}{h}$.

61. If $f(x) = \sqrt{x}$, find $\lim\limits_{h \to 0} \dfrac{f(1+h) - f(1)}{h}$.

9.2 Continuity

You may remember the puzzle in the margin from elementary school: the challenge was to draw the figure without lifting your pencil from the paper or retracing any of the lines. In calculus we are concerned with figures that can be drawn without lifting a pencil from the paper, but we focus our attention on functions. The idea of *continuity* evolved from the notion of a curve "without breaks or jumps" to a rigorous definition given by Karl Weierstrass (1815–1897). Galileo and Leibniz had thought of continuity in terms of the density of points on a curve, but they were in error since the rational numbers have this property of denseness, yet do not form a continuous curve. Another mathematician, J. W. R. Dedekind (1831–1916), took an entirely different approach and concluded that continuity is due to the division of a curve into two parts so that there is one and only one point that makes this division.

As Dedekind wrote, "By this commonplace remark, the secret of continuity is to be revealed."* We begin with a discussion of *continuity at a point*. It may seem strange to talk about continuity *at a point*, but it should seem natural to talk about a curve being "discontinuous at a point," as illustrated by Example 1.

EXAMPLE 1   Which of the following curves appear to have a discontinuity at the point $x = 1$?

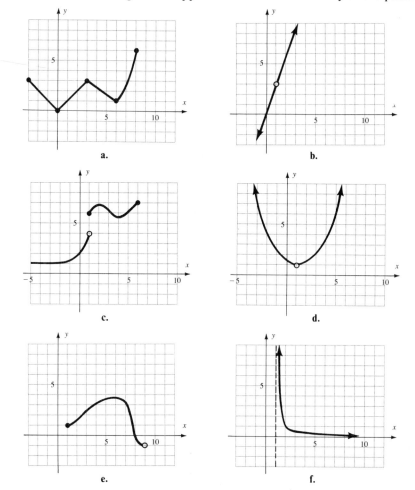

Solution   The functions whose graphs are shown in parts **b**, **c**, **d** and **f** are obviously discontinuous at $x = 1$. If, however, $x = 1$ happens to be the endpoint of an interval (part **e**) we do not say the curve is discontinuous at $x = 1$.   ■

The solution to Example 1 illustrates the idea of continuity from an intuitive standpoint, but we need to define and formalize this idea more completely. There are two essential conditions for a function f to be continuous at a point c. First, $f(c)$ must be defined. For example, the curve in part **e** of Example 1 is not continuous at its right endpoint because it is not defined at $x = 9$ (the open dot indicates an excluded point).

---

* From Carl Boyer, *A History of Mathematics* (New York: Wiley, 1968), p. 607.

EXAMPLE 2 Are any of the given functions continuous at $x = 0$?

a. $f(x) = \dfrac{3x^2 - x + 2}{x}$ **b.** $g(x) = \dfrac{x^2 - 5x}{x}$ **c.** $s(x) = \begin{cases} x + 3 & \text{if } x > 0 \\ x^2 & \text{if } x < 0 \end{cases}$

Solution None of these functions is defined at $x = 0$. ■

A second condition for continuity at a point $x = c$ is that the function makes no jumps at this point. This means that if "x is close to c," then "$f(x)$ must be close to $f(c)$." Looking at Example 1, we see that the graphs in parts **b**, **c**, and **d** jump at the point $x = 1$. We recognize this as the concept of limit and now define the concept of continuity at a point:

Definition of Continuity at a Point

A function f is continuous at a point $x = c$ if

1. $f(c)$ exists and
2. $\lim\limits_{x \to c} f(x) = f(c)$

Test the continuity of each function in Examples 3–6 at the point $x = 1$. If it is not continuous at $x = 1$, tell why.

EXAMPLE 3 $f(x) = \dfrac{x^2 + 2x - 3}{x - 1}$

Solution Not continuous at $x = 1$, since f is not defined at this point. ■

EXAMPLE 4 $g(x) = \dfrac{x^2 + 2x - 3}{x - 1}$ if $x \neq 1$ and $g(x) = 6$ if $x = 1$

Solution Note that g is very similar to f in Example 3 except that g is defined at $x = 1$. Now, to test the second condition of continuity, $g(1) = 6$ and

$$\lim_{x \to 1} g(x) = \lim_{x \to 1} \frac{x^2 + 2x - 3}{x - 1}$$
$$= \lim_{x \to 1} \frac{(x - 1)(x + 3)}{x - 1}$$
$$= \lim_{x \to 1} (x + 3)$$
$$= 4$$

Since $\lim\limits_{x \to 1} g(x) \neq g(1)$, we see that g is not continuous at $x = 1$. ■

EXAMPLE 5 $h(x) = \dfrac{x^2 + 2x - 3}{x - 1}$ if $x \neq 1$ and $h(x) = 4$ if $x = 1$

Solution Compare h with g of Example 4. We see that both conditions of continuity are satisfied, which means that h is continuous at $x = 1$. ■

EXAMPLE 6 $m(x) = 3x^3 + 5x^2 - 4x + 1$

Solution $m(1) = 3(1)^3 + 5(1)^2 - 4(1) + 1 = 5$, so m is defined at $x = 1$. Also, $\lim\limits_{x \to 1} m(x) = m(1)$, so the function is continuous at $x = 1$.   ■

Note that Example 6 is a polynomial function, and from the previous section we know that if f is any polynomial function, then

$$\lim_{x \to c} f(x) = f(c)$$

for any real number c. Thus we immediately have the following result:

Continuity of a Polynomial

> Every polynomial function is continuous at every point in its domain.

If a function is continuous for every point in a given open interval, then we say that the function is continuous over that interval. A function that is not continuous is said to be a **discontinuous function**. For example, polynomials are continuous for all real numbers. On the other hand,

$$f(x) = \frac{x + 1}{x - 1}$$

is continuous for $-1 \le x \le 0$ but not for $0 \le x \le 2$ since f is undefined at $x = 1$. The function h from Example 5,

$$h(x) = \begin{cases} \dfrac{x^2 + 2x - 3}{x - 1} & \text{if } x \ne 1 \\ 4 & \text{if } x = 1 \end{cases}$$

is continuous for all real numbers. As you can see from the above examples, we usually concerned with finding points of discontinuity. The following result summarizes the properties of continuity:

Continuity Theorem

> Let f and g be continuous functions at $x = c$. Then the following functions are also continuous at $x = c$:
>
> **1.** $f + g$ **2.** $f - g$ **3.** fg **4.** f/g $(g \ne 0)$

Since most of the functions we will discuss in this book are continuous over certain intervals, our task will be to look for *points of discontinuity*. These points may be values for which the definition of the function changes or values that cause division by zero. We call such points **suspicious points**. We then check for the continuity at the suspicious points and use the continuity theorem for the rest of the points in the interval.

EXAMPLE 7 Let $f(x) = \dfrac{x^2 + 3x - 10}{x - 2}$. Check continuity.

Solution The suspicious point is $x = 2$ since this is a value for which the function $(x - 2)$ is equal to zero. The function f is not defined at this point, so it is *discontinuous at* $x = 2$. By the continuity theorem (part 4), it is continuous at all other points. ∎

EXAMPLE 8 Let $g(x) = \begin{cases} \dfrac{x^2 + 3x - 10}{x - 2} & \text{if} \quad x \neq 2 \\ 6 & \text{if} \quad x = 2 \end{cases}$ Check continuity.

Solution The suspicious point is $x = 2$ since it is a value for which the definition of the function changes. Check this value:

Step 1 $g(2) = 6$, so the function is defined at $x = 2$.

Step 2 $\displaystyle\lim_{x \to 2} g(x) = \lim_{x \to 2} \dfrac{x^2 + 3x - 10}{x - 2}$

$\qquad\qquad = \displaystyle\lim_{x \to 2} \dfrac{(x + 5)(x - 2)}{x - 2}$

$\qquad\qquad = \displaystyle\lim_{x \to 2} (x + 5)$

$\qquad\qquad = 7$

But $g(2) = 6$ and thus $\displaystyle\lim_{x \to 2} g(x) \neq g(2)$, so the function is *discontinuous at* $x = 2$. ∎

EXAMPLE 9 Let $G(x) = \begin{cases} \dfrac{x^2 + 3x - 10}{x - 2} & \text{if} \quad x \neq 2 \\ 7 & \text{if} \quad x = 2 \end{cases}$ Check continuity.

Solution Note that we have simply redefined the functional value of g of Example 8 at a single point. Thus

Step 1 $G(2) = 7$ and

Step 2 $\displaystyle\lim_{x \to 2} G(x) = 7$ (from Example 8) and $\displaystyle\lim_{x \to 2} G(x) = G(2)$. Thus the function is continuous for all real numbers. ∎

Note in Example 9 that you do not need to check the continuity at other points in the domain. The other points are not suspicious points and you only need to apply the continuity theorem for all of these points.

EXAMPLE 10 Let $f(x) = \begin{cases} 3 - x & \text{if} \quad -5 \leq x < 2 \\ x - 2 & \text{if} \quad 2 \leq x \leq 5 \end{cases}$ Check continuity.

Solution The suspicious point is $x = 2$ since it is a value for which the definition of the function changes. Check continuity by applying the definition: that is, f is defined at $x = 2$ since $f(2) = 0$. In order to find the limit as $x \to 2$ you need to check both the left- and right-hand limits (since the function is defined according to a different function depending on from which direction x approaches 2).

$$\lim_{x \to 2^-} f(x) = \lim_{x \to 2^-} (3 - x) = 1$$

$$\lim_{x \to 2^+} f(x) = \lim_{x \to 2^+} (x - 2) = 0$$

Since the left- and right-hand limits are different, $\lim_{x \to 2} f(x)$ does not exist, so the function is discontinuous at $x = 2$. If you draw the graph of f you can see the discontinuity at $x = 2$. Note that f is continuous over $-5 \le x < 2$ and over $2 \le x \le 5$ but not over $-5 \le x \le 5$.

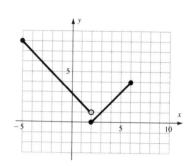

EXAMPLE 11   Let

$$g(x) = \begin{cases} 2 - x & \text{if} \quad -5 \le x < 2 \\ x - 2 & \text{if} \quad 2 \le x < 5 \end{cases} \qquad \text{Check continuity.}$$

Solution   The suspicious point is $x = 2$. g is defined at $x = 2$ since $g(2) = 0$

$$\lim_{x \to 2^-} g(x) = \lim_{x \to 2^-} (2 - x) = 0$$

$$\lim_{x \to 2^+} g(x) = \lim_{x \to 2^+} (x - 2) = 0$$

By looking at the graph of g you can see that it is continuous over $-5 \le x \le 5$. Note that $g(x) = |x - 2|$. Since $\lim_{x \to 2^+} g(x) = \lim_{x \to 2^-} g(x)$, we can write

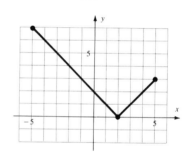

$$\lim_{x \to 2} g(x) = 0$$

The function is continuous at the suspicious point, so we conclude that g is continuous for all real numbers.

Problem Set 9.2

In Problems 1–12 find all suspicious points and tell which of those are points of discontinuity.

1.

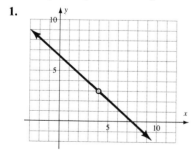

2.

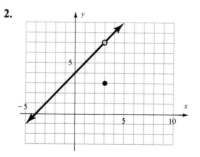

3.

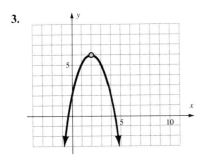

4.

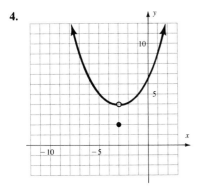

5.

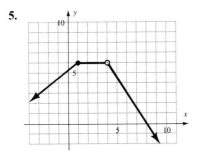

6.

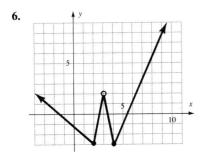

7.

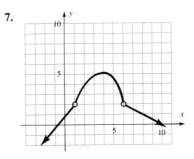

8.

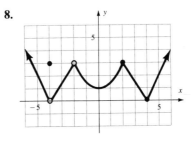

9.

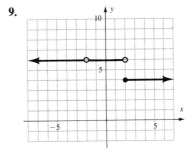

10.

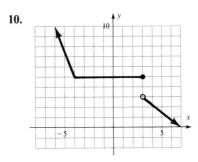

11.

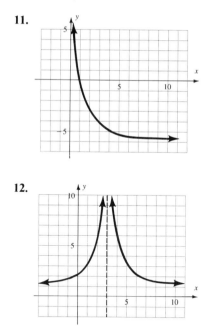

12.

13. Is the graph in Problem 11 continuous at $x = 1$? Explain your answer.

14. Is the graph in Problem 12 continuous at $x = 2.5$? Explain your answer.

15. Is the graph in Problem 11 continuous at $x = 0.001$? Explain your answer.

16. Is the graph in Problem 12 continuous at $x = 2.9999$? Explain your answer.

Which of the functions described in Problems 17–22 represent continuous functions? State the domain, if possible, for each example.

17. The humidity on a specific day at a given location considered as a function of time

18. The temperature on a specific day at a given location considered as a function of time

19. The selling price of IBM stock on a specific day considered as a function of time

20. The number of unemployed people in the United States during December 1989 considered as a function of time

21. The charges for a telephone call from Los Angeles to New York considered as a function of time

22. The charges for a taxi ride across town considered as a function of mileage

In Problems 23–41 a function is defined over a certain domain.

State whether the function is continuous at all points in this domain. Give the points of discontinuity, if any.

23. $f(x) = \dfrac{1}{x^2 + 5}, \quad -5 \le x \le 5$

24. $f(x) = \dfrac{1}{x^2 - 5}, \quad -5 \le x \le 5$

25. $f(x) = \dfrac{1}{x^2 - 9}, \quad 5 \le x \le 10$

26. $f(x) = \dfrac{1}{x^2 - 9}, \quad -5 \le x \le 5$

27. $f(x) = \dfrac{x - 1}{x^2 - 1}, \quad -5 \le x \le 5$

28. $f(x) = \dfrac{3x}{x^3 - x}, \quad -5 \le x \le 5$

29. $f(x) = \dfrac{x + 2}{x^2 - 6x - 16}, \quad -5 \le x \le 5$

30. $f(x) = \dfrac{x + 2}{x^2 - 6x - 16}, \quad -10 \le x \le 10$

31. $f(x) = \begin{cases} \dfrac{1}{x - 3}, & -5 \le x \le 5, x \ne 3 \\ 4, & x = 3 \end{cases}$

32. $f(x) = \dfrac{x^2 - x - 6}{x + 2}, \quad -5 \le x \le 5$

33. $f(x) = \begin{cases} \dfrac{x^2 - x - 6}{x + 2}, & -5 \le x \le 5, x \ne -2 \\ -4, & x = -2 \end{cases}$

34. $f(x) = \begin{cases} \dfrac{x^2 - x - 6}{x + 2}, & -5 \le x \le 5, x \ne -2 \\ -5, & x = -2 \end{cases}$

35. $f(x) = \dfrac{x^2 - 3x - 10}{x + 2}, \quad 0 \le x \le 5$

36. $f(x) = \dfrac{x^2 - 3x - 10}{x + 2}, \quad -5 \le x \le 5$

37. $f(x) = \begin{cases} \dfrac{x^2 - 3x - 10}{x + 2}, & -5 \le x \le 5, x \ne -2 \\ -3, & x = -2 \end{cases}$

38. $f(x) = \begin{cases} \dfrac{x^2 + x + 1}{x^3 - 1}, & 0 \le x \le 5, x \ne 1 \\ 1, & x = 1 \end{cases}$

39. $f(x) = \begin{cases} 1 & \text{if } x \text{ is rational} \\ -1 & \text{if } x \text{ is irrational} \end{cases}$

40. $f(x) = |x|, \quad -5 \le x \le 5$

41. $f(x) = |x - 2|, \quad -5 \le x \le 5$

APPLICATIONS

42. A rental agency will lease trucks at a cost C of $.55 per mile if the annual mileage is under 10,000 miles, but the rate is lowered to $.40 per mile if the mileage is 10,000 or more miles.
   a. Is C a continuous function? What is the domain?
   b. If m is the number of miles we can write $C(m)$ to represent the mileage charge. What is $C(9000)$? What is $C(11,000)$?
   c. Graph the function C.

43. Postal charges are $.25 for the first ounce and $.20 for each additional ounce or fraction thereof. Let c be the cost function for mailing a letter weighing w ounces.
   a. Is c a continuous function? What is the domain?
   b. What is $c(1.9)$? $c(2.01)$? $c(2.89)$?
   c. Graph the function c.

44. On July 10, 1913, the following temperatures were recorded at Death Valley, California:

Time	8	10	12	1	3	5	7
F°	90	115	123	134	130	128	105

Let T represent the temperature in degrees Fahrenheit and assume that T is a function of time (measured on a 24-hour clock) so that $T(8) = 90,\ldots,$ $T(12) = 123,\ldots,$ $T(19) = 105$.
   a. Is T a continuous function? What is the domain?
   b. On the day in question, what is the minimum number of times during the day that the temperature was $100°F$? $110°F$? $80°F$?

45. Give an example of a function defined on $1 \le x \le 3$ that is discontinuous at $x = 2$ and continuous elsewhere on the interval.

46. Give an example of a function defined on $3 \le x \le 7$ that is discontinuous at $x = 4$ and $x = 5$ and continuous elsewhere on the interval.

9.3 Rates of Change

Elementary mathematics focuses on formulas and relationships among variables but does not include the analysis of quantities that are in a state of constant change. For example, the following problem might be found in elementary mathematics. If you drive at 55 miles per hour (mph) for a total of 3 hours, how far did you travel? This example uses a formula, $d = rt$, and the answer is $d = 55(3) = 165$. However, this model does not adequately describe the situation in the real world. You could drive for 3 hours at an *average rate* of 55 mph, but you probably could not drive for 3 hours at a *constant rate* of 55 mph—in reality, the rate would be in a state of constant change. Other applications in which rates may not remain constant quickly come to mind:

> Profits changing with sales
> Population changing with the growth rate
> Property taxes changing with the tax rate
> Tumor sizes changing with chemotherapy
> The speed of falling objects changing over time

The rate at which one quantity changes relative to another is mathematically described by using a concept called a **derivative**. We will see that the rate of change of one quantity relative to another is mathematically determined by finding the slope of a line drawn tangent to a curve.

We begin with a simple example. Consider the speed of a moving object, say, a car. By *speed* we mean the rate at which the distance traveled varies with time. We can measure speed in two ways: *average speed* and *instantaneous speed*. We begin with the average speed of a commuter driving from Santa Rosa, California, to San Francisco, California, along Highway 101.*

---

* Statistics and illustration from Chris Smith, "One Commuter's Morning on the Road," *The Press Democrat*, Santa Rosa, California, March 10, 1985.

Morning on the Road

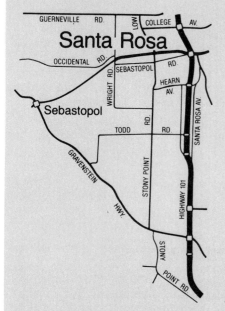

How goes the early morning drive endured daily by Sonoma County residents employed in Marin and San Francisco?

The following is the log of a commute trip made March 5. Interviews with commuters bore out that it was typical for a Tuesday journey.

6:09 A.M.—Trip odometer set at zero. Pull southbound onto 101 at College Avenue. It's dark, although a faint glow appears in the sky to the east. Southbound cars are spaced by 4 to 10 car lengths and move at 55 to 60 mph. Northbound traffic very light.

6:25—16 miles. Washington Boulevard exit, Petaluma. Had to touch the brakes as traffic bunched up.

6:26—17 miles. Traffic thickens on the bridge over the Petaluma River but still moves at the speed limit. (We were lucky. Commuters say there often are backups here as early as 6:15.)

6:30—21 miles. Sailed past the adult-only theater and the Sonoma-Marin county line.

6:34—25 miles. The highway spreads from two lanes in each direction to three just past the thorny-crowned McGraw-Hill building near Novato. Still cruising freely. It's quite light, but nobody's killing their headlights.

6:36—26.7 miles. Glide past the Delong Avenue–Central Novato exit.

Average speed over the 26.7 miles between Santa Rosa and Novato: 60 mph.

6:37—27.5 miles. Big trouble ahead. A long line of cars, brake lights aglow, backed up from the junction with Highway 37, the route from Vallejo and Lakeville Highway.

6:38—28.2 miles. Humph! Stopped dead as the Highway 37 interchange looms just ahead. To the left, northbound cars are zipping gleefully along.

6:39—28.5 miles. It's stop-and-creep. 5 mph, up to 15, then 0. Two thick lanes of traffic from Highway 37 are trying to merge from the right. The newcomers reluctantly weave their way in.

6:46—29.6 miles. Reach the Ignacio Boulevard exit. Took eight minutes to cover the last 1.4 miles.

6:47—30 miles. Speed eases up to 30 mph on the strip in front of Hamilton Field, but in an instant the red lights blaze anew and its stop-and-go again.

7:01—34 miles. Speed varies between 20 and 30 mph. Rocketed past the Freitas Parkway exit in Terre Linda at 25 mph.

7:03—35 miles. The river of cars oozes at 35 mph past the dramatic Marin Civic Center.

Average speed over the 8.4 miles between Novato and central San Rafael: 18 mph.

7:09—38 miles. Crawl at 20 mph past the exit to the Richmond Bridge.

7:11—40 miles. San Francisco Bay, the Larkspur ferry terminal and San Quentin just passed to the left and, at Lucky Drive in Corte Madera, the highway gains a fourth lane. But it's a diamond lane, open from 6 to 9 A.M. only to buses, cars with three or more occupants and vanpools. Traffic in the three right lanes moves at 40 mph; the few and far-between buses, carpools and vanpools in the left lane zip along at 55 or 60.

7:15—43 miles. The left lane ceases to be a diamond lane at the bridge over Richardson Bay and is open to all.

7:18—44 miles. Four solid rows of cars grunt at 25 mph up the Waldo Grade from Sausalito.

7:28—50 miles. We reach the toll plaza, having hit 40 mph between the Waldo Tunnel and the Golden Gate Bridge and traverse the bridge at a steady 25.

Average speed over the 13.2 miles between San Rafael and the toll plaza: 36 mph.

Average speed over the entire 49.8-mile, one-hour and 19-minute commute: 38 mph.

The five-mile drive through San Francisco's Marina district and up Columbus to the financial district took another 15 minutes.

Commuters said the drive home at night is usually worse. We took their word for it.

—By CHRIS SMITH

Figure 9.4 Distance and time for a commute car

To find the average speed we use the formula $d = rt$ or $r = \dfrac{d}{t}$; that is, we divide the distance traveled by the elapsed time:

$$\text{Average speed} = \frac{\text{distance traveled}}{\text{elapsed time}}$$

EXAMPLE 1   Find the average speed using Figure 9.4 for

a. 6:09 A.M. to 6:39 A.M. (0–30 min)   **b.** 6:39 A.M. to 7:09 A.M. (30–60 min)
c. 7:01 A.M. to 7:28 A.M. (52–79 min)

Solution   **a.** 0–30 min: Average speed $= \dfrac{28.5 - 0}{\frac{1}{2} - 0} = 57$ mph   Note: mph is the same as mi/hr

b. 30–60 min: Average speed $= \dfrac{38 - 28.5}{1 - \frac{1}{2}} = 19$ mph

c. 52–79 min: Average speed $= \dfrac{50 - 34}{\frac{79}{60} - \frac{52}{60}} \approx 35.6$ mph   ■

We can generalize the work done in Example 1 by writing the following formula:

$$\text{Average speed} = \frac{d_2 - d_1}{t_2 - t_1} \text{ mph}$$

where the car travels d_1 miles in t_1 hours and d_2 miles in t_2 hours. The relationship is shown graphically in Figure 9.5, where the average speed is the slope of the line joining the points (d_1, t_1) and (d_2, t_2). This line is sometimes called the **secant line**.

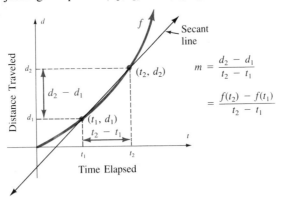

Figure 9.5 Geometrical interpretation of average speed

Average Rate of Change

The *average rate of change* of a function with respect to x between x_1 and x_2 is given by the formula

$$\frac{f(x_2) - f(x_1)}{x_2 - x_1}$$

EXAMPLE 2 Find the average rate of change of f with respect to x between $x = 3$ and 5 if $f(x) = x^2 - 4x + 7$.

Solution Let $x_1 = 3$ and $x_2 = 5$. Then $f(x_1) = f(3) = (3)^2 - 4(3) + 7 = 4$ and $f(x_2) = f(5) = 5^2 - 4(5) + 7 = 12$. Therefore the average rate of change is

$$\frac{f(x_2) - f(x_1)}{x_2 - x_1} = \frac{f(5) - f(3)}{5 - 3}$$

$$= \frac{12 - 4}{2}$$

$$= 4$$

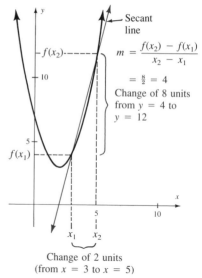

$$m = \frac{f(x_2) - f(x_1)}{x_2 - x_1}$$

$$= \tfrac{8}{2} = 4$$

Change of 8 units from $y = 4$ to $y = 12$

Change of 2 units (from $x = 3$ to $x = 5$)

The slope of the secant line is 4, as shown in Figure 9.6.

Figure 9.6 Average rate of change of f, where $f(x) = x^2 - 4x + 7$ between $x = 3$ and $x = 5$

EXAMPLE 3 Figure 9.7 shows the number of law enforcement officers killed in the line of duty in the United States from 1970 to 1980.

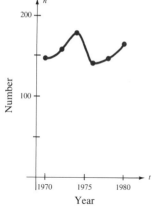

Figure 9.7 Number of law enforcement officers killed in the line of duty

Year	Number
1970	146
1972	157
1974	179
1976	140
1978	146
1980	162

Find the average rate of change of law enforcement officers killed in the line of duty for the following periods of time:

a. 1978–1980   **b.** 1976–1978   **c.** 1974–1978   **d.** 1970–1978

Solution

a. $\dfrac{162 - 146}{1980 - 1978} = \dfrac{16}{2} = 8$

This is the average rate at which police officers were killed between the years 1978 and 1980

b. $\dfrac{146 - 140}{1978 - 1976} = \dfrac{6}{2} = 3$

The number of officers killed between 1976 and 1978 increased by 3 officers per year

c. $\dfrac{146 - 179}{1978 - 1974} = \dfrac{-33}{4} = -8.25$

A negative rate of change indicates a decrease in the rate at which officers were killed between 1974 and 1978

d. $\dfrac{146 - 146}{1978 - 1970} = \dfrac{0}{8} = 0$

There is no change in the rate at which officers were killed between 1970 and 1978

Note in Example 3 that the secant lines for finding the average rate of change can vary greatly and really do not tell us a great deal about the rate *at a particular time*. Imagine that the curve shown in Figure 9.7 is a continuous curve for a roller coaster track with a fixed point at 1978. Also imagine a roller coaster car on the track a distance of h units from the fixed location where $h \to 0$ (that is, the roller coaster car is rolling toward the fixed location).

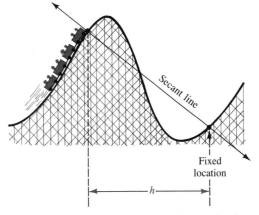

As the car moves along the track we obtain a sequence of secant lines.

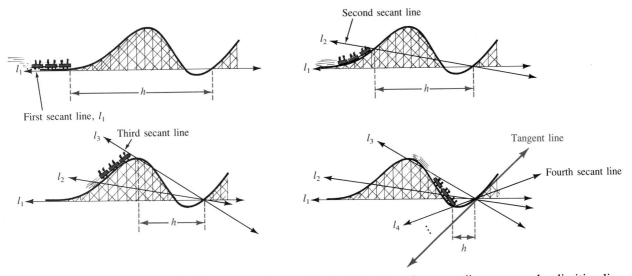

As the car gets close to the fixed location, the secant lines approach a limiting line (shown in color). This limiting line for which $h \to 0$ is called the **tangent line**.

Instantaneous Rate of Change

Given a function f and the graph of $y = f(x)$, the *tangent line* at the point $(x, f(x))$ is the line that passes through this point with slope

$$\lim_{h \to 0} \frac{f(x + h) - f(x)}{h}$$

if this limit exists. The slope of the tangent line is also referred to as the *instantaneous rate of change* of the function f with respect to x.

EXAMPLE 4 Find the instantaneous rate of change of the function $f(x) = 5x^2$ with respect to x.

Solution Find $\lim\limits_{h \to 0} \dfrac{f(x + h) - f(x)}{h}$.

We evaluate this expression in small steps:

1. $f(x)$ is given
2. $f(x + h) = 5(x + h)^2 = 5x^2 + 10xh + 5h^2$
3. $f(x + h) - f(x) = (5x^2 + 10xh + 5h^2) - 5x^2$
   $$= 10xh + 5h^2$$
4. $\dfrac{f(x + h) - f(x)}{h} = \dfrac{10xh + 5h^2}{h} = 10x + 5h$
5. $\lim\limits_{h \to 0} \dfrac{f(x + h) - f(x)}{h} = \lim\limits_{h \to 0}(10x + 5h) = 10x$ ■

Now we will use the concept of instantaneous rate of change in a business application. Suppose the profit, P, for a manufacturer is a function of the number of units produced and behaves according to the model

$$P(x) = 50x - x^2$$

The graph of this profit function is shown in Figure 9.8. Now suppose that present production is 10 units. What is the profit?

$$P(10) = 50(10) - 10^2 = 400$$

Increase in profits for a 5-unit increase in production is not the same.

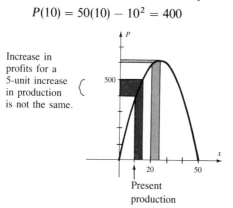

Figure 9.8 Profit function

What is the per unit increase in profit if production is increased from 10 to 20 units?

$$P(20) = 50(20) - 20^2 = 600$$

Increased profit: $P(20) - P(10) = 600 - 400 = 200$

Per unit increase in profit: $\dfrac{P(20) - P(10)}{20 - 10} = \dfrac{200}{10} = 20$

EXAMPLE 5 What is the per unit increase in profit if production is increased from 10 units to:

a. 15 units?   **b.** 11 units?

Solution **a.** $\dfrac{P(15) - P(10)}{15 - 10} = \dfrac{525 - 400}{5} = \dfrac{125}{5} = 25$

b. $\dfrac{P(11) - P(10)}{11 - 10} = \dfrac{429 - 400}{1} = \dfrac{29}{1} = 29$ ■

From Example 5 we see that the per unit increase in profit is different if we increase from 10 to 15 units than if we increase from 10 to 11 units. We would therefore talk about the average per unit increase in profit from 10 to 15 or from 10 to 11 units.

Let us consider this situation in general. An increase in production from x units to $x + h$ units would produce an average per unit increase in profit for a function P as follows:

$$\frac{P(x + h) - P(x)}{x + h - x} = \frac{P(x + h) - P(x)}{h}$$

For the instantaneous rate of change we can let $h \to 0$. *This represents the per unit increase in profit at a production level of x units.* In business this per unit increase is called the **marginal profit** and is defined by

$$\lim_{h \to 0} \frac{P(x + h) - P(x)}{h}$$

EXAMPLE 6 Find the marginal profit for the above model; that is, find

$$P(x) = 50x - x^2$$

Solution 1. $P(x) = 50x - x^2$ is given
2. $P(x + h) = 50(x + h) - (x + h)^2 = 50x + 50h - x^2 - 2xh - h^2$
3. $P(x + h) - P(x) = (50x + 50h - x^2 - 2xh - h^2) - (50x - x^2)$
$$= 50h - 2xh - h^2$$
4. $\dfrac{P(x + h) - P(x)}{h} = \dfrac{(50 - 2x - h)h}{h} = 50 - 2x - h$
5. $\displaystyle\lim_{h \to 0} \frac{P(x + h) - P(x)}{h} = \lim_{h \to 0}(50 - 2x - h) = 50 - 2x$ ∎

We check the results of Example 6 with our previous results:

1. Production level of 10 units where $x = 10$.

$$50 - 2(10) - h = 30 - h$$

2. Increases: From 10 to 20 ($h = 10$): $30 - h = 30 - 10 = 20$
From 10 to 15 ($h = 5$): $30 - h = 30 - 5 = 25$
From 10 to 11 ($h = 1$): $30 - h = 30 - 1 = 29$

3. Marginal profit ($h = 0$): $30 - h = 30$

The final application of the concept of an instantaneous rate of change is **velocity**.

EXAMPLE 7 At an amusement park there is a "free fall" ride called "The Edge." The ride involves falling 100 feet in 2.5 seconds. As you are falling, you pass the 16-foot mark at one second and the 64-foot mark at two seconds. What is the velocity at the *instant* the ride passes the 100-foot mark (measured from the top)?

Solution Let $s(t)$ be the distance from the top t seconds after release. Also assume that $s(t) = 16t^2$.*

---

* This is the formula for free fall in a vacuum; it is sufficiently accurate for our purposes in this problem.

Courtesy of Great America
Theme Park, Santa Clara, CA.

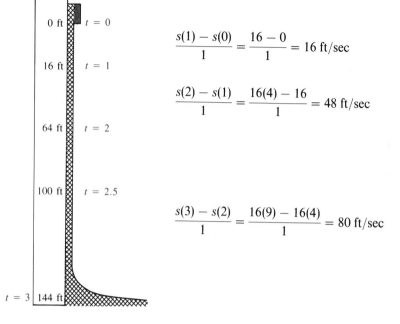

$$\frac{s(1) - s(0)}{1} = \frac{16 - 0}{1} = 16 \text{ ft/sec}$$

$$\frac{s(2) - s(1)}{1} = \frac{16(4) - 16}{1} = 48 \text{ ft/sec}$$

$$\frac{s(3) - s(2)}{1} = \frac{16(9) - 16(4)}{1} = 80 \text{ ft/sec}$$

In general, the average velocity from time $t = t_1$ to $t = t_1 + h$ is given by the following formula:

$$\text{Average velocity} = \frac{\text{change in position}}{\text{change in time}}$$

$$= \frac{s(t_1 + h) - s(t_1)}{h}$$

$$= \frac{16(t + h)^2 - 16t^2}{h}$$

$$= \frac{16t^2 + 32th + 16h^2 - 16t^2}{h}$$

$$= \frac{32th + 16h^2}{h}$$

$$= 32t + 16h$$

Now, to find the velocity at a particular instant in time, we simply need to consider the limit as $h \to 0$:

$$\text{Instantaneous velocity} = \lim_{h \to 0} \frac{s(t_1 + h) - s(t_1)}{h}$$

$$= \lim_{h \to 0} (32t + 16h)$$

$$= 32t$$

We can now use these formulas to find the average and instantaneous velocities:

Time interval	Average velocity $32t + 16h$	Instantaneous velocity $32t$
At t = 0		$32(0) = 0$
From $t = 0$ to $t = 1$ ($h = 1$)	$32(0) + 16(1) = 16$	
At t = 1		$32(1) = 32$
From $t = 1$ to $t = 2$ ($h = 1$)	$32(1) + 16(1) = 48$	
From $t = 1$ to $t = 3$ ($h = 2$)	$32(1) + 16(2) = 64$	
At t = 2		$32(2) = 64$
$\vdots$	$\vdots$	$\vdots$

We are given that the ride passes 100 feet at 2.5 seconds, so the instantaneous velocity at that instant is

$$32(2\tfrac{1}{2}) = 80 \text{ ft/sec}$$

(By the way, 80 ft/sec is approximately 54 mph.) ∎

Problem Set 9.3

APPLICATIONS

The graph shows the height h of a projectile after t seconds. Find the average rate of change of height with respect to the changes in t in Problems 1–4.

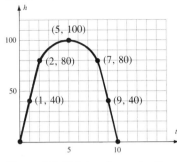

Figure 9.9 Height of a projectile in feet t seconds after fired

1. 1 to 7
3. 1 to 2

2. 1 to 5
4. 2 to 9

The graph shows company output as a function of the number of workers. Find the average rate of change of output for the changes in the number of workers in Problems 5–8.

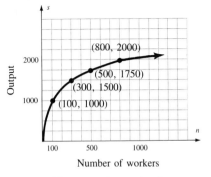

Figure 9.10 Output in production relative to the number of employees at Kampbell Construction

5. 100 to 800
7. 500 to 800

6. 300 to 800
8. 300 to 500

The SAT scores of entering first-year college students are shown in Figure 9.11. Find the average yearly rate of change of the scores for the periods of time in Problems 9–12.

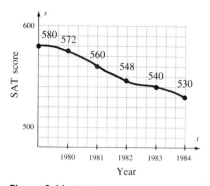

Figure 9.11 SAT scores at Riveria College

9. 1979 to 1984   **10.** 1980 to 1984
11. 1982 to 1984   **12.** 1983 to 1984

Figure 9.12 shows the average rate change of the Gross National Product (GPN) for the years 1978–1984. Find the average yearly rate of change of the GPN for the years in Problems 13–16.

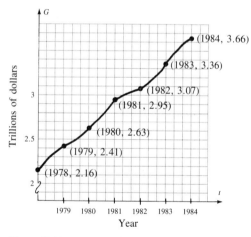

Figure 9.12 Gross National Product in trillions of dollars

13. 1978 to 1984   **14.** 1979 to 1984
15. 1980 to 1984   **16.** 1981 to 1984

The news article in Figure 9.4 gives some average speeds. Verify the accuracy of the following calculations. Answer to the nearest tenth.

17. "Average speed over the 26.7 miles between Santa Rosa and Novato: 60 mph." *Time:* 6:09 A.M. to 6:36 A.M.

18. "Average speed over the 8.4 miles between Novato and central San Rafael: 18 mph." *Time:* 6:36 A.M. to 7:03 A.M.

19. "Average speed over the 13.2 miles between San Rafael and the toll plaza: 36 mph." *Time:* 7:03 A.M. to 7:28 A.M.

20. "Average speed over the entire 49.8-mile, 1-hour and 19-minute commute: 38 mph." *Time:* 6:09 A.M. to 7:28 A.M.

Trace the curves in Problems 21–26 onto your own paper and draw the secant line passing through P and Q. Next, imagine h → 0 and draw the tangent line at P assuming that Q moves along the curve to the point P. Finally, estimate the slope of the curve at P using the slope of the tangent line you have drawn.

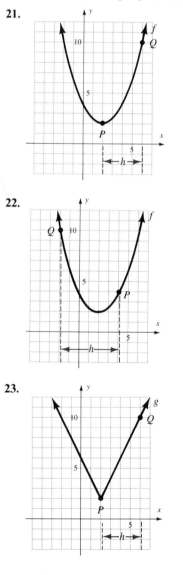

24.

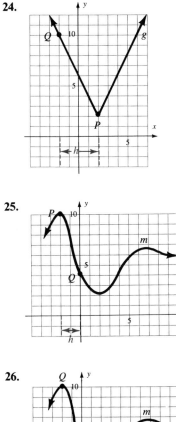

25.

26.

Find the average rate of change for the functions in Problems 27–35.

27. $f(x) = 4 - 3x$ for $x = -3$ to $x = 2$

28. $f(x) = 5x - 1$ for $x = -5$ to $x = -1$

29. $f(x) = 5$ for $x = -3$ to $x = 3$

30. $f(x) = -2$ for $x = 0$ to $x = 5$

31. $f(x) = 3x^2$ for $x = 1$ to $x = 3$

32. $f(x) = x^2 - 3x$ for $x = 0$ to $x = 3$

33. $y = -2x^2 + x + 4$ for $x = 1$ to $x = 4$

34. $y = \sqrt{x}$ for $x = 4$ to $x = 9$

35. $y = \dfrac{-2}{x + 1}$ for $x = 1$ to $x = 5$

Find the instantaneous rate of change for the functions in Problems 36–44. These are the same functions as those given in Problems 27–35.

36. $f(x) = 4 - 3x$ for $x = -3$

37. $f(x) = 5x - 1$ for $x = -5$

38. $f(x) = 5$ for $x = -3$ **39.** $f(x) = -2$ for $x = 0$

40. $f(x) = 3x^2$ for $x = 1$ **41.** $f(x) = x^2 - 3x$ for $x = 0$

42. $y = -2x^2 + x + 4$ for $x = 1$

43. $y = \sqrt{x}$ for $x = 4$ **44.** $y = \dfrac{-2}{x + 1}$ for $x = 1$

APPLICATIONS

Suppose the profit, P, for a manufacturer is a function of the number of units produced and behaves according to the model

$$P(x) = 50x - x^2$$

Also suppose that the present production is 20 units. Use this model for Problems 45–48.

45. What is the per unit increase in profit if production is increased from 20 to 30 units?

46. Repeat Problem 45 for an increase from 20 to 25 units.

47. Repeat Problem 45 for an increase from 20 to 21 units.

48. What is the marginal profit for P at $x = 20$?

The cost, C, in dollars for producing x items is given by

$$C(x) = 30x^2 - 100x$$

Use this model for Problems 49–55.

49. Find the average rate of change of cost as x increases from 100 to 200 items.

50. Repeat Problem 49 for an increase from 100 to 110 items.

51. Repeat Problem 49 for an increase from 100 to 101 items.

52. Repeat Problem 49 for an increase from 100 to $(100 + h)$ items.

53. Repeat Problem 49 for an increase from x to $(x + h)$ items.

54. Find $\displaystyle\lim_{h \to 0} \frac{C(100 + h) - C(100)}{h}$.

55. Find $\displaystyle\lim_{h \to 0} \frac{C(x + h) - C(x)}{h}$.

56. Attach a possible meaning for the result of Problem 54. In business the result of Problem 54 is called the *marginal cost* for the production level of 100. We will discuss this concept later in the text.

57. Attach a possible meaning for the result of Problem 55. In business the result of Problem 55 is called the *marginal cost* for the production level of x. We will discuss this concept later in the text.

9.4 Definition of Derivative

The concept of the derivative is a very powerful mathematical idea, and the variety of applications is almost unlimited. In the last section we investigated the instantaneous rate of change of a function f per unit change in x (marginal profit, instantaneous speed, velocity, and the slope of a line tangent to a curve at a particular point). All of these ideas can be summarized by a single concept, called the **derivative**:

Definition of Derivative

> For a given function f, we define the *derivative of f at x*, denoted by $f'(x)$, to be
>
> $$f'(x) = \lim_{h \to 0} \frac{f(x+h) - f(x)}{h}$$
>
> provided this limit exists.

If the limit does not exist, then we say that f is *not differentiable* at x. We have now developed all of the techniques necessary to apply the definition of derivative to a variety of functions.

EXAMPLE 1 Use the definition to find the derivative of $y = x^2$.

Solution We carry out the five-step process of the last section:

1. $f(x) = x^2$ is given
2. $f(x+h) = (x+h)^2$ Evaluate f at $x+h$
   $$= x^2 + 2xh + h^2$$
3. $f(x+h) - f(x) = (x^2 + 2xh + h^2) - x^2$
   $$= 2xh + h^2$$
4. $\dfrac{f(x+h) - f(x)}{h} = \dfrac{2xh + h^2}{h}$
   $$= \frac{h(2x+h)}{h} = 2x + h$$
5. $\displaystyle\lim_{h \to 0} \frac{f(x+h) - f(x)}{h} = \lim_{h \to 0}(2x + h) = 2x$

The derivative of $y = x^2$ is $2x$. Sometimes we write $y' = 2x$ or $f'(x) = 2x$. ∎

EXAMPLE 2 Find the equation of the line tangent to the curve $y = x^2$ at $x = 2$.

Solution From Example 1, $y' = 2x$, so the slope of the tangent line at any point is $2x$. When $x = 2$, then $y = 2^2$, so we are looking for the tangent line passing through $(2, 4)$ with slope equal to the derivative. Remember, the slope of a curve at a point is the value of the derivative at that point. Thus, at $x = 2$, $f'(2) = 2(2) = 4$. Now we can use the point-slope form:

$$y - 4 = 4(x - 2)$$
$$y = 4x - 4 \quad \text{or, in standard form, } 4x - y - 4 = 0$$

∎

EXAMPLE 3 Find the instantaneous rate of change of profit $P(x) = 80x - 25x^2$, if the present level of production is 50 units.

Solution Evaluate the derivative at $x = 50$ (because that is the present production level). First, find $P'(x)$:

1. $P(x) = 80x - 25x^2$

2. $P(x + h) = 80(x + h) - 25(x + h)^2$
$$= 80x + 80h - 25x^2 - 50xh - 25h^2$$

3. $P(x + h) - P(x) = (80x + 80h - 25x^2 - 50xh - 25h^2) - (80x - 25x^2)$
$$= 80h - 50xh - 25h^2$$

4. $\dfrac{P(x + h) - P(x)}{h} = \dfrac{80h - 50xh - 25h^2}{h}$
$$= 80 - 50x - 25h$$

5. $\lim\limits_{h \to 0} \dfrac{P(x + h) - P(x)}{h} = \lim\limits_{h \to 0}(80 - 50x - 25h) = 80 - 50x$

Thus $P'(x) = 80 - 50x$ and $P'(50) = -2420$. ∎

We use tangent lines to sketch functions in the next chapter. It is worth noting here, though, that if the derivative of a function is positive, then the curve is rising at that point (since the slope of the tangent line is also positive); and if the derivative is negative at a point, then the curve is falling at that point. If the derivative is 0, then the tangent is horizontal. In Example 2, we found the equation of the tangent line of the function $y = x^2$ at $(2, 4)$. This tangent line and others are shown in Figure 9.13.

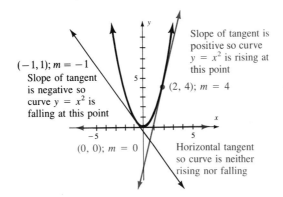

Figure 9.13 Graph of $y = x^2$ with tangent lines

$(-1, 1)$; $m = -1$
Slope of tangent is negative so curve $y = x^2$ is falling at this point

Slope of tangent is positive so curve $y = x^2$ is rising at this point

$(2, 4)$; $m = 4$

$(0, 0)$; $m = 0$

Horizontal tangent so curve is neither rising nor falling

A derivative may not exist at a particular point. If the limit does not exist at $x = a$, then we say that the function is not differentiable at $x = a$. The concept of differentiability is more easily understood if you relate it to tangent lines. Geometrically, a tangent line will not exist when the graph of the function "has a sharp point," as shown in Figure 9.14.

The slope of a curve does not exist at a point for which the derivative is not defined. This means that there might be a vertical tangent line, and when this occurs we also say there is no slope (see Figure 9.14c). A point for which a function is not continuous cannot have a derivative at that point (see Figures 9.14d and e). However, be careful with this statement because the converse is not true. **A function may be continuous at $x = a$ but still not be differentiable there** (see Figures 9.14a–9.14c).

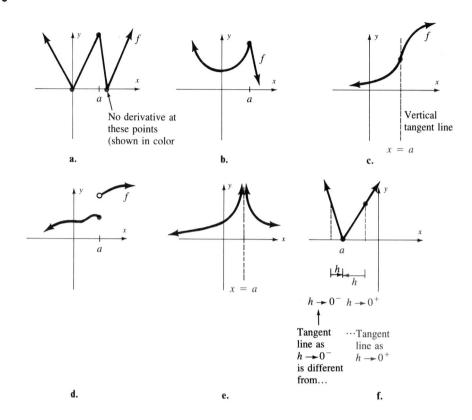

Figure 9.14 Points on a curve for which the derivative is not defined

EXAMPLE 4   Let $y = |x|$. Show that y is continuous at $x = 0$ but not differentiable at $x = 0$.

Solution   First, $f(x) = |x|$ is continuous at $x = 0$ since $f(0) = |0| = 0$ is defined and $\lim\limits_{x \to 0} |x| = 0$. Next, find $f'(x)$:

1. $f(x) = |x|$
2. $f(x + h) = |x + h|$
3. $f(x + h) - f(x) = |x + h| - |x|$
4. $\dfrac{f(x + h) - f(x)}{h} = \dfrac{|x + h| - |x|}{h}$
5. $\lim\limits_{h \to 0} \dfrac{f(x + h) - f(x)}{h} = \lim\limits_{h \to 0} \dfrac{|x + h| - |x|}{h} = \lim\limits_{h \to 0} \dfrac{|h|}{h}$ at $x = 0$

Now consider the left- and right-hand limits:

$$\lim_{h \to 0^+} \frac{|h|}{h} = \lim_{h \to 0^+} \frac{h}{h} = 1 \qquad \text{If } h \to 0^+, \text{ then } h \text{ is positive, so } |h| = h$$

$$\lim_{h \to 0^-} \frac{|h|}{h} = \lim_{h \to 0^-} \frac{-h}{h} = -1 \qquad \text{If } h \to 0^-, \text{ then } h \text{ is negative, so } |h| = -h$$

Since the left- and right-hand limits are not the same, we see that this limit does not exist. Therefore the derivative does not exist. Thus $f(x) = |x|$ is continuous at $x = 0$ but is not differentiable at $x = 0$.   ∎

In the last section *marginal profit* was defined. In business and economics the adjective *marginal* means rate of change. Mathematically this means that it is a derivative. Another example of this usage is **marginal cost**, which means the rate of change in cost per unit change in production at an output level of x units. This idea is illustrated with the following example.

EXAMPLE 5 Suppose the total cost in thousands of dollars for manufacturing x thousand items is described by the following equation:

$$C(x) = 2x^2 + 4x + 2500$$

Also suppose that current production is 50,000 items ($x = 50$). Derive the formula for marginal cost and then find the marginal cost for $x = 50$. Also find the actual cost for producing one additional item at this production level.

Solution The actual cost of producing h additional items is $C(x + h) - C(x)$. The per unit cost is

$$\frac{C(x + h) - C(x)}{h}$$

This represents the slope of the secant line joining $(x, C(x))$ and $(x + h, C(x + h))$. Now the rate of change is found by taking the limit as $h \to 0$:

$$C'(x) = \lim_{h \to 0} \frac{C(x + h) - C(x)}{h}$$

For this example, $C(x) = 2x^2 + 4x + 2500$ and $x = 50$.

$$C(50) = 2(50)^2 + 4(50) + 2500 = 7700$$
$$C(50 + h) = 2(50 + h)^2 + 4(50 + h) + 2500$$
$$= 5000 + 200h + 2h^2 + 200 + 4h + 2500$$
$$= 7700 + 204h + 2h^2$$
$$C'(x) = \lim_{h \to 0} \frac{(7700 + 204h + 2h^2) - 7700}{h}$$
$$= \lim_{x \to 0} \frac{204h + 2h^2}{h}$$
$$= \lim_{h \to 0} (204 + 2h)$$
$$= 204$$

If we want to find the *actual cost* of producing the fifty-first unit we need to find

$$\frac{C(51) - C(50)}{1} = 7906 - 7700 = 206 \qquad \blacksquare$$

The actual cost of producing one unit is very close to the marginal cost. It is usually easier to find the derivative (especially after studying the next two sections) than it is to calculate the actual cost of producing one additional unit. For this reason economists often use marginal cost to approximate the actual cost of producing one additional unit.

Problem Set 9.4

Find all points where the functions in Problems 1–6 do not have derivatives.

1.

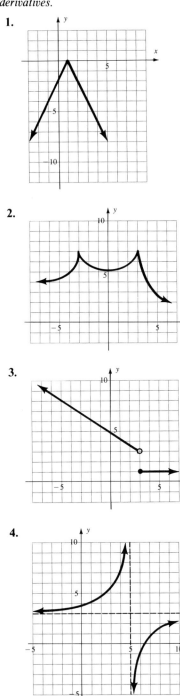

2.

3.

4.

5.

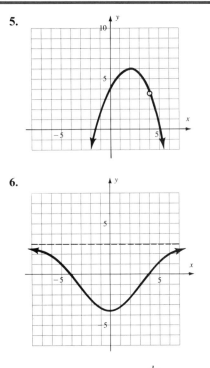

6.

Find the derivative, $f'(x)$ or $\dfrac{dy}{dx}$, of each of the functions in Problems 7–15 by using the derivative definition.

7. $f(x) = 2x^2$ **8.** $f(x) = 5x^2$

9. $f(x) = -3x^2$ **10.** $y = 2x + 1$

11. $y = 4 - 5x$ **12.** $y = 25 - 250x$

13. $y = 3x^2 + 4x$ **14.** $y = 3 + 2x - 3x^2$

15. $y = -3x^2 - 50x + 125$

Find the equation of the line tangent to the curves in Problems 16–21 at the given point.

16. $y = 2x^2$ at $x = 4$

17. $y = 5x^2$ at $x = -3$

18. $y = 2x + 1$ at $x = 3$

19. $y = 4 - 5x$ at $x = -2$

20. $y = 3x^2 + 4x$ at $x = 0$

21. $y = 3 + 2x - 3x^2$ at $x = -1$

APPLICATIONS

In Problems 22–24 let $C(x) = x^2 - 60x + 2500$ be the cost function for producing x items.

22. Derive the formula for the marginal cost.

23. Find the marginal cost for a production level of 20 items.

24. Find the actual cost of producing one additional item if the current production level is 20 items.

In Problems 25–27 suppose the number (in millions) of bacteria present in a culture at time t is given by the formula $N(t) = 2t^2 - 200t + 1000$.

25. Derive the formula for the instantaneous rate of change of the number of bacteria with respect to time.

26. Find the instantaneous rate of change of the number of bacteria with respect to time at time $t = 3$.

27. Find the instantaneous rate of change of the number of bacteria with respect to time at the beginning of this experiment.

28. An object moving in a straight line travels d miles in t hours according to the formula $d(t) = \frac{1}{3}t^2 + 5t$. What is the object's velocity when $t = 3$?

Find the derivative of each of the functions given in Problems 29–34 by using the derivative definition.

29. $g(x) = x^3$

30. $g(x) = 2x^3$

31. $f(x) = -\dfrac{3}{x}$

32. $f(x) = -\dfrac{2}{x+1}$

33. $y = 2\sqrt{x}$

34. $y = \sqrt{3x}$

9.5 Differentiation Techniques, Part I

Differentiation is one of the most powerful and useful concepts in mathematics but it would indeed be cumbersome if we had to apply the definition of derivative as we did in the last section every time we wanted to use this concept. Luckily there are many shortcuts to the process, and these are presented as differentiation techniques in this and the following two sections. In stating and working with these derivative formulas it is helpful to use some alternative notations for derivative. We have already used y', f' and $f'(x)$. Some other notations include

$$\frac{dy}{dx} \qquad \frac{d}{dx}y \qquad \frac{d}{dx}f(x) \qquad D_x y \qquad D_x[f(x)]$$

These are usually pronounced "dee y, dee x," "derivative of y with respect to x," "derivative of f with respect to x," "derivative of y with respect to x," and "derivative of f of x with respect to x," respectively. We will use these notations interchangeably.

In this section we develop five very important derivative formulas, in the next section two more, and in Section 9.7 a derivative formula called the *chain rule*. We begin by finding a formula for the derivative of a power; that is, if $y = x^n$, then what is y'? In the last section (Example 2) we found that if $y = x^2$, then $f'(x) = 2x$. How about $y = x^3$?

EXAMPLE 1   If $f(x) = x^3$, find $f'(x)$.

Solution
1. $f(x) = x^3$
2. $f(x + h) = (x + h)^3 = x^3 + 3x^2h + 3xh^2 + h^3$
3. $f(x + h) - f(x) = x^3 + 3x^2h + 3xh^2 + h^3 - x^3$
$$= 3x^2h + 3xh^2 + h^3$$
4. $\dfrac{f(x + h) - f(x)}{h} = \dfrac{h(3x^2 + 3xh + h^2)}{h}$
$$= 3x^2 + 3xh + h^2$$
5. $\displaystyle\lim_{h \to 0} \dfrac{f(x + h) - f(x)}{h} = \lim_{h \to 0}(3x^2 + 3xh + h^2) = 3x^2$ ∎

If $y = x^4$, then, if we carry out the steps shown in Example 1, we will find $f'(x) = 4x^3$. The details are left as an exercise. Finally, look for a pattern:

$$y = x^2 \qquad \rightarrow \qquad y' = 2x$$
$$y = x^3 \qquad \rightarrow \qquad y' = 3x^2$$
$$y = x^4 \qquad \rightarrow \qquad y' = 4x^3$$
$$y = x^5 \qquad \rightarrow \qquad ?$$

Can you replace the question mark by the pattern you found?

$$y = x^5 \qquad \rightarrow \qquad y' = 5x^4$$
$$y = x^6 \qquad \rightarrow \qquad y' = 6x^5$$
$$\vdots$$
$$y = x^n \qquad \rightarrow \qquad y' = nx^{n-1}$$

We have simply demonstrated the plausibility of the following **power rule**. The general proof for any real number n requires algebra beyond the scope of this course, but we can look at several specific cases to get a feel for the proof.

Power Rule

If f is a differentiable function, and if $f(x) = x^n$, then

$$f'(x) = nx^{n-1}$$

for any real number n.

EXAMPLE 2 If $y = x$, then $n = 1$, and the power rule asserts that $y' = 1 \cdot x^{1-1} = x^0 = 1$. Verify this result by using the definition of a derivative.

Solution
1. $f(x) = x$
2. $f(x + h) = x + h$
3. $f(x + h) - f(x) = h$
4. $\dfrac{f(x + h) - f(x)}{h} = 1$
5. $\displaystyle\lim_{h \to 0} \dfrac{f(x + h) - f(x)}{h} = \lim_{h \to 0} 1 = 1$ ∎

EXAMPLE 3 Find derivatives of the following functions by using the power rule.

   a. x^4 **b.** x^5 **c.** x^6 **d.** x^0 **e.** x^{-1}
   f. x^{-2} **g.** $x^{1/3}$ **h.** $\sqrt{x}$ **i.** $x^{-2.5}$

Solution
   a. If $y = x^4$, then $y' = 4x^3$.
   b. If $y = x^5$, then $y' = 5x^4$.
   c. If $y = x^6$, then $y' = 6x^5$.
   d. If $y = x^0$, then $y' = 0x^{-1} = 0$.
   e. If $y = x^{-1}$, then $y' = (-1)x^{-1-1} = -x^{-2}$ or $\frac{-1}{x^2}$.
   f. If $y = x^{-2}$, then $y' = (-2)x^{-2-1} = -2x^{-3}$ or $\frac{-2}{x^3}$.
   g. If $y = x^{1/3}$, then $y' = (\frac{1}{3})x^{1/3-1} = \frac{1}{3}x^{-2/3}$.

h. If $y = \sqrt{x}$, write (or think of it) as $y = x^{1/2}$; now apply the power rule: $y' = (\frac{1}{2})x^{1/2-1} = \frac{1}{2}x^{-1/2}$. Since the problem was given in radical notation, we state the answer in radical notation: $y' = \dfrac{1}{2\sqrt{x}}$.

i. If $y = x^{-2.5}$, then $y' = -2.5x^{-3.5}$.   ∎

A constant rule can be easily derived using the derivative definition:

1. $f(x) = k$ is given
2. $f(x + h) = k$
3. $f(x + h) - f(x) = 0$
4. $\dfrac{f(x + h) - f(x)}{h} = 0$
5. $\displaystyle\lim_{h \to 0} \dfrac{f(x + h) - f(x)}{h} = \lim_{h \to 0} 0 = 0$

Constant Rule

> If $f(x) = k$ for some constant k, then
> $$f'(x) = 0$$

EXAMPLE 4   Use the constant rule to find the derivatives of the given functions.

a. $y = 5$ **b.** $y = -18$ **c.** $y = \sqrt{3}$ **d.** $y = \pi$

Solution   The derivative of any constant is 0; thus

a. $y' = 0$ **b.** $y' = 0$ **c.** $y' = 0$ **d.** $y' = 0$ ∎

The power rule and the constant rule for derivatives can be combined to prove the following result:

Constant Times a Function

> If $y = kf(x)$ for a differentiable function f, then
> $$y' = kf'(x)$$

This derivative formula for a constant times a function is easy to derive using the definition of a derivative. (The procedure is left as an exercise.) Example 5 illustrates how to use the formula.

EXAMPLE 5   Find the derivatives of the given functions.

a. $y = 3x^2$ **b.** $y = 6x^{-5}$ **c.** $y = -\frac{3}{5}x^{10}$ **d.** $y = -2\sqrt{x}$

Solution   **a.** $y' = 3[(2)x^1]$
           $= 6x$
       c. $y' = -\frac{3}{5}[(10)x^9]$
           $= -6x^9$

b. $y' = 6[(-5)x^{-6}]$
       $= -30x^{-6}$
d. $y' = -2[(\frac{1}{2})x^{1/2-1}]$
       $= -x^{-1/2}$ ∎

The next two derivative formulas are for sums or differences of functions:

Sum Rule

Difference Rule

> Suppose f and g are differentiable functions of x.
> If $y = f + g$, then $y' = f' + g'$.
> If $y = f - g$, then $y' = f' - g'$.

We assume that if f and g are differentiable at a point $x = a$, then the sum rule implies that the sum $f + g$ is also differentiable at $x = a$.

The following proof of the sum formula is optional, and you may skip over to Example 6 if you are not interested in deriving these formulas. We begin by writing y as $u(x)$ so that the sum formula can be written as $u(x) = f(x) + g(x)$. We want to show that $u'(x) = f'(x) + g'(x)$. We use the definition of derivative:

1. $u(x) = f(x) + g(x)$
2. $u(x + h) = f(x + h) + g(x + h)$
3. $u(x + h) - u(x) = [f(x + h) + g(x + h)] - [f(x) + g(x)]$
$$= [f(x + h) - f(x)] + [g(x + h) - g(x)]$$
4. $\dfrac{u(x + h) - u(x)}{h} = \dfrac{[f(x + h) - f(x)] + [g(x + h) - g(x)]}{h}$
$$= \frac{f(x + h) - f(x)}{h} + \frac{g(x + h) - g(x)}{h}$$
5. $\displaystyle\lim_{h \to 0} \frac{u(x + h) - u(x)}{h} = \lim_{h \to 0}\left[\frac{f(x + h) - f(x)}{h} + \frac{g(x + h) - g(x)}{h}\right]$
$$= \lim_{h \to 0}\frac{f(x + h) - f(x)}{h} + \lim_{h \to 0}\frac{g(x + h) - g(x)}{h}$$

Thus $u'(x) = f'(x) + g'(x)$. The proof for the difference formula is identical to the one for the sum formula.

These last two derivative formulas, along with the ones already discussed, allow us to easily find the derivatives of polynomials, as shown in Example 6.

EXAMPLE 6   Find the derivatives of the given functions.
  a. $y = 5x^2 - 3x + 15$ **b.** $y = 9x^3 - 4x^2 + 5x - 12$

Solution   **a.** $y' = 5(2)x - 3(1) + 0$ **b.** $y' = 9(3)x^2 - 4(2)x + 5(1) - 0$
  $= 10x - 3$ $= 27x^2 - 8x + 5$ ∎

EXAMPLE 7   If $f(x) = 6x^8 - 5\sqrt{x} + \dfrac{4}{x}$, find $f'(x)$.

Solution   Write (or think of) this as $f(x) = 6x^8 - 5x^{1/2} + 4x^{-1}$. Then
$$f'(x) = 6(8)x^7 - 5(\tfrac{1}{2})x^{-1/2} + 4(-1)x^{-2}$$
$$= 48x^7 - \frac{5}{2\sqrt{x}} - \frac{4}{x^2}$$ ∎

EXAMPLE 8 Suppose the amount of calcium remaining in a person's bloodstream after a certain test is given by the formula $A = t^{-3/2}$ for $t \geq \frac{1}{2}$. How fast is the body removing calcium from the blood on the second day?

Solution The rate of change (per day) of calcium in the blood is given by the derivative

$$A'(t) = \frac{dA}{dt} = -\frac{3}{2}t^{-5/2}$$

When $t = 2$, this rate is $-\frac{3}{2}(2)^{-5/2}$. This rate can be approximated on a calculator by pressing

$\boxed{2}\ \boxed{x^y}\ \boxed{2.5}\ \boxed{+/-}\ \boxed{\times}\ \boxed{1.5}\ \boxed{+/-}\ \boxed{=}$ -0.2651650429

The amount of calcium in the blood is changing at the rate of about -0.27 units per day when $t = 2$. The negative sign tells us that the amount of calcium is decreasing. ■

Note in Example 8 that the variable is not x and that the symbol used is dA/dt. This symbol is read as "derivative of A with respect to t," or as "dee A, dee tee." Even though the variables x and y are used most of the time, keep in mind that any variables may be substituted for x and y.

The derivative rules of this section are summarized below for easy reference:

Differentiation Formulas

POWER RULE: If $y = x^n$,   then $y' = nx^{(n-1)}$ for any real number n.

CONSTANT RULE: If $y = k$,   then $y' = 0$.

CONSTANT TIMES A FUNCTION RULE: If $y = kf$,   then $y' = kf'$.

SUM RULE: If $y = f + g$,   then $y' = f' + g'$.

DIFFERENCE RULE: If $y = f - g$,   then $y' = f' - g'$.

Problem Set 9.5

Find the derivatives of the functions in Problems 1–30 and simplify them algebraically.

1. $y = x^7$

2. $y = x^{16}$

3. $y = x^{12}$

4. $y = x^{-3}$

5. $y = x^{-5}$

6. $y = 25$

7. $y = -130$

8. $y = 5x^{-4}$

9. $y = -4x^{-8}$

10. $y = 5\sqrt{x}$

11. $y = -\frac{1}{2}\sqrt{x}$

12. $y = \frac{3}{4}\sqrt{x}$

13. $y = 5x^{-8}$

14. $y = -3x^{-2}$

15. $y = 12x^{5/4}$

16. $y = 5x^2 - 9$

17. $y = 3x^2 + x$

18. $y = 5x^3 - 9x^2$

19. $y = 2x^2 - 5x - 6$

20. $y = 5x^2 - 5x + 12$

21. $y = 5x^3 - 5x^2 + 4x - 5$

22. $y = 6x^3 - 25x^2 - 6x + 45$

23. $y = x^{-3} + x^2 + x^{-1}$

24. $y = x^{-5} - x^{-3} - x^{-1}$

25. $y = (x^4 + 2)^2$

26. $y = (x^5 - 1)^2$

27. $y = \dfrac{2}{x} + \dfrac{5}{x^2}$

28. $y = \dfrac{3}{x} - \dfrac{2}{x^3}$

29. $y = -5x^7 + 2\sqrt{x} - 3x^{-1}$

30. $y = 4x^{-3/4} - 5\sqrt{x} + 2x^{-3}$

APPLICATIONS

31. The cost of producing a certain type of boat is $C(x) = 20x^2 + 500x + 250,000$. What is the marginal cost?

32. The profit function for a certain item is $P(x) = 45x - 3x^3$. What is the marginal profit?

33. The relationship between the current and the time on an X-ray machine is given by the formula $m = 400/t$. Find dm/dt.

34. The relationship between the current and the distance on an X-ray machine is given by the formula $m = 256/s^2$. Find dm/ds.

35. The earnings (in thousands of dollars) of Amdex Corporation are a function of the number of years since its founding in 1982 according to the formula $A(t) = 0.05t^2 + 25t + 5$. At what rate are the earnings changing in 1988?

36. In a certain experiment, subjects are found to learn according to the model $N = 25\sqrt{x}$, where N is the number of tasks learned in x hours. How fast are the subjects learning at the end of the second hour?

37. Repeat Problem 36 except now how fast are the subjects learning at the end of the fifth hour?

38. In a wildlife reserve the population P of foxes depends on the population x of rabbits according to the formula $P = 0.0005x + 0.00001x^2$. What is the rate at which the population of foxes is changing when the number of rabbits is 100,000?

39. Repeat Problem 38 for a population of 1 million rabbits.

40. The number of gallons, G, of water pumped into a reservoir after t minutes is given by the formula $g(t) = 2t + \sqrt{t}$. At what rate is water flowing into the reservoir (in gallons per hour) when $t = 10$?

41. Show that if $y = x^4$, then $y' = 4x^3$ by using the derivative definition.

42. Derive the formula for a constant times a function by using the derivative definition.

43. Prove the formula for the difference of functions by using the derivative definition.

9.6 Differentiation Techniques, Part II

There are two additional shortcut differentiation formulas that simplify work with calculus. These are the *product* and *quotient* formulas. Since the sum and difference formulas are easy to remember and are intuitively obvious, students often incorrectly try to generalize to product and quotient formulas. The derivative of a sum or difference is the sum or difference of the derivatives. But this is not true of the product and quotient formulas:

Warning   *The derivative of a product is not the product of the derivatives.* The following simple example will illustrate this fact. Let $f(x) = x^2$ and $g(x) = x^3$. Consider $y = f(x)g(x) = x^5$:

$$f'(x) = 2x \qquad g'(x) = 3x^2 \qquad y' = 5x^4$$

and

$$f'(x)g'(x) = 2x(3x^2) = 6x^3 \neq 5x^4$$

Note that $y' \neq f'g'$. The product formula tells us how to find the derivative of a product without the necessity of first multiplying them together.

Product Rule

If f and g are differentiable functions such that $y = f(x)g(x)$, then

$$y' = f(x)g'(x) + f'(x)g(x)$$

This formula can more easily be remembered in the following form: if $y = fg$, then

$$y' = fg' + f'g \qquad \text{First function times the derivative of the second plus second times the derivative of the first}$$

This differentiation formula is proved in Example 4, but we will first work through a couple of examples to make sure you understand how it works.

EXAMPLE 1   If $f(x) = x^2$, $g(x) = x^3$, and $y = fg$, verify that $y' = fg' + f'g = 5x^4$.

Solution $f'(x) = 2x$ and $g'(x) = 3x^2$, so

$$fg' + f'g = (x^2)(3x^2) + (2x)(x^3)$$
$$= 3x^4 + 2x^4$$
$$= 5x^4$$

Thus $y' = fg' + f'g = 5x^4$.   ∎

EXAMPLE 2   If $y = (2x^3 + 5x - 3)(x^4 - 3x + 1)$, find dy/dx and simplify.

Solution   You could multiply the factors together to obtain a polynomial and then use the results of the last section to find the derivative, or you can use the product rule. This example illustrates the product rule.

$$\frac{dy}{dx} = \frac{d}{dx}[(2x^3 + 5x - 3)(x^4 - 3x + 1)]$$

$$= (2x^3 + 5x - 3)\frac{d}{dx}(x^4 - 3x + 1) + (x^4 - 3x + 1)\frac{d}{dx}(2x^3 + 5x - 3)$$

$$= (2x^3 + 5x - 3)(4x^3 - 3) + (x^4 - 3x + 1)(6x^2 + 5)$$

$$= 8x^6 + 20x^4 - 12x^3 - 6x^3 - 15x + 9 + 6x^6 - 18x^3 + 6x^2 + 5x^4 - 15x + 5$$

$$= 14x^6 + 25x^4 - 36x^3 + 6x^2 - 30x + 14$$ ∎

EXAMPLE 3   If $y = (\sqrt{x} + 1)(2\sqrt{x} - 3)$, find dy/dx and simplify.

Solution $$\frac{dy}{dx} = \frac{d}{dx}[(x^{1/2} + 1)(2x^{1/2} - 3)]$$

$$= (x^{1/2} + 1)\frac{d}{dx}(2x^{1/2} - 3) + (2x^{1/2} - 3)\frac{d}{dx}(x^{1/2} + 1)$$

$$= (x^{1/2} + 1)\left[2\left(\frac{1}{2}\right)x^{-1/2}\right] + (2x^{1/2} - 3)\left[\frac{1}{2}x^{-1/2}\right]$$

$$= (x^{1/2} + 1)x^{-1/2} + (2x^{1/2} - 3)\left[\frac{1}{2}x^{-1/2}\right]$$

$$= x^0 + x^{-1/2} + x^0 - \frac{3}{2}x^{-1/2}$$

$$= 2 - \frac{1}{2}x^{-1/2}$$

$$= 2 - \frac{1}{2\sqrt{x}} \cdot \frac{\sqrt{x}}{\sqrt{x}}$$

$$= 2 - \frac{\sqrt{x}}{2x}$$

$$= \frac{4x - \sqrt{x}}{2x}$$ ∎

Example 3 shows that the product rule is not always the most economical method. Notice that if you first multiply the factors to obtain

$$y = 2x - \sqrt{x} - 3$$

Find the derivative by using the power rule:

$$y' = 2 - \frac{1}{2}x^{-1/2} = 2 - \frac{\sqrt{x}}{2x} = \frac{4x - \sqrt{x}}{2x}$$

EXAMPLE 4   Prove the product rule. That is, if f and g are differentiable functions such that $y = f(x)g(x)$, then prove that

$$y' = f(x)g'(x) + f'(x)g(x)$$

Solution   Use the definition of derivative:

$$y' = \lim_{h \to 0} \frac{f(x + h)g(x + h) - f(x)g(x)}{h}$$

Add and subtract $f(x)g(x + h)$ in the numerator:

$$y' = \lim_{h \to 0} \frac{f(x + h)g(x + h) - f(x)g(x) + f(x)g(x + h) - f(x)g(x + h)}{h}$$

$$= \lim_{h \to 0} \frac{[f(x + h)g(x + h) - f(x)g(x + h)] + [f(x)g(x + h) - f(x)g(x)]}{h}$$

$$= \lim_{h \to 0} \left[g(x + h) \cdot \frac{f(x + h) - f(x)}{h} + f(x) \cdot \frac{g(x + h) - g(x)}{h} \right]$$

$$= \lim_{h \to 0} \left[g(x + h) \cdot \frac{f(x + h) - f(x)}{h} \right] + \lim_{h \to 0} \left[f(x) \cdot \frac{g(x + h) - g(x)}{h} \right]$$

Since $g(x)$ is differentiable it is continuous, so $\lim_{h \to 0} g(x + h) = g(x)$. Using this and the definition of derivative, we have

$$y' = g(x)f'(x) + f(x)g'(x)$$ ∎

The derivative formula for quotients also does *not* follow the pattern of simply taking the quotient of the derivatives. The quotient formula is stated in the following box, and you are asked to prove it in the problems. It is very easy to prove because you can write f/g as fg^{-1} and use the product rule together with the chain rule, which is discussed in the next section.

Quotient Rule

If f and g are differentiable functions such that $y = f(x)/g(x)$ then $y' = [g(x)f'(x) - f(x)g'(x)]/[g(x)]^2$. This formula can be more easily stated and remembered by using the following notation:

If $y = \dfrac{f}{g}$, then

$$y' = \frac{gf' - fg'}{g^2}$$

The *bottom* function times the derivative of the *top* minus the top times the derivative of the bottom, all divided by the bottom squared (*b* comes *before t* alphabetically)

EXAMPLE 5 If $y = \dfrac{2x^3}{3x + 1}$, find $\dfrac{dy}{dx}$ and simplify.

Solution $y' = \dfrac{(3x + 1)(2x^3)' - (2x^3)(3x + 1)'}{(3x + 1)^2}$

$= \dfrac{(3x + 1)(6x^2) - (2x^3)(3)}{(3x + 1)^2}$

$= \dfrac{18x^3 + 6x^2 - 6x^3}{(3x + 1)^2}$

$= \dfrac{12x^3 + 6x^2}{(3x + 1)^2}$ ∎

EXAMPLE 6 If $y = \dfrac{3x^2 + 2}{5x^2 - 1}$, find $\dfrac{dy}{dx}$ and simplify.

Solution $y' = \dfrac{(5x^2 - 1)(6x) - (3x^2 + 2)(10x)}{(5x^2 - 1)^2}$

$= \dfrac{30x^3 - 6x - 30x^3 - 20x}{(5x^2 - 1)^2}$

$= \dfrac{-26x}{(5x^2 - 1)^2}$ ∎

EXAMPLE 7 An apple factory can produce x thousand gallons of apple juice per week at a weekly cost of

$$C(x) = \frac{100(150x + 1)}{x}$$

Show that the cost is always decreasing.

Solution The cost is decreasing when the derivative is negative. Thus we need to show the inequality $C'(x) < 0$. First find $C'(x)$.

Method I: Quotient rule:

$$C'(x) = 100\left[\frac{x(150) - (150x + 1)(1)}{x^2}\right]$$

$$= 100\left(\frac{150x - 150x - 1}{x^2}\right)$$

$$= \frac{-100}{x^2}$$

Method II: The product rule. Write $C(x) = 150{,}000 + 100x^{-1}$:

$$C'(x) = -(1)100x^{-2} = -100x^{-2}$$

Do not forget that quotients can often be written as products and that the problem may be easier to work as a product than as a quotient. However, in either case, you will obtain the same results if you do not make any mistakes. Since $C'(x) = -100x^{-2}$, we see that $C'(x) < 0$ for all x. Thus the cost is always decreasing. ∎

Problem Set 9.6

Find the derivative for each of the functions in Problems 1–28 and simplify.

1. $f(x) = 5x^2(x^2 - 6)$

2. $f(x) = 3x^3(x^2 + 7)$

3. $f(x) = (x + 1)(x - 2)$

4. $f(x) = (2x - 5)(x + 7)$

5. $g(x) = (3x^2 + 5)(2x^2 - 5)$

6. $g(x) = (9x^2 - 8)(2x^2 - 5)$

7. $g(x) = 5x^4(2x^2 - 5x + 1)$

8. $g(x) = 9x^5(3x^2 - 2x + 15)$

9. $y = \dfrac{x}{x - 3}$

10. $y = \dfrac{x}{3x - 1}$

11. $y = \dfrac{x + 5}{x - 3}$

12. $y = \dfrac{3x + 1}{15 - x}$

13. $y = \dfrac{x^2 + 3}{x^2 - 5}$

14. $y = \dfrac{5 - 7x^3}{1 + x^3}$

15. $y = \dfrac{x^2 + 2x - 5}{8x^2 - 3x + 1}$

16. $y = \dfrac{2x^2 - 25x}{3x^2 - 6x + 135}$

17. $f(x) = (2x^2 - 1)(x^3 + 2x^2 - 3)$

18. $f(x) = (3x^2 + 2)(x^3 - 5x^2 + 4)$

19. $f(x) = (4x^2 + x)(x^3 - 3x^2 + 13)$

20. $f(x) = (x^2 + 3x)(x^3 + 4x^2 + 25)$

21. $f(x) = (3x^3 - 2x + 5)(2x^4 + 5x - 9)$

22. $f(x) = (7x^4 - 8x^2 + 144)(5x^3 + 25)$

23. $f(x) = 5x^{2/3}(5x^{-1} + 3x)$

24. $f(x) = 3x^2(x^{-2} + x)$

25. $f(x) = 6\sqrt{x}(2x^2 - 5)$

26. $f(x) = 2(\sqrt{x} + 3x)(\sqrt{x} - x)$

27. $g(x) = \dfrac{2x}{5x^2 - 11x + 3}$

28. $g(x) = \dfrac{5x}{125 - 25x - 3x^2}$

Find the slope of the tangent to the graph of f at the x given in Problems 29–34.

29. $f(x) = \dfrac{5x^2 + 5x}{x - 5}$ at $x = 1$

30. $f(x) = \dfrac{x + 2}{4x - 7x^2}$ at $x = -1$

31. $f(x) = x^{1/2}(x^2 + 3)$ at $x = 4$

32. $f(x) = x^{2/3}(x^2 - 5)$ at $x = 8$

33. $f(x) = \dfrac{x^2 - 5x + 1}{x^2 + 4x + 4}$ at $x = 0$

34. $f(x) = \dfrac{x^3 + 1}{x^3 - 1}$ at $x = 0$

Find an equation of the tangent line to the graph of f at the given value of x in Problems 35–40.

35. $f(x) = \dfrac{5x^2 + 5x}{x - 5}$ at $x = 0$

36. $f(x) = \dfrac{x + 2}{4x - 7x^2}$ at $x = 1$

37. $f(x) = x^{1/2}(x^2 + 3)$ at $x = 4$

38. $f(x) = x^{2/3}(x^2 - 5)$ at $x = 8$

39. $f(x) = \dfrac{x^2 - 5x + 1}{x^2 + 4x + 4}$ at $x = 1$

40. $f(x) = \dfrac{x^3 + 1}{x^3 - 1}$ at $x = 0$

APPLICATIONS

41. Suppose the demand, D, for a commodity in a free market decreases as the price x increases according to the formula

$$D(x) = \frac{100{,}000}{x^2 + 15x + 25} \quad \text{for } x \text{ between } \$10 \text{ and } \$20$$

Find the rate of change of demand with respect to the price change.

42. The price of an item is governed by the formula

$$P(x) = 1500 - \frac{575}{x}$$

Find the rate of change of the price when the demand is $x = 20$. Is the price increasing or decreasing at that point?

43. A drug's effect in the bloodstream can be plotted as a function of the time, t, since it was administered according to the formula

$$f(t) = \frac{t}{0.01 + 0.005t}$$

Find the rate at which the drug's effect is changing relative to time t.

44. Learning a certain task can be measured by counting the number of items, N, learned by a subject in a given amount of time, t, according to the formula

$$N(t) = \frac{\sqrt{t}}{2t^2 + 1}$$

Find the rate of learning after 2 hours.

9.7 The Chain Rule

We now come to the last of the differentiation techniques. These techniques enable us to efficiently find the derivative of almost every function we will encounter. Remember, however, as you progress through this or other courses, that if you encounter a function that does not "fit" any of the differentiation formulas we have developed, you can resort to using the derivative definition.

Consider a function $y = (3x + 1)^2$. How would you find the derivative?

Method I: Expand

$$y = (3x + 1)^2 = 9x^2 + 6x + 1$$
$$y' = 18x + 6 = 6(3x + 1)$$

Method II: Product rule:

$$y = (3x + 1)^2 = (3x + 1)(3x + 1)$$
$$y' = (3x + 1)(3x + 1)' + (3x + 1)'(3x + 1)$$
$$= (3x + 1)(3) + 3(3x + 1) = 6(3x + 1)$$

Method III: Power rule:

$$y = (3x + 1)^2 = u^2 \qquad \text{where} \qquad u = 3x + 1$$

$y' = 2u$ using the power rule, *but this does not agree with the results of methods I and II!* $2u = 2(3x + 1) \neq 6(3x + 1)$

Warning!

There appears to be a "missing factor" of 3 in this result; notice that $u' = 3$ and that the missing factor is precisely the derivative of u are no coincidence. Every time we find the derivative by using the power rule we will need to include a **correction factor**.

$$y' = 2u \cdot u' = 2u(3)$$
$$= 6u \qquad \text{or} \qquad 6(3x + 1)$$

Your reaction might well be, why bother with method III? But suppose that the power were 3, 4, or 5 rather than 2. Expanding and using the product rule is not practical for large powers. With method III and a correction factor:

Derivative
Correction factor

Function	Power rule	$\downarrow$
$(4x + 1)^2$	$2(4x + 1) \cdot$	$4 = 8(4x + 1)$
$(5x + 1)^2$	$2(5x + 1) \cdot$	$5 = 10(5x + 1)$
$(6x + 1)^3$	$3(6x + 1)^2 \cdot$	$6 = 18(6x + 1)^2$
$(2 - 7x)^3$	$3(2 - 7x)^2 \cdot (-7)$	$= -21(2 - 7x)^2$
$(5 - 5x)^9$	$9(5 - 5x)^8 \cdot (-5)$	$= -45(5 - 5x)^8$

These results lead us to the following general result:

Generalized Power Rule

> Let u be some differentiable function of x and let $y = u^n$. Then, for any real number n,
>
> $$y' = nu^{n-1} \cdot u'$$

EXAMPLE 1 Let $y = (2x^2 + 3x - 5)^3$. Find y'.

Solution $y' = \underbrace{3(2x^2 + 3x - 5)^2}(4x + 3)$

Correction factor is $u' = 4x + 3$

Power rule where $u = 2x^2 + 3x - 5$

$= 3(4x + 3)(2x^2 + 3x - 5)^2$ Answers may be left in factored form ∎

EXAMPLE 2 Let $y = 15\sqrt{3x^2 + x}$. Find dy/dx and simplify.

Solution $\dfrac{dy}{dx} = 15\underbrace{\left(\dfrac{1}{2}\right)(3x^2 + x)^{-1/2}}\underbrace{(6x + 1)}$

Correction factor is $u' = 6x + 1$

Power rule where $u = 3x^2 + x$

$$\frac{dy}{dx} = \frac{15(6x + 1)}{2\sqrt{3x^2 + x}}$$ ∎

The product and quotient rules can also be combined along with the power rule.

EXAMPLE 3 Let $y = (x^2 + 5x - 8)(5x + 2)^3$. Find y' and simplify.

Solution This is a product, so begin with the product rule:

$$y' = (x^2 + 5x - 8)[(5x + 2)^3]' + (x^2 + 5x - 8)'(5x + 2)^3$$
$$= (x^2 + 5x - 8)(3)(5x + 2)^2(5) + (2x + 5)(5x + 2)^3$$

Correction factor for $5x + 2$

$$= (5x + 2)^2[15(x^2 + 5x - 8) + (2x + 5)(5x + 2)]$$
$$= (5x + 2)^2(15x^2 + 75x - 120 + 10x^2 + 29x + 10)$$
$$= (5x + 2)^2(25x^2 + 104x - 110)$$ ∎

EXAMPLE 4 Let $y = \dfrac{(5x - 3)^4}{3 - 8x}$. Find dy/dx and simplify.

Solution This is a quotient, so begin with the quotient rule:

$$\frac{dy}{dx} = \frac{(3 - 8x)4(5x - 3)^3(5) - (5x - 3)^4(-8)}{(3 - 8x)^2}$$ Did you notice the correction factor?

$$= \frac{20(3 - 8x)(5x - 3)^3 + 8(5x - 3)^4}{(3 - 8x)^2}$$

$$= \frac{4(5x - 3)^3[5(3 - 8x) + 2(5x - 3)]}{(3 - 8x)^2}$$ Common factor

$$= \frac{4(5x - 3)^3[15 - 40x + 10x - 6]}{(3 - 8x)^2}$$

$$= \frac{4(5x - 3)^3(9 - 30x)}{(3 - 8x)^2} = \frac{12(3 - 10x)(5x - 3)^3}{(3 - 8x)^2}$$ ∎

EXAMPLE 5   Let $y = [(2x - 1)(5x^3 + 2)]^9$. Find dy/dx.

Solution   This is a power, so begin with the power rule:

$$y' = 9[(2x - 1)(5x^3 + 2)]^8[(2x - 1)(5x^3 + 2)]'$$
$$= 9[(2x - 1)(5x^3 + 2)]^8[(2x - 1)(15x^2) + (2)(5x^3 + 2)] \quad \leftarrow \text{Product rule}$$
$$= 9[(2x - 1)(5x^3 + 2)]^8[30x^3 - 15x^2 + 10x^3 + 4] \qquad \text{on the}$$
$$= 9[(2x - 1)(5x^3 + 2)]^8[40x^3 - 15x^2 + 4] \qquad \text{correction}$$
$$= 9(40x^3 - 15x^2 + 4)[(2x - 1)(5x^3 + 2)]^8 \qquad \text{factor} \quad \blacksquare$$

The generalized power rule is a special case of a result called the **chain rule**.

Chain Rule

> Suppose y is a differentiable function of u and u, in turn, is a differentiable function of x. Then
> $$\frac{dy}{dx} = \frac{dy}{du} \cdot \frac{du}{dx}$$

All the examples we can give at this point could also be solved using the chain rule. For example, if $y = u^6$ and $u = 2x^3 - 5x$, then

$$\frac{dy}{dx} = \frac{dy}{du} \cdot \frac{du}{dx}$$
$$= 6u^5(6x^2 - 5)$$
$$= 6(2x^3 - 5x)^5(6x^2 - 5)$$

EXAMPLE 6   Suppose $y = 2u^5$ and $u = 2x^3 - 5x$. Find dy/dx.

Solution   First,

$$\frac{dy}{du} = 10u^4 \qquad \text{and} \qquad \frac{du}{dx} = 6x^2 - 5$$

Now, use the chain rule:

$$\frac{dy}{dx} = \frac{dy}{du} \cdot \frac{du}{dx}$$
$$= 10u^4(6x^2 - 5)$$
$$= 10(2x^3 - 5x)^4(6x^2 - 5) \qquad \text{Write } u \text{ in terms of } x \text{ by substitution} \quad \blacksquare$$

Problem Set 9.7

Find the derivative of the functions in Problems 1–34 and simplify.

1. $f(x) = (3x + 2)^3$

2. $f(x) = (6x + 5)^3$

3. $f(x) = (5x - 1)^4$

4. $f(x) = (4 - 2x)^5$

5. $y = (2x^2 + x)^3$

6. $y = (3x^2 - 2x)^3$

7. $y = (2x^2 - 3x + 2)^2$

8. $y = (3x^2 + 5x - 1)^2$

9. $g(x) = (x^3 + 5x)^4$

10. $g(x) = (2x^3 - 7x^2)^4$

11. $m(x) = (2x^2 - 5x)^{-2}$

12. $m(x) = (6x^2 - 3x)^{-3}$

13. $t(x) = (x^4 + 3x^3)^{-1}$

14. $t(x) = (5x^3 - 2x^2)^{-2}$

15. $y = 5(4x^3 + 3x^2)^3$

16. $y = 6(x^3 + 4x^2)^4$

17. $y = (x^2 - 3x)^{1/4}$

18. $y = (2x^2 + 5x)^{2/3}$

19. $y = \sqrt{x^2 + 16}$

20. $y = \sqrt{x^2 - 25}$

21. $y = 5\sqrt{x^3 + 8}$

22. $y = 9\sqrt{x^4 + 3x}$

23. $y = x\sqrt{3x + 1}$

24. $y = 2x\sqrt{5x - 3}$

25. $f(x) = (2x+1)^2(3x+2)^3$

26. $f(x) = (1 - 3x)^3(4 - x)^2$

27. $y = (5x + 1)^2(4x + 3)^{-1}$

28. $y = (9x + 2)^2(2x - 5)^{-1}$

29. $g(x) = \dfrac{(2x - 5)^2}{5x + 3}$

30. $g(x) = \dfrac{(7x + 11)^3}{x^2 + 3}$

31. $t(x) = \dfrac{(x + 5)^4}{(2x - 5)^2}$

32. $t(x) = \dfrac{(x^2 - 3x + 1)^2}{(5x + 2)^2}$

33. $f(x) = \dfrac{3x}{\sqrt{x^2 + 1}}$

34. $f(x) = \dfrac{-2x}{\sqrt{4 - x^2}}$

APPLICATIONS

35. A colony of bacteria that has been sprayed by a bacterial agent in a controlled situation can be measured in terms of N, the number of viable bacteria remaining after t minutes, according to the formula

$$N(t) = 10^9(50 - 2t)^3$$

What is the rate of decrease at the end of 10 minutes?

36. A psychologist finds that after t minutes $(0 \le t \le 10)$ a person is able to score

$$s(t) = \frac{100t^2}{(2t + 10)^2}$$

on a performance test. Is the person's score increasing or decreasing at time $t = 10$ minutes?

37. Prove the quotient rule.

9.8 Summary and Review

IMPORTANT TERMS

Average rate of change   [9.3]
Chain rule   [9.7]
Constant rule   [9.5]
Constant times a function rule   [9.5]
Continuity   [9.2]
Correction factor   [9.7]
Derivative   [9.3, 9.4]
Difference rule   [9.5]
Discontinuous function   [9.2]
Generalized power rule   [9.7]
Indeterminate form   [9.1]
Instantaneous rate of change   [9.3]

Limit   [9.1]
Marginal cost   [9.4]
Marginal profit   [9.3]
Power rule   [9.5]
Product rule   [9.6]
Quotient rule   [9.6]
Rationalizing the numerator   [9.1]
Secant line   [9.3]
Sum rule   [9.5]
Suspicious point   [9.2]
Tangent line   [9.3]

SAMPLE TEST

For additional practice there are a large number of review problems categorized by objective in the Student Solutions Manual. The following sample test (40 minutes) is intended to review the main ideas of this chapter.

Find the limits in Problems 1–4.

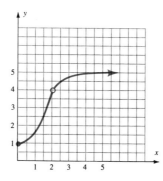

1. $\lim\limits_{x \to 2} f(x)$.

2. $\lim\limits_{x \to 2} \dfrac{2 - 5x + 2x^2}{x - 2}$.

3. $\lim\limits_{x \to 2} \dfrac{6x + 1}{3 - x}$.

4. $\lim\limits_{|x| \to \infty} \dfrac{5x^2 - 3x + 1}{8x^2 - 5x + 4}$.

5. Find the point(s) of discontinuity, if any.

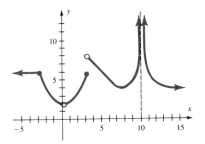

6. Find the point(s) of discontinuity, if any.

$$f(x) = \begin{cases} \dfrac{2x^2 - 7x - 15}{x - 5} & \text{if} \quad -5 \le x \le 5, x \ne 5 \\ 13 & \text{if} \quad x = 5 \end{cases}$$

7. Find the average rate of change for the function $y = 5 + x - 2x^2$ for $x = 1$ to $x = 3$.

8. Find the instantaneous rate of change for the function in Problem 7 at $x = 1$.

9. State the definition of derivative.

10. Use the definition of derivative to find the derivative of $y = 1/x$.

Find dy/dx for the functions in Problems 11–18.

11. $y = \frac{29}{125}$

12. $y = 3x^2 - 5x + 3$

13. $y = 5x^2 - 2x^{-1}$

14. $y = 2(x - 5)(3x + 2)$

15. $y = \dfrac{15}{4x - 5}$

16. $y = (3x^2 - 5x)^{1/4}$

17. $y = \dfrac{(x - 5)^3}{1 - 2x}$

18. $y = \dfrac{5x^3}{\sqrt{x - 5}}$

19. Suppose that the profit, P, for a manufacturer is a function of the number of units produced and behaves according to the model

$$P = \frac{1000 - x^2}{100 - x}$$

If the current level of production is 20 units, what is the per unit change in profit if production is increased from 20 to 25 units?

20. Find the marginal profit for the function given in Problem 19.

Modeling Application 6

Instantaneous Acceleration: A Case Study of the Mazda 626

Consider the performance of the 1982 Mazda 626 Sport Coup, as shown in the graph below:*

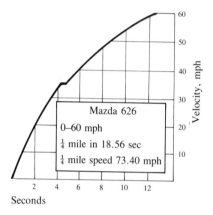

Write a paper discussing the average rate of travel, the distance traveled, the velocity (the instantaneous rate of travel), and the acceleration.** For general guidelines about writing this essay, see Modeling Application 1 on page 130.

---

* From *Time*, January 4, 1982.

** This modeling application is adapted from Peter A. Lindstrom, "A Beginning Calculus Project," UMAP, Vol. 5, No. 3, pp. 271–276.

CHAPTER 10

Additional Derivative Topics

CHAPTER CONTENTS

10.1 Implicit Differentiation

10.2 Differentials

10.3 Business Models Using Differentiation

10.4 Related Rates

10.5 Summary and Review
Important Terms
Sample Test

APPLICATIONS

Management (*Business, Economics, Finance, and Investments*)

Rate of change of price with respect to number of items (10.1, Problems 28–29)
Relationship between sales and advertising cost (10.2, Problem 31)
Changes in revenue and profit (10.2, Problems 37–38)
Demand equation (10.3, Problem 1)
Revenue, cost, or profit functions (10.3, Problems 2–4, 10)
Marginal revenue, cost, or profit (10.3, Problems 5–8, 11, 13–15, 19–20)
Average marginal revenue, cost, or profit (10.3, Problems 9, 12 and 16)
Profit and loss (10.3, Problem 17)
Rate of change of profit (10.4, Problems 10–11)
Positioning of a robot arm in an assembly-line conveyor belt (10.4, Problem 13)
Batching process (10.4, Problems 14–15)
Rate of change of wholesale price of apples (10.4, Problem 19)
Maximize monthly revenue from an apartment complex (10.5, Problem 9)

Life sciences (*Biology, Ecology, Health, and Medicine*)

Concentration of alcohol in the bloodstream (10.2, Problem 33)
Average adult pulse rate as a function of a person's height (10.2, Problem 34)
Area of a circular oil slick (10.2, Problem 35)
Area of a dilated pupil (10.2, Problem 36)
The level of carbon monoxide in a city (10.4, Problem 16)
The rate of rabies as a function of the skunk population (10.4, Problem 20)
Treatment of a stomach disorder (10.4, Problems 21–22)
Cost-benefit model for pollution (10.5, Problem 10)

Social sciences (*Demography, Political Science, Population, Psychology, Society, and Sociology*)

Effect of advertising on voting (10.2, Problem 39)
Effect on learning Spanish vocabulary by changing study time (10.2, Problem 40)

General interest

Seismologist application with concentric circles (10.1, Problem 30)
Distance of one car from another (10.4, Problem 12)
Ripples in a pond (10.4, Problems 17–18)

Modeling application—Publishing: An Economic Model

CHAPTER OVERVIEW   This chapter enhances and amplifies the idea of the derivative. The discussion should strengthen your understanding of the concept of the derivative by emphasizing that it is a rate of change.

PREVIEW   First we are introduced to finding a derivative without actually solving for y; this is called *implicit differentiation*. Then meaning is given to dy and dx in the concept called a *differential*. Differentials, derivatives, and implicit differentiation are used in building some business models. The chapter closes by relating derivatives and rates.

PERSPECTIVE   The concept of a derivative is such an important idea in mathematics that we need time to develop an appreciation of some of the many different places it can be used. This chapter forms a bridge between the definition and manipulation involved in finding a derivative (developed in the last chapter) and the applications of derivatives (introduced in the next chapter).

10.1 Implicit Differentiation

Sometimes we are given a function or an equation in terms of two or more variables, say, x and y. In order to use the derivative formulas developed in Chapter 9 it is necessary to **explicitly** solve for y. For example, consider the equation

$$3x^2 - 2x + 5y - 10 = 0$$

Find the slope of this curve at the point $(4, -6)$. To do this, we need to find the derivative. We solve for y:

$$5y = -3x^2 + 2x + 10$$

$$y = -\frac{3}{5}x^2 + \frac{2}{5}x + 2$$

Therefore

$$y' = -\frac{3}{5}(2)x + \frac{2}{5} = -\frac{6}{5}x + \frac{2}{5}$$

Notice we used y' (which is the same as dy/dx) to denote the derivative. The value at the point $(4, -6)$ is found by substitution of the x and y values ($x = 4, y = -6$ in this example):

$$y' = -\frac{6}{5}(4) + \frac{2}{5} = -\frac{22}{5}$$

Another way of finding the derivative is to use the chain rule to find it **implicitly**; that is, without first solving for y. For this example,

$$3x^2 - 2x + 5y - 10 = 0$$

Think of y as a function of x so that

$$D_x x = 1 \qquad \text{and} \qquad D_x y = y'$$

Now differentiate both sides (with respect to x) to obtain

$$D_x(3x^2 - 2x + 5y - 10) = D_x(0)$$
$$D_x(3x^2) + D_x(-2x) + D_x(5y) + D_x(-10) = D_x(0)$$
$$6x - 2 + 5y' = 0$$

Solve for y':

$$5y' = -6x + 2$$
$$y' = -\frac{6}{5}x + \frac{2}{5}$$

We can now find the slope at $(4, -6)$ as before. You might ask, why would we want to find the derivative implicitly? Why not always solve for y and then find the derivative? Sometimes it is inconvenient, or even impossible to solve for y, but nevertheless we can find the derivative. The idea is to take the derivative of both sides of the equation, thus treating y as a function of x and using the chain rule. Remember, the derivative of x with respect to x is 1, but the derivative of y with respect to x is y'.

EXAMPLE 1   Find y' where $x^5 - y^5 = 211$.

Solution   Take the derivative of both sides with respect to x:

$$5x^4 - \underbrace{5y^4 \cdot y'}_{} = 0$$

Derivative of 211 is 0

Chain rule requires correction factor of y'

Now, solve for y':

$$5x^4 = 5y^4y'$$
$$y' = \frac{x^4}{y^4}$$

∎

Note that y' is stated in terms of both x and y, so if you were asked for the slope of the curve at some point you would need to know both components of the point, not just the x value when we found y' explicitly. In Example 1, the slope of $x^5 - y^5 = 211$ at $(3, 2)$ is

$$y' = \frac{x^4}{y^4} = \frac{81}{16}$$

EXAMPLE 2   Find y' from Example 1 explicitly.

Solution
$$x^5 - y^5 = 211$$
$$y^5 = x^5 - 211$$
$$y = (x^5 - 211)^{1/5}$$
$$y' = \frac{1}{5}(x^5 - 211)^{-4/5}(5x^4) = \frac{x^4}{(x^5 - 211)^{4/5}}$$

∎

Note that this result agrees with

$$y' = \frac{x^4}{y^4}$$

from Example 1 since

$$y = (x^5 - 211)^{1/5} \quad \text{and} \quad y^4 = (x^5 - 211)^{4/5}$$

Warning Remember that all the usual derivative procedures apply when doing implicit differentiation, especially when using the product rule. Consider Example 3 carefully.

EXAMPLE 3 If $x^2 y^3 = 1$, find y'.

Solution We use implicit differentiation; do not forget the product rule:

$$x^2(y^3)' + (x^2)'(y^3) = 0$$
$$x^2(3y^2 \cdot y') + (2x)(y^3) = 0$$
$$3x^2 y^2 y' + 2xy^3 = 0$$
$$3x^2 y^2 y' = -2xy^3$$
$$y' = \frac{-2xy^3}{3x^2 y^2}$$
$$= \frac{-2y}{3x}$$

■

EXAMPLE 4 Find the line tangent to $x^2 + xy + y^2 - 7 = 0$ at $(1, 2)$.

Solution First find the slope of the curve at $(1, 2)$, which means find the derivative at that point. Carry out implicit differentiation:

$$2x + xy' + y + 2yy' + 0 = 0$$
$$xy' + 2yy' = -2x - y$$
$$y'(x + 2y) = -2x - y$$
$$y' = \frac{-2x - y}{x + 2y}$$

At $(1, 2)$:

$$y' = \frac{-2(1) - 2}{1 + 2(2)} = \frac{-4}{5}$$

Finally, the equation of the tangent line is found by using the point-slope form:

$$y - y_1 = m(x - x_1)$$
$$y - 2 = \frac{-4}{5}(x - 1)$$
$$5y - 10 = -4x + 4$$
$$4x + 5y - 14 = 0$$

■

Problem Set 10.1

Suppose that x and y are related by the equations in Problems 1–16 and use implicit differentiation to find dy/dx.

1. $y + 5x^2 + 12 = 0$
2. $3x^3 - y + 15 = 0$
3. $xy = 5$
4. $5xy = x^2$
5. $x^2 - 3xy = 50$
6. $10x = xy$
7. $x^2 + y^2 = 4$
8. $x^3 - y^3 = 9$
9. $2x^4 - 5y^3 = 7$
10. $(2x + 1)^2 + (3y - 5)^2 = 41$
11. $x^2 - xy + y^2 = 1$
12. $x^5y^3 = -1$
13. $3x^2y^3 - 3xy^2 + 5xy = 2$
14. $5x^5y^2 - 5xy^2 - 8x = 100$
15. $x^3 + 2x^2 + xy - 4 = 0$
16. $x^3 - 3x^2 + 2xy + 3 = 0$

Find the standard form of the equation of the line tangent to the curves in Problems 17–27 at the indicated point.

17. $x^2 + y^2 - 4 = 0$ at $(0, 2)$
18. $x^3 - y^3 - 9 = 0$ at $(2, -1)$
19. $2x^4 - 5y^3 - 7 = 0$ at $(-1, -1)$
20. $(2x + 1)^2 + (3y - 5)^2 = 41$ at $(2, 3)$
21. $x^2 + xy + y^2 - 7 = 0$ at $(1, -3)$
22. $x^2 + xy + y^2 - 12 = 0$ at $(2, -4)$
23. $(x - 2)^2 + (y - 1)^2 = 9$ at $x = 2$
24. $(x - 3)^2 + (y - 1)^2 = 16$ at $x = 3$
25. $x^2 - 2xy + y^2 = 0$ at $x = -1$
26. $\dfrac{(x - 1)^2}{9} + \dfrac{(y + 1)^2}{16} = 1$ at $x = 1$
27. $\dfrac{(x + 1)^2}{25} + \dfrac{(y - 1)^2}{4} = 1$ at $x = -1$

APPLICATIONS

28. Find the rate of change of the price, p, in dollars, with respect to the number of items, x, if
$$x = \sqrt{5000 - p^2}$$

29. Find the rate of change of the price, p, in dollars, with respect to the number of items, x, if
$$x = p^2 - 5p + 500$$

30. Seismologists sometimes use equations of circles and their relationships to study the concentric circles around seismological stations. Show that the circles with the equations
$$x^2 + y^2 - 12x - 6y + 25 = 0 \quad \text{and}$$
$$x^2 + y^2 + 2x + y - 10 = 0$$
have tangent lines with the same slope at $x = 2$.

10.2 Differentials

Suppose we consider dx and dy from the derivative dy/dx as two separate quantities. In Chapter 2 we used the symbols Δx and Δy to mean the change in x and the change in y, as follows. Let (x_1, y_1) and (x_2, y_2) be any two points on some curve: then

$$\Delta x = x_2 - x_1 \quad \text{and} \quad \Delta y = y_2 - y_1$$

Now we will let x be fixed and *define dx*, called the **differential of x**, to be an independent variable equal to the change in x. That is, we define dx to be Δx. Then, if f is differentiable at x, we *define dy*, called the **differential of y**, according to the formula

$$dy = f'(x)\,dx$$

Thus, if $dx \neq 0$, then

$$\frac{dy}{dx} = f'(x)$$

There is a very clear geometrical representation of what we have done here, shown in Figure 10.1. Note that $dx = \Delta x$ (the change in x), Δy = the change in y that occurs for a change of Δx, and dy = the rise of the tangent line relative to $\Delta x = dx$.

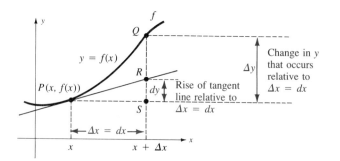

Figure 10.1 Geometrical definition of dx and dy

EXAMPLE 1 Use Figure 10.1 to describe Δx, Δy, dx, and dy.

Solution Since $\Delta x = dx$, we see that these are the length of the segment PS; these quantities represent the change in x. The number Δy represents the change in y, which is the length of the segment QS. Finally, the distance dy is defined to be the rise of the tangent line at the point P, so dy is the length of the segment RS. ∎

EXAMPLE 2 Find Δy and dy for $f(x) = 6x - 2x^2$ when $x = 2$ and $\Delta x = 0.1$.

Solution For this example, $y = f(x)$, so

$$x_1 = 2 \quad \text{and} \quad x_2 = x_1 + \Delta x = 2 + 0.1 = 2.1$$
$$y_1 = f(x_1) = f(2) = 6(2) - 2(2)^2 = 4$$
$$y_2 = f(x_2) = f(2.1) = 6(2.1) - 2(2.1)^2 = 3.78$$

Now

$$dy = f'(x)\,dx$$
$$= (6 - 4x)(0.1) \qquad \text{Since } f'(x) = 6 - 4x \text{ and } \Delta x = 0.1$$

When $x = 2$,

$$dy = [6 - 4(2)](0.1) = -0.2$$

To find Δy we know that

$$\Delta x = x_2 - x_1 \quad \text{and} \quad x_2 = x_1 + \Delta x$$

so that

$$\Delta y = f(x_2) - f(x_1)$$
$$= f(x_1 + \Delta x) - f(x_1) \qquad \text{Substitute}$$
$$= f(2 + 0.1) - f(2)$$
$$= f(2.1) - f(2)$$
$$= 3.78 - 4$$
$$= -0.22$$

∎

Approximations Using Differentials

Note in Example 2 that the numerical values for Δy and dy are almost the same. Since it is generally easier to calculate dy than Δy, it is often convenient to approximate a change in y by using dy. Example 3 illustrates this idea.

EXAMPLE 3 Suppose the demand for a certain product is a function of its price, x, and can be specified according to the formula

$$d(x) = -3x^3 - 2x^2 + 1000$$

How would the demand change as the price changes from \$2 to \$2.10?

Solution The question asks for Δy when $x = 2$ and $\Delta x = 0.10$. Since

$$\Delta y = f(x + \Delta x) - f(x)$$

you want to find

$$\begin{aligned}
\Delta y &= f(2.10) - f(2) \\
&= [-3(2.10)^3 - 2(2.10)^2 + 1000] - [-3(2)^3 - 2(2)^2 + 1000] \\
&= 963.397 - 968 \qquad \text{By calculator} \\
&= -4.603
\end{aligned}$$

This means that you would expect the demand to go down about 5 units. Instead of doing this calculation, however, you might approximate Δy by using dy:

$$\begin{aligned}
dy &= f'(x)\,dx \\
&= (-9x^2 - 4x)\,dx
\end{aligned}$$

Now substitute the values $x = 2$ and $\Delta x = 0.10$:

$$\begin{aligned}
dy &= [-9(2)^2 - 4(2)](0.10) \\
&= -4.4
\end{aligned}$$

You would estimate that the demand would decrease by four units. ∎

In Example 3 the error introduced by using dy instead of Δy is one unit, but there is a lot less "work" involved. Closer examination shows that the error is only about 0.203. Example 4 shows how we can estimate the amount of error introduced in an approximation.

EXAMPLE 4 What is the error introduced by using dy instead of Δy in Example 3 for an increase in price from 2 to $2 + h$ dollars?

Solution For an increase from 2 to $2 + h$,

$$\begin{aligned}
\Delta y &= f(x + \Delta x) - f(x) \\
&= [-3(2 + h)^3 - 2(2 + h)^2 + 1000] - [-3(2)^3 - 2(2)^2 + 1000]
\end{aligned}$$

This is a formidable calculation (the details are left for you). After simplification we find that

$$\Delta y = -44h - 20h^2 - 3h^3$$

However, if we approximate this by dy, we see that

$$dy = f'(x)\,dx$$
$$= (-9x^2 - 4x)\,dx$$

At $x = 2$ and $dx = h$,

$$dy = [-9(2)^2 - 4(2)]h = -44h$$

By comparing these two calculations we see that the error introduced by using dy rather than Δy is

$$-20h^2 - 3h^3$$

If h is small (such as 0.1), we see that the error will be small ($-.203$ in this case). ■

These examples lead us to the following summary:

Differential Approximation

> If $f'(x)$ exists, then, for small Δx,
> $$\Delta y \approx dy$$
> and
> $$f(x + \Delta x) = f(x) + \Delta y$$
> $$\approx f(x) + dy$$
> $$= f(x) + f'(x)\,dx$$

EXAMPLE 5 Colortex manufactures pigments for tinting various paints. Its profit function is given by

$$P(x) = 10x - \frac{x^2}{1000} - 3000$$

where x is the number of tubes of pigment produced. What is the expected change in profit if production is changed from 2000 to 2010 tubes of pigment?

Solution The change in profit is ΔP and the change in production is Δx. The change in production, $\Delta x = 10$, is given, and since it is relatively small we will approximate ΔP by dP:

$$dP = P'(x)\,dx$$
$$= \left(10 - \frac{2x}{1000}\right)dx$$
$$= (10 - 0.002x)\,dx$$

Since $\Delta x = dx = 10$ at a production level of 2000, we have

$$dP = [10 - 0.002(2000)](10) = 60$$

This means that we would expect the profit to increase by $60 if production were increased by 10 tubes of pigment. ■

Differential Formulas

Each of the derivative rules derived in the last chapter has a corresponding differential form. For example, the generalized power rule states that

$$\text{If} \quad y = u^n \quad \text{then} \quad \frac{dy}{dx} = nu^{n-1} \cdot \frac{du}{dx}$$

By multiplying both sides by dx we obtain a differential form of this same derivative formula:

$$\text{If} \quad y = u^n \quad \text{then} \quad dy = nu^{n-1} \cdot du$$

All of the derivative rules previously discussed are now summarized in differential form:

Differential Rules

If u and v are differentiable functions and c is a constant, then

CONSTANT RULE: $dc = 0$

POWER RULE: $du^n = nu^{n-1} \cdot du$

SUM RULE: $d(u + v) = du + dv$

DIFFERENCE RULE: $d(u - v) = du - dv$

PRODUCT RULE: $d(uv) = u\,dv + v\,du$

QUOTIENT RULE: $d\left(\dfrac{u}{v}\right) = \dfrac{v\,du - u\,dv}{v^2}$

EXAMPLE 6   **a.** If $y = \sqrt{x}$, then $\dfrac{dy}{dx} = \dfrac{1}{2}x^{-1/2}$, so, in differential form,

$$dy = \frac{1}{2\sqrt{x}}\,dx$$

b. If $u = x^2 + 1$, then $\dfrac{du}{dx} = 2x$, so, in differential form,

$$du = 2x\,dx$$

∎

Problem Set 10.2

Find dy for the functions in Problems 1–20.

1. $y = 5x^3$

2. $y = 35x^4$

3. $y = 10x^{-1}$

4. $y = 100x^{-2}$

5. $y = 100x^3 - 50x + 10$

6. $y = 500x^5 - 40x^2 + 5000$

7. $y = 3\sqrt{x-1}$

8. $y = 5\sqrt{x^2 + 4}$

9. $y = 5x$

10. $y = 6x$

11. $y = (5x - 3)^2(x^2 - 3)$

12. $y = (2x^2 + 1)(3x - 1)^3$

13. $y = \dfrac{3x + 1}{x - 2}$

14. $y = \dfrac{5x + 2}{4x + 3}$

15. $y = x^2\left(1 - \dfrac{1}{x} + \dfrac{3}{x^2}\right)$

16. $y = x^3\left(1 - \dfrac{5}{x} + \dfrac{10}{x^2}\right)$

17. $y = \left(5 - \dfrac{1}{x^2}\right)\left(1 + \dfrac{1}{x}\right)$

18. $y = \left(6 - \dfrac{10}{x^3}\right)\left(5x - \dfrac{1}{x}\right)$

19. $y = \dfrac{x^2 - 5x + 1}{x^2 + 3x - 2}$

20. $y = \dfrac{2x^2 + 3x - 5}{3x^2 - 2x + 1}$

Evaluate dy and Δy for the values in Problems 21–24.

21. $y = f(x) = x^2 - 2x + 5$, $x = 10, \Delta x = 0.1$

22. $y = f(x) = 2x^2 - 3x$, $x = 5, \Delta x = 0.2$

23. $y = f(x) = \sqrt{5x}$, $x = 5$, $\Delta x = 0.15$

24. $y = f(x) = \sqrt{3x - 2}$, $x = 100$, $\Delta x = 2$

Estimate Δy by using dy in Problems 25–30.

25. $y = \dfrac{2x - 3}{5x + 2}$, $x = 100$, $\Delta x = 3$

26. $y = \dfrac{x^2 + 1}{x - 3}$, $x = 30$, $\Delta x = 1$

27. $y = 20\left(3 - \dfrac{1}{x^2}\right)$, $x = 10$, $\Delta x = 0.02$

28. $y = \dfrac{1 + 3x}{\sqrt{x}}$, $x = 20$, $\Delta x = 0.1$

29. $y = \dfrac{450(x + 300)}{\sqrt{x - 50}}$, $x = 1000$, $\Delta x = 10$

30. $y = \dfrac{1000(20 - 30x)}{\sqrt{x^2 + 20}}$, $x = 50$, $\Delta x = 5$

Use differential approximations in Problems 31–40.

APPLICATIONS

31. Suppose that sales, S, can be expressed as a function of advertising cost (c, in thousands of dollars) according to the formula

$$S(c) = 500c - c^2$$

Estimate the increase in sales that will result by increasing the advertising budget from $100,000 to $110,000.

32. The average cost (in dollars) to manufacture x items is

$$A(x) = 0.05x^3 + 0.1x^2 + 0.5x + 10$$

Approximate the change in the average cost as x changes from 10 to 11.

33. The concentration of alcohol in the bloodstream x hours after drinking 1 ounce of alcohol is approximately

$$A(x) = \dfrac{3x}{100 + x^2}$$

Estimate the change in concentration as x changes from 3 to 3.5.

34. The average adult pulse rate, b, in beats per minute can be expressed as a function of the person's height, h (in inches), according to the formula

$$b = \dfrac{600}{\sqrt{h}}$$

Approximate the change in pulse rate for a change in height from 72 to 73.5 inches.

35. Find the approximate increase in the area of a circular oil slick as its radius increases from 2 to 2.1 miles.

36. The pupil of a patient's eye is nearly circular and will dilate when the patient is given a certain drug. Estimate the increase in the area of a patient's pupil if the radius increases from 4 to 4.1 millimeters.

37. A company manufactures and sells x items per day. If the cost and revenue equations (in dollars) are

$$C(x) = 400 + 30x \qquad \text{and} \qquad R(x) = 60x$$

find the approximate changes in revenue and profit if production is increased from 100 to 110 items.

38. A company produces and sells x items per day. The cost and revenue equations (in dollars) are

$$C(x) = 3x^2 - 40x + 5000 \quad \text{and} \quad R(x) = 500x - \dfrac{x^2}{10}$$

Find the approximate changes in revenue and profit if production is increased from 1000 to 1010 items.

39. The number of people, N, expected to vote in the next election is a function of number of hours, x, of television advertising according to the formula

$$N(x) = 25,000 + \dfrac{x^2}{10} - \dfrac{x^3}{50,000}$$

Find the approximate change in N as x changes from 1000 to 1100 hours.

40. A student learns y Spanish vocabulary words in x hours according to the formula

$$y = 50\sqrt{x}$$

What is the approximate increase in the number of words learned as x changes from 3 to 3.5 hours?

10.3   Business Models Using Differentiation

The application of derivatives, that is, rates of change, to cost, revenue, and profit is an important component in the decision-making process for business executives and managers. The word *marginal* is used in business and economics to mean rate of change. Some of these ideas have appeared in examples in previous parts of this book, but for completeness and easy reference they are summarized below:

Marginal Analysis

Suppose that x is the number of units produced in some time interval and that

$$C(x) = \text{total cost} \quad \text{and} \quad R(x) = \text{total revenue}$$

then

$$P(x) = R(x) - C(x) = \text{TOTAL PROFIT}$$
$$C'(x) = \text{MARGINAL COST}$$
$$R'(x) = \text{MARGINAL REVENUE}$$
$$P'(x) = \text{MARGINAL PROFIT}$$

Marginal can be related to the definition of *derivative* as follows:

Marginal cost is the rate of change in cost per unit change in production at a given output level.

Marginal revenue is the rate of change in revenue per unit change in production at a given output level.

Marginal profit is the rate of change of profit per unit change in production at a given output level.

In practice, the smallest level of change in production is one unit. This was denoted in the last section by $\Delta x = dx = 1$. The actual changes in cost, revenue, and profit are denoted by ΔC, and ΔR, and ΔP, respectively, and are approximated by dC, dR, and dP. Recall from the last section that

$$dC = C'(x)\, dx,$$
$$dR = R'(x)\, dx, \text{ and}$$
$$dP = P'(x)\, dx$$

EXAMPLE 1 Suppose it is known that the cost (in dollars) for a product is given by the equation

$$C(n) = 18,500 + 8.45n$$

where n is the number of items sold. Find the marginal cost.

Solution The marginal cost is $C'(n)$:

$$C'(n) = 8.45$$

This means that it costs an additional $8.45 to produce one more item at all production levels. ∎

The marginal cost is not always a constant. If

$$C(n) = 500 + .045n^2$$

then $C'(n) = 0.09n$, which is a function of n. This means that the marginal cost depends on the production level of n units. For example, if present production is 1000 items and if we want to know the cost of producing one additional item, then $n = 1000$, $\Delta n = 1$, and

$$dC = C'(n)\,dn$$
$$= 0.09(1000)(1)$$
$$= 90$$

This means that to increase production by one unit would lead to an increase in cost of $90.

EXAMPLE 2 Suppose that the market demand for the product in Example 1 is linear and that it has been found that at a list price of $15 a company will sell 7500 items, but at a $25 price sales would drop to 2500 items. Also suppose that the net price is 80% of the selling price. Find the marginal revenue.

Solution Since it is assumed that the market demand is linear, you can find the demand equation:

$$m = \frac{7500 - 2500}{15 - 25}$$
$$= \frac{5000}{-10}$$
$$= -500$$

For the demand equation, use the form $y - y_1 = m(x - x_1)$, where the independent variable x is the price p and the dependent variable y is the number of items n. The demand equation is

$$n - 7500 = -500(p - 15)$$
$$= -500p + 7500$$
$$n = -500p + 15{,}000$$

Next, the revenue function, $R(p)$, is found by

$$\text{REVENUE} = (\text{NET COST})(\text{NUMBER SOLD})$$
$$R(p) = (0.8p)(n)$$
$$= 0.8p(-500p + 15{,}000) \qquad \text{Substitute } n = -500p + 15{,}000$$
$$= -400p^2 + 12{,}000p$$

Finally, the marginal revenue is the derivative of the revenue function, so

$$R'(p) = -800p + 12{,}000$$

This means that the marginal revenue is a function of the price of the item. At a price of $10, the marginal revenue is

$$R'(10) = -800(10) + 12{,}000$$
$$= 4000$$

A $1 increase in the $10 list price would increase revenue about $4,000. ∎

EXAMPLE 3 What are the approximate changes in revenue per unit change in production for the information given in Example 2 at production levels of 10,000, 8000, and 5000 items?

Solution From Example 2,

$$n = -500p + 15,000$$

$$\text{so } p = \frac{15,000 - n}{500}$$

$$\text{For } n = 10,000, \quad p = \frac{15,000 - 10,000}{500} = \$10$$

$$n = 8000, \quad p = \frac{15,000 - 8000}{500} = \$14$$

$$n = 5000, \quad p = \frac{15,000 - 5000}{500} = \$20$$

Therefore, since $R'(p) = -800p + 12,000$

$$R'(10) = \$4,000$$
$$R'(14) = \$800$$
$$R'(20) = -\$4,000$$

A \$1 increase in price would increase revenue by \$4,000 at a production level of 10,000 items and by \$800 at a production level of 8000 items, but at a production level of 5000 items the price increase would produce a *decrease* in revenue of \$4,000. ∎

The **profit function** P is found by

PROFIT = REVENUE − COST

We need to consider profit as a function of the number of items produced, but production is determined by price. Let us look again at the information in Examples 1–3. Profit P is a function of price p and $n = -500p + 15,000$, so from the previous discussion

$$\text{Revenue} = -400p^2 + 12,000p$$
$$\text{Cost} = 18,500 + 8.45n$$
$$= 18,500 + 8.45(-500p + 15,000)$$
$$= 145,250 - 4225p$$

Thus, since

Profit = Revenue − Cost

we have

$$P(p) = (-400p^2 + 12,000p) - (145,250 - 4225p)$$
$$= -400p^2 + 16,225p - 145,250$$

The marginal profit is

$$P'(p) = -800p + 16,225$$

Table 10.1 gives some values of this function.

TABLE 10.1
Marginal profit

List price	Production level	Marginal profit, $
10	10,000	8,225
12	9,000	6,625
14	8,000	5,025
16	7,000	3,425
18	6,000	1,825
20	5,000	225
22	4,000	−1,375
24	3,000	−2,975

Table 10.1 shows that at a production level of 10,000 items ($10 list), a $1 increase in price would increase profit by $8,225; at a production level of 5000 items ($20 list), a $1 increase in price would increase profit by $225; but at a production level of 4000 items ($22 list), a $1 increase in price would *decrease* profit by $1,375.

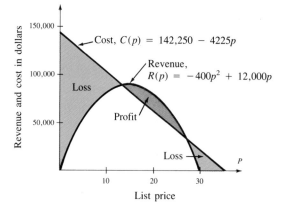

Figure 10.2 Cost, revenue, and profit functions

Sometimes marginal analysis is carried out relative to the average cost, average revenue, or average profit. *Average* here means *cost per unit*, *revenue per unit*, or *profit per unit*. The average of a number x is denoted by $\bar{x}$.

Average Marginal Analysis

If x is the number of units of a product produced in some time interval, then

$$\bar{C}(x) = \frac{C(x)}{x} = \text{cost per unit} = \text{AVERAGE TOTAL COST}$$

$$\bar{C}'(x) = \text{MARGINAL AVERAGE COST}$$

$$\bar{R}(x) = \frac{R(x)}{x} = \text{revenue per unit} = \text{AVERAGE TOTAL REVENUE}$$

$$\bar{R}'(x) = \text{MARGINAL AVERAGE REVENUE}$$

$$\bar{P}(x) = \frac{P(x)}{x} = \text{profit per unit} = \text{AVERAGE TOTAL PROFIT}$$

$$\bar{P}'(x) = \text{MARGINAL AVERAGE PROFIT}$$

EXAMPLE 4 Find the marginal average cost, revenue, and profit for the functions

$$C(n) = 20,000 + 5.45n$$
$$R(n) = -0.0016n^2 - 24n$$
$$P(n) = -0.0016n^2 - 32.45n - 142,250$$

Solution $\bar{C}(n) = \dfrac{C(n)}{n} = \dfrac{20,000 + 5.45n}{n}$

$$= 20,000n^{-1} + 5.45$$
$$\bar{C}'(n) = -20,000n^{-2}$$
$$\bar{R}(n) = \dfrac{R(n)}{n} = \dfrac{-0.0016n^2 - 24n}{n}$$
$$= -0.0016n - 24$$
$$\bar{R}'(n) = -0.0016$$
$$\bar{P}(n) = \dfrac{-0.0016n^2 - 32.45n - 142,250}{n}$$
$$= -0.0016n - 32.45 - 142,250n^{-1}$$
$$\bar{P}'(n) = -0.0016 + 142,250n^{-2}$$

Problem Set 10.3

APPLICATIONS

In Problems 1–7 suppose that the sales of a company are presently 10,000 items per year and that at a list price of $15 the company will sell 20,000 items but at a list price of $25 sales will drop to 5000 items. Also suppose that the net price is 70% of the selling price.

1. What is the demand equation?

2. What is the revenue function?

3. What is the cost function in terms of price if

$$C(n) = 15,000 + 8.5n$$

where n is the number of items sold?

4. What is the profit function in terms of price? (The cost function is given in Problem 3.)

5. Find and interpret the marginal revenue function.

6. Find and interpret the marginal cost function.

7. Find and interpret the marginal profit function.

Suppose the demand equation for a product is

$$x = 5000 - 25p$$

where x is the number of items produced and p is the price. Also suppose the cost equation is

$$C(x) = 50,000 + 50x$$

and that the net price (or net cost) is 70% of the selling price.

Use this information in Problems 8–17.

8. What is the marginal cost?

9. What is the marginal average cost?

10. What is the revenue equation in terms of x?

11. What is the marginal revenue?

12. What is the marginal average revenue?

13. Find $R'(1000)$ and $R'(2500)$ and interpret.

14. What is the marginal profit?

15. Find $P'(1000)$ and $P'(2500)$ and interpret.

16. What is the marginal average profit?

17. Graph the cost and revenue functions on the same coordinate system and show the regions of profit and loss.

18. The total cost to produce x items is

$$C(x) = 1000 + 5x - x^2 + x^3$$

Find the marginal cost.

19. The total revenue for x items is

$$R(x) = 5x - 0.001x^2$$

Find the marginal revenue.

20. The profit in dollars from the sale of x items is

$$P(x) = x^3 - 3x^2 + 5x + 10$$

Find the marginal profit.

10.4   Related Rates

The concept of derivative has been defined and interpreted. We now use the idea that the derivative represents a rate of change of one variable with respect to another variable. That is, we focus on the fact that

$$\frac{dy}{dx} = \text{the rate of change of } y \text{ with respect to } x$$

$$\frac{dP}{dx} = \text{the rate of change of } P \text{ with respect to } x$$

$$\frac{dz}{dt} = \text{the rate of change of } z \text{ with respect to } t$$

and so on.

If a formula is given as an equation, we know that we often solve that equation for an unknown value. However, sometimes the information we have at hand is the *rate* at which the variables are changing. We now explore how to use a given formula to find unknown quantities when we *know* the rates of change. For example, if the profit P is given by the equation

$$P = 75x - \frac{x^3}{50{,}000} - 100{,}000$$

for sales of x items, and if you know the rate at which the profit is changing *with respect to time*, you can transform this profit formula into one involving rates by using implicit differentiation and taking the derivative of both sides with respect to time:

$$\frac{dP}{dt} = \frac{d}{dt}\left(75x - \frac{x^3}{50{,}000} - 100{,}000\right)$$

$$= 75\frac{dx}{dt} - \frac{3}{50{,}000}x^2\frac{dx}{dt}$$

Notice that this formula now relates

$$\frac{dP}{dt} = \text{the rate of change of } P \text{ with respect to time } t$$

$$\frac{dx}{dt} = \text{the rate of change of } x \text{ with respect to time } t$$

Since the formula relates rates, the problem is called a **related rate** problem. When working a related rate problem you must distinguish between the general situation and the specific situation. The *general situation* comprises properties that are true at *every* instant of time, while the *specific situation* refers to those properties that are true only at the *particular* instant of time that the problem investigates. When working related rate problems you should first formulate a model that incorporates all the facts that are true at every instant of time (the general situation). Then substitute into the equation from the general situation the facts that represent the particular instant under investigation (the specific situation). Finally, solve the equation from the specific situation for the desired unknown. This process is illustrated by the following examples.

EXAMPLE 1   Profit, P, is related to the number of items produced, x, by the formula

$$P = 75x - \frac{x^3}{50,000} - 100,000$$

Suppose that production is increasing by 100 units per week. How fast is the profit increasing when production is at 1000 items?

Solution   *The General Situation:*   The formula for the general situation is given. We take the derivative of both sides of the given equation with respect to t in order to transform the equation into one on related rates:

$$\frac{dP}{dt} = 75\frac{dx}{dt} - \frac{3}{50,000}x^2\frac{dx}{dt}$$

This equation is true for every instant of time in this problem.

The Specific Situation: At the instant under consideration in this problem (namely, when production is at 1000 items), we are given that production is increasing at 100 units per week. That is, we are given $dx/dt = 100$. Thus, when production is 1000 items, we have

$$\frac{dP}{dt} = 75(100) - \frac{3}{50,000}(1000^2)(100)$$

$$= 1500$$

This means that the profit is increasing at $1,500 per week when production is 1000 items and when production is increasing at 100 items per week.   ■

EXAMPLE 2   Two conveyor belts are moving away from an assembly point at right angles to one another, as shown in Figure 10.3. If two items simultaneously leave the assembly table, how fast is the distance between them changing after 2 minutes?

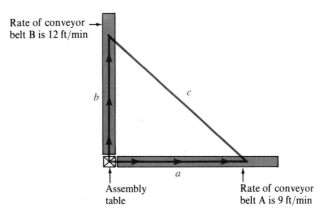

Rate of conveyor belt B is 12 ft/min

b

c

a

Assembly table

Rate of conveyor belt A is 9 ft/min

Figure 10.3 Assembly point configuration

Solution   **The general situation:**   Since a, b, and c in Figure 10.3 represent the sides of a right triangle, they are related by the formula

$$c^2 = a^2 + b^2$$

Convert this formula into one involving rates by taking the derivative of both sides with respect to time t:

$$2c\frac{dc}{dt} = 2a\frac{da}{dt} + 2b\frac{db}{dt}$$

The specific situation: At the instant under consideration (namely at $t = 2$ minutes), you know that $da/dt = 9$ and $db/dt = 12$. Find dc/dt when $t = 2$. When $t = 2$, $a = 18$, $b = 24$, and

$$c^2 = 18^2 + 24^2$$
$$c = 30$$

Substitute these values into the formula relating the rates:

$$2(30)\frac{dc}{dt} = 2(18)(9) + 2(24)(12)$$

$$\frac{dc}{dt} = 15$$

The items are moving apart at 15 feet per minute. ∎

EXAMPLE 3 A cylindrical vat with dimensions given in Figure 10.4 is being filled at the rate of 1000 cubic feet per minute. How fast is the surface rising when the depth is 12 feet?

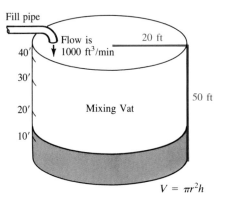

Figure 10.4 Mixing vat

$$V = \pi r^2 h$$

Solution **The general situation:** The necessary relationship from Figure 10.4 requires a geometric formula from your previous work. This is the formula for the volume of a cylinder: $V = \pi r^2 h$. In this problem, $r = 20$ feet (constant), so that the formula we will use is $V = 400\pi h$. Convert this to a formula involving rates:

$$\frac{dV}{dt} = \frac{d}{dt}(400\pi h)$$

$$= 400\pi\frac{dh}{dt}$$

The specific situation: The instant under investigation is when the depth is 12 feet. At this instant, $dV/dt = 1000$ and you want to find dh/dt. Substitute these values into the formula from the general situation:

$$1000 = 400\pi \frac{dh}{dt}$$

$$\frac{dh}{dt} = \frac{1000}{400\pi} \approx 0.796$$

That is, the height is rising at 0.796 feet per minute (this is about $9\frac{1}{2}$ inches per minute). ∎

The procedure for solving related rate problems is summarized below:

Procedure for Solving Related Rate Problems

THE GENERAL SITUATION

1. Draw a figure if appropriate. Use letters to describe the data in the problem. Since rates involve variables and not constants, be careful not to label a quantity with a number unless it *never* changes in the problem.
2. Find a formula relating the variables. (The appropriate formula for most of the problems in this section is given.)
3. Differentiate the equation: usually implicitly and usually with respect to time.

THE SPECIFIC SITUATION

4. List the known quantities; list as unknown the quantity you want to find. If there are other variables or quantities in the problem use the given formula to eliminate those "extra" variables. Substitute in all the values in the formula. The only remaining variable should be the unknown. Solve for the unknown.

EXAMPLE 4   Illustrate steps 3 and 4 of the above procedure to find and interpret dy/dt where $x^3 + 5y^2 = 84$ and $dx/dt = 10$ where $x = 4$. Assume that both x and y are positive.

Solution   **The general situation** (*step 3*):   You are asked to find the rate at which y is changing with respect to time at *that particular instant* when $x = 4$ if you know that x is changing at a rate of 10 units per unit of time. This means that $dx/dt = 10$ is part of the general situation (it does not change throughout the problem). On the other hand, $x = 4$ and dy/dt are part of the specific situation (this is the instant with which we are concerned). Differentiate the formula implicitly:

$$3x^2 \frac{dx}{dt} + 10y \frac{dy}{dt} = 0$$

Known: $dx/dt = 10$ To find: dy/dt

The specific situation (*step 4*):   The formula still involves both an x and a y, but we are interested in finding dy/dt at the instant when $x = 4$ (x is not a constant in this problem), so we use the *given* formula to find y *at that instant*:

$$x^3 + 5y^2 = 84$$
$$4^3 + 5y^2 = 84$$
$$5y^2 = 20$$
$$y^2 = 4$$
$$y = 2, -2 \qquad \text{Reject } y = -2 \text{ since } y \text{ must be positive}$$

Now substitute all values into the formula; the only unknown value should be dy/dt:

$$3x^2\frac{dx}{dt} + 10y\frac{dy}{dt} = 0$$

$$3(4)^2(10) + 10(2)\frac{dy}{dt} = 0$$

$$\frac{dy}{dt} = -24$$

The result says that y is decreasing at the rate of 24 units per unit of time.   ■

Problem Set 10.4

In Problems 1–9 find the indicated rate, given the other information. Assume that both x and y are positive.

1. Find dy/dt where $x^2 + y^2 = 25$ and $dx/dt = 4$ when $x = 3$.

2. Find dx/dt where $x^2 + y^2 = 25$ and $dy/dt = 2$ when $x = 4$.

3. Find dy/dt where $5x^2 - y = 100$ and $dx/dt = 10$ when $x = 10$.

4. Find dx/dt where $4x^2 - y = 100$ and $dy/dt = -6$ when $x = 1$.

5. Find dx/dt where $y = 2\sqrt{x} - 9$ and $dy/dt = 5$ when $x = 9$.

6. Find dy/dt where $y = 5\sqrt{x+9}$ and $dx/dt = 2$ when $x = 7$.

7. Find dy/dt where $xy = 10$ and $dx/dt = -2$ when $x = 5$.

8. Find dy/dt where $5xy^2 = 10$ and $dx/dt = -2$ when $x = 1$.

9. Find dx/dt where $x^2 + xy - y^2 = 10$ and $dy/dt = 5$ when $x = 4$ and $y > 0$.

APPLICATIONS

10. The profit, P, in dollars, is related to the number of items produced, x, by the formula

$$P = 125x - \frac{x^2}{200} - 500 \qquad \text{where } 0 \le x \le 1000$$

Suppose that production is increasing at five units per week. How fast is the profit changing when production is at 200 items per week?

11. Suppose that production in Problem 10 is decreasing by one unit per week. How fast is the profit changing when production is at 200 items?

12. Two cars start driving from the same point. One goes north at a rate of 30 mph and the other west at 40 mph. How fast is the distance between them changing after 5 hours?

13. A product is moving along a conveyor belt at the rate of 3 feet per second. A robot arm is suspended 8 feet above the belt. At what rate is the distance between the product and

the robot arm changing when the product is 9 feet from the base of the arm?

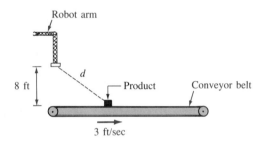

14. How fast is the surface of the vat described in Example 3 rising when the depth is 20 feet?

15. How fast is the surface of the vat described in Example 3 rising if it is being filled at a rate of 500 cubic feet per minute and the other details remain the same?

16. The level of carbon monoxide (in parts per million, ppm) in a city can be predicted by considering it as a function of the number of registered automobiles in that city according to the formula.

$$P = 1 + 0.2x + 0.001x^2$$

If the number of automobiles is increasing at 4000 per year, how is the level of carbon monoxide changing when the city has exactly 25,000 cars?

17. A pebble is thrown into a still pond and the result is a circular ripple. If the radius of this circle is expanding at 1 foot per second, how fast is the area changing when the radius is 10 feet? (Use $A = \pi r^2$.)

18. How fast is the circumference of the circular ripple in Problem 17 changing? (Use $C = 2\pi r$.)

19. The wholesale price of apples is d dollars per ton, and the daily supply x is related to the price by the formula

$$d = \frac{-5x}{x + 1000} + 70$$

Suppose that there are 2000 tons available today and that the supply is decreasing at 200 tons per day. At what rate is the price changing?

20. The number of cases of rabies, r, in an area is related to the number of skunks, s, according to the formula

$$r = 0.0005s^{2/3}$$

A program to eliminate the skunk population is instituted and it is estimated that 200 skunks per day are destroyed and that the present skunk population is 5000. At what rate is the number of cases of rabies changing?

21. A certain medical procedure requires that a spherical balloon be inserted into the stomach and then be inflated. If the radius of the balloon is increasing at the rate of 0.3 centimeter per minute, how fast is the volume changing when the radius is 4 centimeters? (Use $V = \frac{4}{3}\pi r^3$.)

22. How fast is the surface area of the sphere in Problem 21 increasing? (Use $S = 4\pi r^2$.)

23. A weather balloon is rising vertically at the rate of 10 feet per second. An observer is standing on the ground 300 feet from the point where the balloon was released. At what rate is the distance between the observer and the balloon changing when the balloon is 400 feet high?

10.5 Summary and Review

IMPORTANT TERMS

Average marginal analysis   [10.3]
Differential   [10.2]
Explicit differentiation   [10.1]
Implicit differentiation   [10.1]
Marginal analysis   [10.3]

Marginal cost   [10.3]
Marginal profit   [10.3]
Marginal revenue   [10.3]
Profit function   [10.3]

SAMPLE TEST

For additional practice there are a large number of review problems categorized by objective in the Student Solutions Manual. The following sample test (40 minutes) is intended to review the main ideas of this chapter.

Find dy/dx in Problems 1–3 by using implicit differentiation.

1. $y - 10x^2 + 6x - 15 = 0$
2. $6xy = 20x$
3. $x^4 + 5xy - 3x^2y + 9xy^2 - 155 = 0$
4. Find the standard form of the equation of the line tangent to the circle $(x - 2)^2 + y^2 = 25$ at the point $(5, 4)$.

Find the indicated rate, given the information in Problems 5–8.

5. Find dy/dt where $9x^2 + 16y^2 = 145$ and $dx/dt = 4$ when $x = 3$.
6. Find dy/dt where $xy = 9$ and $dx/dt = -8$ when $x = 1$.
7. Find dx/dt where $y = 25x^2$ and $dy/dt = -4$ when $x = 3$.
8. Find dx/dt where $x^2 + 2xy + y^2 = 64$ and $dy/dt = 3$ when $x = 4$.
9. A property management company manages 100 apartments renting for $500 with all the apartments rented. For each $50 per month increase in rent there will be 2 vacancies with no possibility of filling them. If x represents the number of $50 price increases, find the marginal revenue.
10. Suppose that for a company manufacturing lawn chairs, the cost and revenue functions are

$$C = 15,000 + 45x$$
$$R = 100x - \frac{x^2}{2000}$$

where the production output is x chairs per week. If production is increasing at a rate of 50 chairs per week when production output is 2000 chairs per week, find the rate of increase (decrease) in **a.** cost   **b.** revenue   **c.** profit.

Modeling Application 7

Publishing: An Economic Model

Karlin Press sells its *World Dictionary* at a list price of $20 and presently sells 5000 copies per year. Suppose you are being considered for an editorial position and are asked to do an analysis of the price and sales of this book in order to assess your competency for the position. The annual costs associated with this book are summarized in the following table. The dictionary presently has a net cost of 80% of the list price.

Costs associated with publishing *World Dictionary*

Cost	Amount, $
Advertising	1,750.00
Author's royalty	0.00
Binding	1.85 per book
Composition	3,600.00
Computer services	750.00
Investment return	3,000.00
Operating overhead	9,000.00*
Printing	4.90 (per book)
Set-up charges	400.00
Storage	.50 (per book)
Taxes	1.20 (per book)

* Salaries and offices are prorated for this title.

Analyze the price and sales of the dictionary. Your analysis should reach a conclusion about pricing the book for maximum revenue and/or maximum profit. Some consideration should also be given to market demand (which is assumed to be linear). Suppose you do some additional market research to find that at a list price of $15, the company will sell 7500 copies, but at $25 sales would drop to 2500. You might also want to give some thought to inventory and reprint schedule. To this end you find that the annual inventory (for maximum profit) should be 6885, but 7000 copies are allowed because they need some for office and sample copies. For general guidelines about writing this essay, see the commentary accompanying Modeling Application 1 on page 130.

CHAPTER 11
Applications and Differentiation

CHAPTER CONTENTS

11.1 First Derivatives and Graphs

11.2 Second Derivatives and Graphs

11.3 Curve Sketching— Relative Maximums and Minimums

11.4 Absolute Maximum and Minimum

11.5 Summary and Review Important Terms Sample Test Cumulative Review for Chapters 9–11

APPLICATIONS

Management (*Business, Economics, Finance, and Investments*)

Minimal average cost (11.1, Problems 31–32)
Determining when sales are increasing (11.1, Problem 33)
Marginal revenue (11.2, Problem 39; 11.3, Problem 28; 11.5, Problem 9)
Minimize average cost (11.2, Problem 40)
Cost-benefit model (11.3, Problem 27)
Maximum profit (11.4, Problems 19–24)
Tour pricing for group fares (11.4, Problems 27–28)
Largest area in an enclosure (11.4, Problem 30)
Maximize volume of a box (11.4, Problem 31)
Cost of producing a textbook (11.4, Problem 32)

Life sciences (*Biology, Ecology, Health, and Medicine*)

Air pollution (11.4, Problem 25)
Maximize yield per acre (11.4, Problem 29)

Social sciences (*Demography, Political Science, Population, Psychology, Society, and Sociology*)

Learning curve (11.1, Problem 34)
Voting patterns (11.4, Problem 26)

General interest

Largest area in an enclosure (11.4, Problem 30)
Maximum volume of a box (11.4, Problem 31)
Minimize cardboard for a poster (11.5, Problem 10)

Cumulative review—Chapters 9–11

The calculus of Chapters 9 and 10 is put to work in this chapter by discussing some important applications of the derivative. The applications demonstrate some of the power and versatility of differential calculus.

Graphing, aided by the idea of a derivative, leads to curve sketching and the determination of relative maximums and minimums. We then turn to optimization (finding the absolute maximum or minimum), an extremely important concept in the business world.

We continue to explore the usefulness of calculus in a variety of different real world settings. The only reason for introducing the derivative in this book is so it can be applied as a real world model and can be used to derive additional mathematics to use in real world models.

11.1 First Derivatives and Graphs

We continue our study of the derivatives and their applications in this chapter. Many of the phenomena we will study can be better understood by looking at a graph of the relationship between two variables, so our first application is to apply the idea of the derivative to the graph of a function.

A notation called **interval notation** is useful in much of the following discussion. Interval notation uses the idea of an ordered pair listing the left and right endpoints of the interval as the first and second components, respectively. The ordered pair is enclosed in brackets or parentheses. Brackets are used when the endpoint is included in the interval, and parentheses are used when the endpoint is excluded from the interval. The idea is rather simple, as Table 11.1 shows.

TABLE 11.1
Interval notation for line segments

	Interval notation	Inequality notation	Line graph
Closed interval	$[a, b]$	$a \leq x \leq b$	
	$[a, b)$	$a \leq x < b$	
	$(a, b]$	$a < x \leq b$	
Open interval	(a, b)	$a < x < b$	

The smaller number of the pair of numbers a and b must be the first component. The symbols ∞ and $-\infty$ are used to denote rays, as shown in Table 11.2.

TABLE 11.2
Interval notation for rays

Interval notation	Inequality notation	Line graph
$(-\infty, b]$	$x \le b$	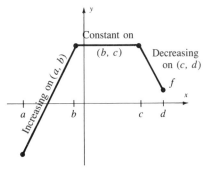
$(-\infty, b)$	$x < b$	
$[a, \infty)$	$x \ge a$	
(a, ∞)	$x > a$	

Note that the open interval notation ( ) is always used with ∞ and $-\infty$ and that [ ] is never used.

We can now apply interval notation to some of the ideas discussed in the last chapter. Consider a function f and its graph, as shown in Figure 11.1.

Figure 11.1 Graph of a function f

As we move from left to right it looks like f is increasing on the interval (a, b); this means that if $x_2 > x_1$ on (a, b), then $f(x_2) > f(x_1)$. The function in Figure 11.1 also looks like the graph falls on the interval (c, d), which means that $f(x_2) < f(x_1)$. The function also looks like it is constant on the interval (b, c). Notice that here we only classify functions as **increasing**, **decreasing**, or **constant** on open intervals. Thus:

Increasing and Decreasing Functions

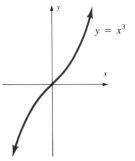

If $f'(x) > 0$, then the function is *increasing*, and if $f'(x) < 0$, then the function is *decreasing*.

Be careful about this statement—it does *not* mean that if the function is increasing at a point then its derivative is positive. There are conditions other than the derivative being positive that can cause a curve to be increasing at a given point. For example, if $y = x^3$, then

$$y' = 3x^2$$

which is 0 at the point $x = 0$. However, the graph of $y = x^3$ shows that the function is increasing at $x = 0$. The correct result is summarized in Table 11.3.

TABLE 11.3
Increasing and decreasing
functions on the interval
(a, b)

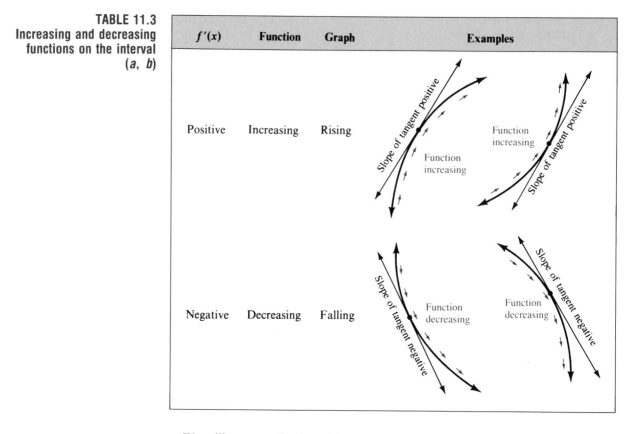

$f'(x)$	Function	Graph	Examples
Positive	Increasing	Rising	
Negative	Decreasing	Falling	

We will now apply these ideas to parabolas whose equations have the form $y = ax^2 + bx + c$. Section 2.5 promised that this type of parabola would be discussed after we learned some calculus techniques. Now recall that parabolas of this form are functions and open upward if $a > 0$ and downward if $a < 0$. Furthermore, the vertex is the extreme point—it is a maximum point if $a < 0$ and a minimum point if $a > 0$. The slope of the parabola at the vertex must also be zero. These possibilities are summarized in Figure 11.2.

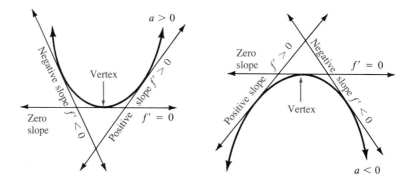

Figure 11.2 Graphs of parabolas showing slope

EXAMPLE 1   Graph $y = 3x^2 + 12x + 9$, and indicate the intervals for which the slope is increasing and for which it is decreasing. Also find the vertex by using the derivative.

Solution   If $y = 3x^2 + 12x + 9$, then $y' = 6x + 12$. Set $y' = 0$ and solve:

$$6x + 12 = 0$$
$$6x = -12$$
$$x = -2$$

We can use the work at the left to decide where the function is increasing and where it is decreasing: decreasing on $(-\infty, -2)$, increasing on $(-2, \infty)$. This is shown below.

Thus, when $x = -2$, $y' = 0$ (horizontal tangent line), so the vertex has a first component of -2. To find the second component substitute into the given equation:

$$y = 3(-2)^2 + 12(-2) + 9$$
$$= 12 - 24 + 9$$
$$= -3$$

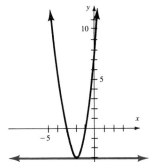

Horizontal tangent

The vertex of the parabola $y = 3x^2 + 12x + 9$ is $(-2, -3)$, and the parabola opens up ($a = 3$, which is greater than zero). You also know (from Chapter 2) that the y-intercept is $(0, 9)$ and that the parabola is symmetric, so it is now easy to graph the parabola, as shown in the margin.

From the graph it is obvious where the function is increasing and where it is decreasing, but you can also determine this by looking at the derivative $f'(x) = 6x + 12$.

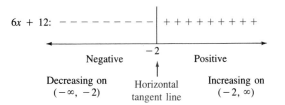

EXAMPLE 2   Graph $y = 10x - x^2$.

Solution   First, find the vertex: $y' = 10 - 2x$
Then, find horizontal tangent: If $y' = 0$, then $10 - 2x = 0$
$$10 = 2x$$
$$5 = x$$

If $x = 5$, then
$$y = 10(5) - (5)^2 = 25$$
The vertex is the point $(5, 25)$.

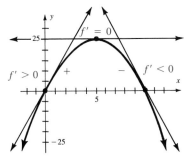

Increasing on $(-\infty, 5)$
Decreasing on $(5, \infty)$

The parabola opens downward ($a = -1$ is less than zero), has the y-intercept at $(0, 0)$, and is symmetric about the vertical line through the vertex. The graph is shown in the margin.

Since we will be graphing functions that are not just parabolas, some new terminology will be helpful. If f is a continuous function on an interval $[a, b]$, then there will be some point c_1 in that interval such that $f(c_1) \geq f(x)$ for all x in that interval. Such a point is called an **absolute maximum** on the interval. Similarly, there is another point, c_2, on the interval (not necessarily different from c_1) such that $f(c_2) \leq f(x)$ for all points x in that interval. This point is called an **absolute**

minimum. In addition to these absolute extremes there may also be points that are higher or lower than the surrounding points; these are called **relative maximums** and **minimums**. Several of these points are shown in Figure 11.3.

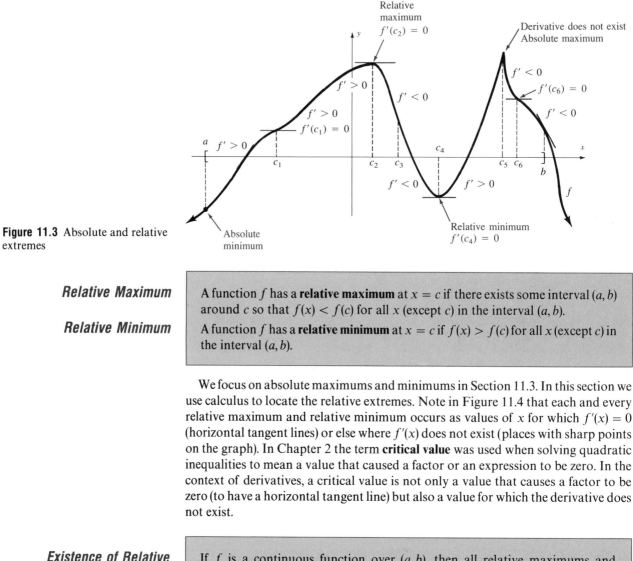

Figure 11.3 Absolute and relative extremes

Relative Maximum	A function f has a **relative maximum** at $x = c$ if there exists some interval (a, b) around c so that $f(x) < f(c)$ for all x (except c) in the interval (a, b).
Relative Minimum	A function f has a **relative minimum** at $x = c$ if $f(x) > f(c)$ for all x (except c) in the interval (a, b).

We focus on absolute maximums and minimums in Section 11.3. In this section we use calculus to locate the relative extremes. Note in Figure 11.4 that each and every relative maximum and relative minimum occurs as values of x for which $f'(x) = 0$ (horizontal tangent lines) or else where $f'(x)$ does not exist (places with sharp points on the graph). In Chapter 2 the term **critical value** was used when solving quadratic inequalities to mean a value that caused a factor or an expression to be zero. In the context of derivatives, a critical value is not only a value that causes a factor to be zero (to have a horizontal tangent line) but also a value for which the derivative does not exist.

Existence of Relative Maximums and Minimums	If f is a continuous function over (a, b), then all relative maximums and relative minimums, if they exist, must occur at critical values of x; that is, at points where $f'(x) = 0$ or where $f'(x)$ does not exist.

Be careful about the way you read this theorem. It does *not* say that if a point is a critical point, then it is a relative maximum or minimum. (Unfortunately, this *incorrect* line of reasoning would provide a very simple test for relative extremes.) We must therefore develop a careful *correct* strategy that will lead us to identify the relative maximums and minimums. We use two derivative tests for this purpose. We first discuss a derivative test that works in all cases. In the next section we discuss an easier, second derivative test that is dependent on the first derivative test since it does not work in all cases.

The first test, appropriately called the **first derivative test**, makes use of the fact that the derivative gives the slope of a tangent line of a curve at a particular point. Suppose that c is a critical value. Let c^- be a point to the left of c (but to the right of any other critical value) and let c^+ be a point to the right of c (but to the left of any other critical value). Then there are four possibilities:

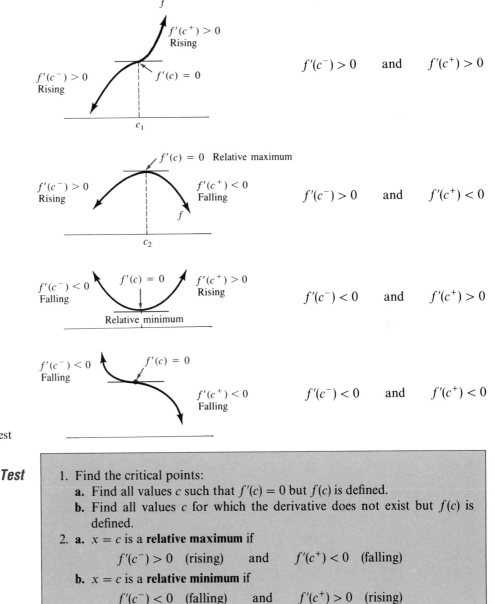

$$f'(c^-) > 0 \quad \text{and} \quad f'(c^+) > 0$$

$$f'(c^-) > 0 \quad \text{and} \quad f'(c^+) < 0$$

$$f'(c^-) < 0 \quad \text{and} \quad f'(c^+) > 0$$

$$f'(c^-) < 0 \quad \text{and} \quad f'(c^+) < 0$$

Figure 11.4 First derivative test

First Derivative Test

1. Find the critical points:
   a. Find all values c such that $f'(c) = 0$ but $f(c)$ is defined.
   b. Find all values c for which the derivative does not exist but $f(c)$ is defined.
2. **a.** $x = c$ is a **relative maximum** if
   $$f'(c^-) > 0 \quad \text{(rising)} \quad \text{and} \quad f'(c^+) < 0 \quad \text{(falling)}$$
   b. $x = c$ is a **relative minimum** if
   $$f'(c^-) < 0 \quad \text{(falling)} \quad \text{and} \quad f'(c^+) > 0 \quad \text{(rising)}$$

EXAMPLE 3 Graph $f(x) = x^2 - 8x + 7$ by using the first derivative test.

Solution $f'(x) = 2x - 8$

1. Find the critical points:

a. $f'(x) = 0$ or $2x - 8 = 0$

$$x = 4$$

The value $x = c_1 = 4$ is a critical value.

b. f is a polynomial function so it has a derivative at all points. This means that f' is defined for all x.

2. Let $c^- = 3$ and $c^+ = 5$:

$$f'(3) = 2(3) - 8 < 0 \quad \text{(falling)}$$
$$f'(5) = 2(5) - 8 > 0 \quad \text{(rising)}$$

Thus, by the first derivative test, the point $x = 4$ is a relative minimum. The graph is shown at the right. Note that a few additional points are needed. When $x = 4$, $f(x) = 4^2 - 8(4) + 7 = -9$ and the y-intercept is 7; the third plotted point is found by symmmetry.

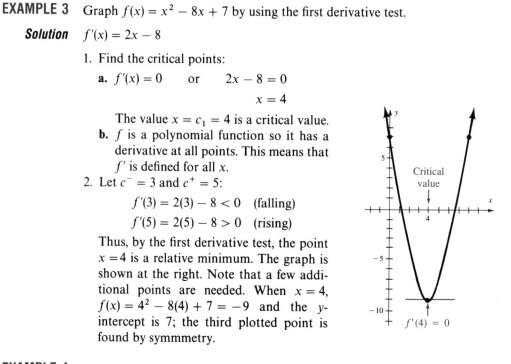

■

EXAMPLE 4 Find the critical values for $f(x) = 5x^3 + 4x^2 - 12x - 25$.

Solution Critical values are values for which $f'(x) = 0$ or $f'(x)$ is not defined.

$$f'(x) = 15x^2 + 8x - 12$$

We see that $f'(x)$ is defined for all x, so we look for x values for which

$$f'(x) = 0$$
$$15x^2 + 8x - 12 = 0$$
$$(3x - 2)(5x + 6) = 0$$

The critical values are $x = \frac{2}{3}$ and $x = -\frac{6}{5}$. ■

If we want to graph the curve whose equation is given in Example 4 we can use the first derivative test. We want to know where $f'(x)$ is positive and where it is negative. Look at the factored form of $f'(x)$:

$3x - 2$:	$- - - - -$	$- - - - -$	$+ + + + + +$
$5x + 6$:	$- - - - -$	$+ + + + +$	$+ + + + + +$
Product:	Positive	Negative	Positive

$$-\frac{6}{5} \qquad \frac{2}{3}$$

Rising   Falling   Rising

Relative   Relative
maximum   minimum

By plotting some points you can graph this curve, as shown in Figure 11.5.

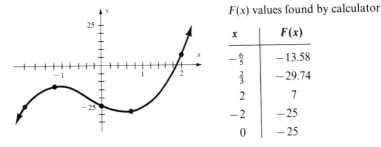

F(x) values found by calculator

x	$F(x)$
$-\frac{6}{5}$	-13.58
$\frac{2}{3}$	-29.74
2	7
-2	-25
0	-25

Figure 11.5 Graph of
$f(x) = 5x^3 + 4x^2 - 12x - 25$

EXAMPLE 5   Find the critical values for $f(x) = |x + 1|$.

Solution   If $x > -1$, then $f(x) = x + 1$ and $f'(x) = 1$.
If $x < -1$, then $f(x) = -x - 1$ and $f'(x) = -1$.
Since these are not the same we say that the derivative does not exist at $x = c = -1$.
Apply the first derivative test. Let

$$f'(c^-) = -1 \qquad \text{since} \qquad c^- < -1 \quad \text{(falling)}$$
$$f'(c^+) = 1 \qquad \text{since} \qquad c^+ > -1 \quad \text{(rising)}$$

Thus $x = -1$ is a relative minimum.   ∎

It is well known in economics that the minimal average cost of a product occurs when the average cost is equal to the marginal cost. Although this principle will not be proved here (see Problem 40 in Problem Set 11.2), it is illustrated below.

EXAMPLE 6   A product has a cost function given by

$$C(x) = 0.125x^2 + 20{,}000$$

where x is the number of units produced. Show that the minimal average cost occurs when the average cost is equal to the marginal cost.

Solution   First, find the average cost: $\bar{C}(x) = \dfrac{C(x)}{x} = 0.125x + 20{,}000x^{-1} \qquad x > 0$

Next, find the minimal average cost: $\bar{C}'(x) = 0.125 - 20{,}000x^{-2}$
Critical values: $\bar{C}'(x)$ does not exist where $x = 0$, but $\bar{C}(x)$ does not exist at $x = 0$, so $x = 0$ is not a critical value.

$$\bar{C}'(x) = 0$$
$$0.125 - 20{,}000x^{-2} = 0$$
$$0.125 = 20{,}000x^{-2}$$
$$0.125x^2 = 20{,}000$$
$$x^2 = 160{,}000$$
$$x = 400, -400$$
$$\uparrow$$
$$\text{Reject since } x > 0$$

There is thus only one critical value, $c = 400$. Use the first derivative test. Let

$$c^- = 390 \quad \text{and} \quad \bar{C}'(c^-) < 0 \quad \text{(curve falling)}$$
$$c^+ = 410 \quad \text{and} \quad \bar{C}'(c^+) > 0 \quad \text{(curve rising)}$$

Thus c is a relative minimum.

Now, find the marginal cost: $C'(x) = 0.25x$. Find the value for which the average cost is equal to the marginal cost:

$$\frac{\bar{C}(x)}{x} = C'(x)$$

Then

$$0.125x + 20,000x^{-1} = 0.25x$$
$$20,000x^{-1} - 0.125x = 0$$
$$20,0000 - 0.125x^2 = 0 \quad \text{Multiply both sides by } x \ (x > 0)$$
$$160,000 - x^2 = 0 \quad \text{Divide both sides by } 0.125$$
$$(400 - x)(400 + x) = 0$$
$$x = 400, \ -400$$
$$\uparrow$$
$$\text{Reject since } x > 0$$

This relationship is shown graphically in Figure 11.6.

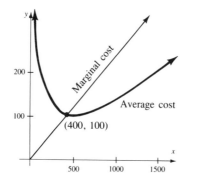

Figure 11.6 The minimal average cost of a product occurs when the average cost is equal to the marginal cost

Problem Set 11.1

Graph the curves in Problems 1–16 by using the derivative to find the vertex.

1. $f(x) = 8x^2$ **2.** $f(x) = -12x^2$

3. $f(x) = -20x^2$ **4.** $2x^2 + 5y = 0$

5. $5y + 15x^2 = 0$ **6.** $5x^2 + 4y = 20$

7. $f(x) = 5x^2 - 20x + 2$ **8.** $f(x) = 9 + 24x - 12x^2$

9. $x^2 + 4y - 3x + 1 = 0$ **10.** $x^2 - 4x + 10y + 13 = 0$

11. $9x^2 + 6x + 18y - 23 = 0$ **12.** $9x^2 + 6y + 18x - 23 = 0$

13. $y = 5x^2 - 3x + 1$ **14.** $y = 2x^2 + 5x - 8$

15. $y = 7x^2 + 2x - 3$ **16.** $y = 2x^2 - 5x + 3$

Identify the intervals for which the functions in Problems 17–22

are increasing, decreasing, or constant. Also identify the places where there is a horizontal tangent.

17.

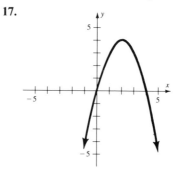

18.

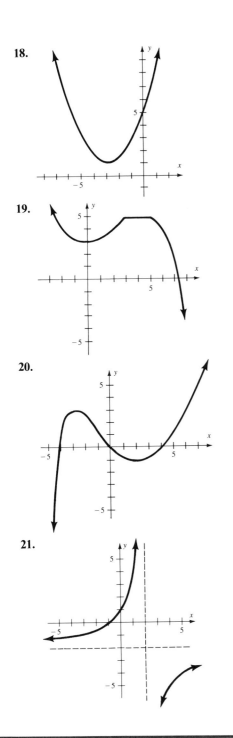

19.

20.

21.

22.

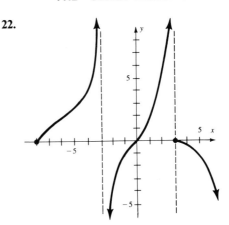

For the functions in Problems 23–30, (a) find the critical values and (b) determine the intervals for which the functions are increasing and decreasing.

23. $f(x) = 5 + 10x - x^2$

24. $f(x) = 10 + 6x - x^2$

25. $f(x) = x^2 - 12x + 6$

26. $f(x) = x^2 + 5x + 6$

27. $g(x) = 2x^3 - 3x^2 - 36x + 4$

28. $g(x) = x^3 + 3x^2 - 24x - 4$

29. $y = x^3 + 5x^2 + 8x - 5$

30. $y = \frac{8}{3}x^3 - 5x^2 - 3x + 10$

APPLICATIONS

31. The cost of producing x units of a product is given by the formula

$$C(x) = 0.25x^2 + 12,100$$

Show that the minimal average cost occurs when the average cost is equal to the marginal cost.

32. Show the result of Problem 31 graphically.

33. If a company has sales of

$$S(x) = 10,000 + 5000x - 25x^2 - x^3$$

where x is the amount spent on advertising, in thousands of dollars, when are the sales increasing?

34. The time t, in minutes, that it takes to learn a list of x items is given by the formula

$$t(x) = 5x\sqrt{x - 10} \qquad \text{for } x \geq 10$$

For what values of x is t increasing?

11.2 Second Derivatives and Graphs

We have now used the derivative to find out where f is increasing and where it is decreasing. Since the derivative of f is also a function we can repeat the process and use the derivative of f' to determine where f' is increasing and where it is decreasing.

The derivative of a derivative is called its **second derivative** and is denoted by f''. Taking still another derivative gives a function f''' called the **third derivative**. Successive derivatives are denoted by

$$f', f'', f''', f^{(4)}, f^{(5)}, \ldots, f^{(n)}$$

EXAMPLE 1 Note the use of parentheses for fourth derivatives and higher. This is to avoid confusion with powers f^4, f^5, and so on.

Find all successive derivatives of $f(x) = 2x^5 - 3x^2 + 5x - 35$.

Solution $f'(x) = 10x^4 - 6x + 5$

$f''(x) = 40x^3 - 6$

$f'''(x) = 120x^2$

$f^{(4)}(x) = 240x$

$f^{(5)}(x) = 240$

$f^{(6)}(x) = f^{(7)}(x) = \cdots = 0$ ■

Another commonly used notation for higher derivatives uses the dy/dx notation as follows:

$$\frac{dy}{dx}, \frac{d^2y}{dx^2}, \frac{d^3y}{dx^3}, \frac{d^4y}{dx^4}, \ldots, \frac{d^ny}{dx^n}$$

EXAMPLE 2 If $y = 8x^{1/2}$, find the first three derivatives.

Solution $\dfrac{dy}{dx} = 8\left(\dfrac{1}{2}\right)x^{1/2-1} = 4x^{-1/2}$

$\dfrac{d^2y}{dx^2} = 4\left(-\dfrac{1}{2}\right)x^{-1/2-1} = -2x^{-3/2}$

$\dfrac{d^3y}{dx^3} = -2\left(-\dfrac{3}{2}\right)x^{-3/2-1} = 3x^{-5/2}$ ■

Now we can use the second derivative to develop a second derivative test to inform us about the shape of a graph. If a curve lies above its tangent line at each point on an interval (a, b), it is **concave upward** over (a, b); if a curve lies below its

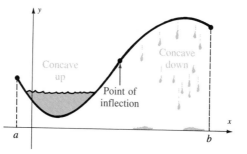

Figure 11.7 Concavity and inflection points

tangent line at each point on the interval (a, b), it is **concave downward**. A point on a graph that separates a concave downward portion of a curve from a concave upward portion is called an **inflection point**. Figure 11.7 illustrates these ideas.

We can now use the second derivative to obtain additional information about a graph. We consider two possibilities: $f''(x) > 0$ and $f''(x) < 0$.

Case I: If $f''(x) > 0$ on (a, b), then f' is increasing on (a, b).
What are the possibilities for f when f' is increasing?

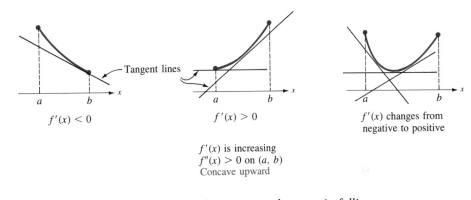

$f'(x) < 0$ $f'(x) > 0$ $f'(x)$ changes from
                                          negative to positive

$f'(x)$ is increasing
$f''(x) > 0$ on (a, b)
Concave upward

f' is increasing and is negative; the curve is falling
f' is increasing and is positive; the curve is rising
f' is increasing and changes from negative to positive

Curves having the shape illustrated by $f''(x) > 0$ are concave upward.

Case II: If $f''(x) < 0$ on (a, b), then f' is decreasing on (a, b).
What are the possibilities for f when f' is decreasing?

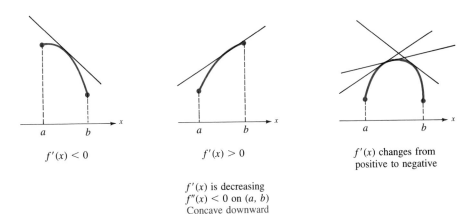

$f'(x) < 0$ $f'(x) > 0$ $f'(x)$ changes from
                                          positive to negative

$f'(x)$ is decreasing
$f''(x) < 0$ on (a, b)
Concave downward

f' is decreasing and negative; the curve is falling
f' is decreasing and positive; the curve is rising
f' is decreasing and changes from positive to negative

Curves having the shape illustrated by $f''(x) < 0$ are concave downward.

EXAMPLE 3 Given $4x^3 - 5x^2 - 8x - 2y + 20 = 0$,

  a. Find the critical values.

  b. Find the intervals for which the function is increasing or decreasing.

  c. Find the interval for which the function is concave upward or concave downward.

Solution   **a.** You could first solve for y and then find the derivative, but instead you find y' implicitly:

$$12x^2 - 10x - 8 - 2y' = 0$$
$$2y' = 12x^2 - 10x - 8$$
$$y' = 6x^2 - 5x - 4$$

Set $y' = 0$ and solve for the critical values.

$$6x^2 - 5x - 4 = 0$$
$$(3x - 4)(2x + 1) = 0$$

The critical values are $x = \frac{4}{3}$ and $x = -\frac{1}{2}$.

  b. The intervals on which the function is increasing or decreasing can be found by finding where the derivative is positive or negative. The procedure is identical to that used when solving inequalities (see Section 1.6).

$$6x^2 + 5x - 4 = (3x - 4)(2x + 1)$$

Plot the critical values and solve on a number line.

$3x - 4$:	$- - - - -$	$- - - - - - - - - -$	$+ + + + + + + + + + +$
$2x + 1$:	$- - - - -$	$+ + + + + + + + + +$	$+ + + + + + + + + + +$

$$-\tfrac{1}{2} \qquad\qquad \tfrac{4}{3}$$

Positive	Negative	Positive
Increasing on	Decreasing on	Increasing on
$(-\infty, -\tfrac{1}{2})$	$(-\tfrac{1}{2}, \tfrac{4}{3})$	$(\tfrac{4}{3}, \infty)$

  c. For the concavity, look at the second derivative

$$y'' = 12x - 5$$

Set $y'' = 0$ and solve $12x - 5 = 0$

$$x = \tfrac{5}{12}$$

The first derivative is not zero at $x = \frac{5}{12}$, so

$12x - 5$:	$- - - - - - -$	$+ + + + + + +$

$$\tfrac{5}{12}$$

Negative	Positive
Concave down on	Concave up on
$(-\infty, \tfrac{5}{12})$	$(\tfrac{5}{12}, \infty)$

∎

EXAMPLE 4   Identify the intervals for which the function is increasing, the intervals for which it is decreasing, where it is concave upward, where it is concave downward, as well as places where there is a horizontal tangent line.

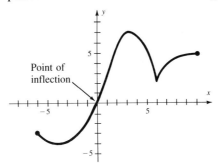

Solution   Function is increasing on $(-4, 3)$ and on $(6, 10)$.
Function is decreasing on $(-6, -4)$ and on $(3, 6)$.
The point of inflection is at $x = 0$; it is the point $(0, 0)$.
The curve is concave up on $(-6, 0)$ and concave down on $(0, 6)$ and $(6, 10)$.
The horizontal tangents are at $x = -4$, $x = 3$, and $x = 10$.
(There is no horizontal tangent at $x = 6$ since there is not even a derivative defined at $x = 6$.) ∎

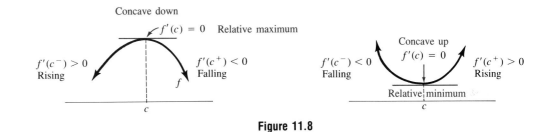

Figure 11.8

For most of your work you will use what is called the **second derivative test**. Figure 11.8 illustrates the two possibilities that give a relative maximum or a relative minimum. If a curve has a horizontal tangent line at $x = c$ and is concave upward ($f'' > 0$) in an interval containing the point of tangency, then $f(c)$ is a relative minimum. On the other hand, if the curve is concave downward ($f'' < 0$), then $f(c)$ is a relative maximum. This observation leads us to the second derivative test:

Second Derivative Test

If f is a continuous function on an interval (a, b),

1. Find the critical points of f. Suppose that c is a critical value of f.
2. **a.** If $f''(c) > 0$, then $x = c$ is a relative minimum.
   Note: Greater than zero, concave *up*; thus a relative minimum.
   b. If $f''(c) < 0$, then $x = c$ is a relative maximum.
   Note: Less than zero, concave *down*; thus a relative maximum.
   c. If $f''(c) = 0$, then the second derivative test fails and gives no information, so the first derivative test must be used.

EXAMPLE 5   Graph $f(x) = x^2 - 8x + 7$ by using the second derivative test.*

Solution $f(x) = x^2 - 8x + 7$

$f'(x) = 2x - 8$; critical value $x = 4$

$f''(x) = 2$; thus, there is a relative minimum
at $x = 4$ since $f''(4) = 2 > 0$

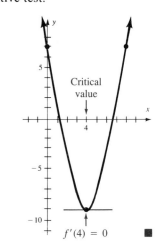

EXAMPLE 6   Apply the second derivative test to $f(x) = 3x^4 + 8x^3 - 6x^2 - 24x + 6$.

Solution $f'(x) = 12x^3 + 24x^2 - 12x - 24$

$f''(x) = 36x^2 + 48x - 12$

1. Critical values: This is a polynomial, so the derivative is defined everywhere.
$f'(x) = 0$ implies

$$12x^3 + 24x^2 - 12x - 24 = 0$$
$$x^3 + 2x^2 - x - 2 = 0$$
$$x^2(x + 2) - (x + 2) = 0$$
$$(x^2 - 1)(x + 2) = 0$$
$$(x - 1)(x + 1)(x + 2) = 0$$

The critical values are $x = 1, -1, -2$.

2. Test $f''(c)$ for $c = 1, -1,$ and -2:

$f''(x) = 36x^2 + 48x - 12$	
$f''(1) = 36(1)^2 + 48(1) - 12 > 0$	Relative minimum
$f''(-1) = 36(-1)^2 + 48(-1) - 12 < 0$	Relative maximum
$f''(-2) = 36(-2)^2 + 48(-2) - 12 > 0$	Relative minimum

If you now want to graph $f(x)$ from Example 6, you need to find the y-components for the critical values:

$x = 1$;	$f(1) = -13$	The point $(1, -13)$ is a relative minimum
$x = -1$;	$f(-1) = 19$	The point $(-1, 19)$ is a relative maximum
$x = -2$;	$f(-2) = 14$	The point $(-2, 14)$ is a relative minimum

---

* Compare this example with Example 3 in Section 11.1, which solved this problem by using the first derivative test.

If there were any other relative maximums or minimums, they would have shown up in the list of critical values (which they did not), so it is easy to complete the graph, as shown in Figure 11.9.

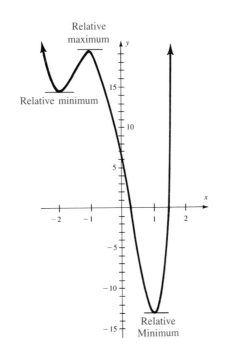

Figure 11.9 Graph of $f(x) = 3x^4 + 8x^3 - 6x^2 - 24x + 6$

EXAMPLE 7 Graph $f(x) = x^4$.

Solution Apply the second derivative test.

$$f'(x) = 4x^3$$ Critical value is $x = 0$

$$f''(x) = 12x^2 \qquad f''(0) = 0$$ The second derivative test fails, so revert to the first derivative test

Since $c = 0$ is the only critical value, let $c^- = -1$ and $c^+ = 1$. Then:

$$f'(-1) < 0$$ The curve is decreasing at $x = -1$

$$f'(1) > 0$$ The curve is increasing at $x = 1$

Therefore $x = 0$ is a relative minimum.

Plot some points and complete the graph, as shown in Figure 11.10:

If $x = 0,$ $\qquad f(0) = 0$

If $x = 1,$ $\qquad f(1) = 1$

If $x = 2,$ $\qquad f(2) = 16$

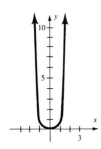

Figure 11.10 Graph of $f(x) = x^4$

$f(-x) = f(x)$, so the curve is symmetric with respect to the x-axis; so it is not necessary to compute negative values of x. ∎

Problem Set 11.2

Find the indicated derivatives for the functions in Problems 1–12.

1. $f(x) = 2x^5 - 3x^4 + x^3 - 5x^2 + 19x - 120$; $f'''(x)$

2. $f(x) = 25 + x - 3x^2 + 4x^3 - x^4 + 3x^5$; $f^{(5)}(x)$

3. $y = \sqrt{5x}$; $\dfrac{d^4y}{dx^4}$

4. $y = 18\sqrt{3x}$; $\dfrac{d^3y}{dx^3}$

5. $g(x) = 4x^3 - 2x^{-1}$; $g^{(4)}(x)$

6. $g(x) = 5x^{-2} - 3x$; $g'''(x)$

7. $y = 3x^{5/3}$; $\dfrac{d^3y}{dx^3}$

8. $y = -4x^{-3/2}$; $\dfrac{d^4y}{dx^4}$

9. $y = \dfrac{3}{x-1}$; $\dfrac{d^3y}{dx^3}$

10. $y = \dfrac{-5}{2x+1}$; $\dfrac{d^3y}{dx^3}$

11. $y = \dfrac{x^2-1}{x+4}$; $\dfrac{d^2y}{dx^2}$

12. $y = \dfrac{3x^2+5}{x-9}$; $\dfrac{d^2y}{dx^2}$

Find all relative maximums and minimums in Problems 13–30 by using either the first or the second derivative test as appropriate. You do not need to graph these functions.

13. $f(x) = (2x+1)^4$

14. $f(x) = (3x-5)^4$

15. $f(x) = \dfrac{x^2+9}{x}$

16. $f(x) = \dfrac{x^2+1}{x}$

17. $f(x) = \sqrt{x}$

18. $f(x) = \sqrt{x^2+4}$

19. $f(x) = 8x^2 - x^4$

20. $f(x) = 2x^2 - x^4$

21. $f(x) = x^3 + 5x^2 - 8x + 10$

22. $f(x) = 6x^3 - 21x^2 - 36x + 15$

23. $y = 4x^3 - 27x^2 - 30x - 6$

24. $y = 2x^3 + 7x^2 - 40x + 5$

25. $g(x) = x + \dfrac{1}{x}$

26. $g(x) = x + \dfrac{4}{x}$

27. $g(x) = \dfrac{x^2}{x-1}$

28. $g(x) = \dfrac{x^2}{x-9}$

29. $f(x) = (x+1)^{2/3}$

30. $f(x) = (x-3)^{2/3}$

For the functions in Problems 31–38, (a) find the critical values; (b) determine the intervals for which the functions are increasing and for which they are decreasing; and (c) determine the intervals for which the functions are concave up or concave down.

31. $f(x) = 3 + 12x - x^2$

32. $f(x) = 8 + 6x - x^2$

33. $g(x) = x^3 - 7x^2 - 5x + 8$

34. $g(x) = x^3 - 5x^2 - 8x + 10$

35. $y = x^3 + 11x^2 - 45x + 125$

36. $y = x^3 - 2x^2 - 15x - 75$

37. $12x^3 - 5x^2 - 4x - 2y + 14 = 0$

38. $2x^3 - 12x^2 - 30x - 6y + 5 = 0$

APPLICATIONS

39. A manufacturer has determined that the price of selling x items is given by the formula

$$P(x) = 5 - \left(\dfrac{x}{100}\right)^2$$

a. Find an expression for the revenue.
b. Find the marginal revenue.
c. Is the marginal revenue increasing or decreasing when $x = 100$ items?

40. Suppose C represents the total cost for a manufacturing process and is expressed as a function of the number of items, x, produced in a given period of time. Also suppose there is a value x in the domain that minimizes the average cost, $\bar{C}$. (That is, $\bar{C}(x) = C(x)/x$.) Show that the average cost is minimized at an x where the average cost equals the marginal cost.

11.3 Curve Sketching—Relative Maximums and Minimums

One of the most important tools in mathematics is the ability to quickly sketch a wide variety of functions. In this section we combine the graphing techniques of calculus with those we used in Chapter 2.*

---

* If you omitted Chapter 2, it is now a good idea to look at Section 2.6.

Table 11.4 presents procedures for graphing functions. It is, of course, not necessary to use all the procedures to graph every function. And, if any of the procedures are too difficult, it is possible to omit that particular procedure. The more you know about a curve, the fewer points you need to plot.

TABLE 11.4
Graphing strategy

Step	Procedure
Simplify	First, simplify, if possible, the function you wish to graph. That is, combine similar terms, reduce fractions, and simplify radical expressions
Second derivative test	Use the second derivative test to find the relative maximum or minimum values: 1. Find the critical values, c 2. $f''(c) > 0$, relative minimum $f''(c) < 0$, relative maximum $f''(c) = 0$, test fails The curve is concave upward if $f''(x) > 0$ The curve is concave downward if $f''(x) < 0$
First derivative test	Use the first derivative test if the second derivative test fails: 1. Relative minimum if $f'(c^-) < 0$ and $f'(c^+) > 0$ 2. Relative maximum if $f'(c^-) > 0$ and $f'(c^+) < 0$ The curve is increasing if $f'(x) > 0$ The curve is decreasing if $f'(x) < 0$
Asymptotes	If $f(x) = P(x)/D(x)$, where $P(x)$ and $D(x)$ are polynomial functions with no common factors, then: 1. *Vertical asymptote:* $x = r$ if $D(r) = 0$ 2. *Horizontal asymptote:* $y = 0$ if the degree of $P(x)$ is less than the degree of $D(x)$; $y = a_n/b_n$ if the degree of $P(x)$ is the same as the degree of $D(x)$; and a_n and b_n are the leading coefficients of $P(x)$ and $D(x)$, respectively
Symmetry	1. *With respect to the x-axis:* Substitute $-y$ for y; if the equation remains unchanged, then it is symmetric with respect to the x-axis 2. *With respect to the y-axis:* Substitute $-x$ for x; if the equation remains unchanged, then it is symmetric with respect to the y-axis 3. *With respect to the origin:* Substitute $-x$ for x and $-y$ for y simultaneously; if the equation remains unchanged, then it is symmetric with respect to the origin
Intercepts	1. *x-intercept:* Set $y = 0$ and solve for x 2. *y-intercept:* Set $x = 0$ and solve for y
Plot points	Plot any additional points necessary to draw the graph

EXAMPLE 1 Graph $f(x) = \dfrac{x^2}{x-2}$.

Solution
$$f'(x) = \frac{(x-2)(2x) - x^2(1)}{(x-2)^2}$$

$$= \frac{2x^2 - 4x - x^2}{(x-2)^2}$$

$$= \frac{x^2 - 4x}{(x-2)^2} = \frac{x(x-4)}{(x-2)^2}$$

$f'(x) = 0$ implies $\dfrac{x(x-4)}{(x-2)^2} = 0$

$$x(x-4) = 0$$

$$x = 0, 4 \qquad \text{Critical values}$$

The derivative is not defined at $x = 2$, and we see that there is an asymptote $x = 2$. Next, we apply the second derivative test:

$$f''(x) = \frac{(x-2)^2(2x-4) - x(x-4)(2)(x-2)}{(x-2)^4}$$

$$= \frac{8}{(x-2)^3} \qquad \text{There were several simplification steps left out; can you fill in the details?}$$

$f''(0) < 0$ There is a relative maximum at $x = 0$; $f(0) = 0$, so the point is $(0, 0)$.

$f''(4) > 0$ There is a relative minimum at $x = 4$; $f(4) = 8$, so the point is $(4, 8)$.

Plot the points $(0, 0)$ and $(4, 8)$ and label them appropriately, as shown in Figure 11.11a. It looks at first glance like they are labeled incorrectly (the top point as a relative minimum and the bottom point as a relative maximum). However, after the asymptote is drawn, you should understand as shown in Figure 11.11b.

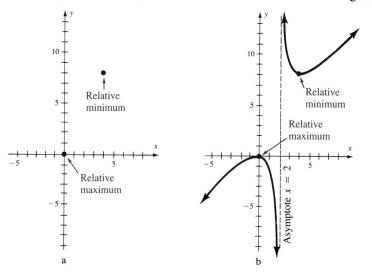

Figure 11.11 Graph of $x^2/(x-2)$

EXAMPLE 2 Graph $f(x) = 12x^{1/2} - 2x^{3/2}$.

Solution *First derivative:*

$$f'(x) = 6x^{-1/2} - 3x^{1/2}$$

Critical values: $x = 0$; $f'(0)$ is not defined but $f(0)$ is defined

$$f'(x) = 0; \qquad 6x^{-1/2} - 3x^{1/2} = 0$$
$$3x^{-1/2}(2 - x) = 0$$
$$2 - x = 0 \qquad \text{Divide both sides by } 3x^{-1/2} \ (x \neq 0)$$
$$x = 2$$

$f(2) = 12(2)^{1/2} - 2(2)^{3/2} = 12\sqrt{2} - 4\sqrt{2} = 8\sqrt{2}$; this is the point $(2, 8\sqrt{2})$.
$f'(x) > 0$ (increasing) on $(0, 2)$ and $f'(x) < 0$ (decreasing) on $(2, \infty)$
Second derivative:

$$f''(x) = -3x^{-3/2} - \frac{3}{2}x^{-1/2}$$

$$f''(2) = -3(2)^{-3/2} - \frac{3}{2}(2)^{-1/2} < 0;$$

thus $x = 2$ is a relative maximum.

$f''(x) < 0$ on $(0, \infty)$, so the curve is concave down.
There are no asymptotes.
There is no symmetry.
Intercepts:
y-intercept $(x = 0)$:

$$12x^{1/2} - 2x^{3/2} = 12(0)^{1/2} - 2(0)^{3/2} = 0; \text{ this is the point } (0, 0).$$

x-intercept $(y = 0)$:

$$12x^{1/2} - 2x^{3/2} = 0$$
$$2x^{1/2}(6 - x) = 0$$
$$x = 6, x = 0$$

These are the points $(6, 0)$ and $(0, 0)$.
The graph is shown in Figure 11.12.

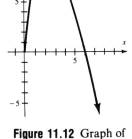

Figure 11.12 Graph of $f(x) = 12x^{1/2} - 2x^{3/2}$

Problem Set 11.3

In Problems 1–6, draw a curve on the interval (a, b) satisfying the stated conditions.

1. $f'(x) > 0$ and $f''(x) > 0$

2. $f'(x) > 0$ and $f''(x) < 0$

3. $f'(x) < 0$ and $f''(x) > 0$

4. $f'(x) < 0$ and $f''(x) < 0$

5. Passes through $(-2, 3)$, $(0, 5)$, and $(2, 7)$; $f'(-2) = 0$ and $f'(2) = 0$; $f''(x) > 0$ if $x < 0$, $f''(x) < 0$ if $x > 0$, and $f''(0) = 0$

6. Passes through $(2, 8)$, $(5, 6)$, and $(8, 4)$; $f'(2) = 0$ and $f'(8) = 0$; $f''(x) < 0$ if $x < 5$, $f''(x) > 0$ if $x > 5$, and $f''(5) = 0$

Graph the functions in Problems 7–26.

7. $y = 6x - x^2$

8. $y = 20x - 5x^2$

9. $y = 2x^2 - 3x + 5$

10. $3x^2 + 4x - 2y + 8 = 0$

11. $y = x^3 + 3x^2 - 9x + 5$

12. $y = x^3 - 3x^2 - 24x + 10$

13. $f(x) = 3x^4 + 8x^3 - 6x^2 - 24x - 5$

14. $f(x) = 3x^4 + 20x^3 - 24x^2 - 240x + 20$

15. $f(x) = x^3 + 1$

16. $f(x) = x^5$

17. $g(x) = x^4 - 1$

18. $g(x) = (x - 1)^4$

19. $y = \dfrac{x^2}{x - 1}$

20. $y = \dfrac{x^2}{x + 1}$

21. $y = \dfrac{x + 1}{x - 1}$

22. $y = \dfrac{2x + 1}{x - 1}$

23. $y = 3x^{2/3}$

24. $y = -3x^{1/3}$

25. $f(x) = 6x^{1/2} - 4x^{3/2}$

26. $f(x) = 6x^{1/2} - 12x^{3/2} + 5$

APPLICATIONS

27. The cost of removing $p\%$ of the pollutants from the atmosphere is given by the formula

$$C(p) = \frac{20{,}000p}{100 - p}$$

Graph $C(p)$.

28. A manufacturer has determined that the price of selling x items is given by the formula

$$P(x) = 5 - \left(\frac{x}{100}\right)^2$$

Sketch the marginal revenue function.

11.4 Absolute Maximum and Minimum

We can now use the derivative in one of its most powerful applications for management, life sciences, and social sciences—that of finding the absolute maximum and absolute minimum of a given function. All of the necessary techniques have now been developed, so the examples in this section illustrate the wide variety of ways these ideas can be applied. Absolute maximum and minimum were defined in the last section, and we only need now to be aware that the **absolute maximum** of a continuous function will occur at the critical value giving the largest value of the function, or else it will occur at an endpoint of a closed interval. On the other hand, the **absolute minimum** will occur at the critical value giving the smallest value of the function or at an endpoint. These situations were illustrated in Figure 11.3, which is repeated here for easy reference.

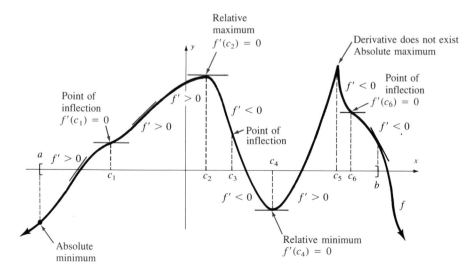

Procedure for Finding the Absolute Maximum and Absolute Minimum

Given a continuous function f over an interval $[a, b]$, there exists an absolute maximum and minimum on the interval. To find these absolute extremes, carry out the following steps:

1. Find the endpoints and the critical values: $a, c_1, c_2, \ldots, c_n, b$. Remember, the critical values are values that cause the derivative to be either zero or undefined.
2. Evaluate $f(a), f(c_1), f(c_2), \ldots, f(b)$.
3. The **absolute maximum** of $f(x)$ is the largest of the values found in step 2.
4. The **absolute minimum** of $f(x)$ is the smallest of the values found in step 2.

EXAMPLE 1 Find the absolute maximum and minimum for

$$f(x) = 3x^5 - 50x^3 + 135x + 20 \qquad \text{on } [-2, 4]$$

Solution 1. $f'(x) = 15x^4 - 150x^2 + 135$
2. Critical values:

$$15x^4 - 150x^2 + 135 = 0$$
$$x^4 - 10x^2 + 9 = 0$$
$$(x^2 - 9)(x^2 - 1) = 0$$
$$(x - 3)(x + 3)(x - 1)(x + 1) = 0$$
$$x = 3, -3, 1, -1$$

┌ Endpoints ┐
↓   ↓
3. List: $-2, \underbrace{-1, 1, 3}, 4$

Critical values
$x = -3$ is not in the interval $[-2, 4]$, so do not test $x = -3$

4. Test each of these values:

$$f(-2) = 3(-2)^5 - 50(-2)^3 + 135(-2) + 20 = 54$$
$$f(-1) = 3(-1)^5 - 50(-1)^3 + 135(-1) + 20 = -68$$
$$f(1) = 3(1)^5 - 50(1)^3 + 135(1) + 20 = 108$$
$$f(3) = 3(3)^5 - 50(3)^3 + 135(3) + 20 = -196 \qquad \text{Absolute minimum}$$
$$f(4) = 3(4)^5 - 50(4)^3 + 135(4) + 20 = 432 \qquad \text{Absolute maximum} \qquad ■$$

As noted, the real value of finding absolute maximum and minimum values is in the applications to which these ideas can be applied. The remainder of this section concerns such applications.

EXAMPLE 2 Westel Corporation manufactures telephones and has developed a new cellular phone. Production analysis shows that its price must not be less than $50; if x units are sold, then the price is given by the formula

$$p(x) = 150 - x$$

The total cost of producing x units is given by the formula

$$C(x) = 2500 + 30x$$

Find the maximum profit, and determine the price that should be charged to maximize the profit.

Solution The profit, $P(x)$, is found by

$$
\begin{aligned}
P(x) &= \text{REVENUE} - \text{COST} \\
&= (\text{number of items})(\text{price per item}) - \text{cost} \\
&= x \cdot p(x) - C(x) \\
&= x(150 - x) - (2500 + 30x) \\
&= 150x - x^2 - 2500 - 30x \\
&= -x^2 + 120x - 2500
\end{aligned}
$$

Find the absolute maximum value of P:

1. Critical points: $P'(x) = -2x + 120$ and $-2x + 120 = 0$ when $x = 60$. There are no values for which the derivative is not defined, so there is one critical value: $x = 60$.

2. Endpoints: Given $p(x) \geq 50$, we see that

$$
\begin{aligned}
150 - x &\geq 50 \\
100 &\geq x \\
x &\leq 100
\end{aligned}
$$

so one endpoint is 100. The other endpoint is merely implied, namely, 0 phones, so that the domain is $[0, 100]$.

3. Find the absolute maximum of P by checking $x = 0$, 60, and 100:

$$
\begin{aligned}
P(0) &= -2500 \\
P(60) &= -(60)^2 + 120(60) - 2500 = 1100 \\
P(100) &= -(100)^2 + 120(100) - 2500 = -500
\end{aligned}
$$

4. The absolute maximum of P thus occurs when $x = 60$.

The answer to the question asked might not be the absolute maximum or minimum value, and you must be careful to answer the question asked. For example, this problem asks for the price. The price needs to be found for the maximum value of x:

$$
\begin{aligned}
p(60) &= 150 - 60 \\
&= 90
\end{aligned}
$$

The phones should be priced at $90 per unit. ∎

The analysis in Example 2 uses the techniques of this chapter, but we were solving this type of problem in Chapter 2. Two additional techniques will simplify the work shown in Example 2. The first relates second-degree problems to the material of this chapter. It is called the **second derivative test for absolute maximum and minimum**:

Second Derivative Test for Absolute Maximum and Minimum

If f is a continuous function with only *one* critical value c and if $f''(c)$ exists and is not zero [that is, if $f''(c) = 0$, then this test fails], and if

$\qquad f''(c) > 0 \qquad$ then $x = c$ is an *absolute minimum*

$\qquad f''(c) < 0 \qquad$ then $x = c$ is an *absolute maximum*

This test tells us in Example 2 that since there was only one critical value at $x = 60$ and since $P''(x) = -2, \quad x = 60$ *must* be the absolute maximum, so we did not need to go through the evaluation of P at 0, 60, and 100.

The second technique involves assuming that the business in question has known cost and revenue functions $C(x)$ and $R(x)$, as shown in Figure 11.13. Profit is positive when $R(x)$ exceeds $C(x)$. The points a and b are the break-even points.

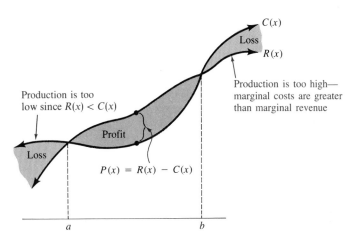

Figure 11.13 Maximum value of a profit function

The maximum profit occurs at a critical point c_1 of $P(x)$. Assume that $P'(x)$ exists for all x in some interval, usually $[0, \infty)$ and that the critical point c_1 occurs at some x in this interval, so that

$\qquad P'(x) = 0$

Then

$\qquad P(x) = R(x) - C(x)$
$\qquad P'(x) = R'(x) - C'(x)$
$\qquad \ \ \ 0 = R'(x) - C'(x)$
$\qquad R'(x) = C'(x)$

Maximum Profit

The maximum profit is achieved when the marginal revenue and marginal cost are equal.

Turning, one final time, to Example 2, we see that

$$R(x) = x(150 - x) = 150x - x^2 \quad \text{and} \quad C(x) = 2500 + 30x$$

so $\quad R'(x) = 150 - 2x \qquad\qquad\qquad C'(x) = 30$

Thus, the maximum profit is found when

$$150 - 2x = 30$$
$$120 = 2x$$
$$60 = x$$

EXAMPLE 3 As more and more industrial areas are constructed, there is a growing need for standards ensuring the control of the pollutants released into the air. Suppose that the air pollution at a particular location is based on the distance from the source of the pollution according to the principle that for distances greater than or equal to 1 mile, the concentration of particulate matter (in parts per million, ppm) decreases as the reciprocal of the distance from the source. This means that if you live 3 miles from a plant emitting 60 ppm, the pollution at your home is $\frac{60}{3} = 20$ ppm. On the other hand, if you live 10 miles from the plant, the pollution at your home is $\frac{60}{10} = 6$ ppm. Suppose now that two plants 10 miles apart are releasing 60 and 240 ppm, respectively, and you want to know the location between them at which the pollution is a minimum.

Solution

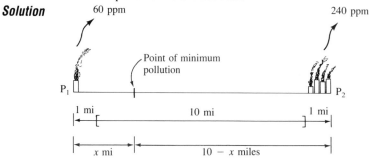

Let x be the distance from plant P_1. The domain is $[1, 9]$ since the given formula implies that you cannot be closer than 1 mile to either plant.

$$\text{Pollution from } P_1 = \frac{60}{x}$$

$$\text{Pollution from } P_2 = \frac{240}{10 - x}$$

The total pollution at any point is the sum of the pollution from the plants:

$$P(x) = \frac{60}{x} + \frac{240}{10 - x}$$

Find the absolute minimum value of P on $[1,9]$:

1. Critical points:

 a. Values for which $P'(x) = 0$:

$$P(x) = 60x^{-1} + 240(10 - x)^{-1}$$
$$P'(x) = -60x^{-2} - 240(10 - x)^{-2}(-1)$$
$$= -60x^{-2} + 240(10 - x)^{-2}$$

 Set $P'(x) = 0$ and solve:

$$\frac{-60}{x^2} + \frac{240}{(10 - x)^2} = 0$$

$-60(10 - x)^2 + 240x^2 = 0$ Multiply by $x^2(10 - x)^2$

$-(100 - 20x + x^2) + 4x^2 = 0$ Divide by 60

$$3x^2 + 20x - 100 = 0$$
$$(3x - 10)(x + 10) = 0$$
$$x = \tfrac{10}{3}, \; -10$$

    Reject this value since it is not in the domain

 b. Values for which $P'(x)$ do not exist: $x = 0$ and $x = 10$.
 Reject these since they are not in the domain $[1,9]$.

2. Endpoints: $x = 1, x = 9$.

3. Find the absolute minimum of P by checking $x = 1, \tfrac{10}{3}$, and 9:

$$P(1) = \frac{60}{1} + \frac{240}{9} \approx 87 \text{ ppm}$$

$$P\left(\frac{10}{3}\right) = \frac{60}{\frac{10}{3}} + \frac{240}{\frac{20}{3}} = 54 \text{ ppm}$$

$$P(9) = \frac{60}{9} + \frac{240}{1} \approx 247 \text{ ppm}$$

The minimum pollution is found at $3\tfrac{1}{3}$ miles from plant P_1.   ■

EXAMPLE 4   The voting patterns of a geographical region show that the percentage, P, of voters in a national election varies according to age, x, and fits the model predicted by the formula.

$$P(x) = 0.00002x^3 - 0.00195x^2 + 0.06x \qquad 18 \le x \le 50$$

Graph P and discuss its possible meaning, and then find the absolute maximum and minimum percent of the population voting in a national election.

Solution $P'(x) = 0.00006x^2 - 0.00390x + 0.06$

Set $P'(x) = 0$ and solve to find the critical values:

$0.00006x^2 - 0.00390x + 0.06 = 0$

$6x^2 - 390x + 6000 = 0$ Multiply by 100,000

$x^2 - 65x + 1000 = 0$ Divide by 6

$(x - 25)(x - 40) = 0$

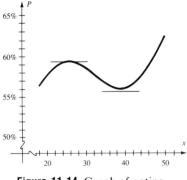

Figure 11.14 Graph of voting patterns

$P''(x) = 0.00012x - 0.00390$, so

$$P''(25) = 0.00012(25) - 0.00390 < 0 \qquad \text{Relative maximum}$$
$$P''(40) = 0.00012(40) - 0.00390 > 0 \qquad \text{Relative minimum}$$

Find some values of P (which are needed both for the graph and for the absolute maximum and absolute minimum):

$$P(25) = 0.59375$$
$$P(40) = 0.56$$
$$P(18) = 0.56484$$
$$P(50) = 0.625$$

From ages 18 to 25 the percent of the population voting is increasing (derivative positive), and from 25 to 40 it is decreasing, at which time it begins to increase again. For the above graph and calculations we see that the absolute maximum percentage of the population voting is at age 50, the endpoint of the interval $[18, 50]$. ∎

Problem Set 11.4

Find the absolute maximum and minimum for the functions in Problems 1–18.

1. $f(x) = 3x^5 - 50x^3 + 135x + 750$ on $[-5, 0]$
2. $f(x) = 3x^5 - 50x^3 + 135x + 750$ on $[-3, 3]$
3. $g(x) = x^3 - 2x^2 - 15x + 42$ on $[0, 5]$
4. $g(x) = x^3 + 5x^2 + 3x + 20$ on $[-5, 0]$
5. $h(x) = 3x^5 - 25x^3 + 60x - 200$ on $[0, 3]$
6. $h(x) = 3x^5 - 25x^3 + 60x - 200$ on $[-3, 3]$
7. $f(x) = 1 + x^{1/3}$ on $[-1, 8]$
8. $g(x) = 1 - x^{2/3}$ on $[-1, 8]$
9. $f(x) = \sqrt{x} + x$ on $[0, 4]$
10. $g(x) = x - \sqrt{x}$ on $[0, 9]$
11. $f(x) = (x + 1)^2(x - 3)$ on $[0, 5]$
12. $g(x) = (x - 1)(x + 4)^3$ on $[-5, 1]$
13. $f(x) = 2(x - 3)^{-1}$ on $[0, 5]$
14. $f(x) = 2x(x - 1)^{-1}$ on $[0, 3]$
15. $f(x) = |x|$ on $[-3, 3]$
16. $f(x) = |x + 2|$ on $[-3, 3]$
17. $f(x) = \dfrac{x - 1}{(x + 1)^2}$ on $[0, 3]$
18. $f(x) = \dfrac{x}{(x - 1)^2}$ on $[-2, 0]$

APPLICATIONS
19. A consulting firm determines that the price equation for a certain product is $p = 1000 - 10x$ dollars, where x is the number of items produced. The cost function is

$C(x) = 30,000 - 500x$. Find the maximum profit, and determine the price that should be charged to make the maximum profit.

20. The price equation for a new product is $50 - x$ dollars, where x is the number of items produced. The cost function is $C(x) = 500 - 20x$. Find the maximum profit, and determine the price that should be charged to make the maximum profit. [Note: Price and cost must both be nonnegative.]

In Problems 21–24 revenue and cost functions are given. Find the number of items, x, which produces a maximum profit.

21. $R(x) = 35x - 0.03x^2$; $\quad C(x) = 6000 + 5x$
22. $R(x) = 2000x - 0.05x^2$; $\quad C(x) = 120,000 + 150x$
23. $R(x) = 5x + x/(x - 64)$; $\quad C(x) = 1000 + x$, where $0 \le x \le 62$
24. $R(x) = 6x + x/(x - 245)$; $\quad C(x) = 5000 + x$, where $0 \le x \le 240$

25. Rework Example 3, except assume that the plants are 20 miles apart.
26. Rework Example 4, except assume that the voting formula is
$$P(x) = 0.00002x^3 - 0.00207x^2 + 0.0648x$$

27. A tour agency is booking a tour and has 100 people signed up. The price of a ticket is $2,000 per person. The agency has booked a plane seating 150 people at a cost of $125,000. Additional costs to the agency are incidental fees of $500 per person. For each $5 that the price is lowered, a new person will sign up. How much should the price be lowered to maximize the profit for the tour agency?

28. Suppose that the tour agency described in Problem 27 is able to obtain a booking for a plane that will seat 225 people at the same price for the charter as the first airline. Answer the question under these conditions.

29. A viticulturist estimates that if 50 grapevines are planted per acre, each grapevine will produce 150 pounds of grapes. For each additional grapevine planted per acre (up to 20), the average yield per vine drops by 2 pounds. How many grapevines should be planted to maximize the yield per acre?

30. A farmer has 1000 feet of fence and wishes to enclose a rectangular plot of land. The land borders a river and no fence is required on that side. What should the length and width of the rectangle be in order to include the largest possible area?

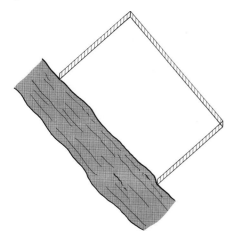

31. A shipping box is to be constructed from a piece of corregated cardboard 10 inches long and 5 inches wide. A square is to be cut out of each corner so that the sides can be folded up to form the box.
What dimensions for the box will maximize the volume of the finished box?

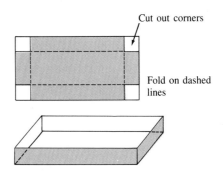

Cut out corners

Fold on dashed lines

32. The cost of producing a certain calculus textbook is $25 per volume. The number of books sold is given by the formula

$$n = \frac{50,000}{x - 20} + 2000$$

where x is the price of the book. The profit $P(x)$ is the number of books sold times the price less $25. That is,

$$P(x) = (x - 25)n$$

In order to maximize the profit, what should be the selling price of the book if the domain for x is $[25, 50]$?

11.5 Summary and Review

IMPORTANT TERMS

Absolute maximum [11.1, 11.4]	Intercepts [11.3]
Absolute minimum [11.1, 11.4]	Interval notation [11.1]
Asymptote [11.3]	Maximum profit [11.4]
Concave downward [11.2]	Relative maximum [11.1]
Concave upward [11.2]	Relative minimum [11.1]
Critical value [11.1].	Second derivative [11.2]
Decreasing function [11.1]	Second derivative test [11.2]
First derivative test [11.1]	Second derivative test for absolute maximum and minimum [11.4]
Increasing function [11.1]	Symmetry [11.3]
Inflection point [11.2]	Third derivative [11.2]

SAMPLE TEST *For additional practice there are a large number of review problems categorized by objective in the Student Solution Manual. The following sample test (40 minutes) is intended to review the main ideas of this chapter.*

1. **a.** Find all derivatives of $y = 1 - 3x^3 + x^5$.
   b. Find the first four derivatives of $g(x) = 3x^5 - 3x^{-1}$.

Graph the functions in Problems 2–5.

2. $x^2 - 6x + 3y - 4 = 0$

3. $g(x) = x^3 + 3x^2 - 9x + 5$

4. $y = 6x^{1/3} - 2x^{1/2}$

5. $y = \dfrac{x - 2}{x + 1}$

Find all the relative maximums and minimums for the functions in Problems 6 and 7.

6. $f(x) = 8x^2 - 2x^4$

7. $g(x) = x + 8/x$

8. Find the absolute maximum and minimum for
$$y = (x - 1)(x + 3)^3 \text{ on } [-5, 5]$$

9. A manufacturer has determined that the price of selling x items is given by the formula
$$p(x) = 50 - (x/1000)^2$$
   Is the marginal revenue increasing or decreasing when $x = 100$ items?

10. A rectangular cardboard poster is to have 208 square inches for printed matter. It is also to have a 3-inch margin at the top and a 2-inch margin at the sides and bottom.

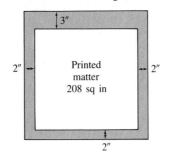

Find the length and width (to the nearest inch) of the poster so that the amount of cardboard used is minimized.

Cumulative Review for Chapters 9–11

Evaluate the limits in Problems 1–5.

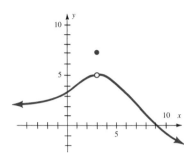

1. Find $\lim\limits_{x \to 3} f(x)$ of graph shown in the margin.

2. Find $\lim\limits_{x \to -4} \dfrac{x + 8}{x - 4}$.

3. Find $\lim\limits_{x \to 5} \dfrac{2x^2 - 7x - 15}{x - 5}$.

4. Find $\lim\limits_{x \to 2} \dfrac{3x^2 - 5x - 2}{x - 2}$.

5. Find $\lim\limits_{x \to \infty} \dfrac{(3x + 1)(5x - 2)}{x^2}$.

Find the points of discontinuity over the given domain for the functions in Problems 6–8.

6. $f(x) = \dfrac{x^2 - 15x + 56}{x - 8}$

7. $f(x) = \begin{cases} \dfrac{x^2 - 15x + 56}{x - 8} & \text{if } 0 \le x \le 10, x \ne 8 \\ 7 & \text{if } x = 8 \end{cases}$

8. $f(x) = \begin{cases} \dfrac{x^2 - 15x + 56}{x - 8} & \text{if } 0 \le x \le 10, x \ne 8 \\ 1 & \text{if } x = 8 \end{cases}$

9. Find the instantaneous rate of change for $y = 4 - \frac{1}{x}$ at $x = 1$.

10. State the definition of derivative.

Find the derivative of the functions in Problems 11–16.

11. $y = 25 - 12x^2 - 5x^3$

12. $y = 5x\sqrt{9 - x}$

13. $y = (2 - 5x)^5$

14. $y = \dfrac{(2x + 5)^8}{x^2 - 5}$

15. $3x^5 y^4 = 10$

16. $2x^2 + xy + 3y^2 = 100$

17. Find the equation of the line tangent to $f(x) = 2x - 6x^3$ at the point $(1, -4)$.

18. Graph $2x^2 + 24x - 3y - 3 = 0$ using calculus.

19. Find all derivatives of $y = 2x^4 - x^3 + 3x^2 + 25$.

20. Find all relative maximums and minimums of $f(x) = 4x^2 - x^4$.

21. Find the absolute maximum and minimum of $y = \sqrt{x} - x$ on $[0, 9]$.

22. Find the maximum profit when $R(x) = 20x + 0.11x^2 - 0.01x^3$ and $C(x) = 10{,}000 + x$.

APPLICATIONS

23. A manufacturer has determined that the price of an item is determined by the number of items sold according to the following price formula for x items sold:

$$p(x) = 25 - \frac{x}{500}$$

Is the marginal revenue increasing or decreasing when $x = 100$ items?

24. Find the marginal cost for the function $C(x) = 20x^3(5x - 100)^2$.

25. The cost-benefit model relating the cost, C, of removing $p\%$ of the pollutants from the atmosphere is

$$C(p) = \frac{50{,}000p}{100 - p}$$

where the domain for p is $[0, 100)$. Find the minimum cost.

CHAPTER 12

Exponential and Logarithmic Functions

CHAPTER CONTENTS

12.1 Exponential Functions

12.2 Logarithmic Functions

12.3 Logarithmic and Exponential Equations

12.4 Derivatives of Logarithmic and Exponential Functions

12.5 Summary and Review
Important Terms
Sample Test

APPLICATIONS

Management (*Business, Economics, Finance, and Investments*)

Finding the future value in an interest-bearing account (12.1, Problems 31–37)
Number of units sold as a function of advertising expenditures (12.2, Problem 37)
Determining the time for an interest-bearing account (12.3, Problems 31–36)
Monthly payment on an installment loan (12.3, Problems 41–42)
Maximum value of a demand equation (12.4, Problem 35)
Marginal cost (12.4, Problem 36)
Rate at which a money supply is increasing (12.4, Problem 39)
Sales growth pattern (12.4, Problem 42)
Price supply equation (12.4, Problem 44)
Amount spent on advertising (12.5, Problem 18)
Equilibrium for supply/demand (12.5, Problem 19)

Life sciences (*Biology, Ecology, Health, and Medicine*)

Determining the pH of a substance (12.2, Problem 38)
Petroleum reserves and time until they are exhausted (12.3, Problems 45–46)
Spread of a disease in a town (12.4, Problems 37–38)
Blood pressure of the aorta artery (12.4, Problem 40)
Amount of drug present in the body after being administered (12.4, Problem 41)

Social sciences (*Demography, Political Science, Population, Psychology, Society, and Sociology*)

Learning curve for a typing test (12.2, Problem 40)
World growth rate (12.3, Problems 37–38)
Forgetting curve in psychology (12.3, Problems 39–40)
Dating an artifact in archaeology (12.3, Problems 43–44; 12.6, Problem 17)
Percent who will hear a presidential announcement (12.5, Problem 20)

General interest

Magnitude of an earthquake on the Richter scale (12.2, Problem 39)

Modeling application—World Running Records

CHAPTER
OVERVIEW

There are many applications involving growth or decay that cannot be modeled with only the algebraic functions considered thus far. Two additional functions—exponential and logarithmic—are now added to our repertoire. Each is defined and graphed, and we learn how to find their derivatives.

PREVIEW

We are first introduced to exponential functions and then to logarithmic functions. We then solve equations involving these functions, and, finally, we learn how to differentiate them.

PERSPECTIVE

Persons needing calculus for management (business, economics, finance, and/or investments), life sciences (biology, ecology, health, and/or medicine), or social sciences (demography, political science, population, psychology, society, and/or sociology) need to understand how things grow (as in money or populations) and how things decay (as in forgetting or inflation). In order to understand these concepts, two of the most useful functions are introduced in this chapter: exponential and logarithmic functions. The material of this chapter will be particularly important in Section 15.4 when solving differential equations.

12.1 Exponential Functions

Linear, quadratic, polynomial, and rational functions are all called **algebraic functions**. An algebraic function is a function that can be expressed in terms of algebraic operations alone. If a function is not algebraic, it is called a **transcendental function**. In this chapter two examples of transcendental functions—*exponential* and *logarithmic* functions—are discussed.

Exponential Function

> The function f is an *exponential function* if
> $$f(x) = b^x$$
> where b is a positive constant other than 1 and x is any real number. The number x is called the **exponent** and b is called the **base**.

Recall that if n is a natural number, then

$$b^n = \underbrace{b \cdot b \cdot b \cdots \cdot b}_{n \text{ factors}}$$

Furthermore, if $b \neq 0$, then $b^0 = 1$, $b^{-n} = 1/b^n$, and $b^{1/n} = \sqrt[n]{b}$. Also, $b^{m/n} = (b^{1/n})^m$.

These definitions are used in conjunction with five **laws of exponents**:

Laws of Exponents

For a and b positive real numbers and rational numbers p and q, there are five rules that govern the use of exponents. The form 0^0 and division by zero, as well as roots of negative numbers, are excluded whenever they occur.

First law: $b^p \cdot b^q = b^{p+q}$

Second law: $\dfrac{b^p}{b^q} = b^{p-q}$

Third law: $(b^p)^q = b^{pq}$

Fourth law: $(ab)^p = a^p b^p$

Fifth law: $\left(\dfrac{a}{b}\right)^p = \dfrac{a^p}{b^p}$

EXAMPLE 1   Simplify the following expressions.

a. $16^{1/2} = (4^2)^{1/2}$
$= 4^1$
$= 4$

b. $-16^{1/2} = -(4^2)^{1/2}$
$= -4$

c. $(-16)^{1/2}$ is not defined since b must be greater than or equal to zero for square roots ($x = \frac{1}{2}$ is a square root)

d. $343^{2/3} = (7^3)^{2/3}$
$= 7^2$
$= 49$

e. $25^{-3/2} = (5^2)^{-3/2}$
$= 5^{-3}$
$= 1/5^3$
$= 1/125$ ∎

EXAMPLE 2   Use the ordinary rules of algebra to simplify the following expressions.

a. $x(x^{2/3} + x^{1/2}) = x^1 x^{2/3} + x^1 x^{1/2}$
$= x^{1+2/3} + x^{1+1/2}$
$= x^{5/3} + x^{3/2}$

b. $(x^{1/2} + y^{1/2})(x^{1/2} - y^{1/2}) = x^{1/2}x^{1/2} - x^{1/2}y^{1/2} + x^{1/2}y^{1/2} - y^{1/2}y^{1/2}$
$= x - y$ ∎

The next step is to enlarge the domain of x to include all real numbers. This is done with the help of the following property:

Squeeze Theorem for Exponents

Suppose b is a real number greater than 1. Then for any real number x there is a unique real number b^x. Moreover, if p and q are any two rational numbers such that $p < x < q$, then
$$b^p < b^x < b^q$$

The squeeze theorem gives meaning to expressions such as $2^{\sqrt{3}}$. Since

$$1.732 < \sqrt{3} < 1.733,$$

the squeeze theorem says that

$$2^{1.732} < 2^{\sqrt{3}} < 2^{2.733}$$

This means that even though only rational exponents are defined, we can handle any real number exponent by using the squeeze theorem to give us any desired degree of accuracy.

We can now consider an exponential function since the definition allows exponents to be any real number.

EXAMPLE 3   Graph the function $f(x) = 2^x$.

Solution   Begin by plotting points, as shown in the table and Figure 12.1.

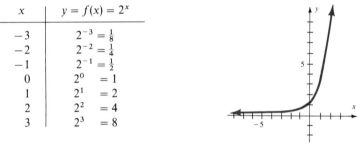

x	$y = f(x) = 2^x$
-3	$2^{-3} = \frac{1}{8}$
-2	$2^{-2} = \frac{1}{4}$
-1	$2^{-1} = \frac{1}{2}$
0	$2^0 \quad = 1$
1	$2^1 \quad = 2$
2	$2^2 \quad = 4$
3	$2^3 \quad = 8$

Figure 12.1 Graph of $y = 2x$

Connect these points with a smooth curve to obtain the graph shown in Figure 12.1. Note that the squeeze theorem allows us to draw this as a smooth curve. ∎

EXAMPLE 4   Sketch $f(x) = \left(\dfrac{1}{2}\right)^x$

Solution   Notice that $y = (1/2)^x = (2^{-1})^x$ or 2^{-x}.

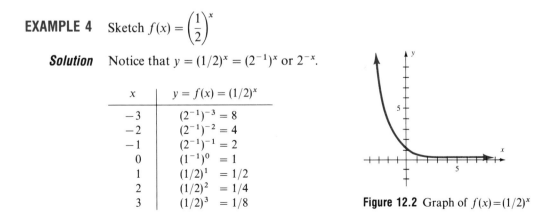

x	$y = f(x) = (1/2)^x$
-3	$(2^{-1})^{-3} = 8$
-2	$(2^{-1})^{-2} = 4$
-1	$(2^{-1})^{-1} = 2$
0	$(1^{-1})^0 \quad = 1$
1	$(1/2)^1 \quad = 1/2$
2	$(1/2)^2 \quad = 1/4$
3	$(1/2)^3 \quad = 1/8$

Figure 12.2 Graph of $f(x)=(1/2)^x$

Connect the points shown in the table with a smooth graph, as shown in Figure 12.2. ∎

The exponential function can be used as a model for certain types of growth or decay. If $b > 1$ (as in Example 3), then the exponential function is called a **growth function**, and if $0 < b < 1$ (as in Example 4), then it is called a **decay function**.

Interest*

One of the most common examples of growth is the way money grows in a bank account. This is a very important concept not only in business but in personal money management. If a sum of money, called the **principal**, is denoted by P and invested at an annual interest rate of r for t years, then the *future amount* of money present is denoted by A and is found by

$$A = P + I$$

where I denotes the amount of interest. **Interest** is an amount of money paid for the use of another's money. **Simple interest** is found by multiplication:

$$I = Prt$$

For example, $1,000 invested for 3 years at 15% simple interest generates interest of

$$I = \$1,000(0.15)(3) = \$450$$

so the future amount in 3 years is

$$A = \$1,000 + \$450 = \$1,450$$

Most businesses, however, pay interest on the interest as well as on the principal, and when this is done, it is called **compound interest**. For example, $1,000 invested at 15% annual interest compounded annually for 3 years can be found as follows:

First year: $A = P + I$

$\qquad\qquad = P \cdot 1 + Pr \qquad\qquad I = Prt$ and $t = 1$

$\qquad\qquad = P(1 + r) \qquad\qquad$ For this example, $A = \$1,000(1 + 0.15)$
$\qquad\qquad\qquad\qquad\qquad\qquad\qquad\qquad\quad = \$1,150$

Second year: $A = P(1 + r) + I \qquad$ The total amount from the first year
$\qquad\qquad\qquad\qquad\qquad\qquad\qquad\quad$ becomes the principal for the second year

$\qquad\qquad = P(1 + r) \cdot 1 + P(1 + r) \cdot r$

$\qquad\qquad = P(1 + r)(1 + r)$

$\qquad\qquad = \mathbf{P(1 + r)^2} \qquad\qquad$ For this example, $A = \$1,000(1 + 0.15)^2$
$\qquad\qquad\qquad\qquad\qquad\qquad\qquad\qquad\quad = \$1,322.50$

Third year: $A = \mathbf{P(1 + r)^2 \cdot 1 + P(1 + r)^2 \cdot r}$

$\qquad\qquad = \mathbf{P(1 + r)^2(1 + r)}$

$\qquad\qquad = \mathbf{P(1 + r)^3} \qquad\qquad$ For this example, $A = \$1,000(1 + 0.15)^3$
$\qquad\qquad\qquad\qquad\qquad\qquad\qquad\qquad\quad = \$1,520.88$

Notice that with simple interest the amount in 3 years is $1,450 as compared with $1,520.88 when interest is compounded annually.

---

* This is a review of material in Chapter 5. It is included here for those who skipped Chapter 5.

The pattern illustrated in Example 3 can be generalized:

Compound Interest or
Future Value Formula

> If a principal of P dollars is invested at an interest rate of i per period for a total of N periods, then the future amount A is given by the formula
> $$A = P(1 + i)^N$$

Compound interest is usually stated in terms of an annual interest rate r and a given number of years t. The frequency of compounding (that is, the number of compoundings per year) is denoted by n where

$n = 1$ if compounded annually
$n = 2$ if compounded semiannually
$n = 4$ if compounded quarterly
$n = 12$ if compounded monthly
$n = 360$ if compounded daily (ordinary interest)
$n = 365$ if compounded daily (exact interest)

Therefore, $i = r/n$ and $N = nt$, as illustrated in Examples 5 and 6.

EXAMPLE 5 If \$12,000 is invested for 5 years at 18% compounded semiannually, what is the amount present at the end of 5 years?

Solution $P = \$12,000; r = 0.18; i = \frac{0.18}{2} = 0.09; t = 5; N = 2(5) = 10$; thus

$$A = P(1 + i)^N$$
$$= \$12,000(1 + 0.09)^{10}$$
$$\approx \$28,408.36 \qquad \text{Use a calculator and round answers involving money to the nearest cent; however, do not round until you are ready to state your } \textit{final} \text{ answer} \qquad \blacksquare$$

EXAMPLE 6 Rework Example 5 except now the interest is compounded monthly.

Solution $P = \$12,000; i = \frac{0.18}{12} = 0.015; N = 5(12) = 60$. Thus

$$A = \$12,000(1 + 0.015)^{60}$$
$$\approx \$29,318.64 \qquad \qquad \blacksquare$$

Examples 5 and 6 show how the amount increases as the number of times it is compounded increases. A reasonable extension is to ask what happens if the interest is compounded even more frequently than monthly. Can we compound daily, hourly, every minute, or every split second? The answer is yes; in fact, money can be compounded **continuously**, which means that at every instant the newly accumulated interest is used as part of the principal for the next instant. In order to understand these concepts consider the following contrived example. Suppose \$1 is invested at 100% interest for 1 year compounded at different intervals. The compound interest formula for this example is

$$A = \left(1 + \frac{1}{n}\right)^n$$

The calculations of this formula for different values of n are shown in Table 12.1.

TABLE 12.1
Effect of compound interest on
a $1 investment

Number of periods	Formula	Amount
Annually, $n = 1$	$(1 + \frac{1}{1})^1$	$2.00
Semiannually, $n = 2$	$(1 + \frac{1}{2})^2$	$2.25
Quarterly, $n = 4$	$(1 + \frac{1}{4})^4$	$2.44
Monthly, $n = 12$	$(1 + \frac{1}{12})^{12}$	$2.61
Daily, $n = 360$	$(1 + \frac{1}{360})^{360}$	$2.715
Hourly, $n = 8640$	$(1 + \frac{1}{8640})^{8640}$	$2.7181

If these calculations are continued for even larger n, you will obtain the following:

$n = 10,000$	the formula yields	2.718145926
$n = 100,000$		2.718268237
$n = 1,000,000$		2.718280469
$n = 10,000,000$		2.718281828
$n = 100,000,000$		2.718281828

The calculator cannot distinguish values of $(1 + 1/n)^n$ for larger n. These values are approaching a particular number. This number, it turns out, is an irrational number, so it does not have a convenient decimal representation. (That is, its decimal representation does not terminate and does not repeat.) Mathematicians have agreed to denote this number by the symbol e, which is defined as a limit:

The Number e

$$\lim_{n \to \infty} (1 + 1/n)^n = e$$

For interest *compounded continuously*, the following formula is used:

$$A = Pe^{rt}$$

You can find e, as well as powers of e, by using Table 10 in Appendix E or by using a calculator.

EXAMPLE 7 Find e, e^2, and e^{-3}.

Solution On a calculator, locate a key labeled e^x. First enter the value for x, then press e^x:

$e \approx 2.718281828$ PRESS: $\boxed{1}\ \boxed{e^x}$

$e^2 \approx 7.389056099$ $\boxed{2}\ \boxed{e^x}$

$e^{-3} \approx 0.0497870684$ $\boxed{3}\ \boxed{+/-}\ \boxed{e^x}$ ■

EXAMPLE 8   Find the future value of the principal in Example 5 if the interest is compounded continuously.

Solution   Since $P = \$12,000$, $r = 0.18$, and $t = 5$,

$$A = \$12,000e^{(0.18)(5)}$$

$$\approx \$29,515.24 \quad \text{PRESS:} \quad \boxed{.18} \boxed{\times} \boxed{5} \boxed{=} \boxed{e^x} \boxed{\times} \boxed{12000} \boxed{=} \quad \blacksquare$$

You should memorize at least the first six digits of e:

$$e \approx 2.71828$$

Problem Set 12.1

Simplify the expressions in Problems 1–18. Eliminate negative exponents from your answers.

1. $25^{1/2}$ **2.** $-25^{1/2}$ **3.** $(-25)^{1/2}$

4. $(-27)^{1/3}$ **5.** $-27^{1/3}$ **6.** $27^{1/3}$

7. $7^{1/3} \cdot 7^{2/3}$ **8.** $8^{4/3} \cdot 8^{-1/3}$ **9.** $1000^{-1/3}$

10. $0.001^{-2/3}$ **11.** $100^{-3/2}$ **12.** $0.01^{-3/2}$

13. $x^{1/2}(x^{1/2} + x^{1/2})$ **14.** $x(x^{1/2} + x^{-1/2})$

15. $x^{2/3}(x^{-2/3} + x^{1/3})$ **16.** $x^{1/4}(x^{3/4} + x^{-1/4})$

17. $(x^{1/2} + y^{1/2})^2$ **18.** $(x^{1/2} - y^{1/2})^2$

Sketch the graph of each function in Problems 19–24.

19. $y = 3^x$ **20.** $y = 4^x$ **21.** $y = (1/3)^x$

22. $y = e^x$ **23.** $y = e^{-x}$ **24.** $y = -e^{-x}$

Evaluate the expressions in Problems 25–30.

25. e^3 **26.** e^{-2} **27.** $e^{0.05}$

28. $e^{-0.2}$ **29.** $e^{0.045}$ **30.** $e^{-4.85}$

APPLICATIONS

31. If $\$1,000$ is invested at 7% compounded annually, how much money will be in the account in 25 years?

32. If $\$1,000$ is invested at 12% compounded semiannually, how much money will there be in 10 years?

33. If $\$1,000$ is invested at 16% compounded continuously, how much money will there be in 25 years?

34. If $\$1,000$ is invested at 14% compounded continuously, how much money will there be in 10 years?

35. If $\$8,500$ is invested at 18% compounded monthly, how much money will there be in 4 years?

36. If $\$3,600$ is invested at 15% compounded daily, how much money will there be in 7 years? (Use a 365-day year; this is *exact interest.*)

37. If $\$9,400$ is invested at 14% compounded daily, how much money will there be in 6 months? (Use a 360-day year; this is *ordinary interest.*)

12.2 Logarithmic Functions

Suppose the exponent for an exponential function is the unknown. For example, the compound interest formula

$$A = P(1 + i)^N$$

was used in the last section to find A. Now suppose that A, P, and i are known, but the length of time it will take for P to grow to A is not known. If the exponent in an equation is the unknown value, the equation is called an **exponential equation**. Consider

$$A = b^x$$

where $b > 1$. How can we solve this exponential equation for x? Notice that

x is the exponent of base b that yields the value A

This can be rewritten as

> x = **exponent of base** b **to get** A

It appears that the equation is now solved for x, but we have simply changed the notation. The expression "exponent of b to get A" is called, for historical reasons, "the log of A to the base b." That is,

> x = **log** A **to base** b

And this phrase is shortened to the notation

> $x = \log_b A$

The term *log* is an abbreviation for **logarithm**.

Logarithmic Function

> The function f defined for $x > 0$ by
> $$f(x) = \log_b x$$
> where $b > 0$, $b \neq 1$, is called the **logarithmic function with base** b.

Remember the definition of logarithm is nothing more than a notational change.

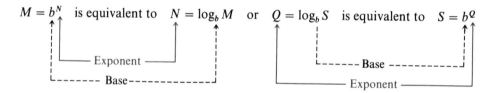

$M = b^N$ is equivalent to $N = \log_b M$ or $Q = \log_b S$ is equivalent to $S = b^Q$

EXAMPLE 1 Change the following expressions from exponential form to logarithmic form.

a. $5^2 = 25$ — The base is 5 and the exponent is 2: $\log_5 25 = 2$.
b. $3^2 = 9$ — This is the same as $\log_3 9 = 2$.
c. $\frac{1}{8} = 2^{-3}$ — In logarithmic form: $\log_2 \frac{1}{8} = -3$.
d. $\sqrt{16} = 4$ — Logarithmic form: $\log_{16} 4 = \frac{1}{2}$. (Remember $\sqrt{16} = 16^{1/2}$.) ■

EXAMPLE 2 Change the following expressions from logarithmic form to exponential form.

a. $\log_{10} 100 = 2$ — The base is 10 and the exponent is 2: $10^2 = 100$.
b. $\log_{10} 1/1000 = -3$ — This is the same as $10^{-3} = 1/1000$.
c. $\log_3 1 = 0$ — Exponential form: $3^0 = 1$. ■

To **evaluate a logarithm** means to find a numerical value for the given logarithm. The first ones you are asked to evaluate use a property of logarithms that says if the bases on two logarithms are the same and the numbers are equal, then the exponents must be equal.

Exponential Property of
Equality

For positive real b,

If $b^x = b^y$, then $x = y$

EXAMPLE 3 Evaluate the given logarithms.

  a. $\log_2 64$ **b.** $\log_3 \frac{1}{9}$ **c.** $\log_9 27$ **d.** $\log_{10} 1$ **e.** $\log_{10} 100$

Solution **a.** Since it is usually necessary to supply a variable to convert to exponential form, we use N in these examples. That is,

$$N = \log_2 64$$

We write this in exponential form: $2^N = 64$

$$2^N = 2^6$$
$$N = 6 \qquad \text{Use the exponential property of equality}$$

Thus $\log_2 64 = 6$.
  b. Let $N = \log_3 \frac{1}{9}$, so $3^N = \frac{1}{9}$

$$3^N = 3^{-2}$$
$$N = -2$$

Thus $\log_3 \frac{1}{9} = -2$.
  c. $N = \log_9 27$, so $9^N = 27$

$$3^{2N} = 3^3$$
$$2N = 3$$
$$N = \frac{3}{2}$$

Thus $\log_9 27 = \frac{3}{2}$.
  d. $\log_{10} 1 = 0$. Can you do this mentally?
  e. $\log_{10} 100 = 2$. ∎

Suppose you cannot use the exponential property of equality. For example, suppose you want to find $\log_{10} 5.03$. Since 5.03 is between 1 and 10 and

$$10^0 = 1$$
$$10^x = 5.03 \qquad \text{You want to find this } x$$
$$10^1 = 10$$

the number x should be between 0 and 1, by the squeeze theorem for exponents. There are tables that show approximations for these exponents. However, calculators have, to a large extent, eliminated the need for extensive log tables. To find x on a calculator that has logarithmic keys, push the keys in the indicated order. For example, $\log_{10} 5.03$ is found by pressing the following keys:

5.03 log   DISPLAY: 0.7015679851

Number first, then log key.   Calculator answers are approximate.

The key for $\log_{10}$ is simply labeled *log*. Base 10 is fairly common, and if the logarithm is to the base 10 it is called a **common logarithm** and is written without the subscript 10.

EXAMPLE 4   Evaluate the following logarithms correct to four decimal places.

    a. log 7.68   **b.** log 852   **c.** log 0.00728

Solution **a.** PRESS: $\boxed{7.68}\boxed{\log}$ DISPLAY: 0.88536122
             To four decimal places, $x = 0.8854$
      b. PRESS: $\boxed{852}\boxed{\log}$ DISPLAY: 2.930439595
             To four decimal places, 2.930.
      c. PRESS: $\boxed{.00728}\boxed{\log}$ DISPLAY: -2.137868621
             To four decimal places, $\log 0.00728 = -2.1379$.   ■

Keep in mind that a logarithm is an exponent. That is, for Example 4a, the answer 0.8854 is the *exponent* on the base of 10 that gives the given number, 7.68, that is:

$$10^{0.8854} \approx 7.68$$

In addition to common logarithms, another logarithm is frequently encountered. This is a logarithm to the base e called a **natural logarithm**. Natural logarithms are denoted by

$$\log_e x = \ln x$$

and this is pronounced as "lon x."

EXAMPLE 5   Find ln 3.49 to two decimal places.

Solution BY CALCULATOR: If your calculator has a $\boxed{\log}$ key, chances are it also has an $\boxed{\ln}$ key:

    $\boxed{3.49}\boxed{\ln}$ DISPLAY: 1.249901736   ■

It is important to realize that any logarithms (whether common or natural) are only as accurate as the input number (3.49 in this example). Thus, rounding to two decimal places gives $\ln 3.49 \approx 1.25$.

EXAMPLE 6   Find ln 0.403.

Solution BY CALCULATOR: $\boxed{.403}\boxed{\ln}$ DISPLAY: -0.908818717
         Thus $\ln 0.403 \approx -0.909$. Always keep in mind that a logarithm is nothing more than an exponent, so this calculator display means that

$$e^{-0.909} \approx 0.403$$

                                                   ■

To graph a logarithmic function you use the definition of logarithm and then construct a table of values.

EXAMPLE 7   Graph $y = \log_2 x$.

Solution   From the definition of logarithm, $2^y = x$; use this equation to construct a table of values:

y	$x = 2^y$
-3	$2^{-3} = \frac{1}{8}$
-2	$2^{-2} = \frac{1}{4}$
-1	$2^{-1} = \frac{1}{2}$
0	$2^0 = 1$
1	$2^1 = 2$
2	$2^2 = 4$
3	$2^3 = 8$

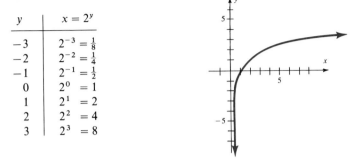

Figure 12.3 Graph of $y = \log_2 x$ ■

Compare your answer with the graph of $y = 2^x$ shown in Figure 12.4.

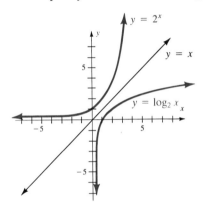

Figure 12.4 Graphs of $y = 2^x$, $y = \log_2 x$, and $y = x$

Functions whose graphs are symmetric with respect to the line $y = x$ as shown in Figure 12.4 are said to be **inverse functions**. This relationship is needed in order to find e on several brands of calculators. If a calculator has

$$\boxed{\ln x} \quad \text{and} \quad \boxed{\text{INV}}$$

keys, but no e^x key, you can still use the calculator to find e^x. Since

$$y = \ln x \quad \text{and} \quad y = e^x$$

are inverse functions, to find e (or e^1) press

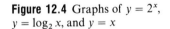   DISPLAY: 2.718281828

These two keys give the inverse of
the $\ln x$ function; that is, they give e^x
when the x value is input just prior
to pressing these keys

If, for example, you need $e^{5.2}$ on such a calculator, press

   DISPLAY: 181.2722419

Problem Set 12.2

Write the equations in Problems 1–6 in logarithmic form.

1. $64 = 2^6$ **2.** $125 = 5^3$ **3.** $9 = (\frac{1}{3})^{-2}$

4. $\frac{1}{2} = 4^{-1/2}$ **5.** $a = b^c$ **6.** $m = n^p$

Write the equations in Problems 7–12 in exponential form.

7. $\log_4 2 = \frac{1}{2}$ **8.** $\log_2 \frac{1}{8} = -3$

9. $\log 0.01 = -2$ **10.** $\log 10{,}000 = 4$

11. $\ln e^2 = 2$ **12.** $\ln x = 0.03$

Use the definition of logarithm or a calculator to evaluate the expressions in Problems 13–30.

13. $\log_b b^2$ **14.** $\log_t t^3$ **15.** $\log_\pi \sqrt{\pi}$

16. $\ln e^4$ **17.** $\log 1000$ **18.** $\log_2 8$

19. $\log_{16} 1$ **20.** $\log 4.27$ **21.** $\log 1.08$

22. $\log 8.43$ **23.** $\log 9760$ **24.** $\log 0.042$

25. $\log 0.321$ **26.** $\log 0.0532$ **27.** $\ln 2.27$

28. $\ln 16.77$ **29.** $\ln 2$ **30.** $\ln 13$

Graph the functions in Problems 31–36.

31. $y = \log_3 x$ **32.** $y = \log_{1/2} x$ **33.** $y = \ln x$

34. $y = \log x$ **35.** $y = \ln \sqrt{x}$ **36.** $y = \log_\pi x$

APPLICATIONS

37. An advertising agency conducts a survey and finds that the number of units sold, N, is related to the amount a spent on advertising (in dollars) according to the following formula:

$$N = 1500 + 300 \ln a \qquad a \geq 1$$

  a. How many units are sold after spending $1,000?

  b. How many units are sold after spending $50,000?

38. The pH of a substance measures its acidity or alkalinity. It is found by the formula

$$pH = -\log[H^+]$$

where $[H^+]$ is the concentration of hydrogen ions in an aqueous solution given in moles per liter.

  a. What is the pH (to the nearest tenth) of a lemon for which $[H^+] = 2.86 \times 10^{-4}$?

  b. What is the pH (to the nearest tenth) of rain water for which $[H^+] = 6.31 \times 10^{-7}$?

39. The Richter scale for measuring earthquakes, developed by Gutenberg and Richter, relates the energy E (in ergs) to the magnitude of the earthquake, M, by the formula

$$M = \frac{\log E - 11.8}{1.5}$$

  a. A small earthquake is one that releases 15^{15} ergs of energy. What is the magnitude of such an earthquake on the Richter scale?

  b. A large earthquake is one that releases 10^{25} ergs of energy. What is the magnitude of such an earthquake on the Richter scale?

40. A learning curve describes the rate at which a person learns certain tasks. If a person sets a goal of typing N words per minute (wpm), the length of time, t (in days), to achieve this goal is given by

$$t = -62.5 \ln(1 - N/80)$$

  a. How long would it take to learn to type 30 wpm?

  b. If we accept this formula, is it possible to learn to type 80 wpm?

41. Prove that $\log_b 1 = 0$ for all $b > 0$, $b \neq 1$.

42. Prove that $\log_b b^x = x$ for all x, where $b > 0$, $b \neq 1$.

43. Prove that $\ln e^x = x$ for all x.

44. Prove that $b^{\log_b x} = x$ for all x, where $b > 0$, $b \neq 1$.

12.3 Logarithmic and Exponential Equations

We now discuss solving two types of equations involving transcendental functions. The first type is called a **logarithmic equation**. All logarithmic equations fall into one of four categories, as illustrated by Examples 1–4.

EXAMPLE 1   Solve $\log_2 \sqrt{2} = x$ for x.

    Solution   *The exponent is the unknown*; apply the definition of logarithm:

$$2^x = \sqrt{2}$$
$$2^x = 2^{1/2}$$

$$x = \frac{1}{2} \qquad \text{Exponential property of equality}$$

■

EXAMPLE 2 Solve $\log_x 25 = 2$.

Solution *The base is the unknown*; apply the definition of logarithm:

$$x^2 = 25$$
$$x = \pm 5$$

Be sure the values you obtain are permissible values for the definition of a logarithm. In this case, $x = -5$ is not a permissible value since a logarithm with a negative base is not defined. Therefore the solution is $x = 5$. ∎

EXAMPLE 3 Solve $\ln x = 5$.

Solution *The power itself is the unknown*; apply the definition of logarithm:

$$e^5 = x$$
$$x \approx 148.41 \qquad \text{PRESS:} \quad \boxed{5}\ \boxed{e^x} \qquad \text{DISPLAY:} \qquad 148.4131591 \qquad ∎$$

The first three categories of logarithmic equation problems all use the definition of logarithm. The last category requires a property that follows from the exponential property of equality.

Log of Both Sides Theorem

> If A, B, and b are positive real numbers with $b \neq 1$, then
> $$\log_b A = \log_b B \qquad \text{is equivalent to} \qquad A = B$$

EXAMPLE 4 Solve $\log_5 x = \log_5 72$.

Solution Use the log of both sides theorem: $x = 72$. ∎

When solving a logarithmic equation, the goal is to write the logarithmic equation with a *single* logarithmic function on either one or both sides of the equation. If the logarithmic equation has the form

$$\log_b A = \log_b B$$

you can use the log of both sides theorem to find the unknown, as Example 4 shows. On the other hand, if the equation has the form

$$\log_b A = N$$

(a log on one side only), then you apply the definition of logarithm to solve the equation, as Examples 1–3 show. In either case, you must first algebraically simplify in order to put the equations into one of these forms. To do this you need some additional theorems about logarithms.

Since logarithms are exponents, we can rewrite the laws of exponents in logarithmic form. For example,

$$b^M \cdot b^N = b^{M+N}$$

This means that if $b^M = A$ and $b^N = B$, then in logarithmic form

$$M = \log_b A \qquad \text{and} \qquad N = \log_b B$$

But if we apply the definition of logarithm to $b^M \cdot b^N = b^{M+N}$, we obtain

$$M + N = \log_b b^M b^N$$

or, by substitution,

$$\log_b A + \log_b B = \log_b AB$$

This is the logarithmic form for the first law of exponents. The second and third laws of exponents can also be written in logarithmic form, as summarized below. You are asked to derive the second and third laws of logarithms in the problem set.

First Law of Logarithms	$\log_b AB = \log_b A + \log_b B$	The log of the product of two numbers is the sum of the logs of those numbers.
Second Law of Logarithms	$\log_b \dfrac{A}{B} = \log_b A - \log_b B$	The log of the quotient of two numbers is the log of the numerator minus the log of the denominator.
Third Law of Logarithms	$\log_b A^p = p \log_b A$	The log of the *p*th power of a number is *p* times the log of that number.

Examples 5 and 6 use these laws of logarithms to solve a logarithmic equation that reduces to the type with a logarithm on both sides of the equation. This means that the last step will be to use the log of both sides theorem.

EXAMPLE 5 Solve $\log_8 3 + (1/2)\log_8 25 = \log_8 x$ for x.

Solution

$$\log_8 3 + \frac{1}{2}\log_8 25 = \log_8 x$$

$$\log_8 3 + \log_8 25^{1/2} = \log_8 x \qquad \text{Third law of logarithms}$$
$$\log_8 3 + \log_8 5 = \log_8 x \qquad 25^{1/2} = \sqrt{25} = 5$$
$$\log_8(3 \cdot 5) = \log_8 x \qquad \text{First law of logarithms}$$
$$15 = x \qquad \text{Log of both sides theorem}$$

The solution is 15. ∎

EXAMPLE 6 Solve $\ln x - (1/2)\ln 2 = (1/2)\ln(x + 4)$ for x.

Solution

$$\ln x - \frac{1}{2}\ln 2 = \frac{1}{2}\ln(x + 4)$$

$$\ln x - \ln 2^{1/2} = \ln(x + 4)^{1/2}$$
$$\ln x - \ln \sqrt{2} = \ln \sqrt{x + 4}$$
$$\ln\left(\frac{x}{\sqrt{2}}\right) = \ln \sqrt{x + 4}$$
$$\frac{x}{\sqrt{2}} = \sqrt{x + 4}$$
$$\frac{x^2}{2} = x + 4$$
$$x^2 - 2x - 8 = 0$$
$$(x - 4)(x + 2) = 0$$
$$x = 4, -2$$

Since $\ln(-2)$ is not defined, $x = -2$ is an extraneous root. Therefore, the solution is 4. ∎

The second type of equations are **exponential equations** and fall into one of three categories.

Base: 10   (common log) e (natural log) b (arbitrary base)

Example: $10^x = 5$ $e^x = 3.456$ $7^x = 3$

Examples 7–9 illustrate each of these types.

EXAMPLE 7   Solve $10^{5x+3} = 195$ for x.

Solution   This is a *common log* problem. Use the definition of logarithm to write it in logarithmic form: $\log 195 = 5x + 3$.

Now solve this equation for x: $\log 195 - 3 = 5x$

$$\frac{\log 195 - 3}{5} = x$$

For an approximate answer use a calculator.

PRESS: | 195 | | log | | − | | 3 | | = | | ÷ | | 5 | | = | DISPLAY: -0.1419930777

■

EXAMPLE 8   Solve $e^{-0.000425t} = \dfrac{1}{2}$ for t.

Solution   This is a natural logarithm problem. Use the definition to write

$$\ln 0.5 = -0.000425t$$

$$t = \frac{\ln .5}{-0.000425}$$

Use a calculator to find $t \approx 1630.934543$; this is about 1600.   ■

Since you do not have tables or calculator keys for bases other than 10 or e, you must proceed differently for an *arbitrary base b*. You should use the log of both sides theorem and the procedure for solving equations. You generally can elect to work with base 10 or base e.

EXAMPLE 9   Solve $7^x = 3$ for x.

Solution   We will work this two ways: base 10 and base e so it is clear that both produce exactly the same answer.

	Base e	*Base 10*	
	$\ln 7^x = \ln 3$	$\log 7^x = \log 3$	Log of both sides
	$x \ln 7 = \ln 3$	$x \log 7 = \log 3$	Third law of exponents
Warning: $\dfrac{\log 3}{\log 7} \neq \log \dfrac{3}{7}$	$x = \dfrac{\ln 3}{\ln 7}$	$x = \dfrac{\log 3}{\log 7}$	Divide both sides by the coefficient of x
	$\approx \dfrac{1.099}{1.946}$	$\approx \dfrac{0.4771}{0.8451}$	Evaluate by calculator
	≈ 0.5646	≈ 0.5646	Divide

■

There is another method for solving the exponential equation of Example 9. This one directly applies the definition of logarithm:

$$7^x = 3 \quad \text{is the same as} \quad \log_7 3 = x$$

If there were a logarithm base 7 key on a calculator or a log base 7 table, you would have the answer. Since there are not, you need one final logarithm theorem that changes logarithms from one base to another.

Change of Base Theorem

$$\log_a x = \frac{\log_b x}{\log_b a}$$

Notice that to change from base a to another (possibly more familiar) base b, you simply change the base on the given logarithm from a to b and then divide by the logarithm to the base b of the old base a. The proof of this theorem is identical to the steps outlined in Example 9. That is, if $a^N = x$, then $N = \log_a x$ and

$$\log_b a^N = \log_b x \qquad \text{Take the } \log_b \text{ of both sides}$$
$$N \log_b a = \log_b x \qquad \text{Third law of exponents}$$
$$N = \frac{\log_b x}{\log_b a} \qquad \text{Divide both sides by } \log_b a$$

$$\text{Thus, } \log_a x = \frac{\log_b x}{\log_b a} \qquad \text{Substitute } N = \log_a x$$

EXAMPLE 10 Change $\log_7 3$ to logarithms with base 10 and evaluate.

Solution $\log_7 3 = \dfrac{\log 3}{\log 7} \approx \dfrac{0.4771}{0.8451} \approx 0.5646$ ∎

EXAMPLE 11 Solve $6^{3x+2} = 200$ for x.

Solution
$$\log_6 200 = 3x + 2 \qquad \text{Use the definition of logarithm}$$
$$\log_6 200 - 2 = 3x \qquad \text{Solve for } x$$
$$x = \frac{\log_6 200 - 2}{3}$$
$$= \frac{\dfrac{\log 200}{\log 6} - 2}{3}$$

By calculator, $x \approx 0.3190157417 \approx 0.32$. ∎

An important application of exponential equations involves the calculation of human population growth. Human populations grow according to the exponential equation

$$P = P_0 e^{rt}$$

where P_0 is the size of the initial population, r is the growth rate, t is the length of time, and P is the size of the population after time t.

EXAMPLE 12 On April 3, 1987 the world population reached 5 billion. If the annual growth rate is 2%, when will the world population reach 6 billion?

Solution The initial population P_0 is 5 (billion), P is 6 (billion), $r = 0.02$, and t is the unknown. Substitute the known values into the population growth formula:

$$6 = 5e^{0.02t}$$

$$\frac{6}{5} = e^{0.02t}$$

$$0.02t = \ln 1.2 \qquad \text{Use the definition of logarithm; note that } \tfrac{6}{5} = 1.2$$

$$t = \frac{\ln 1.2}{0.02}$$

$$\approx 9.11607784$$

We would expect the world's population to reach 6 billion about 9 years after it reached 5 billion; this would be in 1996 (Feb. 9, 1996, if we use the fractional part, but this is really assuming more accuracy than the 2% population growth estimate would allow). ∎

EXAMPLE 13 The 1970 population of San Antonio, Texas, was 654,153, and the 1980 population was 783,296. What was the growth rate of San Antonio for this period?

Solution Since $P_0 = 654{,}153$, $P = 783{,}296$, and $t = 10$, we have

$$P = P_0 e^{rt}$$

$$\frac{P}{P_0} = e^{rt}$$

$$rt = \ln(P/P_0)$$

$$r = \frac{1}{t}\ln\frac{P}{P_0}$$

For this example,

$$r = \frac{1}{10}\ln\frac{783{,}296}{654{,}153}$$

$$\approx 0.0180169389$$

The growth rate is about 1.8%. ∎

Problem Set 12.3

Solve Problems 1–30.

1. $\log_5 25 = x$

2. $\log_2 128 = x$

3. $\log(\tfrac{1}{10}) = x$

4. $\log_x 84 = 2$

5. $\log_x 28 = 2$

6. $\log x = 2$

7. $\ln x = 3$

8. $\ln x = \ln 14$

9. $\ln 9.3 = \ln x$

10. $\log_3 x^2 = \log_3 125$

11. $\ln x^2 = \ln 12$

12. $\log_2 8\sqrt{2} = x$

13. $\log_3 27\sqrt{3} = x$

14. $\log_x 1 = 0$

15. $\log_x 10 = 0$

16. $\log x = 5$

17. $2^x = 128$

18. $8^x = 32$

19. $125^x = 25$

20. $(\tfrac{2}{3})^x = \tfrac{9}{4}$

21. $3^{4x-3} = \tfrac{1}{9}$

22. $27^{2x+1} = 3$

23. $\log_8 5 + \tfrac{1}{2}\log_8 9 = \log_8 x$

24. $\log_7 x - \tfrac{1}{2}\log_7 4 = \tfrac{1}{2}\log_7(2x - 3)$

25. $\ln 10 - \tfrac{1}{2}\ln 25 = \ln x$

26. $\tfrac{1}{2}\ln x = 3\ln 5 - \ln x$

27. $10^{2x-1} = 515$

28. $10^{5-3x} = 0.041$

29. $4^x = 0.82$

30. $5^{-x} = 8$

APPLICATIONS

31. If $1,000 is invested at 12% compounded semiannually, how long will it take (to the nearest half-year) for the money to double?

32. Repeat Problem 31, except compound daily and give your answer to the nearest day (365-day year).

33. Repeat Problem 31, except compound continuously and give your answer to the nearest day (365-day year).

34. How long will it take for $1,000 to triple if it is invested at 12% compounded quarterly? (Give your answer to the nearest quarter.)

35. Repeat Problem 34, except compound continuously and give your answer to the nearest day.

36. If $1,000 is invested at 12% interest compounded quarterly, how long will it take (to the nearest quarter) for the money to reach $2,500?

37. The world population reached 4 billion on March 18, 1976 and 5 billion on April 3, 1987. What was the growth rate for this period of time (to the nearest hundredth of a percent)?

38. Use the date given in Example 12 to estimate the year the population will reach 7 billion.

39. Psychologists are concerned with forgetting. In an experiment, students were asked to remember a set of nonsense syllables, such as "htm." They then had to recall the syllables after t seconds. The model used to describe forgetting in this experiment is

$$R = 80 - 27 \ln t \qquad t > 1$$

where R is the percentage of students who remember the syllables after t seconds.

a. What percentage of the students remembered the syllables after 3 seconds?

b. In how many seconds would only 10% of the students remember the syllables?

40. Solve the forgetting curve formula in Problem 39 for t.

41. If P dollars are borrowed for n months at a monthly interest rate of i, then the monthly payment m is found by the formula

$$m = \frac{Pi}{1 - (1 + i)^{-n}}$$

Use this formula to find the monthly car payment after a down payment of $2,487 on a new car costing $12,487. The car is financed for 4 years at 12%. (*Hint:* $P = \$10,000$ and $i = 0.01$.)

42. A home loan is made for $110,000 at 12% interest for 30 years. What is the monthly payment, and what is the total amount of interest paid? (*Hint:* Use the formula given in Problem 41.)

43. A formula used for carbon-14 dating in archaeology is

$$A = A_0 \left(\frac{1}{2}\right)^{t/5700} \qquad \text{or} \qquad P = \left(\frac{1}{2}\right)^{t/5700}$$

where P is the percentage of carbon-14 present after t years. Solve for t in terms of P.

44. Some bone artifacts found at the Lindemeir site in northeastern Colorado were tested for their carbon-14 content. If 25% of the original carbon-14 was still present, what is the probable age of the artifacts? Use the formula in Problem 43.

45. In 1975 ($t = 0$), the world use of petroleum, P_0, was 19,473 million barrels of oil. If the world reserves are 584,600 million barrels and the growth rate for the use of oil is k, then the total amount A used during a time interval $t > 0$ is given by

$$A = \frac{P_0}{k}(e^{kt} - 1)$$

How long will it be before the world reserves are depleted if $k = 8\%$?

46. Solve the formula in Problem 45 for t.

47. Prove that $\log_b \dfrac{A}{B} = \log_b A - \log_b B$.

48. Prove that $\log_b A^p = p \log_b A$.

12.4 Derivatives of Logarithmic and Exponential Functions

Rates of change and graphs of logarithmic and exponential functions are fairly common applications in management and the life and social sciences. In Chapters 9 and 10 the notion of a derivative as a limit was defined; that is, if $y = f(x)$, then

$$y' = \lim_{h \to 0} \frac{f(x + h) - f(x)}{h}$$

provided this limit exists. We found a variety of derivatives by using this formula and derived some derivative formulas that made the process very efficient, so that in some problems we no longer had to resort to the definition of derivative. Now we will find the derivative of the logarithmic and exponential functions. Our first attempt might be to try to apply some of our derivative formulas, but since the logarithmic and exponential functions are transcendental and not algebraic, none of the derivative formulas applies. Whenever we need to find the derivative of a new class of functions we must go back to the definition of derivative.

We begin by finding the derivative of the natural logarithm; that is, we let $y = \ln x$. We need to find

$$\lim_{h \to 0} \frac{\ln(x + h) - \ln x}{h} = \lim_{h \to 0} \frac{\ln\left(\dfrac{x + h}{x}\right)}{h} \qquad \text{This is a property of logarithms; do you see which one?}$$

$$= \lim_{h \to 0} \frac{1}{h} \ln\left(\frac{x + h}{x}\right)$$

$$= \lim_{h \to 0} \ln\left(\frac{x + h}{x}\right)^{1/h} \qquad \text{Another property of logarithms}$$

$$= \lim_{h \to 0} \ln\left(1 + \frac{h}{x}\right)^{1/h}$$

Let $m = \dfrac{x}{h}$ so that $\dfrac{h}{x} = \dfrac{1}{m}$ and $\dfrac{1}{h} = \dfrac{m}{x}$. Also, as $h \to 0$, $m \to \infty$. Now substitute these changes into the derivative formula:

$$\lim_{h \to 0} \ln\left(1 + \frac{h}{x}\right)^{1/h} = \lim_{m \to \infty} \ln\left(1 + \frac{1}{m}\right)^{m/x}$$

$$= \lim_{m \to \infty} \ln\left[\left(1 + \frac{1}{m}\right)^m\right]^{1/x}$$

$$= \lim_{m \to \infty} \frac{1}{x} \ln\left(1 + \frac{1}{m}\right)^m$$

$$= \frac{1}{x} \lim_{m \to \infty} \ln\left(1 + \frac{1}{m}\right)^m$$

$$= \frac{1}{x} \ln e \qquad \text{Remember the definition of } e$$

$$= \frac{1}{x} \qquad \ln e = 1$$

This turns out to be a very pleasing result because of its simplicity! We can also obtain a more general result by applying the chain rule:

Derivative of Logarithm

If $y = \ln|x|$, then $y' = \dfrac{1}{x}$

If $y = \ln|u|$ where u is a function of x, then $y' = \dfrac{u'}{u}$

The absolute value bars have been added because the domain of $\ln x$ includes only positive values of x. We are not differentiating $|x|$ here but are simply restricting the domain of x.

EXAMPLE 1   If $y = \ln|3x|$, find y'.

Solution $u(x) = 3x$, so $u'(x) = 3$. Thus

$$y' = \frac{3}{3x} = \frac{1}{x}$$

∎

EXAMPLE 2   If $y = \ln|5x^3 + 3x^2 - 4|$, find y'.

Solution $u(x) = 5x^3 + 3x^2 - 4$, so $u'(x) = 15x^2 + 6x$.   (This step is usually done mentally.)

$$y' = \frac{15x^2 + 6x}{5x^3 + 3x^2 - 4}$$

∎

EXAMPLE 3   If $y = (3x^2 + 2x)\ln|5x - 7|$, find y'.

Solution   Use the product rule:

$$y' = (3x^2 + 2x)\left(\frac{5}{5x - 7}\right) + (6x + 2)\ln|5x - 7|$$

$$= \frac{5x(3x + 2)}{5x - 7} + 2(3x + 1)\ln|5x - 7|$$

∎

The notation of logarithms can be confusing. Remember, $\ln x$ means $\log_e x$, $\log x$ means $\log_{10} x$. Also, be sure you distinguish between $\ln x^2$, which means $\ln(xx)$, and $(\ln x)^2$, which means $(\ln x)(\ln x)$.

EXAMPLE 4   The demand equation of a commodity is

$$d = \frac{200\ln(x + 1)}{x}$$

and the cost of producing x units is $C(x) = 2x$. Show that the profit function has a relative maximum and not a minimum point.

Solution   The profit function $P(x)$ is found by $P(x) = R(x) - C(x)$. We are given the cost function, but we need to find $R(x)$. The revenue function is found by multiplying the number of units times the demand, so $R(x) = xd = 200\ln(x + 1)$.

$$P(x) = 200\ln(x + 1) - 2x$$

$$P'(x) = 200\frac{1}{x + 1} - 2$$

$$= 200(x + 1)^{-1} - 2$$

Set $P'(x) = 0$ and solve for x:

$$200(x + 1)^{-1} - 2 = 0$$
$$200 = 2(x + 1)$$
$$99 = x$$

To determine if this is a relative maximum or minimum, check the second derivative:

$$P''(x) = -200(x + 1)^{-2}$$

This function is negative for all values of x, so the profit function has a relative maximum at $x = 99$ and no relative minimum. ■

Now we turn our attention to finding the derivative of the exponential function $y = e^x$. To do this, we use the definition of logarithm to write $x = \ln y$. We can now use the derivative rule for $\ln y$ (found above) by using implicit differentiation on the equation $\ln y = x$:

$$\frac{1}{y} \cdot \frac{dy}{dx} = 1$$

We solve for dy/dx:

$$\frac{dy}{dx} = y$$

Since $y = e^x$, we see that the derivative of e^x is e^x; what could be easier than this! This result shows one of the reasons why base e rather than base 10 is used almost exclusively in more advanced work. Again, this result is generalized using the chain rule:

Derivative of Exponential

> If $y = e^x$, then $y' = e^x$
>
> If $y = e^u$, where u is a function of x, then $y' = u' \cdot e^u$

EXAMPLE 5 If $y = e^{10x}$, then $y' = 10e^{10x}$. ■

EXAMPLE 6 If $y = e^{4x^2 + 5}$, then $y' = 8xe^{4x^2 + 5}$. ■

EXAMPLE 7 If $y = x^3 e^{4x}$, then $y' = x^3(4e^{4x}) + 3x^2 e^{4x}$ Use the product rule first
$$= 4x^3 e^{4x} + 3x^2 e^{4x} \quad \text{or} \quad x^2 e^{4x}(4x + 3)$$ ■

EXAMPLE 8 If $y = \dfrac{e^x}{\ln|x|}$, then $y' = \dfrac{\ln|x| \cdot e^x - e^x \cdot (1/x)}{\ln^2|x|}$ Use the quotient rule first

$$= \frac{\dfrac{e^x}{x}(x \ln|x| - 1)}{\ln^2|x|}$$

$$= \frac{e^x(x \ln|x| - 1)}{x \ln^2|x|}$$ ■

EXAMPLE 9 Suppose you study the spread of a disease introduced into a small town of 2000 persons. Assume that everyone in the town has an equal chance of contracting the disease. A model for predicting the number of people, N, contracting the disease is given by

$$N(t) = \frac{2000}{1 + 1999e^{-0.5t}}$$

where t is the number of days since the disease was introduced into the community. What is the rate at which members of the community are contracting the disease?

Solution The requested function is dN/dt. Write $N(t) = 2000(1 + 1999e^{-0.5t})^{-1}$. Then,

$$\frac{dN}{dt} = 2000(-1)(1 + 1999e^{-0.5t})^{-2}(-0.5)1999e^{-0.5t}$$

$$= \frac{2000(-1)(-0.5)1999e^{-0.5t}}{(1 + 1999e^{-0.5t})^2}$$

$$= \frac{1,999,000e^{-0.5t}}{(1 + 1999e^{-0.5t})^2}$$

This means, for example, that on day 0, $N(0) = 2000/(1 + 1999) = 1$ person has the disease, but dN/dt at $t = 0$ is

$$\frac{1,999,000}{2000^2} = 0.49975$$

so the rate means that nearly one person is contracting the disease every 2 days. On the other hand, after 10 days, $N(10) \approx 138$, so about 138 people have the disease and the rate at which new people are now contracting the disease is dN/dt at $t = 10$:

$$\frac{1,999,000e^{-5}}{(1 + 1999e^{-5})^2} \approx 64.33598929$$

which means that at the time $t = 10$, new people are contracting the disease at the rate of about 64 people per day. ∎

Problem Set 12.4

Find the derivatives of the functions in Problems 1–34.

1. $y = e^{2x}$

2. $y = e^{8x}$

3. $y = e^{-x}$

4. $y = e^{-3x}$

5. $f(x) = xe^x$

6. $f(x) = x^2e^{3x}$

7. $f(x) = (2x^3 + 1)e^{5x^2}$

8. $f(x) = (5x^2 - x)e^{-3x^2}$

9. $y = -2e^{5x}(3x^2 + 5)$

10. $y = (3x + 2e^{5x})^4$

11. $y = \ln|5 - x|$

12. $y = \ln|3 - x^2|$

13. $y = (2x^3 + e^{5x})^4$

14. $y = (e^{6x} - 5x^4)^3$

15. $y = \ln x^4$

16. $y = (\ln x)^4$

17. $y = \ln\sqrt{x}$

18. $y = \sqrt{\ln|x|}$

19. $y = \dfrac{\ln x^3}{e^x}$

20. $y = \dfrac{e^{3x}}{\ln x^2}$

21. $f(x) = \ln\left|\dfrac{x}{x + 2}\right|$

22. $f(x) = \ln\left|\dfrac{x - 3}{x^2}\right|$

23. $f(t) = e^{t^2}(1 - e^{-t})$

24. $f(t) = e^{(t + 1)/t}$

25. $y = (e^x - e^{-x})^2$

26. $y = (e^{-x} + e^x)^3$

27. $y = \dfrac{\ln|x|}{5x + 2}$

28. $y = \dfrac{3x^2 - 7}{\ln|x|}$

29. $f(x) = \dfrac{e^x}{\ln|3x|}$

30. $f(x) = \dfrac{e^{5x}}{\ln|5x|}$

31. $y = \dfrac{2000}{1 + 6e^{0.3x}}$

32. $y = \dfrac{500}{1 - 30e^{0.2x}}$

33. $y = \dfrac{100 \ln|x|}{1 + 40e^{-0.2x}}$

34. $y = \dfrac{900 \ln|x|}{1 - 70e^{-0.3x}}$

APPLICATIONS

35. The demand equation of a commodity is

$$d = \frac{500 \ln(x + 10)}{x^2}$$

where x is the number of units produced ($1 \le x \le 20$). The cost of producing x units is $C(x) = 2x^2$. Find the marginal revenue.

36. Find the marginal cost for the information in Problem 35.

37. The spread of a disease in a town of 5000 people is described by the model

$$N(t) = \frac{5000}{1 + 4999e^{-0.1t}}$$

where N is the number of people contracting the disease after t days.
   a. How many people have the disease on day 0?
   b. What is the rate at which people are getting the disease on day 0?

38. a. How many people in Problem 37 have the disease on the tenth day?
   b. What is the rate at which people are getting the disease on the tenth day?

39. The amount of money P, invested at 12% interest and compounded continuously for t years, has a future value A according to the formula $A = Pe^{0.12t}$. At what rate is the money increasing after 1 year? After 5 years?

40. The blood pressure in the aorta changes between beats according to the formula

$$P(t) = e^{-kt}$$

for an appropriate constant k, where t is the time in milliseconds since the last beat. What is the rate at which the pressure is changing with respect to time after 3 milliseconds if $k = 0.025$?

41. The amount of a drug present in the body is a function of the amount administered and the length of time t (in hours) since it was administered. For 5 milliliters of a certain drug, this behaves according to the formula

$$A(t) = 5e^{-0.03t}$$

What is the rate at which the amount of drug present is changing expressed as a function of time?

42. Sales often follow a growth pattern of rapid initial growth with some leveling off after a period of time. This pattern is described by the formula

$$S(t) = 25,000 - 10,000e^{-0.2t}$$

   a. How many items will be sold initially?
   b. What is the rate of change of sales initially?
   c. What is the rate of change of sales after t years?

43. The price–demand equation for x units of a commodity is given by

$$d(x) = 500e^{-0.1x}$$

Find the marginal revenue.

44. The price–supply equation for x units of the commodity described in Problem 43 is given by

$$s(x) = 50e^{0.05x}$$

Find the marginal supply.

45. If $y = \log x$, show that

$$y' = \frac{\log e}{x}$$

46. If $y = \log_b x$, show that

$$y' = \frac{\log_b e}{x}$$

47. If $y = \log_b u$, where u is a function of x, show that

$$\frac{dy}{dx} = \frac{\log_b e}{u} \cdot \frac{du}{dx}$$

12.5 Summary and Review

IMPORTANT TERMS

Algebraic function [12.1]
Base [12.1]
Change of base theorem [12.3]
Common logarithm [12.2]
Compound interest [12.1]
Continuous interest [12.1]
Decay function [12.1]
Derivative of exponential [12.4]
Derivative of logarithm [12.4]
e [12.1]
Evaluate a logarithm [12.2]
Exponent [12.1]

Exponential equation [12.2]
Exponential function [12.1]
Exponential property of equality [12.2]
Future value formula [12.1]
Growth function [12.1]
Interest [12.1]
Inverse function [12.2]
Laws of exponents [12.1]
Laws of logarithms [12.3]

Log of both sides theorem [12.3]
Logarithm [12.2]
Logarithmic equation [12.3]
Logarithmic function [12.2]
Natural logarithm [12.2]
Principal [12.1]
Simple interest [12.1]
Squeeze theorem for exponents [12.1]
Transcendental function [12.1]

SAMPLE TEST *For additional practice there are a large number of review problems categorized by objective in the Student Solutions Manual. The following sample test (40 minutes) is intended to review the main ideas of this chapter.*

Solve Problems 1–8.

1. $\log_6 36 = x$

2. $\log_x(2x + 15) = 2$

3. $2\log 2 - \dfrac{1}{2}\log 2 = \log \sqrt{x}$

4. $2\ln\dfrac{e}{\sqrt{5}} = 1 - \ln x$

5. $10^{x-1} = 250$

6. $3^{1-2x} = 0.5$

7. $e^{2x+3} = 10$

8. $10^{-x^2} = 0.75$

Find the derivatives in Problems 9–16.

9. $y = 5.5e^{-0.5x^2}$

10. $y = (x^2 - 1)e^{3x}$

11. $y = \ln x^5$

12. $y = (\ln x)^5$

13. $y = \ln|x^2(4 - x^3)|$

14. $y = e^x \ln x$

15. $y = \dfrac{\ln x^2}{4 - x}$

16. $y = \dfrac{1000\ln|x^3|}{2e^x}$

17. The half-life formula for carbon-14 dating is

$$A = 10\left(\frac{1}{2}\right)^{t/5700}$$

where A is the amount present (in milligrams) after t years. Solve this equation for t.

18. An advertising agency conducts a survey that finds that the number of units sold, N, is related to the amount spent on advertising (in dollars) by the formula

$$N = 2500 + 200\ln a \qquad (a \geq 1)$$

What is the rate at which N is changing at the instant that $a = \$20,000$?

19. The price–demand and price–supply equations for x- thousands of units of a commodity are given by

Demand: $d(x) = 200e^{-0.2x}$

Supply: $s(x) = 20e^{0.1x}$

Find the number of units that should be manufactured in order to achieve equilibrium.

20. The president makes a major policy announcement. The percent of the population that will have heard about the announcement is a function of the time t (in days) after the announcement according to the formula

$$N = 1 - e^{-1.5t}$$

where N is the percent written as a decimal. How long will it take for 90% of the population to hear about the announcement?

Modeling Application 8

World Running Records

World records for footraces at all distances have improved consistently ever since records have been kept. For example, if we consider the mile run, the magic 4-minute mile was broken in 1954 by Bannister of the United Kingdom. Since then, the record has decreased steadily to the present time when the world record is under 3 minutes 47 seconds!*

Write a paper that develops a mathematical model to answer the question, Will a 3-minute mile ever be run? Is there an "ultimate" time for a mile run, and, if so, what should we expect that time to be? For general guidelines about writing this essay, see the commentary for Modeling Application 1 on page 130.

---

* This modeling application is from Joseph Brown, "Predicting Future Improvements in Footracing," *MATYC Journal*, Fall 1980, pp. 173–179.

CHAPTER 13
Integration

CHAPTER CONTENTS

13.1 The Antiderivative

13.2 Integration by Substitution

13.3 The Fundamental Theorem of Calculus

13.4 Area between Curves

13.5 Summary and Review
Important Terms
Sample Test

APPLICATIONS

Management (*Business, Economics, Finance, and Investments*)

Finding the cost function given the marginal cost (13.1, Problems 37–40)
Finding the profit function for Turbo Plus, Inc. (13.1, Problem 43)
Finding the cost of producing 100,000 processors (13.1, Problem 44)
Determining the output after *t* hours of work (13.2, Problem 34)
Finding the profit when the marginal profit is known (13.2, Problems 35–36)
Finding the sales (13.2, Problem 37)
Automobile sales from a graph in the *Wall Street Journal* (13.3, Problems 42–43)
Total paper and paperboard production in the U.S. (13.3, Problem 44)
Determining the accumulated sales (13.3, Problems 45–46)
Finding the Gini index to measure money flow (13.3, Problem 47)
Total cost, profit, or production (13.4, Problem 27)
Producer's and consumer's surplus (13.4, Problems 29–31; 13.5, Problem 20)
Yearly profit (13.4, Problem 32; 13.5, Problem 16)
Determining if the eucalyptus tree grows fast enough to make commercial growing a profitable enterprise (13.5, Problem 17)

Life sciences (*Biology, Ecology, Health, and Medicine*)

Infection by a new strain of influenza (13.1, Problems 45–46)
Flu epidemic (13.2, Problem 33)
Predicting oil consumption (petroleum) given present production and rate of consumption (13.2, Problems 40–41; 13.3, Problems 39–41; 13.5, Problem 19)
Amount of pollutants dumped into the Russian River (13.4, Problem 28)
Chemical pollution (13.4, Problem 33)

Social sciences (*Demography, Political Science, Population, Psychology, Society, and Sociology*)

Predicting population given growth rate (13.1, Problems 41–42; 13.2, Problem 39)
Predicting the world population in the year 2000 (13.2, Problem 38)
Growth rate of Chicago, Illinois (13.5, Problem 18)

General interest

Velocity of an object thrown off Hoover Dam (13.1, Problems 47–48)

Modeling application—Computers in Mathematics

CHAPTER OVERVIEW
The last main idea of the book is introduced in this chapter—the idea of an integral and a process called integration.

PREVIEW
The integral is first introduced as an antiderivative, and then an indefinite integral is defined. There are three methods of integration presented in this section: substitution, tables, and an optional method called integration by parts.

PERSPECTIVE
The skills you learn in this chapter will be put to use in the next chapter when applications of integration are considered. As you will see in this chapter, the ability to "reverse" the process of differentiation is important since you quite often know the derivative but need to find the original function. The indefinite integral is used for that purpose.

13.1 The Antiderivative

Calculus is divided into two broad categories. The first, **differential calculus**, involves the definition of derivative and related applications of instantaneous rates of change, marginal costs, profits, and curve sketching, as well as finding maximums and minimums. In this chapter, we begin our study of the second part of calculus, **integral calculus**, which deals with the definition of another limit—the **definite integral**. Integral calculus is used to find areas, volumes, and functions when we have information about the function's rate of change. For example, if we know the present world population and the rate at which it is growing, we can produce a formula (subject to certain assumptions) that predicts the population size at any future time or estimates its size at some time in the past.

EXAMPLE 1
Multiplex Corporation knows that the marginal cost (in thousands of dollars) to produce x items (in thousands) is given by the formula

$$f(x) = 2x + 4$$

Find the cost of producing 50,000 items ($x = 50$) if the fixed costs are $25,000.

Solution
In Section 10.3 marginal cost was defined to be the rate of change in cost per unit change in production at an output level of x units. This means that if F is the cost function, then the marginal cost, f, is a function that can be found by taking the derivative of F at x:

$$F'(x) = f(x)$$

Find a function F so that $F'(x) = f(x)$:

$$F'(x) = 2x + 4$$

By trial and error (we will develop better methods very shortly), we see that

$$F(x) = x^2 + 4x$$

is such a function, but this function is not unique:

$$F_1(x) = x^2 + 4x + 6 \qquad F'_1(x) = 2x + 4$$
$$F_2(x) = x^2 + 4x - 100 \qquad F'_2(x) = 2x + 4$$
$$F_3(x) = x^2 + 4x + 14.8 \qquad F'_3(x) = 2x + 4$$

There are many functions that have $2x + 4$ as a derivative, but they all differ by a constant since the derivative of a constant is 0. Thus

$$F(x) = x^2 + 4x + C$$

for any constant C. The cost of producing 50,000 items is given by $F(50)$, but first we need to find C, the **constant of integration**. The fixed costs are the costs that are present even if $x = 0$. This means that $F(0) = 25$, so

$$F(0) = 0^2 + 4(0) + C = 25$$
$$C = 25$$

Thus

$$F(x) = x^2 + 4x + 25$$

is the cost equation for our original problem, so

$$F(50) = 50^2 + 4(50) + 25 = 2725$$

The cost of producing 50,000 items is $2,725,000. ∎

EXAMPLE 2   Show that $F(x) = x^6$ is an antiderivative of $f(x) = 6x^5$.

Solution   Use the power rule for derivatives:

$$F(x) = x^6$$
$$F'(x) = 6x^5$$ ∎

EXAMPLE 3   Find an antiderivative of $f(x) = x^5$.

Solution   By trial and error (the power rule for derivatives in reverse),

$$F(x) = \frac{x^6}{6} + C$$

Check: $F'(x) = \dfrac{6x^5}{6} + 0 = x^5$ ∎

In Example 3 we went through a (trial and error) process that might be called the process of **antidifferentiation**. A more common name for the process of antidifferentiation is **integration**. If $F(x)$ is an antiderivative of $f(x)$, then $F(x) + C$ is called the **indefinite integral** of $f(x)$. The adjective *indefinite* is used because the constant C is arbitrary or indefinite.

Indefinite Integral

If F is an antiderivative of f, then we write

$$\int f(x)\,dx = F(x) + C$$

and say, "The indefinite integral of $f(x)$ with respect to x is $F(x) + C$." The symbol $\int$ is called the **integral symbol**, $f(x)$ is called the **integrand**, and C is called the **constant of integration**.

We will say more about the symbol dx later, but for now remember that the dx in the indefinite integral identifies the variable of integration, as illustrated by Example 4.

EXAMPLE 4 $\displaystyle\int 3x\,dx$ is read "the indefinite integral of $3x$ with respect to x."

$\displaystyle\int 9y^2\,dy$ is read "the indefinite integral of $9y^2$ with respect to y."

$\displaystyle\int (5z^2 + 3z + 2)\,dz$ is read "the indefinite integral of $5z^2 + 3z + 2$ with respect to z."

$\displaystyle\int x^2 y^3\,dx$ is read "the indefinite integral of $x^2 y^3$ with respect to x."   ■

Integral notation makes it easy to state some integration formulas. For example, the power rule for derivatives says if $f(x) = x^n$, then $f'(x) = nx^{n-1}$, so if this process is reversed we find a corresponding integration formula to be

$$\text{If}\quad f(x) = x^n, \quad\text{then}\quad F(x) = \int x^n\,dx = \frac{x^{n+1}}{n+1} + C \quad n \neq -1$$

To prove this formula, find the derivative of

$$F(x) = \frac{x^{n+1}}{n+1} + C$$

$$F'(x) = \frac{(n+1)x^{n+1-1}}{n+1} + 0$$

$$= x^n$$

EXAMPLE 5   **a.** $\displaystyle\int x^8\,dx = \frac{x^9}{9} + C$

b. $\displaystyle\int x^3\,dx = \frac{x^4}{4} + C$

c. $\displaystyle\int \sqrt{x}\,dx = \int x^{1/2}\,dx$

$$= \frac{x^{3/2}}{\frac{3}{2}} + C = \frac{2}{3}x^{3/2} + C$$

d. $\displaystyle\int dx = \int x^0\,dx = \frac{x^1}{1} + C$

$$= x + C$$

e. $\displaystyle\int \frac{dx}{x^3} = \int x^{-3}\,dx$

$$= \frac{x^{-2}}{-2} + C = -\frac{1}{2}x^{-2} + C$$ ■

If we quickly review the derivative formulas from Chapter 9, we should be able to restate each as an integration formula.

Review of Derivative Formulas

POWER RULE:	If $y = x^n$, then $y' = nx^{n-1}$.
CONSTANT RULE:	If $y = k$, then $y' = 0$.
CONSTANT TIMES A FUNCTION RULE:	If $y = kf$, then $y' = kf'$.
SUM RULE:	If $y = f + g$, then $y' = f' + g'$.
DIFFERENCE RULE:	If $y = f - g$, then $y' = f' - g'$.
LOGARITHMIC RULE:	If $y = \ln|x|$, then $y' = \dfrac{1}{x}$.
EXPONENTIAL RULE:	If $y = e^x$, then $y' = e^x$.

(We will consider the product, quotient, and chain rules later.)

These derivative formulas can be divided into two types: three *direct integration forms* and three *procedural types*. For now, all our examples can be reduced to one of the direct integration types by first using the procedural types.

Integration Formulas

DIRECT INTEGRATION FORMULAS

$$\int x^n \, dx = \frac{x^{n+1}}{n+1} + C \qquad n \neq -1 \qquad \text{Power rule}$$

$$\int x^n \, dx = \ln|x| + C \qquad n = -1 \qquad \text{Logarithmic rule}$$

$$\int e^x \, dx = e^x + C \qquad\qquad\qquad \text{Exponential rule}$$

PROCEDURAL INTEGRATION FORMULAS

$$\int kf(x) \, dx = k \int f(x) \, dx \qquad\qquad \text{The \textbf{integral of a constant} times a function is that constant times the integral}$$

$$\int [f(x) + g(x)] \, dx = \int f(x) \, dx + \int g(x) \, dx \qquad \text{The \textbf{integral of a sum} is the sum of the integrals}$$

$$\int [f(x) - g(x)] \, dx = \int f(x) \, dx - \int g(x) \, dx \qquad \text{The \textbf{integral of a difference} is the difference of the integrals}$$

EXAMPLE 6
$$\int (3x^2 - 4x + 5)\,dx = \int 3x^2\,dx - \int 4x\,dx + \int 5\,dx \qquad \text{Sum and difference formulas}$$

$$= 3\int x^2\,dx - 4\int x\,dx + 5\int dx \qquad \text{Constant formula}$$

$$= 3\left(\frac{x^3}{3} + C_1\right) - 4\left(\frac{x^2}{2} + C_2\right) + 5(x + C_3) \qquad \begin{array}{l}\text{Power} \\ \text{formula}\end{array}$$

$$= x^3 - 2x^2 + 5x + (3C_1 - 4C_2 + C_3)$$

$$= x^3 - 2x^2 + 5x + C \qquad \blacksquare$$

In Example 6 since $3C_1 - 4C_2 + C_3$ represents arbitrary constants, they can be combined into a single constant, C. Also, the steps in Example 6 are shown in great detail so you can relate them to the proper formulas, but in practice your work would look like this:

$$\int (3x^2 - 4x + 5)\,dx = \frac{3x^3}{3} - \frac{4x^2}{2} + 5x + C$$

$$= x^3 - 2x^2 + 5x + C$$

EXAMPLE 7 **a.**
$$\int (5x^4 + 2x^3 + \sqrt[3]{x})\,dx = \frac{5x^5}{5} + \frac{2x^4}{4} + \frac{x^{4/3}}{\frac{4}{3}} + C \qquad \text{Remember, } \sqrt[3]{x} = x^{1/3}$$

$$= x^5 + \frac{x^4}{2} + \frac{3x^{4/3}}{4} + C$$

$$= x^5 + \frac{1}{2}x^4 + \frac{3}{4}x^{4/3} + C$$

b.
$$\int (x - e^x)\,dx = \frac{x^2}{2} - e^x + C$$

c.
$$\int \frac{x^3 + x + 1}{x}\,dx = \int \left(\frac{x^3}{x} + \frac{x}{x} + \frac{1}{x}\right)\,dx \qquad \text{Simplify first, then integrate}$$

$$= \int (x^2 + 1 + x^{-1})\,dx$$

$$= \frac{x^3}{3} + x + \ln|x| + C \qquad \blacksquare$$

As shown in Example 1, it is possible to find C if some value of the original function is known. The most commonly known value is the one when the variable is 0. That is, if

$$F'(x) = f(x)$$

and $f(0)$ is known, then this known value is called the **initial value**.

EXAMPLE 8   The population of Santa Rosa, California, is growing at the rate $3000 + 500t^{1/4}$, where t is measured in years. The present population is 92,500. Predict the population in 5 years.

Solution   The population function for time t is found by knowing

$$P'(t) = 3000 + 500t^{1/4}$$

Thus, by integrating with respect to t,

$$P(t) = \int (3000 + 500t^{1/4})\, dt = 3000t + \frac{500t^{1/4+1}}{\frac{5}{4}} + C$$

$$= 3000t + 400t^{5/4} + C$$

The initial condition ($t = 0$) is 92,500, so

$$P(0) = 3000(0) + 400(0)^{5/4} + C = 92,500$$
$$C = 92,500$$

In 5 years,

$$P(5) = 3000(5) + 400(5)^{5/4} + 92,500$$
$$\approx 110,490.6976$$

The population in 5 years will be about 110,000.   ∎

Problem Set 13.1

Evaluate the indefinite integrals in Problems 1–32. You can check each answer by differentiating your answers.

1. $\int x^7\, dx$

2. $\int x^{10}\, dx$

3. $\int 4x^3\, dx$

4. $\int 12x^5\, dx$

5. $\int 3\, dx$

6. $\int 8\, dx$

7. $\int (5x + 7)\, dx$

8. $\int (5 - 2x)\, dx$

9. $\int dx$

10. $\int (9x^2 - 4x + 3)\, dx$

11. $\int (18x^2 - 6x + 5)\, dx$

12. $\int \frac{dx}{5}$

13. $\int \frac{dx}{9}$

14. $\int \frac{3\, dx}{x}$

15. $\int \frac{5\, dx}{x^2}$

16. $\int -3e^x\, dx$

17. $\int (1 - e^x)\, dx$

18. $\int (x - \sqrt{x})\, dx$

19. $\int (\sqrt[3]{x} + \sqrt{2})\, dx$

20. $\int (\sqrt[3]{5} + \sqrt{2})\, dx$

21. $\int \frac{3x - 1}{\sqrt{x}}\, dx$

22. $\int \frac{5x^2 + 2x + 3}{\sqrt{x}}\, dx$

23. $\int \frac{x}{\sqrt[3]{x}}\, dx$

24. $\int \frac{x^2}{\sqrt[3]{x^2}}\, dx$

25. $\int \frac{x^2 + x + 1}{x}\, dx$

26. $\int \frac{3x^2 + 2x + 1}{x}\, dx$

27. $\int y^3 \sqrt{y}\, dy$

28. $\int (3y^{-2/3} - 4y^{1/2})\, dy$

29. $\int (1 + z^2)^2\, dz$

30. $\int (3z + 1)^2\, dz$

31. $\int (3u^2 - u^{-1} + e^u)\, du$

32. $\int (5u^4 - 3u^{-1} + 4e^u)\, du$

In Problems 33–36 find the antiderivative $F(x)$ for the given function that satisfies the initial conditions.

33. $f(x) = 3x^2 + 4x + 1$; $F(0) = 10$

34. $f(x) = 3x^2 + 4x + 1$; $F(1) = -8$

35. $f(x) = \dfrac{5x + 1}{x}$; $F(1) = 5000$

36. $f(x) = \dfrac{3x^2 + 2x + 1}{x}$; $F(1) = -50$

APPLICATIONS

In Problems 37–40 find the cost function for the given marginal cost functions.

37. $C'(x) = 4x - 8$; fixed cost $= \$5,000$

38. $C'(x) = 128x - 300$; fixed cost $= \$32,000$

39. $C'(x) = 0.009x^2$; fixed cost $= \$18,500$

40. $C'(x) = x + 100e^x$; fixed cost $= \$1,900$

41. The population of New Haven, Connecticut, is growing at the rate of $450 + 600\sqrt{t}$, where t is measured in years, and the present population is 420,000. Predict the population in 5 years.

42. The population of Charlotte, North Carolina, is growing at the rate of $1000 + 500t^{3/4}$, where t is measured in years, and the present population is 330,000. Predict the population in 10 years.

43. The marginal profit of Turbo Plus is

$$P'(x) = 200x - 10,000$$

where x is the sales in thousands of items. Find the profit function if there is a \$50,000 loss ($-\$50,000$ profit) when no items are produced.

44. The marginal cost for producing x new 32-bit microprocessors is

$$C'(x) = 100x^{-2/3} + 0.00001x$$

If the fixed costs are \$120,000, what is the cost of producing 100,000 processors?

45. A new strain of influenza is known to be spreading at the rate of $240t - 3t^2$ cases per day, where t is the number of days measured from the first recorded outbreak. How many people will be affected on the tenth day, if there were 50 cases on the first day?

46. Repeat Problem 45 for the fifth day, if there were five cases on the first day.

47. It is known that the acceleration of a freely falling body is a constant, 32 feet per second per second. Use the formula

$$a(t) = -32$$

to derive a formula for the velocity (at time t) for an object thrown from the top of Hoover Dam (726 feet) with an initial velocity of -72 feet per second. [*Hint:* If v is the velocity function, then $v'(t) = a(t)$; it is negative because the object was thrown downward.]

48. Derive a formula for the height of the object thrown from Hoover Dam in Problem 47. [*Hint:* If h is the height function, then $h'(t) = v(t)$.]

49. The slope of the tangent line to a curve is given by

$$f'(x) = 3x^2 + 5$$

If the point $(1, 2)$ is on the curve, find the equation of the curve.

50. Find the equation of a curve whose slope is $\sqrt{x}$ if the point $(9, 19)$ is on the curve.

13.2 Integration by Substitution

One of the most important integration techniques comes from the *generalized power rule*. For example, consider

$$\int x(x^2 + 1)^2 \, dx$$

We could proceed by writing

$$\int x(x^4 + 2x^2 + 1) \, dx = \int (x^5 + 2x^3 + x) \, dx$$

$$= \int \frac{x^6}{6} + \frac{2x^4}{4} + \frac{x^2}{2} + C$$

$$= \frac{1}{6}x^6 + \frac{1}{2}x^4 + \frac{1}{2}x^2 + C$$

However, it is often not practical (or possible) to multiply [as with $x(x^2 + 1)^{12}$, for example]. We will proceed, instead, by *substitution*. The complicating ingredient is the "correction factor" that we found in Section 9.7. In order to handle this correction factor, we need to use differentials (discussed in Section 10.2).

Let us take another look at the preceding example. This time, instead of multiplying, we will make a substitution. Look at the expression raised to a power, namely, $x^2 + 1$. Let

$$u = x^2 + 1$$

then

$$\frac{du}{dx} = 2x$$

Now, solve for dx:

$$du = 2x\,dx$$
$$dx = \frac{du}{2x}$$

If substitution eliminates the variable x (leaving only the variable u), the equation may be in a simpler, more easily integrable, form.

$$\int x(x^2 + 1)^2\,dx = \int xu^2 \frac{du}{2x}$$

$$= \frac{1}{2}\int u^2\,du \qquad\qquad \textit{Note:} \quad \text{The } x\text{'s cancel and } \frac{1}{2} \text{ is a constant}$$

$$= \frac{1}{2}\frac{u^3}{3} + C \qquad\qquad \text{Use the power rule on } u$$

$$= \frac{(x^2 + 1)^3}{6} + C \qquad\qquad \text{Write } u \text{ in terms of } x \text{ for the final answer}$$

By using substitution we are able to integrate this function without a lot of multiplication. This has tremendous advantages, as shown in Example 1.*

EXAMPLE 1   Evaluate $\int x(x^2 + 1)^{12}\,dx$.

Solution   Let $u = x^2 + 1$; then $\dfrac{du}{dx} = 2x$, so $dx = \dfrac{du}{2x}$

Substitute to obtain

$$\int x\underbrace{(x^2 + 1)}_{u}{}^{12}\underbrace{dx}_{\frac{du}{2x}} = \int xu^{12}\frac{du}{2x} = \frac{1}{2}\int u^{12}\,du$$

$$\underset{\uparrow}{}$$
Ah, ha! This corresponds to the "correction factor" we obtained when using the generalized power rule

$$= \frac{1}{2}\frac{u^{13}}{13} + C = \frac{1}{26}(x^2 + 1)^{13} + C \qquad\qquad \blacksquare$$

---

* If you did expand, you would see that the result agrees with that on page 497:

$$\tfrac{1}{6}(x^6 + 3x^4 + 3x^2 + 1) + C = \tfrac{1}{6}x^6 + \tfrac{1}{2}x^4 + \tfrac{1}{2}x^2 + \tfrac{1}{6} + C$$

EXAMPLE 2   Evaluate $\int x^2 \sqrt{x^3 - 2}\, dx$.

Solution   Let $u = x^3 - 2$; $\dfrac{du}{dx} = 3x^2$, so $dx = \dfrac{du}{3x^2}$. Substitute

$$\int x^2 \sqrt{x^3 - 2}\, dx = \int x^2 \sqrt{u}\, \frac{du}{3x^2}$$

$$= \frac{1}{3} \int u^{1/2}\, du$$

$$= \frac{1}{3} \frac{u^{3/2}}{\frac{3}{2}} + C$$

$$= \frac{2}{9} u^{3/2} + C$$

$$= \frac{2}{9}(x^3 - 2)^{3/2} + C \qquad\blacksquare$$

The procedure is generalized below:

Integration by Substitution

1. Make a choice for u, say, $u = f(x)$ (u is usually inside parentheses, or an exponent, or under a radical).
2. Find $\dfrac{du}{dx} = f'(x)$.
3. Make the substitution

$$u = f(x) \quad \text{and} \quad du = f'(x)\, dx$$

$$\frac{du}{f'(x)} = dx$$

4. *Everything* must be in terms of u with no remaining x's. If this is not possible, try another substitution.
5. Evaluate the integral.
6. Replace u by $f(x)$ so the final answer is in terms of x.

When using the substitution method, make sure that every x is eliminated. If this cannot be done, try a different choice, but remember that not every function *can* be integrated by this technique. This means that *after* substitution and simplification, you *must* have one of the following forms:

$$\int u^n\, du = \begin{cases} \dfrac{u^{n+1}}{n+1} + C & n \neq -1 \\[2mm] \ln|u| + C & n = -1 \end{cases}$$

or

$$\int e^u\, du = e^u + C$$

EXAMPLE 3 Evaluate $\displaystyle\int \frac{x^2\, dx}{(x^3 - 2)^5}$.

Solution Let u be the value in parentheses; that is, let $u = x^3 - 2$. Then

$$du = 3x^2\, dx \qquad \text{so} \qquad dx = \frac{du}{3x^2}$$

$$\int \frac{x^2\, dx}{(x^3 - 2)^5} = \int \frac{x^2}{u^5} \cdot \frac{du}{3x^2} = \int \frac{du}{3u^5}$$

> All x's must be eliminated:
>
> $(x^3 - 2)^5 = u^5$ and $dx = \dfrac{du}{3x^2}$

$$= \frac{1}{3} \int u^{-5}\, du = \frac{1}{3} \frac{u^{-4}}{-4} + C$$

$$= \frac{-(x^3 - 2)^{-4}}{12} + C \qquad \blacksquare$$

EXAMPLE 4 Evaluate $\displaystyle\int e^{3x+5}\, dx$.

Solution Let u be the exponent; that is, $u = 3x + 5$, so

$$du = 3\, dx \qquad \text{and} \qquad dx = \frac{du}{3}$$

Substitute:

$$\int e^{3x+5}\, dx = \int e^u \frac{du}{3}$$

$$= \frac{1}{3} \int e^u\, du$$

$$= \frac{1}{3} e^u + C$$

$$= \frac{1}{3} e^{3x+5} + C \qquad \blacksquare$$

EXAMPLE 5 Evaluate $\displaystyle\int x^3 \sqrt{3x^2 - 1}\, dx$.

Solution Let u be the expression under the radical; that is, $u = 3x^2 - 1$, so

$$du = 6x\, dx \qquad \text{and} \qquad dx = \frac{du}{6x}$$

Substitute:

$$\int x^3 \sqrt{u} \, \frac{du}{6x} = \frac{1}{6} \int x^2 \underbrace{\sqrt{u} \, du}$$

You can handle this

WARNING: x^2 is a variable, so do not try to treat it like a constant. That is, do *not* bring out the x^2 in front of the integral sign

All x's must be eliminated; note:

$$u = 3x^2 - 1$$
$$u + 1 = 3x^2$$
$$\frac{u + 1}{3} = x^2$$

$$= \frac{1}{6} \int \frac{u + 1}{3} \sqrt{u} \, du$$

$$= \frac{1}{18} \int (u^{3/2} + u^{1/2}) \, du$$

$$= \frac{1}{18} \left(\frac{u^{5/2}}{\frac{5}{2}} + \frac{u^{3/2}}{\frac{3}{2}} \right) + C$$

$$= \frac{1}{45} u^{5/2} + \frac{1}{27} u^{3/2} + C$$

$$= \frac{1}{45} (3x^2 - 1)^{5/2} + \frac{1}{27} (3x^2 - 1)^{3/2} + C \qquad \blacksquare$$

Different parts of a problem may require different substitutions.

EXAMPLE 6 Evaluate $\int \left(\dfrac{x^2}{1 - x^3} + xe^{x^2} \right) dx.$

Solution $\displaystyle \int \left(\frac{x^2}{1 - x^3} + xe^{x^2} \right) dx = \int \frac{x^2}{1 - x^3} \, dx + \int xe^{x^2} \, dx$

$$u = 1 - x^3 \qquad\qquad v = x^2$$
$$du = -3x^2 \, dx \qquad\qquad dv = 2x \, dx$$

$$= \int \frac{x^2}{u} \frac{du}{-3x^2} + \int xe^v \frac{dv}{2x}$$

$$= -\frac{1}{3} \int u^{-1} \, du + \frac{1}{2} \int e^v \, dv$$

$$= -\frac{1}{3} \ln|u| + \frac{1}{2} e^v + C$$

$$= -\frac{1}{3} \ln|1 - 3x^2| + \frac{1}{2} e^{x^2} + C \qquad \blacksquare$$

EXAMPLE 7 The TexRite Company has found that the marginal profit for a product is

$$P'(x) = \frac{1600x}{\sqrt[3]{(8x^2 - 33,344)^2}}$$

where x is the number of units sold. If the break-even point is 100 units [that is, $P(100) = 0$], what is the approximate total profit for 1000 items?

Solution $P'(x) = \displaystyle\int \frac{1600x\,dx}{\sqrt[3]{(8x^2 - 33,344)^2}} = \int \frac{1600x}{\sqrt[3]{u^2}} \frac{du}{16x}$

$$= \int 100u^{-2/3}\,du$$

Let $u = 8x^2 - 33,344$
$du = 16x\,dx$

$$= 100\frac{u^{1/3}}{\frac{1}{3}} + C$$

$$= 300(8x^2 - 33,344)^{1/3} + C$$

Now, $P(100) = 0$, so

$$300(8 \cdot 100^2 - 33,344)^{1/3} + C = 0$$
$$C = -10,800$$
$$P(x) = 300(8x^2 - 33,344)^{1/3} - 10,800$$
$$P(1000) \approx 49,116.52389$$

The profit is about $49,000. ∎

Problem Set 13.2

Evaluate the indefinite integrals in Problems 1–32.

1. $\displaystyle\int (5x + 3)^3\,dx$

2. $\displaystyle\int (1 - 5x)^4\,dx$

3. $\displaystyle\int \frac{5}{(5 - x)^4}\,dx$

4. $\displaystyle\int \frac{6}{(x + 8)^2}\,dx$

5. $\displaystyle\int \sqrt{3x + 5}\,dx$

6. $\displaystyle\int \sqrt[5]{9 - 4x}\,dx$

7. $\displaystyle\int 6x(3x^2 + 1)\,dx$

8. $\displaystyle\int 2x(4 - 5x^2)\,dx$

9. $\displaystyle\int \frac{2x + 5}{\sqrt{x^2 + 5x}}\,dx$

10. $\displaystyle\int \frac{2x + 3}{\sqrt{x^2 + 3x}}\,dx$

11. $\displaystyle\int \frac{dx}{6 + 5x}$

12. $\displaystyle\int \frac{dx}{1 - 4x}$

13. $\displaystyle\int \frac{x\,dx}{1 - 3x^2}$

14. $\displaystyle\int \frac{2x\,dx}{x^2 + 5}$

15. $\displaystyle\int e^{5x}\,dx$

16. $\displaystyle\int e^{1-3x}\,dx$

17. $\displaystyle\int 5x^2 e^{4x^3}\,dx$

18. $\displaystyle\int 3xe^{x^2+5}\,dx$

19. $\displaystyle\int \frac{2x - 1}{(4x^2 - 4x)^2}\,dx$

20. $\displaystyle\int \frac{2x - 1}{(x - x^2)^3}\,dx$

21. $\displaystyle\int \frac{4x^3 - 4x}{x^4 - 2x^2 + 3}\,dx$

22. $\displaystyle\int \frac{x^3 - x}{(x^4 - 2x^2 + 3)^2}\,dx$

23. $\displaystyle\int \frac{\ln|x|}{x}\,dx$
Hint: Let $u = \ln|x|$.

24. $\displaystyle\int \frac{\ln|x + 1|}{x + 1}\,dx$
Hint: Let $u = \ln|x + 1|$.

25. $\displaystyle\int \ln e^x\,dx$

26. $\displaystyle\int \ln e^{x^2}\,dx$

27. $\displaystyle\int \left(\frac{3x}{4x^2 + 1} + x^2 e^{x^3}\right)dx$

28. $\displaystyle\int \left(\frac{4x + 6}{\sqrt{x^2 + 3x}} + xe^{3x^2}\right)dx$

29. $\displaystyle\int x\sqrt[3]{(x^2 + 1)^2}\,dx$

30. $\displaystyle\int (x^2 - 4x + 4)^{2/5}\,dx$

31. $\displaystyle\int \frac{x\,dx}{\sqrt{x + 1}}$

32. $\displaystyle\int x^2\sqrt{5 - x}\,dx$

APPLICATIONS

33. A flu epidemic is spreading at the rate of

$$P'(t) = 40t(5 - t^2)^2$$

people per day, where t is the number of days since the first outbreak. If P is a function representing the number of sick people, and if $P(0) = 10$, find P.

34. Let $P(t)$ be the total output after t hours of work. Find P if the rate of production at time t $(1 < t < 40)$ is

$$P'(t) = \frac{50}{50 - t}$$

and production (to the nearest unit) is 71 when $t = 1$.

35. If the marginal profit (in dollars) for a product is

$$P'(x) = \frac{100x}{\sqrt[3]{(x^2 - 36)^2}}$$

where x is the number of units sold, find the profit for 100 items if the break-even point is 10 units.

36. The marginal profit for a product is

$$P'(x) = \frac{20x}{\sqrt{x^2 - 16}}$$

where x is the number of units sold. If the break-even point is 5 units, what is the profit for 25 items?

37. Sales are changing at a rate given by the formula

$$S'(t) = 2000e^{-0.2t}$$

where t is the time in years and $S(t)$ is the sales. What are the sales in 3 years if initial sales are 15,000 units?

38. In 1976 the world population reached 4 billion and was growing at a rate approximated by the formula

$$P'(t) = 0.072e^{0.018t}$$

where t is measured in years since 1976 and $P(t)$ gives the population. What is the predicted population in the year 2000?

39. In 1985 the growth rate of San Antonio, Texas, was given by the formula

$$P'(t) = 14,000e^{0.0175t}$$

where t is measured in years since 1980 and $P(t)$ gives the population. If the population in 1984 was 842,779, predict the population in the year 2000.

40. Between 1980 and 1985 worldwide oil consumption dropped from 2400 million barrels to 1950 million barrels. This rate of consumption is given by the formula

$$R(t) = 78e^{-0.04t}$$

where t is measured in years since 1985. The total consumption since 1985 is given by $T(t)$, where $T'(t) = R(t)$. Find the consumption from 1985 to the year 2000.

41. If we consider worldwide oil consumption from 1925 to 1985, the rate of consumption is given by the formula

$$R(t) = 32.5e^{0.048t}$$

where t is measured in years since 1985. If the worldwide consumption was 1950 million barrels in 1985, predict the consumption from 1985 to the year 2000 if the total consumption is given by $T(t)$, where $T'(t) = R(t)$.

13.3 The Fundamental Theorem of Calculus

In 1986 world oil prices plummeted. The price drop was related to overproduction and declining consumption. Suppose that between 1980 and 1985 the rate of consumption (in billions of barrels) is given by the formula

$$R(t) = 78e^{-0.04t}$$

where t is measured in years after 1985.* If we use this consumption rate, we can predict the amount of oil that will be consumed from 1987 to 1995. The total consumption since 1985 is given by $T(t)$, where $T'(t) = R(t)$. Then

$$T(t) = \int T'(t)\,dt = \int R(t)\,dt$$
$$= \int (78e^{-0.04t})\,dt$$
$$= \frac{78e^{-0.04t}}{-0.04} + C$$
$$= -1950e^{-0.04t} + C$$

* Equations like this are derived in Section 15.4.

We can find C because if $t = 0$, then oil consumption is also zero. Thus

$$T(0) = -1950e^{-0.04t} + C = 0$$
$$-1950 + C = 0$$
$$C = 1950$$

Thus

$$T(2) = -1950e^{-0.04(2)} + 1950 \approx 150 \qquad \text{Total consumption from 1985 to 1987}$$

$$T(10) = -1950e^{-0.04(10)} + 1950 \approx 643 \qquad \text{Total consumption from 1985 to 1995}$$

We can find the total consumption from 1987 to 1995 by subtraction:

$$T(10) - T(2) = 643 - 150 = 493$$

We would predict approximately 493 billion barrels to be consumed, assuming the rate of consumption does not change. However, if prices continue to drop, we would expect the rate of consumption to again turn upward (see Problems 40–41 in Problem Set 13.3).

Let us take a closer look at what we have done. Since $T(t)$ is an antiderivative of $R(t)$ over the interval from 1987 to 1995, we see that $T(10) - T(2)$ is the *net change* of the function T over this interval. In general, if $F(x)$ is any function and a and b are real numbers with $a < b$, then the *net change of $F(x)$ over the interval $[a, b]$* is the number

$$F(b) - F(a)$$

The quantity $F(b) - F(a)$ is often abbreviated by the symbol

$$F(x)\Big|_a^b$$

This discussion suggests the following definition:

Definite Integral

> Let f be a function defined over the interval $[a, b]$.* Then the *definite integral* of f over this interval is denoted by
>
> $$\int_a^b f(x)\,dx$$
>
> and is defined as the net change of an antiderivative of f over that interval. Thus, if $F(x)$ is an antiderivative of $f(x)$, then
>
> $$\int_a^b f(x)\,dx = F(x)\Big|_a^b = F(b) - F(a)$$

---

* When we speak of the closed interval $[a, b]$, it is assumed that $a < b$; that is, we do not allow for $a = b$ or $b < a$.

The function f is called the **integrand** and the constants a and b are called the **limits of integration**. The number x is called a **dummy variable** because the definite integral is a fixed number and not a function of x.

Both the integral sign and the symbol dx are dropped when the antiderivative is found. Also, using the definite integral it is not necessary to take into account a constant of integration. [Remember, in the introductory example, the constant of integration (1950) had no bearing on the final answer since it was first added and then subtracted.] The process of finding the value of a definite integration is called **evaluating the integral**.

EXAMPLE 1 Evaluate the given integrals.

a. $\displaystyle\int_{2}^{3} 6x^2\, dx$ b. $\displaystyle\int_{-2}^{2} x^3\, dx$ c. $\displaystyle\int_{1}^{10} \frac{dx}{x}$

Solution a.

Evaluate at 3 first

$$\int_{2}^{3} 6x^2\, dx = \left.\frac{6x^3}{3}\right|_{2}^{3} = \frac{6(3)^3}{3} - \frac{6(2)^3}{3} = 2(27) - 2(8) = 38$$

Antiderivative of $6x^2$

Subtract the value of the antiderivative at 2

b. $\displaystyle\int_{-2}^{2} x^3\, dx = \left.\frac{x^4}{4}\right|_{-2}^{2} = \frac{2^4}{4} - \frac{(-2)^4}{4} = 0$

c. $\displaystyle\int_{1}^{10} \frac{dx}{x} = \left.\ln|x|\right|_{1}^{10} = \ln|10| - \ln|1| = \ln 10 - 0 \approx 2.3$

Remember, $\ln 1 = 0$ ∎

The most common application of definite integrals involves a very important result called the *fundamental theorem of integral calculus*, which links differentiation, the definite integral, and areas. To understand this theorem we begin with a simple example, *the area under a curve*.

Let f be a continuous nonnegative function on $[a, b]$. Let $A(x)$ be the **area function** that denotes the area of the region of $[a, b]$ bounded by f on top and the x-axis on the bottom and the vertical lines x and a, as shown in Figure 13.1.

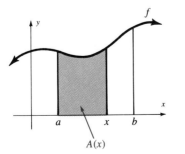

Figure 13.1 The area function

EXAMPLE 2 Let $a = 0$ and find $A(x)$ for each of the given functions.

a. $f(x) = 4$ **b.** $f(x) = 4x$ **c.** $f(x) = 4x + 3$

Solution **a.**

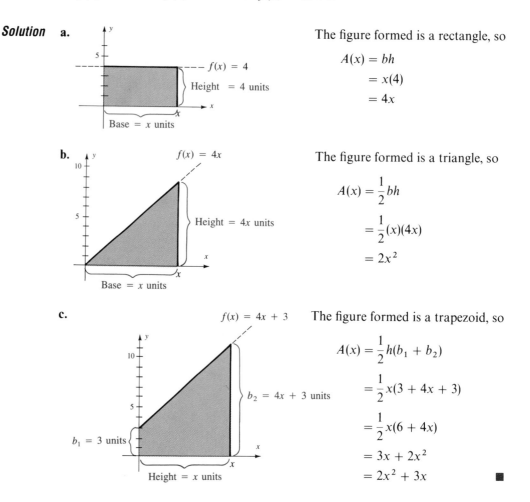

The figure formed is a rectangle, so

$$A(x) = bh$$
$$= x(4)$$
$$= 4x$$

b.

The figure formed is a triangle, so

$$A(x) = \frac{1}{2}bh$$
$$= \frac{1}{2}(x)(4x)$$
$$= 2x^2$$

c.

The figure formed is a trapezoid, so

$$A(x) = \frac{1}{2}h(b_1 + b_2)$$
$$= \frac{1}{2}x(3 + 4x + 3)$$
$$= \frac{1}{2}x(6 + 4x)$$
$$= 3x + 2x^2$$
$$= 2x^2 + 3x$$ ∎

The area function for each part of Example 2 is the antiderivative where $C = 0$. This is true whenever $a = 0$. That is,

Function	Area function	Antiderivative
$f(x) = 4$	$A(x) = 4x$	$F(x) = 4x + C$
$f(x) = 4x$	$A(x) = 2x^2$	$F(x) = 2x^2 + C$
$f(x) = 4x + 3$	$A(x) = 2x^2 + 3x$	$F(x) = 2x^2 + 3x + C$

Now, to find the area bounded by a function f and the x-axis over the entire region from a to b, consider Figure 13.2.

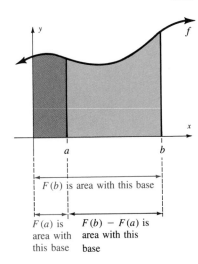

Figure 13.2 $A(x) = F(b) - F(a)$

$F(b)$ is area with this base

$F(a)$ is area with this base $F(b) - F(a)$ is area with this base

From the definition of the definite integral,

$$\int_a^b f(x)\,dx = F(x)\Big|_a^b = F(b) - F(a)$$

we see that the desired area is

$$A(x) = \int_a^b f(x)\,dx$$

This leads us to the **fundamental theorem of integral calculus:**

Fundamental Theorem of Integral Calculus

If f is a continuous nonnegative function on $[a,b]$ and $A(x)$ is the area function defined as the area of the region bounded by the graph of f and the x-axis on the interval $[a, x]$, then $A(x)$ is a differentiable function of x and
$$A'(x) = f(x)$$

The situation described by this theorem is shown in Figure 13.3. Notice that

$$A(x + h) - A(x)$$

is the area of the region shaded in color. This area is approximated by the area of the rectangle with base h and height $f(x)$. Thus

$$A(x + h) - A(x) \approx f(x) \cdot h$$

$$\frac{A(x + h) - A(x)}{h} \approx f(x)$$

If we take the limit as $h \to 0$, we see that

$$\lim_{h \to 0} \frac{A(x + h) - A(x)}{h} = f(x)$$

But this is the definition of derivative, so $A'(x) = f(x)$.

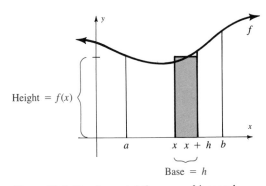

Height $= f(x)$

a x $x + h$ b

Base $= h$

Figure 13.3 Fundamental theorem of integral calculus

EXAMPLE 3 The graph below showing the rate of domestic cars sold is from the March 5, 1986, issue of the *Wall Street Journal*.

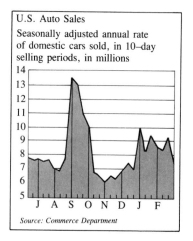

Let f be the rate shown. Then the fundamental theorem of integral calculus tells us that the *number* of cars sold is the area under this curve. (Since f is defined by a graph, the best we can do is to estimate the area. We will use a rectangular approximation here, but will develop other estimation methods in Section 14.3.) Estimate the sales in September.

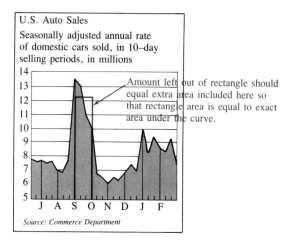

Solution $A = bh$

$$= 1(12.2)$$

$$= 12.2$$

About 12.2 million cars were sold between September 1 and October 1.

Most problems in this text have the function f given as a formula, so we can find the desired area more precisely than we did in Example 3.

The fundamental theorem of integral calculus is restated below:

<table>
<tr>
<td>Restatement of the Fundamental Theorem of Integral Calculus</td>
<td>

The area of a continuous nonnegative function f defined on an interval $[a, b]$ and bounded by the function f, the x-axis, and the vertical lines $x = a$ and $x = b$ is

$$\int_a^b f(x)\, dx = F(b) - F(a)$$

where F is any antiderivative of f.

</td>
</tr>
</table>

EXAMPLE 4   Find the area under the curve $y = x^2$ on $[2, 4]$.

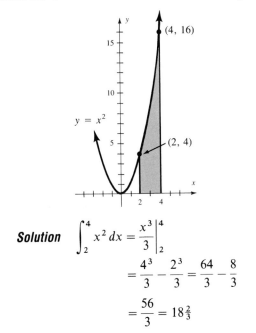

Solution

$$\int_2^4 x^2\, dx = \left. \frac{x^3}{3} \right|_2^4$$

$$= \frac{4^3}{3} - \frac{2^3}{3} = \frac{64}{3} - \frac{8}{3}$$

$$= \frac{56}{3} = 18\tfrac{2}{3} \qquad \blacksquare$$

The fundamental theorem of integral calculus is the most important theorem of calculus. It ties together the ideas of limits, derivatives, areas, and antiderivatives. It is sometimes stated as two separate theorems:

INTEGRALS OF DERIVATIVES: $\qquad \displaystyle\int_a^b F'(t)\, dt = F(b) - F(a)$

DERIVATIVES OF INTEGRALS: $\qquad \displaystyle\frac{d}{dx} \int_0^x f(t)\, dt = f(x)$

In this course we are primarily interested in the first form, which shows how to evaluate integrals by antidifferentiation.

There are many properties of the definite integral, several of which are analogous to those we stated for the indefinite integral:

Properties of the Definite Integral

1. $\displaystyle\int_a^a f(x)\,dx = 0$ where a is in the domain of f

2. $\displaystyle\int_a^b dx = b - a$

3. $\displaystyle\int_a^b f(x)\,dx = -\int_b^a f(x)\,dx$

4. $\displaystyle\int_a^b f(x)\,dx = \int_a^c f(x)\,dx + \int_c^b f(x)\,dx$ for any points a, b, and c on the closed interval $[a, b]$

5. $\displaystyle\int_a^b kf(x)\,dx = k\int_a^b f(x)\,dx$

6. $\displaystyle\int_a^b [f(x) \pm g(x)]\,dx = \int_a^b f(x)\,dx \pm \int_a^b g(x)\,dx$

EXAMPLE 5   Evaluate the given integrals.

a. $\displaystyle\int_{10}^{10} (3x^2 - \sqrt{x})\,dx$ **b.** $\displaystyle\int_4^3 (3x^2 - \sqrt{x})\,dx$

c. $\displaystyle\int_{-2}^{\pi/\sqrt{2}} e^t\,dt + \int_{\pi/\sqrt{2}}^2 e^t\,dt$ correct to the nearest tenth

Solution   **a.** $\displaystyle\int_{10}^{10} (3x^2 - \sqrt{x})\,dx = 0$

b. $\displaystyle\int_4^3 (3x^2 - \sqrt{x})\,dx = -\int_3^4 (3x^2 - \sqrt{x})\,dx$

$$= -\left[\frac{3x^3}{3} - \frac{x^{3/2}}{\frac{3}{2}}\right]_3^4$$

$$= -\left(4^3 - \frac{2}{3}\cdot 4^{3/2}\right) + \left(3^3 - \frac{2}{3}\cdot 3^{3/2}\right)$$

$$= -\left(64 - \frac{16}{3}\right) + (27 - 2\sqrt{3})$$

$$= \frac{16}{3} - 37 - 2\sqrt{3}$$

$$= \frac{-95 - 6\sqrt{3}}{3}$$

c. $\displaystyle\int_{-2}^{\pi/\sqrt{2}} e^t\,dt + \int_{\pi/\sqrt{2}}^2 e^t\,dt = \int_{-2}^2 e^t\,dt = e^t\Big|_{-2}^2 = e^2 - e^{-2} \approx 7.2$ ∎

Always be careful about the evaluation of the variable when you use substitution in order to find the antiderivative.

EXAMPLE 6   Evaluate $\displaystyle\int_1^{\sqrt{5}} \frac{6x\,dx}{\sqrt{1+3x^2}}$. ∎

Solution
$$\int_1^{\sqrt{5}} \frac{6x\,dx}{\sqrt{1+3x^2}} = \int_{x=1}^{x=\sqrt{5}} \frac{du}{\sqrt{u}}$$

Note that we need to say $x = \sqrt{5}$ and $x = 1$ are the limits of integration so that they are not confused with the variable u

$$= \int_{x=1}^{x=\sqrt{5}} u^{-1/2}\,du$$

$$= 2u^{1/2}\Big|_{x=1}^{x=\sqrt{5}}$$

Also notice: $u = 1 + 3x^2$
$$du = 6x\,dx$$

$$= 2\sqrt{1+3x^2}\,\Big|_1^{\sqrt{5}}$$

$$= 2(4) - 2(2)$$

$$= 4$$ ∎

Instead of substituting back to the original variable, we can change the limits of integration. For Example 6, since $u = 1 + 3x^2$, we can substitute $x = 1$ (lower limit):

$$u = 1 + 3(1)^2 = 4$$

and $x = \sqrt{5}$ (upper limit):

$$u = 1 + 3(\sqrt{5})^2 = 16$$

Then we can write

If $x = \sqrt{5}$, then $u = 16$

Note the change in the limits of integration

$$\int_1^{\sqrt{5}} \frac{6x\,dx}{\sqrt{1+3x^2}} = \int_4^{16} u^{-1/2}\,du$$

If $x = 1$, then $u = 4$.

$$= 2u^{1/2}\Big|_4^{16}$$

$$= 2(\sqrt{16} - \sqrt{4})$$

$$= 4$$

Substitution with the Definite Integral

If f is continuous on the set of values taken on by g, and g' is continuous on $[a,b]$, and if f has an antiderivative on that interval, then

$$\int_a^b f[g(x)]g'(x)\,dx = \int_{g(a)}^{g(b)} f(u)\,du \qquad u = g(x),\quad du = g'(x)\,dx$$

provided these integrals exist.

EXAMPLE 7 Evaluate $\displaystyle\int_1^2 e^{-3t}\,dt$ correct to three decimal places.

Solution $\displaystyle\int_1^2 e^{-3t}\,dt = -\frac{1}{3}\int_{-3}^{-6} e^u\,du = -\frac{1}{3}e^u\Big|_{-3}^{-6} = -\frac{1}{3}(e^{-6} - e^{-3}) \approx 0.016$

Let $u = -3t$; if $t = 2$, then $u = -6$
$du = -3dt$; if $t = 1$, then $u = -3$

Problem Set 13.3

Evaluate the definite integrals in Problems 1–20.

1. $\displaystyle\int_2^3 x^2\,dx$
2. $\displaystyle\int_{-2}^3 x^3\,dx$
3. $\displaystyle\int_1^4 \sqrt{x}\,dx$

4. $\displaystyle\int_{-3}^1 \sqrt{1-x}\,dx$
5. $\displaystyle\int_{-1}^2 x(1+x^4)\,dx$

6. $\displaystyle\int_{-1}^1 x^3(1+x^4)\,dx$
7. $\displaystyle\int_3^8 (x-2)^{-1}\,dx$

8. $\displaystyle\int_1^{e^3} \frac{dx}{x}$
9. $\displaystyle\int_1^{32} x^{-2/5}\,dx$

10. $\displaystyle\int_4^9 x^{3/2}\,dx$
11. $\displaystyle\int_0^1 \frac{dx}{(2x-1)^2}$

12. $\displaystyle\int_{-1}^0 \frac{dx}{(1-3x)^3}$
13. $\displaystyle\int_1^2 \left(x^{-2} - \frac{2}{x} + x^3\right)dx$

14. $\displaystyle\int_1^2 \left(x^2 + \frac{5}{x} - \sqrt{x}\right)dx$
15. $\displaystyle\int_6^6 \left(x^2 + \frac{\sqrt{5x}}{3} - 5\right)dx$

16. $\displaystyle\int_{-1}^1 x\sqrt{1+3x^2}\,dx$
17. $\displaystyle\int_{-1}^1 \frac{x\,dx}{\sqrt{1+3x^2}}$

18. $\displaystyle\int_1^e x^{-1}\ln^2 x\,dx$
19. $\displaystyle\int_{-1}^0 5x^2(x^3+1)^{10}\,dx$

20. $\displaystyle\int_{-1}^0 3x^4(1-x^5)^3\,dx$

Find the area bounded by the x-axis, the curve $y = f(x)$, and the given vertical lines in Problems 21–34. Sketch the area you find.

21. $y = 3$, $x = 0$, $x = 4$
22. $y = 5$, $x = 0$, $x = 4$
23. $y = 3x$, $x = 0$, $x = 4$
24. $y = 5x$, $x = 0$, $x = 4$

25. $y = 2x + 1$, $x = 0$, $x = 4$
26. $y = 3x + 2$, $x = 0$, $x = 4$
27. $y = x + 4$, $x = 2$, $x = 5$
28. $y = 5x + 2$, $x = 1$, $x = 3$
29. $y = x^2 + 1$, $x = -2$, $x = 1$
30. $y = x^2 + 2$, $x = -3$, $x = 1$
31. $y = \dfrac{1}{(2x+1)^2}$, $x = 1$, $x = 3$
32. $y = \dfrac{x}{\sqrt{1+3x^2}}$, $x = 0$, $x = \dfrac{1}{2}$
33. $y = e^{5x}$, $x = 0$, $x = 2$
34. $y = \dfrac{\ln x}{x}$, $x = 1$, $x = e^2$

35. If f and g are continuous on $[-3, 2]$ and

$$\int_{-3}^0 f(x)\,dx = 4, \quad \int_0^2 f(x)\,dx = 10,$$

and $\displaystyle\int_{-3}^2 g(x) = -5$, find

a. $\displaystyle\int_{-3}^2 f(x)\,dx$
b. $\displaystyle\int_2^{-3} g(x)\,dx$

c. $\displaystyle\int_{-3}^2 [2f(x) - 3g(x)]\,dx$

36. If f and g are continuous on $[5, 10]$ and

$$\int_5^8 f(x)\,dx = -3, \quad \int_8^{10} f(x)\,dx = 4,$$

and $\displaystyle\int_5^{10} g(x)\,dx = 2$, find

a. $\displaystyle\int_5^{10} f(x)\,dx$ **b.** $\displaystyle\int_7^7 [f(x) + g(x)]\,dx$

c. $\displaystyle\int_{10}^5 g(x)\,dx$

37. If f is continuous on $[1, 5]$ and

$$\int_1^3 f(x)\,dx = 9 \quad \text{and} \quad \int_1^5 f(t)\,dt = 15, \text{ find } \int_3^5 f(y)\,dy.$$

38. If g is continuous on $[-3, 2]$ and

$$\int_{-3}^2 g(x)\,dx = 7 \quad \text{and} \quad \int_0^2 g(t)\,dt = 4, \text{ find } \int_{-3}^0 g(y)\,dy.$$

APPLICATIONS

39. Using the information at the beginning of this section, predict the total number of barrels of oil consumed between 1987 and 2000.

40. Suppose that because of the 1986 crash in oil prices the demand for oil changes so it conforms to the formula

$$R(t) = 32.5e^{0.048t}$$

Use this formula to predict the amount of oil consumed between 1987 and 1995 using 1985 as $t = 0$.

41. Repeat Problem 40 for the years 1987 to 2000.

42. The graph below, from the March 5, 1986, issue of the *Wall Street Journal*, shows the rate of U.S. automobile sales. Estimate the total sales for June–December.

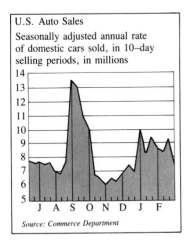

U.S. Auto Sales
Seasonally adjusted annual rate of domestic cars sold, in 10–day selling periods, in millions

Source: Commerce Department

43. Use the graph above to estimate the total automobile sales for the last quarter of the year (Oct. 1 to Dec. 31).

44. The graph below shows the rate of paper and paperboard production in the United States as reported by the American Paper Institute. Estimate the total production for 1985.

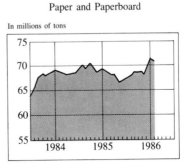

Paper and Paperboard

In millions of tons

45. The sales of Thornton Publishing are continuously growing according to the function

$$S(t) = 50e^t$$

where $S(t)$ is the sales in dollars in the tth year. What are the accumulated sales after 10 years?

46. When will the accumulated sales in Problem 45 pass $1 million?

47. In economics, the Gini index, G, can be used to measure how money is distributed among the population according to the formula

$$G = 1 - \int_0^1 2f(x)\,dx$$

The closer G is to zero, the more the money is spread evenly throughout the population. Find G for the function $f(x) = x^3$.

48. Find the area bounded by the curve $y = 1 - x^2$ and the x-axis.

49. Find the area bounded by the curve $y = 9 - x^2$ and the x-axis.

13.4 Area between Curves

The fundamental theorem of integral calculus relates an area to a definite integral. However, there is an important restriction of the function f; namely, f is a non-negative function. Compare the areas shown by the two graphs in Figure 13.4.

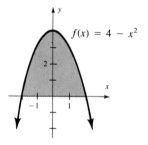

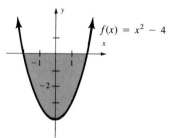

Figure 13.4 Areas bounded by two curves

For the curve bounded by $f(x) = 4 - x^2$, the fundamental theorem applies:

$$\int_{-2}^{2} (4 - x^2)\,dx = 4x - \frac{x^3}{3}\bigg|_{-2}^{2}$$

$$= 8 - \frac{8}{3} - \left(-8 + \frac{8}{3}\right)$$

$$= 16 - \frac{16}{3}$$

$$= \frac{32}{3}$$

This is the area

For the curve bounded by $f(x) = x^2 - 4$, the fundamental theorem does not apply:

$$\int_{-2}^{2} (x^2 - 4)\,dx = \frac{x^3}{3} - 4x\bigg|_{-2}^{2}$$

$$= \frac{8}{3} - 8 - \left(-\frac{8}{3} + 8\right)$$

$$= \frac{16}{3} - 16$$

$$= -\frac{32}{3}$$

This is not the area

It appears that if f is negative over $[a, b]$ then the area is the opposite of the integral.

Figure 13.5 Area defined by a function over $[a, b]$. A_2 and A_4 are positive numbers representing the areas and A_1, A_3, and A_5 are opposites of the enclosed areas (that is, negative areas)

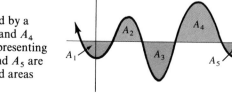

Notice from Figure 13.5,

$$\int_{a}^{b} f(x)\,dx = |A_1| + |A_2| + |A_3| + |A_4| + |A_5|$$
$$= -A_1 + A_2 - A_3 + A_4 - A_5$$

Also, $|A_1| = -A_1$, $|A_3| = -A_3$, and $|A_5| = -A_5$ since these are all negative; and $|A_2| = A_2$ and $|A_4| = A_4$ since these are all positive.

EXAMPLE 1 $\int_{-2}^{2} x^3\,dx = \frac{x^4}{4}\bigg|_{-2}^{2} = \frac{2^4}{4} - \frac{(-2)^4}{4} = 0$

The definite integral is *not* the area.

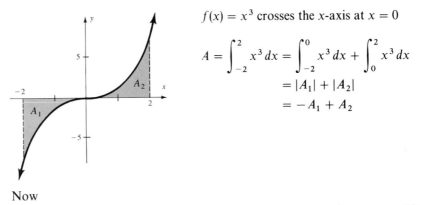

$f(x) = x^3$ crosses the x-axis at $x = 0$

$$A = \int_{-2}^{2} x^3\, dx = \int_{-2}^{0} x^3\, dx + \int_{0}^{2} x^3\, dx$$
$$= |A_1| + |A_2|$$
$$= -A_1 + A_2$$

Now

$$A_1 = \int_{-2}^{0} x^3\, dx = \frac{x^4}{4}\bigg|_{-2}^{0} = 0 - \frac{16}{4} - 0 = -4 \quad \text{and} \quad A_2 = \int_{0}^{2} x^3\, dx = \frac{x^4}{4}\bigg|_{0}^{2} = 4$$

Thus

$$A = -A_1 + A_2 = -(-4) + 4 = 8 \qquad \blacksquare$$

The difficulties encountered when dealing with problems like Example 1 and functions with both positive and negative values (like the function shown in Figure 13.5) can be eased if we find the area *between* two curves.

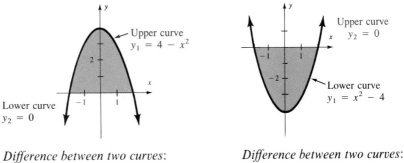

Difference between two curves:

$$y_1 - y_2 = (4 - x^2) - 0 = 4 - x^2$$

Upper curve

$$\text{Area} = \int_{a}^{b} (y_1 - y_2)\, dx$$

Lower curve

Difference between two curves:

$$y_2 - y_1 = 0 - (x^2 - 4) = 4 - x^2$$

Upper curve

$$\text{Area} = \int_{a}^{b} (y_2 - y_1)\, dx$$

Lower curve

Area between Two Curves

Suppose that f and g are continuous functions on $[a, b]$ and that $f(x) \geq g(x)$ for $a \leq x \leq b$.* The area of the region bounded by f above, g below, $x = a$ on the left, and $x = b$ on the right is

$$A = \int_{a}^{b} [f(x) - g(x)]\, dx$$

* This means that the curves f and g do not cross.

Look at Example 1 again. The integral must be split into two parts in order for the upper curve to be listed first. The figures below give several possibilities, along with how each relates to the fundamental theorem of integral calculus. In each case A is the region shaded in color.

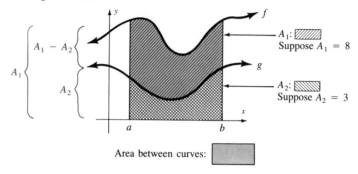

Area between curves: �

Both functions nonnegative

$$A = A_1 - A_2$$

where

$$A = \int_a^b [f(x) - g(x)]\, dx = \overset{\text{Top curve}}{\int_a^b f(x)\, dx} - \underset{\text{Bottom curve}}{\int_a^b g(x)\, dx}$$

$$A = \boxed{\!/\!/\!/} - \boxed{\text{\\\\}} = 8 - 3 = 5$$

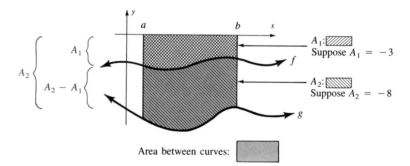

Area between curves: ▨

Both functions nonpositive

$$A = A_1 - A_2$$

where

$$A = \int_a^b [f(x) - g(x)]\, dx = \overset{\text{Top curve}}{\int_a^b f(x)\, dx} - \underset{\text{Bottom curve}}{\int_a^b g(x)\, dx}$$

$$A = -\boxed{\!/\!/\!/} - [-\boxed{\text{\\\\}}]$$

Opposite since the integral is negative

$$= -3 - (-8) = 5$$

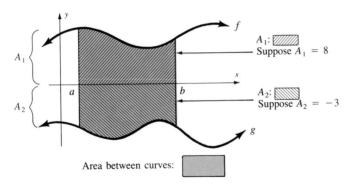

Area between curves:

One positive function and one negative function

$$A = A_1 - A_2$$

where

$$A = \int_a^b [f(x) - g(x)]\, dx$$

Top curve

$$= \int_a^b f(x)\, dx - \int_a^b g(x)\, dx$$

Bottom curve

$$A = \boxed{} - \boxed{}$$
$$= 8 - (-3)$$
$$= 11$$

Curves cross at $x = c$ where $a < c < b$

$$A = R_1 + R_2$$

where

$$R_1 = \int_a^c \underbrace{[g(x) - f(x)]}\, dx$$

Function g is on top in this region

$$R_2 = \int_c^b \underbrace{[f(x) - g(x)]}\, dx$$

Function f is on top in this region

EXAMPLE 2 Find the area between the curves $y = x^2$ and $y = 6 - x$ from $x = -3$ to $x = 2$.

Solution First sketch the curves.

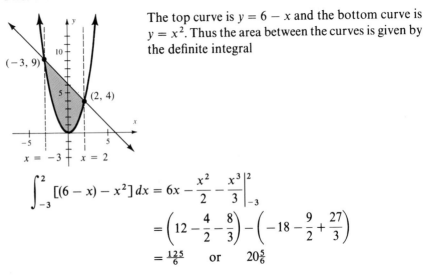

The top curve is $y = 6 - x$ and the bottom curve is $y = x^2$. Thus the area between the curves is given by the definite integral

$$\int_{-3}^2 [(6 - x) - x^2]\, dx = 6x - \frac{x^2}{2} - \frac{x^3}{3}\Big|_{-3}^2$$

$$= \left(12 - \frac{4}{2} - \frac{8}{3}\right) - \left(-18 - \frac{9}{2} + \frac{27}{3}\right)$$

$$= \frac{125}{6} \quad \text{or} \quad 20\frac{5}{6} \qquad \blacksquare$$

Sometimes the limits of integration for problems like Example 2 are implied instead of given. In such cases you need to solve the system of equations formed by the boundaries in the problem:

$$\begin{cases} y = x^2 \\ y = 6 - x \end{cases}$$

By substitution,

$$6 - x = x^2$$
$$x^2 + x - 6 = 0$$
$$(x + 3)(x - 2) = 0$$
$$x = -3, 2$$

The points of intersection for the curves are now found:

If $x = -3$, then $y = 6 - (-3) = 9$;   point: $(-3, 9)$
If $x = 2$, then $y = 6 - 2 = 4$;   point: $(2, 4)$

You would use these values as the limits of integration as shown in Example 2.

EXAMPLE 3   Find the area between the curve $y = x^2 - 9$ and the lines $x = 1$ and $x = 9$.

Solution

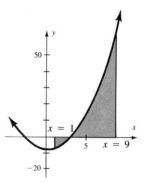

The top curve is not always the same on $[1, 9]$. The curve $y = x^2 - 9$ crosses the x-axis:

$$x^2 - 9 = 0$$
$$(x - 3)(x + 3) = 0$$
$$x = 3, -3$$
$$\quad\quad\uparrow$$
$$\text{Not in interval}$$

The area is found by dividing the interval into two different integrals:

$$\int_1^3 \underbrace{[0 - (x^2 - 9)]}_{\text{Top curve}} dx + \int_3^9 \underbrace{[(x^2 - 9) - 0]}_{\text{Top curve}} dx = \left(-\frac{x^3}{3} + 9x\right)\Big|_1^3 + \left(\frac{x^3}{3} - 9x\right)\Big|_3^9$$

$$= (-9 + 27) - \left(-\frac{1}{3} + 9\right) + (243 - 81) - (9 - 27)$$

$$= \frac{568}{3} \text{ or } 189\frac{1}{3} \qquad\qquad\blacksquare$$

EXAMPLE 4   If the 1986 crash of oil prices brings about a change in the worldwide consumption rate of petroleum products (*Wall Street Journal*, February 11, 1986), then we can measure the effect of that change by finding the area between two curves. For example, suppose that

pre-1985 rate of consumption: $R_1(t) = 78e^{-0.04t}$ $(0 \le t \le 5)$

post-1985 rate of consumption: $R_2(t) = 50.2e^{0.048t}$ $(t \ge 5)$

for a base year of 1980 $(t = 0)$. How much extra oil would be consumed between 1985 and 1990?

Solution

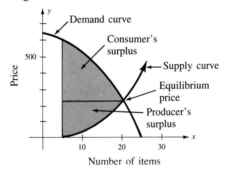

$$\int_5^{10} (50.2e^{0.048t} - 78e^{-0.04t}) \, dt$$

$$= \frac{50.2}{0.048} e^{0.048t} - \frac{78}{-0.04} e^{-0.04t} \Big|_5^{10}$$

$$\approx 2997.27 - 2926.04$$

$$\approx 71.2$$

The additional oil consumed would be about 71.2 billion barrels.   ■

An important business application of the area between curves involves the concepts of **consumer's surplus** and **producer's surplus**. The intersection of the demand and supply functions provides an *equilibrium price* or point, as shown in Figure 13.6.

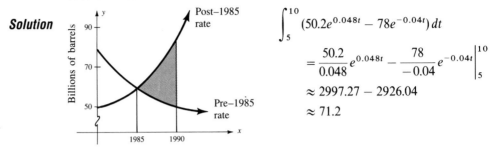

Figure 13.6 Consumer's and producer's surplus

Consumers who would be willing to pay more than the price set by the equilibrium point will benefit by paying the lower price. The area that represents this situation is the region below the demand curve and above the line where the price equals the equilibrium price. This region is referred to as the *consumer's surplus*.

On the other hand, producers who would be willing to sell the product for less than the equilibrium price will benefit when the market price is set at the equilibrium price, so this region is called the *producer's surplus* and is represented as the area below the line where the price equals the equilibrium price and above the supply curve.

If (n, p) is the equilibrium point for a demand function $D(x)$ and a supply function $S(x)$, then

$$\text{Consumer's surplus} = \int_0^n [D(x) - p] \, dx$$

$$\text{Producer's surplus} = \int_0^n [p - S(x)] \, dx$$

EXAMPLE 5 The price, in dollars, for a product is

$$D(x) = 650 - x - x^2$$

where x is the number of items in the domain $[0, 25]$. The supply curve, in dollars, is given by

$$S(x) = x^2 - 9x + 10$$

where x is in the domain $[10, 25]$. Find the consumer's and producer's surplus.

Solution First, find the equilibrium price.

$$D(x) = 650 - x - x^2$$
$$S(x) = x^2 - 9x + 10$$

Find x so that $D(x) = S(x)$:

$$650 - x - x^2 = x^2 - 9x + 10$$
$$2x^2 - 8x - 640 = 0$$
$$x^2 - 4x - 320 = 0$$
$$(x - 20)(x + 16) = 0$$
$$x = 20, -16$$

└── Reject the negative value

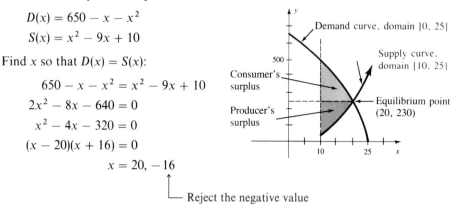

The domain for the surplus is the intersection of the domain for the supply and demand; in this example it is given as $[10, 25]$. For $x = 20$, the price is 230, so the equilibrium point is $(20, 230)$.

$$\text{Consumer's surplus} = \int_{10}^{20} [(650 - x - x^2) - 230] \, dx$$

$$= \int_{10}^{20} (420 - x - x^2) \, dx$$

$$= 420x - \frac{x^2}{2} - \frac{x^3}{3} \Big|_{10}^{20}$$

$$= 8400 - 200 - \frac{8000}{3} - 4200 + 50 + \frac{1000}{3}$$

$$\approx 1716.67$$

$$\text{Producer's surplus} = \int_{10}^{20} [230 - (x^2 - 9x + 10)] \, dx$$

$$= \int_{10}^{20} (220 + 9x - x^2) \, dx$$

$$= 220x + \frac{9x^2}{2} - \frac{x^3}{3} \Big|_{10}^{20}$$

$$\approx 1216.67$$

The consumer's surplus is $1,716.67$, and the producer's surplus is $1,216.67$.

Problem Set 13.4

Write integrals to express the shaded region in Problems 1–8.
Both positive regions (color) and negative regions (gray) should
be included.

1.

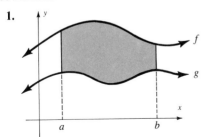

2.

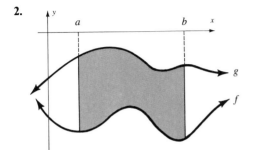

3.

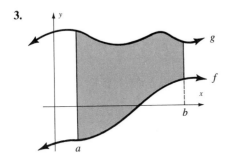

4.

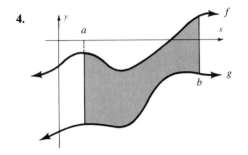

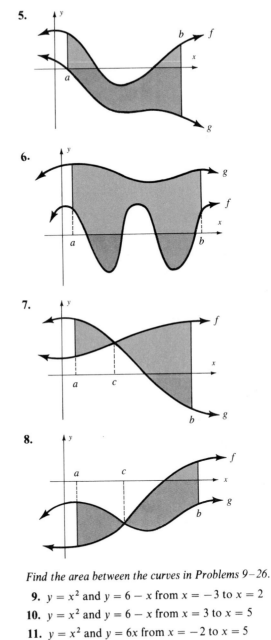

Find the area between the curves in Problems 9–26.

9. $y = x^2$ and $y = 6 - x$ from $x = -3$ to $x = 2$

10. $y = x^2$ and $y = 6 - x$ from $x = 3$ to $x = 5$

11. $y = x^2$ and $y = 6x$ from $x = -2$ to $x = 5$

12. $y = x^2 + 4$ and $y = 2x + 4$ from $x = 0$ to $x = 1$

13. $y = x^2 + 4$ and $y = 2x + 4$ from $x = -1$ to 1

14. $y = x^2 + 4$ and $y = 2x + 4$ from $x = -1$ to 5

15. $y = x^2 - 4$ and the x-axis

16. $y = x^2 - 9$ and the x-axis

17. $y = x^3$ and $y = x^2$ from $x = 0$ to $x = 1$

18. $y = x^3$ and $y = x^2$ from $x = 0$ to $x = 2$

19. $y = x^2 - x$ and $y = 6$ 20. $y = x^2 - 5x$ and $y = 6$

21. $y = x^2$ and $y = \sqrt{x}$ 22. $y = x$ and $y = \sqrt{x}$

23. $y = \sqrt{x}$ and $y = 4$ 24. $y = \sqrt{x}$ and $y = 9$

25. $y = x^3$ and $y = x$ 26. $y = x$ and $y = x^4$

APPLICATIONS

27. The Sorite Corporation has found that production costs are determined by the function

$$C(x) = \frac{100}{\sqrt{x+1}} + 50$$

where $C(x)$ is the cost to produce the xth item. What is the total cost of producing 2000 items?

28. The EPA (Environmental Protection Agency) estimates that pollutants are being dumped into the Russian River according to the formula

$$P(t) = \frac{1000}{\sqrt[3]{(t+5)^2}}$$

where t is the year and $P(t)$ is pollutants in tons. What is the total amount of pollutants that will be dumped into the river in the next 25 years?

29. The price, in cents, for a product is

$$D(x) = 1000 - 15x - x^2$$

where x is the number of items in the domain $[10, 30]$. The supply function is given by

$$S(x) = x^2 - 3x - 40$$

Find the producer's surplus.

30. Find the consumer's surplus for the information in Problem 29.

31. Find the consumer's and producer's surplus if the demand and supply functions are

$$D(x) = \frac{236 - 107x}{(x-3)^2} \quad \text{and} \quad S(x) = x + 20$$

for x on $[0, 3)$, where x is the number of items (in thousands).

32. The marginal revenue and marginal cost (in dollars) per day are given in terms of the tth month by the formulas

$$R'(t) = 500e^{0.01t} \quad \text{and} \quad C'(t) = 50 - 0.1t$$

where $R(0) = C(0) = 0$, and t is the month. Find the total profit for the first year.

33. A chemical spill is known to kill fish according to the formula

$$N(t) = \frac{100,000}{\sqrt{t + 250}}$$

where N is the number of fish killed on the tth day since the spill. What is the total number of fish killed in 30 days?

13.5 Summary and Review

IMPORTANT TERMS

Antidifferentiation   [13.1]
Area between two curves   [13.3]
Area function   [13.3]
Constant of integration   [13.1]
Consumer's surplus   [13.4]
Definite integral   [13.1, 13.3]
Differential calculus   [13.1]
Dummy variable   [13.3]
Evaluate an integral   [13.3]
Fundamental theorem of integral calculus   [13.3]
Indefinite integral   [13.1]

Initial value   [13.1]
Integral calculus   [13.1]
Integral of a constant   [13.1]
Integral of a difference   [13.1]
Integral of a sum   [13.1]
Integral symbol   [13.1]
Integrand   [13.1, 13.3]
Integration   [13.1]
Integration by substitution   [13.2]
Integration formulas   [13.1, 13.3]
Limits of integration   [13.3]
Producer's surplus   [13.4]

SAMPLE TEST   *For additional practice there are a large number of review problems categorized by objective in the Student Solutions Manual. The following sample test (40 minutes) is intended to review the main ideas of this chapter.*

Evaluate the integrals in Problems 1–12. Be sure to include the constant of integration when working with indefinite integrals.

1. $\displaystyle\int x^5\,dx$

2. $\displaystyle\int \frac{dx}{x}$

3. $\displaystyle\int e^u\,du$

4. $\displaystyle\int u^{-4}\,du$

5. $\displaystyle\int (18x^2 - 6x - 3)\,dx$

6. $\displaystyle\int \frac{(x+1)}{2x^2 + 4x + 5}\,dx$

7. $\displaystyle\int \left(e^{3x+1} + \frac{1}{e}\right)dx$

8. $\displaystyle\int \frac{6x^2 - 3x + 2}{x}\,dx$

9. $\displaystyle\int_{-1}^{3} \frac{x^3 + 2}{x^2}\,dx$

10. $\displaystyle\int_{2}^{2\sqrt{3}} x\sqrt{2x^2 + 1}\,dx$

11. $\displaystyle\int_{1}^{8} [x^{1/3}(1 + x^{4/3})^3]\,dx$

12. $\displaystyle\int_{3\sqrt{2}}^{5} x\sqrt{x^2 - 9}\,dx$

13. Find the area bounded by the x-axis, the curve $y = \sqrt{3x + 4}$, and the lines $x = 4$ and $x = \frac{5}{3}$.

14. Find the area bounded by the curves $y = x^2$, $y = 6 - x$, and the lines $x = 0$ and $x = 4$.

15. Find the area bounded by the curves $y = x^3$ and $y = x$.

16. The marginal profit of Campress Press-Ons is

$$P'(x) = 250x - 5000$$

where x is the number of items. Find the profit function if there is a $2,500 loss when no items are produced.

17. One of the fastest growing trees is the eucalyptus. In order to determine whether it is economically feasible to harvest it as a crop for firewood, it must grow at least 6 feet in 5 years ($t = 6$, since it is 1-year-old when planted). If the growth rate is

$$1.2 + 5t^{-4} \qquad t \geq 1$$

feet per year, where t is the time in years, is it economically feasible to plant these trees for commercial harvesting if they are 12 inches tall when planted (that is, when $t = 1$)?

18. In 1986 the growth rate of Chicago, Illinois, was

$$P'(t) = -0.035e^{-0.011t}$$

where t is measured in years after 1980 and $P(t)$ gives the population in millions. If the population in 1984 was 2,992,472, predict the population at the turn of the century.

19. The demand for oil conforms to the formula

$$R(t) = 32.4e^{0.048t}$$

for t years after 1985. If the total oil still left in the earth is estimated to be 670 billion barrels, estimate the length of time before all available oil is consumed if the rate does not change.

20. The price, in dollars, for a product is given by the demand function

$$D(x) = 250 + 10x - x^2$$

where x is the number of items (in thousands) in the domain [0, 10]. The supply function is given by

$$S(x) = x^2 - 10x + 300$$

Find the consumer's and the producer's surplus.

Modeling Application 9

Computers in Mathematics

In the late 1970s, mathematics instruction was forever altered with the advent of powerful, inexpensive hand-held calculators. These calculators allowed both instructors and students to focus on concepts and processes rather than on the tedious and complicated calculations involved in realistic problems. While computers can be programmed to carry out calculator-type simplifications, there are several fundamental stumbling blocks that prohibit them from being used to as great an extent as calculators: they are expensive; they offer little advantage (if any) over the simplification of arithmetic calculations; and they can do only very limited algebraic simplification and manipulation.

This limitation is being overcome by computer languages that do symbolic manipulation. For this modeling application, investigate one of the following computer languages that does symbolic manipulation:

MAXSYMA is a trademark for a language available from Symbolics, Inc., 11 Cambridge Center, Cambridge, MA 02142.

muMath is a copyrighted program marketed by The Soft Warehouse, Box 11174, Honolulu, HI 46828-0174.

PowerMath is a Macintosh program from Brainpower, Inc., 24009 Ventura Boulevard, Suite 250, Calabasas, CA 91302.

TK! Solver is available for the Apple, Macintosh, IBM PC, TRS-80, and DEC.

After your investigation, write a paper describing a symbolic language. For general guidelines about writing this essay, see the commentary for Modeling Application 1 on page 130.

CHAPTER 14

Additional Integration Topics

CHAPTER CONTENTS

14.1 **Integration by Parts**

14.2 **Integration by Table**

14.3 **Numerical Integration**

14.4 **Summary and Review**
     Important Terms
     Sample Test

APPLICATIONS

Management (*Business, Economics, Finance, and Investments*)

Total profit for a 5-year period (14.1, Problem 32)

Marginal profit (14.1, Problem 34)

Maintenance costs on a building (14.1, Problem 35)

Total cost of producing 1000 items given variable production cost (14.2, Problem 66)

Determining the accumulated sales for Simon's Simple Tool Kit (14.2, Problem 67)

Depreciation over a 5-year period when the depreciation rate is known (14.4, Problem 18)

Area under the 10-year Treasury Bond rate curve (1980–1985) as reported in the *Wall Street Journal* (14.4, Problems 19–20)

Life sciences (*Biology, Ecology, Health, and Medicine*)

Determining the amount of contamination in a water supply (14.1, Problem 33)

Amount of pollutants dumped into a river (14.2, Problem 65)

Contour map (14.3, Problem 37)

Social sciences (*Demography, Political Science, Population, Psychology, Society, and Sociology*)

Performance determination by knowing the learning rate of a given task (14.4, Problem 17)

General interest

Determining which flashbulb produces the most light (14.3, Problem 38)

Modeling application—Modeling the Nervous System

<table>
<tr><td>

CHAPTER OVERVIEW

</td><td>

Our study of integration is expanded by the introduction of three new techniques for integration: by parts, by tables, and by numerical integration.

</td></tr>
<tr><td>

PREVIEW

</td><td>

We first learn to find the counterpart of the product rule for differentiation, in a process called *integration by parts*. Next we learn to use tables to integrate fairly complicated functions. Finally, for functions we cannot integrate directly, the process goes full circle back to the fundamental theorem of integral calculus to approximate integrals by using areas, in a process called *numerical integration*.

</td></tr>
<tr><td>

PERSPECTIVE

</td><td>

More and more emphasis today is given to integration by using tables. Although techniques of integration are important, new computer technology has made numerical integration and integration by table much more important than before. A brief table of integrals can be found on the endpapers (the inside covers) of this book. If you need more extensive tables of integration, you can find them at your college bookstore.

</td></tr>
</table>

14.1   Integration by Parts

In Section 13.1 we were able to restate integration formulas corresponding to the power, sum, and difference differentiation rules. In Section 13.2 we used substitution to reverse the chain rule. Now we develop an integration formula for the product rule for differentiation.

Suppose that u and v are differentiable functions of x. Then

$$\frac{d}{dx}(uv) = u\frac{dv}{dx} + v\frac{du}{dx}$$

or, using differentials,

$$d(uv) = u\,dv + v\,du$$

Integrate both sides to obtain

$$\int d(uv) = \int u\,dv + \int v\,du$$

$$uv = \int u\,dv + \int v\,du$$

Solve this equation for $\int u\,dv$ and obtain a formula called **integration by parts**:

Integration by Parts

$$\int u\,dv = uv - \int v\,du$$

EXAMPLE 1   Evaluate $\int xe^x\,dx$ using integration by parts.

Solution To use integration by parts, you must choose u and dv so that the new integral is easier to integrate than the original.

Integrate by parts:

Formula:
$$\int u\,dv = uv - \int v\,du$$

Example 1
$$\int xe^x\,dx = xe^x - \int e^x\,dx = xe^x - e^x + C$$

> Let $u = x$ and $dv = e^x\,dx$; then
>
> $du = dx$ $\qquad v = \int e^x\,dx = e^x$ *See note below

■

Integration by parts is often difficult the first time you try to do it because there is no absolute choice for u and dv. In the previous example, you might have chosen

> $u = e^x$ and $dv = x\,dx$
>
> $du = e^x\,dx$ $\qquad v = \int x\,dx = \dfrac{x^2}{2}$

Then
$$\int xe^x\,dx = e^x\frac{x^2}{2} - \int \frac{x^2}{2}e^x\,dx$$
$$\underset{u\quad v}{\uparrow\ \uparrow}\qquad \underset{v\qquad du}{\underbrace{\quad}\ \underbrace{\quad}}$$
$$= \frac{1}{2}x^2e^x - \frac{1}{2}\int x^2 e^x\,dx$$

---

* *Note:* When we wrote $v = \int e^x\,dx = e^x$ above, we should have written $v = \int e^x\,dx = e^x + K$. Then our work would have looked like

$$\int xe^x\,dx = x(e^x + K) - \int (e^x + K)\,dx$$
$$= xe^x + xK - \int e^x\,dx - \int K\,dx$$
$$= xe^x + xK - \int e^x\,dx - Kx + C_1$$
$$= xe^x - \int e^x\,dx + C_1$$
$$= xe^x - e^x + C_2 + C_1$$
$$= xe^x - e^x + C \qquad \text{where} \qquad C = C_1 + C_2$$

The intermediate constant K in integration by parts will always drop out, so we usually omit the constant when calculating v from dv and include only the constant C at the end.

Note, however, that this choice of u and dv leads to a more complicated form than the original. Therefore, when you are integrating by parts, if you make a choice for u and dv that leads to a more complicated form than when you started, consider going back and making another choice for u and dv.

Sometimes you may have to integrate by parts more than once, as illustrated by Example 2.

EXAMPLE 2 Evaluate $\int 5x^2 e^{3x}\, dx$ using integration by parts.

Solution

Let $u = x^2$ and $dv = e^{3x}\, dx$

$$du = 2x\, dx \qquad v = \int e^{3x}\, dx = \frac{1}{3}e^{3x}$$

Integrate by parts:

$$\int \underset{\substack{\uparrow\quad\uparrow \\ u\quad dv}}{5x^2 e^{3x}\, dx} = 5\left[\underset{\substack{\uparrow\quad\;\uparrow \\ u\quad\; v}}{x^2 \left(\frac{1}{3}e^{3x}\right)} - \int \underset{\substack{\uparrow\quad\;\uparrow \\ v\quad\; du}}{\frac{1}{3}e^{3x} 2x\, dx} \right]$$

$$= \frac{5}{3}x^2 e^{3x} - \frac{10}{3}\int x e^{3x}\, dx$$

$$= \frac{5}{3}x^2 e^{3x} - \frac{10}{3}\int x e^{3x}\, dx \qquad \text{Now use integration by parts again}$$

$$u = x \qquad \text{and} \qquad dv = e^{3x}\, dx$$

$$du = dx \qquad v = \int e^{3x}\, dx = \frac{1}{3}e^{3x}$$

$$= \frac{5}{3}x^2 e^{3x} - \frac{10}{3}\left[x\left(\frac{1}{3}e^{3x}\right) - \int \frac{1}{3}e^{3x}\, dx \right]$$

$$= \frac{5}{3}x^2 e^{3x} - \frac{10}{9}x e^{3x} + \frac{10}{9}\int e^{3x}\, dx$$

$$= \frac{5}{3}x^2 e^{3x} - \frac{10}{9}x e^{3x} + \frac{10}{9}\frac{e^{3x}}{3} + C$$

$$= \frac{5}{3}x^2 e^{3x} - \frac{10}{9}x e^{3x} + \frac{10}{27}e^{3x} + C \qquad\blacksquare$$

EXAMPLE 3 Evaluate $\int \ln x\, dx$ by integrating by parts.

Solution   Let $u = \ln x$ and $dv = dx$

$$du = \frac{1}{x} dx \qquad\qquad v = \int dx = x$$

$$\int \ln x \, dx = x \ln x - \int x \left(\frac{1}{x} \right) dx$$

$$= x \ln x - x + C \qquad\qquad \blacksquare$$

If you use integration by parts with the definite integral, be sure to evaluate the first part at both of the limits of integration, as illustrated by Example 4.

┌WARNING: Do not forget this evaluation

EXAMPLE 4 $\displaystyle\int_0^1 xe^{2x} \, dx = \frac{1}{2} xe^{2x} \Big|_0^1 - \frac{1}{2} \int_0^1 e^{2x} \, dx$

$$u = x \qquad \text{and} \qquad dv = e^{2x} \, dx$$
$$du = dx \qquad\qquad v = (\tfrac{1}{2})e^{2x}$$

┌── It is usually easier to simplify the algebra and then
      do one evaluation here at the end of the problem

$$= \left(\frac{1}{2} xe^{2x} - \frac{1}{4} e^{2x} \right) \Big|_0^1 = \left(\frac{1}{2} e^2 - \frac{1}{4} e^2 \right) - \left(0 - \frac{1}{4} \right) = \frac{1}{4} e^2 + \frac{1}{4} \qquad \blacksquare$$

Remember that integration by parts, as with other integration methods, is not a method that "always works." There are many functions whose integrals cannot be found by the methods described in this book. In fact, there are functions, such as e^{-x^2}, which cannot be integrated at all. For that reason, we develop a method of approximate integration in Section 14.3.

Problem Set 14.1

Evaluate the integrals in Problems 1–28 by parts, if possible.

1. $\displaystyle\int 12xe^{3x} \, dx$

2. $\displaystyle\int 60xe^{5x} \, dx$

9. $\displaystyle\int \ln(x + 1) \, dx$

10. $\displaystyle\int \ln(2x - 1) \, dx$

3. $\displaystyle\int_0^1 xe^{-x} \, dx$

4. $\displaystyle\int_0^{1/2} 9xe^{-2x} \, dx$

11. $\displaystyle\int \frac{x \, dx}{\sqrt{1 - x}}$

12. $\displaystyle\int \frac{x \, dx}{\sqrt{2 - 3x}}$

5. $\displaystyle\int x\sqrt{1 - x} \, dx$

6. $\displaystyle\int 3x\sqrt{1 - 2x} \, dx$

13. $\displaystyle\int 6x^2 e^{2x} \, dx$

14. $\displaystyle\int 8x^2 e^{-2x} \, dx$

7. $\displaystyle\int x(x + 2)^3 \, dx$

8. $\displaystyle\int x(3x - 1)^4 \, dx$

15. $\displaystyle\int x^2 \sqrt{1 - 2x} \, dx$

16. $\displaystyle\int 5x^2 \sqrt{1 - x} \, dx$

17. $\int x^3 \sqrt[3]{1 - x^2}\, dx$

18. $\int x \ln x\, dx$

19. $\int x^2 \ln x\, dx$

20. $\int x^2 \ln 3x\, dx$

21. $\int 6x^2 e^{4x}\, dx$

22. $\int 24x^2 e^{3x}\, dx$

23. $\int \frac{x^3\, dx}{\sqrt{1 - x^2}}$

24. $\int \frac{x^3\, dx}{\sqrt{x^2 + 1}}$

25. $\int \frac{dx}{x \ln x}$

26. $\int \sqrt{x} \ln x\, dx$

27. $\int_1^e (\ln x)^2\, dx$

28. $\int_{1/3}^e 3(\ln 3x)^2\, dx$

29. Derive the formula

$$\int x^m e^x\, dx = x^m e^x - m \int x^{m-1} e^x\, dx$$

30. Derive the formula

$$\int (\ln x)^m\, dx = x(\ln x)^m - m \int (\ln x)^{m-1}\, dx$$

31. a. Evaluate the following integral by parts:

$$\int \frac{x^3}{x^2 - 1}\, dx$$

b. Evaluate the integral in part a by first dividing the integrand.

APPLICATIONS

32. The marginal cost and revenue equations (in millions of dollars) for a product are given by

$$R'(x) = \frac{x}{\sqrt{x - 1}} \quad \text{and} \quad C'(x) = 0.2x$$

where x is the time in years. The area between the graphs of the marginal functions for a time period where $R'(x) > C'(x)$ is the total accumulated profit bounded by two values of x. What is the total profit for years 1 through 5?

33. A contaminated water supply is treated, and the rate of harmful bacteria t days after the treatment is given by

$$N'(t) = \frac{t}{(1 + t)^{1/3}} \qquad 0 \le t \le 10$$

where $N(t)$ is the number of millions of bacteria per liter of water. If the initial count is 10,000, find $N(t)$, and then find the bacteria count after 7 days.

34. If the marginal profit is given by

$$P'(t) = \frac{2t}{(1 + t)^{2/3}}$$

where t is the time in years, find P if the profit at time 0 is $0.

35. Maintenance costs for a certain building can be approximated by

$$M'(x) = 100x^2 + \frac{5000x}{x^2 + 1}$$

where x is the age of the building in years and $M(x)$ is the total accumulated cost of maintenance for x years. Find the total maintenance costs for the period from 5 to 10 years after the building was built.

14.2 Integration by Table

Professional mathematicians and scientists often integrate functions by using tables of integrals. One of the most common sources of integral tables is the Chemical Rubber Company's *Standard Mathematical Tables*, which is much more extensive than what we need for this course. We have, however, a shortened version on the endpapers (the inside covers) of this book.

To use a Brief Integral Table, first classify the integral by type. The types listed in our table are:

> Forms containing $a + bx$
> Forms containing $\sqrt{x^2 \pm a^2}$ or $\sqrt{a^2 - x^2}$
> Logarithmic forms
> Exponential forms

More extensive tables of integrals are available, but these four fundamental types will serve our purposes, and if you learn how to use this integral table you will be able to use more extensive tables.

After deciding which form applies, match the individual type with the problem at hand by making appropriate choices for the constants. More than one form may apply, but the results derived by using different formulas will be the same (except for the constant).

Take a few moments to look at the **Brief Integral Table**. Notice that this integral table has two basic types of **integration formulas**. The first gives a formula that is the antiderivative, while the second simply rewrites the integral in another form. Examples 1 and 2 illustrate each of these types.

EXAMPLE 1 Evaluate $\int x^2(3 - x)^5 dx$.

Solution This is an integral of the form $a + bx$; from the Brief Integral Table, we see that Formula 11 applies, where $a = 3$, $b = -1$, and $n = 5$:

$$\int x^2(3 - x)^5 dx = \frac{1}{(-1)^3}\left[\frac{(3 - x)^{5+3}}{5 + 3} - 2(3)\frac{(3 - x)^{5+2}}{5 + 2} + 3^2\frac{(3 - x)^{5+1}}{5 + 1}\right] + C$$

$$= -\left[\frac{(3 - x)^8}{8} - 6\frac{(3 - x)^7}{7} + 9\frac{(3 - x)^6}{6}\right] + C$$

$$= -\frac{1}{8}(3 - x)^8 + \frac{6}{7}(3 - x)^7 - \frac{3}{2}(3 - x)^6 + C$$

$$= -\frac{1}{112}(3 - x)^6[14(3 - x)^2 - 96(3 - x) + 168] + C$$

$$= -\frac{1}{112}(3 - x)^6[14(9 - 6x + x^2) - 288 + 96x + 168] + C$$

$$= -\frac{1}{112}(3 - x)^6(14x^2 + 12x + 6) + C$$

$$= -\frac{1}{56}(3 - x)^6(7x^2 + 6x + 3)$$ ∎

EXAMPLE 2 Evaluate $\int \ln^4 x \, dx$.

Solution This is an integral in logarithmic form; from the Brief Integral Table we see that Formula 34 applies. Notice that this form involves another integral.

$$\int \ln^4 x \, dx = x \ln^4 x - 4 \int \ln^{4-1} x \, dx$$

$$= x \ln^4 x - 4\left(x \ln^3 x - 3 \int \ln^{3-1} x \, dx\right) \qquad \text{Formula 34, again}$$

$$= x \ln^4 x - 4x \ln^3 x + 12 \int \ln^2 x \, dx$$

Now use Formula 33:

$$= x \ln^4 x - 4x \ln^3 x + 12(x \ln^2 x - 2x \ln x + 2x) + C$$

$$= x \ln^4 x - 4x \ln^3 x + 12x \ln^2 x - 24x \ln x + 24x + C$$ ∎

It is often necessary to make substitutions before using one of the integration formulas, as shown in Example 3.

EXAMPLE 3 Evaluate $\displaystyle\int \frac{x\,dx}{\sqrt{8-5x^2}}$.

Solution This is an integral of the form $\sqrt{a^2-x^2}$, but it does not exactly match any of the Formulas 15–29. Note, however, that, except for the coefficient of 5, it is like Formula 24. Let $u=\sqrt{5}\,x$ (so $u^2=5x^2$); then $du=\sqrt{5}\,dx$:

$$x=\frac{u}{\sqrt{5}}$$
$$dx=\frac{du}{\sqrt{5}}$$

$$\int \frac{x\,dx}{\sqrt{8-5x^2}}=\int \frac{\dfrac{u}{\sqrt{5}}\cdot\dfrac{du}{\sqrt{5}}}{\sqrt{8-u^2}}$$

$$=\frac{1}{5}\int \frac{u\,du}{\sqrt{8-u^2}}$$

Now apply Formula 24, where $a^2=8$:

$$\frac{1}{5}\int \frac{u\,du}{\sqrt{8-u^2}}=\frac{1}{5}(-\sqrt{8-u^2})+C$$

$$=-\frac{1}{5}\sqrt{8-5x^2}+C \qquad\blacksquare$$

As you can see from Example 3, using an integral table is not a trivial task. In fact, other methods of integration are often preferable, if possible. For Example 3, you can let $u=8-5x^2$ and integrate by substitution:

$$u=8-5x^2 \qquad\text{and}\qquad du=-10x\,dx \qquad\text{so}\qquad dx=\frac{du}{-10x}$$

Substitute:

$$\int \frac{x\,dx}{\sqrt{8-5x^2}}=\int \frac{x\cdot\dfrac{du}{-10x}}{\sqrt{u}}$$

$$=\frac{-1}{10}\int u^{-1/2}\,du$$

$$=\frac{-1}{10}2u^{1/2}+C$$

$$=\frac{-1}{5}\sqrt{8-5x^2}+C$$

This answer is the same, of course, as the one we obtained in Example 3. The point of this calculation is to emphasize that you should try simple methods of integration before turning to the table of integrals.

Also, since sometimes more than one formula can be used, it is to our advantage to pick the one that best simplifies the algebra, as shown by Example 4.

EXAMPLE 4 Evaluate $\int \sqrt{1 - 3x} \, dx$.

Solution *Method I:* Substitution (without integral tables). Let $u = 1 - 3x$, then $du = -3 \, dx$, and

$$\int \sqrt{1 - 3x} \, dx = \int u^{1/2}\left(\frac{du}{-3}\right)$$

$$= -\frac{1}{3}\frac{u^{3/2}}{\frac{3}{2}} + C$$

$$= -\frac{2}{9}u^{3/2} + C$$

$$= -\frac{2}{9}(1 - 3x)^{3/2} + C$$

Method II: Use Formula 9, with $a = 1$, $b = -3$, and $n = \frac{1}{2}$:

$$\int \sqrt{1 - 3x} \, dx = \frac{(a + bx)^{n+1}}{(n + 1)b} + C$$

$$= \frac{(1 - 3x)^{3/2}}{(\frac{1}{2} + 1)(-3)} + C$$

$$= -\frac{2}{9}(1 - 3x)^{3/2} + C$$

Method III: Use Formula 13, with $a = 1$, $b = -3$, and $m = 0$:

$$\int \sqrt{1 - 3x} \, dx = \frac{2}{-3(2 \cdot 0 + 3)} \underset{\uparrow}{[x^0} \underset{\uparrow}{\sqrt{(1 - 3x)^3}} - 0] + C$$

$$= -\frac{2}{9}\sqrt{(1 - 3x)^3} + C$$

$$= -\frac{2}{9}(1 - 3x)^{3/2} + C$$

Note: Even though this integral is 0, you must not forget the constant of integration, which was incorporated into the formula ∎

EXAMPLE 5 Evaluate $\int 5x^2 \sqrt{3x^2 + 1} \, dx$.

Solution This is similar to Formula 25, but you must take care of the 5 (using procedural Formula 1) and the 3 (by making a substitution). Let $u = \sqrt{3}\,x$, then $du = \sqrt{3}\,dx$, and

$$\int 5x^2 \sqrt{3x^2 + 1}\, dx = \int 5\left(\frac{u^2}{3}\right)\sqrt{u^2 + 1}\,\frac{du}{\sqrt{3}}$$

$$dx = du/\sqrt{3}$$

$$3x^2 = u^2$$

$$x^2 = u^2/3$$

$$= \frac{5}{3\sqrt{3}} \int u^2 \sqrt{u^2 + 1}\, du$$

$$= \frac{5}{3\sqrt{3}}\left[\frac{u}{4}\sqrt{(u^2+1)^3} - \frac{1}{8}\frac{u}{1}\sqrt{u^2+1} - \frac{1}{8}\ln|u + \sqrt{u^2+1}|\right] + C$$

Note the use of the $\mp$ symbol in conjunction with the $\pm$ symbol in Formula 25. If you use the top symbol one place, then you must use the top symbol throughout

$$= \frac{5u}{12\sqrt{3}}(u^2+1)^{3/2} - \frac{5u}{24\sqrt{3}}(u^2+1)^{1/2} - \frac{5}{24\sqrt{3}}\ln|u + \sqrt{u^2+1}| + C$$

$$= \frac{5\sqrt{3}\,x}{12\sqrt{3}}(3x^2+1)^{3/2} - \frac{5\sqrt{3}\,x}{24\sqrt{3}}(3x^2+1)^{1/2} - \frac{5}{24\sqrt{3}}\ln|\sqrt{3}\,x + \sqrt{3x^2+1}| + C$$

$$= \frac{5x}{12}(3x^2+1)^{3/2} - \frac{5x}{24}(3x^2+1)^{1/2} - \frac{5}{24\sqrt{3}}\ln|\sqrt{3}\,x + \sqrt{3x^2+1}| + C \qquad \blacksquare$$

Problem Set 14.2

Integrate the expressions in Problems 1–12 using the Brief Integral Table. Indicate the formula used.

1. $\displaystyle\int (1 + bx)^{-1}\, dx$

2. $\displaystyle\int (a + bx)^5\, dx$

3. $\displaystyle\int \frac{x\, dx}{\sqrt{x^2 + a^2}}$

4. $\displaystyle\int \frac{x\, dx}{\sqrt{a^2 - x^2}}$

5. $\displaystyle\int \frac{dx}{x^2\sqrt{x^2 - a^2}}$

6. $\displaystyle\int \frac{dx}{x^2\sqrt{a^2 - x^2}}$

7. $\displaystyle\int x \ln x\, dx$

8. $\displaystyle\int x^2 \ln x\, dx$

9. $\displaystyle\int x^{-1} \ln x\, dx$

10. $\displaystyle\int x^5 \ln x\, dx$

11. $\displaystyle\int x e^{ax}\, dx$

12. $\displaystyle\int \frac{dx}{a + be^{2x}}$

Evaluate the integrals in Problems 13–64. If you use the Brief Integral Table, state the formula used.

13. $\displaystyle\int \frac{x^2\, dx}{\sqrt{x^2 + 1}}$

14. $\displaystyle\int \frac{x^2\, dx}{\sqrt{x^3 + 1}}$

15. $\displaystyle\int \frac{dx}{x^2\sqrt{x^2 + 16}}$

16. $\displaystyle\int \frac{dx}{x^2\sqrt{16x^2 + 1}}$

17. $\displaystyle\int \frac{x\, dx}{\sqrt{4x^2 + 1}}$

18. $\displaystyle\int \frac{x^3\, dx}{\sqrt{4x^4 + 1}}$

19. $\displaystyle\int \frac{dx}{x\sqrt{1 - 9x^2}}$

20. $\displaystyle\int \frac{\sqrt{4x^2 + 1}}{x}\, dx$

21. $\displaystyle\int \frac{x\, dx}{\sqrt{x^2 + 4}}$

22. $\displaystyle\int \frac{dx}{x\sqrt{x^2 + 4}}$

23. $\displaystyle\int (1 + x)^3\, dx$

24. $\displaystyle\int (1 + 5x)^3\, dx$

25. $\int x(1 + x)^3 \, dx$

26. $\int 4x(1 - 6x)^3 \, dx$

27. $\int x\sqrt{1 + x} \, dx$

28. $\int x\sqrt{1 + 3x} \, dx$

29. $\int xe^{4x} \, dx$

30. $\int xe^{-5x} \, dx$

31. $\int x \ln 2x \, dx$

32. $\int x \ln 3x \, dx$

33. $\int x^2 \ln 5x \, dx$

34. $\int x^3 \ln x \, dx$

35. $\int \dfrac{dx}{3 + 5e^x}$

36. $\int \dfrac{dx}{1 + e^{5x}}$

37. $\int \dfrac{dx}{1 + e^{2x}}$

38. $\int \dfrac{e^{2x} \, dx}{1 + e^{2x}}$

39. $\int \dfrac{dx}{\sqrt{1 + x}}$

40. $\int \dfrac{x^3 + 2x + 1}{x} \, dx$

41. $\int x^2(1 + x)^3 \, dx$

42. $\int x^3(1 + x)^3 \, dx$

43. $\int 5x^2(1 - x)^3 \, dx$

44. $\int 3x^2(3 - 2x)^3 \, dx$

45. $\int 2x\sqrt{1 - 2x} \, dx$

46. $\int 3x\sqrt{4 - 3x} \, dx$

47. $\int x^2(2 + 3x)^3 \, dx$

48. $\int x^2\sqrt{4x^2 + 1} \, dx$

49. $\int x^2\sqrt{4x^3 + 1} \, dx$

50. $\int x^2 e^x \, dx$

51. $\int x^2 e^{3x} \, dx$

52. $\int \sqrt{2 + 9x^2} \, dx$

53. $\int \dfrac{\sqrt{2 + 9x}}{x^2} \, dx$

54. $\int \dfrac{\sqrt{2 + 9x}}{x} \, dx$

55. $\int \dfrac{\sqrt{2 + 9x^2}}{x} \, dx$

56. $\int 3\sqrt{5 + 4x^2} \, dx$

57. $\int \dfrac{1}{\sqrt{2 + 9x^2}} \, dx$

58. $\int x\sqrt{9x^2 - 5} \, dx$

59. $\int 5x\sqrt{9 - 16x^2} \, dx$

60. $\int \ln^5 x \, dx$

61. $\int \dfrac{\sqrt{x^2 - 1}}{x^2} \, dx$

62. $\int x^3(x^4 + 1)^8 \, dx$

63. $\int x^2\sqrt{3 + 10x} \, dx$

64. $\int (\ln 3x + e^{5x} + \sqrt{x^2 + 1}) \, dx$

APPLICATIONS

65. A local citizens' group estimates that pollutants are being dumped into the Eel River according to the formula

$$P(t) = \frac{5000}{t\sqrt{100 + t^2}} \qquad t \geq 1$$

where t is the month and $P(t)$ is the tons of pollutants. What is the total amount of pollutants that will be dumped into the river in the next 11 months (that is, from $t = 1$ to $t = 12$)?

66. Davis Industries has production costs for a product determined by the function

$$C(x) = \frac{100x}{\sqrt{x^2 + 100}} + 50$$

where $C(x)$ is the cost to produce the xth item. What is the total cost of producing 1000 items?

67. The sales of Simon's Simple Tool Kit are growing according to the function

$$S(t) = te^{0.1t} \qquad t \geq 12$$

where $S(t)$ is the sales in thousands of dollars in the tth month. What are the accumulated sales for 12 months (that is, for $t = 12$ to $t = 24$)?

14.3 Numerical Integration

Calculators or computers are being used more and more to evaluate definite integrals (see the Modeling Application at the end of Chapter 13). **Numerical integration** is the process of finding a numerical approximation for a definite integral *without actually carrying out the antidifferentiation process.* This means that you can easily write a computer or calculator program to evaluate definite integrals if you understand numerical integration. Numerical integration is also useful when it is difficult or impossible to find an antiderivative when carrying out the integration process.

We begin our study of numerical integration by using simple functions. Suppose a function f is continuous on $[a, b]$. Then there are three common approximations to the definite integral:

Approximations for the
Definite Integral

$\int_a^b f(x)\,dx$ can be approximated by

RECTANGLES: $A_n = \Delta x[f(x_0) + f(x_1) + \cdots + f(x_{n-1})]$

TRAPEZOIDS: $T_n = \dfrac{\Delta x}{2}[f(x_0) + 2f(x_1) + 2f(x_2) + \cdots + 2f(x_{n-1}) + f(x_n)]$

PARABOLAS: $P_n = \dfrac{\Delta x}{3}[f(x_0) + 4f(x_1) + 2f(x_2) + \cdots 4f(x_{n-1}) + f(x_n)]$

(true for n an even integer)

where $\Delta x = \dfrac{b-a}{n}$ for some positive integer n. (Δx is read "delta x.")

Even though there are many other approximation schemes, these will suffice for our needs. While we will not prove these approximations, it is instructive to see where these formulas come from.

Rectangular Approximation

Let f be a function continuous on a closed interval $[a, b]$, $a < b$, and suppose further that $f \geq 0$ on the interval, as shown in Figure 14.1.

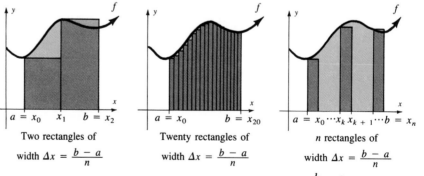

Two rectangles of
width $\Delta x = \dfrac{b-a}{n}$

Twenty rectangles of
width $\Delta x = \dfrac{b-a}{n}$

n rectangles of
width $\Delta x = \dfrac{b-a}{n}$

Figure 14.1 Rectangular approximation of the definite integral

Divide the interval into n subintervals, each of length $\Delta x = \dfrac{b-a}{n}$. The endpoints of these subintervals are

$a, \quad a + \Delta x, \quad a + 2\Delta x, \quad a + 3\Delta x, \quad \ldots, \quad a + n\Delta x$

which are, for convenience, labeled $x_0, x_1, x_2, \ldots, x_n$, respectively. The area of the kth rectangle is

$f(x_k)\,\Delta x$

If you add the areas of all the n rectangles, you clearly obtain A_n.

EXAMPLE 1 Let $f(x) = x^\pi + 1$. Find an approximation for

$$\int_1^3 (x^\pi + 1)\,dx$$

by using a rectangular approximation for $n = 1, 2, 4$, and 8. (We pick, $1, 2, 4, 8, \ldots$ only as a matter of convenience; *any* subdivision will suffice.)

Solution *First approximation, $n = 1$:*

$$A_1 = \int_1^3 (x^\pi + 1)\,dx \approx f(x_0)\,\Delta x$$

$$= f(1)(2)$$
$$= (1^\pi + 1)(2)$$
$$= 4$$

$y = x^\pi + 1$

$f(1) = 1^\pi + 1 = 2$

$a = 1 \quad 2 \quad b = 3$
$x_0 = a \qquad x_1 = b$

$$\Delta x = \frac{3 - 1}{1} = 2$$

Second approximation, $n = 2$:

$y = x^\pi + 1$

$f(1) = 2$
$f(2) = 2^\pi + 1 \approx 9.82$

$a = 1 \quad 2 \quad b = 3$
$x_0 = a \qquad x_2 = b$
$\qquad x_1 = 2$

$$\Delta x = \frac{3 - 1}{2} = 1$$

$$A_2 = \int_1^3 (x^\pi + 1)\,dx \approx f(x_0)\,\Delta x + f(x_1)\,\Delta x$$

$$= f(1)(1) + f(2)(1)$$
$$= (1^\pi + 1) + (2^\pi + 1)$$
$$\approx 2 + 9.82$$
$$= 11.82$$

Third approximation, n = 4:

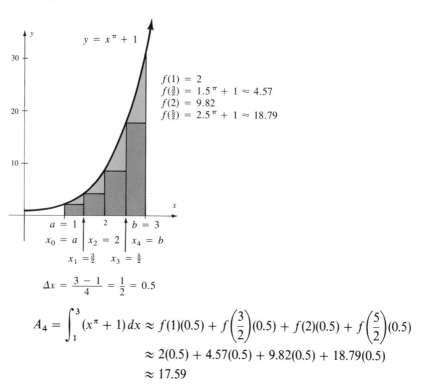

$$f(1) = 2$$
$$f(\tfrac{3}{2}) = 1.5^{\pi} + 1 \approx 4.57$$
$$f(2) = 9.82$$
$$f(\tfrac{5}{2}) = 2.5^{\pi} + 1 \approx 18.79$$

$$\Delta x = \frac{3 - 1}{4} = \frac{1}{2} = 0.5$$

$$A_4 = \int_1^3 (x^{\pi} + 1)\, dx \approx f(1)(0.5) + f\left(\frac{3}{2}\right)(0.5) + f(2)(0.5) + f\left(\frac{5}{2}\right)(0.5)$$
$$\approx 2(0.5) + 4.57(0.5) + 9.82(0.5) + 18.79(0.5)$$
$$\approx 17.59$$

Notice that as $n \to \infty$, the approximate area (gray) is more closely approximating the actual area (color).

Fourth approximation, n = 8:

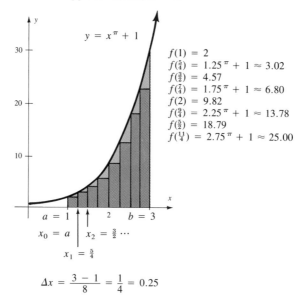

$$f(1) = 2$$
$$f(\tfrac{5}{4}) = 1.25^{\pi} + 1 \approx 3.02$$
$$f(\tfrac{3}{2}) = 4.57$$
$$f(\tfrac{7}{4}) = 1.75^{\pi} + 1 \approx 6.80$$
$$f(2) = 9.82$$
$$f(\tfrac{9}{4}) = 2.25^{\pi} + 1 \approx 13.78$$
$$f(\tfrac{5}{2}) = 18.79$$
$$f(\tfrac{11}{4}) = 2.75^{\pi} + 1 \approx 25.00$$

$$\Delta x = \frac{3 - 1}{8} = \frac{1}{4} = 0.25$$

$$A_8 = \int_1^3 (x^\pi + 1) \, dx \approx f(1)(0.25) + f(1.25)(0.25) + f(1.5)(0.25) + \cdots + f(2.75)(0.25)$$

$$\approx 0.25[f(1) + f(1.25) + f(1.5) + f(1.75) + f(2) + f(2.25) + f(2.5) + f(2.75)]$$

$$\approx 0.25(83.78)$$

$$\approx 20.95 \qquad\blacksquare$$

Trapezoidal Approximation

In order to carry out a trapezoidal approximation, you simply approximate the area represented by the definite integral using trapezoids instead of rectangles, as shown in Figure 14.2. To derive the trapezoidal approximation, use the formula for the area of a trapezoid:

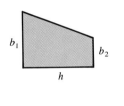

$$A = \left(\frac{b_1 + b_2}{2}\right) h$$

for bases b_1 and b_2 and height h. From this formula you can find the area of the kth trapezoid:

$$\frac{f(x_{k-1}) + f(x_k)}{2} \Delta x \qquad \begin{array}{l} \Delta x = h \text{ in the trapezoidal formula } b_1 = f(x_{k-1}) \text{ and} \\ b_2 = f(x_k) \end{array}$$

Therefore the sum of the areas of all the trapezoids is

$$T_n = \left[\frac{f(x_0) + f(x_1)}{2} + \frac{f(x_1) + f(x_2)}{2} + \cdots + \frac{f(x_{n-1}) + f(x_n)}{2}\right] \Delta x$$

$$= \frac{\Delta x}{2}[f(x_0) + 2f(x_1) + 2f(x_2) + \cdots + 2f(x_{n-1}) + f(x_n)]$$

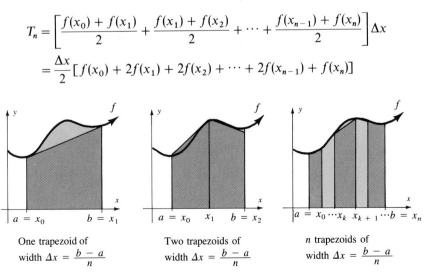

One trapezoid of width $\Delta x = \dfrac{b - a}{n}$

Two trapezoids of width $\Delta x = \dfrac{b - a}{n}$

n trapezoids of width $\Delta x = \dfrac{b - a}{n}$

Figure 14.2 Trapezoidal approximation of the definite integral

EXAMPLE 2 Let $f(x) = x^\pi + 1$. Find an approximation for

$$\int_1^3 (x^\pi + 1)\,dx$$

by using a trapezoidal approximation for $n = 1$ and $n = 4$.

Solution *First approximation, $n = 1$:*

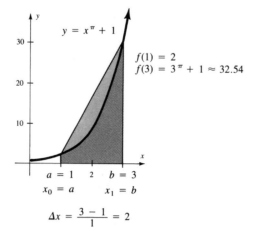

$$f(1) = 2$$
$$f(3) = 3^\pi + 1 \approx 32.54$$

$$\Delta x = \frac{3 - 1}{1} = 2$$

$$T_1 = \int_1^3 (x^\pi + 1)\,dx \approx \frac{f(x_0) + f(x_1)}{2}\,\Delta x$$

$$= \frac{f(1) + f(3)}{2}\,\Delta x$$

$$\approx \frac{2 + 32.54}{2}(2)$$

$$\approx 34.54$$

Second approximation, $n = 4$:

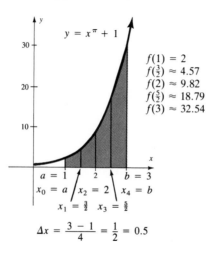

$$f(1) = 2$$
$$f(\tfrac{3}{2}) \approx 4.57$$
$$f(2) \approx 9.82$$
$$f(\tfrac{5}{2}) \approx 18.79$$
$$f(3) \approx 32.54$$

$$\Delta x = \frac{3 - 1}{4} = \frac{1}{2} = 0.5$$

$$T_4 = \int_1^3 (x^\pi + 1)\,dx \approx \frac{0.5}{2}[f(1) + 2f(1.5) + 2f(2) + 2f(2.5) + f(3)]$$
$$\approx 0.25(2 + 9.14 + 19.65 + 37.58 + 32.54)$$
$$\approx 0.25(100.92)$$
$$\approx 25.23$$ ∎

Parabolic Approximation

The parabolic approximation, also known as **Simpson's rule**, uses parabolas to approximate the area under the graph of a function.

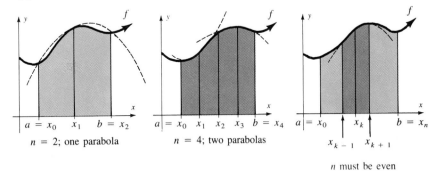

$n = 2$; one parabola $n = 4$; two parabolas n must be even

Figure 14.3 Parabolic approximation, of the definite integral

We will not attempt to derive Simpson's rule, but it is illustrated in Example 3.

EXAMPLE 3   Let $f(x) = x^\pi + 1$. Find an approximation for

$$\int_1^3 (x^\pi + 1)\,dx$$

by using Simpson's approximation for $n = 6$.

Solution   First, gather the necessary values,

$$\Delta x = \frac{3 - 1}{6} = \frac{1}{3}$$

and $f(1) = 2$, $f(\frac{4}{3}) \approx 3.47$, $f(\frac{5}{3}) \approx 5.98$, $f(2) \approx 9.82$, $f(\frac{7}{3}) \approx 15.32$, $f(\frac{8}{3}) \approx 22.79$, $f(3) \approx 32.54$. Thus the integral is approximated by

$$P_6 \approx \frac{\frac{1}{3}}{3}[2 + 4(3.47) + 2(5.98) + 4(9.82) + 2(15.32) + 4(22.79) + 32.54]$$

$$\approx \frac{1}{9}(221.46) \approx 24.60$$ ∎

If you compare Examples 1–3 with the exact value of the definite integral, you will note that each successive approximation is better than the previous and that the larger the value for n the more accurate the results. For the definite integral in Examples 1–3 we see that

$$\int_1^3 (x^\pi + 1)\,dx = \frac{x^{\pi+1}}{\pi + 1} + x \Big|_1^3 \approx 24.60793128$$

EXAMPLE 4 Use Simpson's rule with $n = 4$ to approximate the value of

$$\int_0^1 \frac{dx}{x+1}$$

to four decimal places. Check your answer by finding the exact value of the definite integral.

Solution $\Delta x = \dfrac{b-a}{n} = \dfrac{1-0}{4} = 0.25$

Therefore, since $f(x) = \dfrac{1}{x+1}$,

$$P_4 = \int_0^1 \frac{dx}{x+1} = \frac{0.25}{3}[f(x_0) + 4f(x_1) + 2f(x_2) + 4f(x_3) + f(x_4)]$$

$$= \frac{1}{12}[f(0) + 4f(0.25) + 2f(0.5) + 4f(0.75) + f(1)]$$

$$\approx \frac{1}{12}[1 + 4(0.8) + 2(0.\bar{6}) + 4(0.5714285714) + 0.5]$$

$$\approx \frac{1}{12}(8.319047619)$$

$$\approx 0.6932539683 \qquad \text{or} \qquad 0.6933 \quad \text{(to four decimal places)}$$

The exact value of the definite integral is found as follows:

$$\int_0^1 \frac{dx}{x+1} = \left| \ln(x+1) \right|_0^1 = \ln 2 - \ln 1 = \ln 2$$

(This is 0.6931 to four decimal places.) ■

Problem Set 14.3

Let $\displaystyle\int_1^6 \sqrt{x+3}\,dx$. Then

$$\int_1^6 \sqrt{x+3}\,dx = \frac{2}{3}(x+3)^{3/2}\Big|_1^6 = \frac{2}{3}(27) - \frac{2}{3}(8) = \frac{38}{3}$$

Approximate this integral by finding the requested values in Problems 1–12.

1. A_1	**2.** A_2	**3.** A_3	**4.** A_4
5. T_1	**6.** T_2	**7.** T_3	**8.** T_4
9. P_2	**10.** P_4	**11.** P_6	**12.** T_8

Let $\displaystyle\int_2^4 \frac{x\,dx}{(1+2x)^2}$. *Note that this can be evaluated by using the Brief Integral Table (Formula 10; $a = 1, b = 2, n = -2$):*

$$\frac{1}{b^2}\left[\ln|a+bx| + \frac{a}{a+bx} \right]_2^4 = \frac{1}{4}\left[\ln|1+2x| + \frac{1}{1+2x} \right]_2^4$$

$$= 0.25(\ln 9 + \tfrac{1}{9}) - 0.25(\ln 5 + \tfrac{1}{5})$$

$$\approx 0.5770839221 - 0.4523594781$$

$$\approx 0.124724444$$

Approximate this integral by finding the requested values in Problems 13–24.

13. A_1	**14.** A_2	**15.** A_3	**16.** A_4
17. T_1	**18.** T_2	**19.** T_3	**20.** T_4
21. P_2	**22.** P_4	**23.** P_6	**24.** P_8

Approximate the integrals in Problems 25–28 by using a rectangular approximation. Round your answer to two decimal places using $n = 4$.

25. $\displaystyle\int_3^7 \frac{\sqrt{1+x}}{x^3}\,dx$ **26.** $\displaystyle\int_1^4 \ln x\,dx$

27. $\displaystyle\int_1^3 x^x\,dx$ **28.** $\displaystyle\int_0^2 e^{-x^2}\,dx$

Approximate the integrals in Problems 29–32 by using a trapezoidal approximation. Round your answer to two decimal places using $n = 4$.

29. $\displaystyle\int_3^7 \frac{\sqrt{1+x}}{x^3}\,dx$ **30.** $\displaystyle\int_1^4 \ln x\,dx$

31. $\displaystyle\int_{1}^{3} x^{x}\,dx$ **32.** $\displaystyle\int_{0}^{2} e^{-x^{2}}\,dx$

Approximate the integrals in Problems 33–36 by using Simpson's rule. Round your answer to two decimal places using $n = 4$.

33. $\displaystyle\int_{3}^{7} \frac{\sqrt{1+x}}{x^{3}}\,dx$ **34.** $\displaystyle\int_{1}^{4} \ln x\,dx$

35. $\displaystyle\int_{1}^{3} x^{x}\,dx$ **36.** $\displaystyle\int_{0}^{2} e^{-x^{2}}\,dx$

APPLICATIONS

37. A landscape architect needs to estimate the number of cubic yards of fill necessary to level the area shown below. Use Simpson's rule with $\Delta x = 5$ feet and y values equal to the distances measured in the figure to find the surface area. Assume that the depth averages 3 feet. (*Note:* 1 cubic yard equals 27 cubic feet.)

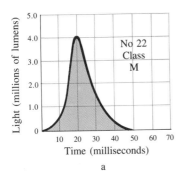

Each contour line is 5 feet from the others

38. The rate at which flashbulbs give off light varies during the flash. For some bulbs, the light output, measured in lumens, reaches a peak and fades quickly (as shown in part a of the figure below) and for others the light output stays level for a relatively longer period of time (shown in part b).

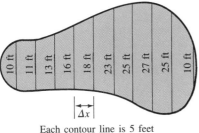

a

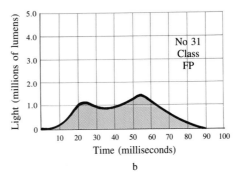

b

The amount A of light emitted by the flashbulb in the interval from 20 to 70 milliseconds after the button is pressed is given by the formula

$$A = \int_{20}^{70} L(t)\,dt \text{ lumens-milliseconds}$$

where $L(t)$ is the lumen output of the bulb as a function of time. Use the trapezoidal rule and the numerical data from the table below to determine which bulb gets more light to the film.*

Light output versus time

Time after ignition (in milliseconds)	Light output (#22 bulb)	Light output (#31 bulb)
0	0	0
5	0.2	0.1
10	0.5	0.3
15	2.6	0.7
20	4.2	1.0
25	3.0	1.2
30	1.7	1.0
35	0.7	0.9
40	0.35	1.0
45	0.2	1.1
50	0	1.3
55		1.4
60		1.3
65		1.0
70		0.8
75		0.6
80		0.3
85		0.2
90		0

* From W. U. Walton et al., *Integration* (Newton, MA: Project CALC, Education Development Center, 1975), p. 83. Data in the table from *Photographic Lamp and Equipment Guide*, P4–15P, General Electric Company, Cleveland, Ohio. The material is quoted from Thomas and Finney, *Calculus and Analytic Geometry*, 6th ed. (Reading, MA: Addison-Wesley Publishing Company, 1984), pp. 311–312.

14.4 Summary and Review

Brief integral table   [14.2]
Integration by parts   [14.1]
Integration formulas   [14.2]
Numerical integration   [14.3]
Parabolic approximation   [14.3]
Rectangular approximation   [14.3]
Simpson's rule   [14.3]
Trapezoidal approximation   [14.3]

SAMPLE TEST   *For additional practice there are a large number of review problems categorized by objective in the Student Solutions Manual. The following sample test (40 minutes) is intended to review the main ideas of this chapter.*

Evaluate the indefinite integrals in Problems 1–9. Use substitution or integration by parts, if possible.

1. $\displaystyle\int (2x - 1)^3\,dx$ **2.** $\displaystyle\int x(2x - 1)^3\,dx$ **3.** $\displaystyle\int x^2(2x - 1)^3\,dx$

4. $\displaystyle\int x(2x^2 - 1)^3\,dx$ **5.** $\displaystyle\int x^2\sqrt{1 - 5x}\,dx$ **6.** $\displaystyle\int \frac{3}{1 - e^{3x}}\,dx$

7. $\displaystyle\int \ln^2 x\,dx$ **8.** $\displaystyle\int \frac{4x^2 + 5x - 3}{x}\,dx$ **9.** $\displaystyle\int \frac{8x + 5}{4x^2 + 5x - 3}\,dx$

State the formula you would use from the Brief Integral Table to approximate the integrals in Problems 10–13. You do not need to actually carry out the integration.

10. $\displaystyle\int x^2(20 - 5x)^4\,dx$ **11.** $\displaystyle\int \frac{\sqrt{100 - x}}{x^5}\,dx$

12. $\displaystyle\int \ln^3 x\,dx$ **13.** $\displaystyle\int e^t t^{-4}\,dt$

Find a decimal approximation of the integral below, where $n = 4$, to two decimal places:

$$\int_0^3 \sqrt{9 - x^2}\,dx$$

14. Rectangular approximation

15. Trapezoidal approximation

16. Simpson's rule

17. The rate of learning a certain task is estimated by the formula

$$\frac{dN}{dt} = 5e^{-0.1t}$$

Find an expression for N if it is known that at time $t = 0$, $N = 0$.

18. The total depreciation at the end of t years is given by $f(t)$ and the depreciation rate is

$$f'(t) = 200x^2\sqrt{10 - x}$$

Find the total depreciation (to the nearest $10) for the first 5 years.

19. The graph below (from the *Wall Street Journal*, February 26, 1986) shows the 10-year Treasury Bond rate for the years 1980–1985. Estimate the area under this curve using a rectangular approximation where $n = 6$.

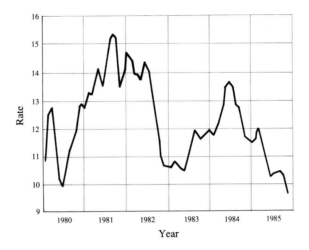

20. How might you obtain a better approximation of the area requested in Problem 19?

Modeling Application 10

Modeling the Nervous System

The nervous system has the ability to respond to a variety of stimuli by generating electrical impulses and transmitting them through nerve cells. Nerve cells vary in length from a few micrometers to more than a meter. Nerve cells that transmit impulses from sense organs to the spinal cord and brain are called *sensory* or *afferent neurons*. Those that transmit messages to the muscles and glands are called motor or efferent neurons. The central nervous system (CNS) consists of the brain and spinal cord; the peripheral nervous system (PNS) consists of the sensory and motor neurons; and the autonomic nervous system (ANS) consists of neurons that regulate processes not under conscious control.

Do some reading in the following sources* and write a paper discussing a model that describes reaction time in the central nervous system. For general guidelines about writing this essay, see the commentary for Modeling Application 1 on page 130.

---

* Horelick, Brindell, Sinan Koont, and Sheldon F. Gottleif. *Modeling the Nervous System: Reaction Time and the Central Nervous System.* ©1983 COMAP, Inc., 271 Lincoln Street, Suite 4, Lexington, MA 02173. COMAP is a nonprofit corporation engaged in research and development in mathematics education.

Erlanger, J., and H. S. Gasser. *Electrical Signs of Nervous Activity* (Philadelphia: University of Pennsylvania Press, 1937).

CHAPTER 15

Applications and Integration

CHAPTER CONTENTS

15.1 Business Models

15.2 Probability Density Functions

15.3 Improper Integrals

15.4 Differential Equations

15.5 Summary and Review
Important Terms
Sample Test

APPLICATIONS

Management (*Business, Economics, Finance, and Investments*)

Money flow (15.1, Problems 7–12; 15.4, Problem 2)
Monthly lease payments (15.1, Problems 27–28)
Time period for a piece of equipment to pay for itself (15.1, Problems 29–30)
Depreciation (15.1, Problems 37–38)
Optimum time to overhaul equipment (15.1, Problem 39)
Net investment flow (15.1, Problem 40)
Capital value (15.3, Problems 19–21; 15.5, Problem 15)
Estimating the life of a product (15.2, Problems 41–43)
Domar's capital expansion model (15.4, Problem 30)
Franchising opportunity for spit-roasted chicken (15.5, Problem 13)

Life sciences (*Biology, Ecology, Health, and Medicine*)

Reaction time of a drug (15.1, Problems 25–26)
Average daily temperature (15.1, Problems 31–32)
Population of a species (15.1, Problems 35–36)
Dispersion of seeds by the wind (15.2, Problem 37)
Probability of rainfall (15.2, Problem 40)
Buildup of radioactive material in the atmosphere (15.3, Problems 22–24)
Oil produced by an oil well (15.3, Problem 30)
Growth rate of bacteria in milk (15.4, Problem 31)

Social sciences (*Demography, Political Science, Population, Psychology, Society, and Sociology*)

Average earth population (15.1, Problem 36)
Survey responses (15.2, Problems 1, 4)
Sibling relationships (15.2, Problem 5)
Learning curve (15.2, Problem 36)
Estimating the crime rate (15.4, Problem 29)

General interest

Life of a battery (15.2, Problem 2; 15.3, Problem 27)
Breaking strength of a rope (15.2, Problem 38)

Cumulative review for Chapters 12–15; Cumulative review for Chapters 9–15

15.1 Business Models

Suppose Wayne Savick introduced a new batching process computer with sales of $S(t)$ units per year t years after it was first introduced. He found that

$$S(t) = 5 + 15t^2$$

for $0 \le t \le 3$. How many computers were sold in the second year? Consider

$$S(2) = 5 + 15(2)^2 = 65$$

How should this be interpreted? Were all 65 computers sold during the second year? Not really; notice that

$$S(1) = 5 + 15(1)^2 = 20$$

Since sales are 20 units when $t = 1$ and 65 units when $t = 2$, it seems reasonable that the number of units sold at different times during the second year is between 20 and 65. Remember, you cannot assume that the annual rate of sales is constant on a yearly basis. The graph of S is shown in Figure 15.1.

A better approximation for finding the total number of computers sold during the entire second year is found when we calculate

$$S\left(\frac{3}{2}\right) = 5 + 15(1.5)^2 = 38.75$$

Thus, during the first half of the second year, the number of units produced should be at least

$$\frac{1}{2}(20) = 10$$

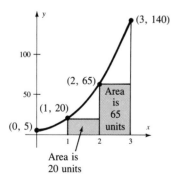

Figure 15.1 Graph of the sales of a new batching process computer

and during the second half of the second year it should be

$$\frac{1}{2}(38.75) = 19.375$$

This approximation is shown in Figure 15.2.

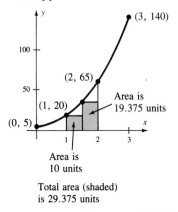

Figure 15.2

Area is 10 units

Total area (shaded) is 29.375 units

The number of square units in the area of the rectangles in Figure 15.2 represents the sales if sales during the first half of the year remain at a constant 20 units and during the second half at a constant 38.75 units. However, sales are not constant. So we now see (by the fundamental theorem of integral calculus) that the desired answer is

$$\int_1^2 (5 + 15t^2)\, dt = 5t + \frac{15t^3}{3}\Big|_1^2 = [10 + 5(2)^3] - [5 + 5(1)^3] = 40$$

EXAMPLE 1 During the first year for which a new product is in the market, $S(t)$ units are sold after t months according to the formula

$$S(t) = 90\sqrt{t} + 150$$

Find the total sales for the first 9 months.

Solution
$$\int_0^9 (90\sqrt{t} + 150)dt = \frac{90t^{3/2}}{\frac{3}{2}} + 150t \Big|_0^9$$

$$= 60t^{3/2} + 150t \Big|_0^9$$

$$= 60(9)^{3/2} + 150(9) = 2970$$

This means that about 2970 units are sold in the first 9 months. ∎

The definite integral we used in Example 1 is called the **total value** over a period of time.

Total Value of a Function over Time

The *total value* of an integrable function f over $[a, b]$ is given by

$$\int_a^b f(x)\, dx$$

A common variation of total value is to determine the length of time it will take for a piece of new equipment to pay for itself.

EXAMPLE 2   A solar heater is being considered as a replacement for an electric heater. The cost of the solar heater is bid at $38,500, and the rate of savings in operating costs by using a solar instead of an electric heater is given by

$$S(x) = 8000x + 3000$$

where $S(x)$ is in dollars and x is the number of years the solar heater will be in use. How long will it take the solar heater to pay for itself?

Solution   Let t be the number of years the solar heater is used until the operating cost savings is equal to the initial replacement cost. Then

$$\int_0^t (8000x + 3000)\, dx = 38{,}500$$

$$\frac{8000x^2}{2} + 3000x \bigg|_0^t = 38{,}500$$

$$\frac{8000t^2}{2} + 3000t = 38{,}500$$

$$4000t^2 + 3000t - 38{,}500 = 0$$

$$8t^2 + 6t - 77 = 0$$

$$(4t - 11)(2t + 7) = 0$$

$$t = \frac{11}{4}, -\frac{7}{2}$$

↑
Reject negative in this application

Thus it would take about $\frac{11}{4}$ years or 2 years and 9 months for the solar heater to pay for itself.   ■

Another application of total value is the **total money flow** over a period of time. You can surely think of several situations where money flows in and out of an account; for example, your checking account or bank receipts or payments for a business. A measure of this flow of money in an account is the **total money flow**. Consider an example where the amount of money is stable; say, you have $100 stored in a box under your bed. We can let $f(x) = 100$ (a constant), which, if

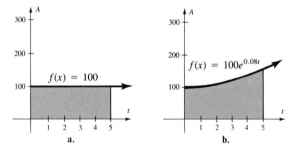

Figure 15.3 Total money flow

graphed, is a horizontal line. If this money is left in the box for 1 year, the money flow (in units of a year) is $100. For 5 years it is $5 \cdot 100 = 500$. Look at the graph in Figure 15.3a and you will see that the money flow is the area under the curve.

The situation is the same if the function f is not a constant. We will now assume that the investment is deposited in a bank (instead of a shoebox) that pays continuous interest. Since the formula for the amount, A, of money given an investment of P dollars at a rate of r percent per year compounded continuously for t years is given by $A = Pe^{rt}$, we see that if the annual rate is 8%, then $f(x) = 100e^{0.08t}$. This graph is shown in Figure 15.3b. The money flow (for $t = 5$, for example) is the area under this curve:

$$\int_0^5 100e^{0.08x}\,dx = 1250e^{0.08x}\Big|_0^5 = 1250e^{0.4} - 1250e^0 = 614.78$$

This discussion leads to the following definition for total money flow:

Total Money Flow

The *total money flow* for P dollars over a period of T years at a rate r is given by the formula

$$F = \int_0^T Pe^{rt}\,dt$$

EXAMPLE 3   If $25,000 is deposited at 9% interest compounded continuously, find the amount present in 10 years and the total money flow over that 10-year period.

Solution $A = 25,000e^{0.09t}$

$\quad\quad = 25,000e^{0.09(10)}$

$\quad\quad \approx 61,490.08$

$F = \int_0^{10} 25,000e^{0.09t}\,dt$

$\quad = \dfrac{25,000e^{0.09t}}{0.09}\Big|_0^{10}$

$\quad \approx \$683,223.09 - \$277,777.78$

$\quad = \$405,445.31$

The amount present in 10 years is $61,490.08 and the total money flow after 10 years is $405,445.31. ∎

Suppose a typist's speed in words per minute is a function of the time of a typing test, as shown in Figure 15.4.

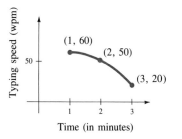

Figure 15.4 Typing speed as a function of the length of time of the typing test

What is the typist's average speed for the three timed tests?

$$\text{Average} = \frac{60 + 50 + 20}{3} = \frac{130}{3} = 43 \text{ wpm}$$

However, this average does not accurately represent the situation because there are infinitely many values for t over the interval $[1, 3]$. If we use the fundamental theorem of integral calculus, we could derive a formula for the **average value** of a function over an interval:

Average Value

The *average value* of a continuous function f over $[a, b]$ is

$$\frac{1}{b-a} \int_a^b f(x)\, dx$$

EXAMPLE 4   If the typist's speed in the above discussion is given by the continuous function

$$w(t) = -10t^2 + 20t + 50 \qquad \text{for } t \text{ in } [1, 3]$$

find the average typing speed in words per minute.

Solution   The average is

$$\frac{1}{3-1} \int_1^3 (-10t^2 + 20t + 50)\, dt = \frac{1}{2}\left[\frac{-10t^3}{3} + \frac{20t^2}{2} + 50t \right]_1^3$$

$$= \frac{1}{2}\left[(-90 + 90 + 150) - \left(-\frac{10}{3} + 10 + 50 \right) \right]$$

$$\approx 47 \text{ wpm}$$

∎

EXAMPLE 5   Given the demand function

$$p(x) = 50e^{0.03x}$$

find the average price (in dollars) over the interval $[0, 100]$.

Solution   Average price $= \dfrac{1}{100 - 0} \displaystyle\int_0^{100} 50e^{0.03x}\, dx$

$$= \frac{1}{2} \int_0^{100} e^{0.03x}\, dx$$

$$= \frac{e^{0.03x}}{0.06} \Big|_0^{100}$$

$$= \frac{50}{3}(e^{0.03(100)} - e^{0.03(0)})$$

$$= \frac{50}{3}(e^3 - 1) \approx 318.09$$

The average price is approximately \$318.

∎

Problem Set 15.1

Find the total value of the functions in Problems 1–6.

1. $f(x) = \sqrt{x} + 100$ on $[4, 9]$

2. $S(x) = 50\sqrt[3]{x} - 30x$ on $[1, 8]$

3. $m(x) = \dfrac{x - 1}{x + 1}$ on $[0, 2]$

4. $t(x) = \dfrac{2x + 1}{x - 2}$ on $[3, 5]$

5. $f(t) = 65e^{-5t}$ on $[0, 3]$

6. $f(t) = 1000e^{0.005t}$ on $[0, 2]$

Determine the total money flow for Problems 7–12.

7. $8,000 at 8% for 6 years

8. $83,000 at 10% for 2 years

9. $2,500 at 5% for 1 year

10. $125,000 at 11% for 15 years

11. $6,500 at 7% for 5 years

12. $4 million at 10% for 8 years

Find the average value for the functions in Problems 13–18.

13. $f(x) = 6x^2 + 3x - 10$ on $[-1, 1]$

14. $f(x) = 2x^3 - 7x^2 + 4x - 11$ on $[1, 3]$

15. $f(x) = \dfrac{x}{x + 1}$ on $[1, 5]$

16. $f(x) = \dfrac{x - 1}{x + 5}$ on $[1, 5]$

17. $f(t) = e^{-5t}$ on $[0, 3]$

18. $f(x) = xe^x$ on $[1, 2]$

APPLICATIONS

19. A new product has sales given by

    $$S(t) = 60 + 24t^2 \qquad \text{for } 0 \le t \le 3$$

    where t is the number of years since the product was introduced. Find the number of items sold in the second year.

20. Repeat Problem 19 for the third year.

21. Repeat Problem 19 for the total 3-year period.

22. During the first year a new commodity has been on the market, y units per month were sold after x months elapsed, according to the formula

    $$y = 300\sqrt[3]{x} + 50 \qquad 0 \le x \le 12 \qquad .$$

    Find the total sales during the first 6 months.

23. Repeat Problem 22 for the second 6 months.

24. Repeat Problem 22 for the entire first year.

25. The rate of reaction to a certain drug is measured in minutes after administration according to the formula

    $$R(t) = x^{-0.5}$$

    Evaluate the total reaction in the first 10 minutes.

26. Repeat Problem 25 for the first hour.

27. An equipment leasing company determines that the rate of maintenance on a piece of new equipment is

    $$M(x) = 150(1 + \sqrt[3]{x^2})$$

    where x is the number of years the equipment has been leased. What should be the total amount charged for a 4-year lease? What amount should be added to the monthly lease payment to pay for the maintenance for a 4-year lease?

28. Repeat Problem 27 for a 3-year lease.

29. Suppose that a new piece of equipment costs $15,000. The rate of operating cost savings is $S(x)$ in dollars and is given by the formula

    $$S(x) = 2400x + 10,500$$

    where x is the number of years the equipment has been used. How long will it take the piece of equipment to pay for itself?

30. If a piece of machinery has a rate of operating cost savings given by

    $$S(x) = 4800x + 22,500$$

    how long it will take for the piece of machinery to pay for itself if x is the number of years the equipment has been used and the original cost was $5,100?

31. The temperature (in degrees Fahrenheit) varies according to the formula

    $$F(t) = 6.44t - 0.23t^2 + 30$$

    Find the average daily temperature if t is the time of day (in hours) measured from midnight.

32. Repeat Problem 31 for the daylight hours. That is, for t on the interval $[6, 20]$.

33. If the demand function is

    $$p = D(x) = \frac{480}{x}$$

    what is the average price (in dollars) for x on $[10, 50]$?

34. If the supply function is

$$p = S(x) = 10(e^{0.03x} - 1)$$

what is the average price (in dollars) over the interval $[10, 50]$?

35. The number of rabbits in a limited geographical area (such as an island) is approximated by

$$P(t) = 500 + t - 0.25t^2$$

for t the number of years on the interval $[0, 5]$. What is the average number of rabbits in the area over the 5-year time period?

36. The world population (in billions) is given by

$$P(t) = 5e^{0.03t} \qquad (t = 0 \text{ in } 1987)$$

What is the average population of the earth during the next 30 years?

37. If the total depreciation at the end of t years is represented by $f(t)$, then the depreciation rate is $f'(t)$ over an interval $[0, t]$. Since $f'(t)$ is usually known, the total depreciation can be found by using the formula

$$f(t) = \int_0^t f'(x)\, dx$$

Suppose a \$38,000 piece of equipment is depreciated over a 10-year period using *straight-line depreciation*. That is, each year the depreciation is

$$\frac{38,000}{10} = 3800$$

So $f'(x) = 3800$. Find the total depreciation for the first 3 years.

38. Many items do not depreciate at a constant rate (see Problem 37). Automobiles or computers, for example, depreciate much more quickly in the early years and more slowly toward the end of the time interval. Suppose an automobile depreciates according to the formula

$$f'(x) = 3000\sqrt{5 - x}$$

Find the total depreciation for the first 3 years.

39. A piece of machinery requires an overhaul after time t. If $E(t)$ represents the expense connected with the equipment, we see that

$$E(t) = C + \text{total depreciation}$$

where C is the cost of overhaul. Now, from Problem 37,

$$E(t) = C + \int_0^t f'(x)\, dx$$

(Also note: $E'(t) = f'(t)$ since the derivative of C is zero.) The average expense, $A(t)$, is found by dividing by the

number of years t:

$$A(t) = \frac{E(t)}{t}$$

If no other factors are involved, the best time to overhaul the equipment is the value of t for which E has a relative minimum:

$$A'(t) = \frac{tE'(t) - E(t)}{t^2}$$

Set $A'(t) = 0$ and solve for $E'(t)$ to find the critical value:

$$\frac{tE'(t) - E(t)}{t^2} = 0$$

$$tE'(t) - E(t) = 0 \qquad \text{Multiply both sides by } t^2$$

$$\underbrace{E'(t)}_{} = \frac{E(t)}{t}$$

$E'(t) = f'(x)$ ↖——— Average expense
is the rate of depreciation

This critical value is a relative minimum (this is left for you to verify), so you see that the best time to overhaul occurs when the rate of depreciation equals the average expense. If a piece of equipment depreciates according to the formula

$$f'(x) = 1000\sqrt{x} \qquad \text{and} \qquad C = 500$$

when should the equipment be overhauled to minimize the average expense?

40. If P dollars is invested at a rate of r percent per year compounded continuously for t years, then

$$A = Pe^{rt}$$

is the future amount. The rate of change of A with respect to time t is called the *net investment flow* and is found by

$$\frac{dA}{dt} = Pe^{rt}(r) = Pre^{rt}$$

Find the net investment flow for the amounts in Example 3.

41. The process by which a corporation increases its accumulated wealth is called *capital formation*. If the net investment flow (see Problem 40) is given by a function $f(x)$, then the increase in capital over the interval $[a, b]$ is

$$\int_a^b f(t)\, dt = F(b) - F(a)$$

Suppose Intel Corporation has a net investment flow approximated by the function $f(t) = \sqrt{t}$, where t is in years and f is in millions of dollars per year. What is the amount of capital formation over the next 5 years?

15.2 Probability Density Functions*

The models we have considered so far have been deterministic models; now we turn to a **probabilistic model**, used for situations that are random in character and that attempts to predict the outcomes of these events with a certain stated or known degree of accuracy. For example, if we toss a coin, it is impossible to predict in advance whether the outcome will be a head or a tail. Our intuition tells us that the outcome is equally likely to be a head or a tail, and somehow we sense that if we repeat the experiment of tossing a coin a large number of times, heads will occur "about half the time." To check this out, I recently flipped a coin 1000 times and obtained 460 heads and 540 tails. The percentage of heads is $\frac{460}{1000} = .46 = 46\%$, which is called the **relative frequency**.

Relative Frequency of a Repeated Experiment

If an experiment is repeated n times and an event occurs m times, then

$$\frac{m}{n}$$

is called the *relative frequency* of the event.

Our task is to create a model that will assign a number p, called the *probability of an event*, which will predict the relative frequency. This means that for a *sufficiently large number of repetitions* of an experiment

$$p \approx \frac{m}{n}$$

Probabilities can be obtained in one of three ways:

1. *Theoretical probabilities* (also called a priori models) are obtained by logical reasoning. For example, the probability of rolling a die and obtaining a 3 is $\frac{1}{6}$ because there are six possible outcomes, each with an equal chance of occurring, so a 3 should appear $\frac{1}{6}$ of the time.
2. *Empirical probabilities* (also called a posteriori models) are obtained from experimental data. For example, an assembly line producing brake assemblies for General Motors produces 1500 brakes per day. The probability of a defective brake can be obtained by experimentation. Suppose the 1500 brakes are tested and 3 are found to be defective. Then the relative frequency, or probability, is

$$\frac{3}{1500} = .002 \text{ or } .2\%$$

3. *Subjective probabilities* are obtained from experience and indicate a measure of "certainty." For example, a TV reporter studies the satellite maps and issues a prediction about tomorrow's weather based on past experience under similar circumstances: "80% chance of rain tomorrow."

A probability measure must conform to these different ways of using the word *probability*. We begin by defining some terms. An **experiment** is the observation of

---

* The material from Chapters 6–8 needed for this section is reviewed here, so that it is not necessary to have studied those chapters.

any physical occurrence. A **sample space** of an experiment is the set of all possible outcomes. An **event** is a subset of the sample space. If an event is the empty set, it is called the **impossible event**; and if it has only one element, it is called a **simple event**.

EXAMPLE 1 List the sample space for each experiment, and then list one event for each of the sample spaces and state whether it is a simple event.

  a. Experiment 1: recording the results of the UCLA football team in a given season
  b. Experiment 2: simultaneously tossing a coin and rolling a die
  c. Experiment 3: installing a computer chip and recording the time to failure

Solution **a.** Experiment 1: $S = \{w, l, t\}$
  where w, l, and t denote win, lose, and tie, respectively. An example of an event E is that "UCLA wins," which is $E = \{w\}$. This is a simple event. Another example is event F, "UCLA does not lose," which is $F = \{w, t\}$; this is not a simple event.
  b. Experiment 2: $S = \{1H, 1T, 2H, 2T, 3H, 3T, 4H, 4T, 5H, 5T, 6H, 6T\}$
  An example of an event is "obtaining an even number or a head," which is $E = \{1H, 2H, 2T, 3H, 4H, 4T, 5H, 6H, 6T\}$. E is not a simple event.
  c. Experiment 3:
  $$S = \{t \mid t \geq 0\}$$
  An example of an event is "the chip lasts more than 1000 hours," which is $E = \{t \mid t > 1000\}$. E is not a simple event. ∎

If X is a variable selected from the sample space in a random fashion, it is called a **random variable**. If S has a finite number of outcomes (as in experiments 1 and 2), then X is called a **discrete random variable**; and if X is not discrete (as in experiment 3), then it is called a **continuous random variable**.

Suppose the heights of 60 students in a mathematics class are recorded (to the nearest inch), as shown in Table 15.1. The results in the table are often referred to as a **frequency distribution** because such results show the distribution of the number of each occurrence.

TABLE 15.1
Frequency distribution of student heights

X = height (in inches)	61 or smaller	62	63	64	65	66	67	68	69	70	71	72	73 or taller
Number of occurrences (frequency)	2	2	2	3	5	9	14	10	5	3	2	1	2
Relative frequency	$\frac{2}{60}$	$\frac{2}{60}$	$\frac{2}{60}$	$\frac{3}{60}$	$\frac{5}{60}$	$\frac{9}{60}$	$\frac{14}{60}$	$\frac{10}{60}$	$\frac{5}{60}$	$\frac{3}{60}$	$\frac{2}{60}$	$\frac{1}{60}$	$\frac{2}{60}$

Probability Distribution

A **probability distribution** is the collection of all values that a random variable assumes along with the probabilities that correspond to these values. Furthermore,

1. $P(X = x_1) + P(X = x_2) + \cdots + P(X = x_n) = 1$
2. $0 \leq P(X = x_i) \leq 1$ for every $1 \leq i \leq n$

It is very useful to display the information in a probability distribution graphically in a bar graph called a **histogram**.

EXAMPLE 2   Represent the information in Table 15.1 in a histogram.

Solution   Use the horizontal axis to delineate the values of the random variable (the heights) and the vertical axis to represent the relative frequencies. Draw each bar so that the width is 1 unit.

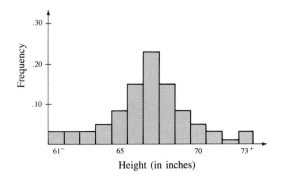

Figure 15.5 Histogram for relative frequencies

There is a very important relationship between the area of the rectangles in a histogram and the probabilities. Since the width of each bar is 1 unit, the **area of each bar is the probability of occurrence for that random variable**. Therefore the **sum of the areas of the rectangles is 1**. This property is very important in the study of probability distributions.

It is possible to record the heights of the students more accurately. In fact, theoretically, the height of a student could be any positive real number in some domain (say, 0 to 100 inches). We can therefore consider the graph of the relative frequencies as a continuous curve, as shown in Figure 15.6.

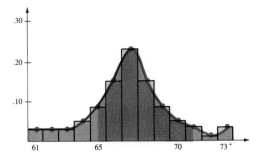

Figure 15.6 Relative frequencies as a continuous curve (this curve is drawn by connecting the points at the top of the bars and smoothing the resulting polygon into a curve)

If we now define the **probability** of an event as the relative frequency of occurrence of the event, and if the random variable is assumed to be continuous, then we can use the area under the curve between two values to find the probability that the random variable is between those values. For example,

$$P(65 \leq X \leq 71)$$

is the probability that a student's height is between 65 and 71 inches and is shown by the shaded region in Figure 15.7.

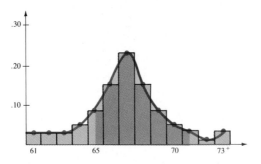

Figure 15.7 $P(65 \leq X \leq 71)$

Since the definite integral can be used to find the area under the graph of $f(x)$ from $x = a$ to $x = b$, if we can find a function f to describe a relative frequency curve, then the probability that x will be between a and b is

$$P(a \leq X \leq b) = \int_a^b f(x)\, dx$$

A function f that can be used to describe a relative frequency curve is called a **probability density function**. Such a function must satisfy certain conditions, as summarized below:

Probability Density Function

If X is a random variable over the interval $[a, b]$, then f is a *probability density function* if

$$\int_a^b f(x)\, dx = 1 \qquad \text{and} \qquad f(x) \geq 0 \qquad \text{for all } x \text{ on } [a, b]$$

EXAMPLE 3 Show that $f(x) = \dfrac{x}{6} - \dfrac{1}{12}$ is a probability density function on $[1, 4]$.

Solution 1. $\displaystyle\int_1^4 \frac{x}{6} - \frac{1}{12}\, dx = \frac{x^2}{12} - \frac{x}{12}\Big|_1^4 = \frac{16}{12} - \frac{4}{12} - \frac{1}{12} + \frac{1}{12} = \frac{12}{12} = 1$

2. If x is between 1 and 4, then f is between $\dfrac{1}{12}$ and $\dfrac{7}{12}$, so $f(x) \geq 0$ for all X on $[a, b]$.

Thus f is a probability density function. ■

EXAMPLE 4 Suppose that the life of a computer chip is described by the probability density function

$$f(x) = \frac{72}{35x^3}$$

where x is the number of months it will function properly. The domain for x is the interval $[1, 6]$. What is the probability that the chip will last longer than 4 months?

Solution The probability that the chip will last longer than 4 months is

$$P(4 \leq X \leq 6) = \int_4^6 \frac{72}{35x^3}\, dx = \frac{72}{35}\frac{x^{-2}}{-2}\Big|_4^6 = -\frac{36}{35x^2}\Big|_4^6 = -\frac{1}{35} + \frac{9}{140} = .036$$

■

Sometimes it is necessary to construct a probability density function. For example, suppose $f(x) = x^2$ on $[1,4]$. Then

$$\int_1^4 x^2 = \frac{x^3}{3}\bigg|_1^4 = \frac{64}{3} - \frac{1}{3} = \frac{63}{3}$$

If you want to construct a probability density function using f, simply multiply by the reciprocal of the value of the integral, in this case, $\frac{3}{63}$, so that the value of the integral will be 1:

$$f(x) = \frac{3}{63} x^2$$

is a probability density function.

One of the principal tasks in the study of probability involves finding a probability density function that describes the experiment under study. Once it has been developed, it is called a **probability distribution function** and its characteristics are summarized so that when similar experiments are studied the distribution function does not have to be derived again. Many mathematical models have been developed for use in most of the common applications. These are divided into two broad categories—**discrete** and **continuous distributions**—as shown in Table 15.2. If you take a course in probability you will derive each of these and find out when to use them.

TABLE 15.2
Common probability distribution functions

Discrete distributions	Common use
Binomial	The experiment has a fixed number of independent trials that can be classified into two categories where the probabilities remain constant for each trial
Geometric	This is a variation of the binomial distribution that gives the probability of getting the *first* success on the nth trial
Poisson	The experiment consists of the arrivals of entities requiring service (such as in a line, gas station, planes arriving at an airport)
Others	Bernoulli, negative binomial, hypergeometric

Continuous distributions	Common use
Uniform	The probability density function is a constant. All simple events are equally likely
Exponential	The probability density function has a graph that is exponential
Normal	The probability density function has a graph that is a bell-shaped curve
Others	Gamma, Cauchy, Student's, chi-square, Rayleigh, Maxwell, F distribution, beta distribution

In this course, we use only the three most common continuous distributions.

Uniform Probability Distribution

If all the simple events in the sample space are equally likely, then an appropriate model is the **uniform probability distribution**. What this means is that the probability density function is a constant (the graph is a horizontal line). Suppose the given interval is $[1, 5]$. Then the length of the interval is $5 - 1 = 4$, so if the probability density function is a constant it must be $\frac{1}{4}$ (so that the area is 1), as shown in Figure 15.8.

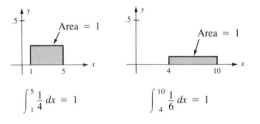

$$\int_1^5 \frac{1}{4}\, dx = 1 \qquad\qquad \int_4^{10} \frac{1}{6}\, dx = 1$$

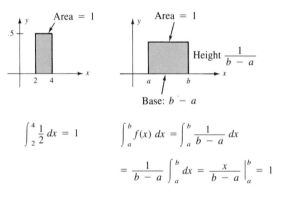

$$\int_2^4 \frac{1}{2}\, dx = 1 \qquad\qquad \int_a^b f(x)\, dx = \int_a^b \frac{1}{b-a}\, dx$$

$$= \frac{1}{b-a} \int_a^b dx = \frac{x}{b-a}\Big|_a^b = 1$$

Figure 15.8 Some uniform distributions

Notice that if the interval is $[4, 10]$, then $f(x) = \frac{1}{6}$; if the interval is $[2, 4]$, then $f(x) = \frac{1}{2}$. This observation leads us to the following definition of the uniform probability distribution:

Uniform Distribution

A continuous random variable x is said to be *uniformly distributed* over an interval $[a, b]$ if it has a probability density function

$$f(x) = \frac{1}{b-a} \qquad \text{for } a \le x \le b$$

The graph is shown in Figure 15.8.

EXAMPLE 5 Cholesterol levels are artificially introduced into blood samples so that the resulting mixtures are uniformly distributed over the interval $[300, 400]$. What is the probability that the cholesterol level is between 240 and 265?

Solution   The probability density function for x is given by

$$f(x) = \frac{1}{400 - 300} = \frac{1}{100} = .01 \qquad \text{for } 300 \le x \le 400$$

Then the desired probability is

$$\int_{240}^{265} .01 \, dx = .01x \Big|_{240}^{265}$$
$$= .01(265 - 240)$$
$$= .25$$

Exponential Probability Distribution

A probability density function will often follow an exponential curve. We find this curve in response to the question, "How long do you need to wait if you are observing a sequence of events occurring in time in order to observe the nth occurrence of the event?"

Exponential Distribution

> A continuous random variable x is said to be *exponentially distributed* over an interval $[0, \infty)$ if it has a probability density function
> $$f(x) = ke^{-kx} \qquad \text{for } k > 0$$
> The graph is shown in Figure 15.9.

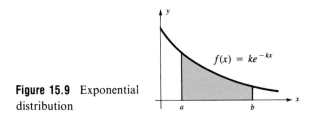

Figure 15.9 Exponential distribution

EXAMPLE 6   The useful life of a machine part is given by

$$f(x) = 0.015e^{-0.015t} \qquad 0 \le t < \infty$$

where t is the number of months to failure. What is the probability that the part will fail in the first year?

Solution $$\int_0^{12} 0.015e^{-0.015t} \, dt = \frac{0.015e^{-0.015t}}{-0.015} \Big|_0^{12}$$
$$= -e^{-0.015(12)} + e^{-0.015(0)}$$

$$\approx .16$$

Normal Probability Distribution

The most common probability distribution is the one associated with the so-called bell-shaped curve. Suppose we survey the results of 100,000 IQ scores and obtain the frequency distribution shown in Figure 15.10.

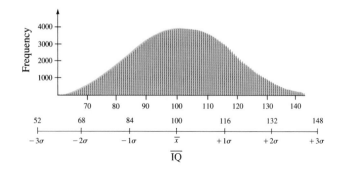

Figure 15.10 Frequencies of IQ scores

If we connect the endpoints of the bars in Figure 15.10 by drawing a smooth curve, we obtain a curve very close to a curve called the *normal distribution curve*, or simply the **normal curve**, as shown in Figure 15.11.

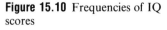

Figure 15.11 Normal distribution curve

$$f(x) = \frac{e^{-(x-\mu)^2/2\sigma^2}}{\sigma\sqrt{2\pi}}$$

Normal Distribution

> A continuous random variable x is said to be **normally distributed** over an interval $(-\infty, \infty)$ if it has a probability density function
>
> $$f(x) = \frac{e^{-(x-\mu)^2/2\sigma^2}}{\sigma\sqrt{2\pi}}$$
>
> where μ and σ are real numbers with $\sigma \geq 0$. The number μ is called the **mean** and the number σ is called the **standard deviation**. The graph is shown in Figure 15.11.

Since the normal distribution is a probability distribution, we know the area under this curve is 1. Therefore we can relate the area to probabilities as follows for a random variable X (see Figure 15.12):

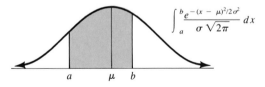

$$\int_a^b \frac{e^{-(x-\mu)^2/2\sigma^2}}{\sigma\sqrt{2\pi}}\,dx$$

Figure 15.12

$P(a \le X \le b)$ is the area under the associated normal curve between a and b.

$P(X > \mu) = P(X < \mu) = \frac{1}{2}$; that is, the curve is symmetric about the mean.

$P(X = x) = 0$ for any real number x. (Since there are infinitely many possibilities, the probability of a particular value is 0.)

$P(X < x) = P(X \le x)$ for any real number x.

$P(X < x) = 1 - P(X \le x)$ for any real number x.

If $\mu = 0$ and $\sigma = 1$, then the curve is called the **standard normal curve**.

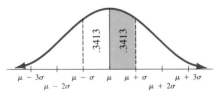 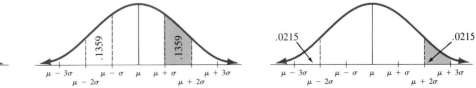

Since the normal distribution is so important for so many applications, and since the calculation of the integral for the probability density function is so complicated, tables of approximate values of the definite integral of the standard normal curve have been calculated. Table 4 in Appendix E is such a table. It contains values of

$$P(a \le x \le b) = \int_a^b \frac{1}{\sqrt{2\pi}} e^{-x^2/2} \, dx$$

EXAMPLE 7   Find $P(X \le .57)$, $P(X > -.13)$, and $P(-.05 < X < .93)$.

Solution $P(X \le .57) = .7157$ From Table 4

$P(X > -.13) = 1 - P(X \le -.13) = 1 - .4483$ From Table 4

$\qquad\qquad\quad = .5517$

$P(-.05 < X < .93) = .8238 - .4801$ From Table 4

$\qquad\qquad\qquad\qquad = .34376$ ■

What if you are not working with a standard normal curve? That is, suppose the curve is a normal curve but does not have a mean of 0. You can then use the information in Figure 15.13, which, for convenience, is summarized in Figure 15.14.

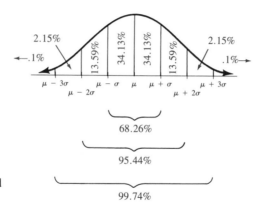

Figure 15.13 Areas under a normal curve. Notice that approximately 34% (.3413) lies between μ and $\mu + \sigma$; 14% (.1359) lies between $\mu + \sigma$ and $\mu + 2\sigma$; and 2% (.0215) lies between $\mu + 2\sigma$ and $\mu + 3\sigma$

EXAMPLE 8   A teacher claims to grade "on a curve." That is, the teacher believes that the scores on a given test are normally distributed. If 200 students take the exam, with mean 73 and standard deviation 9, how would the teacher grade the students?

Solution   First, draw a normal curve with a mean 73 and standard deviation 9, as shown below.

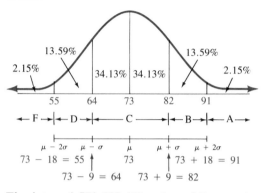

Figure 15.14

The interval $[73, 82]$ $(73 + 1\sigma = 82)$ contains about 34% of the class, and the interval $[82, 91]$ $(73 + 2\sigma = 91)$ contains about 14%. Finally, about 2% of the class will score above 91 or below 54:

Score	Letter grade	Number	Percent
92–100	A	4	2%
83–91	B	28	14%
64–82	C	136	68%
55–63	D	28	14%
0–54	F	4	2%

EXAMPLE 9   The Ridgemont Light Bulb Company tests a new line of light bulbs and finds them to be normally distributed, with a mean life of 98 hours and a standard deviation of 13.

a. What percentage of bulbs will last fewer than 72 hours?
b. What is the probability that a bulb selected at random will last more than 111 hours?

Solution   Draw a normal curve with mean 98 and standard deviation 13.

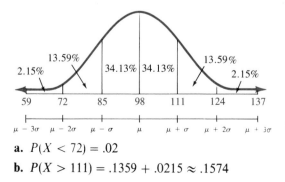

a. $P(X < 72) = .02$
b. $P(X > 111) = .1359 + .0215 \approx .1574$

If a finer division of standard deviations than 1, 2, or 3 is needed, a **z-score** is used. The z-score translates any normal curve into a standard normal curve by using the simple calculation given below:

z-Score

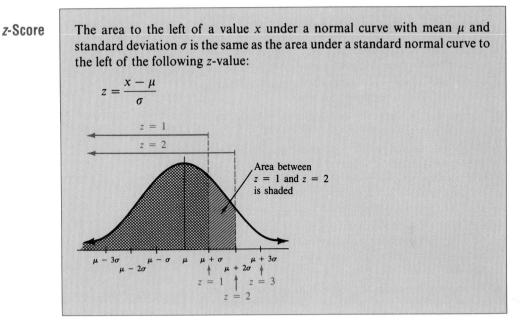

The area to the left of a value x under a normal curve with mean μ and standard deviation σ is the same as the area under a standard normal curve to the left of the following z-value:

$$z = \frac{x - \mu}{\sigma}$$

Area between $z = 1$ and $z = 2$ is shaded

EXAMPLE 10   Find the probability that one of the light bulbs described in Example 9 will last between 110 and 120 hours.

Solution   From Example 9, $\mu = 98$ and $\sigma = 13$. Let $x = 110$; then

$$z = \frac{110 - 98}{13}$$

$$\approx .92$$

Now, look up $z = .92$ in Table 4 and find

$$P(X < 110) = P(z < .92)$$
$$= .8212$$

For $x = 120$,

$$z = \frac{120 - 98}{13} \approx 1.69$$

Again, from Table 4,

$$P(X < 120) = P(z < 1.69)$$
$$= .9545$$

Thus

$$P(110 \leq X \leq 120) = .9545 - .8212$$
$$= .1333.$$

Problem Set 15.2

APPLICATIONS

Describe the sample space for the experiments in Problems 1–6. Define a random variable and characterize it as a discrete or continuous random variable.

1. Ten people are asked if they graduated from college, and the number of people responding yes is recorded.

2. One hundred Everready Long-Life® batteries are tested, and the life of each battery is recorded.

3. The number of words in which an error is made on a typing test consisting of 500 words is recorded.

4. A person is randomly selected for a public opinion poll and asked two questions: Are you male (m) or female (f)? Do you consider yourself a Democrat (d), Republican (r), Independent (i), or are you not registered (n) to vote?

5. A psychologist is studying sibling relationships in families with three children. Each family is asked about the sex of the children in the family and the results are recorded. If a family has a girl, then a boy, and finally another boy, this would be recorded as {gbb}.

6. A sample of five radios is selected from an assembly line and tested. The number of defective radios is recorded.

7. The heights of 30 students are (numbers are rounded to the nearest inch): 66, 68, 64, 70, 67, 67, 68, 64, 65, 66, 64, 70, 72, 71, 69, 64, 63, 70, 71, 63, 68, 67, 65, 69, 65, 67, 66, 69, 69, 67. Give the frequency distribution, define the random variable, and draw a histogram.

8. The wages of employees of a small accounting firm are (in thousands of dollars): 10, 15, 15, 15, 20, 20, 25, 40, 60, 8, 8, 8, 8, 8, 6, 4. Give the frequency distribution, define the random variable, and draw a histogram.

9. The times that a bank teller spent on each transaction were recorded as follows (rounded to the nearest minute): 8, 7, 3, 6, 5, 7, 3, 4, 5, 5, 4, 2, 1, 5, 2, 1, 3, 3, 4, 6, 4, 4, 2, 5, 4, 6, 4, 5, 4, 5. Give the frequency distribution, define the random variable, and draw a histogram.

Determine whether the functions in Problems 10–17 are probability density functions.

10. $f(x) = 6x$ on $[1, 5]$

11. $f(x) = \dfrac{x^2}{72}$ on $[1, 5]$

12. $f(x) = 6x$ on $[0, \sqrt{3}/3]$

13. $f(x) = \frac{1}{4}$ on $[1, 5]$

14. $f(x) = \frac{1}{4}$ on $[3, 6]$

15. $f(x) = 4$ on $[1, 4]$

16. $f(x) = \dfrac{3}{125}x^2$ on $[0, 5]$

17. $f(x) = \dfrac{3}{64}x^2$ on $[1, 4]$

Find k such that each function in Problems 18–27 is a probability density function over the given interval.

18. $f(x) = kx$ on $[2, 6]$

19. $f(x) = kx$ on $[-2, 3]$

20. $f(x) = kx^2$ on $[-1, 3]$

21. $f(x) = kx^2$ on $[0, 5]$

22. $f(x) = 3k$ on $[1, 4]$

23. $f(x) = \dfrac{k}{4}$ on $[1, 7]$

24. $f(x) = kx^{2/3}$ on $[0, 8]$

25. $f(x) = kx^{1/2}$ on $[1, 4]$

26. $f(x) = k/x$ on $[1, 6]$

27. $f(x) = k/x^2$ on $[1, 4]$

Find the area under the standard normal curve satisfying the conditions in Problems 28–33.

28. $X < -2$

29. $X < 1.23$

30. $X > 1$

31. $X > 1.69$

32. $.5 < X < 1.61$

33. $-2.8 \le X \le -.46$

34. The price of an item in dollars is a continuous random variable with a probability density function of $f(x) = 2$ for $1.25 \le x \le 1.75$. What is the probability that the price is more than $1.50?

35. A number is selected at random from the interval $[0, 100]$. The probability density function is $f(x) = .01$. Find the probability that the number selected is between 35 and 75.

36. The number of minutes to learn a certain task is a random variable with a probability density function of

$$f(t) = 0.1e^{-0.1t}$$

Find the probability that the task is learned in less than 10 minutes.

37. The seeds of many plants are dispersed by the wind, and the distance x (in feet) that the seed travels is given by a probability density function

$$d(x) = 0.5e^{-0.5x}$$

Find the probability that the seeds will be dispersed from 5 to 10 feet away.

38. The breaking strength of a rope (in pounds) is normally distributed, with a mean of 100 pounds and a standard deviation of 16. What is the probability that the rope will break with a force of 132 pounds?

39. The diameter of an electric cable is normally distributed, with a mean of 0.9 inch and a standard deviation of 0.01. What is the probability that the diameter will exceed 0.91 inch?

40. The annual rainfall in Ferndale, California, is known to be normally distributed, with a mean of 35.5 inches and a standard deviation of 2.5. What is the probability that the rainfall will exceed 32 inches?

41. If a light bulb is normally distributed with a mean life of 250 hours and a standard deviation of 25 hours, what is the probability that the bulb will burn for less than 210 hours?

42. What is the probability that the bulb described in Problem 41 will burn for more than 270 hours?

43. What is the probability that the bulb described in Problem 41 will burn between 240 and 280 hours?

15.3 Improper Integrals

The graph of the function $f(x) = e^{-x}$ is shown in Figure 15.15. Note that the function is unbounded since $y = 0$ is a horizontal asymptote.*

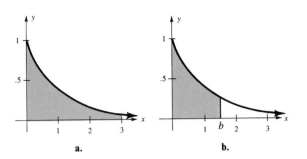

Figure 15.15 Graph of $y = e^{-x}$

a.   **b.**

It is possible to find the area of this region even though it is unbounded. To find the area we need to consider the integral

$$\int_0^\infty e^{-x}\,dx$$

but this integral has not yet been defined. It is called an **improper integral** and is defined by considering a limit. To do this, we consider a vertical line $x = b$, as shown in Figure 15.15b. We find the shaded area:

$$\int_0^b e^{-x}\,dx = -e^{-x}\Big|_0^b$$

$$= -e^{-b} - (-e^0)$$

$$= 1 - e^{-b}$$

Now we let the vertical line $x = b$ slide to the right; that is, suppose $b \to \infty$. The area changes according to

$$\lim_{b \to \infty}(1 - e^{-b}) = 1 \qquad \text{since } \lim_{b \to \infty} e^{-b} = 0$$

This limit is defined to be the area shown in Figure 15.15a, so

$$\int_0^\infty e^{-x}\,dx = 1$$

---

* If you studied Section 15.2 you will see that this function is the exponential probability distribution. This discussion will verify that the function is, indeed, a probability density function.

Improper Integral

If f is continuous over the indicated interval and the limit exists, then the following integrals are called *improper integrals*:

$$\int_a^\infty f(x)\,dx = \lim_{b \to \infty} \int_a^b f(x)\,dx$$

$$\int_{-\infty}^b f(x)\,dx = \lim_{a \to -\infty} \int_a^b f(x)\,dx$$

$$\int_{-\infty}^\infty f(x)\,dx = \int_{-\infty}^c f(x)\,dx + \int_c^\infty f(x)\,dx \qquad \text{where } c \text{ is any point on } (-\infty, \infty) \text{ where both improper integrals on the right exist}$$

If the indicated limit exists, then the improper integral is said to *converge*; if the limit does not exist, then the improper integral is said not to exist or to *diverge*.

EXAMPLE 1   Evaluate the following integrals.

a. $\displaystyle\int_1^\infty 2x^{-3}\,dx$ **b.** $\displaystyle\int_{-\infty}^{-1} \frac{dx}{x}$ **c.** $\displaystyle\int_{-\infty}^\infty xe^{-x^2}\,dx$

Solution   **a.** $\displaystyle\int_1^\infty 2x^{-3}\,dx = \lim_{b \to \infty} \int_1^b 2x^{-3}\,dx = \lim_{b \to \infty} \left(-x^{-2}\right)\Big|_1^b = \lim_{b \to \infty}(-b^{-2} + 1) = 1$

b. $\displaystyle\int_{-\infty}^{-1} \frac{dx}{x} = \lim_{a \to -\infty} \int_a^{-1} x^{-1}\,dx = \lim_{a \to -\infty} (\ln|x|)\Big|_a^{-1} = \lim_{a \to -\infty} (0 - \ln|a|)$

This limit does not exist, so we say that this integral diverges.

c. $\displaystyle\int_{-\infty}^\infty xe^{-x^2}\,dx = \int_{-\infty}^0 xe^{-x^2}\,dx + \int_0^\infty xe^{-x^2}\,dx$

$$= \lim_{a \to -\infty} \int_a^0 xe^{-x^2}\,dx + \lim_{b \to \infty} \int_0^b xe^{-x^2}\,dx$$

$$= \lim_{a \to -\infty} \int_{x=a}^{x=0} \frac{-1}{2} e^u\,du + \lim_{b \to \infty} \int_{x=0}^{x=b} \frac{-1}{2} e^u\,du \qquad \boxed{\begin{array}{l} \text{Let } u = -x^2 \\ du = -2x\,dx \end{array}}$$

$$= \lim_{a \to -\infty} (-0.5e^{-x^2})\Big|_a^0 + \lim_{b \to \infty} (-0.05e^{-x^2})\Big|_0^b$$

$$= \lim_{a \to -\infty} (-0.5 + 0.5e^{-a^2}) + \lim_{b \to \infty} (-0.05e^{-b^2} + 0.5)$$

$$= -0.5 + 0 + 0 + 0.5$$

$$= 0 \qquad\qquad\blacksquare$$

EXAMPLE 2   The **capital value**, V, of a property over T years is given by

$$V = \int_0^T Re^{-rt}\,dt$$

where R is the annual rent, or income, and r is the current interest rate. Suppose the current interest rate is 8% and you have an indeterminant lease (no date of termination) paying $60,000 per year. What is the capital value?

Solution $V = \int_0^\infty 60{,}000e^{-0.08t}\,dt = \lim_{b \to \infty} \int_0^b 60{,}000e^{-0.08t}\,dt = 60{,}000 \lim_{b \to \infty} \int_0^b e^{-0.08t}\,dt$

$= 60{,}000 \lim_{b \to \infty} \frac{e^{-0.08t}}{-0.08}\Big|_0^b = -750{,}000 \lim_{b \to \infty}(e^{-0.08b} - 1)$

$= -750{,}000(0 - 1)$

$= 750{,}000$

The capital value for the lease is $750,000. ■

Problem Set 15.3

Evaluate the integrals in Problems 1–18 that converge.

1. $\int_0^\infty e^{-x}\,dx$

2. $\int_{-\infty}^0 e^{-x}\,dx$

3. $\int_0^\infty e^{-x/2}\,dx$

4. $\int_1^\infty \frac{dx}{x^4}$

5. $\int_1^\infty \frac{dx}{x^3}$

6. $\int_1^\infty \frac{dx}{\sqrt{x}}$

7. $\int_1^\infty \frac{dx}{\sqrt[3]{x}}$

8. $\int_1^\infty \frac{dx}{x^{0.9}}$

9. $\int_1^\infty \frac{dx}{x^{1.1}}$

10. $\int_1^\infty \frac{x\,dx}{\sqrt{x^2 + 2}}$

11. $\int_{-\infty}^0 \frac{2x\,dx}{x^2 + 1}$

12. $\int_1^\infty \frac{x\,dx}{(1 + x^2)^2}$

13. $\int_{-\infty}^\infty x^2\,dx$

14. $\int_{-\infty}^\infty (x^2 + 1)^{-1/2}\,dx$

15. $\int_{-\infty}^\infty \frac{3x\,dx}{(3x^2 + 2)^3}$

16. $\int_e^\infty \frac{dx}{x \ln x}$

17. $\int_e^\infty \frac{dx}{x(\ln x)^2}$

18. $\int_1^\infty \ln x\,dx$

APPLICATIONS

19. Suppose the current interest rate is 6% and you have an indeterminant lease paying $20,200 per year. What is the capital value?

20. Suppose the current interest rate is 11% and you have an indeterminant lease paying $3,600 per year. What is the capital value?

21. Show that the capital value of a rental property with an indeterminant lease paying an annual rent of $R with a current interest rate of r percent is R/r.

22. The amount of radioactive material being released into the atmosphere annually, that is, the amount present at time T, is given by

$$A = \int_0^T Pe^{-rt}\,dt$$

If a recent United Nations publication estimates that

$r = 0.002$ and $P = 200$ millirems, estimate the total future buildup of radioactive material in the atmosphere.

23. Suppose that $r = 0.05$ in Problem 22. Estimate the total future buildup of radioactive material in the atmosphere.

24. Show that the buildup of radioactive material in the atmosphere as reported in Problem 22 approaches a limiting value of P/r.

25. Find the area of the unbounded region between the x-axis and the curve $y = 2/(x - 4)^3$ for $x \geq 6$.

26. Find the area of the unbounded region between the x-axis and the curve $y = \dfrac{2}{(x - 4)^3}$ for $x \leq 2$.

27. For a particular type of battery, the probability density function for t hours to be the life of a battery selected at random is given by

$$f(t) = 0.01e^{-0.01t}$$

What is the probability that the battery will last at least 50 hours?

28. At Caltex Corporation the probability density function for t minutes to be the length of time that a randomly selected telephone customer must wait to be helped is given by

$$f(t) = 0.25e^{-t/4}$$

What is the probability that a customer is kept waiting more than 3 minutes?

29. The probability density function that a new telephone will need servicing x months after it is purchased is given by

$$f(x) = 0.05e^{-x/20}$$

If the phone is guaranteed for a year, what is the probability that a customer selected at random will have a phone that is no longer covered by the guarantee but that needs servicing?

30. Suppose that an oil well produces $P(t)$ thousand barrels of crude oil according to the formula

$$P(t) = 100e^{-0.02t} - 100e^{-0.1t}$$

where t is the number of months the well has been in production. What is the total amount of oil produced by the oil well?

15.4   Differential Equations

Many of the applied integration problems we have worked on have involved knowing the derivative of a function and needing to find the original function. The process of finding this original function is called *antidifferentiation,* and we have used integration in order to find the desired result. More generally, if an equation involves an unknown function (often denoted by y) and one or more of its derivatives, it is called a **differential equation**. If the known derivative is dy/dx, it is called a **first-order differential equation**.

The simplest type of differential equation is one in which there are no y terms. These differential equations can be solved by integration. We use this type of equation to introduce the terminology we will use when solving differential equations. Then we will turn to differential equations with both x and y terms in which the variables can be separated.

The first type of differential equation is solved by antidifferentiation, which is pretty much what we have been doing since Chapter 13.

Differential Equation
Theorem 1

A differential equation of the type

$$\frac{dy}{dx} = f'(x)$$

has the solution

$$y = \int f(x)\,dx + C$$

where C is an arbitrary constant.

This result is easily derived by integrating, as shown in Example 1.

EXAMPLE 1   Solve the differential equation $y' = 6x$.

Solution

$$\frac{dy}{dx} = 6x \qquad \text{First set up the problem by writing } y' \text{ as } \frac{dy}{dx}$$

$$dy = 6x\,dx \qquad \text{Separate variables (multiply both sides by } dx)$$

$$\int dy = \int 6x\,dx \qquad \text{Integrate both sides}$$

$$y + C_1 = \frac{6x^2}{2} + C_2 \qquad \text{Evaluate the integrals}$$

$$y = 3x^2 + (C_2 - C_1) \qquad \text{Combine constants}$$

$$y = 3x^2 + C$$

∎

The last form in Example 1 is called the **general solution**. It combines all the constants into one arbitrary constant. In practice, though, your work for Example 1 should look like this:

$$\frac{dy}{dx} = 6x$$

$$y = \int 6x \, dx + C = 3x^2 + C$$

If we take different values for C, we will obtain **particular solutions**. For example,

$$3x^2 \qquad 3x^2 + 5 \qquad \text{and} \qquad 3x^2 - 10$$

are all particular solutions. Thus, if you know the value of a function at a particular point, you can solve for C, as shown by Example 2.

EXAMPLE 2   Solve $y' = 4x - e^{-x} - \sqrt{x}$ if $y = 50$ when $x = 0$.

Solution $y = \int (4x - e^{-x} - \sqrt{x}) \, dx$

$$= \frac{4x^2}{2} - \frac{e^{-x}}{-1} - \frac{x^{3/2}}{\frac{3}{2}} + C = 2x^2 + e^{-x} - \frac{2}{3}x\sqrt{x} + C$$

Now, if $x = 0$, then $y = 50$, so by substitution you obtain

$$50 = 2(0)^2 + e^{-0} - \frac{2}{3}(0)\sqrt{0} + C$$

$$50 = 1 + C$$

$$49 = C$$

Thus the particular solution is

$$y = 2x^2 + e^{-x} - \frac{2}{3}x\sqrt{x} + 49 \qquad\qquad\qquad\qquad \blacksquare$$

If the known value for the independent variable is 0, then the values $x = 0, y = 50$ are called an **initial condition** or a **boundary condition**.

You can verify that a function is a solution of a differential equation by differentiation. You might wish to check the results of Example 2 in this fashion.

Sometimes the differential equation involves both x and y, and if it is possible to separate the variables, you can use the following theorem.

Differential Equation Theorem 2

If a differential equation can be written in the form

$$g(y) \, dy + f(x) \, dx = 0$$

then the general solution is given by

$$\int g(y) \, dy + \int f(x) \, dx = C$$

where C is an arbitrary constant.

EXAMPLE 3   Find the general solution of

$$\frac{dy}{dx} = x^2 y$$

and check by differentiating.

Solution   Algebraically **separate the variables** by dividing each side by y and multiplying by dx:

$$y^{-1} \, dy = x^2 \, dx$$

$$y^{-1} \, dy - x^2 \, dx = 0$$

$$\int y^{-1} \, dy - \int x^2 \, dx = C$$

$$\ln|y| - \frac{x^3}{3} = C$$

Check: Take the derivative of both sides with respect to x:

$$\frac{1}{y} y' - \frac{3x^2}{3} = 0$$

Solve for $y' = dy/dx$:

$$\frac{y'}{y} = x^2$$

$$y' = x^2 y$$

so it checks.   ■

Note in Example 3 that we did not solve for y. Sometimes, however, we will want to solve for y. For Example 3,

$$\ln|y| = \frac{1}{3} x^2 + C$$

$$y = e^{1/3 x^2 + C} = e^{1/3 x^2} e^C = M e^{1/3 x^2}$$

where the constant e^C is written as the constant M. You can also check by finding the derivative of this result to see if it gives the original differential equation.

The last examples of this section lead to some of the most important applications in growth and decay. Suppose we begin with the assumption that the rate of change is linear.

EXAMPLE 4   Find the general solution of $dy/dx = ky$, where k is a constant.

Solution   Separate the variables:

$$y^{-1} \, dy - k \, dx = 0$$

$$\int y^{-1} \, dy - \int k \, dx = C$$

$$\ln|y| - kx = C$$

$$\ln|y| = kx + C$$

$$y = e^{kx + C} = e^{kx} e^C = M e^{kx}$$

where the constant e^C is written as M.   ■

Example 4 is so common that its status is elevated to that of a theorem:

Differential Equation
Theorem 3

If the rate of change of a variable y with respect to x is linear, then we say that $dy/dx = ky$ for some constant k. That is, the rate of change of y is directly proportional to the quantity itself. The solution of this differential equation is

$$y = Me^{kx}$$

EXAMPLE 5   If the divorce rate has been a constant 5% (1975–1985) and if the number of divorces in 1984 was 1,155,000, estimate the number of divorces in 1990.

Solution   Since the rate is a constant, we can write

$$\frac{dy}{dt} = 0.05y$$

where y is the number of divorces and t is the time. From Theorem 3,

$$y = Me^{0.05t}$$

The initial condition is $y = 1,155,000$ for $t = 0$ (1984):

$$1,155,000 = Me^{0.05(0)}$$
$$1,155,000 = M$$

Thus

$$y = 1,155,000e^{0.05t}$$

Now, for 1990, $t = 6$, so

$$y = 1,155,000e^{0.05(6)}$$
$$\approx 1,559,000 \quad \text{(Calculator: } 1,559,086.923)$$ ■

As you work through the problems below, be aware that we have discussed only the simplest techniques for solving differential equations. You are still not able to solve most differential equations, so the problem set comprises a very select set of differential equations for you to solve. The general solution of differential equations is substantial enough to constitute a separate mathematics course.

Problem Set 15.4

Solve the differential equations in Problems 1–20.

1. $\dfrac{dy}{dx} = x^2$

2. $\dfrac{dy}{dx} = 5$

3. $\dfrac{dy}{dx} = 8x - 10$

4. $\dfrac{dy}{dx} = 5 - x$

5. $y' = 4x^3 - 3x^2 - 5$

6. $y' = 10 - 9x^2 + 2x$

7. $5\dfrac{dy}{dx} = e^x$

8. $8\dfrac{dy}{dx} = \sqrt{x}$

9. $12\dfrac{dy}{dx} = \sqrt{5x + 1}$

10. $\dfrac{3}{2}\dfrac{dy}{dx} = e^{4x - 3}$

11. $yy' = x$

12. $yy' = 5x^2$

13. $y^2\dfrac{dy}{dx} = x^3 - 3$

14. $y\dfrac{dy}{dx} = \sqrt{x} + e^{3x} + 4$

15. $\dfrac{dy}{dx} = 2xy$

16. $\dfrac{dy}{dx} = 4x^2y - 3xy + y$

17. $\dfrac{dP}{dt} = 0.02P$

18. $\dfrac{dP}{dt} = 1250P$

19. $\dfrac{dN}{dt} = 0.001N$

20. $\dfrac{dN}{dt} = 0.15N$

28. $\dfrac{dy}{dx} = \dfrac{1 + x^2}{xy}$ for $y = 4$ when $x = 2$

Solve the differential equations in Problems 21–28 for a particular solution.

21. $\dfrac{dy}{dx} = x^2 y^{-2}$ for $y = 5$ when $x = 0$

22. $x^2 \dfrac{dy}{dx} = y$ for $y = 2$ when $x = 1$

23. $\dfrac{dP}{dt} = 0.02P$ for $P = e^3$ when $t = 0$

24. $\dfrac{dN}{dt} = 0.12N$ for $N = 2.3$ million when $t = 0$ (estimate to two places)

25. $x\dfrac{dy}{dx} - y\sqrt{x} = 0$ for $y = 1$ when $x = 1$

26. $5xy - 3y = \dfrac{dy}{dx}$ for $y = e^2$ when $x = 2$

27. $\dfrac{dy}{dx} = \dfrac{xy}{1 + x^2}$ for $y = 2$ when $x = -1$

APPLICATIONS

29. You read a newspaper story that says that the crime rate in a certain area is increasing at a constant rate of 3% per year. If 3500 major crimes were reported in 1986, estimate the number of major crimes in 1996.

30. Domar's capital expansion model is based on the present value of an investment, P, the investment productivity (a constant h), the marginal productivity to consume (a constant k), and time, t:

$$\dfrac{dP}{dt} = hkP$$

Solve this equation for P where the initial investment is $150,000.

31. Suppose that the growth rate of bacteria in milk is a constant 10% per hour and that the maximum number (in millions) permitted is 1000. How long (to the nearest hour) will the milk be acceptable if we assume that initially there are 12 million bacteria?

15.5 Summary and Review

IMPORTANT TERMS

Average value   [15.1]
Boundary condition   [15.4]
Capital value   [15.3]
Continuous distribution   [15.2]
Continuous random variable   [15.2]
Differential equation   [15.4]
Discrete distribution   [15.2]
Discrete random variable   [15.2]
Event   [15.2]
Experiment   [15.2]
Exponential probability
   distribution   [15.2]
Frequency distribution   [15.2]
General solution   [15.4]
Histogram   [15.2]
Impossible event   [15.2]
Improper integral   [15.3]
Initial condition   [15.4]
Mean   [15.2]
Normal curve   [15.2]
Normal probability
   distribution   [15.2]

Particular solution   [15.4]
Probabilistic model   [15.2]
Probability   [15.2]
Probability density function   [15.2]
Probability distribution   [15.2]
Probability distribution
   function   [15.2]
Random variable   [15.2]
Relative frequency   [15.2]
Sample space   [15.2]
Separation of variables   [15.4]
Simple event   [15.2]
Standard deviation   [15.2]
Standard normal curve   [15.2]
Total money flow   [15.1]
Total value   [15.1]
Uniform probability
   distribution   [15.2]
z-Score   [15.2]

SAMPLE TEST *For additional practice, there are a large number of review problems categorized by objective in the Student Solutions Manual. The following sample test (40 minutes) is intended to review the main ideas of this chapter.*

1. Find the total value of the functions below. Round to the nearest unit.

   a. $f(x) = \sqrt[3]{x + 7}$ on $[1, 20]$ **b.** $f(x) = e^{-0.02x}$ on $[0, 10]$

2. Determine the total money flow for \$35,000 at 12% for 6 years.

3. Find the average value for $f(x) = x\sqrt{25 - x^2}$ on $[0, 3]$.

4. **a.** Give an example of an experiment with a discrete random variable.

   b. Give an example of an experiment with a continuous random variable.

5. Prepare a frequency distribution and draw a histogram for an experiment where you roll a pair of dice 100 times and record the sum of the spots on the tops of the dice.

6. Find the constant k in order to make the following function a probability density function over the given interval.

   $$f(x) = kx \qquad \text{on } [1, 100]$$

7. Find the area under the standard normal curve for

   a. $X > 0$ **b.** $X < -0.1$ **c.** $-0.1 \le X \le 0.2$

8. What is the mean and standard deviation for the standard normal curve?

Evaluate the improper integrals in Problems 9–10 that converge.

9. $\displaystyle\int_{1}^{\infty} 0.5e^{-0.5x}\, dx$

10. $\displaystyle\int_{-\infty}^{1} \frac{x\, dx}{x^2 + 1}$

Solve the differential equations in Problems 11–12.

11. $4y' = 6 + 10x$ 12. $\dfrac{dy}{dx} = \dfrac{xy}{5 - y^2}$ for $y = 1$ when $x = 0$

13. A new spit-roasted chicken franchise estimates growth at the phenomenal rate of 22.5% per year (that is, the rate of change of y with respect to x is linear). If there are 60 outlets at the present time, how many franchises will there be in 10 years?

14. Suppose that a teacher grades an exam on a curve. It is known that the mean is 75 and the standard deviation is 5. There are 45 students in the class.

   a. How many students will receive a C?

   b. How many students will receive an A?

   c. What score would be necessary to obtain an A?

   d. If an exam paper is selected at random, what is the probability that it will be a failing paper?

15. Suppose that the current interest rate is 10% and the British Embassy in Washington, D.C., has an indeterminant lease paying \$84,000 per year. What is the capital value of this lease?

Cumulative Review for Chapters 12–15

1. Graph $y = e^x$

2. Graph $y = \ln \sqrt{x}$

Solve the equations in Problems 3–6.

3. $\log\left(\dfrac{x-5}{6}\right) = 2$

4. $3\ln\dfrac{e}{\sqrt[3]{5}} = 3 - \ln x$

5. $10^{-x} = 0.5$

6. $e^{1-x} = 105$

7. Fill in the blanks to complete the properties of the definite integral.

a. $\displaystyle\int_a^a f(x)\,dx = $ _____

b. $\displaystyle\int_a^b dx = $ _____

c. $\displaystyle\int_a^b f(x)\,dx = $ _____ where $b < a$

d. $\displaystyle\int_a^c f(x)\,dx + \int_c^b f(x)\,dx = $ _____ where $a < c < b$

e. $\displaystyle\int_a^b kf(x)\,dx = $ _____

f. $\displaystyle\int_a^b [f(x) \pm g(x)]\,dx = $ _____

g. $\displaystyle\int_{x=a}^{x=b} u\,dv = $ _____

Evaluate the integrals in Problems 8–15.

8. a. $\displaystyle\int \dfrac{du}{u}$

b. $\displaystyle\int du$

9. a. $\displaystyle\int (5x^4 + 3x^2 + 5)\,dx$

b. $\displaystyle\int e^{2x}\,dx$

10. $\displaystyle\int x^2(5 - 2x^3)^4\,dx$

11. $\displaystyle\int \ln^4 5x\,dx$

12. $\displaystyle\int \dfrac{100\,dx}{5 - e^{-x}}$

13. $\displaystyle\int \dfrac{8x - 3}{4x^2 - 3x + 2}\,dx$

14. $\displaystyle\int_1^\infty x^{-3/2}\,dx$

15. $\displaystyle\int_{-\infty}^\infty \dfrac{x\,dx}{(x^2 + 1)^2}$

16. Write a formula for the area bounded above by the curve $y = 1/\sqrt{x^2 + 4}$, below by the x-axis, on the left by the y-axis, and on the right by the line $x = t$.

17. Find the area bounded by $y = x^2$ and $y = 32 - x^2$.

18. Find $\displaystyle\int_1^2 \ln x^2\,dx$ correct to two decimal places using one (or all) of the following approximations for $n = 4$.

a. Rectangular approximation   **b.** Trapezoidal approximation   **c.** Simpson's rule

APPLICATIONS

19. A piece of replacement equipment costs $48,000. The rate of operating cost savings is $S(x)$ in dollars and is given by the formula

$$S(x) = 5000x + 2000$$

where x is the number of years the piece of equipment will be used. How long will it take the piece of equipment to pay for itself?

20. In the course of any year, the number y of cases of a disease is reduced by 10% according to the growth formula

$$y = P_0 e^{-0.1t}$$

for t years with P_0 cases reported today. If there are 100,000 cases today, how long will it take to reduce the number of cases to less than 10,000?

21. Use the formula $A = P(1 + i)^N$ to find out how long after depositing $1,000 at 8% compounded daily (365-day year) you must wait in order to have $5,000.

22. In 1986 the growth rate of Houston, Texas, was given by the formula

$$P'(x) = 1.6e^{0.025t}$$

where t is measured in years since 1980 and $P(t)$ is the population in millions. If the 1986 population was 1,860,000, predict the population at the turn of the century.

23. A businessperson receives a shipment of Christmas trees on November 20. The sales pattern is such that the inventory moves slowly at the beginning but as Christmas approaches, the demand increases so that x days after November 20 the inventory is y trees, where

$$y = 2450 - 2x^2 \qquad \text{for } 0 \le x \le 35$$

What is the average inventory for the first 30 days?

24. Let x be a normally distributed random variable with mean 55 and standard deviation 10. Find the following probabilities.
    a. $P(X \ge 55)$ b. $P(X > 60)$ c. $P(40 \le X \le 50)$

25. The wait time (in seconds) for a response at a particular terminal in a time-share network is a continuous random variable (x) with a probability density function

$$f(x) = 0.01e^{-x/100}$$

What is the probability that a user must wait more than 1 minute for a response?

Cumulative Review for Chapters 9–15

1. In your own words, discuss the meaning of limit. Use examples and graphs.

2. In your own words, discuss the definition of derivative. Include in your discussion reasons why the derivative is important, as well as some of the principal applications of derivative.

3. In your own words, discuss the definition of integral. Include in your discussion reasons why the integral is important, as well as some of the principal applications of the integral.

4. State the fundamental theorem of integral calculus. Discuss why you think this theorem is fundamental.

5. Solve the following equations.
    a. $10^{x-1} = 25$ b. $e^{x-1} = 11$ c. $8^{x-1} = 5$

Find the derivatives of the functions in Problems 6–8.

6. $y = 5x^{-1} + 3\sqrt{x}$

7. $y = \dfrac{-3}{1 - 2x}$

8. $y = \ln|1 - 11x|$

Evaluate the integrals in Problems 9–11.

9. $\displaystyle\int (3x^2 - 2x)\,dx$

10. $\displaystyle\int_0^2 4x\,dx$

11. $\displaystyle\int_1^3 5x^{-1}\,dx$

APPLICATIONS

12. Changes in oxygen pressure (P_{O_2}) have been recorded on the graph below for time (in seconds) on the interval $[0, 20]$.

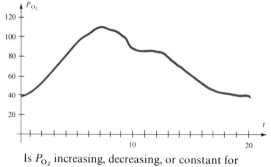

Is P_{O_2} increasing, decreasing, or constant for
a. $t < 3$?   **b.** $10 \le t \le 13$?   **c.** $t \ge 15$?

13. The cost of producing x units of a product is $C(x)$, where

$$C(x) = 5x + 8000$$

The product sells for $130 per unit.
a. What is the break-even point?
b. What revenue will the company receive if it sells just that number of units?

14. A company finds its sales are related to the amount spent on training programs by

$$T(x) = \frac{50 + 25x}{x + 5}$$

where T is the sales in hundreds of thousands of dollars when x thousand dollars are spent on training. Find the rate of change of sales when $x = 5$.

15. Suppose the profit from an item is $P(x) = x^3 + 5x^2 - 8x + 10{,}000$ where x is the price in dollars. Find the maximum possible profit.

16. The rate of sales of a brand of razor blades, in thousands, is given by $S(x) = 3x^2 + 4x$, where x is the time in months that the new product has been on the market. Find the total sales after 10 months.

17. Find the area of the region enclosed by the curves

$$y = 2x - 4 \quad \text{and} \quad y^2 = 4x \quad \text{for } x \ge 1$$

(*Note:* $y^2 = 4x$ can be broken into two functions, $y = 2\sqrt{x}$ and $y = -2\sqrt{x}$.)

18. A manufacturer finds it costs her $x^2 + 5x + 7$ dollars to produce x tons of dulconite. At production levels above 3 tons, she must hire additional workers, and her costs increase by $3x - 9$ dollars on the total production. If the price she receives is $13 per ton regardless of how much she manufactures, and if her plant capacity is 10 tons, what level of output maximizes profits?

19. Graph $f(x) = \frac{1}{4}x^4 - \frac{3}{2}x^2$ showing the relative maximums, relative minimums, and points of inflection.

20. The functional life of a timing device selected at random is determined by the probability density function

$$f(t) = 0.005e^{-0.005t}$$

where t is the number of months it has been in operation. What is the probability that the device will operate for more than 5 years ($t = 60$ months)?

CHAPTER 16
Functions of Several Variables

CHAPTER CONTENTS

16.1 Three-Dimensional Coordinate System

16.2 Partial Derivatives

16.3 Maximum-Minimum Applications

16.4 Lagrange Multipliers

16.5 Multiple Integrals

16.6 Summary and Review
Important Terms
Sample Test

APPLICATIONS

Management (*Business, Economics, Finance, and Investments*)

Cost function (16.1, Problems 43, 45, 46; 16.2, Problem 50; 16.6, Problem 15)
Revenue function (16.1, Problem 44; 16.2, Problem 51)
Profit (16.2, Problem 49)
Future value of an investment (16.2, Problems 55–56)
Rate of change of an annuity (16.2, Problem 57)
Maximize profit (16.3, Problems 19, 21)
Minimize labor cost for a function of two variables (16.3, Problem 20)
Minimize cost of shipping container (16.3, Problem 22)
Least material to construct a shipping container (16.3, Problem 23)
Maximum yield on farm production (16.4, Problem 15)
Minimize cost relative to supply (16.4, Problem 16)
Maximum area for a fenced enclosure, given fixed costs (16.4, Problem 17)
Minimum surface area for a standard size Coke (16.4, Problem 18)
Marginal cost and revenue of a function of two variables (16.4, Problem 20)
Average value of a function (16.5, Problems 49–50)
Cobb-Douglas production function (16.5, Problem 51)

Life sciences (*Biology, Ecology, Health, and Medicine*)

Amount of blood flow as a function of size of the blood vessel (16.1, Problem 47)
Poiseuille's law for blood flow (16.2, Problem 52)
Surface area of a human body (16.2, Problems 53–54)
Supplying a 1000 calorie diet while minimizing the cost (16.4, Problem 19)

Social sciences (*Demography, Political Science, Population, Psychology, Society, and Sociology*)

Intelligence quotient (IQ) (16.1, Problem 42)

General interest

Cost function for finishing a room (16.1, Problems 45–46)
Largest volume that can be mailed in the U.S. (16.3, Problem 24)

Modeling application—The Cobb-Douglas Production Function

CHAPTER OVERVIEW   In order to build realistic mathematical models for real life situations it is necessary to consider functions of more than one variable. This chapter defines and discusses such functions.

PREVIEW   The important concept of this chapter is that of a partial derivative. We use it in maximum-minimum applications and least squares. Another method of maximization, Lagrange multipliers, is introduced in Section 16.4. Multiple integration applications, including finding volumes, are developed in Section 16.5.

PERSPECTIVE   Even though this chapter may be considered optional because of time constraints, it is very important in serious model building. The concept of a function of two, three, or more variables is an easy one; unfortunately, the graphical representation of these functions is not. If you do not have time to consider the ideas of this chapter in class, it is useful to study this chapter on your own after you have completed this course.

16.1   Three-Dimensional Coordinate System

Many real life models involve more than one variable. Suppose, for example, we consider one of the most fundamental applications, that of the cost of producing an item. If Ballad Corporation produces a single record with fixed costs of $2,000 and a unit cost of $.35, then

$$C(x) = 2000 + 0.35x$$

for x records produced. However, if a second record is produced with additional fixed costs of $500 and a unit cost of $.30, then the total cost of producing x records of the first type and y records of the second type requires what we call a **function of two independent variables** x and y:

$$C(x, y) = 2500 + 0.35x + 0.30y$$

Function of Two or More Variables

Suppose D is a collection of ordered n-tuples of real numbers $(x_1, x_2, \ldots, x_n)$. Then a function f with **domain** D is a rule that assigns a number

$$z = f(x_1, x_2, \ldots, x_n)$$

to each n-tuple in D. The function's **range** is the set of z values the function assumes. The symbol z is called the **dependent variable** of f, and f is said to be a **function of the n independent variables** $x_1, x_2, \ldots, x_n$.

You have already considered many examples of functions of several variables, as Example 1 shows.

EXAMPLE 1

Area of a rectangle: $K(l, w) = lw$
Volume of a box: $V(l, w, h) = lwh$
Simple interest: $I(P, r, t) = P(1 + rt)$
Compound interest: $A(P, r, t, n) = P(1 + \frac{r}{n})^{nt}$

Find each of the requested values and interpret your results in terms of what you know about each of these formulas.

a. $K(25, 15)$
b. $V(5, 20, 30)$
c. $I(100{,}000, 0.08, 15)$
d. $A(450{,}000, 0.09, 30, 12)$

Solution

a. $K(25, 15) = 25(15) = 375$; the area of a 25 by 15 rectangle is 375 square units.
b. $V(5, 20, 30) = 5(20)(30) = 3000$; the volume of a 5 by 20 by 30 box is 3000 cubic units.
c. $I(100{,}000, 0.08, 15) = 100{,}000[1 + 0.08(15)] = 100{,}000[2.2] = 222{,}000$; future value of a $100,000 investment at 8% simple interest for 15 years is $220,000.
d. $A(450{,}000, 0.09, 30, 12) = 450{,}000(1 + \frac{0.09}{12})^{30(12)} = 450{,}000(1.0075)^{360} \approx$ 6,628,759.26; the future value of a $450,000 investment at 9% compounded monthly is approximately $6,628,759.26. ■

Even though a function of several variables has been defined for the general case and Example 1 shows functions of several variables, this chapter focuses primarily on functions of two variables. That is, $z = f(x, y)$ is the notation used for z, a function of two independent variables x and y. In order to graph such a function we need to consider **ordered triplets** (x, y, z) and a **three-dimensional coordinate system**, just as we have already considered ordered pairs (x, y) and a two-dimensional coordinate system. We draw a coordinate system with three mutually perpendicular axes, as shown in Figure 16.1.

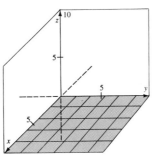

Figure 16.1 Three-dimensional coordinate system

Think of the x-axis and the y-axis as the floor and the z-axis as a line perpendicular to the floor. All of the graphs we have done up to now in this book would now be drawn on the "floor." Example 2 shows how to plot points in three dimensions.

EXAMPLE 2 Graph the following ordered triplets.

Solution **a.** $(10, 20, 10)$
b. $(-12, 6, 12)$
c. $(-12, -18, 6)$
d. $(20, -10, 18)$

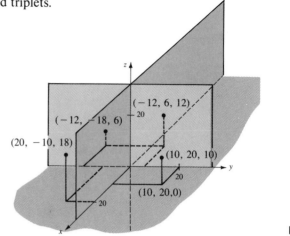

If you orient yourself in a room (your classroom, for example), as shown in Figure 16.2, you will notice certain important planes:

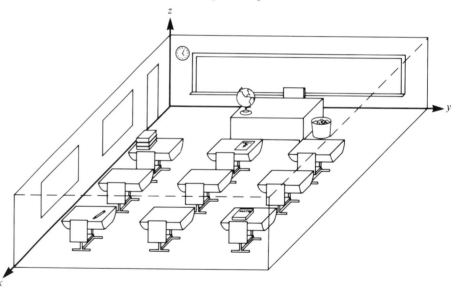

Figure 16.2 A typical classroom; assume the dimensions are 25 by 30 feet with an 8 foot ceiling

Floor: **xy-plane.** Equation is $z = 0$.

Ceiling: a plane parallel to the xy-plane. Equation is $z = 8$.

Front wall: **yz-plane.** Equation is $x = 0$.

Back wall: plane parallel to the yz-plane. Equation is $x = 30$.

Left-side wall: **xz-plane.** Equation is $y = 0$.

Right-side wall: plane parallel to the xz-plane. Equation is $x = 25$.

The xy-, xz-, and yz-planes are called the **coordinate planes**. Name the coordinates of several objects in the figure.

Just as points in the plane are associated with ordered pairs satisfying an equation in two variables, points in space are associated with ordered triplets satisfying an equation. The graph of any function of the form $z = f(x, y)$ is called a **surface**. It is beyond the scope of this course to have you spend a great deal of time graphing three-dimensional surfaces, but you should be aware that computer programs have simplified the task of graphing surfaces, as shown in Figure 16.3.

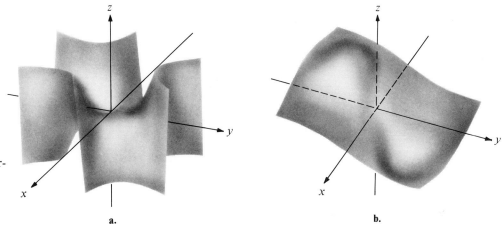

Figure 16.3 Graphs of surfaces in three dimensions:
a. $z = x^3 - 3xy^2$
b. $z = \dfrac{-1}{x^2 + y^2 + 1}$

a.

b.

The remainder of this section is a brief introduction to some of the more common three-dimensional surfaces.

Planes

The graph of $ax + by + cz = d$ is a **plane** if $a, b, c,$ and d are real numbers (not all zero).

EXAMPLE 3   Graph the planes defined by the given equations.

    a. $x + 3y + 2z = 6$ **b.** $y + z = 5$ **c.** $x = 4$

Solution   It is customary to show only the portion of the graph that lies in the **first octant** (that is, where $x, y,$ and z are all positive). To graph a plane, find some ordered triplets satisfying the equation. The best ones to use are often those on one of the coordinate axes.

a. Let $x = 0$ and $y = 0$; then $z = 3$; point is $(0, 0, 3)$; let $x = 0$ and $z = 0$; then $y = 2$; point is $(0, 2, 0)$; let $y = 0$ and $z = 0$; then $x = 6$; point is $(6, 0, 0)$. Plot these points as shown in Figure 16.4a and use them to draw the plane.

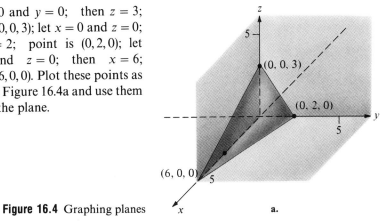

Figure 16.4 Graphing planes

a.

b. When one of the variables is missing from the equation of a plane, then that plane is parallel to the axis corresponding to the missing variable; in this case it is parallel to the x-axis. Draw the line $y + z = 5$ on the yz-plane, and then complete the plane as shown in Figure 16.4b.

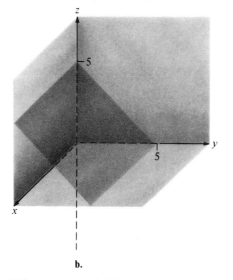

b.

c. When two variables are missing, then the plane is parallel to one of the coordinate planes, as shown in Figure 16.4c.

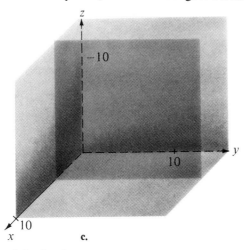

c.

Quadric Surfaces

The graph of the equation

$$Ax^2 + By^2 + Cz^2 + Dxy + Exz + Fyz + Gy + Hy + Iz + J = 0$$

is called a **quadric surface**. The **trace** of a curve is found by setting one of the variables equal to a constant and then graphing the resulting curve. If $x = k$ (k a constant), then the resulting curve is drawn in the plane $x = k$, which is parallel to the yz-plane; similarly, if $y = k$, then the curve is drawn in the plane $y = k$, which is parallel to the xz-plane; and if $z = k$, then the curve is drawn in the plane $z = k$, which is parallel to the xy-plane. Table 16.1 shows the quadric surfaces.

TABLE 16.1 Quadric surfaces

Surface	Description	Surface	Description	Surface	Description
Elliptic cone	The trace in the xy-plane is a point; in planes parallel to the xy-plane it is an ellipse. Traces in the xz- and yz-planes are intersecting lines; in planes parallel to these they are hyperbolas $$z^2 = \frac{x^2}{a^2} + \frac{y^2}{b^2}$$	Elliptic paraboloid	The trace in the xy-plane is a point; in planes parallel to the xy-plane it is an ellipse. Traces in the xz- and yz-planes are parabolas $$z = \frac{x^2}{a^2} + \frac{y^2}{b^2}$$	Ellipsoid or sphere	The traces in the coordinate planes are ellipses. $$\frac{x^2}{a^2} + \frac{y^2}{b^2} + \frac{z^2}{c^2} = 1$$ If $a^2 = b^2 = c^2 = r^2$, then the graph is a sphere $$x^2 + y^2 + z^2 = r^2$$
Hyperboloid of one sheet	The trace in the xy-plane is an ellipse; in the xz- and yz-planes the traces are hyperbolas $$\frac{x^2}{a^2} + \frac{y^2}{b^2} - \frac{z^2}{c^2} = 1$$	Hyperboloid of two sheets	There is no trace in the xy-plane. In planes parallel to the xy-plane, which intersect the surface, the traces are ellipses. Traces in the xz- and yz-planes are the hyperbolas $$\frac{x^2}{a^2} + \frac{y^2}{b^2} - \frac{z^2}{c^2} = -1$$		
Hyperbolic paraboloid	The trace in the xy-plane is two intersecting lines; in planes parallel to the xy-plane the traces are hyperbolas. Traces in the xz- and yz-planes are parabolas $$z = \frac{y^2}{b^2} - \frac{x^2}{a^2}$$				

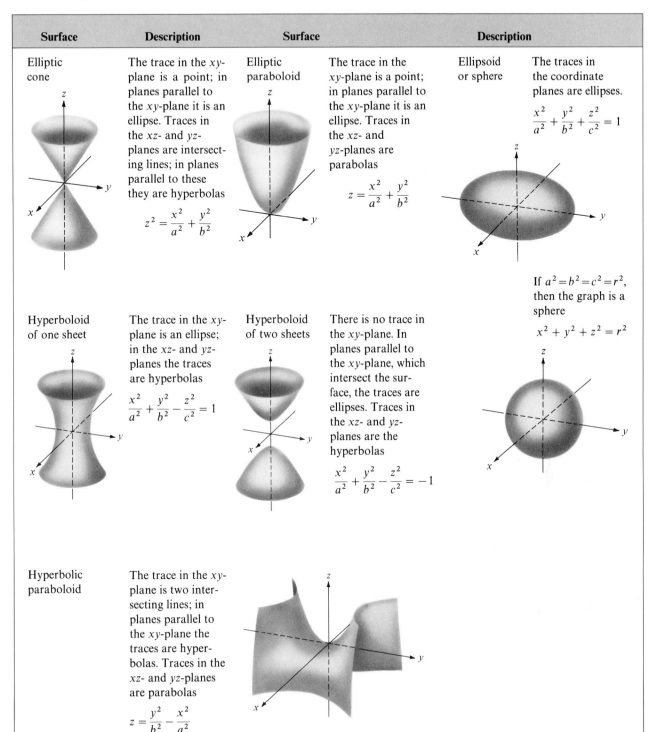

Circular Cylinders

The graphs of

$$y^2 + z^2 = r^2 \qquad x^2 + z^2 = r^2 \qquad \text{and} \qquad x^2 + y^2 = r^2$$

are **right circular cylinders** of radius r, parallel to the x-axis, y-axis, and z-axis, respectively.

EXAMPLE 4   Graph the following equations.

       a. $x^2 + y^2 = 9$ **b.** $y^2 + z^2 = 16$ **c.** $x^2 + z^2 = 25$

Solution   **a.** This is a cylinder parallel to the z-axis (the z variable is missing), as shown in Figure 16.5a.

b. This is a cylinder parallel to the x-axis, as shown in Figure 16.5b.

c. This is a cylinder parallel to the y-axis, as shown in Figure 16.5c.

Figure 16.5 Graphs of right circular cylinders.

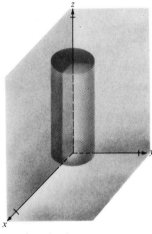

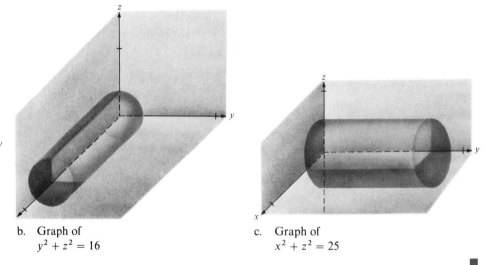

a. Graph of
     $x^2 + y^2 = 9$

b. Graph of
     $y^2 + z^2 = 16$

c. Graph of
     $x^2 + z^2 = 25$

Problem Set 16.1

Evaluate the functions K, V, I, and A from Example 1 in Problems 1–10, and interpret your results.

1. $K(15, 35)$ **2.** $K(45, 90)$

3. $V(3, 5, 8)$ **4.** $V(15, 25, 8)$

5. $I(500, 0.05, 3)$ **6.** $I(1250, 0.08, 12)$

7. $A(2500, 0.12, 6, 4)$ **8.** $A(5500, 0.18, 5, 6)$

9. $A(110,000, 0.09, 30, 12)$ **10.** $A(250,000, 0.15, 15, 12)$

Evaluate $f(x, y) = x^2 - 2xy + y^2$ for the values in Problems 11–14.

11. $f(2, 3)$ **12.** $f(-1, 4)$

13. $f(-2, 5)$ **14.** $f(0, 6)$

Evaluate $g(x, y) = \dfrac{2x - 4y}{x^2 + y^2}$ for the values in Problems 15–18.

15. $g(2, 1)$ **16.** $g(-3, 2)$

17. $g(5, -3)$ **18.** $g(-3, -4)$

Evaluate $h(x, y) = \dfrac{e^{xy}}{\sqrt{x^2 + y^2}}$ for the values in Problems 19–22. Round answers to the nearest hundredth.

19. $h(0, 5)$ **20.** $h(-2, 3)$

21. $h(-2, -3)$ **22.** $h(1, 1)$

Graph the ordered triplets in Problems 23–25.

23. a. $(1, 2, 3)$ **b.** $(-3, 2, 4)$
    c. $(1, -4, 3)$ **d.** $(-5, -9, -8)$

24. a. $(2, 4, 3)$ **b.** $(-3, 2, 4)$
    c. $(10, -20, -5)$ **d.** $(-1, -2, -3)$

25. a. $(10, 5, 20)$ **b.** $(5, -15, -5)$
    c. $(3, 2, -4)$ **d.** $(-5, -1, 3)$

Graph the surfaces in Problems 26–41.

26. $2x + y + 3z = 6$

27. $x + 2y + 5z = 10$

28. $x + y + z = 1$

29. $3x - 2y - z = 12$

30. $z^2 = \dfrac{x^2}{4} + \dfrac{y^2}{9}$

31. $z = \dfrac{x^2}{4} + \dfrac{y^2}{9}$

32. $\dfrac{x^2}{1} + \dfrac{y^2}{4} + \dfrac{z^2}{9} = 1$

33. $\dfrac{x^2}{9} + \dfrac{y^2}{4} + \dfrac{z^2}{25} = 1$

34. $x^2 + y^2 + z^2 = 9$

35. $z = x^2 + y^2$

36. $\dfrac{x^2}{9} - \dfrac{y^2}{1} + \dfrac{z^2}{4} = 1$

37. $\dfrac{x^2}{9} + \dfrac{y^2}{1} - \dfrac{z^2}{4} = -1$

38. $y^2 + z^2 = 25$

39. $x^2 + y^2 = 36$

40. $x^2 + z^2 = 4$

41. $y^2 + z^2 = 20$

APPLICATIONS

42. The intelligence quotient (IQ) is defined as $Q(x, y) = \dfrac{100x}{y}$

where Q is the IQ, x is a person's mental age as measured on a standardized test, and y is a person's chronological age measured in years. Find and interpret
    a. $Q(15, 13)$ **b.** $Q(6, 9)$ **c.** $Q(15, 15)$ **d.** $Q(10.5, 9.8)$

43. A company manufactures two types of golf carts. The first has a fixed cost of $2,500, a variable cost of $800, and x are produced. The second has a fixed cost of $1,200, a variable cost of $550, and y are produced. Write a cost function $C(x, y)$ and find
    a. $C(10, 15)$ **b.** $C(5, 25)$ **c.** $C(15, 10)$ **d.** $C(0, 30)$

44. If the revenue function for the golf carts in Problem 43 is $R(x, y) = 1500x + 900y$ find
    a. $R(10, 15)$ **b.** $R(5, 25)$ **c.** $R(15, 10)$ **d.** $R(0, 30)$

45. If the dimensions of the room in Figure 16.2 are x feet wide, y feet long, and z feet high, and if ceiling material is $2 per square foot, wall material is $.75 per square foot, and floor material is $1.25 per square foot, write a cost function for the ceiling, floor, and wall (assuming no doors or windows).

46. If $C(x, y, z)$ is the cost function for Problem 45, find
    a. $C(25, 30, 8)$ **b.** $C(12, 14, 8)$ **c.** $C(15, 20, 10)$

47. The amount of blood flowing in a blood vessel measured in milliliters is given by $F(l, r) = 0.002l/r^4$ where l is the length of the blood vessel and r is the radius. Find
    a. $F(3.1, 0.002)$ **b.** $F(15.3, 0.001)$ **c.** $F(6, 0.005)$

16.2 Partial Derivatives

One of the most important and useful concepts in mathematics is that of a derivative. In this section we consider the derivative of a function of several variables. We begin with a geometric interpretation and then apply this interpretation to some particular examples.

Consider a surface $z = f(x, y)$, as shown in Figure 16.6a.

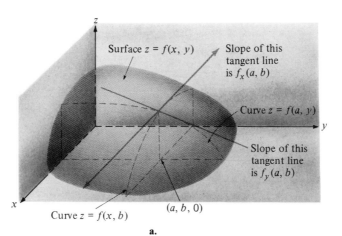

Figure 16.6a

a.

Hold one of the variables constant, say, $y = b$. This gives a curve $z = f(x, b)$, which is the intersection of the plane $y = b$ and the surface. The slope of this curve is called the **partial derivative** of f with respect to x and is denoted by f_x or $\partial z/\partial x$ (see Figure 16.6b). Similarly, if we let $x = a$ (a constant), then the curve that is the intersection of this plane and the surface has slope f_y or $\partial z/\partial y$, which is called the **partial derivative** of f with respect to y (see Figure 16.6c).

Figure 16.6 Geometric interpretation of a partial derivative

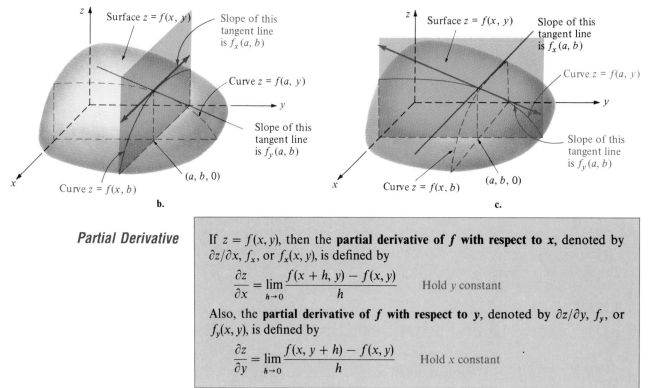

b.

c.

Partial Derivative

If $z = f(x, y)$, then the **partial derivative of f with respect to x**, denoted by $\partial z/\partial x$, f_x, or $f_x(x, y)$, is defined by

$$\frac{\partial z}{\partial x} = \lim_{h \to 0} \frac{f(x + h, y) - f(x, y)}{h} \qquad \text{Hold } y \text{ constant}$$

Also, the **partial derivative of f with respect to y**, denoted by $\partial z/\partial y$, f_y, or $f_y(x, y)$, is defined by

$$\frac{\partial z}{\partial y} = \lim_{h \to 0} \frac{f(x, y + h) - f(x, y)}{h} \qquad \text{Hold } x \text{ constant}$$

This definition says that these are the usual derivatives found by holding y or x, respectively, constant. Similar definitions could be formulated for functions with three or more independent variables.

EXAMPLE 1 Let $f(x, y) = 3xy^2 - 2x^3y + 5x$ and $g(x, y) = (3x - 5y)^4$. Find the requested partial derivatives.

a. f_x **b.** f_y **c.** g_x **d.** g_y

Solution **a.** Treat y as a constant and find the derivative with respect to x:

$$f_x = \frac{\partial}{\partial x} f(x, y) = 3y^2 - 6x^2y + 5$$

b. Treat x as a constant and find the derivative with respect to y:

$$f_y = \frac{\partial}{\partial y} f(x, y) = 6xy - 2x^3$$

c. Treat y as a constant: $g_x = \frac{\partial}{\partial x} g(x, y) = 4(3x - 5y)^3(3) = 12(3x - 5y)^3$

d. Treat x as a constant:

$$g_y = \frac{\partial}{\partial y} g(x, y) = 4(3x - 5y)^3(-5) = -20(3x - 5y)^3 \qquad \blacksquare$$

EXAMPLE 2 Suppose that x is the inventory (in thousands of dollars) and y is the number of employees of a dress store, so that $80 \leq x \leq 250$ and $3 \leq y \leq 8$. Also, suppose the weekly profit function (in dollars) is

$$P(x, y) = 5000 + 10x - 15xy + 10x^2 - 3(y - 8)^2$$

At the present time the inventory is \$105,000 ($x = 105$) and there are four employees ($y = 4$). Approximate the rate of change of P per unit change in x if y remains fixed at 4.

Solution This is the partial derivative of P with respect to x at $(105, 4)$:

$$P_x = 10 - 15y + 20x \bigg|_{(105,4)} = 10 - 15(4) + 20(105) = 2050$$

This says that the instantaneous rate of change of the profit is changing \$2,050 per \$1,000 change in x when y remains fixed at four employees. $\qquad \blacksquare$

It is possible to find the partial derivative of a partial derivative. The idea is very straightforward, but the notation is at first difficult to understand. Study the following definition of **higher-order partial derivatives**.

Higher-Order Partial Derivatives

If $z = f(x, y)$, then

$$\frac{\partial^2 z}{\partial x^2} = \frac{\partial}{\partial x}\left(\frac{\partial z}{\partial x}\right) = f_{xx}(x, y) = f_{xx}$$

$$\frac{\partial^2 z}{\partial y^2} = \frac{\partial}{\partial y}\left(\frac{\partial z}{\partial y}\right) = f_{yy}(x, y) = f_{yy}$$

$$\frac{\partial^2 z}{\partial x \, \partial y} = \frac{\partial}{\partial x}\left(\frac{\partial z}{\partial y}\right) = f_{yx}(x, y) = f_{yx}$$

$$\frac{\partial^2 z}{\partial y \, \partial x} = \frac{\partial}{\partial y}\left(\frac{\partial z}{\partial x}\right) = f_{xy}(x, y) = f_{xy}$$

For the mixed partial derivative $\dfrac{\partial^2 z}{\partial x \, \partial y} = f_{yx}$, start with z and first differentiate with respect to y (keep x constant) and then with respect to x. For $\dfrac{\partial^2 z}{\partial y \, \partial x} = f_{xy}$, start with z and first differentiate with respect to x (keep y constant) and then with respect to y.

EXAMPLE 3 For $z = f(x, y) = 5x^2 - 2xy + 3y^3$, find the requested higher-order partial derivatives. Pay particular attention to the notation—part of what this example is illustrating is the variety in notation that can be used for higher-order partial derivatives.

a. $\dfrac{\partial^2 z}{\partial x\, \partial y}$ **b.** $\dfrac{\partial^2 z}{\partial y\, \partial x}$ **c.** $\dfrac{\partial^2 z}{\partial x^2}$ **d.** $f_{xy}(3, 2)$

e. $\dfrac{\partial^2}{\partial x\, \partial y}(2x^4 - 3x^2 y^2 + 5y^3 + 25)\Big|_{(-2, 5)}$

Solution   **a.** First differentiate with respect to y:

$$\frac{\partial z}{\partial y} = -2x + 9y^2$$

Then differentiate with respect to x:

$$\frac{\partial^2 z}{\partial x\, \partial y} = \frac{\partial}{\partial x}\left(\frac{\partial z}{\partial y}\right)$$

$$= \frac{\partial}{\partial x}(-2x + 9y^2) = -2$$

b. First differentiate with respect to x, then with respect to y:

$$\frac{\partial z}{\partial x} = 10x - 2y \qquad \text{and} \qquad \frac{\partial^2 z}{\partial y\, \partial x} = \frac{\partial}{\partial y}(10x - 2y) = -2$$

c. Differentiate with respect to x twice:

$$\frac{\partial z}{\partial x} = 10x - 2y \qquad \text{and} \qquad \frac{\partial^2 z}{\partial x^2} = \frac{\partial}{\partial x}(10x - 2y) = 10$$

d. Differentiate first with respect to x and then with respect to y; finally, evaluate at $(3, 2)$:

$$f_x(x, y) = 10x - 2y \qquad \text{and} \qquad f_{xy}(x, y) = -2$$

At $(3, 2)$ the value is -2. Notice that since the value is a constant it is -2 at all points.

e. First differentiate with respect to y, then with respect to x; finally, evaluate at $(-2, 5)$:

$$\frac{\partial^2}{\partial x\, \partial y}(2x^4 - 3x^2 y^2 + 5y^3 + 25) = \frac{\partial}{\partial x}(-6x^2 y + 15y^2) = -12xy$$

At the point $(-2, 5)$:

$$\frac{\partial^2}{\partial x\, \partial y}(2x^4 - 3x^2 y^2 + 5y^3 + 25)\Big|_{(-2, 5)} = -12xy\Big|_{(-2, 5)} = -12(-2)(5) = 120 \quad \blacksquare$$

Notice from parts a and b that

$$\frac{\partial^2 z}{\partial x\, \partial y} = \frac{\partial^2 z}{\partial y\, \partial x}$$

but in general *this is not true.* However, for all of the functions in this book it will be true.

Problem Set 16.2

Let

$$z = f(x, y) = 5x^2 - 3x^3y^4 + 2y^3 - 15 \quad and$$
$$w = g(x, y) = (4x - 3y)^5$$

Find the derivatives in Problems 1–20.

1. f_x
2. g_x
3. g_y
4. f_y
5. $f_x(1, 2)$
6. $g_x(2, -1)$
7. $g_y(3, -1)$
8. $f_y(-2, 3)$
9. $\dfrac{\partial^2 z}{\partial x \partial y}$
10. $\dfrac{\partial^2 z}{\partial y \partial x}$
11. $\dfrac{\partial^2 w}{\partial y \partial x}$
12. $\dfrac{\partial^2 w}{\partial x \partial y}$
13. $f_{xx}(0, 2)$
14. $f_{xy}(1, 2)$
15. $f_{yx}(-1, 0)$
16. $f_{yy}(2, -1)$
17. $g_{xx}(0, 2)$
18. $g_{xy}(1, 2)$
19. $g_{yx}(-1, 0)$
20. $g_{yy}(2, -1)$

Let

$$z = f(x, y) = e^{3x + 2y} \quad and$$
$$w = g(x, y) = \sqrt{x^2 - 3y^2}$$

Find the derivatives in Problems 21–40.

21. f_x
22. g_x
23. g_y
24. f_y
25. $f_x(1, 2)$
26. $g_x(-1, 2)$
27. $g_y(3, -2)$
28. $f_y(-2, 3)$
29. $\dfrac{\partial^2 z}{\partial x \partial y}$
30. $\dfrac{\partial^2 z}{\partial y \partial x}$
31. $\dfrac{\partial^2 w}{\partial y \partial x}$
32. $\dfrac{\partial^2 w}{\partial x \partial y}$
33. $f_{xx}(0, 2)$
34. $f_{xy}(1, 2)$
35. $f_{yx}(-1, 0)$
36. $f_{yy}(2, -1)$
37. $g_{xx}(2, 0)$
38. $g_{xy}(2, 1)$
39. $g_{yx}(-1, 0)$
40. $g_{yy}(2, -1)$

Find f_x, f_y, and f_λ for the functions in Problems 41–44.

41. $f(x, y, \lambda) = x + 2xy + \lambda(xy - 10)$
42. $f(x, y, \lambda) = 2x + 2y + \lambda xy$
43. $f(x, y, \lambda) = x^2 + y^2 - \lambda(3x + 2y - 6)$
44. $f(x, y, \lambda) = x^2 - y^2 - \lambda(5x - 3y + 10)$

Find $\partial f/\partial b$ and $\partial f/\partial m$ in Problems 45–48.

45. $f(b, m) = (10m + 5b)^2 + (2m + b)$
46. $f(b, m) = (m + b - 4)^2 + (2m + 2b - 8)^2$
47. $f(b, m) = (m + b + 1)^2 + (2m + 2b + 2)^2 + (3m + 3b + 3)^2$
48. $f(b, m) = (2m - b - 3)^3 + (2m - b - 3)^2 + (2m - b - 3)$

APPLICATIONS

49. Hartwell Corporation sells microwave ovens and finds that its profit is a function of the price, p, of the oven as well as the amount spent on advertising, a, according to the function

$$P(p, a) = 2ap + 50p - 10p^2 - 0.1a^2p - 100$$

Find and interpret $\partial P/\partial a$ as well as $\partial P/\partial p$.

50. Ritetex is producing x units of one item and y units of another. The cost function is

$$C(x, y) = 3700 + 2500x + 550y$$

Find and interpret $C_x(x, y)$ and $C_y(x, y)$.

51. The revenue function for the items produced by Ritetex in Problem 50 is

$$R(x, y) = 1500x + 900y$$

If P is the profit function, find $P_x(5, 10)$ and $P_y(5, 10)$ and interpret what these numbers mean.

52. The amount of blood flowing in a blood vessel (in milliliters) is given by

$$F(l, r) = 0.002l/r^4$$

where l is the length of the blood vessel and r is the radius. (This formula is called *Poiseuille's law*.) Find $\partial F/\partial r$ and $\partial F/\partial l$ and interpret.

53. The number of square feet of a person's body area is a function of the person's height and weight:

$$A(w, h) = 40.5w^{0.425}h^{0.725}$$

where A is the surface area in square feet, w is the weight in pounds, and h is the height in feet. Find your own surface area.

54. Find $A_w(180, 6)$ and $A_h(180, 6)$ for the function A of Problem 53 and interpret the results.

55. If $100 is invested at an annual rate of r for t years, the future value, A, is given by the formula

$$A = 100(1 + r)^t$$

What is the instantaneous rate of change of A per unit change in r if t remains fixed at 5?

56. If $25 is deposited monthly to an account paying an annual rate of r compounded monthly for t years, the

future value, A, is given by the formula

$$A = 25\left[\frac{(1 + \frac{r}{12})^{12t} - 1}{\frac{r}{12}}\right]$$

What is the instantaneous rate of change of A per unit change of r if t remains fixed at 5?

57. If P is the present value of an ordinary annuity of equal payments of \$25 per month for t years at an annual interest rate of r, then

$$P = 25\left[\frac{1 - (1 + \frac{r}{12})^{-12t}}{\frac{r}{12}}\right]$$

What is the instantaneous rate of change of P per unit change in r if t remains fixed at 5?

16.3 Maximum-Minimum Applications

We have seen how important optimization is for functions of a single variable. In this section we consider relative maximums and minimums of functions of two variables; that is, of the type $z = f(x, y)$. In this discussion we assume that all second-order partial derivatives exist. Geometrically, the existence of all second-order partial derivatives guarantees that the surface has no tears, ruptures, sharp points, edges, or corners, as shown in Figure 16.7. We will be looking for bulges or dents, which are called **relative maximums** and **minimums**.

Relative Maximum and Relative Minimum

> Let a function be defined by $f(x, y)$ for each point in some region of the plane containing the x- and y-axes. Let (a, b) be some point in this region. If there exists a circular region with center at (a, b) such that for all (x, y) in that region,
>
> $f(x, y) \le f(a, b)$ then (a, b) is a **relative maximum**
>
> $f(x, y) \ge f(a, b)$ then (a, b) is a **relative minimum**

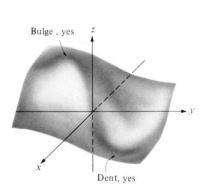

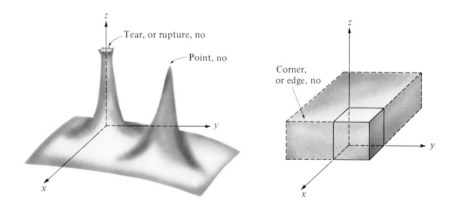

Figure 16.7 Surfaces that have or do not have partial derivatives existing at all points

The top of the bulge in Figure 16.7 is a relative maximum, and the bottom of the dent is a relative minimum. In this section we are not concerned with boundary points or absolute maximum-minimum theory but, nevertheless, we will be able to consider a great many maximum and minimum problems.

Remember when you did maximums and minimums for a function of a single variable? You first found critical values $x = c$; c was not necessarily a maximum or a minimum but just a possibility. You needed to test further to determine if c was a maximum, minimum, or neither by using the first or second derivative tests. We do the same for functions of two variables. We find **critical points** (a, b). These are points for which the partial derivatives f_x and f_y are equal to zero.

Critical Points

> Let $z = f(x, y)$ have a relative maximum or a relative minimum at the point (a, b). If both f_x and f_y exist at (a, b), then
>
> $$f_x(a, b) = 0 \quad \text{and} \quad f_y(a, b) = 0$$
>
> On the other hand, if (a, b) is a point for which
>
> $$f_x(a, b) = 0 \quad \text{and} \quad f_y(a, b) = 0$$
>
> then (a, b) is called a **critical point** for the function f. A critical point is *not necessarily* a relative maximum or a minimum.

If you find a point (a, b) that is a critical point, then there are three possibilities, as shown in Figure 16.8.

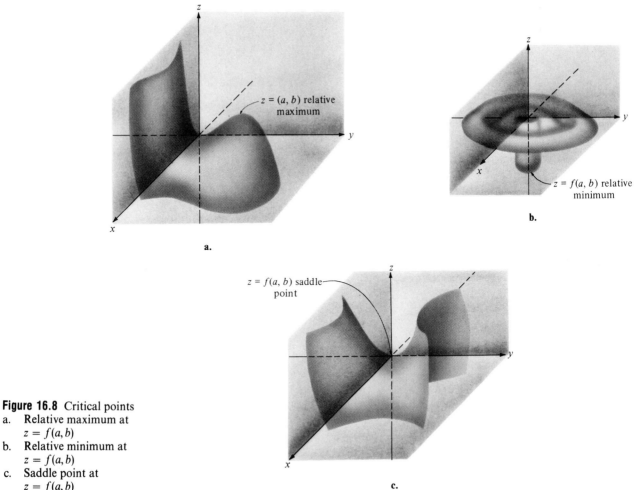

Figure 16.8 Critical points
a. Relative maximum at $z = f(a, b)$
b. Relative minimum at $z = f(a, b)$
c. Saddle point at $z = f(a, b)$

Notice that a saddle point is a point for which the two first partial derivatives are zero but is not itself a relative maximum or a relative minimum. For example, $z = x^2 - y^2$ has both $f_x(0,0) = 0$ and $f_y(0,0) = 0$ and is a saddle point. A **saddle point** is neither a maximum nor a minimum.

EXAMPLE 1 Find critical values for the surface

$$f(x, y) = 3x^2 + 4y^2 - 2xy + 22x + 30$$

Solution First find the partial derivatives:

$$f_x(x, y) = 6x - 2y + 22 \qquad \text{and} \qquad f_y(x, y) = 8y - 2x$$

Next, set these partials equal to zero:

$$6x - 2y + 22 = 0 \qquad \text{and} \qquad 8y - 2x = 0$$

Finally, solve the system

$$3 \begin{cases} 6x - 2y = -22 & \leftarrow \text{This is } 6x - 2y + 22 = 0 \\ -2x + 8y = 0 & \leftarrow \text{This is } 8y - 2x = 0 \end{cases}$$

$$+ \begin{cases} 6x - 2y = -22 \\ -6x + 24y = 0 \end{cases}$$

$$22y = -22$$
$$y = -1$$

To find x, substitute into either of the equations:

$$8y - 2x = 0 \qquad \text{so} \qquad 8(-1) - 2x = 0$$
$$x = -4$$

A critical value is thus $(-4, -1)$. This says that if f has any maximums or minimums they must occur at $(x, y) = (-4, -1)$. ∎

EXAMPLE 2 Find the critical values for the surface

$$f(x, y) = 6xy - 4x^3 - 4y^3 - 10$$

Solution $f_x = 6y - 12x^2 \qquad \text{and} \qquad f_y = 6x - 12y^2$

Set these equal to zero and solve the system:

$$\begin{cases} 6y - 12x^2 = 0 \\ 6x - 12y^2 = 0 \end{cases}$$

Solve by substitution; from the first equation,

$$6y = 12x^2$$
$$y = 2x^2$$

Substitute into the second equation:

$$6x - 12(2x^2)^2 = 0$$
$$6x - 48x^4 = 0$$
$$x - 8x^4 = 0$$

$$x(1 - 8x^3) = 0$$
$$x(1 - 2x)(1 + 2x + 4x^2) = 0$$

Solve by setting each factor equal to zero:

$$x = 0 \qquad 1 - 2x = 0 \qquad 1 + 2x + 4x^2 = 0$$
$$x = \tfrac{1}{2} \qquad \text{No real solution (the discriminant is negative)}$$

Finally, find y:

$$y = 2x^2 \quad \text{so if} \quad x = 0, \quad y = 0$$
$$\text{if} \quad x = \tfrac{1}{2}, \quad y = 2(\tfrac{1}{2})^2 = \tfrac{1}{2}$$

The critical values are $(0, 0)$ and $(\tfrac{1}{2}, \tfrac{1}{2})$. This means that if f has any maximums or minimums they must occur at $(x, y) = (0, 0)$ or $(x, y) = (\tfrac{1}{2}, \tfrac{1}{2})$. ∎

Since there is no guarantee that a critical value is a maximum or a minimum, we need a further test, called a **second derivative test**, to find the relative maximums or minimums.

Second Derivative Test for Functions of Two Variables

If *all* of the following conditions hold:

1. $z = f(x, y)$
2. All second-order partial derivatives exist in some circular region containing (a, b) as the center
3. (a, b) is a critical point [that is, $f_x(a, b) = 0$ and $f_y(a, b) = 0$]
4. $A = f_{xx}(a, b)$, $B = f_{xy}(a, b)$, $C = f_{yy}(a, b)$, $D = B^2 - AC$

Also if:

$D < 0$ and $A < 0$, then $f(a, b)$ is a **relative maximum**

$D < 0$ and $A > 0$, then $f(a, b)$ is a **relative minimum**

$D > 0$, then f has a **saddle point** at (a, b)

$D = 0$, then the *test fails*

EXAMPLE 3 Apply the second derivative test for the critical values found in Examples 1 and 2.

Solution **a.** For Example 1, $A = f_{xx} = 6$, $B = f_{xy} = -2$, and $C = f_{yy} = 8$, so $D = B^2 - AC = 4 - 6(8) < 0$. Thus, $D < 0$ and $A > 0$, so the point $(-4, -1)$ is a relative minimum. Even though you would not be expected to graph this surface, it is shown in Figure 16.9a.

b. For Example 2, point $(0, 0)$:

$$f_{xx}(x, y) = -24x \qquad \text{so} \qquad A = f_{xx}(0, 0) = 0$$
$$f_{xy}(x, y) = 6 \qquad \text{so} \qquad B = f_{xy}(0, 0) = 6$$
$$f_{yy}(x, y) = -24y \qquad \text{so} \qquad C = f_{yy}(0, 0) = 0$$
$$D = B^2 - AC = 36 - 0 > 0$$

Thus $D > 0$, so there is a saddle point at $(0, 0)$. This means that when

$$(x, y) = (0, 0),$$

$$z = f(0, 0) = 6(0)(0) - 4(0)^3 - (0)^3 - 10$$
$$= -10$$

Thus the saddle point is $(0, 0, -10)$.

From now on, it is assumed that you can find the z value for a critical point (a, b), and it will not explicitly be shown. This means that when we refer to the point $(\frac{1}{2}, \frac{1}{2})$ on the surface $z = f(x, y)$ we really mean the point $(\frac{1}{2}, \frac{1}{2}, -\frac{73}{8})$ since

$$f\left(\frac{1}{2}, \frac{1}{2}\right) = 6\left(\frac{1}{2}\right)\left(\frac{1}{2}\right) - 4\left(\frac{1}{2}\right)^3 - \left(\frac{1}{2}\right)^3 - 10$$

$$= -\frac{73}{8}$$

For Example 2, point $(\frac{1}{2}, \frac{1}{2})$:

$$A = f_{xx}(\tfrac{1}{2}, \tfrac{1}{2}) = -12$$
$$B = f_{xy}(\tfrac{1}{2}, \tfrac{1}{2}) = 6$$
$$C = f_{yy}(\tfrac{1}{2}, \tfrac{1}{2}) = -12$$
$$D = B^2 - AC = 36 - 144 < 0$$

Thus $D < 0$ and $A < 0$, so there is a relative maximum at $(\frac{1}{2}, \frac{1}{2})$. Even though you are not expected to graph this surface, it is shown in Figure 16.9 so that you can see the relative maximum and the saddle point.

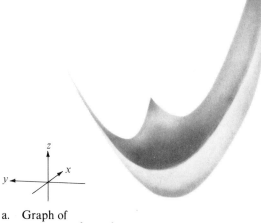

a. Graph of
$f(x, y) = 3x^2 + 4y^2 - 2xy + 22x - 30$

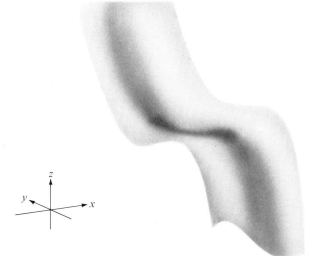

b. Graph of
$f(x, y) = 6xy - 4x^3 - 4y^3 - 10$

Figure 16.9 Graph of surfaces from Examples 1 and 2

$$x(1 - 8x^3) = 0$$
$$x(1 - 2x)(1 + 2x + 4x^2) = 0$$

Solve by setting each factor equal to zero:

$x = 0$	$1 - 2x = 0$	$1 + 2x + 4x^2 = 0$
	$x = \frac{1}{2}$	No real solution (the discriminant is negative)

Finally, find y:

$$y = 2x^2 \quad \text{so if} \quad x = 0, \quad y = 0$$
$$\text{if} \quad x = \tfrac{1}{2}, \quad y = 2(\tfrac{1}{2})^2 = \tfrac{1}{2}$$

The critical values are $(0,0)$ and $(\frac{1}{2}, \frac{1}{2})$. This means that if f has any maximums or minimums they must occur at $(x, y) = (0,0)$ or $(x, y) = (\frac{1}{2}, \frac{1}{2})$.   ■

Since there is no guarantee that a critical value is a maximum or a minimum, we need a further test, called a **second derivative test**, to find the relative maximums or minimums.

Second Derivative Test for Functions of Two Variables

If *all* of the following conditions hold:

1. $z = f(x, y)$
2. All second-order partial derivatives exist in some circular region containing (a, b) as the center
3. (a, b) is a critical point [that is, $f_x(a, b) = 0$ and $f_y(a, b) = 0$]
4. $A = f_{xx}(a, b)$, $B = f_{xy}(a, b)$, $C = f_{yy}(a, b)$, $D = B^2 - AC$

Also if:

$D < 0$ and $A < 0$, then $f(a, b)$ is a **relative maximum**

$D < 0$ and $A > 0$, then $f(a, b)$ is a **relative minimum**

$D > 0$, then f has a **saddle point** at (a, b)

$D = 0$, then the *test fails*

EXAMPLE 3   Apply the second derivative test for the critical values found in Examples 1 and 2.

Solution   **a.** For Example 1, $A = f_{xx} = 6$, $B = f_{xy} = -2$, and $C = f_{yy} = 8$, so $D = B^2 - AC = 4 - 6(8) < 0$. Thus, $D < 0$ and $A > 0$, so the point $(-4, -1)$ is a relative minimum. Even though you would not be expected to graph this surface, it is shown in Figure 16.9a.

b. For Example 2, point $(0, 0)$:

$f_{xx}(x, y) = -24x$	so	$A = f_{xx}(0,0) = 0$
$f_{xy}(x, y) = 6$	so	$B = f_{xy}(0,0) = 6$
$f_{yy}(x, y) = -24y$	so	$C = f_{yy}(0,0) = 0$

$$D = B^2 - AC = 36 - 0 > 0$$

Thus $D > 0$, so there is a saddle point at $(0, 0)$. This means that when

$$(x, y) = (0, 0),$$

$$z = f(0, 0) = 6(0)(0) - 4(0)^3 - (0)^3 - 10$$
$$= -10$$

Thus the saddle point is $(0, 0, -10)$.

From now on, it is assumed that you can find the z value for a critical point (a, b), and it will not explicitly be shown. This means that when we refer to the point $(\frac{1}{2}, \frac{1}{2})$ on the surface $z = f(x, y)$ we really mean the point $(\frac{1}{2}, \frac{1}{2}, -\frac{73}{8})$ since

$$f\left(\frac{1}{2}, \frac{1}{2}\right) = 6\left(\frac{1}{2}\right)\left(\frac{1}{2}\right) - 4\left(\frac{1}{2}\right)^3 - \left(\frac{1}{2}\right)^3 - 10$$

$$= -\frac{73}{8}$$

For Example 2, point $(\frac{1}{2}, \frac{1}{2})$:

$$A = f_{xx}(\tfrac{1}{2}, \tfrac{1}{2}) = -12$$
$$B = f_{xy}(\tfrac{1}{2}, \tfrac{1}{2}) = 6$$
$$C = f_{yy}(\tfrac{1}{2}, \tfrac{1}{2}) = -12$$
$$D = B^2 - AC = 36 - 144 < 0$$

Thus $D < 0$ and $A < 0$, so there is a relative maximum at $(\frac{1}{2}, \frac{1}{2})$. Even though you are not expected to graph this surface, it is shown in Figure 16.9 so that you can see the relative maximum and the saddle point.

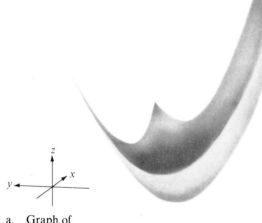

a. Graph of
$f(x, y) = 3x^2 + 4y^2 - 2xy + 22x - 30$

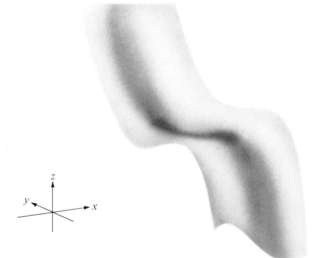

b. Graph of
$f(x, y) = 6xy - 4x^3 - 4y^3 - 10$

Figure 16.9 Graph of surfaces from Examples 1 and 2

EXAMPLE 4 A manufacturer makes two models of widgets, standard and deluxe. Market research shows the following profit equations:

$$\text{Standard} = [44{,}000 + 500(2y - 5x)]x$$
$$\text{Deluxe}\ \ \ = [500(x - y)]y$$

where x is the price of the standard model and y the price of the deluxe model. How should the widgets be priced to maximize the profit?

Solution Let $P = $ profit, then

$$P(x, y) = \text{Profit from standard widget} + \text{profit from deluxe widget}$$
$$= [44{,}000 + 500(2y - 5x)]x + [500(x - y)]y$$
$$= 44{,}000x + 1000xy - 2500x^2 + 500xy - 500y^2$$
$$= 44{,}000x + 1500xy - 2500x^2 - 500y^2$$

$$P_x(x, y) = 44{,}000 + 1500y - 5000x \quad \text{and} \quad P_y(x, y) = 1500x - 1000y$$

Set these equal to zero and solve:

$$y = \tfrac{3}{2}x \qquad \text{From } P_y(x, y) = 0$$

Substitute this into the equation for $P_x(x, y) = 0$:

$$44{,}000 + 1500(\tfrac{3}{2}x) - 5000x = 0$$
$$-2750x = -44{,}000$$
$$x = 16$$

If $x = 16$, then $y = (\tfrac{3}{2})16 = 24$; a critical value is thus $(16, 24)$. Finally, apply the second derivative test:

$$A = P_{xx} = -5000$$
$$B = P_{xy} = 1500$$
$$C = P_{yy} = -1000$$
$$D = B^2 - AC = (1500)^2 - (-5000)(-1000) < 0$$

Thus $D < 0$ and $A < 0$, so $(16, 24)$ is a relative maximum. This means that the profit is maximized if the standard model is priced at \$16 and the deluxe model is priced at \$24. ∎

Problem Set 16.3

Find the relative maximums, relative minimums, and saddle points in Problems 1–18. If the second derivative test fails, simply say so and do not continue with further analysis.

1. $f(x, y) = 3x^2 + 5xy + y^2$
2. $f(x, y) = x^2 + xy - 3x + 2y + 5$
3. $f(x, y) = x^2 + xy + y^2 - 3y$
4. $f(x, y) = x^2 + xy + 2x - 3y + 1$
5. $f(x, y) = x^3 - 3xy - y^3$
6. $f(x, y) = x^3 + y^3 - 3x - 3y$

7. $f(x, y) = 3xy - 5x^2 - y^2 + 3x - 5y - 4$
8. $f(x, y) = 4xy - 6x^2 - y^2 + 2x - 6y - 4$
9. $f(x, y) = xy + 2x - 3y - 4$
10. $f(x, y) = xy - x^2 - 2y^2 + x - y - 5$
11. $f(x, y) = x^2 - y + e^x$
12. $f(x, y) = x^2 + y - e^y$
13. $f(x, y) = xe^y$
14. $f(x, y) = ye^x$
15. $f(x, y) = 4xy - x^4 - y^4$

16. $f(x, y) = xy - x^4 - y^2$

17. $f(x, y) = x^3 + 3xy + y^3$

18. $f(x, y) = x^3 - 3xy + y^3 + 15$

APPLICATIONS

19. Miltex Corporation finds that its profit (in thousands of dollars) is given by the function

$$P(a, n) = -3a^2 - 5n^2 + 34a - 2n + 2an + 40$$

where a is the amount spent on advertising (in thousands of dollars), and n is the number of items (in thousands). Find the maximum value of P and the values of a and n that yield this maximum.

20. The labor cost (per item) for Miltex Corporation is approximated by the function

$$L(x, y) = 0.5x^2 + 5y^2 - 8x - 26y + 3xy + 55$$

where x is the number of hours of machine time, and y is the number of hours of finishing time. Find the minimum labor cost and the values of x and y that yield this minimum.

21. A manufacturer makes two models of class rings, standard and deluxe. The standard costs $50 to manufacture and the deluxe costs $110. Market research shows the follow-

ing income equations:

$$\text{Standard} = [4000 + 50(2y - 3x)]x$$
$$\text{Deluxe} = [50(x - y)]y$$

where x is the number of standard rings and y the number of deluxe rings. How should the rings be priced to maximize the profit?

22. A closed rectangular box with a volume of 8 cubic feet is made from two kinds of material. The top and bottom are made of material costing $.25 per square foot and the sides require special reinforcing that raises the cost to $.50 per square foot. What is the cost and dimensions of the box (to the nearest inch) so that the cost of materials is minimized?

23. What are the dimensions of a rectangular box, open at the top, having a volume of 64 cubic feet, and using the least amount of material for its construction?

24. The U.S. Postal Service states that a package cannot have a combined length and girth (distance around) exceeding 120 inches. What are the dimensions of the largest (in volume) box that can be mailed?

25. The second derivative test does not mention the case where $D < 0$ and $A = 0$. Show that this case is not possible.

16.4 Lagrange Multipliers

In the last section we found the relative maximums and relative minimums of functions of two variables. Sometimes the function to be maximized or minimized has one or more secondary conditions. These conditions are called **constraints**, and the function to be maximized or minimized is called the **objective function**. We begin by considering an example.

EXAMPLE 1   A university extension agricultural service concludes that on a particular farm the yield of wheat per acre (measured in bushels) is a function of the number of acre-feet of water applied, x, and the pounds of fertilizer, y, applied during the growing season according to the formula

$$f(x, y) = 140 - x^2 - 2y^2$$

where f is the yield function. Suppose that water costs $20 per acre-foot, fertilizer costs $12 per pound, and the farmer will invest $236 per acre for water and fertilizer. How much water and fertilizer should the farmer buy to maximize the yield?

Solution   In this problem the objective function is f and the constraint is

$$20x + 12y = 236$$

(Note that a more realistic constraint would be that the farmer would spend no more than $236, in which case this constraint is $20x + 12y \leq 236$; this type of problem was considered in Chapter 4.)

We could use the procedure developed in the last section and solve the equation $20x + 12y = 236$ for y and substitute this into $f(x, y)$ in order to use the second derivative test. Instead, however, we will develop a new method called the method of **Lagrange multipliers**. To use this method we first need to rewrite the constraint as a function $g(x, y) = 0$:

$$g(x, y) = 20x + 12y - 236$$

Next, a function F of three variables is defined as

$$F(x, y, \lambda) = f(x, y) + \lambda \cdot g(x, y)$$

The variable λ is the Greek letter lambda and is called the *Lagrange multiplier*. It is used to find the solution to this problem as outlined below:

Method of Lagrange Multipliers

The relative maximums and minimums of the function $z = f(x, y)$ subject to the constraint $g(x, y) = 0$ will be among those points (x, y) for which (x, y, λ) is a solution to the system

$$\begin{cases} F_x(x, y, \lambda) = 0 \\ F_y(x, y, \lambda) = 0 \\ F_\lambda(x, y, \lambda) = 0 \end{cases}$$

where

$$F(x, y, \lambda) = f(x, y) + \lambda g(x, y)$$

This method can be accomplished in four steps:

Step 1: Formulate the problem in the form of objective and constraint functions.

Maximize: $\quad f(x, y) = 140 - x^2 - 2y^2$

Subject to: $\quad g(x, y) = 20x + 12y - 236 = 0$

Step 2: Write the function $F(x, y, \lambda)$ by introducing the Lagrange multiplier λ.

$$\begin{aligned} F(x, y, \lambda) &= f(x, y) + \lambda g(x, y) \\ &= 140 - x^2 - 2y^2 + \lambda(20x + 12y - 236) \\ &= 140 - x^2 - 2y^2 + 20\lambda x + 12\lambda y - 236\lambda \end{aligned}$$

Step 3: Find the partial derivatives F_x, F_y, and F_λ and solve the system $F_x = 0$, $F_y = 0$, $F_\lambda = 0$. Solutions to this system are called *critical points for F*.

$$\begin{aligned} F_x &= -2x + 20\lambda \\ F_y &= -4y + 12\lambda \\ F_\lambda &= 20x + 12y - 236 \end{aligned}$$

These partial derivatives lead to the system of equations

$$\begin{cases} -2x + 20\lambda = 0 \\ -4y + 12\lambda = 0 \\ 20x + 12y - 236 = 0 \end{cases} \quad \text{or} \quad \begin{cases} x - 10\lambda = 0 \\ y - 3\lambda = 0 \\ 5x + 3y = 59 \end{cases}$$

Begin by solving the first pair of equations simultaneously:

$$
\begin{array}{r}
3 \quad \begin{cases} x - 10\lambda = 0 \\ y - 3\lambda = 0 \end{cases} \qquad + \qquad \begin{cases} 3x - 30\lambda = 0 \\ -10y + 30\lambda = 0 \\ \hline 3x - 10y = 0 \end{cases}
\end{array}
$$

Use this result along with the third equation of the original system to complete the solution:

$$
\begin{array}{r}
-5 \quad \begin{cases} 3x - 10y = 0 \\ 5x + 3y = 59 \end{cases} \qquad + \qquad \begin{cases} -15x + 50y = 0 \\ 15x + 9y = 177 \\ \hline 59y = 177 \\ y = 3 \end{cases}
\end{array}
$$

If $y = 3$, then $3x - 10(3) = 0$ implies that $x = 10$. Also $y - 3\lambda = 0$. This means that $3 - 3\lambda = 0$ or $\lambda = 1$. This system has only one critical point: $(x, y, \lambda) = (10, 3, 1)$.

Step 4: Evaluate $z = f(x, y)$ at each critical point. The maximum or minimum values of $f(x, y)$ will be among these values in the problem. In this example there is only one critical point.

Since $f(x, y) = 140 - x^2 - 2y^2$, we see

$$f(10, 3) = 140 - 10^2 - 2(3)^2 = 22$$

This says that the maximum yield per acre is 22 bushels. ∎

EXAMPLE 2 Minimize $f(x, y) = x^2 + y^2$ subject to $x + y - 1 = 0$.

Solution Step 1: This was done for us in the statement of the problem:

Minimize: $\qquad f(x, y) = x^2 + y^2$

Subject to: $\qquad g(x, y) = x + y - 1 = 0$

Step 2:

$$F(x, y, \lambda) = f(x, y) + \lambda g(x, y) = x^2 + y^2 + \lambda x + \lambda y - \lambda$$

Step 3:

$$F_x = 2x + \lambda$$
$$F_y = 2y + \lambda$$
$$F_\lambda = x + y - 1$$

Solve the system

$$
\begin{cases}
2x + \lambda = 0 \\
2y + \lambda = 0 \\
x + y - 1 = 0
\end{cases}
$$

to find $x = 0.5$, $y = 0.5$, $\lambda = -1$, or critical point: $(0.5, 0.5, -1)$.

Step 4:

$$f(0.5, 0.5) = 0.5^2 + 0.5^2 = 0.5$$

By checking other points near (0.5, 0.5), you can see that this point is a minimum. This situation is illustrated quite nicely in Figure 16.10.

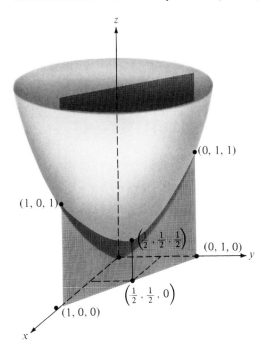

Figure 16.10 Graph of $f(x, y) = x^2 + y^2$ showing that the minimum value on this surface satisfying $x + y = 1$ is 0.5

EXAMPLE 3 Find two numbers whose sum is 100 and whose product is a maximum.

Solution *Step 1:* Let x and y be the two numbers. Then:

Maximize: $f(x, y) = xy$ That is, maximize the product

Subject to: $g(x, y) = x + y - 100 = 0$ This is the constraint; the sum must be 100

Step 2:

$$F(x, y, \lambda) = f(x, y) + \lambda g(x, y)$$
$$= xy + \lambda(x + y - 100)$$
$$= xy + \lambda x + \lambda y - 100\lambda$$

Step 3:

$$F_x = y + \lambda$$
$$F_y = x + \lambda$$
$$F_\lambda = x + y - 100$$

By solving the system

$$\begin{cases} y + \lambda = 0 \\ x + \lambda = 0 \\ x + y = 100 \end{cases}$$

simultaneously, we find that $x = 50$, $y = 50$, and $\lambda = -50$.

Step 4: Since $f(x, y) = xy$, we see that $f(50, 50) = 50(50) = 2500$. Nearby values of x and y yield smaller products, so we conclude that the maximum product is 2500, which occurs for the numbers 50 and 50. ■

The method of Lagrange multipliers can easily be extended to functions of more than two variables, as illustrated by Example 4.

EXAMPLE 4 Suppose the temperature T at any point (x, y, z) in a region of space is given by the formula $T = 5000 - (xy + xz + yz)$. Find the lowest temperature on the plane $x + y + z = 100$.

Solution *Step 1:*

Minimize: $\quad f(x, y, z) = 5000 - xy - xz - yz$

Subject to: $\quad g(x, y, z) = x + y + z - 100 = 0$

Step 2:

$$F(x, y, z, \lambda) = f(x, y, z) + \lambda g(x, y, z)$$
$$= 5000 - xy - xz - yz + \lambda(x + y + z - 100)$$
$$= 5000 - xy - xz - yz + \lambda x + \lambda y + \lambda z - 100\lambda$$

Step 3:

$$F_x = -y - z + \lambda$$
$$F_y = -x - z + \lambda$$
$$F_z = -x - y + \lambda$$
$$F_\lambda = x + y + z - 100$$

Solve the system

$$\begin{cases} y + z - \lambda = 0 \\ x + z - \lambda = 0 \\ x + y - \lambda = 0 \\ x + y + z = 100 \end{cases}$$

From the first two equations: $\quad x - y = 0$
From the first and third equations: $\quad x - z = 0$
From the fourth equation: $\quad z = 100 - x - y$; substitute this result into the equation $x - z = 0$ to find

$$x - (100 - x - y) = 0$$
$$2x + y = 100$$

Finally,

$$\begin{cases} 2x + y = 100 \\ x - y = 0 \end{cases}$$
$$3x = 100$$
$$x = \tfrac{100}{3}$$

By substitution, $y = \tfrac{100}{3}$, $z = \tfrac{100}{3}$, and $\lambda = \tfrac{200}{3}$.

Step 4:

$$f(x, y, z) = 5000 - xy - xz - yz$$

So

$$f\left(\frac{100}{3}, \frac{100}{3}, \frac{100}{3}\right) = 5000 - \left(\frac{100}{3}\right)\left(\frac{100}{3}\right) - \left(\frac{100}{3}\right)\left(\frac{100}{3}\right) - \left(\frac{100}{3}\right)\left(\frac{100}{3}\right)$$

$$= 5000 - 3\left(\frac{100}{3}\right)^2 = 5000 - \frac{10,000}{3} = \frac{5000}{3} \approx 1667$$

By checking other points on the plane, you see that this is a minimum. Thus the minimum temperature is about 1667 degrees. ∎

Problem Set 16.4

Find the relative maximums for $f(x, y)$ in Problems 1–6.

1. $f(x, y) = xy$ subject to $x + y = 20$

2. $f(x, y) = 2xy - 5$ subject to $x + y = 12$

3. $f(x, y) = -2x^2 - 3y^2$ subject to $x + 2y = 24$

4. $f(x, y) = 16 - x^2 - y^2$ subject to $x + 2y = 6$

5. $f(x, y) = x^2 y$ subject to $x + 2y = 14$

6. $f(x, y) = 4xy^2$ subject to $x - 4y = 16$

Find the relative minimums for $f(x, y)$ in Problems 7–12.

7. $f(x, y) = x^2 + y^2$ subject to $x + y = 24$

8. $f(x, y) = x^2 + y^2$ subject to $x + y = 8$

9. $f(x, y) = x^2 + y^2 - xy - 4$ subject to $x + y = 6$

10. $f(x, y) = x^2 + y^2$ subject to $x + y = 9$

11. $f(x, y) = x^2 + y^2$ subject to $x + y = 16$

12. $f(x, y) = x^2 + y^2$ subject to $2x + y = 20$

13. Find two numbers whose sum is 10 and whose product is a maximum.

14. Find two numbers whose sum is 120 and whose product is a maximum.

APPLICATIONS

15. How would the farmer of Example 1 maximize the yield if the amount spent is $100 instead of $236?

16. A wholesaler supplies two types of radio-controlled airplanes, models A and B. Suppose x units of model A and y units of model B can be supplied at a cost of

$$C(x, y) = 6x^2 + 18y^2$$

If the supplier is limited to shipping no more than 100 models, how much of each item should be supplied in order to minimize the cost? (Assume that the number shipped is exactly 100 models.)

17. A rancher needs to build a rectangular fenced enclosure. Because of a difference in terrain, one width costs $3 per foot and the other costs $9 per foot. The length costs $6 per foot. What is the maximum area that can be enclosed if the rancher has $4,000 to spend?

18. A can of Classic Coke® holds about 25 cubic inches. Find the minimum surface area of a can with a volume of 25 cubic inches. Then measure a can of Coke to compare your answer with the size of the actual can.

19. A patient is put on a 1000 calorie diet. Let us suppose (for purposes of this problem) that these calories will be supplied by two food types, meat and vegetables. Let x be the number of ounces of meat costing $.15 per ounce and y the number of ounces of vegetables costing $.03 per ounce. If the two foods produce

$$C(x, y) = 20xy$$

calories, determine what mixture of meat and vegetables should be supplied to minimize the cost.

20. A company buys x items of new equipment and uses y hours of labor at a cost of $C(x, y)$ dollars and a revenue of $R(x, y)$ dollars. The partial derivatives C_x and R_y are the marginal cost and marginal revenue with respect to x, and the partial derivatives C_y and R_y are the marginal cost and marginal revenue with respect to y. Show that if a Lagrange multiplier is used to find the maximum of $R(x, y)$ subject to the constraint $C(x, y) = $ constant, then the Lagrange multiplier is the ratio of the marginal revenue and the marginal cost with respect to each variable at the maximum.

Problems 21–24 involve functions of three variables.

21. Find the maximum value of

$$f(x, y, z) = x - y + z$$

on the sphere

$$x^2 + y^2 + z^2 = 100$$

22. Find the minimum value of

$$f(x, y, z) = x - y + z$$

on the sphere

$$x^2 + y^2 + z^2 = 100$$

23. The temperature T at point (x, y, z) in a region of space is given by the formula

$$T = 100 - xy - xz - yz$$

Find the lowest temperature on the plane

$$x + y + z = 10$$

24. Find the largest product of numbers x, y, and z such that their sum is 24.

16.5 Multiple Integrals

In this chapter we have defined functions with two or more independent variables, defined the derivative of such functions, and then looked at some applications of the derivatives. It seems reasonable to now ask how we integrate functions of two or more variables.

Integration was originally motivated by looking at antiderivatives as the reverse of the process of differentiation. With functions of several variables the question becomes one of reversing the process of partial differentiation. We write $\int f(x, y)\, dx$ to indicate that we are to differentiate with respect to x while holding y fixed, and $\int f(x, y)\, dy$ to indicate that we are to antidifferentiate $f(x, y)$ with respect to y, holding x fixed.

EXAMPLE 1 **a.** Evaluate $\int (12x^2y^3 + 2x - 6y^2)\, dx$.

b. Evaluate $\int (12x^2y^3 + 2x - 6y^2)\, dy$.

Check your results by using partial derivatives.

Solution **a.** $\int (12x^2y^3 + 2x - 6y^2)\, dx = \dfrac{12y^3x^3}{3} + \dfrac{2x^2}{2} - 6y^2x + C(y)$

$$= 4x^3y^3 + x^2 - 6xy^2 + C(y)$$

Integrate with respect to x; note that the constant of integration is a function of y since the derivative of a function of y with respect to x is 0 (remember to think of y as a constant in this context)

Check by taking the partial derivative with respect to x:

$$\frac{\partial}{\partial x}[4x^3y^3 + x^2 - 6xy^2 + C(y)]$$

$$= 4(3)x^2y^3 + 2x - 6y^2 + 0$$
$$= 12x^2y^3 + 2x - 6y^2$$

b. $\int (12x^2y^3 + 2x - 6y^2)\, dy = \dfrac{12x^2y^4}{4} + 2xy - \dfrac{6y^3}{3} + C(x)$

$$= 3x^2y^4 + 2xy - 2y^3 + C(x)$$

This time integrate with respect to y (and treat x as a constant)

Check: $\dfrac{\partial}{\partial y}[3x^2y^4 + 2xy - 2y^3 + C(x)] = 12x^2y^3 + 2x - 6y^2$ ∎

This indefinite integration of a function of two variables extends quite easily to definite integration, as shown in Example 2.

EXAMPLE 2   Evaluate

$$\int_{-1}^{4} (x^2 - xy + y^2)\, dx = \frac{x^3}{3} - \frac{x^2 y}{2} + xy^2 \Big|_{-1}^{4}$$

$$= \left(\frac{64}{3} - 8y + 4y^2 \right) - \left[\frac{-1}{3} - \frac{1}{2} y + (-1)y^2 \right]$$

$$= \frac{65}{3} - \frac{15}{2} y + 5y^2$$ ■

Notice in Example 2 that integrating and evaluating a definite integral, which is a function $f(x, y)$ with respect to x, produces a function of y alone (including constants). This integral, in turn, can be integrated with respect to y:

$$\int_{1}^{3} \left(\frac{65}{3} - \frac{15}{2} y + 5y^2 \right) dy = \frac{65}{3} y - \frac{15}{4} y^2 + \frac{5y^3}{3} \Big|_{1}^{3}$$

$$= \left(65 - \frac{135}{4} + 45 \right) - \left(\frac{65}{3} - \frac{15}{4} + \frac{5}{3} \right)$$

$$= \frac{170}{3}$$

What we have done here is a process called **double integration** that can be summarized using the notation

$$\iint_R f(x, y)\, dA$$

Surface is $z = f(x, y)$

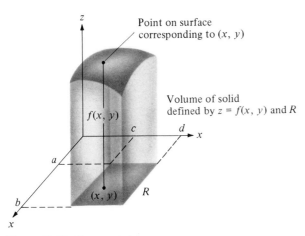

Point on surface corresponding to (x, y)

Volume of solid defined by $z = f(x, y)$ and R

$f(x, y)$

(x, y) R

Figure 16.11 Geometric interpretation of a double integral

where, in this example, $f(x, y) = x^2 - xy + y^2$, dA indicates that this is an integral over a two-dimensional region, and R is a region on which the function is defined—in this example,

$$-1 \le x \le 4$$
$$1 \le y \le 3$$

Geometrically, this process can be viewed as the volume of a solid under the surface $f(x, y)$ (where $f(x, y) \ge 0$) and directly over the rectangle R, as shown in Figure 16.11.

$$V = \iint_R (x^2 - xy + y^2)\, dA$$

Double Integral

The *double integral* of a function $f(x, y)$ over a rectangle R defined by the boundaries

$$a \leq x \leq b \quad \text{and} \quad c \leq y \leq d$$

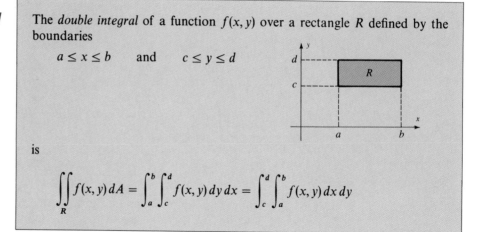

is

$$\iint\limits_{R} f(x, y)\, dA = \int_{a}^{b} \int_{c}^{d} f(x, y)\, dy\, dx = \int_{c}^{d} \int_{a}^{b} f(x, y)\, dx\, dy$$

The function $f(x, y)$ is called the **integrand** and R is called the **region of integration**. The expression dA indicates that it is an integral over a two-dimensional region. The integrals at the right are called **iterated integrals** and the order in which dx and dy are written indicates the order of integration as follows:

$$\int_{c}^{d} \int_{a}^{b} f(x, y)\, dx\, dy = \int_{c}^{d} \left[\int_{a}^{b} f(x, y)\, dx \right] dy$$

Integrate first with respect to x, then with respect to y

$$\int_{a}^{b} \int_{c}^{d} f(x, y)\, dy\, dx = \int_{a}^{b} \left[\int_{c}^{d} f(x, y)\, dy \right] dx$$

Integrate first with respect to y, then with respect to x

EXAMPLE 3 Evaluate

a. $\displaystyle\int_{0}^{3} \int_{0}^{1} (12x^2 y^3 - 4xy)\, dy\, dx$ **b.** $\displaystyle\int_{0}^{1} \int_{0}^{3} (12x^2 y^3 - 4xy)\, dx\, dy$

Solution   **a.** $\displaystyle\int_{0}^{3} \int_{0}^{1} (12x^2 y^3 - 4xy)\, dy\, dx = \int_{0}^{3} \left(\frac{12x^2 y^4}{4} - \frac{4xy^2}{2} \right) \Big|_{0}^{1} dx$ Integrate first with respect to y

$$= \int_{0}^{3} (3x^2 - 2x)\, dx$$

$$= \frac{3x^3}{3} - \frac{2x^2}{2} \Big|_{0}^{3}$$

$$= 27 - 9 = 18$$

b. $\displaystyle\int_{0}^{1} \int_{0}^{3} (12x^2 y^3 - 4xy)\, dx\, dy = \int_{0}^{1} \left(\frac{12x^3 y^3}{3} - \frac{4x^2 y}{2} \right) \Big|_{0}^{3} dy$ Integrate first with respect to x

$$= \int_{0}^{1} (108y^3 - 18y)\, dy$$

$$= \frac{108y^4}{4} - \frac{18y^2}{2} \Big|_{0}^{1}$$

$$= 27 - 9 = 18 \qquad \blacksquare$$

The result in Example 3 is the same regardless of the order of iteration. This is helpful to know when you are setting up a double integral problem, as Example 4 shows.

EXAMPLE 4 Find the volume of the function $z = x + y$ over the region R, where $0 \le x \le 2$ and $1 \le y \le 3$.

Solution We can choose either order of iteration. As a check, we will evaluate both ways.

$$\iint_R (x + y)\, dA$$

Method I:

$$\int_1^3 \int_0^2 (x + y)\, dx\, dy$$

$$= \int_1^3 \left(\frac{x^2}{2} + xy \right) \Big|_0^2 dy$$

$$= \int_1^3 (2 + 2y)\, dy$$

$$= 2y + y^2 \Big|_1^3$$

$$= 6 + 9 - 2 - 1$$

$$= 12$$

Method II:

$$\int_0^2 \int_1^3 (x + y)\, dy\, dx$$

$$= \int_0^2 \left(xy + \frac{y^2}{2} \right) \Big|_1^3 dx$$

$$= \int_0^2 \left(3x + \frac{9}{2} - x - \frac{1}{2} \right) dx$$

$$= \int_0^2 (2x + 4)\, dx$$

$$= x^2 + 4x \Big|_0^2$$

$$= 4 + 8 - 0 - 0$$

$$= 12 \qquad \blacksquare$$

Double Integrals over General Regions

The limits of integration in Examples 1–4 are constant, which corresponds to integration over a rectangular region. We now consider variable limits of integration. There are four types of regions with variable limits:

I. Variable limits of integration for y:

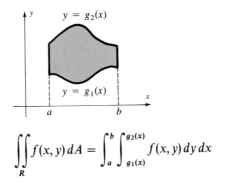

$$\iint_R f(x, y)\, dA = \int_a^b \int_{g_1(x)}^{g_2(x)} f(x, y)\, dy\, dx$$

II. Variable limits of integration for x: $\iint\limits_R f(x, y)\, dA = \int_c^d \int_{h_1(y)}^{h_2(y)} f(x, y)\, dx\, dy$

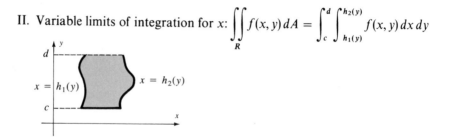

III. Variable limits of integration for either x or y:

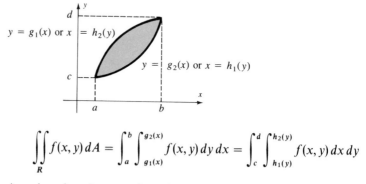

$$\iint\limits_R f(x, y)\, dA = \int_a^b \int_{g_1(x)}^{g_2(x)} f(x, y)\, dy\, dx = \int_c^d \int_{h_1(y)}^{h_2(y)} f(x, y)\, dx\, dy$$

IV. A region that does not have boundaries specified as functions. A region like this, which does not have a boundary that can easily be expressed as a function of x or a function of y, can be broken up into smaller regions that fit one of the previous three types. However, regions such as this are beyond the scope of this book.

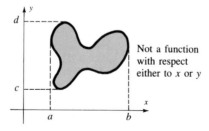

Not a function with respect either to x or y

Double Integrals over Variable Regions

Let $z = f(x, y)$ be a function of two variables. If R is a region defined by $g_1(x) \le y \le g_2(x)$, $a \le x \le b$ (type I), then

$$\iint\limits_R f(x, y)\, dA = \int_a^b \int_{g_1(y)}^{g_2(y)} f(x, y)\, dy\, dx$$

If R is a region defined by $h_1(y) \le x \le h_2(y)$, $c \le y \le d$ (type II), then

$$\iint\limits_R f(x, y)\, dA = \int_c^d \int_{h_1(x)}^{h_2(x)} f(x, y)\, dx\, dy$$

WARNING: When setting up these double integrals, remember that the variable limits of integration are on the inner integral and the constant limits of integration are on the outer integral.

EXAMPLE 5 Evaluate $f(x, y) = 6xy$ over the region shown in the figure.

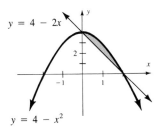

$y = 4 - 2x$

$y = 4 - x^2$

Solution This is a type I with variable limits for y, so we can set up the following integral:

$$\iint_R f(x, y)\, dA = \int_0^2 \int_{4-2x}^{4-x^2} 6xy\, dy\, dx$$

$$= \int_0^2 3xy^2 \Big|_{4-2x}^{4-x^2}\, dx$$

$$= \int_0^2 3x(4 - x^2)^2 - 3x(4 - 2x)^2\, dx$$

$$= \int_0^2 3x(16 - 8x^2 + x^4) - 3x(16 - 16x + 4x^2)\, dx$$

$$= \int_0^2 (3x^5 - 36x^3 + 48x^2)\, dx$$

$$= \left(\frac{3x^6}{6} - \frac{36x^4}{4} + \frac{48x^3}{3}\right)\Big|_0^2$$

$$= 32 - 144 + 128 - 0$$

$$= 16$$

EXAMPLE 6 Evaluate $f(x, y) = x - y$ over the region shown in the figure.

Solution This is a type II with variable limits for x:

$$\int_0^2 \int_{y^2}^{6-y} (x - y)\, dx\, dy = \int_0^2 \left(\frac{x^2}{2} - xy\right)\Big|_{y^2}^{6-y}\, dy$$

$$= \int_0^2 \left[\frac{1}{2}(6 - y)^2 - (6 - y)y - \frac{1}{2}(y^2)^2 + y^3\right] dy$$

$$= \int_0^2 \left(-\frac{1}{2}y^4 + y^3 + \frac{3}{2}y^2 - 12y + 18\right) dy$$

$$= -\frac{y^5}{10} + \frac{y^4}{4} + \frac{3y^3}{6} - \frac{12y^2}{2} + 18y \Big|_0^2$$

$$= -\frac{32}{10} + \frac{16}{4} + \frac{24}{6} - \frac{48}{2} + 36 - 0$$

$$= 16\tfrac{4}{5}$$

$x = 6 - y$

$x = y^2$

One of the hardest parts in evaluating a double integral is deciding which variable to integrate with respect to first, and which second. After that, the next step is deciding the limits of integration. These steps are illustrated for the regions we used in Examples 5 and 6. Example 7, on the other hand, can be integrated both ways.

Problem: Evaluate $\iint\limits_{R} f(x, y)\, dA.$

Examples:

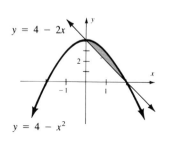

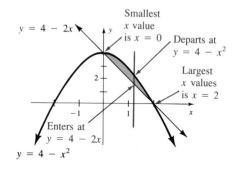

Figure 16.12

Steps: *Type I.* Integrate first with respect to y, then with respect to x:

1: Imagine a vertical line L cutting through R in the direction of increasing y.
2: Integrate from the y value where L enters R to the y value where L departs R (these should be functions of x; that is, solve for y).
3: Choose the x limits that include all the vertical lines that pass through R.

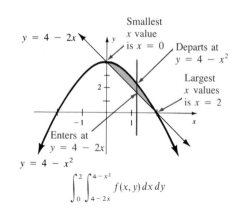

Type II. Integrate first with respect to x, then with respect to y:

1: Imagine a horizontal line L cutting through R in the direction of increasing x.
2: Integrate from the x value where L enters R to the x value where L departs R (these should be functions of y; that is, solve for x).
3: Choose the y limits that include all the horizontal lines that pass through R.

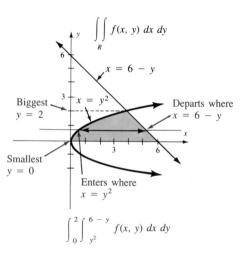

EXAMPLE 7 Evaluate $f(x, y) = x + y$ two different ways over the region shown in the figure.

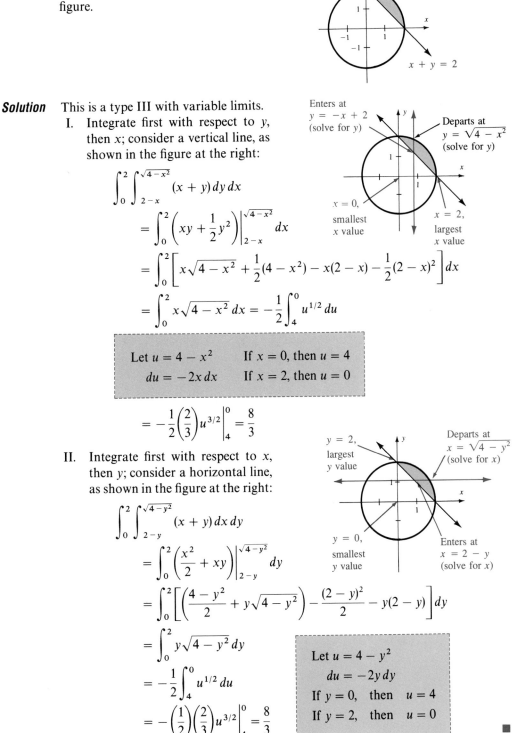

$x^2 + y^2 = 4$

$x + y = 2$

Solution This is a type III with variable limits.

I. Integrate first with respect to y, then x; consider a vertical line, as shown in the figure at the right:

Enters at
$y = -x + 2$
(solve for y)

Departs at
$y = \sqrt{4 - x^2}$
(solve for y)

$x = 0$,
smallest
x value

$x = 2$,
largest
x value

$$\int_0^2 \int_{2-x}^{\sqrt{4-x^2}} (x + y)\, dy\, dx$$

$$= \int_0^2 \left(xy + \frac{1}{2}y^2 \right)\Big|_{2-x}^{\sqrt{4-x^2}} dx$$

$$= \int_0^2 \left[x\sqrt{4 - x^2} + \frac{1}{2}(4 - x^2) - x(2 - x) - \frac{1}{2}(2 - x)^2 \right] dx$$

$$= \int_0^2 x\sqrt{4 - x^2}\, dx = -\frac{1}{2}\int_4^0 u^{1/2}\, du$$

> Let $u = 4 - x^2$ If $x = 0$, then $u = 4$
> $du = -2x\, dx$ If $x = 2$, then $u = 0$

$$= -\frac{1}{2}\left(\frac{2}{3} \right) u^{3/2}\Big|_4^0 = \frac{8}{3}$$

II. Integrate first with respect to x, then y; consider a horizontal line, as shown in the figure at the right:

$y = 2$,
largest
y value

Departs at
$x = \sqrt{4 - y^2}$
(solve for x)

$y = 0$,
smallest
y value

Enters at
$x = 2 - y$
(solve for x)

$$\int_0^2 \int_{2-y}^{\sqrt{4-y^2}} (x + y)\, dx\, dy$$

$$= \int_0^2 \left(\frac{x^2}{2} + xy \right)\Big|_{2-y}^{\sqrt{4-y^2}} dy$$

$$= \int_0^2 \left[\left(\frac{4 - y^2}{2} + y\sqrt{4 - y^2} \right) - \frac{(2 - y)^2}{2} - y(2 - y) \right] dy$$

$$= \int_0^2 y\sqrt{4 - y^2}\, dy$$

$$= -\frac{1}{2}\int_4^0 u^{1/2}\, du$$

> Let $u = 4 - y^2$
> $du = -2y\, dy$
> If $y = 0$, then $u = 4$
> If $y = 2$, then $u = 0$

$$= -\left(\frac{1}{2} \right)\left(\frac{2}{3} \right) u^{3/2}\Big|_4^0 = \frac{8}{3}$$

Reversing the Order of Integration

As Examples 5–7 show, sometimes we integrate first with respect to y (using a vertical line), sometimes with respect to x (using a horizontal line), and other times it is possible to do it either way. If the integration can be done either way, we can change the order of integration. Over a rectangular region, it is simply a matter of interchanging the numerical limits of integration, but when interchanging the order of integration with variable limits, we need to look at the region R more carefully, as Example 8 shows.

EXAMPLE 8   Reverse the order of integration for

$$\int_{2}^{11} \int_{\sqrt{y-2}}^{3} f(x, y)\, dx\, dy$$

Solution   You can see that this integral is integrated first with respect to x (so a horizontal line should be used). This means that to find R, you write

$$x = 3 \qquad \text{and} \qquad x = \sqrt{y-2}$$

and graph these curves as shown.

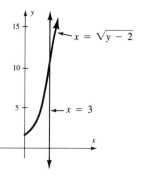

Next, install the limits of integration for y, namely,

$$y = 2 \qquad \text{and} \qquad y = 11$$

Shade R and reverse the order of integration. Then solve each equation for y and draw a vertical line to find the limits of integration.

$$x = \sqrt{y-2}$$
$$x^2 = y - 2$$
$$y = x^2 + 2$$
Depart at
$$y = x^2 + 2$$

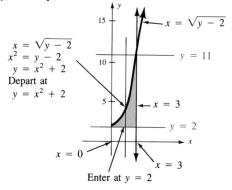

It enters at $y = 2$ and departs at $y = x^2 + 2$; the least value for x is 0 and the largest value for x is 3, so the limits can now be reversed:

$$\int_{0}^{3} \int_{2}^{x^2+2} f(x, y)\, dy\, dx$$

Volume

As stated earlier, a geometrical interpretation of double integration is that of a **volume** under a surface. That idea is now formalized:

Volume Let $z = f(x, y)$ be a nonnegative surface over a region R, as illustrated by Figure 16.13. The volume of the solid under the surface of f and over the region R is

$$\iint_R f(x, y) \, dA$$

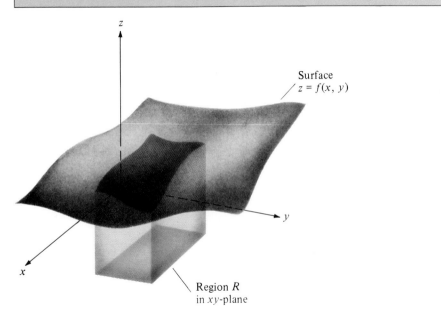

Surface
$z = f(x, y)$

Region R
in xy-plane

Figure 16.13

EXAMPLE 9   Find the volume of the solid bounded on the bottom by the curve with equation $y = x^2$ and the line $2x - y = 0$ and on the top by the plane $z = 10$.

Solution   First draw the region R in two dimensions. (The three-dimensional figure is for reinforcement only; you do not need to draw it for your work.)
We will integrate with respect to y first:

$$\iint_R z \, dA = \int_0^4 \int_{2x}^{x^2} 10 \, dy \, dx = \int_0^4 10y \Big|_{2x}^{x^2} dx$$

$$= \int_0^4 10(x^2 - 2x) \, dx$$

$$= \frac{10x^3}{3} - 10x^2 \Big|_0^4$$

$$= \frac{640}{3} - 160 - 0 = \frac{160}{3}$$

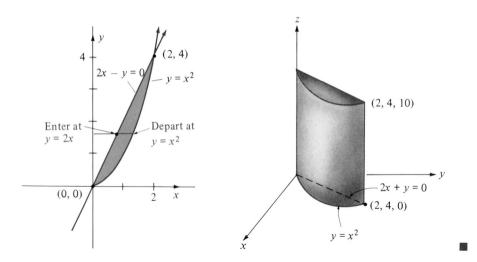

Problem Set 16.5

Evaluate the integrals in Problems 1–12.

1. $\displaystyle\int_0^2 (x^2y + y^2)\,dy$

2. $\displaystyle\int_0^2 (x^2y + y^2)\,dx$

3. $\displaystyle\int_1^4 (xy^2 - x)\,dx$

4. $\displaystyle\int_1^4 (xy^2 - x)\,dy$

5. $\displaystyle\int_0^4 x\sqrt{x^2 + 2y}\,dy$

6. $\displaystyle\int_0^4 x\sqrt{x^2 + 2y}\,dx$

7. $\displaystyle\int_1^2 3e^{x + 2y}\,dx$

8. $\displaystyle\int_1^2 3e^{x + 2y}\,dy$

9. $\displaystyle\int_1^5 xe^{x^2 + 3y}\,dx$

10. $\displaystyle\int_1^5 xe^{x^2 + 3y}\,dy$

11. $\displaystyle\int_1^{e^x} \frac{x}{y}\,dy$

12. $\displaystyle\int_1^{e^x} \frac{x}{y}\,dx$

Evaluate the iterated integrals in Problems 13–24, and draw the region R defined by the double integral.

13. $\displaystyle\int_1^4 \int_0^3 dy\,dx$

14. $\displaystyle\int_{-2}^3 \int_0^4 dy\,dx$

15. $\displaystyle\int_{-3}^2 \int_2^4 dx\,dy$

16. $\displaystyle\int_2^5 \int_{-1}^3 dx\,dy$

17. $\displaystyle\int_1^3 \int_0^y (x + 3y)\,dx\,dy$

18. $\displaystyle\int_0^2 \int_{y^2}^{\frac{1}{2}y + 1} xy\,dx\,dy$

19. $\displaystyle\int_0^2 \int_0^{y^2} xy\,dx\,dy$

20. $\displaystyle\int_0^1 \int_{y^2}^{(y+3)^{1/2}} 2xy\,dx\,dy$

21. $\displaystyle\int_0^5 \int_0^x (2x + y)\,dy\,dx$

22. $\displaystyle\int_1^3 \int_1^{x^2} 3xy\,dy\,dx$

23. $\displaystyle\int_0^1 \int_{x^3}^{x^2} x^2y\,dy\,dx$

24. $\displaystyle\int_{-0.5}^2 \int_{x^2}^{\frac{3}{2}x + 1} 4xy\,dy\,dx$.

Evaluate the integrals in Problems 25–30 by first reversing the order of integration. (Note that these are the odd-numbered Problems 13–23.)

25. $\displaystyle\int_1^4 \int_0^3 dy\,dx$

26. $\displaystyle\int_{-3}^2 \int_2^4 dx\,dy$

27. $\displaystyle\int_1^3 \int_0^y (x + 3y)\,dx\,dy$

28. $\displaystyle\int_0^2 \int_0^{y^2} xy\,dx\,dy$

29. $\displaystyle\int_0^5 \int_0^x (2x + y)\,dy\,dx$

30. $\displaystyle\int_0^1 \int_{x^3}^{x^2} x^2y\,dy\,dx$

Evaluate each double integral over the region R in Problems 31–38.

31. $\displaystyle\iint_R (x^2 + y)\,dx\,dy$ $0 \le x \le 2$
                                       $0 \le y \le 3$

32. $\displaystyle\iint_R (x + 2y)\,dy\,dx$ $-1 \le x \le 2$
                                        $1 \le y \le 3$

33. $\displaystyle\iint_R \frac{dy\,dx}{x}$ $1 \le x \le 3$
                                 $0 \le y \le 1 - x$

34. $\displaystyle\iint_R (3 - 3x^2)\,dy\,dx$ $0 \le x \le 2$
                                     $0 \le y \le -\frac{9}{2}x$

35. $\displaystyle\iint_R y^3 e^{xy}\,dx\,dy$ $0 \le x \le y^2$
                                 $1 \le y \le 2$

36. $\displaystyle\iint_R e^{x+y}\,dx\,dy \qquad \begin{array}{l} 0 \le x \le 2y \\ 0 \le y \le 1 \end{array}$

37. $\displaystyle\iint_R xy\sqrt{x^2+y^2}\,dx\,dy \qquad \begin{array}{l} 1 \le x \le 3 \\ 2 \le y \le 4 \end{array}$

38. $\displaystyle\iint_R xy\sqrt{x^2+y^2}\,dy\,dx \qquad \begin{array}{l} 0 \le x \le 4 \\ 1 \le y \le 3 \end{array}$

Evaluate the integrals in Problems 39–42 two ways.

39. $\displaystyle\iint_R (x+y)\,dA, \quad R$ bounded by $x = y^2$ and $y = x^2$

40. $\displaystyle\iint_R xy\,dA, \quad R$ bounded by $y = x^2$ and $y = 2x$

41. $\displaystyle\iint_R 2xy\,dA, \quad R$ bounded by $x + y = 5, x = 0$, and $y = 0$

42. $\displaystyle\iint_R 3xy^2\,dA, \quad R$ bounded by $y = 2x, y = 3 - x$, and $y = 0$

Find the volume of the solid bounded by the surface $z = f(x, y)$ about the region R as specified in Problems 43–48.

43. $z = 5; \quad R: \quad 1 \le x \le 5, -3 \le y \le 2$

44. $z = x + y; \quad R: \quad -1 \le x \le 2, 0 \le y \le 4$

45. $z = xy\sqrt{x^2+y^2}; \quad R: \quad 0 \le x \le 1, 0 \le y \le 1$

46. $z = x^2y; \quad R: \quad 0 \le x \le 2, 0 \le y \le 4$

47. $z = 5xy; \quad R$ bounded by $x = y^2$ and $y = x^2$

48. $z = 6; \quad R$ bounded by $y = x^2$ and $y = 3x$

49. The *average value* of the function $z = f(x, y)$ over the rectangle $R: \quad a \le x \le b, c \le y \le d$ is

$$\frac{1}{(b-a)(d-c)}\iint_R f(x,y)\,dA$$

Find the average value of the surface $z = f(x, y) = x - 2y$ over the rectangle $0 \le x \le 2, 0 \le y \le 3$.

50. Find the average value of $f(x, y) = 4 - x - y$ over the rectangle $0 \le x \le 2, 0 \le y \le 1$. (See Problem 49 for the definition of average value.)

APPLICATIONS

51. If an industry invests L thousand labor-hours $(0 \le L \le 5)$ and K million dollars $(0 \le K \le 2)$ in the production of P thousand units of a commodity, then P is given by

$$P(L, K) = L^{0.75}K^{0.25}$$

This is called a *Cobb-Douglas production function.** Find the average number of units produced for the indicated ranges of x and y. (See Problem 49 for a definition of average value.)

16.6 Summary and Review

IMPORTANT TERMS

Circular cylinders [16.1]
Constraints [16.4]
Coordinate planes [16.1]
Critical point [16.3]
Dependent variable [16.1]
Domain of a function of several
　variables [16.1]
Double integration [16.5]
First octant [16.1]
Function of n variables [16.1]
Function of two independent
　variables [16.1]
Higher-order partial
　derivatives [16.2]
Integrand [16.5]
Iterated integral [16.5]

Lagrange multiplier [16.4]
Multiple integral [16.5]
Objective function [16.4]
Ordered triplet [16.1]
Partial derivative [16.2]
Plane [16.1]
Quadric surface [16.1]
Range of a function of several
　variables [16.1]
Region of integration [16.5]
Relative maximum [16.3]
Relative minimum [16.3]
Right circular cylinder [16.1]
Saddle point [16.3]
Second derivative test for functions of
　two variables [16.3]

---

* You are asked to do some research on this model in the modeling application at the end of this chapter.

Surface [16.1]   Volume [16.5]
Three-dimensional coordinate xy-Plane [16.1]
   system [16.1] xz-Plane [16.1]
Trace [16.1] yz-Plane [16.1]

SAMPLE TEST   *For additional practice there are a large number of review problems categorized by objective in the Student Solutions Manual. The following sample test (40 minutes) is intended to review the main ideas of this chapter.*

1. Evaluate $P(L, K) = L^{0.4}K^{0.2}$ at the point $(10, 50)$.

2. Plot the following points in three dimensions.
      a. $(0, 3, 0)$ **b.** $(4, -2, 5)$ **c.** $(2, 6, 3)$

3. Graph the surface $2x + y + z = 8$.

4. Graph the surface $y^2 + z^2 - x^2 = 0$.

Find the requested derivatives or integrals in Problems 5–10.

5. f_x for $f(x, y) = (x - 3y)^{10}$

6. f_λ for $f(x, y, \lambda) = 4xy + \lambda(2x + 3y - 50)$

7. f_{xx} for $f(x, y) = 4x^4 - 3x^3y^2 + 2x^2 - 250$

8. $\displaystyle\int_0^1 x^2y^2\sqrt{x^3 + 8}\,dx$

9. $\displaystyle\int_0^9 \int_{\sqrt{x}}^{\sqrt{x}+2} y\,dy\,dx$

10. $\displaystyle\int_0^3 \int_0^{y\sqrt{9-y^2}} dx\,dy$

11. Find the relative maximums, relative minimums, and saddle points of $f(x, y) = 2x^2 - 3xy + y^2$.

12. Use the method of Lagrange multipliers to find two numbers whose sum is 250 and whose product is a maximum.

13. Evaluate

$$\int_1^{e^2} \int_{\ln y}^{2} dx\,dy$$

by reversing the order of integration.

14. Find the volume of the region R bounded by $y = x^2$ and $y = 5x$.

$$\iint_R (x + 2y)\,dA$$

15. A company produces two types of skateboards, standard and competition. The weekly demand and cost equations are:

$$p = 50 - 0.05x + 0.001y$$
$$q = 130 + 0.01x - 0.04y$$

where $\$p$ is the price of the standard skateboard and $\$q$ is the price of the competition model for x, the weekly demand (in hundreds) for standard skateboards, and y, the weekly demand (in hundreds) for competition skateboards. If the cost function is

$$C(x, y) = 90 + 20x + 90y$$

find the revenue function, R, and evaluate $C(5, 8)$ and $R(5, 8)$.

Modeling Application 11

The Cobb-Douglas Production Function

Karlin Corporation manufactures only one product, which is sold at the price P_0. The firm employs a labor force L, which must be paid an average wage p_1. The firm also requires capital K in terms of tools, buildings, and so forth. The cost of using one unit of capital is p_2. Karlin wishes to maximize its profit. Determine what data need to be collected and construct a model that will accomplish Karlin's goal of maximizing profit.

Developing a mathematical model is no easy task. As you have realized by now, there is usually a great deal of work involved. Every good model should include listing the assumptions, translating the assumptions into mathematical notation, building the model, and then checking the model against tabulated data. Read "The Cobb-Douglas Production Function" by Robert Geitz (*The UMAP Journal*, Unit 509. ©1981 Education Development Center, Inc.) This paper is a perfect illustration of the modeling process. After reading it, develop a model to accomplish Karlin Corporation's goal.

APPENDIXES

APPENDIX A   Logic

APPENDIX B   Mathematical Induction

APPENDIX C   Calculators

APPENDIX D   Computers

APPENDIX E   Tables

APPENDIX F   Markov Chains

APPENDIX G   Answers to Odd-Numbered Problems

APPENDIX A

Logic

Logic is a method of reasoning that accepts only inescapable conclusions. This is possible because of the strict way every concept is *defined*, or accepted without definition. So, we begin by defining a crucial term: a **statement**:

Statement

> A *statement* is a meaningful sentence that is either true or false, but not both true and false.

Statements can be either **simple** or **compound**. Simple statements are those without **connectives**—words such as *not, and, or, neither ... nor, if ... then, either ... or, unless, because*, and so on; while compound statements have at least one connective. A compound statement is formed by using connectives to combine simple statements. Because of the definition of a statement, the **truth value** of *any* statement labels it as either true (T) or false (F). The true value of a *compound* statement depends on the truth values of its component parts.

Letters such as $p, q, r, s, \ldots$ are used to denote simple statements, and then certain connectives are defined. There are three **fundamental connectives**: **conjunction** (*and*), **disjunction** (*or*), and **negation** (*not*). These connectives are defined by a **truth table**, which lists all possibilities. The truth values for each connective are chosen to correspond to everyday usage.

Fundamental Connectives

Conjunction:	*p and q*	symbolized by	$p \wedge q$
Disjunction:	*p or q*	symbolized by	$p \vee q$
Negation:	*not p*	symbolized by	$\sim p$

Truth table definition

p	q	$p \wedge q$	$p \vee q$	$\sim p$
T	T	T	T	F
T	F	F	T	F
F	T	F	T	T
F	F	F	F	T

The definition of the fundamental connectives, along with a truth table, are used to determine the truth or falsity of a compound statement.

EXAMPLE 1   Construct a truth table for the compound statement:

*Alfie did **not** come last night **and** he did **not** pick up his money.*

Solution   Let

   p: Alfie came last night
   q: Alfie picked up his money

Then the statement can be written $\sim p \wedge \sim q$. To begin, list all the possible combinations of truth values for the simple statements p and q:

p	*q*	
T	T	
T	F	
F	T	
F	F	

Insert the truth values for $\sim p$ and $\sim q$:

p	*q*	$\sim p$	$\sim q$	
T	T	F	F	
T	F	F	T	
F	T	T	F	
F	F	T	T	

Finally, insert the truth values for $\sim p \wedge \sim q$:

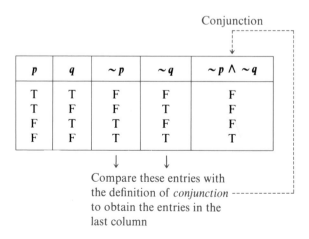

Conjunction

p	*q*	$\sim p$	$\sim q$	$\sim p \wedge \sim q$
T	T	F	F	F
T	F	F	T	F
F	T	T	F	F
F	F	T	T	T

Compare these entries with
the definition of *conjunction*
to obtain the entries in the
last column

The only time the compound statement is true is when *both p* and *q* are false.   ■

EXAMPLE 2   Construct a truth table for $\sim(\sim p)$.

Solution

Negation

p	$\sim p$	$\sim(\sim p)$
T	F	T
F	T	F

Look at this column. Use
the definition of *negation*
for all the truth values in
this column

■

Note that $\sim(\sim p)$ and p have the same truth values. If two statements have the same truth values, one can replace the other in any logical expression. This means that the double negative of a statement is the same as the original statement.

Law of Double Negation

$\sim(\sim p)$ may be replaced by p in any logical expression.

Truth tables can be used to prove certain useful results and to introduce some additional operators. The first one we will consider is called the **conditional**. The statement "if p, then q" is called a *conditional statement*. It is symbolized by $p \to q$; p is called the **antecedent**, and q is called the **consequent**.

Definition of the Conditional

The *conditional* is defined by the following truth table:

p	q	$p \to q$
T	T	T
T	F	F
F	T	T
F	F	T

The *if* part of an implication need not be stated first. All the following statements have the same meaning:

Conditional translation	*Example*
If p, then q	If you are 18, then you can vote.
q, if p	You can vote, if you are 18.
p, only if q	You are 18 only if you can vote.
$p \subseteq q$	p is a subset of q.

Some statements, although not originally written as a conditional, can be put into if–then form. For example, "All ducks are birds" can be rewritten as "If it is a duck, then it is a bird." Thus we add one more form to the list:

All *p* are *q*　　　　　　　　All 18-year-olds can vote.

EXAMPLE 3 Translate the following sentence (which appeared on the 1986 federal income tax form) into symbolic form:

If you received capital gains distributions for the year and you do not need Schedule D to report any other gains or losses, do not file that schedule.

Solution *Step 1:* Isolate the simple statements and assign them variables. Let

c:　You received capital gains distributions for the year

g:　You need Schedule D to report other gains for the year

l:　You need Schedule D to report other losses for the year

f:　You need to file Schedule D

Your choice of variables is, of course, arbitrary, but you must be careful not to let variables represent compound statements

Step 2: Rewrite the sentence, making substitutions for the variables.

If *c* and (~*g* or ~*l*), then ~*f*

Step 3: Complete the translation to symbols:

$$[c \wedge (\sim g \vee \sim l)] \rightarrow \sim f$$ ∎

Related to the conditional *p* → *q* are other statements, which are defined below.

Converse, Inverse, and Contrapositive

Given the conditional *p* → *q*, we define:

Converse:　　　　　*q* → *p*
Inverse:　　　　　~*p* → ~*q*
Contrapositive:　　~*q* → ~*p*

Not all these statements are equivalent, as shown by Table A.1.

TABLE A.1

p	*q*	~*p*	~*q*	**Statement** *p* → *q*	**Converse** *q* → *p*	**Inverse** ~*p* → ~*q*	**Contrapositive** ~*q* → ~*p*
T	T	F	F	T	T	T	T
T	F	F	T	F	T	T	F
F	T	T	F	T	F	F	T
F	F	T	T	T	T	T	T

Note that, if the conditional is true, the converse and inverse are not necessarily true. However, the contrapositive is always true if the original conditional is true.

Law of Contraposition

> A conditional may always be replaced by its contrapositive without having its truth value affected.

EXAMPLE 4   Consider the following:

    p: You obey the law.
    q: You will go to jail.

Given $p \to \sim q$, write the converse, inverse, and contrapositive.

Solution

Statement:	$p \to \sim q$	*Converse*:	$\sim q \to p$
Inverse:	$\sim p \to q$	*Contrapositive*:	$q \to \sim p$

The double negative $\sim(\sim q)$ is replaced by q in the inverse and contrapositive. Make this simplification whenever possible

Statement:	$p \to \sim q$	If you obey the law, then you will not go to jail.
Contrapositive:	$q \to \sim p$	If you go to jail, then you did not obey the law.
Converse:	$\sim q \to p$	If you do not go to jail, then you obey the law.
Inverse:	$\sim p \to q$	If you do not obey the law, then you will go to jail.

       The power of logic is in drawing conclusions not explicitly stated or known. There are three main types of reasoning that allow us to draw such conclusions. These are *direct reasoning, indirect reasoning*, and *transitivity*.

       Direct reasoning is the drawing of a conclusion from two premises. The following example, called a **syllogism**, is an illustration:

$p \to q$	If you receive an A on the final, then you will pass the course.
p	You receive an A on the final.
q	You pass the course.

This argument consists of two *premises*, or *hypotheses*, and a *conclusion*; the argument is valid if

$$[(p \to q) \land p] \to q$$

is always true, as shown in Table A.2.

TABLE A.2
Truth table for direct reasoning

p	q	$p \to q$	$(p \to q) \land p$	$[(p \to q) \land p] \to q$
T	T	T	T	T
T	F	F	F	T
F	T	T	F	T
F	F	T	F	T

The pattern of argument is illustrated below:

Direct Reasoning

Major premise:	$p \to q$	
Minor premise:	p	
Conclusion:	$\therefore q$	Three dots ($\therefore$) are used to mean *therefore*

This reasoning is also sometimes called *modus ponens, law of detachment*, or *assuming the antecedent*.

EXAMPLE 5

If you play chess, then you are intelligent.
You play chess.

$\therefore$ You are intelligent. ∎

EXAMPLE 6

If you are a logical person, then you will understand this example.
You are a logical person.

$\therefore$ You understand this example. ∎

We say that these arguments are *valid* since we recognize them as direct reasoning. The following syllogism illustrates what we call **indirect reasoning**:

$p \to q$	If you receive an A on the final, then you will pass the course.
$\sim q$	You did not pass the course.
$\therefore \sim p$	$\therefore$ You did not receive an A on the final.

We can prove this result is valid by direct reasoning as follows:

$$[(\sim q \to \sim p) \land \sim q] \to \sim p$$

Also, since $\sim q \to \sim p$ is logically equivalent to the contrapositive $p \to q$, we have

$$[(p \to q) \land \sim q] \to \sim p$$

We can also prove that indirect reasoning is valid by using a truth table, and you are asked to do this in the problem set. The pattern of argument is illustrated below:

Indirect Reasoning

Major premise:	$p \to q$	
Minor premise:	$\sim q$	
Conclusion:	$\therefore \sim p$	

This method is also known as reasoning by *denying the consequent* or as *modus tollens*.

EXAMPLE 7

If the cat takes the rat, then the rat will take the cheese.
The rat does not take the cheese.

$\therefore$ The cat does not take the rat. ∎

EXAMPLE 8 If you received an A on the test, then I am Napoleon.

I am not Napoleon.

∴ You did not receive an A on the test. ■

Sometimes we must consider some extended arguments. **Transitivity** allows us to reason through several premises to some conclusion. The argument form is as follows:

Transitivity

Premise: $p \to q$

Premise: $q \to r$

Conclusion: $\therefore p \to r$

Transitivity is proved by a truth table, as Table A.3 shows.

TABLE A.3
Truth table for transitivity

p	q	r	$p \to q$	$q \to r$	$p \to r$	$(p \to q) \wedge (q \to r)$	$[(p \to q) \wedge (q \to r)] \to (p \to r)$
T	T	T	T	T	T	T	T
T	T	F	T	F	F	F	T
T	F	T	F	T	T	F	T
T	F	F	F	T	F	F	T
F	T	T	T	T	T	T	T
F	T	F	T	F	T	F	T
F	F	T	T	T	T	T	T
F	F	F	T	T	T	T	T

EXAMPLE 9 If you attend class, then you will pass the course. $p \to q$

If you pass the course, then you will graduate. $q \to r$

∴ If you attend class, then you will graduate. $\therefore p \to r$

Transitivity can be extended so that a chain of several if–then sentences is connected together. For example, we could continue:

If you graduate, then you will get a good job.

If you get a good job, then you will meet the right people.

If you meet the right people, then you will become well-known.

∴ If you attend class, then you will become well-known. ■

Several of these argument forms may be combined into one argument. Remember:

Procedure for Drawing
Logical Conclusions

1. Translate into symbols.
2. Simplify the symbolic argument. Replace a statement by its contrapositive, use direct or indirect reasoning, or use transitivity.
3. Translate the conclusion back into words.

EXAMPLE 10 Form a valid conclusion using all these statements:

1. If I receive a check for $500, then we will go on vacation.
2. If the car breaks down, then we will not go on vacation.
3. The car breaks down.

Solution First, change into symbolic form:

1. $c \to v$	where	c = I receive a $500 check
2. $b \to \sim v$		v = We will go on vacation
3. b		b = Car breaks down

Next, simplify the symbolic argument. In this example the premises must be rearranged:

2. $b \to \sim v$	Second premise
1. $c \to v$	First premise
3. b	

Now $c \to v$ is the same as $\sim v \to \sim c$ (replace the statement by its contrapositive). That is, if 1 is replaced by $1'$ $\sim v \to \sim c$, the following argument is obtained:

2.	$b \to \sim v$	Second premise
$1'$.	$\sim v \to \sim c$	Contrapositive of first premise
	$\therefore b \to \sim c$	Transitive
3.	b	Third premise
	$\therefore \sim c$	Direct reasoning

Finally, translate the conclusion back into words: "I did not receive a check for $500." ∎

Problem Set A

According to the definition, which of the examples in Problems 1–4 are statements?

1. **a.** Hickory, Dickory, Dock, the mouse ran up the clock.
   b. $3 + 5 = 9$ **c.** Is John ugly?
   d. John has a wart on the end of his nose.

2. **a.** March 13, 1984, is Monday. .
   b. Division by zero is impossible.
   c. Logic is not as difficult as I had anticipated.
   d. $4 - 6 = 2$

3. **a.** $6 + 9 \neq 7 + 8$
   b. Thomas Jefferson was the twenty-third president.
   c. Sit down and be quiet!
   d. If wages continue to rise, then prices will also rise.

4. **a.** John and Mary were married on August 3, 1979.
   b. $6 + 12 \neq 10 + 8$
   c. Do not read this sentence.
   d. Do you have a cold?

Translate the statements in Problems 5–15 into symbols. For each simple statement, be sure to indicate the meanings of the symbols you use. Answers are not unique.

5. W. C. Fields is eating, drinking, and having a good time.

6. Sam will not seek and will not accept the nomination.

7. Jack will not go tonight and Rosamond will not go tomorrow.

8. Fat Albert lives to eat and does not eat to live.

9. The decision will depend on judgment or intuition, and not on who paid the most.

10. Everything happens to everybody sooner or later if there is time enough. (G. B. Shaw)

11. We are not weak if we make a proper use of those means which the God of Nature has placed in our power. (Patrick Henry)

12. A useless life is an early death. (Goethe)

13. All work is noble. (Thomas Carlyle)

14. Everything's got a moral if only you can find it. (Lewis Carroll)

15. You don't have to itemize deductions on Schedule A or complete the worksheet if you have earned income of $2,300 or more. (1983 tax form)

Construct a truth table for the statements in Problems 16–39.

16. $\sim p \lor q$ **17.** $\sim p \land \sim q$ **18.** $\sim(p \land q)$

19. $\sim r \lor \sim s$ **20.** $\sim(\sim r)$ **21.** $(r \land s) \lor \sim s$

22. $p \land \sim q$ **23.** $\sim p \lor \sim q$

24. $(\sim p \land q) \lor \sim q$ **25.** $(p \land \sim q) \land p$

26. $(\sim p \lor q) \land (q \land p)$ **27.** $(p \lor q) \lor (p \land \sim q)$

28. $p \lor (p \to q)$ **29.** $p \to (\sim p \to q)$

30. $(p \land q) \to p$ **31.** $\sim p \to \sim (p \land q)$

32. $(p \land q) \land (p \to \sim q)$ **33.** $(p \to p) \to (q \to \sim q)$

34. $(p \to \sim q) \to (q \to \sim p)$ **35.** $(p \to q) \to (\sim q \to \sim p)$

36. $[p \land (p \lor q)] \to p$ **37.** $(p \land q) \land \sim r$

38. $[(p \lor q) \land \sim r] \land r$ **39.** $[p \land (q \lor \sim p)] \lor r$

40. Prove indirect reasoning by using a truth table.

Name the type of reasoning illustrated by the arguments in Problems 41–48.

41. If I inherit \$1,000, I will buy you a cookie.
I inherit \$1,000.
Therefore, I will buy you a cookie.

42. All snarks are fribbles.
All fribbles are ugly.
Therefore, all snarks are ugly.

43. If you understand a problem, it is easy.
The problem is not easy.
Therefore, you do not understand the problem.

44. If I do not get a raise in pay, I will quit.
I do not get a raise.
Therefore, I quit.

45. If Fermat's last theorem is ever proved, then my life is complete.
My life is not complete.
Therefore, Fermat's last theorem is not proved.

46. All mathematicians are eccentrics.
All eccentrics are rich.
Therefore, all mathematicians are rich.

47. No students are enthusiastic.
You are enthusiastic.
Therefore, you are not a student.

48. If you like beer, you will like Bud.
You do not like Bud.
Therefore, you do not like beer.

In Problems 49–65 form a valid conclusion using all the statements for each argument.

49. If you can learn mathematics, then you are intelligent.
If you are intelligent, then you understand human nature.

50. If I am idle, then I become lazy.
I am idle.

51. All trebbles are frebbles.
All frebbles are expensive.

52. If we interfere with the publication of false information, we are guilty of suppressing the freedom of others.
We are not guilty of suppressing the freedom of others.

53. If a nail is lost, then a shoe is lost.
If a shoe is lost, then a horse is lost.
If a horse is lost, then a rider is lost.
If a rider is lost, then a battle is lost.
If a battle is lost, then a kingdom is lost.

54. If you climb the highest mountain, you will feel great.
If you feel great, then you are happy.

55. If $b = 0$, then $a = 0$.
$a \neq 0$.

56. If $a \cdot b = 0$, then $a = 0$ or $b = 0$.
$a \cdot b = 0$.

57. If I eat that piece of pie, I will get fat.
I will not get fat.

58. If we win first prize, we will go to Europe.
If we are ingenious, we will win first prize.
We are ingenious.

59. If I can earn enough money this summer, I will attend college in the fall.
If I do not participate in student demonstrations, then I will not attend college.
I earned enough money this summer.

60. If I am tired, then I cannot finish my homework.
If I understand the material, then I can finish my homework.

61. If you go to college, then you will get a good job.
If you get a good job, then you will make a lot of money.
If you do not obey the law, then you will not make a lot of money.
You go to college.

62. All puppies are nice.
This animal is a puppy.
No nice creatures are dangerous.

63. Babies are illogical.
Nobody is despised who can manage a crocodile.
Illogical persons are despised.

64. Everyone who is sane can do logic.
No lunatics are fit to serve on a jury.
None of your sons can do logic.

65. No ducks waltz.
No officers ever decline to waltz.
All my poultry are ducks.

APPENDIX B

Mathematical Induction

Mathematical induction is an important method of proof in mathematics, allowing us to prove results involving the set of positive integers. Mathematical induction is not the scientific method or the inductive logic used in the experimental sciences; it is a form of deductive logic in which conclusions are inescapable.

The first step in establishing a result by mathematical induction is often the observation of a pattern. Let us begin with a simple example. Suppose we wish to know the sum of the first n odd integers. We could begin by looking for a pattern:

$$
\begin{aligned}
1 &= 1 \\
1 + 3 &= 4 \\
1 + 3 + 5 &= 9 \\
1 + 3 + 5 + 7 &= 16 \\
1 + 3 + 5 + 7 + 9 &= 25
\end{aligned}
$$

Do you see a pattern? It appears that the sum of the first n odd numbers is n^2 since the sum of the first three odd numbers is 3^2, of the first four odd numbers is 4^2, and so on. We now wish to *prove* deductively that

$$
1 + 3 + 5 + \cdots + \underbrace{(2n - 1)}_{\uparrow} = n^2
$$
$$
n\text{th odd number}
$$

is true for all positive integers n. How can we proceed? We prove certain propositions about the positive integers. The proposition is denoted by $P(n)$. For example, in the above problem we let

$$
P(n) = 1 + 3 + 5 + \cdots + (2n - 1) = n^2
$$

This means that

$$
\begin{aligned}
P(1): \quad & 1 = 1^2 \\
P(2): \quad & 1 + 3 = 2^2 \\
P(3): \quad & 1 + 3 + 5 = 3^2 \\
P(4): \quad & 1 + 3 + 5 + 7 = 4^2 \\
& \quad\quad \vdots \\
P(100): \quad & 1 + 3 + 5 + \cdots + 199 = 100^2 \\
& \quad\quad \vdots \\
P(x - 1): \quad & 1 + 3 + 5 + \cdots + (2x - 3) = (x - 1)^2 \\
P(x): \quad & 1 + 3 + 5 + \cdots + (2x - 1) = x^2 \\
P(x + 1): \quad & 1 + 3 + 5 + \cdots + (2x + 1) = (x + 1)^2
\end{aligned}
$$

Now we need to show that $P(n)$ is true for all n (n a positive integer).

Principle of Mathematical
Induction (PMI)

If a given proposition $P(n)$ is true for $P(1)$ and if the truth of $P(k)$ implies the truth of the proposition for $P(k + 1)$, then $P(n)$ is true for all positive integers.

Thus, for proof by mathematical induction, we need to

1. Prove $P(1)$ is true.
2. Assume $P(k)$ is true.
3. Prove $P(k + 1)$ is true.
4. Conclude that $P(n)$ is true for all positive integers n.

Students often have a certain uneasiness when they first use the principle of mathematical induction as a method of proof. Suppose we use this principle with a stack of dominoes, as shown in the cartoon below.

Reproduced by special permission of PLAYBOY magazine; copyright © 1961 by Playboy.

How can the man in the cartoon be certain of knocking over all the dominoes? He would have to be able to knock over the first one. He would have to have the dominoes arranged so that *if* the kth domino falls, then the next one, the $(k + 1)$st, will also fall. That is, each domino is set up so that if it falls, it causes the next one to fall. This is a kind of "chain reaction." The first domino falls; this knocks over the next one (the second domino); the second one knocks over the next one (the third domino); the third one knocks over the next one; this continues until all the dominoes are knocked over.

Let us now return to the example to prove (for all positive integers n)

$$1 + 3 + 5 + \cdots + (2n - 1) = n^2$$

Step 1:   Prove $P(1)$ true: $1 = 1^2$ is true.

Step 2:   Assume $P(k)$ true: $1 + 3 + 5 + \cdots + (2k - 1) = k^2$

Step 3: Prove $P(k + 1)$ true.

TO PROVE: $1 + 3 + 5 + \cdots + [2(k + 1) - 1] = (k + 1)^2$

This is found by substituting $(k + 1)$ for n in the original statement we want to prove. Next, we simplify. This is so we will know when we are finished with step 3 of the proof.

$$1 + 3 + 5 + \cdots + [2(k + 1) - 1] = (k + 1)^2$$
$$1 + 3 + 5 + \cdots + (2k + 2 - 1) \quad = (k + 1)^2$$
$$1 + 3 + 5 + \cdots + (2k + 1) \qquad = (k + 1)^2$$

The procedure for step 3 is to begin with the hypotheses (from step 2) and *prove*

$$1 + 3 + 5 + \cdots + (2k + 1) = (k + 1)^2$$

Statements		Reasons
1. $1 + 3 + 5 + \cdots + (2k - 1) \qquad\qquad = k^2$	1.	By hypothesis (step 2)
2. $1 + 3 + 5 + \cdots + (2k - 1) + (2k + 1) = k^2 + (2k + 1)$	2.	Add $(2k + 1)$ to both sides
3. $1 + 3 + 5 + \cdots + (2k - 1) + (2k + 1) = k^2 + 2k + 1$	3.	Associative
4. $1 + 3 + 5 + \cdots + (2k - 1) + (2k + 1) = (k + 1)^2$	4.	Factoring (distributive)

Step 4: The proposition is true for all positive integers by the principle of mathematical induction (PMI).

EXAMPLE 1 Prove or disprove: $2 + 4 + 6 + \cdots + 2n = n(n + 1)$.

Proof *Step 1:* Prove $P(1)$ true:

$2 \overset{?}{=} 1(1 + 1)$

$2 = 2$

True.

Step 2: Assume $P(k)$. HYPOTHESIS: $2 + 4 + 6 + \cdots + 2k = k(k + 1)$

Step 3: Prove $P(k + 1)$. TO PROVE: $2 + 4 + 6 + \cdots + 2(k + 1) = (k + 1)(k + 2)$

Statements		Reasons
1. $2 + 4 + 6 + \cdots + 2k \qquad\qquad = k(k + 1)$	1.	Hypothesis (step 2)
2. $2 + 4 + 6 + \cdots + 2k + 2(k + 1) = k(k + 1) + 2(k + 1)$	2.	Add $2(k + 1)$ to both sides
3. $\qquad\qquad\qquad\qquad\qquad = (k + 1)(k + 2)$	3.	Factor

Step 4: The proposition is true for all positive integers by PMI. ∎

EXAMPLE 2 Prove or disprove: $n^3 + 2n$ is divisible by 3.

Proof *Step 1:* Prove $P(1)$: $1^3 + 2 \cdot 1 = 3$, which is divisible by 3.

Step 2: Assume $P(k)$. HYPOTHESIS: $k^3 + 2k$ is divisible by 3.

Step 3:   Prove $P(k + 1)$. TO PROVE: $(k + 1)^3 + 2(k + 1)$ is divisible by 3.

Statements		Reasons
1. $(k + 1)^3 + 2(k + 1) = k^3 + 3k^2 + 3k + 1 + 2k + 2$		1. Distributive, associative, and commutative axioms
2. $ = (3k^2 + 3k + 3) + (k^3 + 2k)$		2. Commutative and associative axioms
3. $ = 3(k^2 + k + 1) + (k^3 + 2k)$		3. Distributive axiom
4. $3(k^2 + k + 1)$ is divisible by 3		4. Definition of divisibility by 3
5. $k^3 + 2k$ is divisible by 3		5. Hypothesis
6. $(k + 1)^3 + 2(k + 1)$ is divisible by 3		6. Both terms are divisible by 3 and therefore the sum is divisible by 3

Step 4:   The proposition is true for all positive integers n by PMI.   ■

Example 3 shows that *even though* we make an assumption in step 2, it is not going to help if the proposition is not true.

EXAMPLE 3   Prove or disprove: $n + 1$ is prime.

Proof   *Step 1:*   Prove $P(1)$: $1 + 1 = 2$ is a prime.

Step 2:   Assume $P(k)$. HYPOTHESIS: $k + 1$ is a prime.

Step 3:   Prove $P(k + 1)$. TO PROVE: $(k + 1) + 1$ is a prime. This is not possible since $(k + 1) + 1 = k + 2$, which is not prime whenever k is an even positive integer.

Step 4:   Any conclusions? We cannot conclude that the statement is false, only that induction does not work. But this statement is, in fact, false, and a counterexample is $n = 3$ since $n + 1 = 4$ is not prime.   ■

EXAMPLE 4   Prove or disprove: $1 \cdot 2 \cdot 3 \cdot 4 \cdots \cdot n < 0$.

Proof   Students often slip into the habit of skipping either the first or second step in a proof by mathematical induction. This is dangerous, and it is important to check every step. Suppose you do not verify the first step.

Step 2:   Assume $P(k)$. HYPOTHESIS: $1 \cdot 2 \cdot 3 \cdots \cdot k < 0$

Step 3:   Prove $P(k + 1)$. TO PROVE: $1 \cdot 2 \cdot 3 \cdots \cdot k \cdot (k + 1) < 0$
$1 \cdot 2 \cdot 3 \cdots \cdot k < 0$ by hypothesis.
$k + 1$ is positive since k is a positive integer. Then, since we know that the product of a negative and a positive is negative,

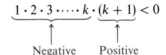

Step 3 is proved.

Step 4:   The proposition is not true for all positive integers since the first step, $1 < 0$, does not hold.   ■

Problem Set B

1. Prove: $1 + 2 + 3 + \cdots + n = \dfrac{n(n+1)}{2}$

for all positive integers n.

2. Prove: $1^2 + 2^2 + 3^2 + \cdots + n^2 = \dfrac{n(n+1)(2n+1)}{6}$

for all positive integers n.

3. Prove:

$1^2 + 3^2 + 5^2 + \cdots + (2n-1)^2 = \dfrac{n(2n-1)(2n+1)}{3}$

for all positive integers n.

4. Prove: $1^3 + 2^3 + 3^3 + \cdots + n^3 = \dfrac{n^2(n+1)^2}{4}$

for all positive integers n.

5. Prove:

$2^2 + 4^2 + 6^2 + \cdots + (2n)^2 = \dfrac{2n(n+1)(2n+1)}{3}$

for all positive integers n.

6. Prove:

$1 \cdot 2 + 2 \cdot 3 + 3 \cdot 4 + \cdots + n(n+1) = \dfrac{n(n+1)(n+2)}{3}$

for all positive integers n.

7. Prove:

$1 \cdot 3 + 2 \cdot 4 + 3 \cdot 5 + \cdots + n(n+2) = \dfrac{n(n+1)(2n+7)}{6}$

for all positive integers n.

8. Prove: $1 + r + r^2 + \cdots + r^n = \dfrac{r^{n+1} - 1}{r - 1}$

for all positive integers n.

9. Prove: $n^5 - n$ is divisible by 5 for all positive integers n.

10. Prove: $n(n+1)(n+2)$ is divisible by 6 for all positive integers n.

11. Prove: $(1 + n)^2 \geq 1 + n^2$ for all positive integers n.

12.
$$1^3 = 1^2$$
$$1^3 + 2^3 = 3^2$$
$$1^3 + 2^3 + 3^3 = 6^2$$
$$1^3 + 2^3 + 3^3 + 4^3 = 10^2$$

Make a conjecture based on the above pattern and then prove or disprove your conjecture.

13.
$$1 = 1$$
$$1 + 4 = 5$$
$$1 + 4 + 7 = 12$$
$$1 + 4 + 7 + 10 = 22$$

Make a conjecture based on the above pattern and then prove or disprove your conjecture.

14. Prove: $\dbinom{k}{r} + \dbinom{k}{r-1} = \dbinom{k+1}{r}$.

Use the formula $\dbinom{n}{m} = \dfrac{n!}{m!(n-m)!}$ and not induction.
You will need this result in Problem 15.

15. The binomial theorem can be proved for any positive integer n by using mathematical induction. Fill in the missing steps and reasons.

TO PROVE:

$$(a + b)^n = \sum_{j=0}^{n} \binom{n}{j} a^{n-j} b^j$$

$$= \binom{n}{0} a^n + \binom{n}{1} a^{n-1} b + \binom{n}{2} a^{n-2} b^2 + \cdots$$

$$+ \binom{n}{r} a^{n-r} b^r + \cdots + \binom{n}{n-1} ab^{n-1} + \binom{n}{n} b^n$$

Step 1: Prove the binomial theorem is true for $n = 1$.
    a. Fill in these details.

Step 2: Assume the theorem is true for $n = k$.
    b. Fill in the statement of the hypothesis.

Step 3: Prove the theorem is true for $n = k + 1$.

TO PROVE: $(a + b)^{k+1} = \sum_{j=0}^{k+1} \binom{k+1}{j} a^{k+1-j} b^j$

BY HYPOTHESIS:
    c. Fill in the statement of the hypothesis.
    d. Fill in the details; the final simplified form is

$(a + b)^{k+1} = a^{k+1} + \cdots$

$$+ \left[\binom{k}{r} + \binom{k}{r-1} \right] a^{k-r+1} b^r + \cdots + b^{k+1}$$

    e. Use Problem 14 to complete the proof.

APPENDIX C

Calculators

This appendix provides a brief introduction to calculators and their use. In the last few years, pocket calculators have been one of the fastest selling items in the United States. This is probably because most people (including mathematicians!) do not like to do arithmetic and a good calculator can be purchased for less than $20.

This book was written with the assumption that you have or will have a calculator. This appendix is included to help you choose a calculator and understand the calculator comments in this book.

Calculators are classified by the types of problems they are equipped to handle, as well as by the type of logic for which they are programmed. The problem of selecting a calculator is compounded by the multiplicity of brands from which to choose.

The different types of calculators are distinguished primarily by their price.

1. *Four-function calculators* (under $10). These calculators have a keyboard consisting of the numerals and the four arithmetic operations, or functions: addition $\boxed{+}$, subtraction $\boxed{-}$, multiplication $\boxed{\times}$, and division $\boxed{\div}$.
2. *Four-function calculators with memory* ($10–$20). Usually these are no more expensive than four-function calculators, and offer a memory register: $\boxed{M}$, $\boxed{STO}$, or $\boxed{M^+}$. The more expensive models may have more than one memory register. Memory registers allow you to store partial calculations for later recall. Some models will even remember the total when they are turned off.
3. *Scientific calculators* ($20–$50). These calculators add additional mathematical functions, such as square root $\boxed{\sqrt{\ }}$, trigonometric $\boxed{\sin}$, $\boxed{\cos}$, and $\boxed{\tan}$, and logarithmic $\boxed{\log}$ and $\boxed{\exp}$. Depending on the particular brand, a scientific model may have other keys as well.
4. *Special-purpose calculators* ($40–$400). Special-use calculators for business, statistics, surveying, medicine, or even gambling and chess are available.
5. *Programmable calculators* ($50–$600). With these calculators you can enter a *sequence* of steps for the calculator to repeat on your command. Some of these calculators allow the insertion of different cards that "remember" the sequence of steps for complex calculations.

For most nonscientific purposes, a four-function calculator with memory is sufficient for everyday usage. **For this book you need a scientific calculator**. Three types of logic are used by scientific calculators: *arithmetic*, *algebraic*, and *RPN*. You need to know the type of logic used by your calculator. To determine the type of logic used by a particular calculator, try this test problem:

$$\boxed{2}\ \boxed{+}\ \boxed{3}\ \boxed{\times}\ \boxed{4}\ \boxed{=}$$

If the answer shown is 20, it is an arithmetic-logic calculator. If the answer is 14 (the correct answer), then it is an algebraic-logic calculator. If the calculator has no equal key $=$ but has an ENTER or SAVE key, then it is an RPN-logic calculator. An RPN-logic calculator will give the answer as 14. In algebra you learn to perform multiplication before addition, so that the correct value for

$$2 + 3 \times 4$$

is 14 (multiply first). An algebraic calculator will "know" this fact and will give the correct answer, whereas an arithmetic calculator will simply work from left to right to obtain the incorrect answer, 20. Therefore, if you have an arithmetic-logic calculator, you will need to be careful about the order of operations. Some arithmetic-logic calculators provide parentheses ( ) so that operations can be grouped as in

2 + (3 × 4) =

but then you must remember to insert the parentheses.

With an RPN calculator, the operation symbol is entered after the numbers have been entered. These three types of logic can be illustrated by the problem $2 + 3 \times 4$:

Figure C.1

a. The Sharp calculator is an example of a calculator with arithmetic logic.

b. This Texas Instruments SR-51-II is an example of a calculator with algebraic logic.

c. The Hewlett-Packard calculator uses RPN logic.

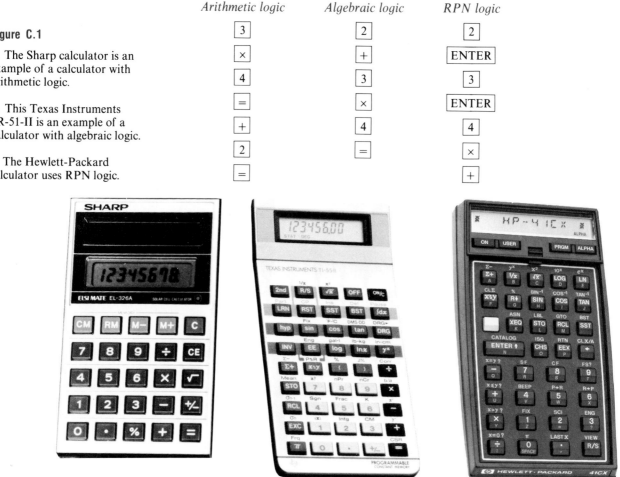

In this book the examples use algebraic- and RPN-logic calculators. The keys to be pressed are indicated by boxes drawn around the numbers and operational signs, as shown above. Numerals for calculator display are illustrated as

$$0 \; 1 \; 2 \; 3 \; 4 \; 5 \; 6 \; 7 \; 8 \; 9$$

Regardless of the type of logic your calculator uses, it is a good idea to check your owner's manual for each type of problem illustrated in the text because there are many different brands of calculators on the market, and many have slight variations in keyboards.

There is also a limit to the accuracy of your calculator. You may have a calculator with a 6-, 8-, or 10-digit display. Test its accuracy with the following example.

EXAMPLE 1 Find $2 \div 3$.

Solution   *Algebraic:* $\boxed{2}$ $\boxed{\div}$ $\boxed{3}$ $\boxed{=}$

RPN: $\boxed{2}$ $\boxed{\text{ENTER}}$ $\boxed{3}$ $\boxed{\div}$

Display: 2   2.   3   .6666666667   ∎

There may be a discrepancy between this answer and the one you obtain on your calculator. Some machines will not round the answer as shown here but will show the display

.6666666666

Others will show a display such as

6.6666 -01

This is a number in scientific notation and should be interpreted as

$$6.6666 \times 10^{-1} \quad \text{or} \quad .66666$$

Most calculators will also use scientific notation when the numbers become larger than that allowed by their display register. Example 2 shows how your calculator handles large numbers. Some calculators simply show an overflow and will not accept larger numbers. You need a calculator that will accept and handle large and small numbers.

EXAMPLE 2 Find 50^6.

Solution   Check your owner's manual to find out how to use an exponent key. Most calculators will work as shown below.

Algebraic: $\boxed{50}$ $\boxed{y^x}$ $\boxed{6}$

RPN: $\boxed{50}$ $\boxed{\text{ENTER}}$ $\boxed{6}$ $\boxed{y^x}$

When the maximum size of the display has been reached, the calculator should automatically switch to scientific notation. The point at which a calculator will do this varies from one type or brand to another. The answer for this example is 1.5625 10, which means

$$1.5625 \times 10^{10} = 15,625,000,000 \quad ∎$$

Problem Set C

Use a calculator to evaluate each of the expressions in Problems 1–31.

1. $(14)(351)$

2. $(218)(263)$

3. $(4158)(0.00456)$

4. $(3.00)^3(182)$

5. $(2.00)^4(1245)(277)$

6. $(6.00)^5(1456)(288)$

7. $\dfrac{(1979)(1356)}{452}$

8. $\dfrac{(515)(20,600)}{200}$

9. $\dfrac{(618)(460)(125)}{650}$

10. $[0.14 + (197)(25.08)]19$

11. $(990)(1117)(342) - 89$

12. $\dfrac{1.00}{0.005 + 0.020}$

13. $\dfrac{1.500 \times 10^4 + (7.000)(67.00)}{20,000}$

14. $(6.28)^{1/2}(4.85)$

15. $(8.23)^{1/2}(6.14)$

16. $\dfrac{1.00}{\sqrt{8.48} - \sqrt{21.3}}$

17. $\dfrac{1.00}{\sqrt{4.83} + \sqrt{2.51}}$

18. $[(4.083)^2(4.283)^3]^{-2/3}$

19. $[(6.128)^4(3.412)^2]^{-1/2}$

20. $\dfrac{16^2 + 25^2 - 9^2}{(2.0)(16)(25)}$

21. $\dfrac{216^2 + 418^2 - 315^2}{(2.00)(216)(418)}$

22. $\dfrac{4.82^2 + 6.14^2 - 9.13^2}{(2.00)(4.82)(6.14)}$

23. $\sqrt{\dfrac{(51)(36)}{212}}$

24. $\sqrt{\dfrac{25 + 49}{1 + 4(51)}}$

25. $\sqrt{\dfrac{45 + 156}{2 + 51(19)}}$

26. $\dfrac{18.361^2 + 15.215^2 - 13.815^2}{(2.0000)(18.361)(15.215)}$

27. $\dfrac{17.813^2 + 13.451^2 - 19.435^2}{2(17.813)(13.451)}$

28. $\dfrac{(2.51)^2 + (6.48)^2 - (2.51)(6.48)(.3462)}{(2.51)(6.48)}$

29. $\dfrac{241^2 + 568^2 - (241)(568)(0.5213)}{(241)(568)}$

30. $1500\left(1 + \dfrac{0.105}{12}\right)^{8(12)}$

APPENDIX D

Computers

Microcomputers are changing the curriculum today as much as calculators changed the curriculum in the last decade. The personal computer is accessible to a great many students, and more and more schools and colleges have computer labs in which programs tailored for specific classes can be used.

For this reason, *Computer Aided Finite Mathematics* has been written as a companion book to this text and is available at your book store. It explains how to use a microcomputer to help you with computationally difficult problems. The programs do not simply "do calculations," but, rather, help you through a process so that you understand each and every step. Additional examples and practice problems are also provided.

Computer Aided Finite Mathematics will be particularly useful in the sections of the book listed in Table D.1.

TABLE D.1
Programs on disk accompanying this book

Text Section	Options	Program number/name
2.3, 4.1	Graph a line	1. Straight Line
2.2, 3.1, 4.1	Print out a table of values	Plotter
	Graph a set of lines	
3.2	Manual product	2. Matrix Product
	(instructional option)	
	Automatic product	
3.3	Manual reduction	3. Row Reduction of
	(instructional option)	a System
	1. Interchange rows	
	2. Multiply a row by a constant	
	3. Divide a row by a constant	
	4. Add multiple of one row to another	
	Automatic reduction	
3.5	Manual inverse	4. Matrix Inverse by
	(instructional option)	Reduction
	Automatic reduction	
3.4, 3.5, 3.6	Solve a system of equations by finding the inverse	5. Matrix Equations
4.5, 4.6, 4.7	Manual reduction	6. Simplex Method
	(instructional option)	

Text Section	Options	Program number/name
	1. Interchange rows 2. Multiply a row by a constant 3. Divide a row by a constant 4. Add multiple of one row to another Automatic reduction	
Chapter 6 Review	Program generates a random set of questions involving the fundamental counting principle, permutations, distinguishable permutations, and combinations.	7. Counting Problems
Chapter 7 Review	Program generates a random set of questions involving probability problems.	8. Probability Problems
Chapter 7 Review	Program generates a random set of questions involving union, intersection, conditional probability, and independent events.	9. Probability Properties
8.2	Mean, median, standard deviation, frequency distribution, and histogram for random or input data	10. Statistics
6.4, 8.3	Binomial coefficients Single binomial coefficient Binomial probabilities Binomial probability histogram	11. Binomial Coefficients and Probabilities
8.6	Coefficient of correlation Least squares line Random or input data	12. Correlation and Regression
	Raise a transition matrix to a power Probability vector Market share after n transitions (instructional option) Automatic reduction	13. Markov Chains
	Plays a two-person zero-sum game a large number of times when operator enters a strategy. The computer then plays either a random or intelligent strategy.	14. Two-Person Zero-Sum Game
5.2–5.5	Lump sum; future value or present value Ordinary annuity or sinking fund Annuity due Present value of an annuity or installment payment	15. Formulas for Finance
Chapter 5 Review	Program generates a random set of questions involving finance problems.	16. Finance Problems
5.4, General Interest	Prints out an amortization table for paying off a loan.	17. Loan Amortization

Additional Available Software

Calculus

Calculus Blackboard. Less Hill (Wadsworth Advanced Books and Software. 511 Forest Lodge Road, Pacific Grove, CA 93950. $40; Apple). Can be used for instructor demonstrations and for graphics.

Calculus I and II (PLATO; Control Data Publishing, Higher Education Marketing Division, MNB 04 A. 3601 West 77th Street, Bloomington, MN 55435. $1,150 each; IBM PC). Nice calculus standalone, but cost is prohibitive.

Introductory Calculus Series (Microphys. 1737 West 2nd Street, Brooklyn, NY 11223. $200; Apple, IBM PC, TRS-80). Includes drill, practice, and a problem generator.

Graphing

Cactusplot. John Losse (Cactusplot. 1442 North McAllister, Tempe, AZ 85281. $60; Apple, IBM PC). Includes demonstrations and graphics.

Electronic Blackboard: Function Plotter. Richard O'Farrell et al. (Brooks/Cole. 511 Forest Lodge Road, Pacific Grove, CA 93950. $50; Apple). Includes demonstrations.

SURF. Mark Bridger (Bridge Software. 31 Champa Street, Newton Upper Falls, MA 02164. $35; IBM PC). Does three-dimensional graphing.

Surfaces for Multivariate Calculus. Roy E. Myers (CONDUIT. University of Iowa, Oakdale Campus, Iowa City, IA 52242. $65; Apple). Includes demonstrations and graphics.

Finite Mathematics

Computer-Aided Finite Mathematics. Chris Avery and Charles Barker (Brooks/Cole. 511 Forest Lodge Road, Pacific Grove, CA 93950. $16.95; Apple). Specifically designed for the finite mathematics portion of this book, with a very attractive price.

Logic

Reasoning: The Logical Process. Connie Ouding (MCE. 157 South Kalamazoo Mall, Suite 250, Kalamazoo, MI 49007. $54.95; Apple). A tutorial that includes problem solving.

Statistics

Advanced Simulation and Statistics Package. P. Lewis, E. Orav, and L. Uribe (Wadsworth Advanced Books and Software. 511 Forest Lodge Road, Pacific Grove, CA 93950. $59.95; IBM PC). Includes graphics and simulation.

KEYSTAT. H. R. Strang and A. H. Innes (Brooks/Cole. 511 Forest Lodge Road, Pacific Grove, CA 93950. $75.75; Apple, CP/M). Computational.

MacSpin (Wadsworth Advanced Books and Software. 511 Forest Lodge Road, Pacific Grove, CA 93950. $79.95; Macintosh). Graphical data analysis package.

Micro Statistics Package, The. E. A. Keller and P. Marsh (Keller, Marsh Associates. 1414 Smith Court, San Luis Obispo, CA 93401. $49.50; IBM PC, HP 110/150).

Symbolic Mathematics

muMATH, muSIMP. The Soft Warehouse (Microsoft. 10700 Northrup Way, Box 97200, Bellevue, WA 98009. $225; Apple, MS-DOS, CP/M).

PowerMath (Brainpower. 24009 Ventura Boulevard, Suite 250, Calabasas, CA 91302. $99.95; Macintosh).

TK! Solver (Lotus, 55 Cambridge Parkway, Cambridge, MA 02142. $399; Apple, Macintosh, IBM PC, TRS-80, DEC).

APPENDIX E

Tables

TABLE 1
Pascal's triangle—combinatorics

n	$\binom{n}{0}$	$\binom{n}{1}$	$\binom{n}{2}$	$\binom{n}{3}$	$\binom{n}{4}$	$\binom{n}{5}$	$\binom{n}{6}$	$\binom{n}{7}$	$\binom{n}{8}$	$\binom{n}{9}$	$\binom{n}{10}$
0	1										
1	1	1									
2	1	2	1								
3	1	3	3	1							
4	1	4	6	4	1						
5	1	5	10	10	5	1					
6	1	6	15	20	15	6	1				
7	1	7	21	35	35	21	7	1			
8	1	8	28	56	70	56	28	8	1		
9	1	9	36	84	126	126	84	36	9	1	
10	1	10	45	120	210	252	210	120	45	10	1
11	1	11	55	165	330	462	462	330	165	55	11
12	1	12	66	220	495	792	924	792	495	220	66
13	1	13	78	286	715	1287	1716	1716	1287	715	286
14	1	14	91	364	1001	2002	3003	3432	3003	2002	1001
15	1	15	105	455	1365	3003	5005	6435	6435	5005	3003
16	1	16	120	560	1820	4368	8008	11440	12870	11440	8008
17	1	17	136	680	2380	6188	12376	19448	24310	24310	19448
18	1	18	153	816	3060	8568	18564	31824	43758	48620	43758
19	1	19	171	969	3876	11628	27132	50388	75582	92378	92378
20	1	20	190	1140	4845	15504	38760	77520	125970	167960	184756

Note: $\binom{n}{m} = \dfrac{n(n-1)(n-2)\cdots\cdots(n-m+1)}{m(m-1)(m-2)\cdots\cdots 3 \cdot 2 \cdot 1}$; $\binom{n}{0} = 1$; $\binom{n}{1} = n$

For coefficients missing from the above table, use the relation

$$\binom{n}{m} = \binom{n}{n-m}$$

For example,

$$\binom{20}{11} = \binom{20}{9} = 167,960$$

TABLE 2
Squares and square roots

n	n^2	$\sqrt{n}$	$\sqrt{10n}$	n	n^2	$\sqrt{n}$	$\sqrt{10n}$
1	1	1.000	3.162	51	2601	7.141	22.583
2	4	1.414	4.472	52	2704	7.211	22.804
3	9	1.732	5.477	53	2809	7.280	23.022
4	16	2.000	6.325	54	2916	7.348	23.238
5	25	2.236	7.071	55	3025	7.416	23.452
6	36	2.449	7.746	56	3136	7.483	23.664
7	49	2.646	8.367	57	3249	7.550	23.875
8	64	2.828	8.944	58	3364	7.616	24.083
9	81	3.000	9.487	59	3481	7.681	24.290
10	100	3.162	10.000	60	3600	7.746	24.495
11	121	3.317	10.488	61	3721	7.810	24.698
12	144	3.464	10.954	62	3844	7.874	24.900
13	169	3.606	11.402	63	3969	7.937	25.100
14	196	3.742	11.832	64	4096	8.000	25.298
15	225	3.873	12.247	65	4225	8.062	25.495
16	256	4.000	12.649	66	4356	8.124	25.690
17	289	4.123	13.038	67	4489	8.185	25.884
18	324	4.243	13.416	68	4624	8.246	26.077
19	361	4.359	13.784	69	4761	8.307	26.268
20	400	4.472	14.142	70	4900	8.367	26.458
21	441	4.583	14.491	71	5041	8.426	26.646
22	484	4.690	14.832	72	5184	8.485	26.833
23	529	4.796	15.166	73	5329	8.544	27.019
24	576	4.899	15.492	74	5476	8.602	27.203
25	625	5.000	15.811	75	5625	8.660	27.386
26	676	5.099	16.125	76	5776	8.718	27.568
27	729	5.196	16.432	77	5929	8.775	27.749
28	784	5.292	16.733	78	6084	8.832	27.928
29	841	5.385	17.029	79	6241	8.888	28.107
30	900	5.477	17.321	80	6400	8.944	28.284
31	961	5.568	17.607	81	6561	9.000	28.460
32	1024	5.657	17.889	82	6724	9.055	28.636
33	1089	5.745	18.166	83	6889	9.110	28.810
34	1156	5.831	18.439	84	7056	9.165	28.983
35	1225	5.916	18.708	85	7225	9.220	29.155
36	1296	6.000	18.974	86	7396	9.274	29.326
37	1369	6.083	19.235	87	7569	9.327	29.496
38	1444	6.164	19.494	88	7744	9.381	29.665
39	1521	6.245	19.748	89	7921	9.434	29.833
40	1600	6.325	20.000	90	8100	9.487	30.000
41	1681	6.403	20.248	91	8281	9.539	30.166
42	1764	6.481	20.494	92	8464	9.592	30.332
43	1849	6.557	20.736	93	8649	9.644	30.496
44	1936	6.633	20.976	94	8836	9.695	30.659
45	2025	6.708	21.213	95	9025	9.747	30.822
46	2116	6.782	21.448	96	9216	9.798	30.984
47	2209	6.856	21.679	97	9409	9.849	31.145
48	2304	6.928	21.909	98	9604	9.899	31.305
49	2401	7.000	22.136	99	9801	9.950	31.464
50	2500	7.071	22.361	100	10000	10.000	31.623

TABLE 3
Binomial probabilities

This table computes the probability of exactly k successes in n independent binomial trials with the probability of success on a single trial equal to p. It thus contains the individual terms for specified choices of k, n, and p.* For entries $0+$, the probability is less than $.0005$, but greater than 0.

$$\binom{n}{k} p^k (1-p)^{n-k}$$

p

n	k	.01	.05	.10	.15	.20	.25	.30	.35	.40	.45	.50	k
2	0	.980	.903	.810	.723	.640	.563	.490	.423	.360	.303	.250	0
	1	.020	.095	.180	.255	.320	.375	.420	.455	.480	.495	.500	1
	2	0+	.003	.010	.023	.040	.063	.090	.122	.160	.202	.250	2
3	0	.970	.857	.729	.614	.512	.422	.343	.275	.216	.166	.125	0
	1	.029	.135	.243	.325	.384	.422	.441	.444	.432	.408	.375	1
	2	0+	.007	.027	.057	.096	.141	.189	.239	.288	.334	.375	2
	3	0+	0+	.001	.003	.008	.016	.027	.043	.064	.091	.125	3
4	0	.961	.815	.656	.522	.410	.316	.240	.179	.130	.092	.063	0
	1	.039	.171	.292	.368	.410	.422	.412	.384	.346	.299	.250	1
	2	.001	.014	.049	.098	.154	.211	.265	.311	.346	.368	.375	2
	3	0+	0+	.004	.011	.026	.047	.076	.111	.154	.200	.250	3
	4	0+	0+	0+	.001	.002	.004	.008	.015	.026	.041	.062	4
5	0	.951	.774	.590	.444	.328	.237	.168	.116	.078	.050	.031	0
	1	.048	.204	.328	.392	.410	.396	.360	.312	.259	.206	.156	1
	2	.001	.021	.073	.138	.205	.264	.309	.336	.346	.337	.313	2
	3	0+	.001	.008	.024	.051	.088	.132	.181	.230	.276	.312	3
	4	0+	0+	0+	.002	.006	.015	.028	.049	.077	.113	.156	4
	5	0+	0+	0+	0+	0+	.001	.002	.005	.010	.018	.031	5
6	0	.941	.735	.531	.377	.262	.178	.118	.075	.047	.028	.016	0
	1	.057	.232	.354	.399	.393	.356	.303	.244	.187	.136	.094	1
	2	.001	.031	.098	.176	.246	.297	.324	.328	.311	.278	.234	2
	3	0+	.002	.015	.041	.082	.132	.185	.235	.276	.303	.313	3
	4	0+	0+	.001	.005	.015	.033	.060	.095	.138	.186	.234	4
	5	0+	0+	0+	0+	.002	.004	.010	.020	.037	.061	.094	5
	6	0+	0+	0+	0+	0+	0+	.001	.002	.004	.008	.016	6
7	0	.932	.698	.478	.321	.210	.133	.082	.049	.028	.015	.008	0
	1	.066	.257	.372	.396	.367	.311	.247	.185	.131	.087	.055	1
	2	.002	.041	.124	.210	.275	.311	.318	.298	.261	.214	.164	2
	3	0+	.004	.023	.062	.115	.173	.227	.268	.290	.292	.273	3
	4	0+	0+	.003	.011	.029	.058	.097	.144	.194	.239	.273	4
	5	0+	0+	0+	.001	.004	.012	.025	.047	.077	.117	.164	5
	6	0+	0+	0+	0+	0+	.001	.004	.008	.017	.032	.055	6
	7	0+	0+	0+	0+	0+	0+	0+	.001	.002	.004	.008	7
8	0	.923	.663	.430	.272	.168	.100	.058	.032	.017	.008	.004	0
	1	.075	.279	.383	.385	.336	.267	.198	.137	.090	.055	.031	1
	2	.003	.051	.149	.238	.294	.311	.296	.259	.209	.157	.109	2
	3	0+	.005	.033	.084	.147	.208	.254	.279	.279	.257	.219	3
	4	0+	0+	.005	.018	.046	.087	.136	.188	.232	.263	.273	4
	5	0+	0+	0+	.003	.009	.023	.047	.081	.124	.172	.219	5
	6	0+	0+	0+	0+	.001	.004	.010	.022	.041	.070	.109	6
	7	0+	0+	0+	0+	0+	0+	.001	.003	.008	.016	.031	7
	8	0+	0+	0+	0+	0+	0+	0+	0+	.001	.002	.004	8

* For $p > .50$, the value of $\binom{n}{k} p^k (1-p)^{n-k}$ is found by using the table entry for $\binom{n}{n-k}(1-p)^{n-k} p^k$.

TABLE 3 (Continued)

p

n	k	.01	.05	.10	.15	.20	.25	.30	.35	.40	.45	.50	k
9	0	.914	.630	.387	.232	.134	.075	.040	.021	.010	.005	.002	0
	1	.083	.299	.387	.368	.302	.225	.156	.100	.060	.034	.018	1
	2	.003	.063	.172	.260	.302	.300	.267	.216	.161	.111	.070	2
	3	0+	.008	.045	.107	.176	.234	.267	.272	.251	.212	.164	3
	4	0+	.001	.007	.028	.066	.117	.172	.219	.251	.260	.246	4
	5	0+	0+	.001	.005	.017	.039	.074	.118	.167	.213	.246	5
	6	0+	0+	0+	.001	.003	.009	.021	.042	.074	.116	.164	6
	7	0+	0+	0+	0+	0+	.001	.004	.010	.021	.041	.070	7
	8	0+	0+	0+	0+	0+	0+	0+	.001	.004	.008	.018	8
	9	0+	0+	0+	0+	0+	0+	0+	0+	0+	.001	.002	9
10	0	.904	.599	.349	.197	.107	.056	.028	.013	.006	.003	.001	0
	1	.091	.315	.387	.347	.268	.188	.121	.072	.040	.021	.010	1
	2	.004	.075	.194	.276	.302	.282	.233	.176	.121	.076	.044	2
	3	0+	.010	.057	.130	.201	.250	.267	.252	.215	.166	.117	3
	4	0+	.001	.011	.040	.088	.146	.200	.238	.251	.238	.205	4
	5	0+	0+	.001	.008	.026	.058	.103	.154	.201	.234	.246	5
	6	0+	0+	0+	.001	.006	.016	.037	.069	.111	.160	.205	6
	7	0+	0+	0+	0+	.001	.003	.009	.021	.042	.075	.117	7
	8	0+	0+	0+	0+	0+	0+	.001	.004	.011	.023	.044	8
	9	0+	0+	0+	0+	0+	0+	0+	.001	.002	.004	.010	9
	10	0+	0+	0+	0+	0+	0+	0+	0+	0+	0+	.001	10
11	0	.895	.569	.314	.167	.086	.042	.020	.009	.004	.001	0+	0
	1	.099	.329	.384	.325	.236	.155	.093	.052	.027	.013	.005	1
	2	.005	.087	.213	.287	.295	.258	.200	.140	.089	.051	.027	2
	3	0+	.014	.071	.152	.221	.258	.257	.225	.177	.126	.081	3
	4	0+	.001	.016	.054	.111	.172	.220	.243	.236	.206	.161	4
	5	0+	0+	.002	.013	.039	.080	.132	.183	.221	.236	.226	5
	6	0+	0+	0+	.002	.010	.027	.057	.099	.147	.193	.226	6
	7	0+	0+	0+	0+	.002	.006	.017	.038	.070	.113	.161	7
	8	0+	0+	0+	0+	0+	.001	.004	.010	.023	.046	.081	8
	9	0+	0+	0+	0+	0+	0+	.001	.002	.005	.013	.027	9
	10	0+	0+	0+	0+	0+	0+	0+	0+	.001	.002	.005	10
	11	0+	0+	0+	0+	0+	0+	0+	0+	0+	0+	0+	11
12	0	.886	.540	.282	.142	.069	.032	.014	.006	.002	.001	0+	0
	1	.107	.341	.377	.301	.206	.127	.071	.037	.017	.008	.003	1
	2	.006	.099	.230	.292	.283	.232	.168	.109	.064	.034	.016	2
	3	0+	.017	.085	.172	.236	.258	.240	.195	.142	.092	.054	3
	4	0+	.002	.021	.068	.133	.194	.231	.237	.213	.170	.121	4
	5	0+	0+	.004	.019	.053	.103	.158	.204	.227	.222	.193	5
	6	0+	0+	0+	.004	.016	.040	.079	.128	.177	.212	.226	6
	7	0+	0+	0+	.001	.003	.011	.029	.059	.101	.149	.193	7
	8	0+	0+	0+	0+	.001	.002	.008	.020	.042	.076	.121	8
	9	0+	0+	0+	0+	0+	0+	.001	.005	.012	.028	.054	9
	10	0+	0+	0+	0+	0+	0+	0+	.001	.002	.007	.016	10
	11	0+	0+	0+	0+	0+	0+	0+	0+	0+	.001	.003	11
	12	0+	0+	0+	0+	0+	0+	0+	0+	0+	0+	0+	12
13	0	.878	.513	.254	.121	.055	.024	.010	.004	.001	0+	0+	0
	1	.115	.351	.367	.277	.179	.103	.054	.026	.011	.004	.002	1
	2	.007	.111	.245	.294	.268	.206	.139	.084	.045	.022	.010	2
	3	0+	.021	.100	.190	.246	.252	.218	.165	.111	.066	.035	3
	4	0+	.003	.028	.084	.154	.210	.234	.222	.184	.135	.087	4
	5	0+	0+	.006	.027	.069	.126	.180	.215	.221	.199	.157	5
	6	0+	0+	.001	.006	.023	.056	.103	.155	.197	.217	.209	6
	7	0+	0+	0+	.001	.006	.019	.044	.083	.131	.177	.209	7
	8	0+	0+	0+	0+	.001	.005	.014	.034	.066	.109	.157	8
	9	0+	0+	0+	0+	0+	.001	.003	.010	.024	.050	.087	9
	10	0+	0+	0+	0+	0+	0+	.001	.002	.006	.016	.035	10
	11	0+	0+	0+	0+	0+	0+	0+	0+	.001	.004	.010	11
	12	0+	0+	0+	0+	0+	0+	0+	0+	0+	0+	.002	12
	13	0+	0+	0+	0+	0+	0+	0+	0+	0+	0+	0+	13

TABLE 3 (Continued)

n	k	.01	.05	.10	.15	.20	.25	.30	.35	.40	.45	.50	k
14	0	.869	.488	.229	.103	.044	.018	.007	.002	.001	0+	0+	0
	1	.123	.359	.356	.254	.154	.083	.041	.018	.007	.003	.001	1
	2	.008	.123	.257	.291	.250	.180	.113	.063	.032	.014	.006	2
	3	0+	.026	.114	.206	.250	.240	.194	.137	.085	.046	.022	3
	4	0+	.004	.035	.100	.172	.220	.229	.202	.155	.104	.061	4
	5	0+	0+	.008	.035	.086	.147	.196	.218	.207	.170	.122	5
	6	0+	0+	.001	.009	.032	.073	.126	.176	.207	.209	.183	6
	7	0+	0+	0+	.002	.009	.028	.062	.108	.157	.195	.209	7
	8	0+	0+	0+	0+	.002	.008	.023	.051	.092	.140	.183	8
	9	0+	0+	0+	0+	0+	.002	.007	.018	.041	.076	.122	9
	10	0+	0+	0+	0+	0+	0+	.001	.005	.014	.031	.061	10
	11	0+	0+	0+	0+	0+	0+	0+	.001	.003	.009	.022	11
	12	0+	0+	0+	0+	0+	0+	0+	0+	.001	.002	.006	12
	13	0+	0+	0+	0+	0+	0+	0+	0+	0+	0+	.001	13
	14	0+	0+	0+	0+	0+	0+	0+	0+	0+	0+	0+	14
15	0	.860	.463	.206	.087	.035	.013	.005	.002	0+	0+	0+	0
	1	.130	.366	.343	.231	.132	.067	.031	.013	.005	.002	0+	1
	2	.009	.135	.267	.286	.231	.156	.092	.048	.022	.009	.003	2
	3	0+	.031	.129	.218	.250	.225	.170	.111	.063	.032	.014	3
	4	0+	.005	.043	.116	.188	.225	.219	.179	.127	.078	.042	4
	5	0+	.001	.010	.045	.103	.165	.206	.212	.186	.140	.092	5
	6	0+	0+	.002	.013	.043	.092	.147	.191	.207	.191	.153	6
	7	0+	0+	0+	.003	.014	.039	.081	.132	.177	.201	.196	7
	8	0+	0+	0+	.001	.003	.013	.035	.071	.118	.165	.196	8
	9	0+	0+	0+	0+	.001	.003	.012	.030	.061	.105	.153	9
	10	0+	0+	0+	0+	0+	.001	.003	.010	.024	.051	.092	10
	11	0+	0+	0+	0+	0+	0+	.001	.002	.007	.019	.042	11
	12	0+	0+	0+	0+	0+	0+	0+	0+	.002	.005	.014	12
	13	0+	0+	0+	0+	0+	0+	0+	0+	0+	.001	.003	13
	14	0+	0+	0+	0+	0+	0+	0+	0+	0+	0+	0+	14
	15	0+	0+	0+	0+	0+	0+	0+	0+	0+	0+	0+	15
16	0	.851	.440	.185	.074	.028	.010	.003	.001	0+	0+	0+	0
	1	.138	.371	.329	.210	.113	.053	.023	.009	.003	.001	0+	1
	2	.010	.146	.275	.277	.211	.134	.073	.035	.015	.006	.002	2
	3	0+	.036	.142	.229	.246	.208	.146	.089	.047	.022	.009	3
	4	0+	.006	.051	.131	.200	.225	.204	.155	.101	.057	.028	4
	5	0+	.001	.014	.056	.120	.180	.210	.201	.162	.112	.067	5
	6	0+	0+	.003	.018	.055	.110	.165	.198	.198	.168	.122	6
	7	0+	0+	0+	.005	.020	.052	.101	.152	.189	.197	.175	7
	8	0+	0+	0+	.001	.006	.020	.049	.092	.142	.181	.196	8
	9	0+	0+	0+	0+	.001	.006	.019	.044	.084	.132	.175	9
	10	0+	0+	0+	0+	0+	.001	.006	.017	.039	.075	.122	10
	11	0+	0+	0+	0+	0+	0+	.001	.005	.014	.034	.067	11
	12	0+	0+	0+	0+	0+	0+	0+	.001	.004	.011	.028	12
	13	0+	0+	0+	0+	0+	0+	0+	0+	.001	.003	.009	13
	14	0+	0+	0+	0+	0+	0+	0+	0+	0+	.001	.002	14
	15	0+	0+	0+	0+	0+	0+	0+	0+	0+	0+	0+	15
	16	0+	0+	0+	0+	0+	0+	0+	0+	0+	0+	0+	16
17	0	.843	.418	.167	.063	.023	.008	.002	.001	0+	0+	0+	0
	1	.145	.374	.315	.189	.096	.043	.017	.006	.002	.001	0+	1
	2	.012	.158	.280	.267	.191	.114	.058	.026	.010	.004	.001	2
	3	.001	.041	.156	.236	.239	.189	.125	.070	.034	.014	.005	3
	4	0+	.008	.060	.146	.209	.221	.187	.132	.080	.041	.018	4
	5	0+	.001	.017	.067	.136	.191	.208	.185	.138	.087	.047	5
	6	0+	0+	.004	.024	.068	.128	.178	.199	.184	.143	.094	6
	7	0+	0+	.001	.007	.027	.067	.120	.168	.193	.184	.148	7
	8	0+	0+	0+	.001	.008	.028	.064	.113	.161	.188	.185	8
	9	0+	0+	0+	0+	.002	.009	.028	.061	.107	.154	.185	9
	10	0+	0+	0+	0+	0+	.002	.009	.026	.057	.101	.148	10
	11	0+	0+	0+	0+	0+	.001	.003	.009	.024	.052	.094	11
	12	0+	0+	0+	0+	0+	0+	.001	.002	.008	.021	.047	12
	13	0+	0+	0+	0+	0+	0+	0+	.001	.002	.007	.018	13
	14	0+	0+	0+	0+	0+	0+	0+	0+	0+	.002	.005	14
	15	0+	0+	0+	0+	0+	0+	0+	0+	0+	0+	.001	15
	16	0+	0+	0+	0+	0+	0+	0+	0+	0+	0+	0+	16
	17	0+	0+	0+	0+	0+	0+	0+	0+	0+	0+	0+	17

TABLE 3 (Continued)

n	k	.01	.05	.10	.15	.20	.25	.30	.35	.40	.45	.50	k
18	0	.835	.397	.150	.054	.018	.006	.002	0+	0+	0+	0+	0
	1	.152	.376	.300	.170	.081	.034	.013	.004	.001	0+	0+	1
	2	.013	.168	.284	.256	.172	.096	.046	.019	.007	.002	.001	2
	3	.001	.047	.168	.241	.230	.170	.105	.055	.025	.009	.003	3
	4	0+	.009	.070	.159	.215	.213	.168	.110	.061	.029	.012	4
	5	0+	.001	.022	.079	.151	.199	.202	.166	.115	.067	.033	5
	6	0+	0+	.005	.030	.082	.144	.187	.194	.166	.118	.071	6
	7	0+	0+	.001	.009	.035	.082	.138	.179	.189	.166	.121	7
	8	0+	0+	0+	.002	.012	.038	.081	.133	.173	.186	.167	8
	9	0+	0+	0+	0+	.003	.014	.039	.079	.128	.169	.185	9
	10	0+	0+	0+	0+	.001	.004	.015	.038	.077	.125	.167	10
	11	0+	0+	0+	0+	0+	.001	.005	.015	.037	.074	.121	11
	12	0+	0+	0+	0+	0+	0+	.001	.005	.015	.035	.071	12
	13	0+	0+	0+	0+	0+	0+	0+	.001	.004	.013	.033	13
	14	0+	0+	0+	0+	0+	0+	0+	0+	.001	.004	.012	14
	15	0+	0+	0+	0+	0+	0+	0+	0+	0+	.001	.003	15
	16	0+	0+	0+	0+	0+	0+	0+	0+	0+	0+	.001	16
	17	0+	0+	0+	0+	0+	0+	0+	0+	0+	0+	0+	17
	18	0+	0+	0+	0+	0+	0+	0+	0+	0+	0+	0+	18
19	0	.826	.377	.135	.046	.014	.004	.001	0+	0+	0+	0+	0
	1	.159	.377	.285	.153	.068	.027	.009	.003	.001	0+	0+	1
	2	.014	.179	.285	.243	.154	.080	.036	.014	.005	.001	0+	2
	3	.001	.053	.180	.243	.218	.152	.087	.042	.017	.006	.002	3
	4	0+	.011	.080	.171	.218	.202	.149	.091	.047	.020	.007	4
	5	0+	.002	.027	.091	.164	.202	.192	.147	.093	.050	.022	5
	6	0+	0+	.007	.037	.095	.157	.192	.184	.145	.095	.052	6
	7	0+	0+	.001	.012	.044	.097	.153	.184	.180	.144	.096	7
	8	0+	0+	0+	.003	.017	.049	.098	.149	.180	.177	.144	8
	9	0+	0+	0+	.001	.005	.020	.051	.098	.146	.177	.176	9
	10	0+	0+	0+	0+	.001	.007	.022	.053	.098	.145	.176	10
	11	0+	0+	0+	0+	0+	.002	.008	.023	.053	.097	.144	11
	12	0+	0+	0+	0+	0+	0+	.002	.008	.024	.053	.096	12
	13	0+	0+	0+	0+	0+	0+	.001	.002	.008	.023	.052	13
	14	0+	0+	0+	0+	0+	0+	0+	.001	.002	.008	.022	14
	15	0+	0+	0+	0+	0+	0+	0+	0+	.001	.002	.007	15
	16	0+	0+	0+	0+	0+	0+	0+	0+	0+	0+	.002	16
	17	0+	0+	0+	0+	0+	0+	0+	0+	0+	0+	0+	17
	18	0+	0+	0+	0+	0+	0+	0+	0+	0+	0+	0+	18
	19	0+	0+	0+	0+	0+	0+	0+	0+	0+	0+	0+	19
20	0	.818	.358	.122	.039	.012	.003	.001	0+	0+	0+	0+	0
	1	.165	.377	.270	.137	.058	.021	.007	.002	0+	0+	0+	1
	2	.016	.189	.285	.229	.137	.067	.028	.010	.003	.001	0+	2
	3	.001	.060	.190	.243	.205	.134	.072	.032	.012	.004	.001	3
	4	0+	.013	.090	.182	.218	.190	.130	.074	.035	.014	.005	4
	5	0+	.002	.032	.103	.175	.202	.179	.127	.075	.036	.015	5
	6	0+	0+	.009	.045	.109	.169	.192	.171	.124	.075	.037	6
	7	0+	0+	.002	.016	.055	.112	.164	.184	.166	.122	.074	7
	8	0+	0+	0+	.005	.022	.061	.114	.161	.180	.162	.120	8
	9	0+	0+	0+	.001	.007	.027	.065	.116	.160	.177	.160	9
	10	0+	0+	0+	0+	.002	.010	.031	.069	.117	.159	.176	10
	11	0+	0+	0+	0+	0+	.003	.012	.034	.071	.119	.160	11
	12	0+	0+	0+	0+	0+	.001	.004	.014	.035	.073	.120	12
	13	0+	0+	0+	0+	0+	0+	.001	.004	.015	.037	.074	13
	14	0+	0+	0+	0+	0+	0+	0+	.001	.005	.015	.037	14
	15	0+	0+	0+	0+	0+	0+	0+	0+	.001	.005	.015	15
	16	0+	0+	0+	0+	0+	0+	0+	0+	0+	.001	.005	16
	17	0+	0+	0+	0+	0+	0+	0+	0+	0+	0+	.001	17
	18	0+	0+	0+	0+	0+	0+	0+	0+	0+	0+	0+	18
	19	0+	0+	0+	0+	0+	0+	0+	0+	0+	0+	0+	19
	20	0+	0+	0+	0+	0+	0+	0+	0+	0+	0+	0+	20

TABLE 4
Standard normal cumulative distribution

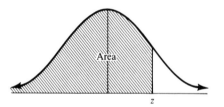

z	0.00	0.01	0.02	0.03	0.04	0.05	0.06	0.07	0.08	0.09
−3.4	0.0003	0.0003	0.0003	0.0003	0.0003	0.0003	0.0003	0.0003	0.0003	0.0002
−3.3	0.0005	0.0005	0.0005	0.0004	0.0004	0.0004	0.0004	0.0004	0.0004	0.0003
−3.2	0.0007	0.0007	0.0006	0.0006	0.0006	0.0006	0.0006	0.0005	0.0005	0.0005
−3.1	0.0010	0.0009	0.0009	0.0009	0.0008	0.0008	0.0008	0.0008	0.0007	0.0007
−3.0	0.0013	0.0013	0.0013	0.0012	0.0012	0.0011	0.0011	0.0011	0.0010	0.0010
−2.9	0.0019	0.0018	0.0017	0.0017	0.0016	0.0016	0.0015	0.0015	0.0014	0.0014
−2.8	0.0026	0.0025	0.0024	0.0023	0.0023	0.0022	0.0021	0.0021	0.0020	0.0019
−2.7	0.0035	0.0034	0.0033	0.0032	0.0031	0.0030	0.0029	0.0028	0.0027	0.0026
−2.6	0.0047	0.0045	0.0044	0.0043	0.0041	0.0040	0.0039	0.0038	0.0037	0.0036
−2.5	0.0062	0.0060	0.0059	0.0057	0.0055	0.0054	0.0052	0.0051	0.0049	0.0048
−2.4	0.0082	0.0080	0.0078	0.0075	0.0073	0.0071	0.0069	0.0068	0.0066	0.0064
−2.3	0.0107	0.0104	0.0102	0.0099	0.0096	0.0094	0.0091	0.0089	0.0087	0.0084
−2.2	0.0139	0.0136	0.0132	0.0129	0.0125	0.0122	0.0119	0.0116	0.0113	0.0110
−2.1	0.0179	0.0174	0.0170	0.0166	0.0162	0.0158	0.0154	0.0150	0.0146	0.0143
−2.0	0.0228	0.0222	0.0217	0.0212	0.0207	0.0202	0.0197	0.0192	0.0188	0.0183
−1.9	0.0287	0.0281	0.0274	0.0268	0.0262	0.0256	0.0250	0.0244	0.0239	0.0233
−1.8	0.0359	0.0352	0.0344	0.0336	0.0329	0.0322	0.0314	0.0307	0.0301	0.0294
−1.7	0.0446	0.0436	0.0427	0.0418	0.0409	0.0401	0.0392	0.0384	0.0375	0.0367
−1.6	0.0548	0.0537	0.0526	0.0516	0.0505	0.0495	0.0485	0.0475	0.0465	0.0455
−1.5	0.0668	0.0655	0.0643	0.0630	0.0618	0.0606	0.0594	0.0582	0.0571	0.0559
−1.4	0.0808	0.0793	0.0778	0.0764	0.0749	0.0735	0.0722	0.0708	0.0694	0.0681
−1.3	0.0968	0.0951	0.0934	0.0918	0.0901	0.0885	0.0869	0.0853	0.0838	0.0823
−1.2	0.1151	0.1131	0.1112	0.1093	0.1075	0.1056	0.1038	0.1020	0.1003	0.0985
−1.1	0.1357	0.1335	0.1314	0.1292	0.1271	0.1251	0.1230	0.1210	0.1190	0.1170
−1.0	0.1587	0.1562	0.1539	0.1515	0.1492	0.1469	0.1446	0.1423	0.1401	0.1379
−0.9	0.1841	0.1814	0.1788	0.1762	0.1736	0.1711	0.1685	0.1660	0.1635	0.1611
−0.8	0.2119	0.2090	0.2061	0.2033	0.2005	0.1977	0.1949	0.1922	0.1894	0.1867
−0.7	0.2420	0.2389	0.2358	0.2327	0.2296	0.2266	0.2236	0.2206	0.2177	0.2148
−0.6	0.2743	0.2709	0.2676	0.2643	0.2611	0.2578	0.2546	0.2514	0.2483	0.2451
−0.5	0.3085	0.3050	0.3015	0.2981	0.2946	0.2912	0.2877	0.2843	0.2810	0.2776
−0.4	0.3446	0.3409	0.3372	0.3336	0.3300	0.3264	0.3228	0.3192	0.3156	0.3121
−0.3	0.3821	0.3783	0.3745	0.3707	0.3669	0.3632	0.3594	0.3557	0.3520	0.3483
−0.2	0.4207	0.4168	0.4129	0.4090	0.4052	0.4013	0.3974	0.3936	0.3897	0.3859
−0.1	0.4602	0.4562	0.4522	0.4483	0.4443	0.4404	0.4364	0.4325	0.4286	0.4247
−0.0	0.5000	0.4960	0.4920	0.4880	0.4840	0.4801	0.4761	0.4721	0.4681	0.4641

TABLE 4 (Continued)

z	0.00	0.01	0.02	0.03	0.04	0.05	0.06	0.07	0.08	0.09
0.0	0.5000	0.5040	0.5080	0.5120	0.5160	0.5199	0.5239	0.5279	0.5319	0.5359
0.1	0.5398	0.5438	0.5478	0.5517	0.5557	0.5596	0.5636	0.5675	0.5714	0.5753
0.2	0.5793	0.5832	0.5871	0.5910	0.5948	0.5987	0.6026	0.6064	0.6103	0.6141
0.3	0.6179	0.6217	0.6255	0.6293	0.6331	0.6368	0.6406	0.6443	0.6480	0.6517
0.4	0.6554	0.6591	0.6628	0.6664	0.6700	0.6736	0.6772	0.6808	0.6844	0.6879
0.5	0.6915	0.6950	0.6985	0.7019	0.7054	0.7088	0.7123	0.7157	0.7190	0.7224
0.6	0.7257	0.7291	0.7324	0.7357	0.7389	0.7422	0.7454	0.7486	0.7517	0.7549
0.7	0.7580	0.7611	0.7642	0.7673	0.7704	0.7734	0.7764	0.7794	0.7823	0.7852
0.8	0.7881	0.7910	0.7939	0.7967	0.7995	0.8023	0.8051	0.8078	0.8106	0.8133
0.9	0.8159	0.8186	0.8212	0.8238	0.8264	0.8289	0.8315	0.8340	0.8365	0.8389
1.0	0.8413	0.8438	0.8461	0.8485	0.8508	0.8531	0.8554	0.8577	0.8599	0.8621
1.1	0.8643	0.8665	0.8686	0.8708	0.8729	0.8749	0.8770	0.8790	0.8810	0.8830
1.2	0.8849	0.8869	0.8888	0.8907	0.8925	0.8944	0.8962	0.8980	0.8997	0.9015
1.3	0.9032	0.9049	0.9066	0.9082	0.9099	0.9115	0.9131	0.9147	0.9162	0.9177
1.4	0.9192	0.9207	0.9222	0.9236	0.9251	0.9265	0.9278	0.9292	0.9306	0.9319
1.5	0.9332	0.9345	0.9357	0.9370	0.9382	0.9394	0.9406	0.9418	0.9429	0.9441
1.6	0.9452	0.9463	0.9474	0.9484	0.9495	0.9505	0.9515	0.9525	0.9535	0.9545
1.7	0.9554	0.9564	0.9573	0.9582	0.9591	0.9599	0.9608	0.9616	0.9625	0.9633
1.8	0.9641	0.9649	0.9656	0.9664	0.9671	0.9678	0.9686	0.9693	0.9699	0.9706
1.9	0.9713	0.9719	0.9726	0.9732	0.9738	0.9744	0.9750	0.9756	0.9761	0.9767
2.0	0.9772	0.9778	0.9783	0.9788	0.9793	0.9798	0.9803	0.9808	0.9812	0.9817
2.1	0.9821	0.9826	0.9830	0.9834	0.9838	0.9842	0.9846	0.9850	0.9854	0.9857
2.2	0.9861	0.9864	0.9868	0.9871	0.9875	0.9878	0.9881	0.9884	0.9887	0.9890
2.3	0.9893	0.9896	0.9898	0.9901	0.9904	0.9906	0.9909	0.9911	0.9913	0.9916
2.4	0.9918	0.9920	0.9922	0.9925	0.9927	0.9929	0.9931	0.9932	0.9934	0.9936
2.5	0.9938	0.9940	0.9941	0.9943	0.9945	0.9946	0.9948	0.9949	0.9951	0.9952
2.6	0.9953	0.9955	0.9956	0.9957	0.9959	0.9960	0.9961	0.9962	0.9963	0.9964
2.7	0.9965	0.9966	0.9967	0.9968	0.9969	0.9970	0.9971	0.9972	0.9973	0.9974
2.8	0.9974	0.9975	0.9976	0.9977	0.9977	0.9978	0.9979	0.9979	0.9980	0.9981
2.9	0.9981	0.9982	0.9982	0.9983	0.9984	0.9984	0.9985	0.9985	0.9986	0.9986
3.0	0.9987	0.9987	0.9987	0.9988	0.9988	0.9989	0.9989	0.9989	0.9990	0.9990
3.1	0.9990	0.9991	0.9991	0.9991	0.9992	0.9992	0.9992	0.9992	0.9993	0.9993
3.2	0.9993	0.9993	0.9994	0.9994	0.9994	0.9994	0.9994	0.9995	0.9995	0.9995
3.3	0.9995	0.9995	0.9995	0.9996	0.9996	0.9996	0.9996	0.9996	0.9996	0.9997
3.4	0.9997	0.9997	0.9997	0.9997	0.9997	0.9997	0.9997	0.9997	0.9997	0.9998

TABLE 5
Compound interest

Compounded amount of $1 for N periods at i% per period.

N	1%	1½%	2%	2½%	3%	3½%	4%	N
1	1.010000	1.015000	1.020000	1.025000	1.030000	1.035000	1.040000	1
2	1.020100	1.030225	1.040400	1.050625	1.060900	1.071225	1.081600	2
3	1.030301	1.045678	1.061208	1.076891	1.092727	1.108718	1.124864	3
4	1.040604	1.061364	1.082432	1.103813	1.125509	1.147523	1.169859	4
5	1.051010	1.077284	1.104081	1.131408	1.159274	1.187686	1.216653	5
6	1.061520	1.093443	1.126162	1.159693	1.194052	1.229255	1.265319	6
7	1.072135	1.109845	1.148686	1.188686	1.229874	1.272279	1.315932	7
8	1.082857	1.126493	1.171659	1.218403	1.266770	1.316809	1.368569	8
9	1.093685	1.143390	1.195093	1.248863	1.304773	1.362897	1.423312	9
10	1.104622	1.160541	1.218994	1.280085	1.343916	1.410599	1.480244	10
11	1.115668	1.177949	1.243374	1.312087	1.384234	1.459970	1.539454	11
12	1.126825	1.195618	1.268242	1.344889	1.425761	1.511069	1.601032	12
13	1.138093	1.213552	1.293607	1.378511	1.468534	1.563956	1.665074	13
14	1.149474	1.231756	1.319479	1.412974	1.512590	1.618695	1.731676	14
15	1.169069	1.250232	1.345868	1.448298	1.557967	1.675349	1.800944	15
16	1.172579	1.268986	1.372786	1.484506	1.604706	1.733986	1.872981	16
17	1.184304	1.288020	1.400241	1.521618	1.652848	1.794676	1.947900	17
18	1.196147	1.307341	1.428246	1.559659	1.702433	1.857489	2.025817	18
19	1.208109	1.326951	1.456811	1.598650	1.753506	1.922501	2.106849	19
20	1.220190	1.346855	1.485947	1.638616	1.806111	1.989789	2.191123	20
25	1.282432	1.450945	1.640606	1.853944	2.093778	2.363245	2.665836	25
30	1.347849	1.563080	1.811362	2.097568	2.427262	2.806794	3.243398	30
35	1.416603	1.683881	1.999890	2.373205	2.813862	3.333590	3.946089	35
40	1.488864	1.814018	2.208040	2.685064	3.262038	3.959260	4.801021	40
45	1.564811	1.954213	2.437854	3.037903	3.781596	4.702359	5.841176	45
50	1.644632	2.105242	2.691588	3.437109	4.383906	5.584927	7.106683	50
55	1.728525	2.267944	2.971731	3.888773	5.082149	6.633141	8.646367	55
60	1.816697	2.443220	3.281031	4.399790	5.891603	7.878091	10.519627	60
65	1.909366	2.632042	3.622523	4.977958	6.829983	9.356701	12.798735	65
70	2.006763	2.835456	3.999558	5.632103	7.917822	11.112825	15.571618	70
75	2.109128	3.054592	4.415835	6.372207	9.178926	13.198550	18.945255	75
80	2.216715	3.290663	4.875439	7.209568	10.640891	15.675738	23.049799	80
85	2.329790	3.544978	5.382879	8.156964	12.335709	18.617859	28.043605	85
90	2.448633	3.818949	5.943133	9.228856	14.300467	22.112176	34.119333	90
95	2.573538	4.114092	6.561699	10.441604	16.578161	26.262329	41.511386	95
100	2.704814	4.432046	7.244646	11.813716	19.218632	31.191408	50.504948	100

N	4½%	5%	5½%	6%	6½%	7%	7½%	N
1	1.045000	1.050000	1.055000	1.060000	1.065000	1.070000	1.075000	1
2	1.092025	1.102500	1.113025	1.123600	1.134225	1.144900	1.155625	2
3	1.141166	1.157625	1.174241	1.191016	1.207950	1.225043	1.242297	3
4	1.192519	1.215506	1.238825	1.262477	1.286466	1.310796	1.335469	4
5	1.246182	1.276282	1.306960	1.338226	1.370087	1.402552	1.435629	5
6	1.302260	1.340096	1.378843	1.418519	1.459142	1.500730	1.543302	6
7	1.360862	1.407100	1.454679	1.503630	1.553987	1.605781	1.659049	7
8	1.422101	1.477455	1.534687	1.593848	1.654996	1.718186	1.783478	8
9	1.486095	1.551328	1.619094	1.689479	1.762570	1.838459	1.917239	9
10	1.552969	1.628895	1.708144	1.790848	1.877137	1.967151	2.061032	10
11	1.622853	1.710339	1.802092	1.898299	1.999151	2.104852	2.215609	11
12	1.695881	1.795856	1.901207	2.012196	2.129096	2.252192	2.381780	12
13	1.772196	1.885649	2.005774	2.132928	2.267487	2.409845	2.560413	13
14	1.851945	1.979932	2.116091	2.260904	2.414874	2.578534	2.752444	14
15	1.935282	2.078928	2.232476	2.396558	2.571841	2.759032	2.958877	15
16	2.022370	2.182875	2.355263	2.540352	2.739011	2.952164	3.180793	16
17	2.113377	2.292018	2.484802	2.692773	2.917046	3.158815	3.419353	17
18	2.208479	2.406619	2.621466	2.854339	3.106654	3.379932	3.675804	18
19	2.307860	2.526950	2.765647	3.025600	3.308587	3.616528	3.951489	19
20	2.411714	2.653298	2.917757	3.207135	3.523645	3.869684	4.247851	20
25	3.005434	3.386355	3.813392	4.291871	4.827699	5.427433	6.098340	25
30	3.745318	4.321942	4.983951	5.743491	6.614366	7.612255	8.754955	30
35	4.667348	5.516015	6.513825	7.686087	9.062255	10.676581	12.568870	35
40	5.816365	7.039989	8.513309	10.285718	12.416075	14.974458	18.044239	40
45	7.248248	8.985008	11.126554	13.764611	17.011098	21.002452	25.904839	45
50	9.032636	11.467400	14.541961	18.420154	23.306679	29.457025	37.189746	50
55	11.256308	14.635631	19.005762	24.650322	31.932170	41.315001	53.390690	55
60	14.027408	18.679186	24.839770	32.987691	43.749840	57.946427	76.649240	60
65	17.480702	23.839901	32.464587	44.144972	59.941072	81.272861	110.039897	65
70	21.784136	30.426426	42.429916	59.075930	82.124463	113.989392	157.976504	70
75	27.146996	38.832686	55.454204	79.056921	112.517632	159.876019	226.795701	75
80	33.830096	49.561441	72.476426	105.795993	154.158907	224.234388	325.594560	80
85	42.158455	63.254353	94.723791	141.578904	211.211062	314.500328	467.433099	85
90	52.537105	80.730365	123.800206	189.464511	289.377460	441.102980	671.060665	90
95	65.470792	103.034676	161.801918	253.546255	396.472198	618.669748	963.394370	95
100	81.588518	131.501258	211.468686	339.302084	543.201271	867.716326	1383.077210	100

TABLE 5 (Continued)
Future and present value

Use this table to find the future value (multiply principle by table number) and to find present value (divide future value by table number).

N	8%	9%	10%	11%	12%	13%	14%	N
1	1.080000	1.090000	1.100000	1.110000	1.120000	1.130000	1.140000	1
2	1.166400	1.188100	1.210000	1.232100	1.254400	1.276900	1.299600	2
3	1.259712	1.295029	1.331000	1.367631	1.404928	1.442897	1.481544	3
4	1.360489	1.411582	1.464100	1.518070	1.573519	1.630474	1.688960	4
5	1.469328	1.538624	1.610510	1.685058	1.762342	1.842435	1.925415	5
6	1.586874	1.677100	1.771561	1.870415	1.973823	2.081952	2.194973	6
7	1.713824	1.828039	1.948717	2.076160	2.210681	2.352605	2.502269	7
8	1.850930	1.992563	2.143589	2.304538	2.475963	2.658444	2.852586	8
9	1.999005	2.171893	2.357948	2.558037	2.773079	3.004042	3.251949	9
10	2.158925	2.367364	2.593742	2.839421	3.105848	3.394567	3.707221	10
11	2.331639	2.580426	2.853117	3.151757	3.478550	3.835861	4.226232	11
12	2.518170	2.812665	3.138428	3.498451	3.895976	4.334523	4.817905	12
13	2.719624	3.065805	3.452271	3.883280	4.363493	4.898011	5.492411	13
14	2.937194	3.341727	3.797498	4.310441	4.887112	5.534753	6.261349	14
15	3.172169	3.642482	4.177248	4.784589	5.473566	6.254270	7.137938	15
16	3.425943	3.970306	4.594973	5.310894	6.130394	7.067326	8.137249	16
17	3.700018	4.327633	5.054470	5.895093	6.866041	7.986078	9.276464	17
18	3.996019	4.717120	5.559917	6.543553	7.689966	9.024268	10.575169	18
19	4.315701	5.141661	6.115909	7.263344	8.612762	10.197423	12.055693	19
20	4.660957	5.604411	6.727500	8.062312	9.646293	11.523088	13.743490	20
25	6.848475	8.623081	10.834706	13.585464	17.000064	21.230542	26.461916	25
30	10.062657	13.267678	17.449402	22.892297	29.959922	39.115898	50.950159	30
35	14.785344	20.413968	28.102437	38.574851	52.799620	72.068506	98.100178	35
40	21.724521	31.409420	45.259256	65.000867	93.050970	132.781552	188.883514	40
45	31.920449	48.327286	72.890484	109.530242	163.987604	244.641402	363.679072	45
50	46.901613	74.357520	117.390853	184.564827	289.002190	450.735925	700.232988	50
55	68.913856	114.408262	189.059142	311.002466	509.320606	830.451725	1348.238807	55
60	101.257064	176.031292	304.481640	524.057242	897.596933	1530.053473	2595.918660	60
65	148.779847	270.845963	490.370725	883.066930	1581.872491	2819.024345	4998.219642	65
70	218.606406	416.730086	789.746957	1488.019132	2787.799828	5193.869624	9623.644985	70
75	321.204530	641.190893	1271.895371	2507.398773	4913.055841	9569.368113	18529.506390	75
80	471.954834	986.551668	2048.400215	4225.112750	8658.483100	17630.940454	35676.981807	80
85	693.456489	1517.932029	3298.969030	7119.560696	15259.205681	32483.864937	68692.981028	85
90	1018.915089	2335.526582	5313.022612	11996.873812	26891.934223	59849.415520	132262.467379	90
95	1497.120549	3593.497147	8556.676047	20215.430053	47392.776624	110268.668614	254660.083396	95
100	2199.761256	5529.040792	13780.612340	34064.175270	83522.265727	203162.874228	490326.238126	100

N	15%	16%	17%	18%	19%	20%	N
1	1.150000	1.160000	1.170000	1.180000	1.190000	1.200000	1
2	1.322500	1.345600	1.368900	1.392400	1.416100	1.440000	2
3	1.520875	1.560896	1.601613	1.643032	1.685159	1.728000	3
4	1.749006	1.810639	1.873887	1.938778	2.005339	2.073600	4
5	2.011357	2.100342	2.192448	2.287758	2.386354	2.488320	5
6	2.313061	2.436396	2.565164	2.699554	2.839761	2.985984	6
7	2.660020	2.826220	3.001242	3.185474	3.379315	3.583181	7
8	3.059023	3.278415	3.511453	3.758859	4.021385	4.299817	8
9	3.517876	3.802961	4.108400	4.435454	4.785449	5.159780	9
10	4.045558	4.411435	4.806828	5.233836	5.694684	6.191736	10
11	4.652391	5.117265	5.623989	6.175996	6.776674	7.430084	11
12	5.350250	5.936027	6.580067	7.287593	8.064242	8.916100	12
13	6.152788	6.885791	7.698679	8.599359	9.596448	10.699321	13
14	7.075706	7.987518	9.007454	10.147244	11.419773	12.839185	14
15	8.137062	9.265521	10.538721	11.973748	13.589530	15.407022	15
16	9.357621	10.748004	12.330304	14.129023	16.171540	18.488426	16
17	10.761264	12.467685	14.426456	16.672247	19.244133	22.186111	17
18	12.375454	14.462514	16.878953	19.673251	22.900518	26.623333	18
19	14.231772	16.776517	19.748375	23.214436	27.251616	31.948000	19
20	16.366537	19.460759	23.105599	27.393035	32.429423	38.337600	20
25	32.918953	40.874244	50.657826	62.668627	77.388073	95.396217	25
30	66.211772	85.849877	111.064650	143.370638	184.675312	237.376314	30
35	133.175523	180.314073	243.503474	327.997290	440.700607	590.668229	35
40	267.863546	378.721158	533.868713	750.378345	1051.667507	1469.771568	40
45	538.769269	795.443826	1170.479411	1716.683879	2509.650603	3657.261988	45
50	1083.657442	1670.703804	2566.215284	3927.356860	5988.913902	9100.438150	50
55	2179.622184	3509.048796	5626.293659	8984.841120	14291.666609	22644.802257	55
60	4383.998746	7370.201365	12335.356482	20555.139966	34104.970919	56347.514353	60
65	8817.787387	15479.940952	27044.628088	47025.180900	81386.522174	140210.646915	65
70	17735.720039	32513.164839	59293.941729	107582.222368	194217.025056	348888.956932	70
75	35672.867976	68288.754533	129998.886072	246122.063716	463470.508558	868147.369314	75
80	71750.879401	143429.715890	285015.802412	563067.660386	1106004.544354	2160228.462010	80
85	144316.646994	301251.407222	624882.336142	1288162.407650	2639317.992285	5375339.686589	85
90	290272.325206	632730.879999	1370022.050417	2947003.540121	6298346.150529	13375565.248934	90
95	583841.327636	1328551.025313	3003702.153303	6742030.208228	15030081.387632	33282686.520228	95
100	1174313.450700	2791651.199375	6585460.885837	15424131.905453	35867089.727971	82817974.522015	100

TABLE 6
Ordinary annuity

Amount of ordinary annuity of $1 at compound interest. Remember, an ordinary annuity requires payment at the end of the period.

i / N	1%	1½%	2%	2½%	3%	3½%	4%	N
1	1.000000	1.000000	1.000000	1.000000	1.000000	1.000000	1.000000	1
2	2.010000	2.015000	2.020000	2.025000	2.030000	2.035000	2.040000	2
3	3.030100	3.045225	3.060400	3.075625	3.090900	3.106225	3.121600	3
4	4.060401	4.090903	4.121608	4.152516	4.183627	4.214943	4.246464	4
5	5.101005	5.152267	5.204040	5.256329	5.309136	5.362466	5.416323	5
6	6.152015	6.229551	6.308121	6.387737	6.468410	6.550152	6.632975	6
7	7.213535	7.322994	7.434283	7.547430	7.662462	7.779408	7.898294	7
8	8.285671	8.432839	8.582969	8.736116	8.892336	9.051687	9.214226	8
9	9.368527	9.559332	9.754628	9.954519	10.159106	10.368496	10.582795	9
10	10.462213	10.702722	10.949721	11.203382	11.463879	11.731393	12.006107	10
11	11.566835	11.863262	12.168715	12.483466	12.807796	13.141992	13.486351	11
12	12.682503	13.041211	13.412090	13.795553	14.192030	14.601962	15.025805	12
13	13.809328	14.236830	14.680332	15.140442	15.617790	16.113030	16.626838	13
14	14.947421	15.450382	15.973938	16.518953	17.086324	17.676986	18.291911	14
15	16.096896	16.682138	17.293417	17.931927	18.598914	19.295681	20.023588	15
16	17.257864	17.932370	18.639285	19.380225	20.156881	20.971030	21.824531	16
17	18.430443	19.201355	20.012071	20.864730	21.761588	22.705016	23.697512	17
18	19.614748	20.489376	21.412312	22.386349	23.414435	24.499691	25.645413	18
19	20.810895	21.796716	22.840559	23.946007	25.116868	26.357180	27.671229	19
20	22.019004	23.123667	24.297370	25.544658	26.870374	28.279682	29.778079	20
25	28.243200	30.063024	32.030300	34.157764	36.459264	38.949857	41.645908	25
30	34.784892	37.538681	40.568079	43.902703	47.575416	51.622677	56.084938	30
35	41.660276	45.592088	49.994478	54.928207	60.462082	66.674013	73.652225	35
40	48.886373	54.267894	60.401983	67.402554	75.401260	84.550278	95.025516	40
45	56.481075	63.614201	71.892710	81.516131	92.719861	105.781673	121.029392	45
50	64.463182	73.682828	84.579401	97.484349	112.796867	130.997910	152.667084	50
55	72.852457	84.529599	98.586534	115.550921	136.071620	160.946890	191.159173	55
60	81.669670	96.214652	114.051539	135.991590	163.053437	196.516883	237.990685	60
65	90.936649	108.802772	131.126155	159.118330	194.332758	238.762876	294.968380	65
70	100.676337	122.363753	149.977911	185.284114	230.594064	288.937865	364.290459	70
75	110.912847	136.972781	170.791773	214.888297	272.630856	348.530011	448.631367	75
80	121.671522	152.710852	193.771958	248.382713	321.363019	419.306787	551.244977	80
85	132.978997	169.665226	219.143939	286.278570	377.856952	503.367394	676.000123	85
90	144.863267	187.929900	247.156656	329.154253	443.348904	603.205027	827.983334	90
95	157.353755	207.606142	278.084960	377.664154	519.272026	721.780816	1012.784648	95
100	170.481383	228.803043	312.232306	432.548654	607.287733	862.611657	1237.623705	100

i / N	4½%	5%	5½%	6%	6½%	7%	7½%	N
1	1.000000	1.000000	1.000000	1.000000	1.000000	1.000000	1.000000	1
2	2.045000	2.050000	2.055000	2.060000	2.065000	2.070000	2.075000	2
3	3.137025	3.152500	3.168025	3.183600	3.199225	3.214900	3.230625	3
4	4.278191	4.310125	4.342266	4.374616	4.407175	4.439943	4.472922	4
5	5.470710	5.525631	5.581091	5.637093	5.693641	5.750739	5.808391	5
6	6.716892	6.801913	6.888051	6.975319	7.063728	7.153291	7.244020	6
7	8.019152	8.142008	8.266894	8.393838	8.522870	8.654021	8.787322	7
8	9.380014	9.549109	9.721573	9.897468	10.076856	10.259803	10.446371	8
9	10.802114	11.026564	11.256260	11.491316	11.731852	11.977989	12.229849	9
10	12.288209	12.577893	12.875354	13.180795	13.494423	13.816448	14.147087	10
11	13.841179	14.206787	14.583498	14.971643	15.371560	15.783599	16.208119	11
12	15.464032	15.917127	16.385591	16.869941	17.370711	17.888451	18.423728	12
13	17.159913	17.712983	18.286798	18.882138	19.499808	20.140643	20.805508	13
14	18.932109	19.598632	20.292572	21.015066	21.767295	22.550488	23.365921	14
15	20.784054	21.578564	22.408663	23.275970	24.182169	25.129022	26.118365	15
16	22.719337	23.657492	24.641140	25.672528	26.754010	27.888054	29.077242	16
17	24.741707	25.840366	26.996403	28.212880	29.493021	30.840217	32.258035	17
18	26.855084	28.132385	29.481205	30.905653	32.410067	33.999033	35.677388	18
19	29.063562	30.539004	32.102671	33.759992	35.516722	37.378965	39.353192	19
20	31.371423	33.065954	34.868318	36.785591	38.825309	40.995492	43.304681	20
25	44.565210	47.727099	51.152588	54.864512	58.887679	63.249038	67.977862	25
30	61.007070	66.438848	72.435478	79.058186	86.374864	94.460786	103.399403	30
35	81.496618	90.320307	100.251364	111.434780	124.034690	138.236878	154.251606	35
40	107.030323	120.799774	136.605614	154.761966	175.631916	199.635112	227.256020	40
45	138.849965	159.700156	184.119165	212.743514	246.324587	285.749311	332.064515	45
50	178.503028	209.347996	246.217476	290.335905	343.179672	406.528929	482.529947	50
55	227.917959	272.712618	327.377486	394.172027	475.879533	575.928593	698.542534	55
60	289.497954	353.583718	433.450372	533.128181	657.689842	813.520383	1008.656538	60
65	366.237831	456.798011	572.083392	719.082861	906.785722	1146.755161	1453.865297	65
70	461.869680	588.528511	753.271204	967.932170	1248.068666	1614.134174	2093.020048	70
75	581.044362	756.653718	990.076429	1300.948680	1715.655875	2269.657419	3010.609352	75
80	729.557699	971.228821	1299.571387	1746.599891	2356.290874	3189.062680	4327.927467	80
85	914.632336	1245.087069	1704.068919	2342.981741	3234.016343	4478.576120	6219.107984	85
90	1145.269007	1594.607301	2232.731017	3141.075187	4436.576302	6287.185427	8934.142195	90
95	1432.684259	2040.693529	2923.671235	4209.104250	6084.187663	8823.853541	12831.924930	95
100	1790.855956	2610.025157	3826.702467	5638.368059	8341.558016	12381.661794	18427.696132	100

TABLE 6 (Continued)

N	8%	9%	10%	11%	12%	13%	14%	N
1	1.000000	1.000000	1.000000	1.000000	1.000000	1.000000	1.000000	1
2	2.080000	2.090000	2.100000	2.110000	2.120000	2.130000	2.140000	2
3	3.246400	3.278100	3.310000	3.342100	3.374400	3.406900	3.439600	3
4	4.506112	4.573129	4.641000	4.709731	4.779328	4.849797	4.921144	4
5	5.866601	5.984711	6.105100	6.227801	6.352847	6.480271	6.610104	5
6	7.335929	7.523335	7.715610	7.912860	8.115189	8.322706	8.535519	6
7	8.922803	9.200435	9.487171	9.783274	10.089012	10.404658	10.730491	7
8	10.636628	11.028474	11.435888	11.859434	12.299693	12.757263	13.232760	8
9	12.487558	13.021036	13.579477	14.163972	14.775656	15.415707	16.085347	9
10	14.486562	15.192930	15.937425	16.722009	17.548735	18.419749	19.337295	10
11	16.645487	17.560293	18.531167	19.561430	20.654583	21.814317	23.044516	11
12	18.977126	20.140720	21.384284	22.713187	24.133133	25.650178	27.270749	12
13	21.495297	22.953385	24.522712	26.211638	28.029109	29.984701	32.088654	13
14	24.214920	26.019189	27.974983	30.094918	32.392602	34.882712	37.581065	14
15	27.152114	29.360916	31.772482	34.405359	37.279715	40.417464	43.842414	15
16	30.324283	33.003399	35.949730	39.189948	42.753280	46.671735	50.980352	16
17	33.750226	36.973705	40.544703	44.500843	48.883674	53.739060	59.117601	17
18	37.450244	41.301338	45.599173	50.395936	55.749715	61.725138	68.394066	18
19	41.446263	46.018458	51.159090	56.939488	63.439681	70.749406	78.969235	19
20	45.761964	51.160120	57.274999	64.202832	72.052442	80.946829	91.024928	20
25	73.105940	84.700896	98.347059	114.413307	133.333870	155.619556	181.870827	25
30	113.283211	136.307539	164.494023	199.020878	241.332684	293.199215	356.786847	30
35	172.316804	215.710755	271.024368	341.589555	431.663496	546.680819	693.572702	35
40	259.056519	337.882445	442.592556	581.826066	767.091420	1013.704243	1342.025099	40
45	386.505617	525.858734	718.904837	986.638559	1358.230032	1874.164630	2590.564800	45
50	573.770156	815.083556	1163.908529	1668.771152	2400.018249	3459.507117	4994.521346	50
55	848.923201	1260.091796	1880.591425	2818.204240	4236.005047	6380.397885	9623.134336	55
60	1253.213296	1944.792133	3034.816395	4755.065839	7471.641112	11761.949792	18535.133283	60
65	1847.248083	2998.288474	4893.707253	8018.790272	13173.937422	21677.110345	35694.426015	65
70	2720.080074	4619.223180	7687.469566	13518.355744	23223.331897	39945.150956	68733.178463	70
75	4002.556624	7113.232148	12708.953714	22785.443391	40933.798673	73602.831635	132346.474212	75
80	5886.935428	10950.574090	20474.002146	38401.025004	72145.692501	135614.926571	254828.441480	80
85	8655.706112	16854.800326	32979.690296	64714.188149	127151.714005	249868.191823	490657.007341	85
90	12723.938616	25939.184247	53120.226118	109053.398293	224091.118528	460372.427073	944724.766995	90
95	18701.506857	39916.634964	85556.760466	183767.545936	394931.471864	848212.835490	1818993.452831	95
100	27484.515704	61422.675465	137796.123398	309665.229724	696010.547721	1562783.647911	3502323.129475	100

N	15%	16%	17%	18%	19%	20%	N
1	1.000000	1.000000	1.000000	1.000000	1.000000	1.000000	1
2	2.150000	2.160000	2.170000	2.180000	2.190000	2.200000	2
3	3.472500	3.505600	3.538900	3.572400	3.606100	3.640000	3
4	4.993375	5.066496	5.140513	5.215432	5.291259	5.368000	4
5	6.742381	6.877135	7.014400	7.154210	7.296598	7.441600	5
6	8.753738	8.977477	9.206848	9.441968	9.682952	9.929920	6
7	11.066799	11.413873	11.772012	12.141522	12.522713	12.915904	7
8	13.726819	14.240093	14.773255	15.326996	15.902028	16.499085	8
9	16.785842	17.518508	18.284708	19.085855	19.923413	20.798902	9
10	20.303718	21.321469	22.393108	23.521308	24.708862	25.958682	10
11	24.349276	25.732904	27.199937	28.755144	30.403546	32.150419	11
12	29.001667	30.850169	32.823926	34.931070	37.180220	39.580502	12
13	34.351917	36.786196	39.403993	42.218663	45.244461	48.496603	13
14	40.504705	43.671987	47.102672	50.818022	54.840909	59.195923	14
15	47.580411	51.659505	56.110126	60.965266	66.260682	72.035108	15
16	55.717472	60.925026	66.648848	72.939014	79.850211	87.442129	16
17	65.075093	71.673030	78.979152	87.068036	96.021751	105.930555	17
18	75.836357	84.140715	93.405608	103.740283	115.265884	128.116666	18
19	88.211811	98.603230	110.284561	123.413534	138.166402	154.740000	19
20	102.443583	115.379747	130.032936	146.627970	165.418018	186.688000	20
25	212.793017	249.214024	292.104856	342.603486	402.042491	471.981083	25
30	434.745146	530.311731	647.439118	790.947991	966.712169	1181.881569	30
35	881.170156	1120.712955	1426.491022	1816.651612	2314.213721	2948.341146	35
40	1779.090308	2360.757241	3134.521839	4163.213027	5529.828982	7343.857840	40
45	3585.128460	4965.273911	6879.290650	9531.577105	13203.424228	18281.309940	45
50	7217.716277	10435.648773	15089.501673	21813.093666	31515.336327	45497.190750	50
55	14524.147893	21925.304976	33089.962703	49910.228445	75214.034786	113219.011287	55
60	29219.991638	46057.508533	72555.038129	114189.666478	179494.583786	281732.571766	60
65	58778.582580	96743.380952	159080.165226	261245.449442	428344.853547	701048.234576	65
70	118231.466926	203201.030246	348782.010169	597673.457599	1022189.605560	1744439.784661	70
75	237812.453171	426798.465828	764693.447483	1367339.242866	2439313.202939	4340731.846568	75
80	478332.529343	896429.474315	1676557.661247	3128148.113254	5821071.286073	10801137.310052	80
85	962104.313290	1882815.045139	3675772.565539	7156452.264725	13891142.064658	26876693.432947	85
90	1935142.168042	3954561.749997	8058947.355395	16372236.334003	33149185.002785	66877821.244672	90
95	3892268.850907	8305937.658205	17668830.313546	37455717.823488	79105686.250695	166413427.601142	95
100	7828749.671335	17445313.746092	38737999.328452	85689616.141407	188774151.199846	414089867.610073	100

TABLE 7
Annuity due

Amount of annuity due for $1 at compound interest. Remember, annuity due requires payment at the beginning of the period.

N	1%	1½%	2%	2½%	3%	3½%	4%	N
1	1.010000	1.015000	1.020000	1.025000	1.030000	1.035000	1.040000	1
2	2.030100	2.045225	2.060400	2.075625	2.090900	2.106225	2.121600	2
3	3.060401	3.090903	3.121608	3.152516	3.183627	3.214943	3.246464	3
4	4.101005	4.152267	4.204040	4.256329	4.309136	4.362466	4.416323	4
5	5.152015	5.229551	5.308121	5.387737	5.468410	5.550152	5.632975	5
6	6.213535	6.322994	6.434283	6.547430	6.662462	6.779408	6.898294	6
7	7.285671	7.432839	7.582969	7.736116	7.892336	8.051687	8.214226	7
8	8.368527	8.559332	8.754628	8.954519	9.159106	9.368496	9.582795	8
9	9.462213	9.702722	9.949721	10.203382	10.463879	10.731393	11.006107	9
10	10.566835	10.863262	11.168715	11.483466	11.807796	12.141992	12.486351	10
11	11.682503	12.041211	12.412090	12.795553	13.192030	13.601962	14.025805	11
12	12.809328	13.236830	13.680332	14.140442	14.617790	15.113030	15.626838	12
13	13.947421	14.450382	14.973938	15.518953	16.086324	16.676986	17.291911	13
14	15.096896	15.682138	16.293417	16.931927	17.598914	18.295681	19.023588	14
15	16.257864	16.932370	17.639285	18.380225	19.156881	19.971030	20.824531	15
16	17.430443	18.201355	19.012071	19.864730	20.761588	21.705016	22.697512	16
17	18.614748	19.489376	20.412312	21.386349	22.414435	23.499691	24.645413	17
18	19.810895	20.796716	21.840559	22.946007	24.116868	25.357180	26.671229	18
19	21.019004	22.123667	23.297370	24.544658	25.870374	27.279682	28.778079	19
20	22.239194	23.470522	24.783317	26.183274	27.676486	29.269471	30.969202	20
25	28.525631	30.513969	32.670906	35.011708	37.553042	40.313102	43.311745	25
30	35.132740	38.101762	41.379441	45.000271	49.002678	53.429471	58.328335	30
35	42.076878	46.275969	50.994360	56.301413	62.275944	69.007603	76.598314	35
40	49.375237	55.081912	61.610023	69.087617	77.663298	87.509537	98.826536	40
45	57.045885	64.568414	73.330564	83.554034	95.501457	109.484031	125.870568	45
50	65.107814	74.788070	86.270989	99.921458	116.180773	135.582837	158.773767	50
55	73.580982	85.797543	100.558264	118.439694	140.153768	166.580031	198.805540	55
60	82.486367	97.657871	116.332570	139.391380	167.945040	203.394974	247.510313	60
65	91.846015	110.434814	133.748679	163.096289	200.162741	247.119577	306.767116	65
70	101.683100	124.199209	152.977469	189.916217	237.511886	299.050690	378.862077	70
75	112.021975	139.027372	174.207608	220.260504	280.809781	360.728561	466.576621	75
80	122.888237	155.001515	197.647397	254.592280	331.003909	433.982524	573.294776	80
85	134.308787	172.210204	223.526818	293.435534	389.192660	520.985253	703.133728	85
90	146.311900	190.748849	252.099789	337.383110	456.649371	624.317203	861.102667	90
95	158.927293	210.720235	283.646659	387.105758	534.850186	747.043145	1053.296034	95
100	172.186197	232.235089	318.476952	443.362370	625.506365	892.803065	1287.128653	100

N	4½%	5%	5½%	6%	6½%	7%	7½%	N
1	1.045000	1.050000	1.055000	1.060000	1.065000	1.070000	1.075000	1
2	2.137025	2.152500	2.168025	2.183600	2.199225	2.214900	2.230625	2
3	3.278191	3.310125	3.342266	3.374616	3.407175	3.439943	3.472922	3
4	4.470710	4.525631	4.581091	4.637093	4.693641	4.750739	4.808391	4
5	5.716892	5.801913	5.888051	5.975319	6.063728	6.153291	6.244020	5
6	7.019152	7.142008	7.266894	7.393838	7.522870	7.654021	7.787322	6
7	8.380014	8.549109	8.721573	8.897468	9.076856	9.259803	9.446371	7
8	9.802114	10.026564	10.256260	10.491316	10.731852	10.977989	11.229849	8
9	11.288209	11.577893	11.875354	12.180795	12.494423	12.816448	13.147097	9
10	12.841179	13.206787	13.583498	13.971643	14.371560	14.783599	15.208119	10
11	14.464032	14.917127	15.385591	15.869941	16.370711	16.888451	17.423728	11
12	16.159913	16.712983	17.286798	17.882138	18.499808	19.140643	19.805508	12
13	17.932109	18.598632	19.292572	20.015066	20.767295	21.550488	22.365921	13
14	19.784054	20.578564	21.408663	22.275970	23.182169	24.129022	25.118365	14
15	21.719337	22.657492	23.641140	24.672528	25.754010	26.888054	28.077242	15
16	23.741707	24.840366	25.996403	27.212880	28.493021	29.840217	31.258035	16
17	25.855084	27.132385	28.481205	29.905653	31.410067	32.999033	34.677388	17
18	28.063562	29.539004	31.102671	32.759992	34.516722	36.378965	38.353192	18
19	30.371423	32.065954	33.868318	35.785591	37.825309	39.995492	42.304681	19
20	32.783137	34.719252	36.786076	38.992727	41.348954	43.865177	46.552532	20
25	46.570645	50.113454	53.965981	58.156383	62.715378	67.676470	73.076201	25
30	63.752388	69.760790	76.419429	83.801677	91.989230	101.073041	111.154358	30
35	85.163966	94.836323	105.765189	118.120867	132.096945	147.913460	165.820476	35
40	111.846688	126.839763	144.118923	164.047684	187.047990	213.609570	244.300759	40
45	145.098214	167.685164	194.245719	225.508125	262.335685	305.751763	356.969354	45
50	186.535665	219.815396	259.759438	307.756059	365.486351	434.985955	518.719693	50
55	238.174268	286.348249	345.383247	417.822348	506.811702	616.243594	750.933224	55
60	302.525362	371.262904	457.290142	565.115872	700.439682	870.466810	1084.305779	60
65	382.718533	479.637912	603.547978	762.227832	965.726794	1227.028022	1562.905195	65
70	482.653815	617.954936	794.701120	1026.008100	1329.193129	1727.123566	2249.996552	70
75	607.191358	794.486404	1044.530633	1379.005601	1827.173507	2428.533438	3236.405054	75
80	762.387795	1019.790262	1371.047813	1851.395885	2509.449781	3412.297067	4652.522027	80
85	955.790791	1307.341422	1797.792719	2483.560646	3444.227405	4792.076448	6685.541082	85
90	1196.806112	1674.337666	2355.531223	3329.539698	4724.953761	6727.288407	9604.202860	90
95	1497.155051	2142.728205	3084.473153	4461.650505	6479.659861	9441.523288	13794.319300	95
100	1871.444474	2740.526415	4037.171102	5976.670142	8883.759287	13248.378119	19809.773342	100

TABLE 7 (Continued)

N	8%	9%	10%	11%	12%	13%	14%	N
1	1.080000	1.090000	1.100000	1.110000	1.120000	1.130000	1.140000	1
2	2.246400	2.278100	2.310000	2.342100	2.374400	2.406900	2.439600	2
3	3.506112	3.573129	3.641000	3.709731	3.779328	3.849797	3.921144	3
4	4.866601	4.984511	5.105100	5.227801	5.352847	5.480271	5.610104	4
5	6.335929	6.523335	6.715610	6.912860	7.115189	7.322706	7.535519	5
6	7.922803	8.200435	8.487171	8.783274	9.089012	9.404658	9.730491	6
7	9.636628	10.028474	10.435888	10.859434	11.299693	11.757263	12.232760	7
8	11.487558	12.021036	12.579477	13.163972	13.775656	14.415707	15.085347	8
9	13.486562	14.192930	14.937425	15.722009	16.548735	17.419749	18.337295	9
10	15.645487	16.560293	17.531167	18.561430	19.654585	20.814317	22.044516	10
11	17.977126	19.140720	20.384284	21.713187	23.133133	24.650178	26.270749	11
12	20.495297	21.953385	23.522712	25.211638	27.029109	28.984701	31.088654	12
13	23.214920	25.019189	26.974983	29.094918	31.392602	33.882712	36.581065	13
14	26.152114	28.360916	30.772482	33.405359	36.279715	39.417464	42.842414	14
15	29.324283	32.003390	34.949730	38.189948	41.753280	45.671735	49.980352	15
16	32.750226	35.973705	39.544703	43.500843	47.883674	52.739060	58.117601	16
17	36.450244	40.301338	44.599173	49.395936	54.749715	60.725138	67.394066	17
18	40.446263	45.018458	50.159090	55.939488	62.439681	69.749406	77.969235	18
19	44.761964	50.160120	56.274999	63.202832	71.052442	79.946829	90.024928	19
20	49.422921	55.764530	63.002490	71.265144	80.698736	91.469917	103.768418	20
25	78.954415	92.323977	108.181765	126.998771	149.333934	175.850098	207.332743	25
30	122.345868	148.575217	180.943425	220.913174	270.292606	331.315113	406.737006	30
35	186.102148	235.124723	298.126805	379.164406	483.463116	617.749325	790.672881	35
40	279.781040	368.291865	486.851811	645.826934	859.142391	1145.485795	1529.908613	40
45	417.426067	573.186021	790.795321	1095.168801	1521.217636	2117.806032	2953.243872	45
50	619.671769	888.441076	1280.299382	1852.335979	2688.020438	3909.243042	5693.754335	50
55	916.837058	1373.500057	2068.650567	3128.206707	4744.325653	7209.849610	10970.373143	55
60	1353.470360	2119.823425	3338.298035	5278.123082	8368.238046	13291.003265	21130.051943	60
65	1995.027929	3268.134436	5383.077978	8900.857202	14754.809912	24495.134690	40691.645657	65
70	2937.686480	5034.953266	8676.216525	15005.374875	26010.131725	45138.020581	78355.823448	70
75	4322.761154	7753.423041	13979.849085	25291.842164	45845.854514	83171.199747	150874.980601	75
80	6357.890263	11936.125758	22521.402360	42625.137755	80803.175601	153244.867025	290504.423288	80
85	9348.162601	18371.732355	36277.659326	71832.748846	142409.919685	282351.056760	559348.988369	85
90	13741.853705	28273.710829	58432.248730	121049.272105	250982.052751	520220.842593	1076986.234374	90
95	20197.627405	43509.132110	94112.436513	203981.975989	442323.248488	958480.504103	2073652.536227	95
100	29683.276961	66950.716257	151575.735738	343728.404994	779531.813448	1765945.522139	3992648.367601	100

N	15%	16%	17%	18%	19%	20%	N
1	1.150000	1.160000	1.170000	1.180000	1.190000	1.200000	1
2	2.472500	2.505600	2.538900	2.572400	2.606100	2.640000	2
3	3.993375	4.066496	4.140513	4.215432	4.291259	4.368000	3
4	5.742381	5.877135	6.014400	6.154210	6.296598	6.441600	4
5	7.753738	7.977477	8.206848	8.441968	8.682952	8.929280	5
6	10.066799	10.413873	10.772012	11.141522	11.522713	11.915904	6
7	12.726819	13.240093	13.773255	14.326996	14.902028	15.499085	7
8	15.785842	16.518508	17.284708	18.085855	18.923413	19.798942	8
9	19.303718	20.321469	21.393108	22.521309	23.708862	24.958682	9
10	23.349276	24.732904	26.199937	27.755144	29.403546	31.150419	10
11	28.001667	29.850169	31.823926	33.931070	36.180220	38.580502	11
12	33.351917	35.786196	38.403993	41.218663	44.244461	47.496603	12
13	39.504705	42.671987	46.102672	49.818022	53.840909	58.195923	13
14	46.580411	50.659505	55.110126	59.965266	65.260682	71.035108	14
15	54.717472	59.925026	65.648848	71.939014	78.850211	86.442129	15
16	64.075093	70.673030	77.979152	86.068036	95.021751	104.930555	16
17	74.836357	83.140715	92.405608	102.740283	114.265884	127.116666	17
18	87.211811	97.603230	109.284561	122.413534	137.166402	153.740000	18
19	101.443583	114.379747	129.032936	145.627970	164.418018	185.688000	19
20	117.810120	133.840506	152.138535	173.021005	196.847442	224.025600	20
25	244.711970	289.088267	341.762681	404.272113	478.430565	566.377300	25
30	499.956918	615.161608	757.503768	933.318630	1150.387481	1418.257883	30
35	1013.345680	1300.027028	1668.994496	2143.648902	2753.914328	3538.009375	35
40	2045.953854	2738.478399	3667.390552	4912.591372	6580.496488	8812.629408	40
45	4122.897729	5759.717737	8048.770061	11247.260984	15712.074831	21937.571928	45
50	8300.373719	12105.352576	17654.716957	25739.450526	37503.250230	54596.628900	50
55	16702.770077	25433.353773	38715.256362	58894.069565	89504.701395	135862.813544	55
60	33602.990383	53426.709898	84889.394611	134743.806444	213598.554705	338079.086119	60
65	67595.369968	112222.321904	186123.793315	308269.630342	509730.375721	841257.881492	65
70	135966.186964	235713.195085	408074.951898	705254.679967	1216405.630617	2093327.741593	70
75	273484.321146	495086.220361	894691.333555	1613460.306582	2902782.711497	5208878.215881	75
80	550082.408744	1039858.190205	1961572.463659	3691214.773639	6927074.830427	12961364.772062	80
85	1106419.960284	2184065.452361	4300653.901680	8444613.672375	16530459.056942	32252032.119537	85
90	2225413.493248	4587291.629996	9428968.405813	19319238.874124	39447530.153314	80253385.493606	90
95	4476109.178543	9634887.683518	20672531.466849	44197747.031716	94135766.638327	199696113.121370	95
100	9003062.122036	20236563.945467	45323459.214289	101113747.046860	224641239.927817	496907841.132087	100

TABLE 8
Present value of an annuity

Amount needed to deposit today to equal the future value of the annuity; it is the amount you can borrow with a given monthly payment on an installment loan.

N	1%	1½%	2%	2½%	3%	3½%	4%	N
1	.990099	.985222	.980392	.975610	.970874	.966184	.961538	1
2	1.970395	1.955883	1.941561	1.927424	1.913470	1.899694	1.886095	2
3	2.940985	2.912200	2.883883	2.856024	2.828611	2.801637	2.775091	3
4	3.901966	3.854385	3.807729	3.761974	3.717098	3.673079	3.629895	4
5	4.853431	4.782645	4.713460	4.645828	4.579707	4.515052	4.451822	5
6	5.795476	5.697187	5.601431	5.508125	5.417191	5.328553	5.242137	6
7	6.728195	6.598214	6.471991	6.349391	6.230283	6.114544	6.002055	7
8	7.651678	7.485925	7.325481	7.170137	7.019692	6.873956	6.732745	8
9	8.566018	8.360517	8.162237	7.970866	7.786109	7.607687	7.435332	9
10	9.471305	9.222185	8.982585	8.752064	8.530203	8.316605	8.110896	10
11	10.367628	10.071118	9.786848	9.514209	9.252624	9.001551	8.760477	11
12	11.255077	10.907505	10.575341	10.257765	9.954004	9.663334	9.385074	12
13	12.133740	11.731532	11.348374	10.983185	10.634955	10.302738	9.985648	13
14	13.003703	12.543382	12.106249	11.690912	11.296973	10.920520	10.563123	14
15	13.865053	13.343233	12.849264	12.381378	11.937935	11.517411	11.118387	15
16	14.717874	14.131264	13.577709	13.055003	12.561102	12.094117	11.652296	16
17	15.562251	14.907640	14.291872	13.712198	13.166118	12.651321	12.165669	17
18	16.398269	15.672561	14.992031	14.353364	13.753513	13.189682	12.659297	18
19	17.226008	16.426168	15.678462	14.978891	14.323799	13.709837	13.133939	19
20	18.045553	17.168639	16.351433	15.589162	14.877475	14.212403	13.590326	20
25	22.023156	20.719611	19.523456	18.424376	17.413148	16.481515	15.622080	25
30	25.807708	24.015838	22.396456	20.930293	19.600441	18.392045	17.292033	30
35	29.408580	27.075595	24.998619	23.145157	21.487220	20.000661	18.664613	35
40	32.834686	29.915845	27.355479	25.102775	23.114772	21.355072	19.792774	40
45	36.094508	32.552337	29.490160	26.833024	24.518713	22.495450	20.720040	45
50	39.196118	34.999688	31.423606	28.362312	25.729764	23.455618	21.482185	50
55	42.147192	37.271467	33.174788	29.713979	26.774428	24.264053	22.108612	55
60	44.955038	39.380269	34.760887	30.908656	27.675564	24.944734	22.623490	60
65	47.626608	41.337786	36.197466	31.964577	28.452892	25.517849	23.046682	65
70	50.168514	43.154872	37.498619	32.897857	29.123421	26.000397	23.394515	70
75	52.587051	44.841600	38.677114	33.722740	29.701826	26.406689	23.680408	75
80	54.888206	46.407323	39.744514	34.451817	30.200763	26.748776	23.915392	80
85	57.077676	47.860722	40.711290	35.096215	30.631151	27.036804	24.108531	85
90	59.160881	49.209855	41.586929	35.665768	31.002407	27.279316	24.267278	90
95	61.142980	50.462201	42.380023	36.169171	31.322656	27.483504	24.397756	95
100	63.028879	51.624704	43.098352	36.614105	31.598905	27.655425	24.504999	100

N	4½%	5%	5½%	6%	6½%	7%	7½%	N
1	.956938	.952381	.947867	.943396	.938967	.934579	.930233	1
2	1.872668	1.859410	1.846320	1.833393	1.820626	1.808018	1.795565	2
3	2.748964	2.723248	2.697933	2.673012	2.648476	2.624316	2.600526	3
4	3.587526	3.545951	3.505150	3.465106	3.425799	3.387211	3.349326	4
5	4.389977	4.329477	4.270284	4.212364	4.155679	4.100197	4.045885	5
6	5.157872	5.075692	4.995530	4.917324	4.841014	4.766540	4.693846	6
7	5.892701	5.786373	5.682967	5.582381	5.484520	5.389289	5.296601	7
8	6.595886	6.463213	6.334566	6.209794	6.088751	5.971299	5.857304	8
9	7.268790	7.107822	6.952195	6.801692	6.656104	6.515232	6.378887	9
10	7.912718	7.721735	7.537626	7.360087	7.188830	7.023582	6.864081	10
11	8.528917	8.306414	8.092536	7.886875	7.689042	7.498674	7.315424	11
12	9.118581	8.863252	8.618518	8.383844	8.158725	7.942686	7.735278	12
13	9.682852	9.393573	9.117079	8.852683	8.599742	8.357651	8.125640	13
14	10.222825	9.898641	9.589648	9.294984	9.013842	8.745468	8.489154	14
15	10.739546	10.379658	10.037581	9.712249	9.402669	9.107914	8.827120	15
16	11.234015	10.837770	10.462162	10.105895	9.767764	9.446649	9.141507	16
17	11.707191	11.274066	10.864609	10.477260	10.110577	9.763223	9.433960	17
18	12.159992	11.689587	11.246074	10.827603	10.432466	10.059087	9.706009	18
19	12.593294	12.085321	11.607654	11.158116	10.734710	10.335595	9.959078	19
20	13.007936	12.462210	11.950382	11.469921	11.018507	10.594014	10.194491	20
25	14.828209	14.093945	13.413933	12.783356	12.197877	11.653583	11.146946	25
30	16.288889	15.372451	14.533745	13.764831	13.058676	12.409041	11.810386	30
35	17.461012	16.374194	15.390552	14.498246	13.686957	12.947672	12.272511	35
40	18.401584	17.159086	16.046125	15.046297	14.145527	13.331709	12.594409	40
45	19.156347	17.774070	16.547726	15.455832	14.480228	13.605522	12.818629	45
50	19.762008	18.255925	16.931518	15.761861	14.724521	13.800746	12.974812	50
55	20.248021	18.633472	17.225170	15.990543	14.902825	13.939939	13.083602	55
60	20.638022	18.929290	17.449854	16.161428	15.032966	14.039181	13.159381	60
65	20.950979	19.161070	17.621767	16.289123	15.127953	14.109940	13.212165	65
70	21.202112	19.342677	17.753304	16.384544	15.197282	14.160389	13.248933	70
75	21.403634	19.484970	17.853947	16.455848	15.247885	14.196359	13.274543	75
80	21.565345	19.596460	17.930953	16.509131	15.284818	14.222005	13.292383	80
85	21.695110	19.683816	17.989873	16.548947	15.311775	14.240291	13.304809	85
90	21.799241	19.752262	18.034954	16.578699	15.331451	14.253328	13.313464	90
95	21.882800	19.805891	18.069447	16.600932	15.345812	14.262623	13.319493	95
100	21.949853	19.847910	18.095839	16.617546	15.356293	14.269251	13.323693	100

TABLE 8 (Continued)

N	8%	9%	10%	11%	12%	13%	14%	N
1	.925926	.917431	.909091	.900901	.892857	.884956	.677193	1
2	1.783265	1.759111	1.735537	1.712523	1.690051	1.668102	1.646661	2
3	2.577097	2.531295	2.486852	2.443715	2.401831	2.361153	2.321632	3
4	3.312127	3.239720	3.169865	3.102446	3.037349	2.974471	2.913712	4
5	3.992710	3.889651	3.790787	3.695897	3.604776	3.517231	3.433081	5
6	4.622880	4.485919	4.355261	4.230538	4.111407	3.997550	3.888668	6
7	5.206370	5.032953	4.868419	4.712196	4.563757	4.422610	4.288305	7
8	5.746639	5.534819	5.334926	5.146123	4.967640	4.798770	4.638864	8
9	6.246888	5.995247	5.759024	5.537048	5.328250	5.131655	4.946372	9
10	6.710081	6.417658	6.144567	5.889232	5.650223	5.426243	5.216116	10
11	7.138964	6.805191	6.495061	6.206515	5.937699	5.686941	5.452733	11
12	7.536078	7.160725	6.813692	6.492356	6.194374	5.917647	5.660292	12
13	7.903776	7.486904	7.103356	6.749870	6.423548	6.121812	5.842362	13
14	8.244237	7.786150	7.366687	6.981865	6.628168	6.302488	6.002072	14
15	8.559479	8.060688	7.606080	7.190870	6.810864	6.462379	6.142168	15
16	8.851369	8.312558	7.823709	7.379162	6.973986	6.603875	6.265060	16
17	9.121638	8.543631	8.021553	7.548794	7.119630	6.729093	6.372859	17
18	9.371887	8.755625	8.201412	7.701617	7.249670	6.839905	6.467420	18
19	9.603599	8.950115	8.364920	7.839294	7.365777	6.937969	6.550369	19
20	9.818147	9.128546	8.513564	7.963328	7.469444	7.024752	6.623131	20
25	10.674776	9.822580	9.077040	8.421745	7.843139	7.329985	6.872927	25
30	11.257783	10.273654	9.426914	8.693793	8.055184	7.495653	7.002664	30
35	11.654568	10.566821	9.644159	8.855240	8.175504	7.585572	7.070045	35
40	11.924613	10.757360	9.779051	8.951051	8.243777	7.634376	7.105041	40
45	12.108402	10.881197	9.862808	9.007910	8.282516	7.660864	7.123217	45
50	12.233485	10.961683	9.914814	9.041653	8.304498	7.675242	7.132656	50
55	12.318614	11.013993	9.947106	9.061678	8.316972	7.683045	7.137559	55
60	12.376552	11.047991	9.967157	9.073562	8.324049	7.687280	7.140106	60
65	12.415983	11.070087	9.979607	9.080614	8.328065	7.689579	7.141428	65
70	12.442820	11.084449	9.987338	9.084800	8.330344	7.690827	7.142115	70
75	12.461084	11.093782	9.992138	9.087283	8.331637	7.691504	7.142472	75
80	12.473514	11.099849	9.995118	9.088757	8.332371	7.691871	7.142657	80
85	12.481974	11.103791	9.996969	9.089632	8.332787	7.692071	7.142753	85
90	12.487732	11.106354	9.998118	9.090151	8.333023	7.692179	7.142803	90
95	12.491651	11.108019	9.998831	9.090459	8.333157	7.692238	7.142829	95
100	12.494318	11.109102	9.999274	9.090642	8.333234	7.692270	7.142843	100

N	15%	16%	17%	18%	19%	20%	N
1	.869565	.862069	.854701	.847458	.840336	.833333	1
2	1.625709	1.605232	1.585214	1.565642	1.546501	1.527778	2
3	2.283225	2.245890	2.209585	2.174273	2.139917	2.106481	3
4	2.854978	2.798181	2.743235	2.690062	2.638586	2.588735	4
5	3.352155	3.274294	3.199346	3.127171	3.057635	2.990612	5
6	3.784483	3.684594	3.589185	3.497603	3.409777	3.325510	6
7	4.160420	4.038565	3.922380	3.811528	3.705695	3.604592	7
8	4.487322	4.343591	4.207163	4.077566	3.954366	3.837160	8
9	4.771584	4.606544	4.450566	4.303022	4.163332	4.030967	9
10	5.018769	4.833227	4.658604	4.494086	4.338935	4.192472	10
11	5.233712	5.028644	4.836413	4.656005	4.486500	4.327060	11
12	5.420619	5.197107	4.988387	4.793225	4.610504	4.439217	12
13	5.583147	5.342334	5.118280	4.909513	4.714709	4.532681	13
14	5.724476	5.467529	5.229299	5.008062	4.802277	4.610567	14
15	5.847370	5.575456	5.324187	5.091578	4.875863	4.675473	15
16	5.954235	5.668497	5.405288	5.162354	4.937700	4.729561	16
17	6.047161	5.748704	5.474605	5.222334	4.989664	4.774634	17
18	6.127966	5.817848	5.533851	5.273164	5.033331	4.812195	18
19	6.198231	5.877455	5.584488	5.316241	5.070026	4.843496	19
20	6.259331	5.928841	5.627767	5.352746	5.100862	4.869580	20
25	6.464149	6.097092	5.766234	5.466906	5.195148	4.947587	25
30	6.565980	6.177198	5.829390	5.516806	5.234658	4.978936	30
35	6.616607	6.215338	5.858196	5.538618	5.251215	4.991535	35
40	6.641778	6.233497	5.871335	5.548152	5.258153	4.996598	40
45	6.654293	6.242143	5.877327	5.552319	5.261061	4.998633	45
50	6.660515	6.246259	5.880061	5.554141	5.262279	4.999451	50
55	6.663608	6.248219	5.881307	5.554937	5.262790	4.999779	55
60	6.665146	6.249152	5.881876	5.555285	5.263004	4.999911	60
65	6.665911	6.249596	5.882135	5.555437	5.263093	4.999964	65
70	6.666291	6.249808	5.882254	5.555504	5.263131	4.999986	70
75	6.666480	6.249908	5.882308	5.555533	5.263147	4.999994	75
80	6.666574	6.249956	5.882332	5.555546	5.263153	4.999998	80
85	6.666620	6.249979	5.882344	5.555551	5.263156	4.999999	85
90	6.666644	6.249990	5.882349	5.555554	5.263157	5.000000	90
95	6.666655	6.249995	5.882351	5.555555	5.263158	5.000000	95
100	6.666661	6.249998	5.882352	5.555555	5.263158	5.000000	100

TABLE 9
Mortality table based on
100,000 persons living at age 0
l_x = number of living;
d_x = number of deaths;
p_x = probability of living;
q_x = probability of dying, for
age x from 0 to 99

x	l_x	d_x	p_x	q_x	x	l_x	d_x	p_x	q_x
0	100,000	708	.9929	.0071	50	87,624	729	.9917	.0083
1	99,292	175	.9982	.0018	51	86,895	792	.9909	.0091
2	99,117	151	.9985	.0015	52	86,103	858	.9900	.0100
3	98,966	144	.9986	.0015	53	85,245	928	.9891	.0109
4	98,822	138	.9986	.0014	54	84,317	1,003	.9881	.0119
5	98,684	133	.9987	.0014	55	83,314	1,083	.9870	.0130
6	98,551	128	.9987	.0013	56	82,231	1,168	.9858	.0142
7	98,423	124	.9987	.0013	57	81,063	1,260	.9845	.0156
8	98,299	121	.9988	.0012	58	79,803	1,357	.9830	.0170
9	98,178	119	.9988	.0012	59	78,446	1,458	.9814	.0186
10	98,059	119	.9988	.0012	60	76,988	1,566	.9797	.0204
11	97,940	120	.9988	.0012	61	75,422	1,677	.9778	.0222
12	97,820	123	.9988	.0013	62	73,745	1,793	.9757	.0243
13	97,697	129	.9987	.0013	63	71,952	1,912	.9734	.0266
14	97,568	136	.9986	.0014	64	70,040	2,034	.9710	.0291
15	97,432	142	.9986	.0015	65	68,006	2,159	.9683	.0318
16	97,290	150	.9985	.0016	66	65,847	2,287	.9653	.0347
17	97,140	157	.9984	.0016	67	63,560	2,418	.9620	.0381
18	96,983	164	.9983	.0017	68	61,142	2,548	.9583	.0417
19	96,819	168	.9983	.0017	69	58,594	2,672	.9544	.0456
20	96,651	173	.9982	.0018	70	55,922	2,784	.9502	.0498
21	96,478	177	.9982	.0018	71	53,138	2,877	.9459	.0542
22	96,301	179	.9982	.0019	72	50,261	2,948	.9414	.0587
23	96,122	182	.9981	.0019	73	47,313	2,993	.9368	.0633
24	95,940	183	.9981	.0019	74	44,320	3,019	.9319	.0681
25	95,757	185	.9981	.0019	75	41,301	3,030	.9266	.0734
26	95,572	187	.9981	.0020	76	38,271	3,030	.9208	.0792
27	95,385	190	.9980	.0020	77	35,241	3,020	.9143	.0857
28	95,195	193	.9980	.0020	78	32,221	2,998	.9070	.0931
29	95,002	198	.9979	.0021	79	29,223	2,957	.8988	.1012
30	94,804	202	.9979	.0021	80	26,266	2,888	.8901	.1100
31	94,602	207	.9978	.0022	81	23,378	2,790	.8807	.1194
32	94,395	212	.9978	.0023	82	20,588	2,659	.8709	.1292
33	94,183	218	.9977	.0023	83	17,929	2,499	.8606	.1394
34	93,965	226	.9976	.0024	84	15,430	2,314	.8500	.1500
35	93,739	235	.9975	.0025	85	13,116	2,113	.8389	.1611
36	93,504	247	.9974	.0027	86	11,003	1,901	.8272	.1728
37	93,257	261	.9972	.0028	87	9,102	1,685	.8149	.1851
38	92,996	280	.9970	.0030	88	7,417	1,470	.8018	.1982
39	92,716	301	.9968	.0033	89	5,947	1,263	.7876	.2124
40	92,415	326	.9965	.0035	90	4,684	1,068	.7720	.2280
41	92,089	354	.9962	.0039	91	3,616	888	.7544	.2456
42	91,735	383	.9958	.0042	92	2,728	725	.7342	.2658
43	91,352	414	.9955	.0045	93	2,003	579	.7109	.2891
44	90,938	447	.9951	.0049	94	1,424	450	.6840	.3160
45	90,491	484	.9947	.0054	95	974	341	.6499	.3501
46	90,007	525	.9942	.0058	96	633	253	.6003	.3997
47	89,482	569	.9937	.0064	97	380	185	.5132	.4869
48	88,913	618	.9931	.0070	98	195	129	.3385	.6615
49	88,295	671	.9924	.0076	99	66	66	.0000	1.0000

Based on the 1958 CSO Mortality Table prepared in cooperation with the National Association of Insurance Commissioner. Courtesy of the Society of Actuaries, Chicago, Illinois.

TABLE 10
Powers of *e*

x	e^x	e^{-x}	x	e^x	e^{-x}	x	e^x	e^{-x}
0.00	1.000	1.000	0.50	1.649	0.607	1.00	2.718	0.368
0.01	1.010	0.990	0.51	1.665	0.600	1.01	2.746	0.364
0.02	1.020	0.980	0.52	1.682	0.595	1.02	2.773	0.361
0.03	1.031	0.970	0.53	1.699	0.589	1.03	2.801	0.357
0.04	1.041	0.961	0.54	1.716	0.583	1.04	2.829	0.353
0.05	1.051	0.951	0.55	1.733	0.577	1.05	2.858	0.350
0.06	1.062	0.942	0.56	1.751	0.571	1.06	2.886	0.346
0.07	1.073	0.932	0.57	1.768	0.566	1.07	2.915	0.343
0.08	1.083	0.923	0.58	1.786	0.560	1.08	2.945	0.340
0.09	1.094	0.914	0.59	1.804	0.554	1.09	2.974	0.336
0.10	1.105	0.905	0.60	1.822	0.549	1.10	3.004	0.333
0.11	1.116	0.896	0.61	1.840	0.543	1.11	3.034	0.330
0.12	1.127	0.887	0.62	1.859	0.538	1.12	3.065	0.326
0.13	1.139	0.878	0.63	1.878	0.533	1.13	3.096	0.323
0.14	1.150	0.869	0.64	1.896	0.527	1.14	3.127	0.320
0.15	1.162	0.861	0.65	1.916	0.522	1.15	3.158	0.317
0.16	1.174	0.852	0.66	1.935	0.517	1.16	3.190	0.313
0.17	1.185	0.844	0.67	1.954	0.512	1.17	3.222	0.310
0.18	1.197	0.835	0.68	1.974	0.507	1.18	3.254	0.307
0.19	1.209	0.827	0.69	1.994	0.502	1.19	3.287	0.304
0.20	1.221	0.819	0.70	2.014	0.497	1.20	3.320	0.301
0.21	1.234	0.811	0.71	2.034	0.492	1.21	3.353	0.298
0.22	1.246	0.803	0.72	2.054	0.487	1.22	3.387	0.295
0.23	1.259	0.795	0.73	2.075	0.482	1.23	3.421	0.292
0.24	1.271	0.787	0.74	2.096	0.477	1.24	3.456	0.289
0.25	1.284	0.779	0.75	2.117	0.472	1.25	3.490	0.287
0.26	1.297	0.771	0.76	2.138	0.468	1.26	3.525	0.284
0.27	1.310	0.763	0.77	2.160	0.463	1.27	3.561	0.281
0.28	1.323	0.756	0.78	2.182	0.458	1.28	3.597	0.278
0.29	1.336	0.748	0.79	2.203	0.454	1.29	3.633	0.275
0.30	1.350	0.741	0.80	2.226	0.449	1.30	3.669	0.273
0.31	1.363	0.733	0.81	2.248	0.445	1.31	3.706	0.270
0.32	1.377	0.726	0.82	2.270	0.440	1.32	3.743	0.267
0.33	1.391	0.719	0.83	2.293	0.436	1.33	3.781	0.264
0.34	1.405	0.712	0.84	2.316	0.432	1.34	3.819	0.262
0.35	1.419	0.705	0.85	2.340	0.427	1.35	3.857	0.259
0.36	1.433	0.698	0.86	2.363	0.423	1.36	3.896	0.257
0.37	1.448	0.691	0.87	2.387	0.419	1.37	3.935	0.254
0.38	1.462	0.684	0.88	2.441	0.415	1.38	3.975	0.252
0.39	1.477	0.677	0.89	2.435	0.411	1.39	4.015	0.249
0.40	1.492	0.670	0.90	2.460	0.407	1.40	4.055	0.247
0.41	1.507	0.664	0.91	2.484	0.403	1.41	4.096	0.244
0.42	1.522	0.657	0.92	2.509	0.399	1.42	4.137	0.242
0.43	1.537	0.651	0.93	2.535	0.395	1.43	4.179	0.239
0.44	1.553	0.644	0.94	2.560	0.391	1.44	4.221	0.237
0.45	1.568	0.638	0.95	2.586	0.387	1.45	4.263	0.235
0.46	1.584	0.631	0.96	2.612	0.383	1.46	4.306	0.232
0.47	1.600	0.625	0.97	2.638	0.379	1.47	4.349	0.230
0.48	1.616	0.619	0.98	2.664	0.375	1.48	4.393	0.228
0.49	1.632	0.613	0.99	2.691	0.372	1.49	4.437	0.225

TABLE 10
(Continued)

x	e^x	e^{-x}	x	e^x	e^{-x}	x	e^x	e^{-x}
1.50	4.482	0.223	2.00	7.389	0.135	2.50	12.182	0.082
1.51	4.527	0.221	2.01	7.463	0.134	2.51	12.305	0.081
1.52	4.572	0.219	2.02	7.538	0.133	2.52	12.429	0.080
1.53	4.618	0.217	2.03	7.614	0.131	2.53	12.554	0.080
1.54	4.665	0.214	2.04	7.691	0.130	2.54	12.680	0.079
1.55	4.712	0.212	2.05	7.768	0.129	2.55	12.807	0.078
1.56	4.759	0.210	2.06	7.846	0.127	2.56	12.936	0.077
1.57	4.807	0.208	2.07	7.925	0.126	2.57	13.066	0.077
1.58	4.855	0.206	2.08	8.004	0.125	2.58	13.197	0.076
1.59	4.904	0.204	2.09	8.085	0.124	2.59	13.330	0.075
1.60	4.953	0.202	2.10	8.166	0.122	2.60	13.464	0.074
1.61	5.003	0.200	2.11	8.248	0.121	2.61	13.599	0.074
1.62	5.053	0.198	2.12	8.331	0.120	2.62	13.736	0.073
1.63	5.104	0.196	2.13	8.415	0.119	2.63	13.874	0.072
1.64	5.155	0.194	2.14	8.499	0.118	2.64	14.013	0.071
1.65	5.207	0.192	2.15	8.585	0.116	2.65	14.154	0.071
1.66	5.259	0.190	2.17	8.671	0.115	2.66	14.296	0.070
1.67	5.312	0.188	2.17	8.758	0.114	2.67	14.440	0.069
1.68	5.366	0.186	2.18	8.846	0.113	2.68	14.585	0.069
1.69	5.420	0.185	2.19	8.935	0.112	2.69	14.732	0.068
1.70	5.474	0.183	2.20	9.025	0.111	2.70	14.880	0.067
1.71	5.529	0.181	2.21	9.116	0.110	2.71	15.029	0.067
1.72	5.585	0.179	2.22	9.207	0.109	2.72	15.180	0.066
1.73	5.641	0.177	2.23	9.300	0.108	2.73	15.333	0.065
1.74	5.697	0.176	2.24	9.393	0.106	2.74	15.487	0.065
1.75	5.755	0.174	2.25	9.488	0.105	2.75	15.643	0.064
1.76	5.812	0.172	2.26	9.583	0.104	2.76	15.800	0.063
1.77	5.871	0.170	2.27	9.679	0.103	2.77	15.959	0.063
1.78	5.930	0.169	2.28	9.777	0.102	2.78	16.119	0.062
1.79	5.989	0.167	2.29	9.875	0.101	2.79	16.281	0.061
1.80	6.050	0.165	2.30	9.974	0.100	2.80	16.445	0.061
1.81	6.110	0.164	2.31	10.074	0.099	2.81	16.610	0.060
1.82	6.172	0.162	2.32	10.176	0.098	2.82	16.777	0.060
1.83	6.234	0.160	2.33	10.278	0.097	2.83	16.945	0.059
1.84	6.297	0.159	2.34	10.381	0.096	2.84	17.116	0.058
1.85	6.360	0.157	2.35	10.486	0.095	2.85	17.288	0.058
1.86	6.424	0.156	2.36	10.591	0.094	2.86	17.462	0.057
1.87	6.488	0.154	2.37	10.697	0.093	2.87	17.637	0.057
1.88	6.553	0.153	2.38	10.805	0.093	2.88	17.814	0.056
1.89	6.619	0.151	2.39	10.913	0.092	2.89	17.993	0.056
1.90	6.686	0.150	2.40	11.023	0.091	2.90	18.174	0.055
1.91	6.753	0.148	2.41	11.134	0.090	2.91	18.357	0.054
1.92	6.821	0.147	2.42	11.246	0.089	2.92	18.541	0.054
1.93	6.890	0.145	2.43	11.359	0.088	2.93	18.728	0.053
1.94	6.959	0.144	2.44	11.473	0.087	2.94	18.916	0.053
1.95	7.029	0.142	2.45	11.588	0.086	2.95	19.016	0.052
1.96	7.099	0.141	2.46	11.705	0.085	2.96	19.298	0.052
1.97	7.171	0.139	2.47	11.822	0.085	2.97	19.492	0.051
1.98	7.243	0.138	2.48	11.941	0.084	2.98	19.688	0.051
1.99	7.316	0.137	2.49	12.061	0.083	2.99	19.886	0.050
						3.00	20.086	0.050

TABLE 11
Common logarithms

N	0	1	2	3	4	5	6	7	8	9
1.0	.0000	.0043	.0086	.0128	.0170	.0212	.0253	.0294	.0334	.0374
1.1	.0414	.0453	.0492	.0531	.0569	.0607	.0645	.0682	.0719	.0755
1.2	.0792	.0828	.0864	.0899	.0934	.0969	.1004	.1038	.1072	.1106
1.3	.1139	.1173	.1206	.1239	.1271	.1303	.1335	.1367	.1399	.1430
1.4	.1461	.1492	.1523	.1553	.1584	.1614	.1644	.1673	.1703	.1732
1.5	.1761	.1790	.1818	.1847	.1875	.1903	.1931	.1959	.1987	.2014
1.6	.2041	.2068	.2095	.2122	.2148	.2175	.2201	.2227	.2253	.2279
1.7	.2304	.2330	.2355	.2380	.2405	.2430	.2455	.2480	.2504	.2529
1.8	.2553	.2577	.2601	.2625	.2648	.2672	.2695	.2718	.2742	.2765
1.9	.2788	.2810	.2833	.2856	.2878	.2900	.2923	.2945	.2967	.2989
2.0	.3010	.3032	.3054	.3075	.3096	.3118	.3139	.3160	.3181	.3201
2.1	.3222	.3243	.3263	.3284	.3304	.3324	.3345	.3365	.3385	.3404
2.2	.3424	.3444	.3464	.3483	.3502	.3522	.3541	.3560	.3579	.3598
2.3	.3617	.3636	.3655	.3674	.3692	.3711	.3729	.3747	.3766	.3784
2.4	.3802	.3820	.3838	.3856	.3874	.3892	.3909	.3927	.3945	.3962
2.5	.3979	.3997	.4014	.4031	.4048	.4065	.4082	.4099	.4116	.4133
2.6	.4150	.4166	.4183	.4200	.4216	.4232	.4249	.4265	.4281	.4298
2.7	.4314	.4330	.4346	.4362	.4378	.4393	.4409	.4425	.4440	.4456
2.8	.4472	.4487	.4502	.4518	.4533	.4548	.4564	.4579	.4594	.4609
2.9	.4624	.4639	.4654	.4669	.4683	.4698	.4713	.4728	.4742	.4757
3.0	.4771	.4786	.4800	.4814	.4829	.4843	.4857	.4871	.4886	.4900
3.1	.4914	.4928	.4942	.4955	.4969	.4983	.4997	.5011	.5024	.5038
3.2	.5051	.5065	.5079	.5092	.5105	.5119	.5132	.5145	.5159	.5172
3.3	.5185	.5198	.5211	.5224	.5237	.5250	.5263	.5276	.5289	.5302
3.4	.5315	.5328	.5340	.5353	.5366	.5378	.5391	.5403	.5416	.5428
3.5	.5441	.5453	.5465	.5478	.5490	.5502	.5514	.5527	.5539	.5551
3.6	.5563	.5575	.5587	.5599	.5611	.5623	.5635	.5647	.5658	.5670
3.7	.5682	.5694	.5705	.5717	.5729	.5740	.5752	.5763	.5775	.5786
3.8	.5798	.5809	.5821	.5832	.5843	.5855	.5866	.5877	.5888	.5899
3.9	.5911	.5922	.5933	.5944	.5955	.5966	.5977	.5988	.5999	.6010
4.0	.6021	.6031	.6042	.6053	.6064	.6075	.6085	.6096	.6107	.6117
4.1	.6128	.6138	.6149	.6160	.6170	.6180	.6191	.6201	.6212	.6222
4.2	.6232	.6243	.6253	.6263	.6274	.6284	.6294	.6304	.6314	.6325
4.3	.6335	.6345	.6355	.6365	.6375	.6385	.6395	.6405	.6415	.6425
4.4	.6435	.6444	.6454	.6464	.6474	.6484	.6493	.6503	.6513	.6522
4.5	.6532	.6542	.6551	.6561	.6571	.6580	.6590	.6599	.6609	.6618
4.6	.6628	.6637	.6646	.6656	.6665	.6675	.6684	.6693	.6702	.6712
4.7	.6721	.6730	.6739	.6749	.6758	.6767	.6776	.6785	.6794	.6803
4.8	.6812	.6821	.6830	.6839	.6848	.6857	.6866	.6875	.6884	.6893
4.9	.6902	.6911	.6920	.6928	.6937	.6946	.6955	.6964	.6972	.6981
5.0	.6990	.6998	.7007	.7016	.7024	.7033	.7042	.7050	.7059	.7067
5.1	.7076	.7084	.7093	.7101	.7110	.7118	.7126	.7135	.7143	.7152
5.2	.7160	.7168	.7177	.7185	.7193	.7202	.7210	.7218	.7226	.7235
5.3	.7243	.7251	.7259	.7267	.7275	.7284	.7292	.7300	.7308	.7316
5.4	.7324	.7332	.7340	.7348	.7356	.7364	.7273	.7380	.7388	.7396
N	0	1	2	3	4	5	6	7	8	9

TABLE 11
(Continued)

N	0	1	2	3	4	5	6	7	8	9
5.5	.7404	.7412	.7419	.7427	.7435	.7443	.7451	.7459	.7466	.7474
5.6	.7482	.7490	.7497	.7505	.7513	.7520	.7528	.7536	.7543	.7551
5.7	.7559	.7566	.7574	.7582	.7589	.7597	.7604	.7612	.7619	.7627
5.8	.7634	.7642	.7649	.7657	.7664	.7672	.7679	.7686	.7694	.7701
5.9	.7709	.7716	.7723	.7731	.7738	.7745	.7752	.7760	.7767	.7774
6.0	.7782	.7789	.7796	.7803	.7810	.7818	.7825	.7832	.7839	.7846
6.1	.7853	.7860	.7868	.7875	.7882	.7889	.7896	.7903	.7910	.7917
6.2	.7924	.7931	.7938	.7945	.7952	.7959	.7966	.7973	.7980	.7987
6.3	.7993	.8000	.8007	.8014	.8021	.8028	.8035	.8041	.8048	.8055
6.4	.8062	.8069	.8075	.8082	.8089	.8096	.8102	.8109	.8116	.8122
6.5	.8129	.8136	.8142	.8149	.8156	.8162	.8169	.8176	.8182	.8189
6.6	.8195	.8202	.8209	.8215	.8222	.8228	.8235	.8241	.8248	.8254
6.7	.8261	.8267	.8274	.8280	.8287	.8293	.8299	.8306	.8312	.8319
6.8	.8325	.8331	.8338	.8344	.8351	.8357	.8363	.8370	.8376	.8382
6.9	.8388	.8395	.8401	.8407	.8414	.8420	.8426	.8432	.8439	.8445
7.0	.8451	.8457	.8463	.8470	.8476	.8482	.8488	.8494	.8500	.8506
7.1	.8513	.8519	.8525	.8531	.8537	.8543	.8549	.8555	.8561	.8567
7.2	.8573	.8579	.8585	.8591	.8597	.8603	.8609	.8615	.8621	.8627
7.3	.8633	.8639	.8645	.8651	.8657	.8663	.8669	.8675	.8681	.8686
7.4	.8692	.8698	.8704	.8710	.8716	.8722	.8727	.8733	.8739	.8745
7.5	.8751	.8756	.8762	.8768	.8774	.8779	.8785	.8791	.8797	.8802
7.6	.8808	.8814	.8820	.8825	.8831	.8837	.8842	.8848	.8854	.8859
7.7	.8865	.8871	.8876	.8882	.8887	.8893	.8899	.8904	.8910	.8915
7.8	.8921	.8927	.8932	.8938	.8943	.8949	.8954	.8960	.8965	.8971
7.9	.8976	.8982	.8987	.8993	.8998	.9004	.9009	.9015	.9020	.9025
8.0	.9031	.9036	.9042	.9047	.9053	.9058	.9063	.9069	.9074	.9079
8.1	.9085	.9090	.9096	.9101	.9106	.9112	.9117	.9122	.9128	.9133
8.2	.9138	.9143	.9149	.9154	.9159	.9165	.9170	.9175	.9180	.9186
8.3	.9191	.9196	.9201	.9206	.9212	.9217	.9222	.9227	.9232	.9238
8.4	.9243	.9248	.9253	.9258	.9263	.9269	.9274	.9279	.9284	.9289
8.5	.9294	.9299	.9304	.9309	.9315	.9320	.9325	.9330	.9335	.9340
8.6	.9345	.9350	.9355	.9360	.9365	.9370	.9375	.9380	.9385	.9390
8.7	.9395	.9400	.9405	.9410	.9415	.9420	.9425	.9430	.9435	.9440
8.8	.9445	.9450	.9455	.9460	.9465	.9469	.9474	.9479	.9484	.9489
8.9	.9494	.9499	.9504	.9509	.9513	.9518	.9523	.9528	.9533	.9538
9.0	.9542	.9547	.9552	.9557	.9562	.9566	.9571	.9576	.9581	.9586
9.1	.9590	.9595	.9600	.9605	.9609	.9614	.9619	.9624	.9628	.9633
9.2	.9638	.9643	.9647	.9652	.9657	.9661	.9666	.9671	.9675	.9680
9.3	.9685	.9689	.9694	.9699	.9703	.9708	.9713	.9717	.9722	.9727
9.4	.9731	.9736	.9741	.9745	.9750	.9754	.9759	.9763	.9768	.9773
9.5	.9777	.9782	.9786	.9791	.9795	.9800	.9805	.9809	.9814	.9818
9.6	.9823	.9827	.9832	.9836	.9841	.9845	.9850	.9854	.9859	.9863
9.7	.9868	.9872	.9877	.9881	.9886	.9890	.9894	.9899	.9903	.9908
9.8	.9912	.9917	.9921	.9926	.9930	.9934	.9939	.9943	.9948	.9952
9.9	.9956	.9961	.9965	.9969	.9974	.9978	.9983	.9987	.9991	.9996
N	0	1	2	3	4	5	6	7	8	9

APPENDIX F

Markov Chains

---

1 Introduction to Markov Chains

In this appendix we build a probability model using a process called a **Markov chain**. A Markov chain process is used when a series of events or experiments consists of a finite number of trials, each with a finite number of possible outcomes having a fixed probability of occurrence. The result of each event or experiment depends only on the result of the immediately preceding experiment. For example, suppose you own 100 shares of CBS stock and are keeping a record of its progress at the close of each trading day. Each day there are three possibilities: up, down, or unchanged. After a lengthy analysis you determine that the following probabilities apply:

<table>
<tr>
<td>If the stock increases one day, the probabilities for the following day are:</td>
<td rowspan="3">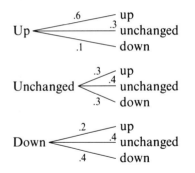</td>
</tr>
<tr>
<td>If the stock remains unchanged one day, the probabilities for the following day are:</td>
</tr>
<tr>
<td>If the stock decreases one day, the probabilities for the following day are:</td>
</tr>
</table>

All of these probabilities can be easily summarized in matrix form:

$$
\begin{array}{c}
\\
\\
\text{Present state}
\end{array}
\begin{array}{c}
\\
\text{Up} \\
\text{Unchanged} \\
\text{Down}
\end{array}
\begin{array}{ccc}
\text{Up} & \text{Unchanged} & \text{Down} \\
\begin{bmatrix} .6 & .3 & .1 \\ .3 & .4 & .3 \\ .2 & .4 & .4 \end{bmatrix}
\end{array}
$$

The entries can each be defined as a conditional probability. For example:

$P(\text{up} \mid \text{up}) = .6$ This is the probability that the stock increases on the day following a day that it had increased

$P(\text{unchanged} \mid \text{up}) = .3$ This is the probability that the stock is unchanged on the day following a day that it had increased

$P(\text{down} \mid \text{up}) = .1$

$P(\text{up} \mid \text{unchanged}) = .3,$ $P(\text{unchanged} \mid \text{unchanged}) = .4,$
$P(\text{down} \mid \text{unchanged}) = .3$

$P(\text{up} \mid \text{down}) = .2,$ $P(\text{unchanged} \mid \text{down}) = .4,$ $P(\text{down} \mid \text{down}) = .4$

The matrix form for these probabilities is called a **transition matrix** because it gives the probability of moving from a present state to the next state. A transition matrix must have the following properties:

1. *It is square* because all possible states are used both as rows and columns.
2. *All entries are between 0 and 1 inclusive* because all entries represent probabilities.
3. *The sum of the entries in any row is 1* because the numbers in the row give the probability of changing from the state listed at the left to one of the states listed across the top.

Probability Vector

A **probability vector** is a $1 \times n$ matrix for which the sum of the n entries is 1. If $s_1, s_2, \ldots, s_n$ are the states of a Markov chain and p_{ij} is the probability that an experiment will be in state s_j if it is in state s_i now, then the row matrices

$$
\begin{array}{cccc}
[p_{11} & p_{12} & \cdots & p_{1n}] \\
[p_{21} & p_{22} & \cdots & p_{2n}] \\
& \vdots & & \\
[p_{n1} & p_{n2} & \cdots & p_{nn}]
\end{array}
$$

Transition Matrix

are the probability vectors, and the matrix T formed by these probability vectors is called the **transition matrix**:

$$
T = \begin{bmatrix}
p_{11} & p_{12} & \cdots & p_{1n} \\
p_{21} & p_{22} & \cdots & p_{2n} \\
& \vdots & & \\
p_{n1} & p_{n2} & \cdots & p_{nn}
\end{bmatrix}
$$

EXAMPLE 1   A mathematics instructor gives surprise quizzes. She never gives a quiz 2 days in a row, but if she does not give a quiz one day, she is just as likely to give a quiz the following day as she is not to give a quiz. This is an example of a Markov chain with two *states:* s_1 = quiz and s_2 = no quiz. The probability of a quiz the next day depends on the present state. We can summarize the probabilities in matrix form:

$$
\begin{array}{cc}
 & \begin{array}{cc} \text{Quiz} & \text{No quiz} \\ \text{next day} & \text{next day} \end{array} \\
\begin{array}{c} \text{Quiz one day} \\ \text{No quiz one day} \end{array} &
\left[\begin{array}{cc} 0 & 1 \\ \frac{1}{2} & \frac{1}{2} \end{array} \right]
\end{array}
$$ ■

Suppose the mathematics instructor in Example 1 uses the same rules for giving quizzes day after day. If today is Monday, what are the possibilities for Wednesday if the class meets daily? Let Q = quiz given and let N = no quiz given. Then we can draw a tree diagram, as shown. The probabilities for each outcome are found by multiplying along each branch of the tree; $P(NNN) = \frac{1}{4}$ means that there is a probability of $\frac{1}{4}$ that no quiz will be given on any of the 3 days.

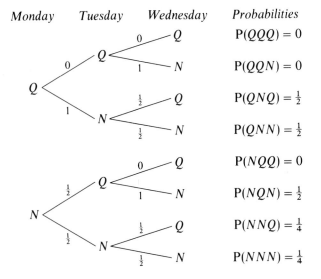

EXAMPLE 2   What is the probability that the instructor described in Example 1 will give a quiz on Wednesday, given the following information?

a. There is a quiz on Monday.
b. There is no quiz on Monday.

What is the probability that there is no quiz on Wednesday, given the following information?

c. There is a quiz on Monday.
d. There is no quiz on Monday.

Solution   We use the tree diagram above and the addition principle for probability.

a. $P(Q \text{ on Wednesday} \mid Q \text{ on Monday}) = P(QQQ) + P(QNQ) = 0 + \frac{1}{2} = \frac{1}{2}$
b. $P(Q \text{ on Wednesday} \mid N \text{ on Monday}) = P(NQQ) + P(NNQ) = 0 + \frac{1}{4} = \frac{1}{4}$

c. $P(N \text{ on Wednesday} \mid Q \text{ on Monday}) = P(QQN) + P(QNN) = 0 + \frac{1}{2} = \frac{1}{2}$

d. $P(N \text{ on Wednesday} \mid N \text{ on Monday}) = P(NQN) + P(NNN) = \frac{1}{2} + \frac{1}{4} = \frac{3}{4}$ ■

EXAMPLE 3 If T is the transition matrix given in Example 1, find T^2.

Solution
$$T^2 = \begin{bmatrix} 0 & 1 \\ \frac{1}{2} & \frac{1}{2} \end{bmatrix}\begin{bmatrix} 0 & 1 \\ \frac{1}{2} & \frac{1}{2} \end{bmatrix}$$

$$= \begin{bmatrix} 0 + \frac{1}{2} & 0 + \frac{1}{2} \\ 0 + \frac{1}{4} & \frac{1}{2} + \frac{1}{4} \end{bmatrix}$$

$$= \begin{bmatrix} \frac{1}{2} & \frac{1}{2} \\ \frac{1}{4} & \frac{3}{4} \end{bmatrix}$$
■

Now compare the results of Examples 2 and 3:

$$P(Q \text{ on Wed.} \mid Q \text{ on Mon.}) = \frac{1}{2} \qquad P(N \text{ on Wed.} \mid Q \text{ on Mon.}) = \frac{1}{2}$$
$$P(Q \text{ on Wed.} \mid N \text{ on Mon.}) = \frac{1}{4} \qquad P(N \text{ on Wed.} \mid N \text{ on Mon.}) = \frac{3}{4}$$

$$\begin{array}{c} \\ \text{Monday} \end{array} \begin{array}{c} \\ Q \\ N \end{array} \overset{\overset{\text{Wednesday}}{\begin{array}{cc} Q & N \end{array}}}{\begin{bmatrix} \frac{1}{2} & \frac{1}{2} \\ \frac{1}{4} & \frac{3}{4} \end{bmatrix}}$$

EXAMPLE 4 What are the probabilities that the instructor described in Example 1 will or will not give a quiz on Friday (4 days after Monday)?

Solution If the above result can be applied twice, we need to find T^4.

$$T^4 = \begin{bmatrix} 0 & 1 \\ \frac{1}{2} & \frac{1}{2} \end{bmatrix}^4 = \begin{bmatrix} 0 & 1 \\ \frac{1}{2} & \frac{1}{2} \end{bmatrix}^2\begin{bmatrix} 0 & 1 \\ \frac{1}{2} & \frac{1}{2} \end{bmatrix}^2 = \begin{bmatrix} \frac{1}{2} & \frac{1}{2} \\ \frac{1}{4} & \frac{3}{4} \end{bmatrix}\begin{bmatrix} \frac{1}{2} & \frac{1}{2} \\ \frac{1}{4} & \frac{3}{4} \end{bmatrix}$$

$$= \begin{bmatrix} \frac{1}{4} + \frac{1}{8} & \frac{1}{4} + \frac{3}{8} \\ \frac{1}{8} + \frac{3}{16} & \frac{1}{8} + \frac{9}{16} \end{bmatrix} = \begin{bmatrix} \frac{3}{8} & \frac{5}{8} \\ \frac{5}{16} & \frac{11}{16} \end{bmatrix}$$

The row 1, column 1 entry in T^4 signifies that the probability of a quiz on Friday, given a quiz on Monday, is $\frac{3}{8}$. ■

Eight (school) days after the initial Monday the probabilities are found by considering T^8:

$$T^8 = \begin{bmatrix} 0 & 1 \\ \frac{1}{2} & \frac{1}{2} \end{bmatrix}^4\begin{bmatrix} 0 & 1 \\ \frac{1}{2} & \frac{1}{2} \end{bmatrix}^4 = \begin{bmatrix} \frac{3}{8} & \frac{5}{8} \\ \frac{5}{16} & \frac{11}{16} \end{bmatrix}\begin{bmatrix} \frac{3}{8} & \frac{5}{8} \\ \frac{5}{16} & \frac{11}{16} \end{bmatrix}$$

$$= \begin{bmatrix} \frac{9}{64} + \frac{25}{128} & \frac{15}{64} + \frac{55}{128} \\ \frac{15}{128} + \frac{55}{256} & \frac{25}{128} + \frac{121}{256} \end{bmatrix}$$

$$= \begin{bmatrix} \frac{43}{128} & \frac{85}{128} \\ \frac{85}{256} & \frac{171}{256} \end{bmatrix} \approx \begin{bmatrix} .3359 & .6641 \\ .3320 & .6680 \end{bmatrix}$$

This suggests that as we take higher and higher powers of T, the result gets closer and closer to

$$\begin{bmatrix} \frac{1}{3} & \frac{2}{3} \\ \frac{1}{3} & \frac{2}{3} \end{bmatrix}$$

Note that the same probability vector appears in both rows. Suppose we multiply this vector $\begin{bmatrix} \frac{1}{3} & \frac{2}{3} \end{bmatrix}$ times the original transition matrix:

$$\begin{bmatrix} \frac{1}{3} & \frac{2}{3} \end{bmatrix} \begin{bmatrix} 0 & 1 \\ \frac{1}{2} & \frac{1}{2} \end{bmatrix} = [0 + \frac{1}{3} \quad \frac{1}{3} + \frac{1}{3}] = \begin{bmatrix} \frac{1}{3} & \frac{2}{3} \end{bmatrix}$$

The answer is still $\begin{bmatrix} \frac{1}{3} & \frac{2}{3} \end{bmatrix}$! This is no coincidence, and if $T, T^2, T^3, \ldots, T^n$, approach a matrix all of whose rows are V, then $VT = V$. For this reason we call V a **fixed probability vector**. We discuss the process for finding a fixed probability vector in the next section.

COMPUTER APPLICATION   Program 13 on the computer disk accompanying this book raises a transition matrix to higher and higher powers so you can "see" it moving toward a stable state.

Problem Set 1

Are the matrices in Problems 1–3 probability vectors?

1. a. $\begin{bmatrix} \frac{1}{2} & \frac{1}{2} & \frac{1}{2} \end{bmatrix}$ **b.** $\begin{bmatrix} \frac{1}{2} & \frac{1}{2} \end{bmatrix}$
   c. $\begin{bmatrix} \frac{2}{3} & \frac{1}{5} \end{bmatrix}$

2. a. $[1 \quad 0 \quad 0]$ **b.** $[.6 \quad .3 \quad .1]$
   c. $[.1 \quad .1 \quad .1]$

3. a. $[.5 \quad .6 \quad -.1]$
   b. $[.05 \quad .15 \quad .63 \quad .17]$
   c. $[0 \quad 0 \quad 0]$

Are the matrices in Problems 4–6 transition matrices?

4. a. $\begin{bmatrix} .3 & .2 & .5 \\ 0 & .1 & .9 \\ .8 & .1 & .1 \end{bmatrix}$ **b.** $\begin{bmatrix} .5 & 0 & .5 \\ .1 & .1 & .1 \\ 1 & 0 & 0 \end{bmatrix}$

5. a. $\begin{bmatrix} \frac{1}{3} & \frac{2}{3} \\ \frac{1}{5} & \frac{4}{5} \end{bmatrix}$ **b.** $\begin{bmatrix} \frac{2}{3} & \frac{4}{5} \\ \frac{5}{6} & \frac{1}{6} \end{bmatrix}$

6. a. $\begin{bmatrix} \frac{1}{4} & \frac{1}{2} & \frac{1}{4} \\ \frac{1}{5} & \frac{3}{5} & \frac{2}{5} \\ 0 & 0 & 0 \end{bmatrix}$ **b.** $\begin{bmatrix} 1 & 0 & 0 \\ 0 & 1 & 0 \\ 0 & 0 & 1 \end{bmatrix}$

If T is the matrix in Problems 7–12, find T^2 and T^3.

7. $\begin{bmatrix} \frac{1}{2} & \frac{1}{2} \\ \frac{1}{4} & \frac{3}{4} \end{bmatrix}$ **8.** $\begin{bmatrix} .1 & .2 & .7 \\ .1 & .8 & .1 \\ .2 & .3 & .5 \end{bmatrix}$

9. $\begin{bmatrix} .6 & .4 \\ .23 & .77 \end{bmatrix}$ **10.** $\begin{bmatrix} 1 & 0 & 0 \\ 0 & 1 & 0 \\ .4 & .3 & .3 \end{bmatrix}$

11. $\begin{bmatrix} .15 & .25 & .6 \\ .5 & .35 & .15 \\ .1 & .8 & .1 \end{bmatrix}$ **12.** $\begin{bmatrix} .29 & .53 & .18 \\ .01 & .99 & 0 \\ .6 & .3 & .1 \end{bmatrix}$

13. If a certain experiment has two states 0 and 1 with the probabilities $P(0|0) = .3$, $P(1|0) = .7$, $P(0|1) = .6$, and $P(1|1) = .4$, write the transition matrix.

14. If a certain experiment has three states 0, 1, and 2 with the probabilities $P(0|0) = .4$, $P(1|0) = .3$, $P(2|0) = .3$, $P(1|0) = .1$, $P(1|1) = .5$, $P(2|1) = .4$, $P(0|2) = .7$, $P(1|2) = .2$, $P(2|2) = .1$, write the transition matrix.

APPLICATIONS

15. A stock broker has been watching the price changes of a particular stock. Over the past year, the broker has found that on a day the stock is traded,

P(increase | increase on previous day) $= .3$
P(decrease | increase on previous day) $= .7$
P(increase | decrease on previous day) $= .8$
P(decrease | decrease on previous day) $= .2$

Write a transition matrix for this information.

16. A city planner has studied the living patterns within a certain greater metropolitan area. Over the last 10 years the study shows that

P(move to city | lived in suburbs previous year) = .1

P(stay in suburbs | lived in suburbs previous year) = .6

P(leave area | lived in suburbs previous year) = .3

P(stay in city | lived in city previous year) = .8

P(move to suburbs | lived in city previous year) = .1

P(leave area | lived in city previous year) = .1

P(move to city | lived out of the area previous year) = .4

P(move to suburbs | lived out of the area previous year) = .6

Write a transition matrix for this information.

17. San Francisco is served by three airports, San Francisco International (SFO), Oakland (OAK), and San Jose (SJC), and a car rental company plans to begin operations in San Francisco by setting up rental and car storage facilities at these three airports. The probabilities that a person will return the car to a particular location is given by the following matrix, T:

Returned to

		SFO	OAK	SJC
	SFO	.8	.1	.1
Rented from	OAK	.2	.7	.1
	SJC	.3	.2	.6

a. Is this a transition matrix? Why or why not?
b. What does the entry .8 signify?
c. Which airport has the lowest percent of returns?

18. A survey of the residents in the greater Dallas–Fort Worth area shows that they purchased their homeowners insurance from one of three companies. The probabilities that a resident will purchase homeowners insurance from one of these companies in the subsequent year are summarized by the following matrix:

Company the
subsequent year

		A	B	C
	A	.4	.4	.2
Present company	B	.1	.8	.1
	C	.1	.2	.7

a. Is this a transition matrix? Why or why not?
b. What does the entry .8 signify?

c. Which company looks like it has the least percentage of satisfied customers?

19. Find T^2 for the matrix in Problem 17.

20. Find T^4 for the matrix in Problem 17.

21. Find T^2 for the matrix in Problem 18.

22. Find T^4 for the matrix in Problem 18.

In Problems 23–30 suppose that the transition matrix in Example 1 is changed to

Next day

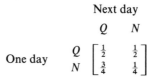

Assume that the instructor does not give a quiz on the first day of class.

23. What is the probability of a quiz on the second day of class?

24. What is the probability of a quiz on the third day of class?

25. What is the probability of a quiz on the fourth day of class?

26. What is P(Q on Wednesday | Q on Monday)?

27. What is P(Q on Wednesday | N on Monday)?

28. What is P(N on Wednesday | Q on Monday)?

29. What is P(N on Wednesday | N on Monday)?

30. Write a matrix that will answer Problems 26–29.

A computer program is called a "learning program" if it can learn by making mistakes and then correcting those mistakes. Suppose the probability that the computer will be functioning properly on a given day depends on whether it was functioning properly on the previous day. These probabilities are summarized by the following transition matrix, where C means functioning properly and E means malfunctioning:

State next day

		C	E
Present state	C	.99	.01
	E	.9	.1

In Problems 31–38 assume that the present state of the program is that it is operating properly and give answers correct to four decimal places.

31. What is the probability that the computer will be in a correct state on the second day (i.e., after two transitions)?

32. What is the probability that the computer will be in a correct state on the third day (i.e., after three transitions)?

33. What is the probability that the computer will be in a correct state on the fourth day (i.e., after four transitions)?

34. What is P(C on Wednesday | E on Monday)?

35. What is P(C on Wednesday | C on Monday)?

36. What is P(E on Wednesday | E on Monday)?

37. What is P(E on Wednesday | C on Monday)?

38. Write a matrix that will answer Problems 34–37.

2 Regular Markov Chains

In Section 1 we discussed transition matrices that summarized the probabilities for changing from one state to another. We also were introduced to the idea of a *fixed probability vector*, which is now defined:

Fixed Probability Vector

> A probability vector V such that $VT = V$ is called a **fixed probability vector** for the matrix T.

To understand this concept, consider a marketing analysis for Tide®. Suppose the following transition matrix is given:

$$
\begin{array}{c}
 \\
\text{Present purchase}
\end{array}
\begin{array}{cc}
 & \text{Next purchase} \\
 & \begin{array}{cc} \text{Tide} & \text{Brand X} \end{array} \\
\begin{array}{c} \text{Tide} \\ \text{Brand X} \end{array} &
\begin{bmatrix} .8 & .2 \\ .4 & .6 \end{bmatrix}
\end{array}
$$

Now, suppose that there are 10 million people who purchase a detergent each month and that 25% of the consumers purchase Tide and 75% purchase Brand X (all other brands lumped together). Also, suppose that every purchaser makes one purchase per month. What share of the market will use Tide in the next month, given these assumptions?

$$
\begin{array}{cc} \text{Tide} & \text{Brand X} \\ [\,.25 & .75\,] \end{array}
\begin{bmatrix} .8 & .2 \\ .4 & .6 \end{bmatrix} = [.5 \quad .5]
$$

Second month:

$$
[.5 \quad .5]\begin{bmatrix} .8 & .2 \\ .4 & .6 \end{bmatrix} = [.6 \quad .4]
$$

Third month:

$$
[.6 \quad .4]\begin{bmatrix} .8 & .2 \\ .4 & .6 \end{bmatrix} = [.64 \quad .36]
$$

Fourth month:

$$
[.64 \quad .36]\begin{bmatrix} .8 & .2 \\ .4 & .6 \end{bmatrix} = [.656 \quad .344]
$$

This means that at the end of the fourth month, Tide could expect 65.6% of the market, or 6.56 million customers. If we continue this process, will the market share for Tide and Brand X stabilize at some specific share of the market for each? If so, this share is represented by the fixed probability vector.

Let $V = \begin{bmatrix} v_1 & v_2 \end{bmatrix}$. Now find V so that

$$\begin{bmatrix} v_1 & v_2 \end{bmatrix}\begin{bmatrix} .8 & .2 \\ .4 & .6 \end{bmatrix} = \begin{bmatrix} v_1 & v_2 \end{bmatrix}$$

$$\begin{cases} .8v_1 + .4v_2 = v_1 \\ .2v_1 + .6v_2 = v_2 \end{cases} \quad \text{or} \quad \begin{cases} .2v_1 - .4v_2 = 0 \\ .2v_1 - .4v_2 = 0 \end{cases}$$

In addition, $\begin{bmatrix} v_1 & v_2 \end{bmatrix}$ is a probability vector, so $v_1 + v_2 = 1$, giving the system

$$\begin{cases} .2v_1 - .4v_2 = 0 \\ .2v_1 - .4v_2 = 0 \\ v_1 + v_2 = 1 \end{cases}$$

$$\begin{bmatrix} .2 & -.4 & \vdots & 0 \\ .2 & -.4 & \vdots & 0 \\ 1 & 1 & \vdots & 1 \end{bmatrix} \rightarrow \begin{bmatrix} 1 & 1 & \vdots & 1 \\ 0 & -.6 & \vdots & -.2 \\ 0 & 0 & \vdots & 0 \end{bmatrix} \rightarrow \begin{bmatrix} 1 & 1 & \vdots & 1 \\ 0 & 1 & \vdots & \frac{1}{3} \\ 0 & 0 & \vdots & 0 \end{bmatrix} \rightarrow \begin{bmatrix} 1 & 0 & \vdots & \frac{2}{3} \\ 0 & 1 & \vdots & \frac{1}{3} \\ 0 & 0 & \vdots & 0 \end{bmatrix}$$

Thus $v_1 = \frac{2}{3}, v_2 = \frac{1}{3}$, and $\begin{bmatrix} \frac{2}{3} & \frac{1}{3} \end{bmatrix}$ is the fixed probability vector. This means that, in the long run, Tide can expect to capture $\frac{2}{3}$ of the market.

The next example ties together the concepts of a fixed probability vector and the successive powers of a transition matrix.

EXAMPLE 1 Suppose a professor gives quizzes (Q) or no quizzes (N) according to the transition matrix

$$\begin{array}{c} \\ Q \\ N \end{array}\begin{array}{c} Q \qquad N \\ \begin{bmatrix} 0 & 1 \\ \frac{1}{2} & \frac{1}{2} \end{bmatrix} \end{array}$$

Find the fixed probability vector $V = \begin{bmatrix} v_1 & v_2 \end{bmatrix}$.

Solution

$$\begin{bmatrix} v_1 & v_2 \end{bmatrix}\begin{bmatrix} 0 & 1 \\ \frac{1}{2} & \frac{1}{2} \end{bmatrix} = \begin{bmatrix} v_1 & v_2 \end{bmatrix}$$

We solve this equation for v_1 and v_2:

$$\begin{bmatrix} v_1 & v_2 \end{bmatrix}\begin{bmatrix} 0 & 1 \\ \frac{1}{2} & \frac{1}{2} \end{bmatrix} = \begin{bmatrix} 0 + \frac{1}{2}v_2 & v_1 + \frac{1}{2}v_2 \end{bmatrix} = \begin{bmatrix} \frac{1}{2}v_2 & v_1 + \frac{1}{2}v_2 \end{bmatrix}$$

If $\begin{bmatrix} \frac{1}{2}v_2 & v_1 + \frac{1}{2}v_2 \end{bmatrix} = \begin{bmatrix} v_1 & v_2 \end{bmatrix}$, the corresponding components are equal:

$$\begin{cases} \frac{1}{2}v_2 = v_1 \\ v_1 + \frac{1}{2}v_2 = v_2 \end{cases}$$

Using this relationship, along with the fact that $v_1 + v_2 = 1$, gives the matrix system

$$\begin{bmatrix} 1 & 1 & \vdots & 1 \\ -1 & \frac{1}{2} & \vdots & 0 \\ 1 & -\frac{1}{2} & \vdots & 0 \end{bmatrix} \rightarrow \begin{bmatrix} 1 & 1 & \vdots & 1 \\ 0 & \frac{3}{2} & \vdots & 1 \\ 0 & -\frac{3}{2} & \vdots & -1 \end{bmatrix} \rightarrow \begin{bmatrix} 1 & 1 & \vdots & 1 \\ 0 & 1 & \vdots & \frac{2}{3} \\ 0 & 0 & \vdots & 0 \end{bmatrix} \rightarrow \begin{bmatrix} 1 & 0 & \vdots & \frac{1}{3} \\ 0 & 1 & \vdots & \frac{2}{3} \\ 0 & 0 & \vdots & 0 \end{bmatrix}$$

Thus $v_1 = \frac{1}{3}$ and $v_2 = \frac{2}{3}$, or $V = \begin{bmatrix} \frac{1}{3} & \frac{2}{3} \end{bmatrix}$. ∎

Now let us compare the solution of Example 1 with the powers of T found in Section 1:

$$T^2 = \begin{bmatrix} \frac{1}{2} & \frac{1}{2} \\ \frac{1}{4} & \frac{3}{4} \end{bmatrix} \qquad T^4 = \begin{bmatrix} \frac{3}{8} & \frac{5}{8} \\ \frac{5}{16} & \frac{11}{16} \end{bmatrix} \qquad T^8 \approx \begin{bmatrix} .3359 & .6641 \\ .3320 & .6680 \end{bmatrix}$$

It is no coincidence that the higher powers of T approach the matrix consisting of the fixed probability vector $\begin{bmatrix} \frac{1}{3} & \frac{2}{3} \end{bmatrix}$:

$$\begin{bmatrix} \frac{1}{3} & \frac{2}{3} \\ \frac{1}{3} & \frac{2}{3} \end{bmatrix}$$

This relationship between the higher powers of T and the fixed probability vector are summarized in the **transition matrix theorem**. However, before we can state this theorem, we need to consider some terminology that will help us describe when the transition matrix theorem applies.

Consider the matrix

$$\begin{bmatrix} 1 & 0 \\ \frac{1}{5} & \frac{4}{5} \end{bmatrix}$$

The powers of this matrix will always have $p_{12} = 0$. We can see this by considering

$$\begin{bmatrix} a & 0 \\ b & c \end{bmatrix} \begin{bmatrix} a & 0 \\ b & c \end{bmatrix}$$

The entry in the first row, second column of the product is

$$a \cdot 0 + 0 \cdot c = 0$$

Thus, regardless of the other entries, the powers of this matrix will always have $p_{12} = 0$. This means that the powers of this matrix will never approach a matrix whose rows are all a probability vector V. However, if for some power of T all the entries are positive, then the powers of the matrix will approach a matrix whose rows are all a probability vector. If this is the case, we call the transition matrix **regular**.

Regular Transition Matrix

> A transition matrix is *regular* if there exists some power of it that contains only positive elements.

EXAMPLE 2   Is $\begin{bmatrix} \frac{1}{5} & \frac{4}{5} \\ 1 & 0 \end{bmatrix}$ regular?

Solution $\begin{bmatrix} \frac{1}{5} & \frac{4}{5} \\ 1 & 0 \end{bmatrix} \begin{bmatrix} \frac{1}{5} & \frac{4}{5} \\ 1 & 0 \end{bmatrix} = \begin{bmatrix} \frac{1}{25} + \frac{4}{5} & \frac{4}{25} + 0 \\ \frac{1}{5} + 0 & \frac{4}{5} + 0 \end{bmatrix}$

$$= \begin{bmatrix} \frac{21}{25} & \frac{4}{25} \\ \frac{1}{5} & \frac{4}{5} \end{bmatrix}$$

Since the second power of the matrix has all positive elements, we see that it is regular.   ∎

If a transition matrix T is regular, then the following theorem can be proved:

Transition Matrix Theorem

If T is a regular transition matrix, then:
1. T has a unique fixed probability vector, V, whose elements are all positive.
2. $T, T^2, T^3, \ldots, T^n$ approach the matrix all of whose rows are V.
3. If W is any probability vector, then $WT, WT^2, WT^3, \ldots, WT^n$ approach the fixed probability vector V.

COMPUTER APPLICATION   If you have access to a computer, you can verify this theorem using Program 13 on the computer disk accompanying this text. In this program, $a(n)$ is used as follows:

$$a(1) = WT, a(2) = WT^2, a(3) = WT^3, \ldots, a(n) = WT^n$$

EXAMPLE 3   Suppose that General Motors (GM), Ford (F), and Chrysler (C) each introduce a front-wheel-drive compact with better than 50 mpg. Assume that each company initially captures about one-third of the market. During the year:

1. General Motors keeps 85% of its customers but loses 10% to Ford and 5% to Chrysler.
2. Ford keeps 80% of its customers but loses 10% to General Motors and 10% to Chrysler.
3. Chrysler keeps 60% of its customers but loses 25% to General Motors and 15% to Ford.

If this trend continues, what share of the market is each likely to have at the end of next year, and what is the long-run prediction if this trend continues?

Solution   The transition matrix is

$$
\begin{array}{c}
 \\
\text{GM} \\
\text{F} \\
\text{C}
\end{array}
\begin{array}{ccc}
\text{GM} & \text{F} & \text{C} \\
\left[\begin{array}{ccc}
\frac{17}{20} & \frac{1}{10} & \frac{1}{20} \\
\frac{1}{10} & \frac{4}{5} & \frac{1}{10} \\
\frac{1}{4} & \frac{3}{20} & \frac{3}{5}
\end{array}\right]
\end{array}
\qquad
\begin{array}{l}
\textit{Note:} \quad 85\% = \frac{85}{100} = \frac{17}{20}; \text{ do the same for} \\
\text{all the percents}
\end{array}
$$

Since the entries are all positive, the matrix is a regular transition matrix and the transition matrix theorem applies. To determine what share of the market each company will have at the end of next year we multiply the original share vector $[\frac{1}{3} \quad \frac{1}{3} \quad \frac{1}{3}]$ by T:

$$
[\tfrac{1}{3} \quad \tfrac{1}{3} \quad \tfrac{1}{3}]
\begin{bmatrix}
\frac{17}{20} & \frac{1}{10} & \frac{1}{20} \\
\frac{1}{10} & \frac{4}{5} & \frac{1}{10} \\
\frac{1}{4} & \frac{3}{20} & \frac{3}{5}
\end{bmatrix}
$$

$$
= [\tfrac{17}{60} + \tfrac{1}{30} + \tfrac{1}{12} \quad \tfrac{1}{30} + \tfrac{4}{15} + \tfrac{1}{20} \quad \tfrac{1}{60} + \tfrac{1}{30} + \tfrac{1}{5}]
$$

$$
= [\tfrac{2}{5} \quad \tfrac{7}{20} \quad \tfrac{1}{4}]
$$

We see that at the end of next year GM will have $\frac{2}{5}$ or 40% of the market, Ford will have 35% of the market, and Chrysler will have 25% of the market.

For the long-range prediction we find the fixed probability vector:

$$[v_2 \quad v_2 \quad v_3]\begin{bmatrix} \frac{17}{20} & \frac{1}{10} & \frac{1}{20} \\ \frac{1}{10} & \frac{4}{5} & \frac{1}{10} \\ \frac{1}{4} & \frac{3}{20} & \frac{3}{5} \end{bmatrix} = [v_1 \quad v_2 \quad v_3]$$

By the third part of the transition matrix theorem, any probability vector $[\frac{1}{3} \quad \frac{1}{3} \quad \frac{1}{3}]$ T approaches this fixed probability vector. Multiplying, we get

$$[\tfrac{17}{20}v_1 + \tfrac{1}{10}v_2 + \tfrac{1}{4}v_3 \quad \tfrac{1}{10}v_1 + \tfrac{4}{5}v_2 + \tfrac{3}{20}v_3 \quad \tfrac{1}{20}v_1 + \tfrac{1}{10}v_2 + \tfrac{3}{5}v_3] = [v_1 \quad v_2 \quad v_3]$$

By the equality of vectors, we obtain the following system:

$$\begin{cases} \tfrac{17}{20}v_1 + \tfrac{1}{10}v_2 + \tfrac{1}{4}v_3 = v_1 \\ \tfrac{1}{10}v_1 + \tfrac{4}{5}v_2 + \tfrac{3}{20}v_3 = v_2 \\ \tfrac{1}{20}v_1 + \tfrac{1}{10}v_2 + \tfrac{3}{5}v_3 = v_3 \end{cases} \quad \text{or} \quad \begin{cases} 3v_1 - 2v_2 - 5v_3 = 0 \\ 2v_1 - 4v_2 + 3v_3 = 0 \\ v_1 + 2v_2 - 8v_3 = 0 \end{cases}$$

If you solve this system of equations, you will find that it is a dependent system. Add to this system the fact that V is a probability vector, namely, $v_1 + v_2 + v_3 = 1$, to obtain the matrix:

$$\begin{bmatrix} 3 & -2 & -5 & | & 0 \\ 2 & -4 & 3 & | & 0 \\ 1 & 2 & -8 & | & 0 \\ 1 & 1 & 1 & | & 1 \end{bmatrix} \rightarrow \begin{bmatrix} 1 & 1 & 1 & | & 1 \\ 0 & -5 & -8 & | & -3 \\ 0 & -6 & 1 & | & -2 \\ 0 & 1 & -9 & | & -1 \end{bmatrix} \rightarrow \begin{bmatrix} 1 & 1 & 1 & | & 1 \\ 0 & 1 & -9 & | & -1 \\ 0 & 0 & -53 & | & -8 \\ 0 & 0 & -53 & | & -8 \end{bmatrix}$$

$$\rightarrow \begin{bmatrix} 1 & 1 & 1 & | & 1 \\ 0 & 1 & -9 & | & -1 \\ 0 & 0 & 1 & | & \frac{8}{53} \\ 0 & 0 & 0 & | & 0 \end{bmatrix} \rightarrow \begin{bmatrix} 1 & 0 & 0 & | & \frac{26}{53} \\ 0 & 1 & 0 & | & \frac{19}{53} \\ 0 & 0 & 1 & | & \frac{8}{53} \\ 0 & 0 & 0 & | & 0 \end{bmatrix}$$

Therefore, $v_1 = \frac{26}{53}$, $v_2 = \frac{19}{53}$, and $v_3 = \frac{8}{53}$. Thus the fixed probability vector is

$$[\tfrac{26}{53} \quad \tfrac{19}{53} \quad \tfrac{8}{53}] \approx [.49 \quad .36 \quad .15]$$

This means that the long-range prediction is that General Motors will have about 49% of the market, Ford 36%, and Chrysler 15%. Note that the transition matrix theorem tells us that this is the long-term prediction *regardless of the original market share*. Even if GM started with 10% of the market, Ford 20%, and Chrysler 70%, the long-term prediction, if the same trend continues, will still be 49% for GM, 36% for Ford, and 15% for Chrysler. ∎

Problem Set 2

In Problems 1–12 determine whether the matrix is regular. For each matrix that is regular, find the matrix that it approaches as it is raised to higher and higher powers.

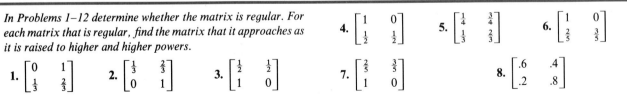

9. $\begin{bmatrix} 0 & 1 & 0 \\ \frac{1}{2} & \frac{1}{4} & \frac{1}{4} \\ 0 & \frac{1}{3} & \frac{2}{3} \end{bmatrix}$ **10.** $\begin{bmatrix} \frac{1}{3} & \frac{1}{3} & \frac{1}{3} \\ 0 & 0 & 1 \\ \frac{3}{10} & \frac{1}{2} & \frac{1}{5} \end{bmatrix}$

11. $\begin{bmatrix} .6 & .2 & .2 \\ .1 & .8 & .1 \\ .1 & .2 & .7 \end{bmatrix}$ **12.** $\begin{bmatrix} 0 & .3 & .7 \\ .4 & 0 & .6 \\ .5 & .5 & 0 \end{bmatrix}$

13. a. Is the following matrix regular?

$$\begin{bmatrix} 0 & 1 \\ 1 & 0 \end{bmatrix}$$

b. Does this matrix have a unique fixed probability vector?

c. Does the transition matrix theorem apply? Why or why not?

14. a. Is the following matrix regular?

$$\begin{bmatrix} 1 & 0 \\ \frac{1}{2} & \frac{1}{2} \end{bmatrix}$$

b. Does this matrix have a unique fixed probability vector?

c. Does the transition matrix theorem apply? Why or why not?

15. Consider the following matrix:

$$\begin{bmatrix} a & 1-a \\ 1-a & a \end{bmatrix}$$

a. What conditions on a are necessary if this is to be a transition matrix?

b. Find the fixed probability vector (assuming that the conditions you found in part a apply).

16. Let

$$T = \begin{bmatrix} 1 & 0 & 0 \\ \frac{1}{2} & 0 & \frac{1}{2} \\ 0 & 0 & 1 \end{bmatrix}$$

a. Find the matrix that T^n approaches.

b. Does T have more than one fixed probability vector?

APPLICATIONS

17. A housewife buys three kinds of dogfood: canned, packaged, and dry. Her dog likes variety so she never buys the same kind on successive days. If she buys canned food one day, then the next day she buys packaged food. But if she buys packaged or dry food, then the next day she is twice as likely to buy canned food as the other types. Find the

transition matrix. In the long run, how often does she buy each of the three varieties?

18. A publishing company has three books to promote. It is company policy to take out an advertisement for one of the books each month in a national journal. The author of book A is the brother-in-law of an editor and has a 60% chance of being picked for advertising each month. The other books will not be picked two months in a row, and if book A is not chosen, then books B and C have an equal chance of being picked. Find the transition matrix. In the long run, how often will each book be advertised?

19. A retailer stocks three brands of coffee. Each brand has about one-third of the market. A survey is taken and finds that each month the following occurs:

a. Alpha Coffee Concern keeps 70% of its customers but loses 10% to Better Bean Coffee and 20% to Carolyn's Certified Coffee.

b. Better Bean Coffee keeps 80% of its customers but loses 10% to Alpha Coffee Concern and 10% to Carolyn's Certified Coffee.

c. Carolyn's Certified Coffee keeps 50% of its customers but loses 10% to Alpha Coffee and 40% to Better Bean Coffee.

Write the transition matrix for this information.

20. What is the market share of each brand of coffee in Problem 19 after 1 month? After 2 months?

21. What is the market share of each brand of coffee in Problem 19 in the long run?

22. What is the market share of each brand of coffee in Problem 19 in the long run if Alpha Coffee starts with 10%, Better Bean with 30%, and Carolyn's Certified Coffee with 60% of the market? What is the market share after 1 month?

23. In an eastern state a survey finds that if a person's father is a Democrat, the person will vote Democratic 70% of the time and Independent the rest of the time. If a person's father is a Republican, the person will vote Republican 50% of the time, Democratic 40% of the time, and Independent 10% of the time. Of the children of Independents, 40% of the time they will vote Independent, 40% Democratic, and 20% Republican. What is the long-range voting pattern for each party?

24. What is the probability that the grandchild of a Democrat will vote Independent in the eastern state described in Problem 23?

25. On Niels' cattle ranch it has been shown that the probability that a Jersey cow will have a Jersey calf is .6, that it will have a Guernsey calf is .2, or that it will have a white-faced calf is .2. The probabilities that a Guernsey cow will have a Jersey, Guernsey, or white-faced calf are .1, .7, and .2, respectively; and the probabilities that a white-faced cow will have a Jersey, Guernsey, or white-faced calf are .1, .1, and .8, respectively. Assume that the herd is now 30% Jersey, 20% Guernsey, and 50% white-faced. What will the distribution of the herd be in two generations?

26. What will the distribution of the herd of cattle described in Problem 25 be after a large number of generations?

27. Two companies compete against each other, and the transition matrix for people changing from one company to another each month is given by the following transition matrix:

$$
\begin{array}{c}
\text{To company} \\
\begin{array}{cc}
\quad\text{I} & \text{II} \\
\end{array} \\
\text{From company} \;
\begin{array}{c}
\text{I} \\
\text{II}
\end{array}
\begin{bmatrix}
.7 & .3 \\
.2 & .8
\end{bmatrix}
\end{array}
$$

What is the long-range expectation for the companies if this trend continues?

28. In a recent close presidential race a poll finds that the incumbent would have to carry California, New York, and Texas to be reelected, so the campaign manager decides to devote the last week of the campaign to these states. She also decides that the incumbent will not campaign in the same state 2 days in a row and that if he campaigns in California, he will be twice as likely to campaign in New York the next day. Also, if he campaigns in New York, he will be twice as likely to campaign in California the next day. If he campaigns in Texas, he will be equally likely to campaign in California or New York the next day. What is the transition matrix? Is the transition matrix regular?

29. If the incumbent of Problem 28 spoke in Texas on Wednesday, what is the probability that he will be in California on Friday?

30. If the incumbent of Problem 28 spoke in California on Wednesday, what is the probability that he will be in California again on Friday?

3 Absorbing Markov Chains

Under certain conditions a stochastic process will give rise to a transition matrix for a Markov chain with the property that once a given state is reached it is impossible to move out of that state. For example, suppose two people each have two quarters and decide to flip one coin each and call "match" or "no match." The winner takes both coins. If they play long enough, common sense tells us that eventually one player will go broke and the other player will end up with all four coins to put an end to the game. The states of having no or four coins are called **absorbing states**.

Absorbing States

> A state i of a Markov chain is an *absorbing state* if once the system reaches state i on same trial, the system remains in that state on all future trials. This will occur whenever $p_{ii} = 1$.

EXAMPLE 1 Let

$$
T = \begin{array}{c}
\begin{array}{ccc}
\quad 1 & 2 & 3
\end{array} \\
\begin{array}{c}
1 \\
2 \\
3
\end{array}
\begin{bmatrix}
.1 & .2 & .7 \\
0 & 1 & 0 \\
.5 & .3 & .2
\end{bmatrix}
\end{array}
$$

Then $p_{12} = .2$ is the probability of going from state 1 to state 2. Since $p_{22} = 1$, the probability of going from state 2 to state 2 (that is, remaining in state 2) is 1. Note also that if one entry in any row is 1, then the other entries must be 0s, since each row of a transition matrix is a probability vector. ∎

EXAMPLE 2 Suppose two people each have two coins and decide to play match–no match. Write the transition matrix for the first player. The position p_{ij} is the probability of changing from state i to state j. In this example, state i means having i coins.

$$
\begin{array}{c}
\text{Number of ending coins} \\
\text{of player 1}
\end{array}
$$

$$
\text{Number of beginning} \atop \text{coins of player 1}
\begin{array}{c}
\\ 0 \\ 1 \\ 2 \\ 3 \\ 4
\end{array}
\begin{array}{ccccc}
0 & 1 & 2 & 3 & 4 \\
\left[\begin{array}{ccccc}
1 & 0 & 0 & 0 & 0 \\
\frac{1}{2} & 0 & \frac{1}{2} & 0 & 0 \\
0 & \frac{1}{2} & 0 & \frac{1}{2} & 0 \\
0 & 0 & \frac{1}{2} & 0 & \frac{1}{2} \\
0 & 0 & 0 & 0 & 1
\end{array}\right]
\end{array}
$$

Note that states 0 and 4 are absorbing. Also note that at each state the player's bankroll will increase or decrease by one coin, each with a probability of $\frac{1}{2}$. ■

Suppose we want to examine the long-term trend for a matrix with one or more absorbing states. Let

$$
T = \begin{array}{c} 1 \\ 2 \\ 3 \end{array}
\begin{array}{ccc}
1 & 2 & 3 \\
\left[\begin{array}{ccc}
1 & 0 & 0 \\
.1 & .5 & .4 \\
0 & 0 & 1
\end{array}\right]
\end{array}
$$

Both states 1 and 3 are absorbing states—once these states are entered, the system will remain in that state. If the Markov chain begins in state 2, it too must eventually end up in either state 1 or state 3. Now consider the powers of T (correct to two decimal places):

$$
T^2 = \left[\begin{array}{ccc}
1 & 0 & 0 \\
.15 & .25 & .6 \\
0 & 0 & 1
\end{array}\right]
\qquad
T^4 = \left[\begin{array}{ccc}
1 & 0 & 0 \\
.19 & .06 & .75 \\
0 & 0 & 1
\end{array}\right]
$$

$$
T^8 = \left[\begin{array}{ccc}
1 & 0 & 0 \\
.20 & .00 & .80 \\
0 & 0 & 1
\end{array}\right]
$$

It appears that if the system begins in state 2, the probability that it will end in absorbing state 1 is $\frac{1}{5}$ (entry p_{21}) and in absorbing state 3, $\frac{4}{5}$ (entry p_{23}). Now remember that with regular Markov chains the final result is independent of the initial state. This is not true for absorbing Markov chains, as we can see from this example.

Absorbing Markov Chain
Theorem

If T is a transition matrix for an absorbing Markov chain, then:
1. $T, T^2, T^3, \ldots$ approach some particular matrix.
2. In a finite number of steps the chain will enter an absorbing state and stay there.
3. The long-term trend depends on the initial state.

The next step is to learn a procedure for finding the probabilities that a particular initial nonabsorbing state will lead to a particular absorbing state. The following theorem outlines this procedure:

Transition Theorem for
Absorbing Markov Chains

Let T be a transition matrix for an absorbing Markov chain.

Step 1   Rearrange the rows and columns of T so that the absorbing states come first:

$$T = \left[\begin{array}{c|c} I_n & 0 \\ \hline P & Q \end{array}\right]$$

where I_n is an identity matrix and 0 is a matrix consisting of all 0s. Matrices P and Q are the matrices consisting of the remaining entries shown in the indicated positions.

Step 2   Let $F = [I_m - Q]^{-1}$, where I_m is an identity matrix of the same order as Q. Remember, $[I_m - Q]^{-1}$ is the inverse of the matrix $[I_m - Q]$. Call F the **fundamental matrix**.

Step 3 FP is a matrix whose entries p_{ij} represent the probability that nonabsorbing state i will lead to absorbing state j.

EXAMPLE 3   Let

$$T = \begin{array}{c} \\ 1 \\ 2 \\ 3 \end{array} \begin{array}{ccc} 1 & 2 & 3 \\ \left[\begin{array}{ccc} 1 & 0 & 0 \\ .1 & .5 & .4 \\ 0 & 0 & 1 \end{array}\right] \end{array}$$

Find and interpret FP.

Solution   *Step 1*

$$\begin{array}{c} 1 \\ 3 \\ 2 \end{array} \begin{array}{ccc} 1 & 2 & 3 \\ \left[\begin{array}{ccc} 1 & 0 & 0 \\ 0 & 0 & 1 \\ \hdashline .1 & .5 & .4 \end{array}\right] \end{array} \qquad \begin{array}{c} 1 \\ 3 \\ 2 \end{array} \begin{array}{ccc} 1 & 3 & 2 \\ \left[\begin{array}{cc|c} 1 & 0 & 0 \\ 0 & 1 & 0 \\ \hline .1 & .4 & .5 \end{array}\right] \end{array}$$

$$I = \left[\begin{array}{cc} 1 & 0 \\ 0 & 1 \end{array}\right] \qquad 0 = \left[\begin{array}{c} 0 \\ 0 \end{array}\right] \qquad P = [.1 \quad .4] \qquad Q = [.5]$$

Step 2 $F = [I_1 - Q]^{-1} = [1 - .5]^{-1} = [.5]^{-1} = [2]$

Step 3 $FP = [2][.1 \quad .4] = [.2 \quad .8] \quad \begin{array}{c} \\ 2 \end{array}$ with column headers $\begin{array}{cc} 1 & 3 \end{array}$

This means that if the system begins with nonabsorbing state 2, the probability that it will end up in absorbing state 1 is .2 and in absorbing state 3 is .8. This is consistent with the arithmetic we did by finding T^2, T^4, and T^8 earlier.   ∎

EXAMPLE 4 Find and interpret FP for the game of match–no match and the matrix

$$
T = \begin{array}{c} \\ 0 \\ 1 \\ 2 \\ 3 \\ 4 \end{array}
\begin{array}{c} \begin{array}{ccccc} 0 & 1 & 2 & 3 & 4 \end{array} \\
\left[\begin{array}{ccccc}
1 & 0 & 0 & 0 & 0 \\
\frac{1}{2} & 0 & \frac{1}{2} & 0 & 0 \\
0 & \frac{1}{2} & 0 & \frac{1}{2} & 0 \\
0 & 0 & \frac{1}{2} & 0 & \frac{1}{2} \\
0 & 0 & 0 & 0 & 1
\end{array} \right]
\end{array}
$$

Solution

$$
\begin{array}{c} \\ 0 \\ 4 \\ 1 \\ 2 \\ 3 \end{array}
\begin{array}{c} \begin{array}{ccccc} 0 & 1 & 2 & 3 & 4 \end{array} \\
\left[\begin{array}{ccccc}
1 & 0 & 0 & 0 & 0 \\
0 & 0 & 0 & 0 & 1 \\
\frac{1}{2} & 0 & \frac{1}{2} & 0 & 0 \\
0 & \frac{1}{2} & 0 & \frac{1}{2} & 0 \\
0 & 0 & \frac{1}{2} & 0 & \frac{1}{2}
\end{array} \right]
\end{array}
\qquad
\begin{array}{c} \\ 0 \\ 4 \\ 1 \\ 2 \\ 3 \end{array}
\begin{array}{c} \begin{array}{ccccc} 0 & 4 & 1 & 2 & 3 \end{array} \\
\left[\begin{array}{ccccc}
1 & 0 & 0 & 0 & 0 \\
0 & 1 & 0 & 0 & 0 \\
\frac{1}{2} & 0 & 0 & \frac{1}{2} & 0 \\
0 & 0 & \frac{1}{2} & 0 & \frac{1}{2} \\
0 & \frac{1}{2} & 0 & \frac{1}{2} & 0
\end{array} \right]
\end{array}
$$

$$
P = \left[\begin{array}{cc}
\frac{1}{2} & 0 \\
0 & 0 \\
0 & \frac{1}{2}
\end{array} \right]
\qquad
Q = \left[\begin{array}{ccc}
0 & \frac{1}{2} & 0 \\
\frac{1}{2} & 0 & \frac{1}{2} \\
0 & \frac{1}{2} & 0
\end{array} \right]
$$

$$
I - Q = \left[\begin{array}{ccc}
1 & 0 & 0 \\
0 & 1 & 0 \\
0 & 0 & 1
\end{array} \right] - \left[\begin{array}{ccc}
0 & \frac{1}{2} & 0 \\
\frac{1}{2} & 0 & \frac{1}{2} \\
0 & \frac{1}{2} & 0
\end{array} \right] = \left[\begin{array}{ccc}
1 & -\frac{1}{2} & 0 \\
-\frac{1}{2} & 1 & -\frac{1}{2} \\
0 & -\frac{1}{2} & 1
\end{array} \right]
$$

$$
F = (I - Q)^{-1} = \left[\begin{array}{ccc}
\frac{3}{2} & 1 & \frac{1}{2} \\
1 & 2 & 1 \\
\frac{1}{2} & 1 & \frac{3}{2}
\end{array} \right]
$$

Note: See Section 3.4 for the method for finding the inverse of a matrix. A computer program for finding the inverse is also available (see Appendix D). Just the final result is shown here.

$$
FP = \left[\begin{array}{ccc}
\frac{3}{2} & 1 & \frac{1}{2} \\
1 & 2 & 1 \\
\frac{1}{2} & 1 & \frac{3}{2}
\end{array} \right]
\left[\begin{array}{cc}
\frac{1}{2} & 0 \\
0 & 0 \\
0 & \frac{1}{2}
\end{array} \right]
$$

$$
= \begin{array}{c} \\ 1 \\ 2 \\ 3 \end{array}
\begin{array}{c} \begin{array}{cc} 0 & 4 \end{array} \\
\left[\begin{array}{cc}
\frac{3}{4} & \frac{1}{4} \\
\frac{1}{2} & \frac{1}{2} \\
\frac{1}{4} & \frac{3}{4}
\end{array} \right]
\end{array}
$$

This means that if the system was originally in state 1 (player 1 having one coin), then the probability of financial ruin (0 coins) is $\frac{3}{4}$ and the probability that all four coins would be obtained is $\frac{1}{4}$. If the system began in state 2 (player 1 having two coins), then there is a probability of $\frac{1}{2}$ of either winning or losing both coins. If the system began in state 3 (player 1 having three coins), then there is a probability of $\frac{1}{4}$ of losing all three coins and $\frac{3}{4}$ of winning all four coins. ∎

Problem Set 3

Determine whether each matrix in Problems 1–12 is a transition matrix for an absorbing Markov chain. If it is, find the absorbing states.

1. $\begin{bmatrix} .3 & .5 & .2 \\ .1 & .7 & .2 \\ .2 & .6 & .2 \end{bmatrix}$
2. $\begin{bmatrix} .1 & .6 & .3 \\ 1 & 0 & 0 \\ .4 & .3 & .3 \end{bmatrix}$

3. $\begin{bmatrix} \frac{1}{2} & \frac{1}{3} & \frac{1}{6} \\ \frac{2}{3} & \frac{1}{6} & \frac{1}{12} \\ \frac{1}{6} & \frac{1}{3} & \frac{1}{2} \end{bmatrix}$
4. $\begin{bmatrix} \frac{1}{3} & \frac{1}{3} & \frac{1}{3} \\ 0 & 1 & 0 \\ \frac{1}{6} & \frac{1}{3} & \frac{1}{2} \end{bmatrix}$

5. $\begin{bmatrix} 0 & 1 & 0 \\ .3 & .2 & .5 \\ .5 & .5 & 0 \end{bmatrix}$
6. $\begin{bmatrix} 1 & 0 & 0 \\ .3 & .2 & .5 \\ .5 & .5 & 0 \end{bmatrix}$

7. $\begin{bmatrix} 1 & 0 & 0 \\ 0 & 1 & 0 \\ 0 & 0 & 1 \end{bmatrix}$
8. $\begin{bmatrix} 0 & 1 & 0 \\ 1 & 0 & 0 \\ 0 & 0 & 0 \end{bmatrix}$

9. $\begin{bmatrix} 1 & 0 & 0 \\ 0 & 1 & 0 \\ .41 & .34 & .25 \end{bmatrix}$
10. $\begin{bmatrix} \frac{1}{2} & \frac{1}{3} & \frac{1}{4} & 0 \\ 0 & 1 & 0 & 0 \\ \frac{1}{4} & \frac{1}{4} & \frac{1}{4} & \frac{1}{4} \\ 0 & 0 & 0 & 1 \end{bmatrix}$

11. $\begin{bmatrix} \frac{1}{2} & \frac{1}{4} & \frac{1}{5} & \frac{1}{20} \\ \frac{1}{3} & 0 & \frac{1}{3} & \frac{1}{3} \\ 0 & 0 & 1 & 0 \\ \frac{1}{5} & \frac{1}{5} & \frac{1}{5} & \frac{2}{5} \end{bmatrix}$
12. $\begin{bmatrix} 0 & 1 & 0 & 0 \\ 0 & 0 & 1 & 0 \\ 0 & 0 & 0 & 1 \\ 1 & 0 & 0 & 0 \end{bmatrix}$

Find and interpret FP for the matrices in Problems 13–24. If approximate answers are necessary, give them correct to two decimal places.

13. $\begin{bmatrix} .2 & 3 & .5 \\ 0 & 1 & 0 \\ 0 & 0 & 1 \end{bmatrix}$
14. $\begin{bmatrix} 1 & 0 & 0 \\ 1 & .8 & 1 \\ 0 & 0 & 1 \end{bmatrix}$

15. $\begin{bmatrix} 1 & 0 & 0 \\ 0 & 1 & 0 \\ .15 & .35 & .5 \end{bmatrix}$
16. $\begin{bmatrix} \frac{1}{3} & \frac{1}{3} & \frac{1}{3} \\ \frac{1}{2} & \frac{1}{4} & \frac{1}{4} \\ 0 & 0 & 1 \end{bmatrix}$

17. $\begin{bmatrix} 1 & 0 & 0 \\ .1 & .9 & 0 \\ .3 & .6 & .1 \end{bmatrix}$
18. $\begin{bmatrix} \frac{1}{4} & \frac{1}{3} & \frac{5}{12} \\ 0 & 1 & 0 \\ \frac{1}{2} & 0 & \frac{1}{2} \end{bmatrix}$

19. $\begin{bmatrix} 1 & 0 & 0 & 0 \\ \frac{1}{3} & \frac{1}{3} & 0 & \frac{1}{3} \\ \frac{1}{4} & \frac{1}{4} & \frac{1}{4} & \frac{1}{4} \\ 0 & 0 & 0 & 1 \end{bmatrix}$

20. $\begin{bmatrix} \frac{1}{2} & \frac{1}{2} & 0 & 0 \\ 0 & 1 & 0 & 0 \\ 0 & 0 & 1 & 0 \\ 0 & \frac{1}{3} & \frac{1}{3} & \frac{1}{3} \end{bmatrix}$

21. $\begin{bmatrix} 1 & 0 & 0 & 0 \\ .4 & .2 & .1 & .3 \\ 0 & 0 & 1 & 0 \\ .1 & .5 & .4 & 0 \end{bmatrix}$

22. $\begin{bmatrix} \frac{1}{5} & \frac{1}{5} & \frac{1}{5} & \frac{1}{5} & \frac{1}{5} \\ 0 & 1 & 0 & 0 & 0 \\ \frac{1}{3} & \frac{1}{3} & \frac{1}{3} & 0 & 0 \\ 0 & 0 & 0 & 1 & 0 \\ 0 & 0 & 0 & 0 & 1 \end{bmatrix}$

23. $\begin{bmatrix} 1 & 0 & 0 & 0 & 0 \\ .1 & .2 & .3 & .3 & .1 \\ .2 & .2 & .2 & .2 & .2 \\ 0 & 0 & 1 & 0 & 0 \\ 0 & 0 & 0 & 0 & 1 \end{bmatrix}$

24. $\begin{bmatrix} 1 & 0 & 0 & 0 & 0 \\ \frac{1}{3} & 0 & \frac{2}{3} & 0 & 0 \\ 0 & \frac{1}{3} & 0 & \frac{2}{3} & 0 \\ 0 & 0 & \frac{1}{3} & 0 & \frac{2}{3} \\ 0 & 0 & 0 & 0 & 1 \end{bmatrix}$

APPLICATIONS

25. a. Write a transition matrix for a match–no match game for two players, each with a single coin.
   b. Find FP.
   c. What is the probability of financial ruin for the first player?

26. a. Write a transition matrix for a match–no match game for two players with a total of six coins between them.
   b. Find FP.
   c. What is the probability of financial ruin if the first player begins with three coins?

d. What is the probability of financial ruin if the first player begins with two coins?

e. What is the probability of financial ruin if the first player begins with four coins?

27. Noxin, Inc., produces color computer monitors. Each monitor is tested and categorized before it is shipped. If it passes the test, it is shipped; if it needs minor repairs, it is returned to the assembly room for repair and retesting; if it then fails the test, it is destroyed. The probabilities for these events are summarized in the following transition matrix:

	Subsequent test		
	Failed	Returned	Passed
Failed	1	0	0
First test Returned	.1	.2	.7
Passed	0	0	1

What is the probability that a returned item will eventually be passed?

28. Rennolds Department Store issues credit cards and has a policy of revoking any cards with delinquent accounts for 3 months. The auditor calculates the probabilities for the following transition matrix by looking at the records of 5000 credit card customers:

	Subsequent month			
	Paid up	Delinquent 1 month	Delinquent 2 months	Revoke card
Paid up	.65	.35	0	0
Delinquent 1 month	.70	0	.30	0
First month Delinquent 2 months	.3	.4	0	.3
Revoke card	0	0	0	1

a. What is the probability that a paid-up customer will eventually have a revoked card?

b. What is the probability that a customer who is 1 month delinquent will eventually have a revoked card?

29. A rat is placed at random in one of the compartments of the maze shown here. When the rat is in some room, there are equal probabilities that it will choose any door in that room. When the rat reaches room 1, it is rewarded with food, and it no longer leaves that room. Room 3 has one-way doors that do not permit the rat to leave the room once it enters. What is the probability that the rat will end up in room 1 if it was originally placed in room 5?

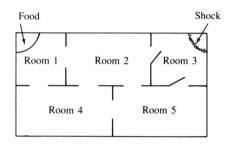

30. The experiment in Problem 29 is repeated. What is the probability that the rat will end up in room 1 if it is placed in room 2?

APPENDIX G

Answers to Odd-Numbered Problems

PROBLEM SET 1.1, PAGES 2–3

Part I. **1.** E **3.** A **5.** C **7.** D **9.** D **11.** C **13.** B **15.** D **17.** False **19.** False **21.** False **23.** False **25.** True **27.** True **29.** True

Part II. **1.** B **3.** C **5.** C **7.** A **9.** B **11.** D **13.** C **15.** A **17.** D

PROBLEM SET 1.2, PAGES 7–8

1. a. I, Q, R **b.** Q', R **3. a.** N, W, I, Q, R **b.** N, W, I, Q, R **5. a.** Q, R **b.** Q, R **7. a.** Q, R **b.** Q', R
9. a. Undefined **b.** W, I, Q, R **11.** 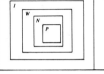 **13.**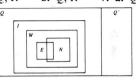

15. False **17.** False **19.** True **21.** True **23.** True **25.** False **27.** True **29.** True **31.** True
33. False **35.** False **37.** False **39.** True **41.** 9 **43.** 19 **45.** $-\pi$ **47.** $\pi - 2$ **49.** $10 - \pi$
51. $\sqrt{20} - 4$ **53.** $8 - \sqrt{50}$ **55.** $7 - 2\pi$ **57.** $-(x + 3)$ **59.** $y - 5$ **61.** $-(4 + 3t)$ **63.** 10 **65.** 11
67. $\pi - 2$ **69.** $2 - \sqrt{3}$

PROBLEM SET 1.3, PAGES 13–14

1. 13 **3.** 11 **5.** 36 **7.** -25 **9.** -49 **11.** 33 **13.** 6 **15.** 9 **17.** 4 **19.** 1 **21.** 0 **23.** 42
25. -71 **27.** -1 **29. a.** $x^2 + 4x - 12$ **b.** $x^2 + x - 20$ **31. a.** $x^2 - 8x + 15$ **b.** $x^2 - x - 12$
33. a. $2y^2 - y - 1$ **b.** $2y^2 - 5y + 3$ **35. a.** $6y^2 + 5y - 6$ **b.** $6y^2 + 13y + 6$ **37. a.** $x^2 - y^2$ **b.** $a^2 - b^2$
39. a. $x^2 + 4x + 4$ **b.** $x^2 - 4x + 4$ **41. a.** $a^2 + 2ab + b^2$ **b.** $a^2 - 2ab + b^2$ **43.** $3x - 4z$ **45.** $4x - 2y + z$
47. $-2y$ **49.** $2x + 2y$ **51.** $-x + 4y - 9$ **53.** $4x^2 - 8x - 1$ **55.** $-3x^2 + 3x - 9$ **57.** $2x^2 - x - 3$
59. $3x^2 - 4x - 11$ **61.** $x^3 + 3x^2 + 3x + 1$ **63.** $x^3 - x^2 - 4x - 2$ **65.** $3x^2 - 10x + 1$ **67.** $15x^3 - 22x^2 + 5x + 2$

PROBLEM SET 1.4, PAGE 18

1. $4x(5y - 3)$ **3.** $2(3x - 1)$ **5.** $x(y + z^2 + 3)$ **7.** $a^2 + b^2$ **9.** $(m - n)^2$ **11.** $(4x - 1)(x + 3)$
13. $(2x - 3)(x + 5)$ **15.** $(3x + 1)(x - 2)$ **17.** $2(x - 8)(x + 3)$ **19.** $x^2(3x + 5)(4x - 3)$
21. $(2x + 1)(2x - 1)(x + 2)(x - 2)$ **23.** $(x - y + 1)(x - y - 1)$ **25.** $5(a - 1)(5a + 1)$ **27.** $-\frac{1}{25}(3x + 10)(7x + 10)$
29. $(a + b + x + y)(a + b - x - y)$ **31.** $(2x - 3)(x + 2)$ **33.** $(6x - 1)(x + 8)$ **35.** $(6x + 1)(x + 8)$
37. $(4x - 3)(x + 4)$ **39.** $(9x - 2)(x - 6)$ **41.** $-(x - 3)(x + 3)(x - 1)(x + 1)$ **43.** $\frac{1}{36}(2x + 1)(2x - 1)(3x + 1)(3x - 1)$
45. $(x - 4)(x + 1)(x^2 - 3x - 8)$ **47.** $(2x + 2y + a + b)(x + y - 3a - 3b)$

PROBLEM SET 1.5, PAGES 21–22

1. $x = 2$ **3.** $x = -5$ **5.** $x = -3$ **7.** $x = -3$ **9.** $x = -2$ **11.** $x = 2$ **13.** $x = 3$ **15.** $x = 2$
17. $x \le 3$ **19.** $x < -5$ **21.** $x \ge -4$ **23.** $x > -3$ **25.** $5 < x < 9$ **27.** $-1 < x < 4$ **29.** $-7 < x < -4$
31. $x = 3$ **33.** $x < -1$ **35.** $x \le 10$ **37.** $x < -2$ **39.** $x \le 1$ **41.** $x = 1$ **43.** $\varnothing$
45. $x = 0$ **47.** identity

PROBLEM SET 1.6, PAGE 27

1. $x = 3, -5$ **3.** $x = 2, -9$ **5.** $x = \frac{4}{5}, -\frac{1}{2}$ **7.** $x = 1, -6$ **9.** $x = 5$ **11.** $x = \frac{1}{4}, -\frac{2}{3}$ **13.** No real solutions
15. $x = \frac{\sqrt{5}}{2}, \frac{-\sqrt{5}}{2}$ **17.** $x = 0, \frac{7}{3}$ **19.** $x = 2, -\frac{1}{3}$ **21.** $x \le 2$ or $x \ge 6$ **23.** $-3 < x < 0$ **25.** $x \le -2$ or $x \ge 8$
27. $x < \frac{1}{3}$ or $x > 4$ **29.** $x \le -3$ or $x \ge 3$ **31.** No real solutions   **33.** $x < -2$ or $x > 3$ **35.** $2 \le x \le 3$
37. $-5 \le x \le 1$ **39.** $1 - \sqrt{3} < x < 1 + \sqrt{3}$

PROBLEM SET 1.7, PAGES 30–31

1. $\dfrac{3x + 3y + 2}{x + y}$ **3.** $\dfrac{x - y}{2}$ **5.** $\dfrac{3x + 2}{x + 2}$ **7.** $\dfrac{11}{2(x + y)}$ **9.** 1 **11.** $3x^2 - 17x - 6$ **13.** $\dfrac{x^2 + 2x + 3}{x^2}$ **15.** $\dfrac{3}{x - 1}$
17. $\dfrac{-x^2 + 5x + 3}{x^2}$ **19.** $\dfrac{x^2 + 2xy + y^2}{xy}$ **21.** $\dfrac{x^2 - 2xy + y^2}{xy}$ **23.** $\dfrac{x^2 + 3x^4y^3 + 3y}{3x^3y^2}$ **25.** $\dfrac{-2}{x + 7}$ **27.** $\dfrac{1 - x^2}{x^2 + 1}$
29. $\dfrac{4(x - 1)}{(x + 2)(x - 2)}$

PROBLEM SET 1.8, PAGE 32

1. 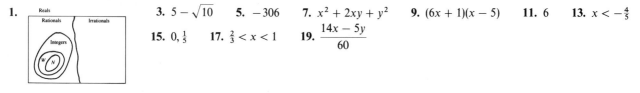   **3.** $5 - \sqrt{10}$ **5.** -306 **7.** $x^2 + 2xy + y^2$ **9.** $(6x + 1)(x - 5)$ **11.** 6 **13.** $x < -\frac{4}{5}$
15. $0, \frac{1}{5}$ **17.** $\frac{2}{3} < x < 1$ **19.** $\dfrac{14x - 5y}{60}$

CHAPTER 2
PROBLEM SET 2.1, PAGES 34–35

1. A   **3.** C   **5.** C   **7.** B   **9.** C   **11.** A   **13.** A   **15.** B   **17.** B   **19.** B

PROBLEM SET 2.2, PAGES 40–43

1. (Answers vary.) A mathematical model is a body of mathematics based on certain assumptions and data about the real world and used to predict future occurrences.
3. (Answers vary.) A function of a variable is a rule that assigns to each value of x exactly one value for $f(x)$.
5. a. y is a function of x since for each year x there will be one and only one closing price y **b.** y is not a function of x since for each closing price x we may not be able to uniquely determine the year y
7. a. \$.29   **b.** \$.53   **9. a.** $f(8) = 11$ **b.** $f(-3) = -11$ **11. a.** $f(2) = 7$ **b.** $f(10) = 47$ **c.** $f(-15) = -78$
d. $f(100) = 497$ **13. a.** $h(0) = 6$ **b.** $h(8) = -26$ **c.** $h(-7) = 34$ **d.** $h(100) = -394$ **15. a.** $m(0) = 1$
b. $m(1) = -1$ **c.** $m(2) = -1$ **d.** $m(3) = 1$ **17. a.** All real numbers   **b.** All real numbers   **19. a.** $x \ne -5$
b. $x \ge \frac{1}{2}$ **21. a.** $f(t) = 5t - 2$ **b.** $f(w) = 5w - 2$ **23. a.** $f(t + h) = 5(t + h) - 2$ **b.** $f(s + t) = 5(s + t) - 2$
25. $f(t + h + 8) = 5(t + h + 8) - 2 = 5t + 5h + 38$ **27.** $g(3 + h) = 2h^2 + 8h + 1$ **29.** $f(2x^2) = 10x^2 - 2$
31. $g(2x^2 - 4x) = 2(2x^2 - 4x)^2 - 4(2x^2 - 4x) - 5 = 8x^4 - 32x^3 + 24x^2 + 16x - 5$ **33.** $g(5x) = 50x^2 - 20x - 5$
35. $g(f(x)) = g(5x - 2) = 50x^2 - 60x + 11$ **37.** $\dfrac{2(x + h) - 2x}{h} = 2$ **39.** $4x + 2h$
41. $\dfrac{2(x + h)^2 - 3 - (2x^2 - 3)}{h} = 4x + 2h$ **43.** $2x + h - 2$ **45. a.** $r(1954) = \$.92$ **b.** $m(1954) = \$.45$
47. $1.48 - 0.34 = \$1.14$ **49. a.** $1.15 - 0.64 = \$.51$ **b.** $e(1984) - e(1944)$ **51. a.** \$.03
b. The average increase in the price of gasoline per year from 1944 to 1984   **53. a.** $0.018 = \dfrac{s(1954) - s(1944)}{10}$
b. $0.0125 = \dfrac{s(1964) - s(1944)}{20}$ **c.** $0.0247 \approx \dfrac{s(1974) - s(1944)}{30}$ **d.** $0.0285 = \dfrac{s(1984) - s(1944)}{40}$ **e.** $\dfrac{s(1944 + h) - s(1944)}{h}$
55. 0.2% **57.** In 2001 the population will be about 342 million   **59. a.** $63{,}800$
b. The average increase of marriages per year from the year 1977 to the year $1977 + h$

PROBLEM SET 2.3, PAGES 45–47

1. A picture of the ordered pairs or solutions for the function

3. The x-intercepts are the points where the graph crosses (or touches) the x-axis. The y-intercept is the point where the graph crosses the y-axis

5. $(2, f(2))$ **7.** $(x_0, f(x_0))$ **9.** $(x_0, g(x_0))$ **11.** $(3, h(3))$ **13.** $(x_0 + t, h(x_0 + t))$

15. The domain is $x = -3$. The range is all real numbers. The x-intercept is $(-3, 0)$, and there is no y-intercept. The graph is *not* a function

17. The domain is $-2 \leq x \leq 5$. The range is $-5 \leq y \leq 3$. The x-intercepts are $(-\frac{3}{2}, 0)$ and $(\frac{15}{4}, 0)$. The y-intercept is $(0, 3)$. This relation is a function

19. The domain and range both are the set of real numbers. The x-intercepts are $(-3, 0), (-1, 0)$, and $(2, 0)$. The y-intercept is $(0, \frac{5}{2})$. This relation is a function

21.   **23.**   **25.**   **27.**

29.   **31.**   **33.**

35.   **37.**   **39.**

41.   **43.**   **45.**

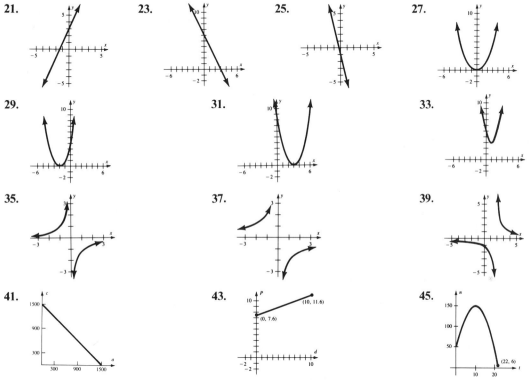

PROBLEM SET 2.4, PAGES 56–57

1. x-intercept is $(-2, 0)$; y-intercept is $(0, 4)$ **3.** x-intercept is $(-1, 0)$; y-intercept is $(0, -\frac{4}{3})$

5. x-intercept is $(-5, 0)$; y-intercept is $(0, 2)$ **7.** No x-intercept; y-intercept is $(0, -2)$ **9.** $m = \frac{1}{3}$ **11.** $m = \frac{5}{7}$

13. $m = 1$ **15.** The slope is 2 and the y-intercept is 4   **17.** The slope is 9 and the y-intercept is 1

19. The slope is $\frac{2}{3}$ and the y-intercept is $\frac{5}{3}$ **21.** The slope is 0 and the y-intercept is 5

23. No slope or y-intercept exist for this line   **25.** Slope $= 2$; y-intercept is -5 **27.** Slope $= -\frac{1}{4}$; y-intercept is 2

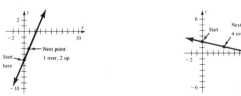

29. Slope $= \frac{3}{5}$; y-intercept is $\frac{2}{5}$ **31.** Slope is $-\frac{1}{3}$; y-intercept is 3   **33.** Slope $= \frac{3}{2}$; y-intercept is 0

35. (No slope) $x = -\frac{5}{2}$ **37.**   **39.**   **41.**

43.   **45.**   **46.**   **47.**

48.   **49.**   **50.**   **51.**

52.   **53.**   **54.**   **55.**

56.   **57.**   **59.** $2x + y + 3 = 0$ **61.** $y = 4$ **63.** $x + y - 2 = 0$

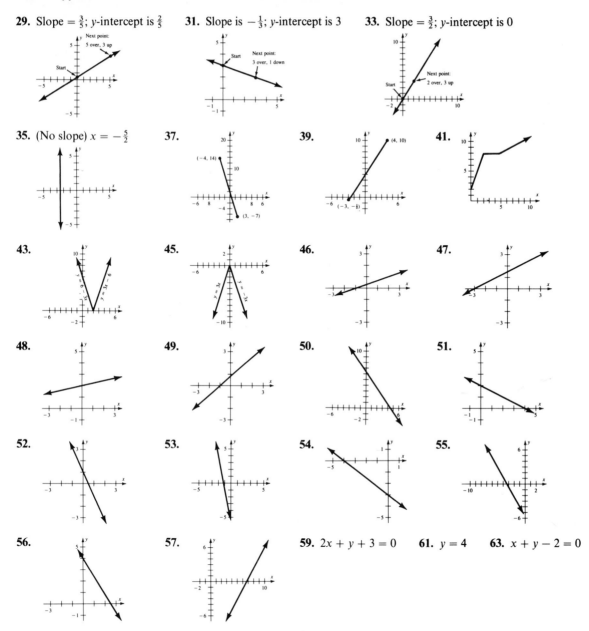

65. $3x - 5y - 27 = 0$ **67.** $2x - y - 4 = 0$ **69.** $y = 6$ **71.** Data points: $(5, 10)$ and $(10, 20)$; equation: $2x - y = 0$
73. Equation: $0.3x - y + 8.2 = 0$. In 1990 the population will be about 17.2 million
75. Let y be the price and x the number of miles. Equation: $y = 60$

77.   **79.** $T = 0.24(I - 15,000) + 2097$ **81.** $\begin{cases} T = 0.32(I - 23,500) + 4349 & \text{if } I \le 28,800 \\ T = 0.36(I - 28,800) + 6045 & \text{if } I > 28,800 \end{cases}$

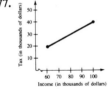

PROBLEM SET 2.5, PAGES 63–64

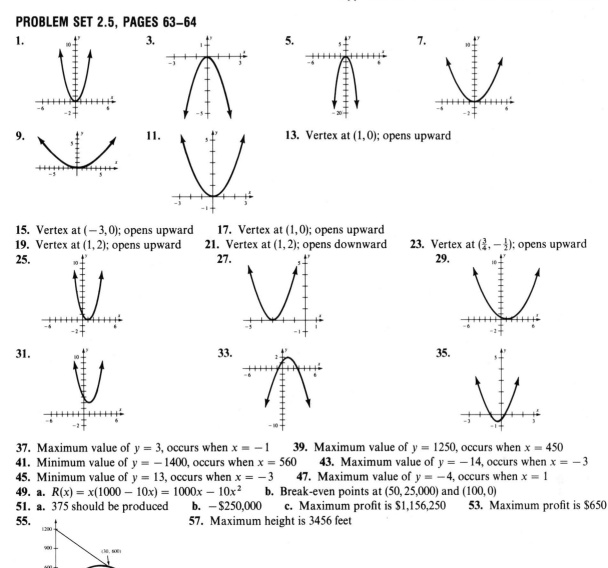

13. Vertex at $(1, 0)$; opens upward

15. Vertex at $(-3, 0)$; opens upward   **17.** Vertex at $(1, 0)$; opens upward
19. Vertex at $(1, 2)$; opens upward   **21.** Vertex at $(1, 2)$; opens downward   **23.** Vertex at $(\frac{3}{4}, -\frac{1}{2})$; opens upward

37. Maximum value of $y = 3$, occurs when $x = -1$ **39.** Maximum value of $y = 1250$, occurs when $x = 450$
41. Minimum value of $y = -1400$, occurs when $x = 560$ **43.** Maximum value of $y = -14$, occurs when $x = -3$
45. Minimum value of $y = 13$, occurs when $x = -3$ **47.** Maximum value of $y = -4$, occurs when $x = 1$
49. a. $R(x) = x(1000 - 10x) = 1000x - 10x^2$ **b.** Break-even points at $(50, 25,000)$ and $(100, 0)$
51. a. 375 should be produced   **b.** $-\$250,000$ **c.** Maximum profit is $\$1,156,250$ **53.** Maximum profit is $\$650$
55.   **57.** Maximum height is 3456 feet

PROBLEM SET 2.6, PAGES 70–71

1. A rational function is the quotient of two polynomials
3. A horizontal asymptote is a horizontal line that the function approaches as $|x|$ gets larger. A horizontal asymptote will be present if the degrees of numerator and denominator are equal (in which case the ratio of leading coefficients gives the value of the asymptote) or if the degree of the numerator is less than that of the denominator (in which case $y = 0$ will be the asymptote)
5. Vertical at $x = \frac{3}{2}$; horizontal at $y = 0$ **7.** Vertical at $x = -3$ and $x = 1$; horizontal at $y = 0$
9. Vertical at $x = \frac{2}{3}$ and $x = -\frac{3}{2}$; horizontal at $y = 0$ **11.** Vertical at $x = -1 + \frac{\sqrt{2}}{2}$ and $x = -1 - \frac{\sqrt{2}}{2}$; horizontal at $y = \frac{5}{2}$
13. No vertical asymptote; horizontal at $y = 0$ **15.** Vertical at $x = -2$ and $x = 1$; horizontal at $y = 0$

17. 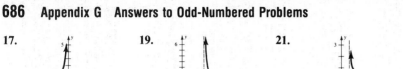 **19.** **21.** **23.**

25. **27.** **29.** $106,667   **31.** **33.** $72,000

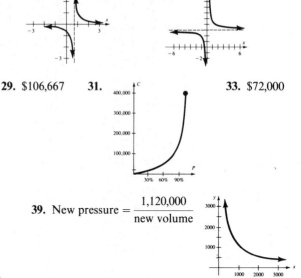

35. **37.** **39.** New pressure $= \dfrac{1{,}120{,}000}{\text{new volume}}$

PROBLEM SET 2.7, PAGE 72

1. 14   **3.** $-4t - 3$ **5.** -4 **7.**   **9.** -8 **11.**

13. $5x + y - 170 = 0$ **15.** $x + 3 = 0$ **17.** 550   **19.** The cost is $4,500

CHAPTER 3
PROBLEM SET 3.1, PAGES 82–84

1.   **3.**   **5.** Dependent system

The answers for variables and constants in Problems 7–17 may vary, but the solutions to the problems will not vary.

7. $a_{11} = 1, a_{12} = -3, b_1 = -7$
$a_{21} = 1, a_{22} = 2, b_2 = 8$
Solution: $(a, b) = (2, 3)$

9. $a_{11} = 3, a_{12} = -2, b_1 = -21$
$a_{21} = 3, a_{22} = 2, b_2 = 3$
Solution: $(m, n) = (-3, 6)$

11. $a_{11} = 2, a_{12} = -3, b_1 = 9$
$a_{21} = -2, a_{22} = 3, b_2 = -9$
Dependent system

13. $a_{11} = 1, a_{12} = 1, b_1 = 2$
$a_{21} = 2, a_{22} = -1, b_2 = 1$
Solution: $(c, d) = (1, 1)$

15. $a_{11} = 3, a_{12} = -4, b_1 = 3$
$a_{21} = 5, a_{22} = 3, b_2 = 5$
Solution: $(q_1, q_2) = (1, 0)$

17. $a_{11} = 7, a_{12} = 1, b_1 = 5$
$a_{21} = 14, a_{22} = -2, b_2 = -2$
Solution: $(\frac{2}{7}, 3)$

19. $(\frac{3}{5}, \frac{1}{2})$ **21.** $(-6, 2)$ **23.** $(-\frac{8}{5}, -\frac{21}{5})$ **25.** $(31, 19)$ **27.** $\left(\dfrac{a}{a^2 - b^2}, \dfrac{-b}{a^2 - b^2} \right)$ **29.** $(1800, 200)$

31. $A(0,0)$, $B(\frac{7}{2},0)$, $C(\frac{5}{2},2)$, $D(0,3)$ **33.** $A(0,0)$, $B(\frac{8}{3},0)$, $C(\frac{5}{3},\frac{5}{3})$, $D(0,\frac{8}{3})$ **35.** $A(0,0)$, $B(2,0)$, $C(\frac{7}{4},\frac{1}{2})$, $D(1,\frac{5}{4})$, $E(0,\frac{3}{2})$
37. 50 miles **39.** 6 hours **41.** \$15 **43. a.** 125 supplied; 75 demanded **b.** \$200 **c.** \$400
d. \$233.33 **e.** 83
45. The line through (1, 800) and (7, 0) intersects the line through (1, 200) and (7, 600) at (\$4, 400 items)
47. If 1000 pairs are produced, both cost and revenue will be \$2,000
49. If 10,000 disks are produced, both cost and revenue will be \$30,000
51. If 3733 cards are produced, both cost and revenue will be \$5,600
53. If 2240 cards are produced, both cost and revenue will be \$4,480
55. 500 shares of Standard Oil, 200 shares of Xerox **57.** 60 units grade A, 20 units grade B

PROBLEM SET 3.2, PAGES 96–99

1. a. 3×3 (square matrix) **b.** 3×1 (column matrix) **c.** 3×2 **d.** 4×1 (column matrix) **e.** 2×3
f. 1×4 (row matrix) **g.** 1×2 (row matrix) **3. a.** $x = 2$, $y = 4$ **b.** $x = 4$, $y = -9$, $z = 6$
c. $a = 1$, $b = 0$, $c = 0$, $d = 1$ **5.**

	Part 1	Part 2	Part 3	Part 4
Factory A	25	42	193	0
Factory B	16	39	150	0
Factory C	50	50	50	50
Factory D	0	0	0	320

7. a. 3×3, c_{12} **b.** 3×3, c_{32} **c.** 3×3, c_{21} **d.** 2×2, c_{11} **e.** 2×2, c_{21} **f.** 2×2, c_{12}
9. a. $[3w + 5x + 8y + 9z]$, 1×1 **b.** $[2x - 3x + 4y + 5z]$, 1×1 **c.** $[2a + 3b + 5c \quad 2d + 3e + 5f]$, 1×2

d. $[3a - 2c + e \quad 3b - 2d + f]$, 1×2 **11. a.** $\begin{bmatrix} 8w & -w & 3w & -2w \\ 8x & -x & 3x & -2x \\ 8y & -y & 3y & -2y \\ 8z & -z & 3z & -2z \end{bmatrix}$, 4×4 **b.** $\begin{bmatrix} 6w & -w & 3w & 2w \\ 6x & -x & 3x & 2x \\ 6y & -y & 3y & 2y \\ 6z & -z & 3z & 2z \end{bmatrix}$, 4×4

c. $\begin{bmatrix} 12 & -24 & 18 \\ 6 & -12 & 9 \\ -2 & 4 & -3 \end{bmatrix}$, 3×3 **d.** $\begin{bmatrix} 10 & 30 & 20 \\ -4 & -12 & -8 \\ -2 & -6 & -4 \end{bmatrix}$, 3×3 **13.** $\begin{bmatrix} 2 & 4 & 2 \\ 6 & -2 & 4 \\ 2 & 2 & 5 \end{bmatrix}$ **15.** $\begin{bmatrix} 0 & 4 & -2 \\ 0 & 0 & 0 \\ -6 & 0 & 5 \end{bmatrix}$

17. Not conformable **19.** Not conformable **21.** Not conformable **23.** Not conformable **25.** $\begin{bmatrix} -1 & -16 & 6 \\ -3 & 1 & -2 \\ 20 & -1 & -20 \end{bmatrix}$

27. Not conformable **29.** Not conformable **31.** $\begin{bmatrix} 19 & 14 & 12 \\ 15 & 7 & 20 \\ 2 & 21 & 19 \end{bmatrix}$ **33.** $\begin{bmatrix} 10 & 5 & 8 \\ 9 & 3 & 11 \\ 6 & 11 & 7 \end{bmatrix}$

35. Not conformable **37.** $\begin{bmatrix} 13 & -4 & 10 \\ 8 & 3 & 4 \\ 21 & 4 & -2 \end{bmatrix}$ **39.** $\begin{bmatrix} 35 & 5 & 18 \\ 46 & 2 & 16 \\ 39 & -7 & 26 \end{bmatrix}$ **41.** $\begin{bmatrix} -1 & 37 & 32 \\ 4 & 14 & 45 \\ 27 & 9 & 28 \end{bmatrix}$

43. $\begin{bmatrix} 34 & 117 & 44 \\ 45 & 143 & 97 \\ 109 & 100 & 151 \end{bmatrix}$ **45.** $\begin{bmatrix} 132 & 83 & 70 \\ 89 & 59 & 77 \\ 172 & 23 & 150 \end{bmatrix}$ **47.** $\begin{bmatrix} 14 & 153 & 100 \\ 20 & 198 & 84 \\ -34 & 189 & 116 \end{bmatrix}$ **49.** $\begin{bmatrix} 13 & 25 & 20 \\ 25 & 31 & 23 \\ 42 & 24 & 33 \end{bmatrix}$

51. $\begin{bmatrix} 13 & 25 & 20 \\ 25 & 31 & 23 \\ 42 & 24 & 33 \end{bmatrix}$ **53.** Not conformable **55.** $\begin{bmatrix} 1 & 2 \\ 4 & 0 \\ -1 & 3 \\ 2 & 1 \end{bmatrix}$ **57.** Not conformable

59. Not conformable (for multiplication)   **61.** Not conformable   **63.** $\begin{bmatrix} 42 & 70 \\ -35 & 7 \end{bmatrix}$

65. a.

	SF	D	A	KC
SF	0	1	0	1
D	1	0	1	0
A	0	1	0	1
KC	1	0	1	0

b. There is no way to go from Kansas City to San Francisco making exactly one stop

c.

Origin	First Stop	Second Stop	Destination	
SF	Dallas	SF	KC	
SF	Dallas	Atlanta	KC	
SF	KC	Atlanta	KC	4 ways
SF	KC	SF	KC	

d.

	SF	D	A	KC
SF	2	0	2	0
D	0	2	0	2
A	2	0	2	0
KC	0	2	0	2

e.

	SF	D	A	KC
SF	0	4	0	4
D	4	0	4	0
A	0	4	0	4
KC	4	0	4	0

67.

US	USSR	Cuba	Mexico	
2	4	1	3	US
4	2	3	4	USSR
1	3	0	1	Cuba
3	4	1	2	Mexico

So, using *two* intermediaries, the US can communicate with:
the USSR in four ways, Cuba in only one way, and Mexico in three ways.

69. False   **71.** True   **73.** True

PROBLEM SET 3.3, PAGES 111–113

1. a. $\begin{bmatrix} 4 & 5 & | & -16 \\ 3 & 2 & | & 5 \end{bmatrix}$ **b.** $\begin{bmatrix} 1 & 1 & 1 & | & 4 \\ 3 & 2 & 1 & | & 7 \\ 1 & -3 & 2 & | & 0 \end{bmatrix}$ **c.** $\begin{bmatrix} 1 & 3 & 1 & 1 & | & 3 \\ 1 & 0 & -2 & 2 & | & 0 \\ 0 & 0 & 1 & 5 & | & -14 \\ 0 & 1 & -3 & -1 & | & 2 \end{bmatrix}$

3. a. $\begin{cases} 2x_1 + x_2 + 4x_3 = 3 \\ 6x_1 + 2x_2 - x_3 = -4 \\ -3x_1 - x_2 = 1 \end{cases}$ **b.** $\begin{cases} x_1 = 5 \\ x_2 = -3 \\ x_3 = 4 \end{cases}$ **c.** $\begin{cases} x_1 = 3 \\ x_2 = 2 \\ x_3 = -8 \\ 0 = 1 \end{cases}$ **5. a.** $\begin{bmatrix} 1 & 5 & 3 & | & -2 \\ -1 & 3 & -2 & | & 3 \\ -2 & 2 & 4 & | & 8 \end{bmatrix}$

b. $\begin{bmatrix} -2 & 2 & 4 & | & 8 \\ -2 & 6 & -4 & | & 6 \\ 1 & 5 & 3 & | & -2 \end{bmatrix}$ **c.** $\begin{bmatrix} -1 & 1 & 2 & | & 4 \\ -1 & 3 & -2 & | & 3 \\ 1 & 5 & 3 & | & -2 \end{bmatrix}$ **d.** $\begin{bmatrix} 0 & 12 & 10 & | & 4 \\ -1 & 3 & -2 & | & 3 \\ 1 & 5 & 3 & | & -2 \end{bmatrix}$ **e.** $\begin{bmatrix} -2 & 2 & 4 & | & 8 \\ -2 & -2 & -5 & | & 5 \\ 1 & 5 & 3 & | & -2 \end{bmatrix}$

7. $(1, 2)$ **9.** $(-6, -4)$ **11.** $(\frac{10}{3} + 2t, 3t)$ **13.** Inconsistent system   **15.** Inconsistent system
17. Inconsistent system   **19.** $(1, 2, 3)$ **21.** $(3, -1, 2)$ **23.** $(3, 2, 5)$ **25.** $(2, -1, 0)$ **27.** $(6, 2, -9)$ **29.** $(1, 1, 2)$
31. $(\frac{8}{5} - 7t, -\frac{19}{5} + t, 5t)$ **33.** Inconsistent system   **35.** $(-7 + 11t, -2 + 7t, 3t)$
37. $(w, x, y, z) = (\frac{18}{5} + 3s - t, \frac{7}{5} + 7s - 14t, 5s, 5t)$ **39.** $(w, x, y, z) = (\frac{11}{2} - 6s, 2 - 4s + t, 3s, 3t)$
41. $(-\frac{3}{2} - 2t, 1, -\frac{1}{4} + 3t, 12t)$ **43.** $(4, -1, 1, -3)$ **45.** $(1, -1, 1, 2)$ **47.** $(\frac{3}{2} + 51t, \frac{1}{3} - 94t, -\frac{1}{3} - 62t, -\frac{3}{2} + 33t, 6t)$
49. Three containers of spray I and four containers of spray II are needed
51. Four units of candy I, five units of candy II, and six units of candy III   **53.** No. The large scoop is the best buy

PROBLEM SET 3.4, PAGES 120–121

1. Yes   **3.** Yes   **5.** Yes   **7.** No   **9.** Yes   **11.** No   **13.** $\begin{bmatrix} 2 & 7 \\ 1 & 4 \end{bmatrix}$ **15.** $\begin{bmatrix} 2 & -5 \\ -1 & 3 \end{bmatrix}$

17. $\dfrac{1}{6}\begin{bmatrix} 0 & 3 \\ 2 & -1 \end{bmatrix}$ **19.** $\dfrac{1}{22}\begin{bmatrix} 2 & -3 \\ 1 & 4 \end{bmatrix}$ **21.** $\begin{bmatrix} 4 & 3 \\ 2 & 2 \end{bmatrix}$ **23.** $\begin{bmatrix} 3 & -17 & -20 \\ 3 & -18 & -20 \\ -1 & 6 & 7 \end{bmatrix}$ **25.** $\begin{bmatrix} 3 & 3 & -1 \\ -2 & -2 & 1 \\ -4 & -5 & 2 \end{bmatrix}$

27. $\dfrac{1}{12}\begin{bmatrix} -2 & 2 & 2 \\ 8 & -8 & 4 \\ 7 & -1 & -1 \end{bmatrix}$ **29.** $\dfrac{1}{10}\begin{bmatrix} -12 & 0 & 8 & 1 \\ 6 & 0 & -4 & 2 \\ 2 & 0 & 2 & -1 \\ 0 & 10 & 0 & 0 \end{bmatrix}$ **31.** $(3,2)$ **33.** $(5,-4)$ **35.** $(-1,4)$ **37.** $(3,1)$

39. $(-2,2)$ **41.** $(23,-30)$ **43.** $(5,6,1)$ **45.** $(1,-2,3)$ **47.** $(5,4,-1)$ **49.** $(1,2,3)$ **51.** $(-3,2,7)$
53. $(w,x,y,z) = (3,-2,1,-4)$ **55.** Six bags of grain I, three bags of grain II, and one bag of grain III
57. Twenty-two bags of grain I, eight bags of grain II, and eleven bags of grain III **59.** Ten oz of each food

PROBLEM SET 3.5, PAGES 126–127

1. $x = t, y = 3t, z = 2t$ **3.** $x = -6t, y = 2t, z = 3t$ **5.** $x_1 = 154t, x_2 = 89t, x_3 = 30t$
7. Answers vary. One possible solution is $x_1 = \$50{,}000, x_2 = \$47{,}500, x_3 = \$30{,}000$
9. Answers vary. One possible solution is $x_1(\text{farmer}) = \$1{,}000, x_2(\text{builder}) = \$1{,}250, x_3(\text{tailor}) = \$1{,}250, x_4(\text{merchant}) = \$1{,}000$
11. The required output in 3 years is about \$136 for farming, \$139 for construction, and \$155 for clothing
13. Farming output = \$400, construction output = \$658, clothing output = \$382

15. $\begin{bmatrix} 1.41 & 0.14 & 0.34 \\ 0.33 & 1.41 & 0.39 \\ 0.17 & 0.06 & 1.16 \end{bmatrix}$ **17.** $347, 251, 214$ **19.** $\begin{bmatrix} 1.41 & 0.00 & 0.00 \\ 0.04 & 1.26 & 0.03 \\ 0.08 & 0.02 & 1.28 \end{bmatrix}$

PROBLEM SET 3.6, PAGES 128–129

1. a. $(-\frac{25}{7}, \frac{64}{21})$; addition **b.** $(3,2)$; substitution **c.** $(5,8)$; graphing **3. a.** $\begin{bmatrix} 8 & 3 & 7 \\ -2 & 16 & 17 \end{bmatrix}$

b. Not conformable **5. a.** $\begin{bmatrix} 11 & 8 & -3 \\ 3 & 8 & -6 \\ 11 & -6 & 14 \end{bmatrix}$ **b.** $\begin{bmatrix} 5 & 2 & 0 \\ 3 & 0 & 0 \\ 2 & 3 & 5 \end{bmatrix}$ **7. a.** $(3,-2)$ **b.** $(2,-5)$

c. $(\frac{7}{5} + 2t, -\frac{24}{5} + 11t, 5t)$ **9. a.** $A^{-1} = \begin{bmatrix} 1 & -1 & 1 \\ 0 & 2 & -1 \\ 2 & 3 & 0 \end{bmatrix}$ **b.** $\begin{bmatrix} 1 & -1 & 1 \\ 0 & 2 & -1 \\ 2 & 3 & 0 \end{bmatrix}\begin{bmatrix} 8 \\ -1 \\ 3 \end{bmatrix} = \begin{bmatrix} 12 \\ -5 \\ 13 \end{bmatrix}$ $x = 12, y = -5, z = 13$

CHAPTER 4
PROBLEM SET 4.1, PAGES 136–137

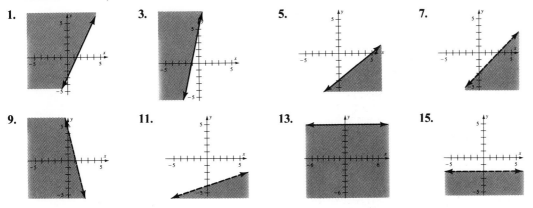

1. **3.** **5.** **7.**

9. **11.** **13.** **15.**

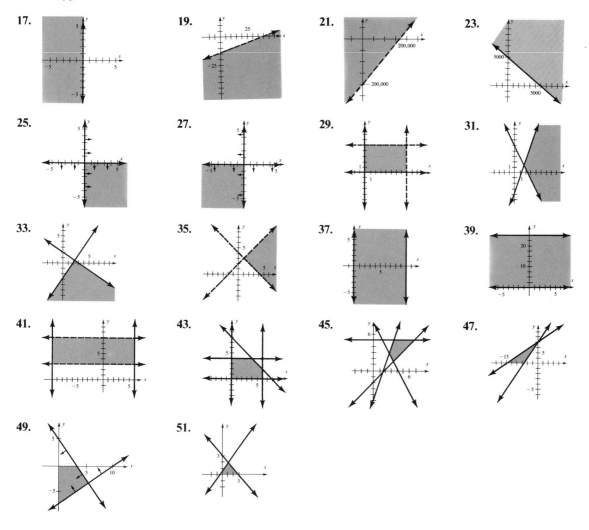

17. 19. 21. 23.

25. 27. 29. 31.

33. 35. 37. 39.

41. 43. 45. 47.

49. 51.

PROBLEM SET 4.2, PAGES 144–147

1. Let x = Number of acres of corn
y = Number of acres of wheat
Maximize: $P = (1.20)(100x) + (2.50)(40y)$
Subject to: $\begin{cases} 120x + 60y \le 24{,}000 \\ 100x + 40y \ge 20{,}000 \\ x + y \le 500 \\ x \ge 0, \quad y \ge 0 \end{cases}$

3. Let x = Number of standard items produced
y = Number of economy items produced
P = Profit
Maximize: $P = 45x + 30y$
Subject to: $\begin{cases} 3x + 3y \le 1500 \\ 2x + 0y \le 800 \\ x \ge 0, \quad y \ge 0 \end{cases}$

5. Let x = Ounces of Corn Flakes
y = Ounces of Honeycombs
Minimize: $C = 0.07x + 0.19y$
Subject to: $\begin{cases} 23x + 14y \ge 322 \\ 7x + 17y \ge 119 \\ x \ge 0, \quad y \ge 0 \end{cases}$

7. Let x = Amount invested in Pertec stock
y = Amount invested in Campbell Municipal Bonds
Maximize: $R = 0.20x + 0.10y$
Subject to: $\begin{cases} x + y \le 100{,}000 \\ x \le 70{,}000 \\ y \ge 20{,}000 \\ y \le 3x \\ x \ge 0, \quad y \ge 0 \end{cases}$

9. Let $x =$ Number of Glassbelt tires produced
$y =$ Number of Rainbelt tires produced
Maximize: $R = 52x + 36y$
Subject to: $\begin{cases} 3x + y \le 200 \\ 2x + 2y \le 400 \\ x \ge 0, \quad y \ge 0 \end{cases}$

11. Let $x =$ Number of commercial guests
$y =$ Number of other guests
Maximize: $P = 4.50x + 3.50y$
Subject to: $\begin{cases} 0.40x + 0.20y \le 50 \\ x + y \le 200 \\ x \ge 0, \quad y \ge 0 \end{cases}$

13. Let $x =$ Number of Alpha units produced
$y =$ Number of Beta units produced
Maximize: $P = 5x + 8y$
Subject to: $\begin{cases} x \le 700 \\ x + 3y \le 1200 \\ x + 2y \le 1000 \\ x \ge 0, \quad y \ge 0 \end{cases}$

15. Let $x =$ Number of A1 animals
$y =$ Number of A2 animals
$z =$ Number of A3 animals
Maximize: $T = x + y + z$
Subject to: $\begin{cases} 12x + 15y + 20z \le 8000 \\ 6x + 20y + 10z \le 10{,}000 \\ 16x + 5y + 10z \le 6000 \\ x \ge 0, \quad y \ge 0, \quad z \ge 0 \end{cases}$

17. Let $x =$ Number of tables manufactured
$y =$ Number of shelves manufactured
$P =$ Profit
Maximize: $P = 4x + 2y$
Subject to: $\begin{cases} x \ge 30, \quad y \ge 25 \\ 2x + y \le 200 \\ 2x + 2y \le 240 \\ 2x + 3y \le 300 \\ x \ge 0, \quad y \ge 0 \end{cases}$

19. Let $x =$ Number of hours of jogging
$y =$ Number of hours of handball
$z =$ Number of hours of dancing
$E =$ Total exercise time
Minimize: $E = x + y + z$
Subject to: $\begin{cases} x \ge 3 \\ y \ge 2 \\ z \ge 5 \\ x \ge y \\ 900x + 600y + 800z \ge 9000 \\ x \ge 0, \quad y \ge 0, \quad z \ge 0 \end{cases}$

21. Let $x =$ Number of units on day shift
$y =$ Number of units on swing shift
$z =$ Number of units on graveyard shift
$C =$ Labor costs
Minimize: $C = 100x + 150y + 180z$
Subject to: $\begin{cases} x \ge 20, \quad x \le 50, \quad y \le 50, \quad z \le 50 \\ y + z \ge 50 \\ 60x \le 1800 \\ 60y \le 1500 \\ 60z \le 3000 \\ x \ge 0, \quad y \ge 0, \quad z \ge 0 \end{cases}$

23. Let $x =$ Number of cases shipped from LA to Chicago
$y =$ Number of cases shipped from LA to Dallas
$z =$ Number of cases shipped from Seattle to Chicago
$w =$ Number of cases shipped from Seattle to Dallas
$C =$ Shipping cost
Minimize: $C = 9x + 7y + 7z + 8w$
Subject to: $\begin{cases} x + y \le 90 \\ z + w \le 130 \\ x + z \ge 80 \\ y + w \ge 110 \\ x \ge 0, \quad y \ge 0, \quad z \ge 0, \quad w \ge 0 \end{cases}$

PROBLEM SET 4.3, PAGES 155–157

1. Maximum at D, minimum at A **3.** No maximum, minimum at C
5. $(0, \frac{9}{2}), (6, 0), (5, 2), (0, 0)$ **7.** $(0, 0), (0, 4), (4, 0), (2, 3)$ **9.** $(6, 0), (6, 2), (4, 4), (0, 4), (0, 0)$

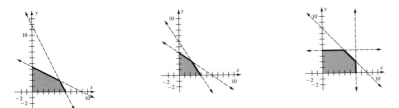

11. (*Note: This is Example 2, Section 4.2.*) $(0, \frac{8}{5}), (0, 4), (\frac{24}{13}, \frac{16}{13})$

13. (*Note: This is Example 3, Section 4.2.*) $(0, 40), (8, 24), (\frac{200}{7}, \frac{60}{7}), (50, 0)$

15. $(3, 2), (5, 5), (7, 5), (\frac{10}{3}, \frac{4}{3})$

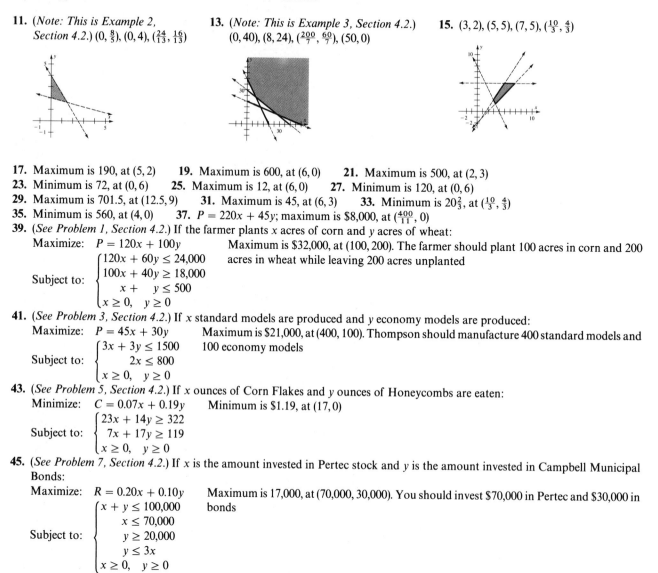

17. Maximum is 190, at $(5, 2)$ **19.** Maximum is 600, at $(6, 0)$ **21.** Maximum is 500, at $(2, 3)$

23. Minimum is 72, at $(0, 6)$ **25.** Maximum is 12, at $(6, 0)$ **27.** Minimum is 120, at $(0, 6)$

29. Maximum is 701.5, at $(12.5, 9)$ **31.** Maximum is 45, at $(6, 3)$ **33.** Minimum is $20\frac{2}{3}$, at $(\frac{10}{3}, \frac{4}{3})$

35. Minimum is 560, at $(4, 0)$ **37.** $P = 220x + 45y$; maximum is \$8,000, at $(\frac{400}{11}, 0)$

39. (*See Problem 1, Section 4.2.*) If the farmer plants x acres of corn and y acres of wheat:

Maximize: $P = 120x + 100y$

Subject to: $\begin{cases} 120x + 60y \le 24,000 \\ 100x + 40y \ge 18,000 \\ x + y \le 500 \\ x \ge 0, \quad y \ge 0 \end{cases}$

Maximum is \$32,000, at $(100, 200)$. The farmer should plant 100 acres in corn and 200 acres in wheat while leaving 200 acres unplanted

41. (*See Problem 3, Section 4.2.*) If x standard models are produced and y economy models are produced:

Maximize: $P = 45x + 30y$

Subject to: $\begin{cases} 3x + 3y \le 1500 \\ 2x \le 800 \\ x \ge 0, \quad y \ge 0 \end{cases}$

Maximum is \$21,000, at $(400, 100)$. Thompson should manufacture 400 standard models and 100 economy models

43. (*See Problem 5, Section 4.2.*) If x ounces of Corn Flakes and y ounces of Honeycombs are eaten:

Minimize: $C = 0.07x + 0.19y$

Subject to: $\begin{cases} 23x + 14y \ge 322 \\ 7x + 17y \ge 119 \\ x \ge 0, \quad y \ge 0 \end{cases}$

Minimum is \$1.19, at $(17, 0)$

45. (*See Problem 7, Section 4.2.*) If x is the amount invested in Pertec stock and y is the amount invested in Campbell Municipal Bonds:

Maximize: $R = 0.20x + 0.10y$

Subject to: $\begin{cases} x + y \le 100,000 \\ x \le 70,000 \\ y \ge 20,000 \\ y \le 3x \\ x \ge 0, \quad y \ge 0 \end{cases}$

Maximum is 17,000, at $(70,000, 30,000)$. You should invest \$70,000 in Pertec and \$30,000 in bonds

PROBLEM SET 4.4, PAGES 162–163

1. Maximize: $z = 30x_1 + 20x_2$

Subject to: $\begin{cases} 2x_1 + x_2 \le 12 \\ 5x_1 + 8x_2 \le 40 \\ x_1 \ge 0, \quad x_2 \ge 0 \end{cases}$

3. Maximize: $z = 100x_1 + 100x_2$

Subject to: $\begin{cases} 3x_1 + 2x_2 \le 12 \\ x_1 + 2x_2 \le 8 \\ x_1 \ge 0, \quad x_2 \ge 0 \end{cases}$

5. Conditions 1 and 3 are violated

7. Conditions 2 and 3 are violated   **9.** Condition 3 is violated   **11.** Condition 2 is violated

13.

x_1	x_2	y_1	y_2	z	
3	1	1	0	0	300
2	2	0	1	0	400
-2	-3	0	0	1	0

15.

x_1	x_2	y_1	y_2	z	
1	1	1	0	0	200
4	2	0	1	0	500
-45	-35	0	0	1	0

17.
x_1	x_2	y_1	y_2	y_3	z	
12	150	1	0	0	0	1200
6	200	0	1	0	0	1200
16	50	0	0	1	0	800
−1	−1	0	0	0	1	0

19.
x_1	x_2	x_3	y_1	y_2	z	
8	5	3	1	0	0	1000
5	1	3	0	1	0	800
−12	−7	−5	0	0	1	0

21.
x_1	x_2	x_3	x_4	y_1	y_2	y_3	y_4	z	
1	1	0	0	1	0	0	0	0	90
0	0	1	1	0	1	0	0	0	130
1	0	1	0	0	0	1	0	0	80
0	1	0	1	0	0	0	1	0	110
−9	−7	−7	−8	0	0	0	0	1	0

23.
x_1	x_2	y_1	y_2	z	
1	0	$\frac{1}{3}$	0	0	30
0	2	−2	1	1	−162
0	−6	4	0	1	360

25.
x_1	x_2	y_1	y_2	z	
1	0	$\frac{1}{3}$	0	0	30
0	1	−1	$\frac{1}{2}$	0	−81
0	0	−2	3	1	−126

27.
x_1	x_2	y_1	y_2	z	
2	1	$\frac{1}{4}$	0	0	10
0	0	$-\frac{3}{2}$	1	0	540
12	0	$\frac{5}{2}$	0	1	100

29.
x_1	x_2	y_1	y_2	y_3	z	
0	−3	1	−1	0	0	100
1	3	0	$\frac{1}{2}$	0	0	50
0	−7	0	$-\frac{3}{2}$	1	0	150
0	25	0	5	0	1	500

31.
x_1	x_2	y_1	y_2	y_3	z	
$\frac{5}{2}$	0	1	$\frac{5}{2}$	0	0	350
$\frac{1}{2}$	1	0	$\frac{1}{2}$	0	0	10
−4	0	0	−6	1	0	100
2	0	0	8	0	1	340

33.
x_1	x_2	x_3	y_1	y_2	y_3	z	
−15	−6	0	1	0	−2	0	60
−11	−2	0	0	1	$-\frac{3}{2}$	0	45
4	2	1	0	0	$\frac{1}{2}$	0	5
14	5	0	0	0	2	1	20

PROBLEM SET 4.5, PAGES 172–174

1. a. $x_1 = 0, x_2 = 0, y_1 = 30, y_2 = 50, z = 0$ **b.** $x_1 = 0, x_2 = 0, y_1 = 120, y_2 = 180, z = 0$
3. a. $x_1 = 0, x_2 = 0, x_3 = 0, y_1 = 60, y_2 = 30, y_3 = 40, z = 0$ **b.** $x_1 = 0, x_2 = 0, x_3 = 0, y_1 = 80, y_2 = 50, y_3 = 60, y_4 = 90, z = 0$
5. a. $x_1 = 10, y_1 = 12, y_3 = 20, x_2 = y_2 = 0, z = 32$. Final tableau
  b. $x_1 = 120, x_3 = 80, y_1 = 20, x_2 = y_2 = y_3 = 0, z = 360$. Not the final tableau

7.
x_1	x_2	y_1	y_2	z	
$\frac{9}{2}$	0	1	$-\frac{3}{4}$	0	17
$\frac{1}{2}$	1	0	$\frac{1}{4}$	0	1
6	0	0	5	1	20

9.
x_1	x_2	y_1	y_2	y_3	z	
1	0	1	$-\frac{1}{5}$	0	0	48
$\frac{1}{3}$	1	0	$\frac{1}{15}$	0	0	$\frac{2}{3}$
3	0	0	$-\frac{1}{5}$	1	0	68
$\frac{5}{3}$	0	0	$\frac{4}{3}$	0	1	$\frac{40}{3}$

11.
x_1	x_2	x_3	y_1	y_2	y_3	z	
0	−16	−48	1	0	−5	0	10
0	−5	−9	0	1	−2	0	6
1	3	9	0	0	1	0	2
0	7	27	0	0	4	1	8

13. Maximum is $z = 600$, when $x_1 = 0, x_2 = 200$
15. Maximum is 480, when $x_1 = 0$ and $x_2 = 30$
17. Maximum is $z = 7500$, when $x_1 = 50, x_2 = 150$
19. Maximum is 4700, when $x_1 = 700, x_2 = 150$
21. Maximum is $z = 50$, when $x_1 = 50, x_2 = 0$
23. Maximum is 400, when $x_1 = 100, x_2 = 0$
27. Maximum is $z = 1600$, when $x_1 = 80, x_4 = 110$
25. Maximum is $z = 1600$, when $x_1 = 0, x_2 = 50, x_3 = 250$
29. The company should manufacture 700 Alpha products and 150 Beta products. Then the (maximum) profit will be $4,700
31. Maximum profit is $1,600, when $x_1 = 0, x_2 = 50, x_3 = 250$

PROBLEM SET 4.6, PAGES 181–182

1. Maximum is 400, when $x_1 = 0$ and $x_2 = 10$ **3.** Maximum is 600, when $x_1 = 20$ and $x_2 = 0$
5. Maximum is $\frac{405}{2}$, when $x_1 = \frac{15}{4}$ and $x_2 = -\frac{9}{2}$ **7.** Maximum is $\frac{1840}{13}$, when $x_1 = \frac{24}{13}$ and $x_2 = \frac{16}{13}$
9. Maximum is 12, when $x_1 = 0$ and $x_2 = 4$ **11.** Minimum is -40, when $x_1 = 2$ and $x_2 = 0$
13. Minimum is 125, when $x_1 = 3$ and $x_2 = 2$ **15.** Maximum is 80, when $x_1 = 15$, $x_2 = 0$, and $x_3 = 10$
17. Minimum cost is $\frac{115}{2}$ or $57.50, when no units of A, $\frac{115}{2}$ units of B, and no units of C are mixed
19. Any ordered pair on the line segment with endpoints (80, 40) and (87.5, 25) will yield the maximum, $400, when substituted
into the objective function. Since we desire integral solutions, we could start at (80, 40) and use the idea of the slope of the line
segment, which is $-\frac{2}{1}$, to generate other integral solutions: (81, 38), (82, 36), (83, 34), (84, 32), (85, 30), (86, 28), (87, 26)
21. Answers vary, but the key idea is that the simplex method always gives the maximum, *but only one* set of values where it
occurs. The occurrence of a maximum will not be unique if and only if the objective function has the same slope as one of the
sides of the (polygonal) feasibility region. In this case we might have noticed that the objective function and the first constraint
both have slopes of -2. Perhaps in such a situation, one should resort to the graphing method when possible

PROBLEM SET 4.7, PAGES 189–191

1. a. $A^T = \begin{bmatrix} 6 & 4 \\ 9 & 8 \end{bmatrix}$ **b.** $B^T = \begin{bmatrix} 5 & 3 \\ 6 & 8 \end{bmatrix}$ **c.** $C^T = \begin{bmatrix} 4 & 6 \\ 9 & 1 \\ 1 & 4 \end{bmatrix}$ **3. a.** $G^T = \begin{bmatrix} 1 \\ 3 \\ 5 \end{bmatrix}$ **b.** $H^T = \begin{bmatrix} 4 & 9 & 6 \end{bmatrix}$

c. $J^T = \begin{bmatrix} 1 & 0 & 3 & 2 \end{bmatrix}$ **5.** Maximize: $z' = 10y_1 + 30y_2$ **7.** Maximize: $z' = 10y_1 + 2y_2 + y_3 + 15y_4$

Subject to: $\begin{cases} 2y_1 + 3y_2 \le 3 \\ 8y_1 + 5y_2 \le 4 \\ y_1 \ge 0, \quad y_2 \ge 0 \end{cases}$ Subject to: $\begin{cases} y_1 + 2y_2 + 3y_4 \le 3 \\ y_1 + y_3 \le 2 \\ y_1 + 3y_2 + 2y_3 + 5y_4 \le 5 \\ y_1 \ge 0, \quad y_2 \ge 0, \quad y_3 \ge 0, \quad y_4 \ge 0 \end{cases}$

9. Minimum value is 24, when $x_1 = 0$ and $x_2 = 6$ **11.** Minimum value is 25, when $x_1 = 5$, $x_2 = 5$, $x_3 = 0$
13. Minimum is $\frac{2000}{3}$, when $x_1 = 0$, $x_2 = 0$, $x_3 = \frac{100}{3}$ **15.** Minimum is 72, when $x_1 = 0$, $x_2 = 6$
17. Minimum is 120, when $x_1 = 0$, $x_2 = 6$ **19.** Minimum is $\frac{62}{3}$, when $x_1 = \frac{10}{3}$, $x_2 = \frac{4}{3}$
21. Minimum is 560, when $x_1 = 4$, $x_2 = 0$
23. The Gainesville plant should operate 30 days and the Sacramento plant should operate 20 days. The minimum cost is
$900,000
25. She should spend $\frac{38}{9}$ hours (about 4 hours, 13 minutes) jogging, 2 hours playing handball, and 5 hours dancing per week
27. The day shift should produce 30 units, the swing shift should produce 25 units, and the graveyard shift should produce
50 units
29. The dealer should ship 30 sets from Burlingame to Hillsborough and 35 sets from San Jose to Palo Alto
31. The dealer should ship 15 refrigerators from Dallas to Fort Worth, 10 from Dallas to Houston, and 25 from San Antonio to
Houston

PROBLEM SET 4.8, PAGES 192–193

1.

	x_1	x_2	x_3	y_1	y_2	y_3	z	
	2	0	1	1	0	0	0	50
	0	④	1	0	1	0	0	90
	3	4	0	0	0	1	0	100
	-6	-25	-3	0	0	0	1	0

First pivot is circled. For extra practice, you can solve this problem; the maximum is 582.5, which occurs when $x_1 = \frac{10}{3}$, $x_2 = \frac{45}{2}$,
and $x_3 = 0$.

3.

	y_1	y_2	y_3	x_1	x_2	z^1	
	9	3	2	1	0	0	50
	1	⑫	3	0	1	0	80
	-18	-36	-30	0	0	1	0

First pivot is circled. For extra practice, you can solve this problem;
the minimum is 750, which occurs when $x_1 = 15$ and $x_2 = 0$.

5. C **7.** C **9.** C

CHAPTER 5
PROBLEM SET 5.1, PAGES 203–204

1. 6, 16, 26, 36   **3.** 1, 4, 16, 64   **5.** 10, -50, 250, -1250 **7.** 4, 11, 25, 53   **9.** 1, 5, 13, 29   **11.** 3, 13, 63, 313
13. 32   **15.** 81   **17.** 10   **19.** 312   **21.** 143   **23.** $\frac{5}{8}$ **25.** $x_n = 5 + 25n$ **27.** $x_n = 4 - 2n$ **29.** $x_n = 4n$
31. $x_n = 3^n$ **33.** $x_n = 2^n$ **35.** $x_n = 3 - 3 \cdot 2^n$ **37.** $x_n = -\frac{9}{4} + \frac{17}{4} \cdot 5^n$ **39.** $x_n = -\frac{2}{3} + \frac{5}{3} \cdot 4^n$ **41.** 20
43. 49   **45.** 11   **47.** 80   **49.** \$32,610.26   **51.** $x_8 \approx 188,470$ **53.** $x_{24} = 2^{24} = 16,777,216$

PROBLEM SET 5.2, PAGES 212–214

1. \$400 simple interest; \$469.33 compounded annually   **3.** \$720 simple interest; \$809.86 compounded annually
5. \$12,000 simple interest; \$43,231.47 compounded annually   **7.** \$1,400 simple interest; \$1,469.33 compounded annually
9. \$2,720 simple interest; \$2,809.86 compounded annually   **11.** \$17,000 simple interest; \$48,231.47 compounded annually
13. 9%; 5; 1.538624; \$1,538.62; \$538.62   **15.** 8%; 3; 1.259712; \$629.86; \$129.86   **17.** 2%; 12; 1.268242; \$634.12; \$134.12
19. 4.5%; 40; 5.816365; \$29,081.83; \$24,081.83   **21.** 5%; 40; 7.039989; \$35,199.95; \$30,199.95
23. 2%; 60; 3.281031; \$13,124.12; \$9,124.12   **25.** 4%; 5; 1.216653; \$1,520.82; \$270.82   **27.** \$24,067.32
29. \$19,898.24   **31.** \$29,960.20   **33.** 6.17%   **35.** 12.75%   **37.** \$16,889.10   **39.** \$153,278.42   **41.** \$5
43. \$12   **45.** \$2,691   **47.** \$188   **49.** \$168,187   **51.** \$773,662   **53.** \$3,655.96   **55.** \$131,444.49
57. \$2,218.89   **59.** Present value is \$50,000 compared to \$50,734.56; best to take \$10,000 now and \$45,000 in 1 year

PROBLEM SET 5.3, PAGE 217

1. \$56,642   **3.** \$59,498   **5.** \$22,620   **7.** \$61,173   **9.** \$61,878   **11.** \$23,299   **13.** \$4,883   **15.** \$220,498
17. \$2,930   **19.** \$72,433   **21.** \$19,075   **23.** \$413,755   **25.** \$24,297   **27.** \$1,374,583   **29.** Answers vary

PROBLEM SET 5.4, PAGES 221–222

1. \$5,628.89   **3.** \$5,655.87   **5.** \$6,934.43   **7.** \$4,313.07   **9.** \$1,181.41   **11.** \$5,250.32   **13.** \$44.42
15. \$272.19   **17.** \$111.52   **19.** \$133.78   **21.** \$887.30   **23.** \$4,802.66   **25.** \$131,022   **27.** \$1,780.28
29. The present value of the annuity is \$25,906.15, so it is the most valuable

PROBLEM SET 5.5, PAGES 224–226

1. \$1,193.20   **3.** \$1,896.70   **5.** \$4,485.45   **7.** \$6,631.19   **9.** \$2,076.83   **11.** \$351.35   **13.** \$5,426.39
15. \$220,000 + \$35,310 = \$255,310   **17. a.** Present value   **b.** \$56,255   **19. a.** Future value   **b.** \$140,522
21. a. Present value   **b.** \$131,444   **23. a.** Future value   **b.** \$292   **25. a.** Present value   **b.** \$165,135
27. a. Installment payments   **b.** \$1,317   **29. a.** Ordinary annuity   **b.** \$175,611

PROBLEM SET 5.6, PAGES 227–228

1. $1000(1.01)^{12(2.5)} = \$1,347.85$ **3.** $\dfrac{10,000}{(1.03)^{20}} = \$5,536.76$ **5.** $2\dfrac{(10,000)(0.09)}{1.09^5 - 1} = \$1,670.92$

7. $10,000(1.0625)^5 = \$13,540.81$ **9.** $450\left(\dfrac{1.11^{26} - 1}{0.11} - 1\right) = \$57,149.45$ **11.** $\dfrac{1,000,000}{(1.12)^{37}} = \$15,098.48$

13. $\dfrac{5,000(0.115)}{(1.115)^{15} - 1} = \$139.62$ **15.** $1,000,000\left[\dfrac{1 - (1.14)^{-5}}{0.14}\right] = \$3,433,081$

17. $(150,000,000)(0.09)(1) + \dfrac{(150,000,000)(0.10)}{(1.10)^{50} - 1} = \$13,628,876$ **19.** C

CHAPTER 6
PROBLEM SET 6.1, PAGES 240–241

1. (Answers vary.) An *element* of a set is a member of the set, and not a subset of the set. A *subset* of a set is not a member of the set, and is a set. For example, 4 is an element (not a subset) of the set of even integers, {4} is a subset (not an element) of the set of even integers

3. (Answers vary.) The universal set is the set of all things (or elements) under consideration and is usually nonempty. The empty set contains no elements

5. $\{m\}, \{y\}, \{m, y\}, \{\ \}$ **7.** $\{y\}, \{o\}, \{u\}, \{y, o\}, \{y, u\}, \{o, u\}, \{y, o, u\}, \{\ \}$ **9.** $\{\ \}, \{3\}, \{6\}, \{9\}, \{3, 6\}, \{3, 9\}, \{6, 9\}, \{3, 6, 9\}$

11. $\{\ \}, \{1\}, \{2\}, \{3\}, \{4\}, \{1, 2\}, \{1, 3\}, \{1, 4\}, \{2, 3\}, \{2, 4\}, \{3, 4\}, \{1, 2, 3\}, \{1, 2, 4\}, \{2, 3, 4\}, \{1, 3, 4\}, \{1, 2, 3, 4\}$

13. $\{2, 6, 8, 10\}$ **15.** $\{3, 4, 5\}$ **17.** $\{2, 3, 5, 6, 8, 9\}$ **19.** $\{1, 3, 4, 5, 6, 7, 10\}$ **21.** $\{1, 2, 3, 4, 5, 6, 7, 9, 10\}$

23. $\{1, 2, 3, 4, 5, 6\}$ **25.** $\{1, 2, 3, 5, 6, 7\}$ **27.** $\{3\}$ **29.** $\{5, 6, 7\}$ **31.** $\{1, 2, 4, 6\}$ **33.** $\{1, 2, 3, 4, 5, 6, 7\}$

35. $\{1, 2, 3, 4, 5\}$ **37.** $\{5, 6, 7\}$ **39. a.** False **b.** True **c.** True **d.** True **e.** False

41. **43.** **45.** **47.**

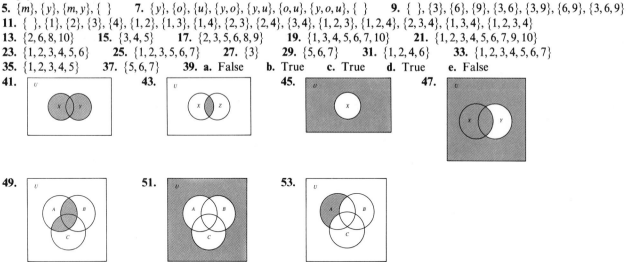

49. **51.** **53.**

55. a. $A \cup B = \{$Bob Wisner, Joan Marsh, Craig Barth, Phyllis Niklas, Shannon Smith, Christy Anton$\}$; union
b. $A \cap B = \{$Phyllis Niklas, Craig Barth$\}$; intersection **57. a.** Answers vary. **b.** False **c.** True **d.** True

e. False **f.** True **g.** True **h.** False **i.** False

59. Let M be the set of tires with defective materials and W be the set of tires with defective workmanship.
Then: $n(M \cup W) = n(M) + n(W) - n(M \cap W) = 72 + 89 - 17 = 144$ **61.** 70

63. **65. a.** 12 **b.** 7 **c.**

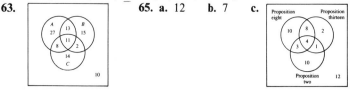

67. a. B **b.** A **c.** U **d.** $\varnothing$ **69.** 1,001,101 **71.** Answers vary

PROBLEM SET 6.2, PAGES 249–250

1. 696 **3.** 40,200 **5.** 24 **7.** 144 **9.** 3,628,800 **11.** 72 **13.** 11,880 **15.** 126 **17.** 330 **19. a.** 9

b. 72 **c.** 504 **d.** 3024 **e.** 1 **21. a.** 95,040 **b.** 60 **c.** 1680 **d.** 1 **e.** $\dfrac{g!}{(g - h)!}$ **23. a.** 210 **b.** 120

c. $\dfrac{50!}{2}$ **d.** 25 **e.** $\dfrac{m!}{(m - 3)!}$ or $m(m - 1)(m - 2)$ **25.** 7! or 5040 **27.** $\dfrac{6!}{2!} = 360$ **29.** $\dfrac{11!}{4!4!2!} = 34,650$

31. $\dfrac{11!}{2!2!2!} = 4,989,600$ **33.** $\dfrac{11!}{3!2!2!2!} = 831,600$ **35.** 210 **37.** 15 **39.** 40,320 **41.** 47,045,520 **43.** 60

45. 175,760,000 **47.** 2,300,000 **49.** 17,331,000 **51.** 1,976,000 **53.** Answers vary

PROBLEM SET 6.3, PAGES 257–258

1. a. 9   **b.** 36   **c.** 84   **d.** 126   **e.** 1   **3. a.** 35   **b.** 1   **c.** 1225   **d.** 25   **e.** $\dfrac{g!}{h!(g-h)!}$ **5. a.** 2520

b. 1   **c.** 45   **d.** $\dfrac{n!}{(n-4)!}$ **e.** $\dfrac{n!}{4!(n-4)!}$ **7. a.** 1   **b.** 23,426   **c.** 1   **d.** 5040   **e.** $\dfrac{n!}{(n-5)!}$ **9. a.** 20

b. 420   **11. a.** 2520   **b.** 1680   **13.** $\dbinom{100}{5} = 75{,}287{,}520$ ways   **15.** $\dbinom{4}{2} = 6$ ways   **17.** $\dbinom{13}{5} = 1287$ ways

19. Permutation; $_8P_5 = 6720$ **21.** Permutation; $_5P_5 = 120$ **23.** Combination; $\dbinom{100}{6} = 1{,}192{,}052{,}400$

25. Permutation; $_6P_6 = 720$ **27.** None; $3 \cdot 3 \cdot 2 \cdot 2 \cdot 1 \cdot 1 = 36$ ways   **29.** Combination; $\dbinom{5}{3} = 10$

31. Permutation; $_6P_5 = 720$ **33.** $\dbinom{5}{2}\dbinom{25}{5}\dbinom{18}{5} = 4{,}552{,}178{,}400$ **35.** $\dfrac{50!}{15!10!25!}$

37. $\dbinom{50}{7,5,38} \approx 96{,}149{,}000{,}000{,}000$ **39.** $\dbinom{10}{4,5,1} = 1260$ **41. a.** $\dbinom{10}{3} = 120$ **b.** $\dbinom{11}{2,8,1} = 495$

c. $\dbinom{10}{1}\dbinom{11}{2} = 550$ **d.** $\dbinom{10}{1}\dbinom{6}{1}\dbinom{5}{1} = 300$ **e.** $\dbinom{6}{2}\dbinom{5}{1} = 75$ **43.** $\dbinom{10}{3}\dbinom{12}{3} = 26{,}400$

45. $\dbinom{n}{k-1} + \dbinom{n}{k} = \dfrac{n!}{(k-1)!(n-k+1)!} + \dfrac{n!}{k!(n-k)!} = \dfrac{k \cdot n!}{k!(n-k+1)!} + \dfrac{(n-k+1) \cdot n!}{k!(n-k+1)!} = \dfrac{(n+1)n!}{k!(n-k+1)!} = \dbinom{n+1}{k}$

PROBLEM SET 6.4, PAGES 263–264

1. 8   **3.** 28   **5.** 56   **7.** 12   **9.** 1   **11.** 153   **13.** $x^5 + 5x^4y + 10x^3y^2 + 10x^2y^3 + 5xy^4 + y^5$
15. $x^4 + 4x^3y + 6x^2y^2 + 4xy^3 + y^4$ **17.** $x^5 - 5x^4y + 10x^3y^2 - 10x^2y^3 + 5xy^4 - y^5$
19. $x^5 + 10x^4 + 40x^3 + 80x^2 + 80x + 32$ **21.** $16x^4 + 96x^3y + 216x^2y^2 + 216xy^3 + 81y^4$
23. $1 - 10x + 45x^2 - 120x^3 + 210x^4 - 252x^5 + 210x^6 - 120x^7 + 45x^8 - 10x^9 + x^{10}$
25. $\dbinom{15}{0}x^{15} - \dbinom{15}{1}x^{14}y + \cdots + (-1)^r\dbinom{15}{r}x^{15-r}y^r + \cdots + \dbinom{15}{14}xy^{14} - \dbinom{15}{15}y^{15}$

27. $\dbinom{20}{0}x^{20} + \dbinom{20}{1}x^{19}y + \cdots + \dbinom{20}{r}x^{20-r}y^r + \cdots + \dbinom{20}{19}xy^{19} + \dbinom{20}{20}y^{20}$ **29.** -330 **31.** 81,081   **33.** -16

35. 2^{100} **37.** 210   **39.** Look at the expansion of $(x+y)^n$ with $x = 1$, $y = 1$ **41.** $\dbinom{n}{r} = \dfrac{n!}{r!(n-r)!}$, $\dbinom{n}{n-r} = \dfrac{n!}{(n-r)!r!}$

PROBLEM SET 6.5, PAGES 264–265

1. $\{2,3,4,8,9,10,12\}$ **3.** $\{5,6\}$ **5. a.** 5040   **b.** 5016   **c.** 6   **d.** 9900   **7. a.** 28   **b.** 168   **9.** 42
11. $\dbinom{4}{2}19 \cdot 18 \cdot 17 \cdot 16 \cdot \left(\dfrac{32!}{6!26!}\right) \approx 5.06 \times 10^{11}$ **13.** $2^{10} = 1024$ **15. a.** 125   **b.** 5   **c.** 25   **d.** 10   **e.** 60

CHAPTER 7
PROBLEM SET 7.1, PAGES 272–273

1. $S = \{2,3,4,5,6,7,8,9,10,11,12\}$ **3.** $\{0,1,2,3,4,5,6,7,8,9,10\}$ **5.** $\{4,5,6,7\}$
7. {Nonnegative integers less than or equal to 500}   **9.** {bbb, bbg, bgb, bgg, gbb, gbg, ggb, ggg}
11. a. $E = \{2,4,6,8,10,12\}$ **b.** $F = \{2,3,5,7,11\}$; not mutually exclusive   **13. a.** $G = \{2\}$
b. $H = \{3,4,5,6,7,8,9,10,11,12\}$; mutually exclusive   **15. a.** $G = \{\text{HHH, HHT, HTH, HTT}\}$
b. $H = \{\text{HHH, HHT, THH, THT}\}$; not mutually exclusive   **17. a.** $F = \{\text{fd, fr, fi, fn}\}$
b. $D = \{\text{md, fd}\}$; not mutually exclusive   **19. a.** $E = \{\text{fd, fr, fi, fn, mr}\}$ **b.** $R = \{\text{md, mr, mi, fd, fr, fi}\}$; not mutually exclusive
21. $L \cup E = \{\text{Roll a number less than 3 or roll an even number}\} = \{1,2,4,6\}$
23. $E \cap L = \{\text{Roll a number less than 3 and even}\} = \{2\}$ **25.** $\bar{L} = \{\text{Roll a 3 or larger}\} = \{3,4,5,6\}$

27. $\bar{E} \cap \bar{L} = \{$Roll an odd number that is also at least a 3$\} = \{3, 5\}$ **29.** Yes   **31.** $\frac{3}{11}$ **33.** $\frac{5}{11}$ **35.** 0   **37.** $\frac{8}{11}$
39. Yes   **41.** $\frac{1}{9}$ **43.** $\frac{5}{9}$ **45.** 0   **47.** $\frac{8}{9}$

PROBLEM SET 7.2, PAGES 279–280

1. $\frac{3}{8}$ **3.** $\frac{2}{8} = \frac{1}{4}$ **5.** $\frac{6}{20} = \frac{3}{10}$ **7.** $\frac{13}{20}$ **9.** 0   **11.** $\frac{4}{52} = \frac{1}{13}$ **13.** $\frac{1}{52}$ **15.** $\frac{16}{52} = \frac{4}{13}$ **17.** .05   **19.** .19
21. Results may vary; yes   **23.** $\frac{4}{36} = \frac{1}{9}$ **25.** $\frac{6}{36} = \frac{1}{6}$ **27.** $\frac{4}{36} = \frac{1}{9}$ **29.** $\frac{2}{36} = \frac{1}{18}$ **31.** $\frac{7}{36}$ **33.** $\frac{18}{36} = \frac{1}{2}$
35. $\frac{8}{36} = \frac{2}{9}$ **37.** $\frac{1}{36}$ **39. a.** $\frac{1}{16}$ **b.** $\frac{2}{16} = \frac{1}{8}$ **c.** $\frac{3}{16}$

41.   **43. a.** $\frac{9}{196}$ **b.** $\frac{1}{196}$ **c.** $\frac{1}{14}$ **d.** $\frac{1}{49}$

	8	7	6	5	4	3	2	1
8	8,8	7,8	6,8	5,8	4,8	3,8	2,8	1,8
7	8,7	7,7	6,7	5,7	4,7	3,7	2,7	1,7
6	8,6	7,6	6,6	5,6	4,6	3,6	2,6	1,6
5	8,5	7,5	6,5	5,5	4,5	3,5	2,5	1,5
4	8,4	7,4	6,4	5,4	4,4	3,4	2,4	1,4
3	8,3	7,3	6,3	5,3	4,3	3,3	2,3	1,3
2	8,2	7,2	6,2	5,2	4,2	3,2	2,2	1,2
1	8,1	7,1	6,1	5,1	4,1	3,1	2,1	1,1

PROBLEM SET 7.3, PAGES 284–285

1. $\frac{10}{16} = \frac{5}{8}$ **3.** $\frac{4}{16} = \frac{1}{4}$ **5.** $\frac{12}{16} = \frac{3}{4}$ **7.** $\frac{6}{13}$ **9.** $\frac{3}{13}$ **11.** $\frac{10}{13}$ **13.** 1 to 3   **15.** $\frac{9}{10}$ **17.** 1 to 12   **19.** $\frac{1}{4}$
21. 0.008   **23.** $\frac{2}{17}$ **25.** $\frac{1}{4}$ **27.** $\frac{3}{8}$ **29.** $\frac{5}{16}$ **31.** $\frac{5}{16}$ **33.** 2 to 7   **35.** $\frac{3}{8} \approx .38$ **37.** $\frac{1}{20} \approx .05$
39. $\frac{1}{15} \approx .07$ **41.** .08   **43.** .31   **45.** .23   **47.** .01   **49.** .33   **51.** .14   **53.** .01   **55.** .21   **57.** .06
59. .00000154   **61.** .42256903

PROBLEM SET 7.4, PAGES 291–292

1. $\frac{3}{8}$ **3.** $\frac{1}{8}$ **5.** $\frac{3}{7}$ **7.** $\frac{7}{8}$ **9.** $\frac{1}{4}$ **11.** $\frac{1}{4}$ **13.** $\frac{2}{5}$ **15.** $\frac{1}{6}$ **17.** $\frac{2}{11}$ **19.** .36   **21.** .19   **23.** .53   **25.** .32
27. .59   **29.** $P(E \mid F) = .50, P(F \mid E) = .20$ **31.** $P(E \mid F) = \frac{1}{3}, P(F \mid E) = \frac{1}{2}$ **33.** $P(E \mid F) = .5, P(F \mid E) = .8$ **35.** 60%
37. $\frac{1}{2}$ **39.** Independent   **41.** Independent   **43.** Not independent   **45.** Independent   **47.** Independent
49. Not independent   **51.** Not independent   **53.** Not independent   **55.** Answers vary

PROBLEM SET 7.5, PAGE 296

1. $\frac{1}{2}$ **3.** $\frac{5}{6}$ **5.** $\frac{1}{12}$ **7.** $\frac{2}{3}$ **9.** $\frac{4}{9}$ **11.** $\frac{11}{12}$ **13.** $\frac{1}{3}$ **15.** $\frac{5}{9}$ **17.** $\frac{1}{4}$ **19.** 0   **21.** $\frac{1}{2}$ **23. a.** $\frac{1}{3}$ **b.** $\frac{5}{12}$
c. $\frac{5}{9}$ **25. a.** $\frac{7}{12}$ **b.** $\frac{1}{6}$ **c.** $\frac{1}{18}$ **27.** $\frac{1}{12}$ **29.** $\frac{1}{6}$ **31.** $\frac{3}{4}$ **33.** $\frac{1}{2}$ **35.** $\frac{1}{3}$ **37.** $\frac{1}{2}$ **39.** $\frac{25}{64}$ **41.** $\frac{15}{32}$
43. $\frac{49}{64}$ **45.** $\frac{3}{28}$ **47.** $\frac{15}{56}$

PROBLEM SET 7.6, PAGES 300–302

1. $.83   **3.** The paper boy's expectation is about 9¢, but the most likely tip is a nickel   **5.** 2¢   **7.** -50¢   **9.** About 2¢
11. $12   **13.** $318   **15.** $85   **17.** -5¢   **19.** -5¢   **21.** -5¢   **23.** -8¢   **25.** -5¢
27. No; the expectation is $-$8,125   **29.** Five wells should be tried   **31.** The optimal number of cars is 14

PROBLEM SET 7.7, PAGES 307–308

1. .05   **3.** .04   **5.** .05   **7.** $\frac{1}{4}$ **9.** $\frac{4}{17}$ **11.** $\frac{1}{4}$ **13.** .04   **15.** .04   **17.** .05   **19.** $\frac{1}{4}$ **21.** .44   **23.** .59
25. $\frac{2}{3}$ **27. a.** .22   **b.** .09   **c.** .70   **29.** $\frac{5}{12}$

PROBLEM SET 7.8, PAGES 309–310

1. a. For a sample space S and event E, a real number $P(E)$ is associated, so that: (i) $0 \leq P(E) \leq 1$); (ii) $P(S) = 1$;
(iii) and, if E and F are mutually exclusive events, then $P(E \cup F) = P(E) + P(F)$ **b.** Their intersection is empty
c. One event does not affect the probability of the other   **3.** No. The probabilities of A, B, and C are $\frac{1}{4}, \frac{1}{4}$, and $\frac{1}{2}$, respectively
5. $\frac{18}{38} \approx .47$ **7. a.** $\frac{125}{375} \approx .333$ **b.** $\frac{125}{600} \approx .208$ **c.** $\frac{150}{625} = .24$ **d.** $\frac{250}{400} = .625$ **9. a.** .33   **b.** .27   **c.** .73

CHAPTER 8
PROBLEM SET 8.1, PAGES 316–317

1.

Inches	Tally	Frequency
63	||	2
64	||||	4
65	|||	3
66	|||	3
67	卌	5
68	|||	3
69	||||	4
70	|||	3
71	||	2
72	|	1

Let x = students' heights (rounded to the nearest inch), where $x = 63, 64, \ldots, 72$

3.

Number of spots	Tally	Frequency
2	|	1
3	|||	3
4	卌	5
5	||	2
6	卌 ||||	9
7	卌 |||	8
8	卌 |	6
9	卌 |	6
10	||||	4
11	卌	5
12	|	1

Let x = sum of dice, where $x = 2, 3, 4, \ldots, 11, 12$

5.

Number of years	Frequency	Relative frequency
13	1	$\frac{1}{25} = .04$
14	2	$\frac{2}{25} = .08$
15	1	$\frac{1}{25} = .04$
16	3	$\frac{3}{25} = .12$
17	3	$\frac{3}{25} = .12$
18	0	$\frac{0}{25} = 0$
19	3	$\frac{3}{25} = .12$
20	3	$\frac{3}{25} = .12$
21	2	$\frac{2}{25} = .08$
22	4	$\frac{4}{25} = .16$
23	1	$\frac{1}{25} = .04$
24	1	$\frac{1}{25} = .04$
25	1	$\frac{1}{25} = .04$
	$\overline{25}$	$\overline{1}$

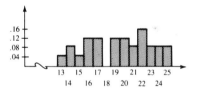

7.

Number waiting	Frequency	Relative frequency
2	20	$\frac{20}{50} = .40$
3	15	$\frac{15}{50} = .30$
4	7	$\frac{7}{50} = .14$
5	5	$\frac{5}{50} = .10$
6	2	$\frac{2}{50} = .04$
7	1	$\frac{1}{50} = .02$
	$\overline{50}$	$\overline{1}$

9.

Number of heads	Frequency	Relative frequency
3	18	$\frac{18}{150} = .12$
2	56	$\frac{56}{150} = .373$
1	59	$\frac{59}{150} = .393$
0	17	$\frac{17}{150} = .113$
	150	1

11.

Number of heads	Relative frequency
0	$\frac{1}{16} = .0625$
1	$\frac{4}{16} = .25$
2	$\frac{6}{16} = .375$
3	$\frac{4}{16} = .25$
4	$\frac{1}{16} = .0625$
	1

13.

Number of acres	Relative frequency
0	$\frac{188}{221} = .851$
1	$\frac{32}{221} = .145$
2	$\frac{1}{221} = .005$
	1

15.

Number of defects	Relative frequency
0	$\frac{64}{125} = .512$
1	$\frac{48}{125} = .384$
2	$\frac{12}{125} = .096$
3	$\frac{1}{125} = .008$
	1

17.

Number of cures	Relative frequency
4	.1296
3	.3456
2	.3456
1	.1536
0	.0256
	1

19.

Number of Almond Joys	Relative frequency
0	$\frac{17}{35} \approx .486$
1	$\frac{3}{7} \approx .429$
2	$\frac{3}{35} \approx .086$
	1

21. **23.**

25. **27.** **29.** **31.**

PROBLEM SET 8.2, PAGES 325–326

	Mean	Median	Mode	Range	Variance	Standard deviation
1.	3	3	None	4	2.5	1.58
3.	105	105	None	4	2.5	1.58
5.	6	7	7	8	6.7	2.58
7.	10	8	None	18	52.0	7.21
9.	91	95	95	17	50.5	7.11
11.	3	3	3	4	1.7	1.29

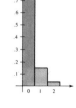

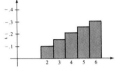

13. Answers vary   **15.** Mean = $15,800; median = $16,000; mode = $10,000
17. Mean = 68.09; median = 70; mode = 70   **19.** Mean = 56; median = 65; mode = 70; range = 90
21. Variance = $141\frac{2}{3}$; standard deviation = 11.90   **23.** Variance = $41,288,888.89; standard deviation = $6,425.64
25. Mean = .75; variance = .6875; standard deviation = .83   **27.** Mean = 3.4; variance = .84; standard deviation = .92
29. Mean = 2.85; variance = 2.3275; standard deviation = 1.53   **31.** Answers vary; all data are the same
33. Answers vary; first student 70, 70, 70, 70, 70, 70; second student 80, 70, 60, 90, 50, 70
35. Answers vary   **37.** Answers vary   **39.** 9.1596   **41.** 9.5736   **43.** 10.4198

PROBLEM SET 8.3, PAGES 332–333

1. $P(X = 0) = \binom{3}{0}(.2)^0(.8)^3 = .512$ **3.** $P(X = 2) = \binom{3}{2}(.2)^2(.8) = .096$ **5.** $1 - P(X = 0) = .488$

7. $P(X = 4) = \binom{4}{4}(.9)^4(.1)^0 = .6561$ **9.** $P(X = 0) = \binom{4}{0}(.9)^0(.1)^4 = .0001$ **11.** .948   **13.** .132

15. .1281   **17.** .016   **19.** .0002

	Mean	Variance	Standard deviation
21.	1.5	1.05	1.02
23.	7.8	2.73	1.65
25.	3.0	1.50	1.22
27.	0.7	0.63	0.79
29.	6.0	2.4	1.55

31. .4812   **33.** .5443   **35.** .484   **37.** 16
39. Mean = 1.2; standard deviation = .98
41. Mean = 30; standard deviation = 3.46   **43. a.** .109375
b. .1143   **45.** .3125   **47.** $n2^{1-n}$

PROBLEM SET 8.4, PAGE 338

1. .0228   **3.** .8413   **5.** .2709   **7.** .9772   **9.** .8413   **11.** .0455   **13.** .2548   **15.** .7926   **17.** 34 persons
19. 8 persons   **21.** 25   **23.** 1 student   **25.** .0228   **27.** .978   **29.** 40.5 inches   **31.** $P(X > 250) = .5$
33. $P(X < 220) = P(z < -1.2) = .1151$ **35.** $P(220 \le X \le 320) = .88$ **37.** .0228   **39.** $P(X > .43) = .0668$

PROBLEM SET 8.5, PAGE 343

1. Yes   **3.** No   **5.** Yes   **7.** Yes   **9.** No   **11.** $P(z < 1.29) = .9015$ **13.** $P(z > 2.84) = .002$ **15.** .2206
17. .567   **19.** .092   **21.** .0005   **23.** $P(z < .28) = .6103$ **25.** $P(z \ge -1.41) = .9207$ **27.** $P(z > -.57) = .7157$
29. $P(-1.41 < z < -.57) = .2050$ **31.** $P(-2.31 < z < -2.22) = .0028$ **33.** $P(-.61 < z < -.52) = .0306$
35. $P(z < 3.5) \approx .9998$ **37.** .5   **39.** $P(z > -1.17) = .8790$ **41.** $P(-8.16 < z < 3.5) = .9998$
43. $P(-2.39 < z < -2.27) = .0032$ **45.** $P(1.11 < z < 1.22) = .0223$
47. $np < 5, P(X > 6) = P(X = 7) + P(X = 8) + P(X = 9) + P(X = 10) = .0035$
49. $np < 5, 1 - P(X = 0) - P(X = 1) - P(X = 2) = .068$ **51.** $P(X > 300) = P(z > 2.12) = .017$

PROBLEM SET 8.6, PAGES 350–351

1. Yes   **3.** No   **5.** Yes   **7.** No   **9.** No   **11.** No   **13.** $r = .995$, at 5%   **15.** $r = -.984$, at 1%

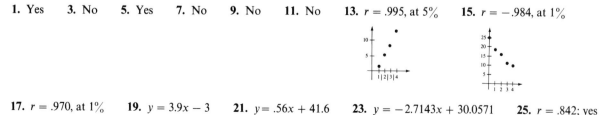

17. $r = .970$, at 1%   **19.** $y = 3.9x - 3$ **21.** $y = .56x + 41.6$ **23.** $y = -2.7143x + 30.0571$ **25.** $r = .842$; yes

27. $y = 2.4545x + 6.090$ **29.** $r = -.732$; not significant at 1% or 5%   **31.** $r = -.750$; not significant at 1% or 5%
33. $r = .924$; significant at 1%   **35.** $y = .00087x + 2.83518$

PROBLEM SET 8.7, PAGES 352–353

1.

Roll	2	3	4	5	6	7	8	9	10	11	12
Frequency	1	3	4	3	8	9	6	6	4	5	1

3.

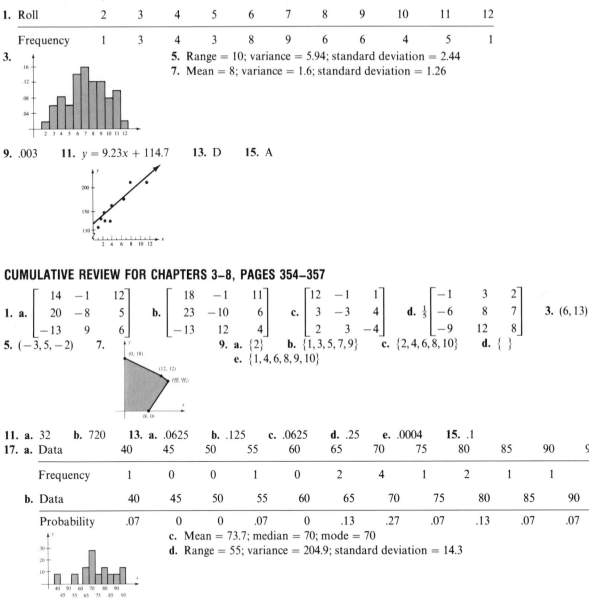

5. Range = 10; variance = 5.94; standard deviation = 2.44
7. Mean = 8; variance = 1.6; standard deviation = 1.26

9. .003   **11.** $y = 9.23x + 114.7$ **13.** D   **15.** A

CUMULATIVE REVIEW FOR CHAPTERS 3–8, PAGES 354–357

1. a. $\begin{bmatrix} 14 & -1 & 12 \\ 20 & -8 & 5 \\ -13 & 9 & 6 \end{bmatrix}$ **b.** $\begin{bmatrix} 18 & -1 & 11 \\ 23 & -10 & 6 \\ -13 & 12 & 4 \end{bmatrix}$ **c.** $\begin{bmatrix} 12 & -1 & 1 \\ 3 & -3 & 4 \\ 2 & 3 & -4 \end{bmatrix}$ **d.** $\frac{1}{5}\begin{bmatrix} -1 & 3 & 2 \\ -6 & 8 & 7 \\ -9 & 12 & 8 \end{bmatrix}$ **3.** $(6, 13)$

5. $(-3, 5, -2)$ **7.**

9. a. $\{2\}$ **b.** $\{1, 3, 5, 7, 9\}$ **c.** $\{2, 4, 6, 8, 10\}$ **d.** $\{\ \}$
e. $\{1, 4, 6, 8, 9, 10\}$

11. a. 32   **b.** 720   **13. a.** .0625   **b.** .125   **c.** .0625   **d.** .25   **e.** .0004   **15.** .1

17. a.

Data	40	45	50	55	60	65	70	75	80	85	90	95
Frequency	1	0	0	1	0	2	4	1	2	1	1	2

b.

Data	40	45	50	55	60	65	70	75	80	85	90	95
Probability	.07	0	0	.07	0	.13	.27	.07	.13	.07	.07	.13

c. Mean = 73.7; median = 70; mode = 70
d. Range = 55; variance = 204.9; standard deviation = 14.3

19. a. $r = -.6013$, no significant correlation at 1% or 5%   **b.** $y = -.16x + 41.6$
21. Maximum number of animals is 400 (all species C)   **23.** "Mixture" should consist of 2 pounds of peanuts   **25.** C

PROBLEM SET 9.1, PAGES 369–371

1. 0   **3.** 8   **5.** 7   **7.** 6   **9.** 2   **11.** 2   **13.** 15   **15.** 358   **17.** −5   **19.** The limit does not exist.
21. The limit does not exist.   **23.** 0   **25.** $-\frac{1}{2}$ **27.** Limit does not exist   **29.** 0   **31.** Limit does not exist

33. 4   **35.** 8   **37.** $\frac{1}{6}$ **39.** 0   **41.** Limit does not exist   **43.** 0   **45.** $\frac{1}{4}$ **47.** 2   **49.** $\frac{3}{2}$ **51.** 5
53. Limit does not exist (grows without bound)   **55.** $\frac{4}{3}$ **57. a.** $7   **b.** $12   **c.** $7   **d.** Does not exist
59. a. Does not exist   **b.** 80%   **c.** 70%   **d.** 50%   **61.** $\frac{1}{2}$

PROBLEM SET 9.2, PAGES 376–379

1. Discontinuity at $x = 4$ **3.** Discontinuity at $x = 2$ **5.** Discontinuity at $x = 4$; additional suspicious point at $x = 1$
7. Discontinuities at $x = 1$ and $x = 6$ **9.** Discontinuities at $x = -2$ and $x = 2$ **11.** Continuous over $x > 0$
13. Yes   **15.** Yes   **17.** Continuous; domain is $0 \le x < 24$, where x is the hour of the day
19. Not continuous; domain is $0 \le t < 24$, where t is the time of the day
21. Not continuous; domain is all positive numbers t, where t is the number of minutes   **23.** Continuous throughout
25. Continuous throughout   **27.** Discontinuities at $x = 1$ and $x = -1$ **29.** Discontinuity at $x = -2$
31. Discontinuity at $x = 3$ **33.** Discontinuity at $x = -2$ **35.** Continuous throughout   **37.** Discontinuity at $x = -2$
39. Discontinuity at all points   **41.** Continuous throughout   **43. a.** No, domain is $w > 0$
b. $c(1.9) = \$.45; c(2.01) = \$.65; c(2.89) = \$.65$ **c.**   **45.** Answers vary

PROBLEM SET 9.3, PAGES 387–389

1. $\frac{20}{3}$ **3.** 40   **5.** $\frac{10}{7}$ **7.** $\frac{5}{6}$ **9.** -10 **11.** -9 **13.** $\frac{1}{4}$ **15.** 0.2575   **17.** 59.3 mph   **19.** 36 mph
21. 0   **23.** 2   **25.** 0   **27.** -3 **29.** 0   **31.** 12   **33.** -9 **35.** $\frac{1}{6}$ **37.** 5   **39.** 0
                                        41. -3 **43.** $\frac{1}{4}$ **45.** 0   **47.** 9   **49.** 8900   **51.** 5930   **53.** $60x + 30h - 100$
                                        55. $60x - 100$ **57.** The instantaneous rate of change of cost at any given x

PROBLEM SET 9.4, PAGES 394–395

1. At $x = 1$ **3.** At $x = 3$ **5.** At $x = 4$ **7.** $\lim\limits_{h \to 0} \dfrac{2(x + h)^2 - 2x^2}{h} = \lim\limits_{h \to 0}(4x + 2h) = 4x$

9. $\lim\limits_{h \to 0} \dfrac{[-3(x + h)^2] - [-3x^2]}{h} = -6x$ **11.** $\lim\limits_{h \to 0} \dfrac{[4 - 5(x + h)] - [4 - 5x]}{h} = -5$

13. $\lim\limits_{h \to 0} \dfrac{[3(x + h)^2 + 4(x + h)] - [3x^2 + 4x]}{h} = 6x + 4$ **15.** $-6x - 50$ **17.** $30x + y + 45 = 0$ **19.** $5x + y - 4 = 0$

21. $8x - y + 6 = 0$ **23.** -20 **25.** $4t - 200$ **27.** -200 **29.** $3x^2$ **31.** $\dfrac{3}{x^2}$ **33.** $\dfrac{1}{\sqrt{x}}$

PROBLEM SET 9.5, PAGES 399–400

1. $y' = 7x^6$ **3.** $y' = 12x^{11}$ **5.** $y' = -5x^{-6}$ **7.** $y' = 0$ **9.** $y' = 32x^{-9}$ **11.** $y' = \dfrac{-1}{4\sqrt{x}} = \dfrac{-\sqrt{x}}{4x}$

13. $y' = -40x^{-9}$ **15.** $y' = 15x^{1/4}$ **17.** $y' = 6x + 1$ **19.** $y' = 4x - 5$ **21.** $y' = 15x^2 - 10x + 4$

23. $y' = -3x^{-4} + 2x - x^{-2} = \dfrac{2x^5 - x^2 - 3}{x^4}$ **25.** $y' = 8x^7 + 16x^3$ **27.** $y' = -2x^{-2} - 10x^{-3} = \dfrac{-2x - 10}{x^3}$

29. $y' = -35x^6 + x^{-1/2} + 3x^{-2} = \dfrac{-35x^8 + x\sqrt{x} + 3}{x^2}$ **31.** $C'(x) = 40x + 500$ **33.** $\dfrac{dm}{dt} = \dfrac{-400}{t^2}$

35. $A'(6) = 0.1(6) + 25 = 25.6$ (thousand dollars)   **37.** $\dfrac{25}{2\sqrt{5}} \approx 6$ tasks per hour

39. $P'(1,000,000) = 0.0005 + 0.00002(1,000,000) \approx 20$ foxes per rabbit

41. $\displaystyle\lim_{h\to 0}\frac{(x+h)^4 - x^4}{h} = \lim_{h\to 0}\frac{x^4 + 4x^3h + 6x^2h^2 + 4xh^3 + h^4 - x^4}{h}$

$$= \lim_{h\to 0}(4x^3 + 6x^2h + 4xh^2 + h^3) = 4x^3$$

43. $\displaystyle\lim_{h\to 0}\frac{[f(x+h) - g(x+h)] - [f(x) - g(x)]}{h} = \lim_{h\to 0}\frac{f(x+h) - f(x)}{h} - \frac{g(x+h) - g(x)}{h}$

$$= \lim_{h\to 0}\frac{f(x+h) - f(x)}{h} - \lim_{h\to 0}\frac{g(x+h) - g(x)}{h}$$

$$= f'(x) - g'(x)$$

PROBLEM SET 9.6, PAGE 404

1. $f'(x) = 20x^3 - 60x$ **3.** $f'(x) = 2x - 1$ **5.** $f'(x) = 24x^3 - 10x$ **7.** $g'(x) = 60x^5 - 125x^4 + 20x^3$

9. $y' = \dfrac{-3}{(x-3)^2}$ **11.** $y' = \dfrac{-8}{(x-3)^2}$ **13.** $y' = \dfrac{-16x}{(x^2-5)^2}$ **15.** $y' = \dfrac{-19x^2 + 82x - 13}{(8x^2 - 3x + 1)^2}$

17. $f'(x) = 10x^4 + 16x^3 - 3x^2 - 16x$ **19.** $f'(x) = 20x^4 - 44x^3 - 9x^2 + 104x + 13$

21. $f'(x) = 42x^6 - 20x^4 + 100x^3 - 81x^2 - 20x + 43$ **23.** $f'(x) = -\frac{25}{3}x^{-4/3} + 25x^{2/3}$

25. $f'(x) = 30x^{3/2} - 15x^{-1/2} = \dfrac{15\sqrt{x}(2x^2 - 1)}{x}$ **27.** $g'(x) = \dfrac{-10x^2 + 6}{(5x^2 - 11x + 3)^2}$ **29.** $f'(1) = \dfrac{5(1)^2 - 50(1) - 25}{(1-5)^2} = -\dfrac{35}{8}$

31. $f'(4) = \dfrac{5}{2}\cdot 4^{3/2} + \left(\dfrac{3}{2}\cdot 4^{-1/2}\right) = \dfrac{83}{4}$ **33.** $f'(0) = -\dfrac{3}{2}$ **35.** $x + y = 0$

37. $y - 38 = \dfrac{83}{4}(x - 4)$ or $83x - 4y - 180 = 0$ **39.** $x + 9y + 2 = 0$ **41.** $D'(x) = \dfrac{-200,000x - 1,500,000}{(x^2 + 15x + 25)^2}$

43. $f'(t) = \dfrac{0.01}{(0.01 + 0.005t)^2}$

PROBLEM SET 9.7, PAGES 407–408

1. $9(3x+2)^2$ **3.** $20(5x-1)^3$ **5.** $3(2x^2 + x)^2(4x + 1)$ or $3x^2(4x+1)(2x+1)^2$ **7.** $2(2x^2 - 3x + 2)(4x - 3)$

9. $4(x^3 + 5x)^3(3x^2 + 5)$ or $4x^3(3x^2 + 5)(x^2 + 5)^3$ **11.** $-2(2x^2 - 5x)^{-3}(4x - 5)$ or $-2x^{-3}(2x-5)^{-3}(4x-5)$

13. $-x^{-4}(4x + 9)(x + 3)^{-2}$ **15.** $90x^5(2x + 1)(4x + 3)^2$ **17.** $\dfrac{1}{4}(2x - 3)(x^2 - 3x)^{-3/4} = \dfrac{2x - 3}{4(x^2 - 3x)^{3/4}}$

19. $x(x^2 + 16)^{-1/2} = \dfrac{x\sqrt{x^2 + 16}}{x^2 + 16}$ **21.** $\dfrac{15}{2}x^2(x^3 + 8)^{-1/2}$ **23.** $\dfrac{1}{2}(9x + 2)(3x + 1)^{-1/2}$ **25.** $(2x + 1)(3x + 2)^2(30x + 17)$

27. $2(5x + 1)(4x + 3)^{-2}(10x + 13)$ **29.** $\dfrac{(2x - 5)(10x + 37)}{(5x - 3)^2}$ **31.** $\dfrac{4(x + 5)^3(x - 10)}{(2x - 5)^3}$ **33.** $3(x^2 + 1)^{-3/2}$

35. The rate of decrease is 5,400,000,000,000 bacteria per minute.   **37.** Answers vary

PROBLEM SET 9.8, PAGES 408–409

1. 4   **3.** 13   **5.** $x = 0$, $x = 3$, and $x = 10$ **7.** -7

9. The derivative of a function f at x is $\displaystyle\lim_{h\to 0}\frac{f(x+h) - f(x)}{h}$ provided this limit exists   **11.** $y' = 0$ **13.** $\dfrac{dy}{dx} = 10x + 2x^{-2}$

15. $\dfrac{dy}{dx} = \dfrac{-60}{(4x - 5)^2}$ **17.** $\dfrac{-(x-5)^2(4x+7)}{(1-2x)^2}$ **19.** The profit is decreasing by $.50 per unit

CHAPTER 10
PROBLEM SET 10.1, PAGE 415

1. $y' = -10x$ **3.** $y' = -\dfrac{y}{x}$ **5.** $y' = \dfrac{2x - 3y}{3x}$ **7.** $y' = -\dfrac{x}{y}$ **9.** $y' = \dfrac{8x^3}{15y^2}$ **11.** $y' = \dfrac{y - 2x}{2y - x}$

13. $y' = \dfrac{3y^2 - 6xy^3 - 5y}{9x^2y^2 - 6xy + 5x}$ **15.** $y' = \dfrac{-3x^2 - 4x - y}{x}$ **17.** $y - 2 = 0$ **19.** $8x + 15y + 23 = 0$

21. $x + 5y + 14 = 0$ **23.** $y - 4 = 0$ or $y + 2 = 0$ **25.** $x - y = 0$ **27.** $y + 1 = 0$ or $y - 3 = 0$ **29.** $p' = \dfrac{1}{2p - 5}$

PROBLEM SET 10.2, PAGES 419–420

1. $dy = 15x^2\,dx$ **3.** $dy = -10x^{-2}\,dx$ **5.** $dy = (300x^2 - 50)\,dx$ **7.** $dy = \frac{3}{2}(x - 1)^{-1/2}\,dx$ **9.** $dy = 5dx$
11. $dy = 2(5x - 3)(10x^2 - 3x - 15)\,dx$ **13.** $dy = -7(x - 2)^{-2}\,dx$ **15.** $dy = (2x - 1)\,dx$

17. $dy = (-5x^{-2} + 2x^{-3} + 3x^{-4})\,dx$ **19.** $dy = \dfrac{8x^2 - 6x + 7}{(x^2 + 3x - 2)^2}\,dx$ **21.** $\Delta y = 1.81; dy = 1.8$

23. $\Delta y = 0.0744457825; dy = 0.075$ **25.** $dy = 0.0002261869$ **27.** $dy = 0.0008$ **29.** $dy = 46.10503548$
31. The sales will increase by 3000 units **33.** The concentration of alcohol will increase by 0.01% **35.** $dA = 1.256637061$
37. The change in revenue is about $600, and the change in profit is $300 **39.** There would be about 14,000 additional votes

PROBLEM SET 10.3, PAGE 425

1. $1500p + n - 42,500 = 0$ **3.** $C(p) = 376,250 - 12,750p$ **5.** $R'(p) = -2100p + 29,750$ **7.** $P'(p) = -2100p + 42,500$
9. $\bar{C}'(x) = -50,000x^{-2}$ **11.** $R'(x) = 140 - 0.056x$
13. $R'(1000) = 84$; at a production level of 1000 items, the revenue is increasing at a rate of $84 per item. $R'(2500) = 0$; at a production level of 2500 items, the revenue does not change per unit change in the number of items
15. $P'(1000) = 34$; at a production level of 1000 items, the profit is increasing at $34 per item. $P'(2500) = -50$; at a production level of 2500 item, the profit is decreasing at $50 per item
17.

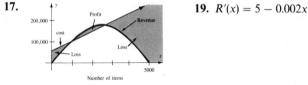

19. $R'(x) = 5 - 0.002x$

PROBLEM SET 10.4, PAGES 430–431

1. $\dfrac{dy}{dt} = -3$ **3.** $\dfrac{dy}{dt} = 1000$ **5.** $\dfrac{dx}{dt} = 15$ **7.** $\dfrac{dy}{dt} = \dfrac{4}{5}$ **9.** $\dfrac{dx}{dt} = \dfrac{10}{9}(\sqrt{10} - 1)$ **11.** $-$$123 per week

13. $\dfrac{27\sqrt{145}}{145} \approx 2.2422$ ft/sec **15.** $\dfrac{500}{400\pi} \approx 0.398$ ft/min. This is about $4\frac{3}{4}$ inches per minute **17.** $20\pi \approx 62.8$ ft^2/sec
19. $.11 per day **21.** 60.3 cm^3/min **23.** 8 ft/sec

PROBLEM SET 10.5, PAGE 431

1. $y' = 20x - 6$ **3.** $y' = \dfrac{4x^3 - 6xy + 9y^2 + 5y}{3x^2 - 18xy - 5x}$ **5.** $\dfrac{dy}{dt} = \pm\dfrac{27}{8}$ **7.** $\dfrac{dx}{dt} = -\dfrac{2}{75}$
9. $R'(x) = 4000 - 200x$

CHAPTER 11
PROBLEM SET 11.1, PAGES 442–443

1. **3.** **5.** **7.**

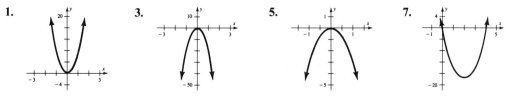

9. **11.** **13.** **15.**

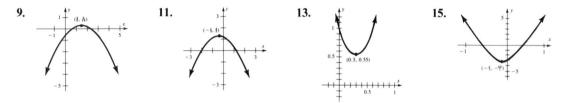

17. Increasing on $(-\infty, 2)$ and decreasing on $(2, \infty)$; horizontal tangent at $(2, 4)$

19. Increasing on $(0, 3)$, decreasing on $(-\infty, 0)$ and $(5, \infty)$ and constant on $(3, 5)$; horizontal tangent at $(0, 3)$ and for all x on $(3, 5)$

21. Increasing on $(-\infty, 2)$ and $(2, \infty)$; no horizontal tangents   **23. a.** $x = 5$

b. Increasing on $(-\infty, 5)$ and decreasing on $(5, \infty)$ **25. a.** $x = 6$ **b.** Increasing on $(6, \infty)$ and decreasing on $(-\infty, 6)$

27. a. $x = -2, x = 3$ **b.** Increasing on $(-\infty, -2)$ and $(3, \infty)$ and decreasing on $(-2, 3)$ **29. a.** $x = -2, x = -\frac{4}{3}$

b. Increasing on $(-\infty, -2)$ and $(-\frac{4}{3}, \infty)$ and decreasing on $(-2, -\frac{4}{3})$

31. Show that the minimum value of $0.25x + 12.000x^{-1}$ is the same as the solution of the equation:

Average cost = marginal cost; that is, the solution of $0.25x + 12,100x^{-1} = 0.5x$

33. Sales are increasing when the amount spent on advertising is less than \$33,333

PROBLEM SET 11.2, PAGE 450

1. $f'''(x) = 120x^2 - 72x + 6$ **3.** $\dfrac{d^4y}{dx^4} = \dfrac{-15\sqrt{5x}}{16x^4}$ **5.** $g^{(4)}(x) = -48x^{-5}$ **7.** $\dfrac{d^3y}{dx^3} = \dfrac{-10\sqrt[3]{x^2}}{9x^2}$ **9.** $\dfrac{d^3y}{dx^3} = -18(x-1)^{-4}$

11. $\dfrac{d^2y}{dx^2} = \dfrac{30}{(x+4)^3}$ **13.** Relative minimum at $(-\frac{1}{2}, 0)$ **15.** Relative maximum at $(-3, -6)$; relative minimum at $(3, 6)$

17. Relative minimum at $(0, 0)$ **19.** Relative maximum at $(-2, 16)$ and $(2, 16)$; relative minimum at $(0, 0)$ **21.** Relative

minimum at $(\frac{2}{3}, \frac{194}{27})$; relative maximum at $(-4, 58)$ **23.** Relative maximum at $(-\frac{1}{2}, \frac{7}{4})$; relative minimum at $(5, -331)$

25. Relative maximum at $(-1, -2)$; relative minimum at $(1, 2)$ **27.** Relative maximum at $(0, 0)$; relative minimum at $(2, 4)$

29. Relative minimum at $(-1, 0)$ **31. a.** Critical value at $x = 6$ **b.** Increasing on $(-\infty, 6)$; decreasing on $(6, \infty)$

c. Concave down throughout   **33. a.** Critical values at $x = -\dfrac{1}{3}$ and $x = 5$ **b.** Increasing on $(-\infty, -\frac{1}{3})$ and $(5, \infty)$;

decreasing on $(-\frac{1}{3}, 5)$ **c.** Concave down on $(-\infty, \frac{7}{3})$; concave up on $(\frac{7}{3}, \infty)$ **35. a.** Critical values at $x = \frac{5}{3}$ and $x = -9$

b. Increasing on $(-\infty, -9)$ and $(\frac{5}{3}, \infty)$; decreasing on $(-9, \frac{5}{3})$ **c.** Concave down on $(-\infty, -\frac{11}{3})$; concave up on $(-\frac{11}{3}, \infty)$

37. a. Critical values at $x = \frac{1}{2}$ and $x = -\frac{2}{9}$ **b.** Increasing on $(-\infty, -\frac{2}{9})$ and $(\frac{1}{2}, \infty)$; decreasing on $(-\frac{2}{9}, \frac{1}{2})$

c. Concave down on $(-\infty, \frac{5}{36})$; concave up on $(\frac{5}{36}, \infty)$ **39. a.** $R(x) = xP(x) = 5x - \dfrac{x^3}{100^2}$ **b.** $R'(x) = 5 - \dfrac{3x^2}{100^2}$

c. Look at the derivative of marginal revenue to see if it is increasing or decreasing: $R''(x) = -\dfrac{6x}{100^2}$. $R''(100) < 0$ falling

so the marginal revenue is decreasing when $x = 100$

PROBLEM SET 11.3, PAGES 453–454

1. Answers vary   **3.** Answers vary   **5.** Answers vary

7.   **9.**   **11.**   **13.**

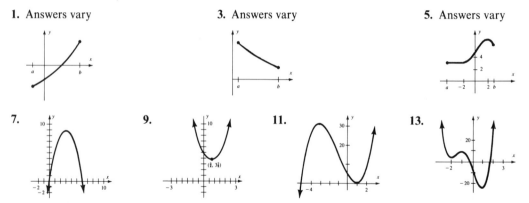

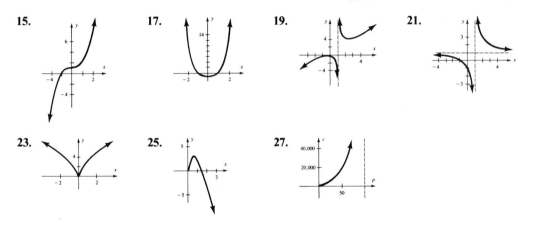

15.   **17.**   **19.**   **21.**

23.   **25.**   **27.**

PROBLEM SET 11.4, PAGES 460–461

1. Maximum = 966 occurs at $x = -3$; minimum = -3050 occurs at $x = -5$
3. Maximum = 42 occurs at $x = 0$ and $x = 5$; minimum = 6 occurs at $x = 3$
5. Maximum = 34 occurs at $x = 3$; minimum = -200 occurs at $x = 0$
7. Maximum = 3 occurs at $x = 8$; minimum = 0 occurs at $x = -1$
9. Maximum = 6 occurs at $x = 4$; minimum = 0 occurs at $x = 0$
11. Maximum = 72 occurs at $x = 5$; minimum = $-\dfrac{256}{27}$ occurs at $x = \dfrac{5}{3}$ **13.** No maximum or minimum exist on $[0, 5]$
15. Maximum = 3 occurs at $x = 3$ or -3; minimum = 0 occurs at $x = 0$
17. Maximum = $\frac{1}{8}$ occurs when $x = 3$; minimum = -1 occurs when $x = 0$
19. At a price of \$250, a maximum profit of \$26,250 will result   **21.** 500 items should be produced
23. 60 items should be produced
25. $P(x) = 60x^{-1} + 240(20 - x)^{-1}$; domain is $[1, 19]$. Minimum pollution is $6\frac{2}{3}$ miles from plant P_1
27. The profit = $P(x) = -5x^2 + 1000x + 25{,}000$, where x is the number of new sign-ups. Profit is maximized when $x = 50$, or when the price is \$1,750
29. 62 grapevines should be planted (remember, $\frac{1}{2}$ grapevine cannot be planted)
31. Width = 2.89, length = 7.89, height = 1.06 (inches); for all practical purposes this would be a box $3'' \times 8'' \times 1''$ with a volume of 24 cu. in.

PROBLEM SET 11.5, PAGE 462

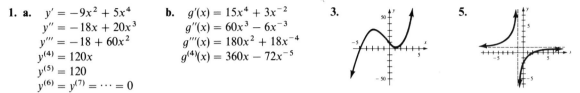

1. a. $y' = -9x^2 + 5x^4$ **b.** $g'(x) = 15x^4 + 3x^{-2}$ **3.**   **5.**
$\qquad$ $y'' = -18x + 20x^3$ $\qquad$ $g''(x) = 60x^3 - 6x^{-3}$
$\qquad$ $y''' = -18 + 60x^2$ $\qquad$ $g'''(x) = 180x^2 + 18x^{-4}$
$\qquad$ $y^{(4)} = 120x$ $\qquad$ $g^{(4)}(x) = 360x - 72x^{-5}$
$\qquad$ $y^{(5)} = 120$
$\qquad$ $y^{(6)} = y^{(7)} = \cdots = 0$
7. The relative maximum is at $x = -2\sqrt{2}$ and the relative minimum is at $x = 2\sqrt{2}$ **9.** Marginal revenue is decreasing when $x = 100$

CUMULATIVE REVIEW FOR CHAPTERS 9–11, PAGES 462–463

1. 5   **3.** 13   **5.** 15   **7.** $x = 8$ **9.** 1   **11.** $y' = -24x - 15x^2$ **13.** $y' = -25(2 - 5x)^4$ **15.** $y' = \dfrac{-5y}{4x}$

17. $16x + y - 12 = 0$ **19.** $y' = 8x^3 - 3x^2 + 6x$; $y'' = 24x^2 - 6x + 6$; $y''' = 48x - 6$; $y^{(4)} = 48$; $y^{(5)} = \cdots = 0$
21. Maximum = $\frac{1}{4}$ occurs at $x = \frac{1}{4}$; minimum = -6 occurs at $x = 9$ **23.** Decreasing
25. Minimum = 0 occurs when $p = 0$

CHAPTER 12
PROBLEM SET 12.1, PAGE 471

1. 5 **3.** Not a real number **5.** -3 **7.** 7 **9.** $\frac{1}{10}$ **11.** $\frac{1}{1000}$ **13.** $2x$ **15.** $1 + x$ **17.** $x + 2x^{1/2}y^{1/2} + y$
19. **21.** **23.** **25.** 20.0855 **27.** 1.0513 **29.** 1.046

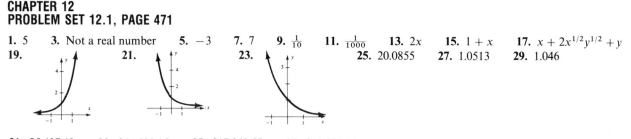

31. \$5,427.43 **33.** \$54,598.15 **35.** \$17,369.57 **37.** \$10,081.44

PROBLEM SET 12.2, PAGE 476

1. $\log_2 64 = 6$ **3.** $\log_{1/3} 9 = -2$ **5.** $\log_b a = c$ **7.** $2 = 4^{1/2}$ **9.** $0.01 = 10^{-2}$ **11.** $e^2 = e^2$ **13.** 2 **15.** $\frac{1}{2}$
17. 3 **19.** 0 **21.** 0.0334 **23.** 3.989 **25.** -0.4935 **27.** 0.8198 **29.** 0.6931
31. **33.** **35.** **37. a.** 3572 **b.** 4746 **39. a.** 3.9 **b.** 8.8

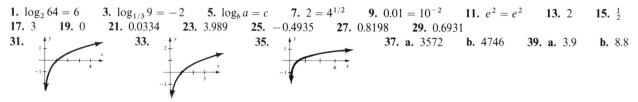

41. $b^0 = 1 \Rightarrow \log_b 1 = 0$ **43.** If $y = e^x$, then $\ln y = x$; $\ln e^x = \ln y = x$

PROBLEM SET 12.3, PAGES 481–482

1. $x = 2$ **3.** $x = -1$ **5.** $x = \sqrt{28} = 2\sqrt{7}$ **7.** $x = e^3 \approx 20.0855$ **9.** $x = 9.3$ **11.** $x = \pm\sqrt{12} = \pm 2\sqrt{3}$ **13.** $x = \frac{7}{2}$
15. There is no solution **17.** $x = 7$ **19.** $x = \frac{2}{3}$ **21.** $x = \frac{1}{4}$ **23.** $x = 15$ **25.** $x = 2$ **27.** $x = 1.8559036$
29. $x = \log_4 0.82 = -0.14315209$ **31.** 6 years **33.** 2108 days (5 years 283 days) **35.** 3342 days (9 years 57 days)
37. 2.02% **39. a.** 50% **b.** 13 seconds **41.** \$263.34 **43.** $t = 5700 \log_{1/2} P$ **45.** About 15.3 years
47. Let $A = b^x$ and $B = b^y$. Then $\log_b A = x$ and $\log_b B = y$. Hence $\log_b \dfrac{A}{B} = \log_b \dfrac{b^x}{b^y} = \log_b b^{(x-y)} = x - y = \log_b A - \log_b B$

PROBLEM SET 12.4, PAGES 486–487

1. $y' = 2e^{2x}$ **3.** $y' = -e^{-x}$ **5.** $f'(x) = xe^x + e^x$ **7.** $f'(x) = 2xe^{5x^2}(10x^3 + 3x + 5)$ **9.** $y' = -2e^{5x}(15x^2 + 6x + 25)$
11. $y' = \dfrac{-1}{5-x} = \dfrac{1}{x-5}$ **13.** $y' = 4(2x^3 + e^{5x})^3(6x^2 + 5e^{5x})$ **15.** $y' = \dfrac{4}{x}$ **17.** $y' = \dfrac{1}{2x}$ **19.** $y' = \dfrac{3(1 - x\ln x)}{xe^x}$
21. $f'(x) = \dfrac{2}{x(x+2)}$ **23.** $f'(t) = e^{t^2-t}(2te^t - 2t + 1)$ **25.** $y' = 2(e^x - e^{-x})(e^x + e^{-x}) = 2(e^{2x} - e^{-2x})$
27. $y' = \dfrac{5x + 2 - 5x\ln|x|}{x(5x+2)^2}$ **29.** $f'(x) = \dfrac{e^x(x\ln|3x| - 1)}{x\ln^2|3x|}$ **31.** $y' = \dfrac{-3600e^{-3x}}{(1 + 6e^{-3x})^2}$ **33.** $y' = \dfrac{100(e^{0.2x} + 40 + 8x\ln|x|)}{xe^{0.2x}(1 + 40e^{-0.2x})^2}$
35. $\left(\dfrac{500\ln(x+10)}{x}\right)' = \dfrac{500x - 500(x+10)\ln(x+10)}{x^2(x+10)}$ **37. a.** 1 **b.** 0.099978 (people per day)
39. $A'(1) \approx 0.1353P$ dollars per year, or about 13.5% of P; $A'(5) \approx 0.21865P$ dollars per year, or about 21.9% of P
41. The amount is changing by $A'(t) = -0.15e^{-0.03t}$ **43.** The marginal revenue is $500e^{-0.1x} - 50xe^{-0.1x}$
45. Answers vary **47.** Answers vary

PROBLEM SET 12.5, PAGE 488

1. $x = 2$ **3.** $x = 8$ **5.** $x \approx 3.4$ **7.** $x \approx -0.35$ **9.** $y' = -5.5e^{-0.5x^2}$ **11.** $y' = \dfrac{5}{x}$ **13.** $y' = \dfrac{8 - 5x^3}{x(4 - x^3)}$

15. $y' = \dfrac{8 - 2x + x\ln x^2}{x(4-x)^2}$ **17.** $t = 5700\log_{1/2}\left(\dfrac{A}{10}\right)$ **19.** 7675 units

CHAPTER 13
PROBLEM SET 13.1, PAGES 496–497

1. $\dfrac{x^8}{8} + C$ **3.** $x^4 + C$ **5.** $3x + C$ **7.** $\dfrac{5x^2}{2} + 7x + C$ **9.** $x + C$ **11.** $6x^3 - 3x^2 + 5x + C$ **13.** $\frac{1}{9}x + C$

15. $\dfrac{-5}{x} + C$ **17.** $x - e^x + C$ **19.** $\frac{3}{4}x^{4/3} + \sqrt{2}x + C$ **21.** $2x^{3/2} - 2x^{1/2} + C$ **23.** $\frac{3}{5}x^{5/3} + C$

25. $\dfrac{x^2}{2} + x + \ln|x| + C$ **27.** $\frac{2}{9}y^{9/2} + C$ **29.** $z + \frac{2}{3}z^3 + \frac{1}{2}z^5 + C$ **31.** $u^3 - \ln|u| + e^u + C$

33. $F(x) = x^3 + 2x^2 + x + 10$ **35.** $F(x) = 5x + \ln|x| + 4995$ **37.** $C(x) = 2x^2 - 8x + 5000$
39. $C(x) = 0.003x^3 + 18,500$ **41.** 426,722 people   **43.** $P(x) = 100x^2 - 10,000x - 50,000$ **45.** 10,931 people
47. $v(t) = -32t - 72$ **49.** $f(x) = x^3 + 5x - 4$

PROBLEM SET 13.2, PAGES 502–503

1. $\dfrac{(5x+3)^4}{20} + C$ **3.** $\dfrac{5}{3(5-x)^3} + C$ **5.** $\dfrac{2(3x+5)^{3/2}}{9} + C$ **7.** $\dfrac{(3x^2+1)^2}{2} + C$ **9.** $2\sqrt{x^2+5x} + C$

11. $\frac{1}{5}\ln|6+5x| + C$ **13.** $-\frac{1}{6}\ln|1-3x^2| + C$ **15.** $\frac{1}{5}e^{5x} + C$ **17.** $\frac{5}{12}e^{4x^3} + C$ **19.** $\dfrac{-1}{4(4x^2-4x)} + C$

21. $\ln|x^4 - 2x^2 + 3| + C$ **23.** $\frac{1}{2}\ln^2|x| + C$ **25.** $\dfrac{x^2}{2} + C$ **27.** $\frac{3}{8}\ln(4x^2+1) + \frac{1}{3}e^{x^3} + C$ **29.** $\frac{3}{10}(x^2+1)^{5/3} + C$

31. $\frac{2}{3}(x+1)^{3/2} - 2\sqrt{x+1} + C$ **33.** $P(t) = \frac{20}{3}(t^2-5)^3 + \frac{2530}{3}$ **35.** $P(100) = 150(100^2-36)^{1/3} - 600 = \$2,627.77$
37. $S(3) = -10000e^{-0.2(3)} + 25,000 \approx \$19,511.88$ **39.** $P(20) \approx 1,120,026$ people   **41.** 714 million barrels

PROBLEM SET 13.3, PAGES 512–513

1. $\frac{19}{3}$ **3.** $\frac{14}{3}$ **5.** 12   **7.** $\ln 6$ **9.** $\frac{35}{3}$ **11.** -1 **13.** $\frac{17}{4} - \ln 4$ **15.** 0   **17.** 0   **19.** $\frac{5}{33}$
21. Area $= 12$ **23.** Area $= 24$ **25.** Area $= 20$ **27.** Area $= 22\frac{1}{2}$ **29.** Area $= 6$ **31.** Area $= \frac{2}{21}$

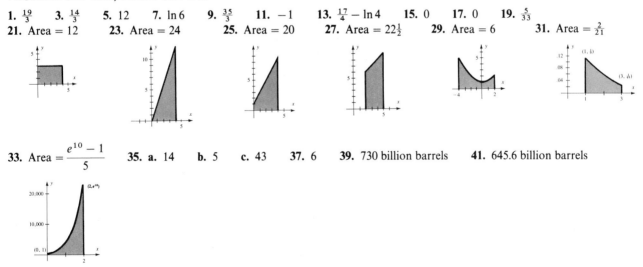

33. Area $= \dfrac{e^{10}-1}{5}$ **35. a.** 14   **b.** 5   **c.** 43   **37.** 6   **39.** 730 billion barrels   **41.** 645.6 billion barrels

43. About 20–25 million automobiles   **45.** 1,101,273.29   **47.** $G = \frac{1}{2}$ **49.** Area $= 36$

PROBLEM SET 13.4, PAGES 521–522

1. $\displaystyle\int_a^b [f(x) - g(x)]\,dx$ **3.** $\displaystyle\int_a^b [g(x) - f(x)]\,dx$ **5.** $\displaystyle\int_a^b [f(x) - g(x)]\,dx$ **7.** $\displaystyle\int_a^c [g(x) - f(x)]\,dx + \int_c^b [f(x) - g(x)]\,dx$
9. $20\frac{5}{6}$ **11.** 48   **13.** 2   **15.** $10\frac{2}{3}$ **17.** $\frac{1}{12}$ **19.** $20\frac{5}{6}$ **21.** $\frac{1}{3}$ **23.** $21\frac{1}{3}$ **25.** $\frac{1}{2}$ **27.** \$108,747
29. \$1516.67   **31.** Consumer's surplus $= \$16.88$; producer's surplus $= \$2$ **33.** 184,362

PROBLEM SET 13.5, PAGE 523

1. $\dfrac{x^6}{6} + C$ **3.** $e^u + C$ **5.** $6x^3 - 3x^2 - 3x + C$ **7.** $\frac{1}{3}e^{3x+1} + e^{-1}x + C$ **9.** $\frac{4}{3}$ **11.** $\frac{3}{16}(17^4 - 16) \approx 15{,}657$

13. $\frac{74}{9}$ **15.** $\frac{1}{2}$ **17.** Yes   **19.** About 14–15 years

CHAPTER 14
PROBLEM SET 14.1, PAGES 529–530

1. $4xe^{3x} - \frac{4}{3}e^{3x} + C$ **3.** $1 - 2e^{-1}$ **5.** $-\frac{2}{3}x(1-x)^{3/2} - \frac{4}{15}(1-x)^{5/2} + C$ **7.** $\dfrac{x}{4}(x+2)^4 - \dfrac{(x+2)^5}{20} + C$

9. $(x+1)\ln(x+1) - x + C$ **11.** $-2x(1-x)^{1/2} - \frac{4}{3}(1-x)^{3/2} + C$ **13.** $3x^2e^{2x} - 3xe^{2x} + \frac{3}{2}e^{2x} + C$

15. $\dfrac{-x^2(1-2x)^{3/2}}{3} - \dfrac{2x(1-2x)^{5/2}}{15} - \dfrac{2(1-2x)^{7/2}}{105} + C$ **17.** $\dfrac{-3x^2(1-x^2)^{4/3}}{8} + \dfrac{9(1-x^2)^{7/3}}{56} + C$ **19.** $\dfrac{x^3}{3}\ln x - \dfrac{x^3}{9} + C$

21. $\frac{3}{2}x^2e^{4x} - \dfrac{3xe^{4x}}{4} + \dfrac{3}{16}e^{4x} + C$ **23.** $-(1-x^2)^{1/2} + \frac{1}{3}(1-x^2)^{3/2} + C$ **25.** $\ln|\ln x| + C$ **27.** $e - 2$

29. Answers vary   **31.** Both answers should give $\dfrac{x^2}{2} + \dfrac{1}{2}\ln|x^2 - 1| + C$ **33.** 14.1 million bacteria in 7 days

35. $32,559.23

PROBLEM SET 14.2, PAGES 534–535

1. Formula 9; $\dfrac{1}{b}\ln|1 + bx| + C$ **3.** Formula 23; $\sqrt{x^2 + a^2} + C$ **5.** Formula 26; $\dfrac{\sqrt{x^2 - a^2}}{a^2x} + C$

7. Formula 30; $\dfrac{x^2}{2}\ln x - \dfrac{x^2}{4} + C$ **9.** Formula 32; $\frac{1}{2}\ln^2 x + C$ **11.** Formula 35; $\dfrac{e^{ax}}{a^2}(ax - 1) + C$

13. Formula 29; $\dfrac{x}{2}\sqrt{x^2 + 1} - \dfrac{1}{2}\ln|x + \sqrt{x^2 + 1}| + C$ **15.** Formula 26; $-\dfrac{\sqrt{x^2 + 16}}{16x} + C$

17. Formula 2; let $u = 4x^2 + 1$; $\dfrac{1}{4}\sqrt{4x^2 + 1} + C$ **19.** Formula 20; $-\ln\left|\dfrac{1 + \sqrt{1 - 9x^2}}{3x}\right| + C$

21. Formula 23; $\sqrt{x^2 + 4} + C$ **23.** Formula 9; let $u = 1 + x$; $\dfrac{(1+x)^4}{4} + C$

25. Formula 10; $\dfrac{(1+x)^5}{5} - \dfrac{(1+x)^4}{4} + C$ **27.** Formula 13; $\dfrac{-2(2 - 3x)\sqrt{(1+x)^3}}{15} + C$

29. Formula 35; $\dfrac{e^{4x}}{16}(4x - 1) + C$ **31.** Formula 30; $\dfrac{x^2}{2}\ln 2x - \dfrac{x^2}{4} + C$ **33.** Formula 31; $\dfrac{x^3}{3}\ln 5x - \dfrac{x^3}{9} + C$

35. Formula 39; $\dfrac{x}{3} - \dfrac{1}{3}\ln|3 + 5e^x| + C$ **37.** Formula 39; $x - \dfrac{1}{2}\ln|1 + e^{2x}| + C$

39. Let $u = 1 + x$; Formula 2; $2\sqrt{1 + x} + C$ **41.** Formula 11; $\dfrac{(1+x)^6}{6} - \dfrac{2(1+x)^5}{5} + \dfrac{(1+x)^4}{4} + C$

43. Formula 11; $\dfrac{-5(1-x)^6}{6} + 2(1-x)^5 - \dfrac{5(1-x)^4}{4} + C$

45. Formula 13; $\dfrac{-2(1 + 3x)\sqrt{(1 - 2x)^3}}{15} + C$ **47.** Formula 11; $\dfrac{1}{27}\left[\dfrac{(2 + 3x)^6}{6} - \dfrac{4(2 + 3x)^5}{5} + (2 + 3x)^4\right] + C$

49. Formula 2; let $u = 4x^3 + 1$; $\dfrac{\sqrt{(4x^3 + 1)^3}}{18} + C$ **51.** Formulas 35 and 36; $\dfrac{e^{3x}}{27}(9x^2 - 6x + 2) + C$

53. Formula 14; $\dfrac{-\sqrt{(2 + 9x)^3}}{2x} + \dfrac{9}{4}\left(2\sqrt{2 + 9x} + \dfrac{1}{\sqrt{2}}\ln\left|\dfrac{\sqrt{2 + 9x} - \sqrt{2}}{\sqrt{2 + 9x} + \sqrt{2}}\right|\right) + C$

55. Formula 21; $\sqrt{9x^2 + 2} - \sqrt{2}\ln\left|\dfrac{\sqrt{2} + \sqrt{9x^2 + 2}}{3x}\right| + C$ **57.** Formula 16; $\frac{1}{3}\ln|3x + \sqrt{9x^2 + 2}| + C$

59. Formula 2; let $u = 9 - 16x^2$; $-\dfrac{5}{48}(9 - 16x^2)^{3/2} + C$ **61.** Formula 28; $-\dfrac{\sqrt{x^2 - 1}}{x} + \ln|x + \sqrt{x^2 - 1}| + C$

63. Formula 11; $\dfrac{(3 + 10x)^{7/2}}{3500} - \dfrac{3(3 + 10x)^{5/2}}{1250} + \dfrac{3(3 + 10x)^{3/2}}{500} + C$ **65.** About 1120 tons of pollutants   **67.** About \$1,480,000

PROBLEM SET 14.3, PAGES 542–543

1. 10   **3.** 11.81415   **5.** 12.5   **7.** 12.64748   **9.** 12.66503   **11.** 12.66664   **13.** 0.16   **15.** 0.13546
17. .12938   **19.** 0.12526   **21.** 0.12476   **23.** 0.12473   **25.** 0.14   **27.** 8.36   **29.** 0.11   **31.** 14.86   **33.** 0.10
35. 13.81   **37.** 100 cubic yards

PROBLEM SET 14.4, PAGES 544–545

1. $\dfrac{(2x - 1)^4}{8} + C$ **3.** $\dfrac{x^2}{8}(2x - 1)^4 - \dfrac{x}{40}(2x - 1)^5 + \dfrac{1}{480}(2x - 1)^6 + C$

5. $-\dfrac{2}{15}x^2(1 - 5x)^{3/2} - \dfrac{8}{375}x(1 - 5x)^{5/2} - \dfrac{16}{13,125}(1 - 5x)^{7/2} + C$ **7.** $x\ln^2 x - 2x\ln x + 2x + C$ (Formula 33)
9. $\ln|4x^2 + 5x - 3| + C$ **11.** Formula 14   **13.** Formula 37   **15.** 6.74   **17.** $50(1 - e^{-0.1t})$
19. Answers vary; about 72 units

CHAPTER 15
PROBLEM SET 15.1, PAGES 553–554

1. $512\tfrac{2}{3}$ **3.** $2 - 2\ln 3 \approx -0.197$ **5.** 13   **7.** \$61,607.44   **9.** \$2,563.55   **11.** \$38,913.42   **13.** -8

15. $1 - \tfrac{1}{4}\ln 3 \approx 0.7253$ **17.** $\dfrac{1 - e^{-15}}{15} \approx 0.0667$ **19.** 116   **21.** 396   **23.** \$4,028.34   **25.** $2\sqrt{10} \approx 6.32$

27. \$1,507.14; \$31.40 per month   **29.** $1\tfrac{1}{4}$ years   **31.** $63°$ **33.** $12\ln 5 \approx \$19.31$ **35.** 500   **37.** \$11,400
39. In about 1.31 years   **41.** \$7,453,560

PROBLEM SET 15.2, PAGES 566–567

1. $\{0, 1, 2, 3, \ldots, 9, 10\}$; $x =$ number of people who responded yes; discrete
3. $\{0, 1, 2, 3, \ldots, 499,500\}$; $x =$ number of words with an error; discrete
5. $\{bbb, bbg, bgb, bgg, gbb, gbg, ggb, ggg\}$; x could be the number of boys or the number of girls; discrete

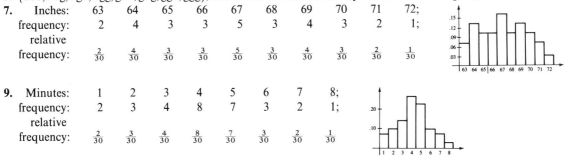

7.

Inches:	63	64	65	66	67	68	69	70	71	72;
frequency:	2	4	3	3	5	3	4	3	2	1;
relative frequency:	$\frac{2}{30}$	$\frac{4}{30}$	$\frac{3}{30}$	$\frac{3}{30}$	$\frac{5}{30}$	$\frac{3}{30}$	$\frac{4}{30}$	$\frac{3}{30}$	$\frac{2}{30}$	$\frac{1}{30}$

9.

Minutes:	1	2	3	4	5	6	7	8;
frequency:	2	3	4	8	7	3	2	1;
relative frequency:	$\frac{2}{30}$	$\frac{3}{30}$	$\frac{4}{30}$	$\frac{8}{30}$	$\frac{7}{30}$	$\frac{3}{30}$	$\frac{2}{30}$	$\frac{1}{30}$

11. Not a probability density function   **13.** Probability density function   **15.** Not a probability density function
17. Not a probability density function   **19.** No value exists   **21.** $\frac{3}{125}$ **23.** $\frac{2}{3}$ **25.** $\frac{3}{14}$ **27.** $\frac{4}{3}$ **29.** .8907
31. .0455   **33.** .3202   **35.** .40   **37.** .075   **39.** .1587   **41.** .0548   **43.** .5403

PROBLEM SET 15.3, PAGES 569–570

1. 1   **3.** 2   **5.** $\frac{1}{2}$ **7.** Diverges   **9.** 10   **11.** Diverges   **13.** Diverges   **15.** 0   **17.** 1   **19.** \$336,666.67
21. $\displaystyle\int_0^\infty Re^{-rt}\,dt = R\left[\dfrac{e^{-rt}}{-r}\right]_0^\infty = R\left[0 - \dfrac{e^0}{-r}\right] = \dfrac{R}{r}$ **23.** 4000 millirems   **25.** $\frac{1}{4}$ **27.** $\dfrac{1}{\sqrt{e}} \approx .6065$ **29.** $\dfrac{1}{e^{0.6}} \approx .5488$

PROBLEM SET 15.4, PAGES 573–574

1. $y = \dfrac{x^3}{3} + C$ **3.** $y = 4x^2 - 10x + C$ **5.** $y = x^4 - x^3 - 5x + C$ **7.** $y = \dfrac{e^x}{5} + C$ **9.** $y = \frac{1}{90}(5x + 1)^{3/2} + C$

11. $y^2 = x^2 + C$ **13.** $y^3 = \frac{3}{4}x^4 - 9x + C$ **15.** $|y| = Me^{x^2}$ **17.** $|P| = Me^{.02t}$ **19.** $|N| = Me^{.001t}$

21. $y^3 = x^3 + 125$ **23.** $\ln|P| = 0.02t + 3$ **25.** $\ln|y| = 2x^{1/2} - 2$ **27.** $\ln|y| = \frac{1}{2}\ln(1 + x^2) + \frac{1}{2}\ln 2$ **29.** 4725

31. about 44 hours

PROBLEM SET 15.5, PAGE 575

1. a. 49   **b.** 9   **3.** $\frac{61}{9}$ **5.** Answers vary   **7. a.** 0.5   **b.** 0.4602   **c.** 0.1191 ‡ **9.** $e^{-.5}$ **11.** $\frac{5}{4}x^2 + \frac{3}{2}x + C$

13. 569   **15.** $840,000

CUMULATIVE REVIEW FOR CHAPTERS 12–15, PAGES 575–577

1.   **3.** 605   **5.** $x = -\log 0.5 \approx 0.301$ **7. a.** 0   **b.** $b - a$ **c.** $-\displaystyle\int_b^a f(x)\,dx$

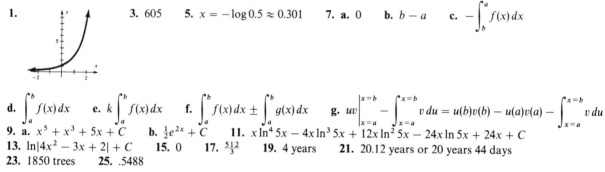

d. $\displaystyle\int_a^b f(x)\,dx$ **e.** $k\displaystyle\int_a^b f(x)\,dx$ **f.** $\displaystyle\int_a^b f(x)\,dx \pm \int_a^b g(x)\,dx$ **g.** $uv\Big|_{x=a}^{x=b} - \displaystyle\int_{x=a}^{x=b} v\,du = u(b)v(b) - u(a)v(a) - \int_{x=a}^{x=b} v\,du$

9. a. $x^5 + x^3 + 5x + C$ **b.** $\frac{1}{2}e^{2x} + C$ **11.** $x\ln^4 5x - 4x\ln^3 5x + 12x\ln^2 5x - 24x\ln 5x + 24x + C$

13. $\ln|4x^2 - 3x + 2| + C$ **15.** 0   **17.** $\frac{512}{3}$ **19.** 4 years   **21.** 20.12 years or 20 years 44 days

23. 1850 trees   **25.** .5488

CUMULATIVE REVIEW FOR CHAPTERS 9–15, PAGES 577–578

1. Answers vary   **3.** Answers vary   **5. a.** $x = 1 + \log 25 \approx 2.3979$ **b.** $1 + \ln 11 \approx 3.3979$ **c.** $1 + \log_8 5 \approx 1.774$

7. $y' = \dfrac{-6}{(1 - 2x)^2}$ **9.** $y' = x^3 - x^2 + C$ **11.** $5\ln 3 \approx 5.493$ **13. a.** 64 units   **b.** $8,320

15. No maximum, profit increases as x increases   **17.** $\frac{19}{3}$

19. Relative minimums at $A(-\sqrt{3}, -2.25)$ and $B(\sqrt{3}, -2.25)$; relative maximum at $C(0,0)$; points of inflection at $D(-1, -1.25)$ and $E(1, -1.25)$

CHAPTER 16
PROBLEM SET 16.1, PAGES 586–587

1. The area of rectangle with dimensions 15 by 35 is $K(15, 35) = 525$ square units

3. The volume of a box 3 by 5 by 8 is $V(3, 5, 8) = 120$ cubic units

5. The simple interest from a $500 investment at 5% for 3 years is $I(500, 0.05, 3) = 575

7. The future value of a $2,500 investment at 12% compounded quarterly for 6 years is $A(2,500, 0.12, 6, 4) = $5,081.99$

9. The future value of a $110,000 investment at 9% compounded monthly for 30 years is $A(110000, 0.09, 30, 12) = $1,620,363$

11. 1   **13.** 49   **15.** 0   **17.** $\frac{11}{17}$ **19.** 0.2   **21.** 111.89

23.

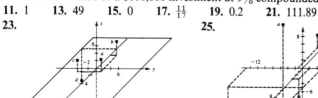

25.

27. Plane with x-, y-, and z-intercepts of $(10, 0, 0)$, $(0, 5, 0)$, and $(0, 0, 2)$, respectively. The portion of the plane in the first octant is pictured

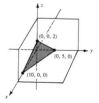

29. Plane with x-, y-, and z-intercepts of $(4, 0, 0)$, $(0, -6, 0)$, and $(0, 0, -12)$, respectively

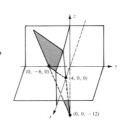

31. Elliptic paraboloid. Looking down the z-axis, the pictured traces are concentric ellipses

33. Ellipsoid with x-, y-, and z-intercepts of $(\pm 3, 0, 0)$, $(0, \pm 2, 0)$, and $(0, 0, \pm 5)$, respectively

35. Circular paraboloid. Looking down the z-axis, the pictured traces are concentric circles

37. Hyperboloid of two sheets with z-intercepts $(0, 0, \pm 2)$. Looking down the z-axis, the pictured traces are concentric ellipses

39. Circular cylinder with x- and y-intercepts of $(\pm 6, 0, 0)$ and $(0, \pm 6, 0)$, respectively

41. Circular cylinder with y- and z-intercepts of $(0, \pm 2\sqrt{5}, 0)$ and $(0, 0, \pm 2\sqrt{5})$, respectively

43. $C(x, y) = 3700 + 800x + 550y$ **a.** \$19,950   **b.** \$21,450
  c. \$21,200   **d.** \$20,200
45. $C(x, y, z) = 2xy + 1.50xz + 1.50yz + 1.25xy$
47. a. $3.875 \times 10^8 \ ml$ **b.** $3.06 \times 10^{10} \ ml$ **c.** $1.92 \times 10^7 \ ml$

PROBLEM SET 16.2, PAGES 591–592

1. $10x - 9x^2y^4$ **3.** $-15(4x - 3y)^4$ **5.** -134 **7.** $-759,375$ **9.** $-36x^2y^3$ **11.** $-240(4x - 3y)^3$ **13.** 10

15. 0 **17.** $-69,120$ **19.** $15,360$ **21.** $3e^{3x+2y}$ **23.** $\dfrac{-3y}{\sqrt{x^2 - 3y^2}}$ **25.** $3e^7$ **27.** Does not exist   **29.** $6e^{3x+2y}$

31. $\dfrac{3xy}{(\sqrt{x^2 - 3y^2})^3}$ **33.** $9e^4$ **35.** $\dfrac{6}{e^3}$ **37.** 0 **39.** 0 **41.** $f_x = 1 + 2y + \lambda y; \ f_y = 2x + \lambda x; \ f_\lambda = xy - 10$

43. $f_x = 2x - 3\lambda; \ f_y = 2y - 2\lambda; \ f_\lambda = -3x - 2y + 6$ **45.** $\dfrac{\partial f}{\partial b} = 100m + 50b + 1; \ \dfrac{\partial f}{\partial m} = 200m + 100b + 2$

47. $\dfrac{\partial f}{\partial b} = 28(m + b + 1); \ \dfrac{\partial f}{\partial m} = 28(m + b + 1)$

49. At a fixed level of advertising, the rate of change of profit per unit change in price is $\dfrac{\partial P}{\partial p} = 2a + 50 - 20p - 0.1a^2$.

At a fixed price, the rate of change of profit per unit change in advertising spent is $\dfrac{\partial P}{\partial a} = 2p - 0.2ap$

51. $P_x(5, 10) = -1000$; the rate of decrease in profit is \$1,000 per unit increase in x, assuming that y stays constant.
$P_y(5, 10) = 350$; the rate of increase in profit is \$350 per unit increase in y, assuming that x stays constant
53. For most people answers will vary between 900 and 1700 square feet   **55.** $500(1 + r)^4$

57. $P_r = -\dfrac{300}{r^2}\left[1 - \left(1 + \dfrac{r}{12}\right)^{-60} - 5r\left(1 + \dfrac{r}{12}\right)^{-61}\right]$

PROBLEM SET 16.3, PAGES 597–598

1. Saddle point at $(0,0,0)$ **3.** Relative minimum at $(-1,2,-3)$ **5.** Saddle point at $(0,0,0)$; relative maximum at $(-1,1,1)$
7. Relative maximum at $(-\frac{9}{11},-\frac{41}{11},\frac{45}{11})$ **9.** Saddle point at $(3,-2,2)$ **11.** No critical values
13. No relative maximums, minimums, or saddle points
15. Relative maximums at $(1,1,2)$ and $(-1,-1,2)$; saddle point at $(0,0,0)$
17. Relative maximum at $(-1,-1,1)$; saddle point at $(0,0,0)$ **19.** $P(6,1) = 141$ (thousand dollars)
21. 50 regular rings and 75 deluxe rings
23. The box should have a square base $2^{7/3}$ by $2^{7/3}$ (about 5.04 by 5.04) feet and a height of $2^{4/3}$ (about 2.52) feet
25. $D = B^2 - AC$; if $A = 0$, then $D = B^2 > 0$

PROBLEM SET 16.4, PAGES 603–604

1. $f(10,10) = 100$ **3.** $f(\frac{72}{11},\frac{96}{11}) = -\frac{3456}{11}$ **5.** $f(\frac{28}{3},\frac{7}{3}) = \frac{5488}{27}$ **7.** $f(12,12) = 288$ **9.** $f(3,3) = 5$ **11.** $f(8,8) = 128$
13. Both numbers are 5   **15.** By applying $\frac{250}{59}$ acre-feet of water and $\frac{75}{59}$ pounds of fertilizer
17. Maximum area is $(\frac{500}{3})^2 \approx 27{,}778$ square feet   **19.** $\sqrt{10} \approx 3.16$ ounces of meat and $5\sqrt{10} \approx 15.81$ ounces of vegetables
21. $f\left(\dfrac{10\sqrt{3}}{3}, \dfrac{-10\sqrt{3}}{3}, \dfrac{10\sqrt{3}}{3}\right) = 10\sqrt{3}$ **23.** $T(\frac{10}{3},\frac{10}{3},\frac{10}{3}) = \frac{200}{3}$

PROBLEM SET 16.5, PAGES 614–615

1. $2x^2 + \frac{8}{3}$ **3.** $\frac{15}{2}(y^2 - 1)$ **5.** $\dfrac{x}{3}(x^2 + 8)^{3/2} - \dfrac{x^4}{3}$ **7.** $3e^{2y+2} - 3e^{2y+1}$ **9.** $\dfrac{e^{3y+25}}{2} - \dfrac{e^{3y+1}}{2}$ **11.** x^2

13. 9   **15.** 10   **17.** $\frac{91}{3}$ **19.** $\frac{16}{3}$ **21.** $\frac{625}{6}$

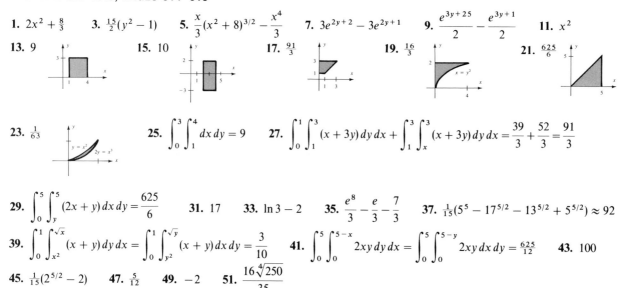

23. $\frac{1}{63}$ **25.** $\displaystyle\int_0^3\int_1^4 dx\,dy = 9$ **27.** $\displaystyle\int_0^1\int_1^3 (x+3y)\,dy\,dx + \int_1^3\int_x^3 (x+3y)\,dy\,dx = \dfrac{39}{3} + \dfrac{52}{3} = \dfrac{91}{3}$

29. $\displaystyle\int_0^5\int_y^5 (2x+y)\,dx\,dy = \dfrac{625}{6}$ **31.** 17   **33.** $\ln 3 - 2$ **35.** $\dfrac{e^8}{3} - \dfrac{e}{3} - \dfrac{7}{3}$ **37.** $\frac{1}{15}(5^5 - 17^{5/2} - 13^{5/2} + 5^{5/2}) \approx 92$
39. $\displaystyle\int_0^1\int_{x^2}^{\sqrt{x}} (x+y)\,dy\,dx = \int_0^1\int_{y^2}^{\sqrt{y}} (x+y)\,dx\,dy = \dfrac{3}{10}$ **41.** $\displaystyle\int_0^5\int_0^{5-x} 2xy\,dy\,dx = \int_0^5\int_0^{5-y} 2xy\,dx\,dy = \frac{625}{12}$ **43.** 100
45. $\frac{1}{15}(2^{5/2} - 2)$ **47.** $\frac{5}{12}$ **49.** -2 **51.** $\dfrac{16\sqrt[4]{250}}{35}$

PROBLEM SET 16.6, PAGE 616

1. 5.49   **3.**   **5.** $10(x - 3y)^9$ **7.** $48x^2 - 18xy^2 + 4$ **9.** 54

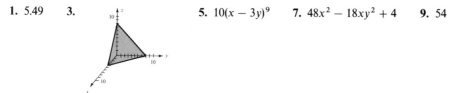

11. Saddle point at $(0,0)$ and no relative maximums or minimums   **13.** $e^2 - 3$
15. $R(x,y) = -0.05x^2 + 50x + 130y - .04y^2 + 0.011xy$; $C = (5,8) = 910$; $R(5,8) = 1286.63$

APPENDIX A, PAGES 626–627

1. a, b, and d are statements   **3.** a, b, and d are statements

5. $e \wedge d \wedge g$; where e = W. C. Fields is eating, d = W. C. Fields is drinking, g = W. C. Fields is having a good time

7. $\sim J \wedge \sim R$; where J = Jack will go tonight, R = Rosamond will go tomorrow

9. $(J \vee I) \wedge \sim P$; where J = The decision will depend on judgment, I = The decision will depend on intuition, P = The decision will depend on who paid the most

11. $p \rightarrow \sim w$; p = We make a proper use of those means which the God of Nature has placed in our power, w = We are weak

13. $w \rightarrow n$; where w = It is work, n = It is noble

15. $(t \vee m) \rightarrow (\sim i \vee w)$; i = You must itemize deductions on Schedule A; w = You must complete the worksheet; t = You have earned an income of $2,300; m = You have earned an income of more than $2,300

17.

p	q	$\sim p$	$\sim q$	$\sim p \wedge \sim q$
T	T	F	F	F
T	F	F	T	F
F	T	T	F	F
F	F	T	T	T

19.

r	s	$\sim r$	$\sim s$	$\sim r \vee \sim s$
T	T	F	F	F
T	F	F	T	T
F	T	T	F	T
F	F	T	T	T

21.

r	s	$r \wedge s$	$\sim s$	$(r \wedge s) \vee \sim s$
T	T	T	F	T
T	F	F	T	T
F	T	F	F	F
F	F	F	T	T

23.

p	q	$\sim p$	$\sim q$	$\sim p \vee \sim q$
T	T	F	F	F
T	F	F	T	T
F	T	T	F	T
F	F	T	T	T

25.

p	q	$\sim q$	$p \wedge \sim q$	$(p \wedge \sim q) \wedge p$
T	T	F	F	F
T	F	T	T	T
F	T	F	F	F
F	F	T	F	F

27.

p	q	$\sim q$	$p \vee q$	$p \wedge \sim q$	$(p \vee q) \vee (p \wedge \sim q)$
T	T	F	T	F	T
T	F	T	T	T	T
F	T	F	T	F	T
F	F	T	F	F	F

29.

p	q	$\sim p$	$\sim p \rightarrow q$	$p \rightarrow (\sim p \rightarrow q)$
T	T	F	T	T
T	F	F	T	T
F	T	T	T	T
F	F	T	F	T

31.

p	q	$\sim p$	$p \wedge q$	$\sim (p \wedge q)$	$\sim p \rightarrow \sim (p \wedge q)$
T	T	F	T	F	T
T	F	F	F	T	T
F	T	T	F	T	T
F	F	T	F	T	T

33.

p	q	$\sim q$	$p \rightarrow p$	$q \rightarrow \sim q$	$(p \rightarrow p) \rightarrow (q \rightarrow \sim q)$
T	T	F	T	F	F
T	F	T	T	T	T
F	T	F	T	F	F
F	F	T	T	T	T

35.

p	q	$\sim p$	$\sim q$	$p \rightarrow q$	$\sim q \rightarrow \sim p$	$(p \rightarrow q) \rightarrow (\sim q \rightarrow \sim p)$
T	T	F	F	T	T	T
T	F	F	T	F	F	T
F	T	T	F	T	T	T
F	F	T	T	T	T	T

37.

p	q	r	~r	p ∧ q	(p ∧ q) ∧ ~r
T	T	T	F	T	F
T	T	F	T	T	T
T	F	T	F	F	F
T	F	F	T	F	F
F	T	T	F	F	F
F	T	F	T	F	F
F	F	T	F	F	F
F	F	F	T	F	F

39.

p	q	r	~p	q ∨ ~p	p ∧ (q ∨ ~p)	[p ∧ (q ∨ ~p)] ∨ r
T	T	T	F	T	T	T
T	T	F	F	T	T	T
T	F	T	F	F	F	T
T	F	F	F	F	F	F
F	T	T	T	T	F	T
F	T	F	T	T	F	F
F	F	T	T	T	F	T
F	F	F	T	T	F	F

41. Direct **43.** Indirect **45.** Indirect **47.** Indirect
49. If you can learn mathematics, then you understand human nature **51.** All trebbles are expensive
53. If a nail is lost, then the kingdom is lost **55.** $b \neq 0$ **57.** I will not eat that piece of pie
59. I will participate in student demonstrations **61.** You obey the law **63.** Babies cannot manage crocodiles
65. If they are my poultry, then they are not officers

APPENDIX B, PAGE 632

Answers vary.

APPENDIX C, PAGE 636

1. 4914 **3.** 18.96048 **5.** 5,517,840 **7.** 5937 **9.** 54,669.23077 **11.** 378,193,771 **13.** 0.77345
15. 17.61441762 **17.** 0.2644086835 **19.** 0.0078046506 **21.** 0.6764741715 **23.** 2.942851909 **25.** .454975922
27. 0.251485404 **29.** 2.259842248

APPENDIX F
PROBLEM SET 1, PAGES 667–669

1. a. No **b.** Yes **c.** No **3. a.** No **b.** Yes **c.** No **5. a.** Yes **b.** No

7. $T^2 = \begin{bmatrix} \frac{3}{8} & \frac{5}{8} \\ \frac{5}{16} & \frac{11}{16} \end{bmatrix}$; $T^3 = \begin{bmatrix} \frac{11}{32} & \frac{21}{32} \\ \frac{21}{64} & \frac{43}{64} \end{bmatrix}$ **9.** $T^2 = \begin{bmatrix} .452 & .548 \\ .3151 & .6849 \end{bmatrix}$; $T^3 = \begin{bmatrix} .39724 & .60276 \\ .346587 & .653413 \end{bmatrix}$

11. $T^2 = \begin{bmatrix} .2075 & .605 & .1875 \\ .265 & .3675 & .3675 \\ .425 & .385 & .19 \end{bmatrix}$; $T^3 = \begin{bmatrix} .3524 & .4136 & .234 \\ .2603 & .4889 & .2509 \\ .2753 & .393 & .3318 \end{bmatrix}$ **13.** $\begin{matrix} & 0 & 1 \\ 0 \\ 1 \end{matrix} \begin{bmatrix} .3 & .7 \\ .6 & .4 \end{bmatrix}$

15.

		Next Day	
		Increase	Decrease
Present	Increase	.3	.7
	Decrease	.8	.2

17. a. No; the bottom row is not a probability vector
b. The probability that a person who rented a car from SFO will return it to SFO is .8 **c.** SJC

19. $T^2 = \begin{bmatrix} .69 & .17 & .15 \\ .33 & .53 & .15 \\ .46 & .29 & .41 \end{bmatrix}$ **21.** $T^2 = \begin{bmatrix} .22 & .52 & .26 \\ .13 & .70 & .17 \\ .13 & .34 & .53 \end{bmatrix}$ **23.** $\frac{3}{4}$ **25.** $\frac{39}{64}$ **27.** $\frac{9}{16}$ **29.** $\frac{7}{16}$ **31.** .99

33. .9890   **35.** .9891   **37.** .0109

PROBLEM SET 2, PAGES 673–675

1. Regular; approaches $\begin{bmatrix} \frac{1}{4} & \frac{3}{4} \\ \frac{1}{4} & \frac{3}{4} \end{bmatrix}$ **3.** Regular; approaches $\begin{bmatrix} \frac{2}{3} & \frac{1}{3} \\ \frac{2}{3} & \frac{1}{3} \end{bmatrix}$ **5.** Regular; approaches $\begin{bmatrix} \frac{4}{13} & \frac{9}{13} \\ \frac{4}{13} & \frac{9}{13} \end{bmatrix}$

7. Regular; approaches $\begin{bmatrix} \frac{5}{8} & \frac{3}{8} \\ \frac{5}{8} & \frac{3}{8} \end{bmatrix}$ **9.** Regular; approaches $\begin{bmatrix} \frac{2}{9} & \frac{4}{9} & \frac{1}{3} \\ \frac{2}{9} & \frac{4}{9} & \frac{1}{3} \\ \frac{2}{9} & \frac{4}{9} & \frac{1}{3} \end{bmatrix}$ **11.** Regular; approaches $\begin{bmatrix} \frac{1}{5} & \frac{1}{2} & \frac{3}{10} \\ \frac{1}{5} & \frac{1}{2} & \frac{3}{10} \\ \frac{1}{5} & \frac{1}{2} & \frac{3}{10} \end{bmatrix}$

13. a. No   **b.** Yes; $[\frac{1}{2} \quad \frac{1}{2}]$ **c.** No   **15. a.** $0 \le a \le 1$ **b.** $[\frac{1}{2} \quad \frac{1}{2}]$

17. $\begin{array}{c} \\ C \\ P \\ D \end{array} \begin{array}{ccc} C & P & D \\ \begin{bmatrix} 0 & 1 & 0 \\ \frac{2}{3} & 0 & \frac{1}{3} \\ \frac{2}{3} & \frac{1}{3} & 0 \end{bmatrix} \end{array}$ approaches $\begin{array}{c} \\ C \\ P \\ D \end{array} \begin{array}{ccc} C & P & D \\ \begin{bmatrix} \frac{2}{5} & \frac{9}{20} & \frac{3}{20} \\ \frac{2}{5} & \frac{9}{20} & \frac{3}{20} \\ \frac{2}{5} & \frac{9}{20} & \frac{3}{20} \end{bmatrix} \end{array}$

In the long run, she will buy canned food 40% of the time, packaged food 45% of the time, and dry food 15% of the time

19. $\begin{array}{c} \\ A \\ B \\ C \end{array} \begin{array}{ccc} A & B & C \\ \begin{bmatrix} .7 & .1 & .2 \\ .1 & .8 & .1 \\ .1 & .4 & .5 \end{bmatrix} \end{array}$ **21.** 25% $(\frac{1}{4})$ for Alpha, 54% $(\frac{13}{24})$ for Better Bean, 21% $(\frac{5}{24})$ for Carolyn's

23. 57% $(\frac{4}{7})$ Democrat, 12% $(\frac{6}{49})$ Republican, 31% $(\frac{15}{49})$ Independent   **25.** 22.5% Jersey, 27.5% Guernsey, 50% white-faced
27. The first company can expect 40% of the market and the second company can expect 60% of the market
29. $\frac{1}{3}$

PROBLEM SET 3, PAGES 679–680

1. Not absorbing   **3.** Not absorbing   **5.** Not absorbing   **7.** All three states are absorbing
9. States 1 and 2 are absorbing   **11.** State 3 is absorbing
13. $FP = [\frac{5}{4}][\frac{3}{10} \quad \frac{1}{2}] = [\frac{3}{8} \quad \frac{5}{8}]$; if the system begins in state 1, the probability that it will end up in absorbing state 2 is $\frac{3}{8}$, and the probability that it will end up in absorbing state 3 is $\frac{5}{8}$
15. $FP = [2][.15 \quad .35] = [.30 \quad .70]$; if the system begins in state 3, the probability that it will end up in absorbing state 1 is .30, and the probability that it will end up in absorbing state 2 is .70

17. $FP = \begin{bmatrix} 10 & 0 \\ \frac{20}{3} & \frac{10}{9} \end{bmatrix} \begin{bmatrix} \frac{1}{10} \\ \frac{3}{10} \end{bmatrix} = \begin{bmatrix} 1 \\ 1 \end{bmatrix}$; since there is only one absorbing state, the probability of ending in that state is 100% regardless of the beginning state

19. $FP = \begin{bmatrix} \frac{3}{2} & 0 \\ \frac{1}{2} & \frac{4}{3} \end{bmatrix} \begin{bmatrix} \frac{1}{3} & \frac{1}{3} \\ \frac{1}{4} & \frac{1}{4} \end{bmatrix} \approx \begin{bmatrix} \frac{1}{2} & \frac{1}{2} \\ \frac{1}{2} & \frac{1}{2} \end{bmatrix}$; if the system begins in state 2, the probability that it will end up in absorbing state 1 is $\frac{1}{2}$, and the probability that it will end up in absorbing state 4 is also $\frac{1}{2}$. If the system begins in state 3, the situation is identical.

21. $FP = \begin{bmatrix} \frac{20}{13} & \frac{6}{13} \\ \frac{10}{13} & \frac{16}{13} \end{bmatrix} \begin{bmatrix} .4 & .1 \\ .1 & .4 \end{bmatrix} = \begin{bmatrix} \frac{43}{65} & \frac{22}{65} \\ \frac{28}{65} & \frac{37}{65} \end{bmatrix}$; if the system is in state 2, there is a probability of $\frac{43}{65}$ that it will end up in absorbing state 1 and a probability of $\frac{22}{65}$ that it will end up in absorbing state 3. If the system begins in state 4, there are probabilities of $\frac{28}{65}$ and $\frac{37}{65}$ that it will end up in absorbing states 1 and 3, respectively

23. $FP = \begin{bmatrix} \frac{5}{3} & \frac{5}{3} & \frac{5}{6} \\ \frac{5}{9} & \frac{20}{9} & \frac{11}{18} \\ \frac{5}{9} & \frac{20}{9} & \frac{29}{18} \end{bmatrix}\begin{bmatrix} .1 & .1 \\ .2 & .2 \\ 0 & 0 \end{bmatrix} = \begin{matrix} 2 \\ 3 \\ 4 \end{matrix}\begin{matrix} 1 & 5 \\ \begin{bmatrix} \frac{1}{2} & \frac{1}{2} \\ \frac{1}{2} & \frac{1}{2} \\ \frac{1}{2} & \frac{1}{2} \end{bmatrix} \end{matrix}$; if the system begins in state 2, 3, or 4, there is an equal chance of ending up in either absorbing state 1 or absorbing state 5

25. a. $\begin{matrix} & 0 & 1 & 2 \\ 0 \\ 1 \\ 2 \end{matrix}\begin{bmatrix} 1 & 0 & 0 \\ \frac{1}{2} & 0 & \frac{1}{2} \\ 0 & 0 & 1 \end{bmatrix}$ **b.** $FP = [1][\frac{1}{2} \quad \frac{1}{2}] = [\frac{1}{2} \quad \frac{1}{2}]$ **c.** 50%   **27.** $\frac{7}{8}$ **29.** $\frac{3}{8}$

Index

Abscissa, 42
Absolute inequality, 19
Absolute maximum, 437, 454
  procedure for finding, 455
  second derivative test, 457
Absolute minimum, 438, 454
  procedure for finding, 455
  second derivative test, 457
Absolute value:
  definition, 5
  function, 51
Acceleration, 410
Accuracy, calculator, 635
Ace, 274
Acidity, 476
Actuary exam questions, 146, 353
Addition:
  matrices, 87
  polynomials, 10
  principle, 271
  property of equations, 19
  property of inequalities, 20
  rational expressions, 29
Additive inverse, 113
Akalinity, 476
Algebraic function, 465
Algebra pretest, 2
Algebra review, 1
Allocation of resources, 137
  example in manufacturing, 140
  example in production, 138, 152
Altered sample space, 286
American Institute of Certified Public
    Accountants, 356
Amortization, 217, 220
  computer program, 638
Analysis of data, 317
And (conjunction), 293
Annual compounding, 205, 469
Annual interest rate, 224

Annual yield, 209
Annuity, 214
  definition, 215
  due, 216
    table, 654
  ordinary, 215
    table, 652
  present value, 217, 592
    table, 656
  summary, 225
Answers, 663
Antecedent, 621
Antiderivative, 491
Antidifferentiation, 492
Anton, Cristy, 240
A posteriori model, 269, 555
Approximations for the definite integral,
    536
A priori model, 268, 555
Archimedes, 362
Area:
  between curves, 513
  function, 505
  histogram, 315
  probability distribution, 557
  procedure for finding, 515
  rectangle, 506
  trapezoid, 506, 539
  triangle, 506
  under a curve, 505
  under a normal curve, 335
Arithmetic mean, 318
Arithmetic sequence, 197
  definition, 198
Arrangement, 242
Assuming the antecedent, 624
Asymptote, 451
  procedure for finding, 68
  vertical, 65, 66, 67, 68
Augmented matrix, 99

Average, 318
  cost, minimal, 441
  expense, 554
  marginal analysis, 424
  rate, 379
  rate of change, 381
  speed, 379, 381
  total cost, 424
  total profit, 424
  total revenue, 424
  value, 552, 615
  winnings, 297
Axis of abscissas, 42
Axis of ordinates, 42

Barth, Craig, 240
Base, 9, 465, 472
Basic feasible solution, 165
Bayes, Thomas, 305
Bayes' Theorem, 305
Bayes tree, 306
Bernoulli, Jacob, 328
Bernoulli distribution, 559
Bernoulli trial, 328
Best-fitting line, 344
Beta distribution, 559
Bimodal distribution, 319
Binomial, 9
Binomial coefficients, computer
    program, 638
Binomial distribution, 327, 559
  compared to normal distribution, 340
  mean, 331
  standard deviation, 331
  theorem, 328
  variance, 331
Binomial experiment, 327
  definition, 327
Binomial multiplication, 13
Binomial probabilities, table of, 644

Binomial random variable, 327
Binomial theorem, 259
    proof of, 632
    statement of, 261
Birthday problem, 292
Blood types, 265
Bond, 223
Boundary, 132
Boundary condition, 571
Boyer, Carl, 372
Brace, 4
Break-even analysis, 62, 82
Break-even linear equation, 134
Break-even point, 134
Brief Integral Table, front and back
    covers
Brown, Joseph, 489
Business models, 420, 548

Calculator, 633
    accuracy, 635
    display, 635
    requirements for this book, 633
    scientific notation, 635
Calculus, 359
    differential, 491
    fundamental theorem, 503
    integral, 491
    software sources, 639
Cancer mortality, 352
Cantor, Georg, 231, 250
Capital expansion model, 574
Capital formation, 554
Capital value, 568
Carbon 14 dating, 482, 488
Cards, deck of, 274
Carlyle, Thomas, 626
Carroll, Lewis, 626
Cartesian coordinate system, 42
Cartesian plane, 42
Cause-and-effect relationship, 348
Cayley, Arthur, 99
Chain rule, 405
    statement of, 407
Change, rate of, 379
Change of base theorem, 480
Charlotte, NC, 497
Chicago, IL, 523
Chi-square distribution, 559
Circular cylinder, 586
Closed half-plane, 132
Closed interval, 434
Closed model, 122
Club, 274
Cobb-Douglas production function, 615,
    617
Coefficients, fractional, 15

Column, 85
Column matrix, 86
Combination, 250
    compared with other counting
        schemes, 255
    compared with permutation, 250
    definition, 251
    formula, 251
    notation, 251
Combinatorics, 230
    table of, 642
Common factor, 14
    definition, 15
Common logarithm, 474
    table of, 661
Common ratio, 199
Common term, 198
Communication matrix, 86
Commutative property, matrices, 93
Comparison, property of, 6
Comparison symbols, 5
Competitive edge, modeling application,
    130
Complement, 235
Complement, Venn diagram, 235
Complementary events, 282
Completely factored, 15
Complex number, 4
Component, 42
Composite, 7
Compound inequality, 21
Compounding period, 469
    annual, 205
    daily, 207
    monthly, 207
    quarterly, 207
    semiannual, 207
Compound interest, 205, 468
    compared with simple, 208
    comparison of compounding period,
        470
    formula, 469
    present value, 210
    table of, 650
Compound statement, 619
Computer, 637
    list of programs, 637
Computer applications:
    binomial probabilities, 330
    binomial theorem, 262
    correlation, 349
    counting problems, 256
    histogram, 325
    least squares, 349
    matrix inverse, 119
    matrix multiplication, 89
    mean, 325

Computer applications (*continued*)
    median, 325
    mode, 325
    pivoting process, 161
    probability problems, 308
    probability properties, 308
    row reduction, 104
    simplex method, 169
    simplex method and rounding, 188
    standard deviation, 325
    systems by inverses, 119
Computers in mathematics, 524
Concave downward, 59, 444, 445, 451
    second derivative test, 447
Concave upward, 59, 444, 451
    second derivative test, 447
Conclusion, 623
    procedure for drawing, 625
Conditional, definition, 621
Conditional equation, 18
Conditional inequality, 20
Conditional probability, 285
    formula for, 288
    notation for, 286
Conformable matrices, 87
Conjunction, 619
Connective, 619
Consequent, 621
Consistent systems, 77
Constant, 8
Constant function, 47, 435
Constant of integration, 492
Constant of proportionality, 202
Constant rule, 397, 399
    differential form, 419
Constant times a function, 397, 399
Constraint, 138, 147, 598
    superfluous, 153
Consumers' surplus, 519
Continuity, 371
    at a point, 373
    polynomial, 374
    Theorem, 374
Continuous compounding, 469
Continuous distribution, 334, 340, 559
Continuous mathematics, 361
Continuous random variable, 313, 556
Contract rate, 223
Contradiction, 20
Contraposition, law of, 623
Contrapositive, 622
Converge, 568
Converse, 622
Convex set, 150
Coordinate axes, 42
Coordinate of a point, 5, 42
Coordinate plane, 42, 582

Coordinate system:
  Cartesian, 42
  one-dimensional, 4
  three-dimensional, 580
  two-dimensional, 42
Corbet, J., 311
Corner point, 150
  procedure for finding, 80
Correction factor, 405
Correlation, 343
  computer program, 638
  negative, 345
  positive, 345
Correlation coefficient:
  formula, 346
  Pearson, 347
Cost, 421, 424
Cost analysis, 76
Cost-benefit model, 69
Cost function, 62
Counting number, 4
Counting problems, computer program,
  638
CPA examination, 192, 193, 228, 357
Craps, 286
Critical point, 439
  functions of several variables, 593
  Lagrange multipliers, 599
Critical value, 25
  correlation, 347
  derivative tests, 438
Cumulative distribution, table of, 648
Cumulative Reviews:
  Chapters 3–8 (finite mathematics),
    354
  Chapters 9–11 (differential calculus),
    462
  Chapters 12–15 (integral calculus),
    576
  Chapters 9–15 (calculus), 577
Curves:
  falling, 445
  rising, 445
  sketching, 450
    procedure, 451
Cylinder, right circular, 586

Daily compounding, 207, 469
Danzig, George, 157
Data, analysis of, 317
Dear Abby, 352
Decay, 572
Decay function, 468
Deck of cards, 274
Decreasing curve, 451
Decreasing function, 435
Dedekind, J., 371

Definite integral, 491
  approximations, 536
  definition, 504
  properties of, 510
  substitution method, 511
Degree, 9
Deleted point, 67
Delta x, 48, 415
Delta y, 48, 415
Demand curve, 81
Demand function, 62
Demand matrix, 124
DeMere, Chevalier, 285
DeMoivre, Abraham, 339
Density function, 558
Denying the consequent, 624
Departing variable, 168
Dependent systems, 77, 108
Dependent variable, 54
  function of two or more variables, 580
Depreciation, 554
Derivative, 360, 379
  definition, 390
  differential approximation, 417
  exponential, 485
  formulas, summary of, 494
  higher, 444
  higher order partial, 589
  of an integral, 509
  logarithm, 483
  notation for, 395
  partial, 588
  second, 444
  tests:
    first, 439
    second, 447
Descartes, Rene, 42
Description method, 231
Diagonal form, 101
Diamond, 274
DiCarlucci, Joseph, 266
Dice, 286
  craps, 286
  sample space, 278
Die (singular of dice), 278
Difference:
  matrices, 88
  rule, 398, 399
    differential form, 419
  of squares, 14, 15
    definition, 16
Difference equation, 196, 197
  general solution, 199, 200
Differential, 415
  approximation, 417
    statement of procedure, 418
  calculus, 491

Differential (*continued*)
  equation, 570
    general solution, 571
    particular solution, 571
    theorems, 570, 571, 573
  formulas, 419
  geometrical definition, 416
  of x, 415
  of y, 415
Differentiation:
  business models, 420
  explicit, 412
  implicit, 412
  techniques, 395, 400
Direct integration formulas, 494
Direct reasoning, 623
  definition, 624
  truth table, 623
Direct variable, *see* Proportionality, 202
Discontinuity, points of, 374
Discontinuous function, 374
Discrete distribution, 334, 340, 559
Discrete mathematics, 361
Discrete random variable, 313, 556
Discriminant, 24
Disjoint sets, 233
Disjunction, 619
Dispersion, 321
Distance, one-dimensional, 7
Distance on a number line, 7
Distinguishable partition, 254
Distinguishable permutation, 248
  general formula, 249
Distributions:
  continuous, 559
  discrete, 559
  exponential definition, 561
  frequency, 556
  normal definition, 562
  probability, 556
  uniform definition, 560
Distributive multiplication, 12
Distributive property, 11
Diverge, 568
Division:
  property of equations, 19
  property of inequalities, 20
  rational expressions, 29
Domain:
  of a function, 35, 37, 42
  of two or more variables, 580
  of a variable, 8
Domar's capital expansion model, 574
Double integral, 605, 607
  definition, 606
  geometric interpretation, 605
  over a variable region, 608

Double negation, 621
Double subscript, 77
Dual, 183
Duality, 182
Dummy variable, 505

e, table of powers, 659
Earthquake, 476
  epicenter, 266
Echelon form, 103
Edge, amusement ride, 385
Eel River, 535
Effective annual yield, 209
Effective rate, 209, 223
Element, 4, 232
Elementary row operations, 101, 104
Eliminating the parameter, 55
Ellipsoid, 585
Elliptic cone, 585
Elliptic paraboloid, 585
Empirical probability, 269, 555
Empty set, 4, 231
  probability of, 272
Entering variable, 168
Entry, 86
Epicenter, 266
Equal:
  matrices, 86
  sets, 233
  symbol, 5
Equal, rational expressions, 29
Equally likely, 273
Equation:
  conditional, 18
  contradiction, 20
  equivalent, 18
  identity, 18
  linear, 18
  open, 18
  properties of, 19
  quadratic, 22
Equations, properties of, 19
Equilibrium point, 71, 81, 519
Equilibrium price, 519
Equivalent matrices, 101
Equivalent systems, 78
Erlanger, J., 546
Essential amino acids, 356
Euler, Leonhard, 233
Euler circles, 233
Evaluate:
  an expression, 10
  a function, 38
  integral, 505
  logarithm, 472
Even number, 7

Event, 269, 556
  probability of, 276
Exact interest, 207, 469
Expectation, 296
Expected value, 296
Experiment, 269, 555
  without replacement, 294
  with replacement, 294
Explicit differentiation, 412
Exponent, 8, 9, 465, 472
  definition, 9
  laws of, 12, 29, 466
  squeeze theorem, 466
Exponential distribution, 559
  definition, 561
  equation, 471, 479
Exponential function:
  definition, 465
  derivative, 485
Exponential property of equality, 473
Exponential rule, 494
Expression, 8
Extremum, *see* Maximum *or* Minimum, 438

Face value of a bond, 223
Factor, 8, 14
Factorial, 245
Factoring, 14
  trinomial, 17
  types of, 14
  over the set of integers, 15
Failure, 327
Fair coin, 273
Fair die, 274
Fair game, 296, 298
False positive result, 311
F-distribution, 559
Feasible region, 148
Feasible solution, 147
Fermat, Pierre, 285
Ferndale, CA, 338, 566
Finance, 195
Financial formulas, computer program, 638
Financial problems:
  summary, 224
  variables, 224
Financial tables, which to use, 224
Finite distribution, 334
Finite mathematics, 361
  definition, 73
  software sources, 639
Finite set, 4
First component, 42
First derivative test, 439, 451
First octant, 583

First-order differential equation, 570
First-order linear difference equation, 197
Fixed cost, 62
Flush, 282, 285
FOIL, 13
  factoring, 16
  multiplication, 14
Foley, Jack, 217, 626
Foley, Rosamond, 217, 249, 332, 626
Fractional coefficients, 15
Frequency, 555
Frequency distribution, 314, 556
  computer program, 638
Full house, 285
Function, 33
  algebraic, 465
  decay, 468
  defined by a graph, 36
  defined by an algebraic formula, 37
  defined by a table, 36
  defined by a verbal rule, 36
  definition, 35
  domain, 37
  exponential, 465
  falling, 436
  growth, 468
  linear, 47
  logarithmic, 472
  n independent variables, 580
  polynomial, 63
  pretest, 34
  quadratic, 58
  rational, 65
  rising, 436
  transcendental, 465
  two independent variables, 580
Fundamental connective, 619
Fundamental counting principle, 243, 244
  compared with other counting procedures, 255
Fundamental property of rational expressions, 29
Fundamental theorem of calculus, 503, 505
  restatement of, 509
  statement of, 507
Fundamental theorem of linear programming, 150
Future value, 208, 224, 468, 591
  compound interest, 207
  formula, 469
  simple interest, 205
  summary, 225
  table of, 650

Galileo, 371
Gamma distribution, 559
Gasser, H., 546
Gauss, Karl, 103
Gaussian elimination, 103
Gauss-Jordan elimination, 99
  procedure summarized, 105
Generalized power rule, 405
General situation, related rates, 426, 429
General solution, differential equation, 571
General term, 197
Geometric distribution, 559
Geometric mean, 326
Geometric sequence, 199
  definition, 199
Gini index, 513
GNP, 388
Goethe, 626
Golden Gate Bridge, 380
Gottlief, Sheldon F., 546
Grade on a curve, 336, 564
Graphing, 33
  first derivative test, 434
  functions, 42, 44
  lines, 50
  method for solving systems, 78
  numbers, 5
  relations, 42
  software sources, 639
  strategy, 451
Greater than, 5
Greatest common factor, 15
Gross National Product, 388
Growth, 572
  function, 468
  human population, 480
Gutenberg, 476

Half-life, 488
Half-plane, 132
Hanford, Washington, 352
Harmonic mean, 326
Heart, 274
Henry, Patrick, 626
Higher derivatives, 444
Higher-order partial derivative, 589
Histogram, 315, 557
  area of, 315
  binomial, 339
  computer program, 638
Homogeneous systems, 123
Hoover Dam, 497
Horizontal asymptote, 67, 68, 451
Horizontal change, 48
Horizontal line, 47, 56

Horizontal tangent line, 437
Houston, TX, 577
Human population growth, 480
Hyperbolic paraboloid, 585
Hyperboloid of one sheet, 585
Hyperboloid of two sheets, 585
Hypergeometric distribution, 559
Hypotheses, 623

Identity:
  addition for matrices, 114
  addition for reals, 113
  multiplication for matrices, 114
  multiplication for reals, 113
Implicit differentiation, 412
Impossible event, 269, 556
Improper integral, 567
  definition, 568
Improper subset, 233
Incidence matrix, 87
Inconsistent systems, 77, 107
Increasing curve, 451
Increasing function, 435
Indefinite integral, 492
Independent events, 289
  test for, 290
Independent systems, 77
Independent variable, 54
Indeterminate form, 367
Index of summation, 203
Indirect reasoning, definition, 624
Induction, mathematical, 626
Inequality:
  absolute, 19
  compound, 21
  conditional, 20
  notation for rays, 435
  notation for segments, 434
  polynomial, 26
  properties of, 20
  quadratic, 25
  symbols, 5
Infinite set, 4
Infinity, limit at, 368
Inflation, 211
Inflection point, 444, 445, 454
Initial basic solution, 165
Initial condition, 571
Initial simplex tableau, 160
Initial value, 196, 495
Input-output matrix, 122
Input-output model, 122
Installment loans, 220
Installment payments:
  formula, 220
  summary, 225
Instantaneous rate of change, 383

Instantaneous speed, 379
Integer, 4, 7
Integral:
  constant, 494
  converge, 568
  of a derivative, 509
  difference, 494
  diverge, 568
  double, definition of, 606
  improper, 567
  indefinite, 492
  iterated, 606
  sum, 494
  symbol, 492
Integral calculus, 491
Integrand, 492, 505, 606
Integration:
  constant of, 492
  exponential rule, 494
  formulas, front cover
  limits of, 505
  logarithmic rule, 494
  numerical, 535
  by parts, 526
  power rule, 494
  procedural formulas, 494
  region of, 606
  reversing the order, 612
  substitution, 497
    procedure, 499
  summary, 494
  by table, 530
  variable limits of, 607
Intercept, 45, 451
Interest, 224, 468
  comparison of simple and compound, 208
  compound, 205
  compounding periods, 469
  continuous, 469
  definition, 204
  exact, 207, 469
  ordinary, 207, 469
  rate, 204
  simple formula, 204
Intermediary messages, communication matrices, 96
Intersection, 235, 293
  probability of, 293
  Venn diagram, 235
Interval notation, 434
  for rays, 435
  for segments, 434
Inventory control, 86
Inverse, 622
  addition for reals, 113
  functions, 475

Inverse (*continued*)
  matrices, 113, 114
    procedure for finding, 117
  matrix computer program, 637
  multiplication for reals, 113
Investment flow, 554
Investment modeling application, 229
Irrational number, 4, 7
Irreducible over the set of integers, 16
Israeli economy model, 127
Iterated integral, 606

Jordan, Camille, 103

Knievel, Evel, 64
Kohn, R., 194
Koont, Sinan, 546
Kronecker, Leopold, 250

Lagrange multiplier, 598
  method of, 599
Lambda, 599
Larsen, Richard, 352
Law of contraposition, 623
Law of detachment, 624
Law of double negation, 621
Least squares line, 348
  formula, 349
Least squares method, 344
Left-hand limit, 364
Leibniz, Gottfried, 359, 371
Leontief, Wassily, 122, 127
Leontief models, 122
Life insurance policies, 301
Limit, 361
  constant, 364
  infinity, 368
  of integration, 505
  intuitive notion, 363
  left-hand, 364
  notation, 363
  polynomial, 366
  right-hand, 364
  of *x* as *x* approaches *a*, 362
Limits of integration, 505
Lindstrom, Peter, 410
Line, 43
Linear combination method, 79
Linear correlation coefficient, 346
Linear equation, 18, 47
  computer program, 637
  summary of forms, 56
Linear function, 47
Linear inequality, 132
Linear programming:
  examples of models, 137
  formulating the problem, 137

Linear programming (*continued*)
  fundamental theorem of, 150
  graphical method, 131
  graphical solution, 147
  nonstandard problem, 174
  procedure for solving, 151
Loaded die, 274
Logarithm:
  change of base, 480
  common, 474
  derivative, 483
  equation solving, 476
  evaluate, 472
  laws of, 478
  natural, 474
Logarithmic equation, 476
Logarithmic function, definition, 472
Logarithmic rule, 494
Logic, 619
  algebraic, 633
  arithmetic, 633
  conclusions, procedure for reaching,
    625
  RPN, 633
  software sources, 639
Log of both sides theorem, 477
Loss, 62
Lower limit of summation, 203
Lump sum, 225

Management accounting examination,
  228
Marginal, 421
  analysis, 421
  average costs, 424
  average profit, 424
  average revenue, 424
  cost, 389, 393
  definition, 393
  profit, 385
Market rate, 223
Market share, computer program, 638
Marsh, Joan, 240
Mathematical expectation, 296, 298
Mathematical induction, 626
  principal of, 629
Matrices (plural of matrix), 85
Matrix, 84, 85
  addition, 87
  augmented, 99
  column, 86
  communication, 87
  conformable, 87
  definition, 85
  elementary row operations, 101
  entry, 86
  equal, 86

Matrix (*continued*)
  equation, computer method of
    solution, 637
  equivalent, 101
  incidence, 87
  input-output, 122
  inverse, 113, 114
    procedure for finding, 117
  multiplication, 89, 92
    computer program for, 89, 637
    definition, 92
  noncommutative, 93
  nonsingular, 115
  product, computer program, 637
  row, 86
  scalar multiplication, 89
  subtraction, 88
  theory, 99
  zero, 86
Maximization, 164
Maximize:
  production, 137
  profit, 137
  return, 137
  revenue, 137
Maximum:
  absolute, 437, 454
  existence of, 438
  method of Lagrange multipliers, 599
  profit, 457, 458
  relative, 438, 451
  relative for functions of several
    variable, 592
  second derivative test, 457
  value, 61, 150
MAXSYMA, 524
Maxwell distribution, 559
Mean, 318, 562
  binomial distribution, 331
  computer program, 638
  geometric, 326
  harmonic, 326
  notation for, 321
  quadratic, 326
  weighted, 321
Measure of central tendency, 318
Median, 318
  computer program, 638
Medical diagnosis, 311
Member, 4, 231
Mikalson, Tom, 211
Milton, J., 311
Minimal average cost, 441
Minimization problem, 178
Minimize:
  calories, 137
  cost, 137

Minimize (*continued*)
  expense, 137
  shipping costs, 137
Minimum:
  absolute, 438, 454
  existence of, 438
  method of Lagrange multipliers, 599
  relative, 438, 451
  relative for functions of several
    variable, 592
  second derivative test, 457
  value, 61, 150
Mode, 318
Model:
  a posteriori, 555
  a priori, 555
  business, 548
  probability, 555
Modeling, steps in, 617
Modeling applications:
  air pollution control, 194
  Cobb-Douglas production function,
    617
  computers in mathematics, 524
  earthquake epicenter, 266
  gaining a competitive edge, 130
  instantaneous acceleration, 410
  investment opportunity, 229
  medical diagnosis, 311
  nervous system, 546
  publishing, 432
  world running records, 489
Modus ponens, 624
Modus tollens, 624
Money flow, 550, 551
Monomial, 9
Monthly compounding, 207, 469
Monthly payment, 482
Mortality table, 658
Multiple integral, 604
Multiplication:
  inverse, 113
  matrices, 89
  polynomials, 11
  property of equations, 19
  property of factorial, 246
  property of inequalities, 20
  rational expressions, 29
Multivariate function, 580
MuMath, 524
Mutually exclusive, 269

Natural logarithm, 474
  table of, 659
Natural number, 4, 7
Negation (*not*), 619
  law of double, 621

Negative binomial distribution, 559
Negative correlation, 345
Negative number, 5
Neither, 619
Nervous system, model for, 546
Net investment flow, 554
Net price, 422
New Haven, CT, 497
Newton, Isaac, 359
Niklas, Phyllis, 240
Nonconformable matrices, 90
Nonsingular matrix, 115
Nonstandard linear programming
    problem, 174
Normal approximation to the binomial,
    338, 340
  minimum sample required, 342
Normal cumulative distribution, table
    of, 648
Normal curve, 333, 562
  associated areas, 335
  standard, 563
  standard form, 334
  z-score, 565
Normal distribution, 333, 559
  compared to binomial distribution,
    340
  definition, 562
  distribution curve, 334
Normally distributed data, 562
No slope, 49, 54
Not (negation), 619
Not equal to symbol, 5
Novato, CA, 380
Null set, 4
Number:
  comparison of types, 7
  complex, 4
  composite, 7
  counting, 4
  even, 7
  integer, 4
  irrational, 4
  negative, 5
  odd, 7
  positive, 5
  prime, 7
  rational, 4
  real, 4
  relationship among types of, 7
  whole, 4
Number line, 4
Numerical coefficient, 8
Numerical integration, 535

Objective function, 139, 147, 598
Octant, 583

Odd number, 7
Odds, 283
Oil consumption, 503
Oil prices, 503
  crash of, 518
  demand, 523
  reserves, 523
One-dimensional coordinate system, 4
Open equation, 18
Open half-plane, 132
Open interval, 434
Open model, 122, 124
Operations research, 299
Opposite, 113
Optimal land use, 137
Optimal solution, 167
Optimum value, 186
Or (disjunction), 293
Order, 85
Ordered four-tuple, 242
Ordered pair, 42, 242
Ordered partition, 252, 254
  compared with other counting
    schemes, 255
  formula, 255
Ordered triplet, 242, 581
Order of operations, 9
  sets, 236
Ordinary annuity, 215
  summary, 225
  table of, 652
Ordinary interest, 207, 469
Ordinate, 42
Origin, 5, 42
Overhaul time, 554
Overlapping sets, 233

Pair, 285
Parabola, 43, 59, 61
  graph of, 436
Parabolic approximation, 541
Parameter, 55, 108
Parametric equations, 55
Parametric form, 55, 56
Parentheses, order of operations with
    Venn Diagram, 235
Partial derivative, 587
  definition, 588
  geometric interpretation, 588
  higher order, 589
Particular solution, differential equation,
    571
Partition, 252, 253
  compared with other counting
    schemes, 255
Partitioned event, probability of, 304
Pascal, Blaise, 264, 285

Pascal's triangle, 259, 264
  table of coefficients, 642
Pearson correlation coefficient, 347
Perfect square, 17
Period, 224
Periodic payment, 224
Periodic payments, 225
Permutation, 242
  compared with combination, 250
  compared with other counting
      schemes, 255
  definition, 242
  distinguishable, 248
  formula, 247
  general formula, 249
  notation, 243
Petaluma, CA, 380
Petroleum:
  prices, 503, 518
  prices, demand, 523
  reserves, 523
  world use, 482
pH, 476
pi, 362
Piecewise linear function, 51
Pigeonhole procedure, 244
Pivot, 103
  column, 167, 169
  element, 168
  row, 102, 168, 169
Pivoting, 103
  operation, 161
  process, definition, 167
Plane, 583
  coordinate, 42
  in three dimensions, 582
  xy, 42, 582
  xz, 582
  yz, 582
Plot, a point, 42
PMI, 629
Point of inflection, 334, 444, 454
Point-slope form, 53, 56
Poiseuille' law, 591
Poisson distribution, 559
Polynomial:
  addition, 10
  classified, 9
  function, 63
  general form, 9
  inequality, 26
  multiplication, 11
  subtraction, 11
Population, 321
  growth or decay, 480
Positive correlation, 345
Positive number, 5

Posteriori model, 269
Power, 9
PowerMath, 524
Power rule, 396, 399, 494
  differential form, 419
  generalized, 405
Premise, 623
Present value, 204, 210, 224
  annuity, 217, 592
    formula, 218
    table of, 656
  summary, 225
  table of, 650
Pretest:
  algebra, 2
  functions and graphs, 34
Price, selling vs. net, 422
Primary waves, 266
Prime, 7
Principal, 204, 468
Priori model, 268
Probabilistic model, 268, 555
Probability, 267, 312
  binomial distribution, 328
  calculated, 280
  complementary events, 282
  computer program, 638
  conditional, 285, 288
  definition, 557
  density function, 558
  distribution, 315, 556
  distribution function, 559
  empirical, 269, 555
  empty set, 272
  equally likely events, 273
  event, 268, 555
  exponential distribution, 561
  given odds, 284
  intersection, 293
  measure, 270
  models, 268
  mutually exclusive and equally likely,
      275
  normal distribution, 333, 334, 562
  partitioned event, 304
  procedure for finding, 276
  related to set notation, 270
  relative frequency, 555
  subjective, 269, 555
  summary of formulas, 295
  theoretical, 268, 555
  uniform distribution, 560
  uniform model, 274
  union, 293
Producer's surplus, 519
Production function, 617
Product of matrices, 92

Product rule, 400
  differential form, 419
Profit, 62, 421, 424
  function, 423
  maximum, 457
Progression, 196, 198
Proper subset, 233
Proportionality, 202
Publishing, an economic model, 432

Quadrant, 42
Quadratic equation, 22
  solution by factoring, 23
  solution by formula, 24
Quadratic formula, 24
Quadratic function, 58
Quadratic inequalities, 25
Quadratic mean, 326
Quadric surface, 584
  summary, 585
Quarterly compounding, 207, 469
Quotient rule, 402
  differential form, 419

Radioactive contamination, 352
Random variable, 313, 556
  binomial, 327
Range:
  of a function, 35, 42
  of a function of two or more
      variables, 580
  of a set of data, 321
Rate:
  average, 379, 381
  bond, 223
  of change, 379
    linear, 572
  constant, 379
  interest, 204
  related, 426
Rational expression:
  definition, 27
  properties of, 29
  simplify, 28
Rational function, 44, 65
Rationalize the denominator, 367
Rationalize the numerator, 367
Rational number, 4, 7
Rayleigh distribution, 559
Real number, 4
  negative, 5
  positive, 5
  relationship of subsets, 7
Reciprocal, 113
Rectangular approximation, 536
Reduced row-echelon form, 103
Region of integration, 606

Regression, 343
    analysis, 344
    computer program, 638
    line, 348
        formula, 349
Related rates, 426
    procedure for solving, 429
Relative frequency, 268, 314, 555
Relative maximum, 438, 451
    first derivative test, 439
    functions of several variables, 592
    graph of, 593
    procedure for finding with functions
        of several variables, 595
    second derivative test, 447
Relative minimum, 438, 451
    first derivative test, 439
    functions of several variables, 592
    graph of, 593
    procedure for finding with functions
        of several variables, 595
    second derivative test, 447
Repetition, 247
Replacement, with and without, 290, 294
Revenue, 421, 424
Revenue function, 62
Reversing the order of integration, 612
Reviews:
    Chapter 1, 31
    Chapter 2, 71
    Chapter 3, 127
    Chapter 4, 191
    Chapter 5, 226
    Chapter 6, 264
    Chapter 7, 308
    Chapter 8, 351
    Chapter 9, 408
    Chapter 10, 431
    Chapter 11, 461
    Chapter 12, 487
    Chapter 13, 522
    Chapter 14, 544
    Chapter 15, 574
    Chapter 16, 615
    cumulative, Chapters 3–8 (finite
        mathematics), 354
    cumulative, Chapters 9–11
        (differential calculus), 462
    cumulative, Chapters 12–15 (integral
        calculus), 576
    cumulative, Chapters 9–15 (calculus),
        577
Richter scale, 476
Right circular cylinder, 586
Right-hand limit, 364
Rise, 48
Roller coaster, 383

Root, 18
Root mean square, 326
Roster method, 231
Roulette, 292, 301
Row, 85
    matrix, 86
    operations, 101, 104
    reduction, computer program, 637
Row-echelon form, 103
Royal flush, 285
Run, 48

Saddle point, 594
    graph of, 593
    procedure for finding, 595
Sample scores, 321
Sample space, 269, 556
    altered, 286
San Antonio, TX, 503
San Francisco, CA, 379
San Rafael, CA, 380
Santa Rosa, CA, 379, 380, 496
Satisfy an equation, 18, 42
SAT scores, 388
Sausalito, CA, 380
Savick, Wayne, 548
Scalar multiplication, 89
Scatter diagram, 344
Schedule X (U.S. tax rates), 57
Scientific calculator, 633
Scientific notation, calculator, 635
Sebastopol, CA, 380
Secant line, 381
Secondary waves, 266
Second component, 42
Second derivative, 444
Second derivative test, 447, 451
    absolute maximum and minimum,
        457
    functions of two variables, 595
Selling price, 422
Semiannual compounding, 207, 469
Separate the variables, differential
        equation, 572
Sequence, 196
    arithmetic, 197
        definition, 199
    geometric, 199
        definition, 199
Set-builder notation, 37, 231
Sets, 4, 231
    disjoint, 233
    empty, 4, 231
    equal, 233
    four general sets, 241
    null, 4
    number of elements, 238

Sets (*continued*)
    numbers, 4
    operations, 231
    overlapping, 233
    solution, 18
    three general sets, 237
    two general sets, 234
    universal, 231
    Venn diagrams, 233
Shaw, G., 626
Sigma, 202
Sigma notation, 202, 318
Significance level, 347
Similar terms, 10
Simple event, 269, 556
Simple interest, 468
    compared with compound, 208
    formula, 204
    present value, 210
Simple statement, 619
Simplex method, 157, 164
    computer program, 637
    definition, 169
    flowchart for, 169
    removing a negative element in
        rightmost column, 177
Simpson's rule, 541
Simultaneous solution, 77
Sinking fund, 222
    summary, 225
Slack variable, 157, 158
    definition, 159
Slope, of a line, 48
Slope-intercept form, 49, 56
Smith, Chris, 379, 380
Smith, Shannon, 240, 241
Social Security, 217
Society of Actuaries, 353
Software, sources for, 639
Solution:
    of an equation, 18
    linear difference equation, 197
    set, 18
Spade, 274
Specific situation, related rates, 426, 429
Speed, 379
Sphere, 585
Square, perfect, 17
Square root:
    definition, 24
    property, 23
    table, 643
Squares table, 643
Squeeze theorem for exponents, 466
Standard deviation, 322, 324, 562
    binomial distribution, 331
    computer program, 638

Standard form, 53, 56, 158
  definition, 54
Standard linear programming problem,
  158
Standard normal cumulative
  distribution, table of, 648
Standard normal curve, 334, 563
Standard position parabola, 58
Statement, 619
Statistical linear relationship, 348
Statistics, 312
  computer program, 638
  software sources, 639
Steepness of a line, 48
Stochastic process, 302
Stopping distance of an automobile, 64
Straight-line depreciation, 554
String of inequalities, 21
Stroup, Donna, 352
Student's distribution, 559
Subjective probability, 269, 555
Subset, 232
  as a combination, 250
  number of a finite set, 263
Substitution method for integration, 497
  definite integration, 511
Substitution method for solving systems,
  79
Subtraction:
  matrices, 88
  polynomials, 10, 11
  property of equations, 19
  property of inequalities, 20
  rational expressions, 29
Success, 327
Summation notation, 202, 318
Sum of matrices, 87
Sum rule, 398, 399
  differential form, 419
Superfluous constraint, 153
Supply and demand, 81
Supply curve, 81
Surface, 583
  quadric, 584
Survey problems, 237, 238
Suspicious points, 374
Syllogism, 623
Symbolic language, 524
Symbolic logic, 241
Symbolic manipulation, 524
Symbolic mathematics, software sources,
  640
Symmetric, 59
Symmetry, 451
Systems:
  dependent, 77, 108
  equations, 76, 77

Systems (*continued*)
  Gauss-Jordan elimination, 105
  homogeneous, 123
  inconsistent, 77, 107
  linear inequalities, 132, 135
  matrix inverse, 119
  m equations with n unknowns, 101
  more equations than unknowns, 107
  more unknowns than equations, 109
  solution by graphing, 77
  solution by linear combinations, 79
  solution by substitution, 79

Tangent line, 383
Target row, 102
Tax rate schedule, 57
Term, 8
  of a sequence, 196
Terms:
  Chapter 1, 31
  Chapter 2, 71
  Chapter 3, 127
  Chapter 4, 191
  Chapter 5, 226
  Chapter 6, 264
  Chapter 7, 308
  Chapter 8, 351
  Chapter 9, 408
  Chapter 10, 431
  Chapter 11, 461
  Chapter 12, 487
  Chapter 13, 522
  Chapter 14, 544
  Chapter 15, 574
  Chapter 16, 615
Test for maximality, 169
Test point, 132
Tests:
  Chapter 1, 31
  Chapter 2, 71
  Chapter 3, 128
  Chapter 4, 192
  Chapter 5, 227
  Chapter 6, 264
  Chapter 7, 308
  Chapter 8, 352
  Chapter 9, 408
  Chapter 10, 431
  Chapter 11, 462
  Chapter 12, 487
  Chapter 13, 522
  Chapter 14, 544
  Chapter 15, 575
  Chapter 16, 616
  cumulative reviews:
    Chapters 3–8 (finite mathematics),
      354

Tests (*continued*)
  Chapters 9–11 (differential
    calculus), 462
  Chapters 12–15 (integral calculus),
    575
  Chapters 9–15 (calculus), 577
Theoretical probability, 268, 555
Third derivative, 444
Three-dimensional coordinate system,
  581
Three dimensions, 580
Time for a loan or deposit, 204
TK! Solver, 524
Total depreciation, 554
Total money flow, 550, 551
Total value, 549
Trace, 584
Transcendental function, 465
Transitivity, 625
Translation, 59
Transportation problem:
  formulation, 143
  solution of, 178
Transpose, definition, 185
Trapezoid, 539
Trapezoidal approximation, 539
Treasury Bond rate, 545
Tree diagram, 232, 243
Trichotomy property, 6
Trinomial, 9
  factoring, 16, 17
Triola, Mario, 348
Truth table, 619
Truth value, 619
Two-dimensional coordinate system, 42

Uniform distribution, 559
  definition, 560
Uniform probability model, 274
Union, 234, 293
  number of elements in, 238
  probability of, 293
  Venn diagram, 234
Universal set, 231
Unless, 619
Upper limit of summation, 203
Utilization of staff, 137

Vallejo, CA, 380
Value of a function, 35
Variable, 8
Variable limits of integration, 607
Variance, 321, 322
  binomial distribution, 331
Velocity, 385
Venn, John, 233, 241

Venn diagram, 233
   complement, 235
   intersection, 235
   number of elements in union, 238
   union, 234
Vertex, 59
Vertical asymptote, 65, 68, 451
Vertical change, 48
Vertical line, 47, 56
Vertical line test, 44, 47
Volume, 613
Von Neumann, John, 182
Von Neumann's duality principle, 186

Walton, W., 543
Weierstrass, Karl, 371

Weighted mean, 321
Whole number, 4, 7
Wisner, Bob, 240
Without repetition, 242
Without replacement, 290, 294
   general agreement, 295
With replacement, 290, 294
World population, 481
World running records, 489

x-axis, 42
x-intercept, 451
   definition, 45
   procedure for finding, 47
xy-plane, 42, 582
xz-plane, 582

y-axis, 42
y-intercept, 451
   definition, 45
   procedure for finding, 47
yz-plane, 582

z-score, 337, 565
Zero, property of, 22
Zero matrix, 86
Zero of a function, 45
Zero slope, 49, 54

Forms Containing $\sqrt{x^2 \pm a^2}$, $\sqrt{a^2 - x^2}$

15. $\displaystyle\int \sqrt{x^2 \pm a^2}\, dx = \frac{x}{2}\sqrt{x^2 \pm a^2} \pm \frac{a^2}{2}\ln|x + \sqrt{x^2 \pm a^2}| + C$

16. $\displaystyle\int \frac{dx}{\sqrt{x^2 \pm a^2}} = \ln|x + \sqrt{x^2 \pm a^2}| + C$

17. $\displaystyle\int x\sqrt{x^2 \pm a^2}\, dx = \frac{1}{3}\sqrt{(x^2 \pm a^2)^3} + C$

18. $\displaystyle\int x\sqrt{a^2 - x^2}\, dx = -\frac{1}{3}\sqrt{(a^2 - x^2)^3} + C$

19. $\displaystyle\int \frac{dx}{x\sqrt{x^2 + a^2}} = -\frac{1}{a}\ln\left|\frac{a + \sqrt{x^2 + a^2}}{x}\right| + C$

20. $\displaystyle\int \frac{dx}{x\sqrt{a^2 - x^2}} = -\frac{1}{a}\ln\left|\frac{a + \sqrt{a^2 - x^2}}{x}\right| + C$

21. $\displaystyle\int \frac{\sqrt{x^2 + a^2}\, dx}{x} = \sqrt{x^2 + a^2} - a\ln\left|\frac{a + \sqrt{x^2 + a^2}}{x}\right| + C$

22. $\displaystyle\int \frac{\sqrt{a^2 - x^2}\, dx}{x} = \sqrt{a^2 - x^2} - a\ln\left|\frac{a + \sqrt{a^2 - x^2}}{x}\right| + C$

23. $\displaystyle\int \frac{x\, dx}{\sqrt{x^2 \pm a^2}} = \sqrt{x^2 \pm a^2} + C$

24. $\displaystyle\int \frac{x\, dx}{\sqrt{a^2 - x^2}} = -\sqrt{a^2 - x^2} + C$

25. $\displaystyle\int x^2\sqrt{x^2 \pm a^2}\, dx = \frac{x}{4}\sqrt{(x^2 + a^2)^3} \mp \frac{a^2 x}{8}\sqrt{x^2 \pm a^2} - \frac{a^4}{8}\ln|x + \sqrt{x^2 \pm a^2}| + C$

26. $\displaystyle\int \frac{dx}{x^2\sqrt{x^2 \pm a^2}} = \mp \frac{\sqrt{x^2 \pm a^2}}{a^2 x} + C$

27. $\displaystyle\int \frac{dx}{x^2\sqrt{a^2 - x^2}} = -\frac{\sqrt{a^2 - x^2}}{a^2 x} + C$

28. $\displaystyle\int \frac{\sqrt{x^2 \pm a^2}}{x^2}\, dx = -\frac{\sqrt{x^2 \pm a^2}}{x} + \ln|x + \sqrt{x^2 \pm a^2}| + C$

29. $\displaystyle\int \frac{x^2\, dx}{\sqrt{x^2 \pm a^2}} = \frac{x}{2}\sqrt{x^2 \pm a^2} \mp \frac{a^2}{2}\ln|x + \sqrt{x^2 \pm a^2}| + C$